FOUNDRY TECHNOLOGY

Source Book
A collection of outstanding articles
from the technical literature

Compiled by
Consulting Editor

PAUL J. MIKELONIS
Corporate Technical Director
General Casting Corporation

American Society for Metals
Metals Park, Ohio 44073

American Foundrymen's Society
Des Plaines, Illinois 60016

Library of Congress Card No: 82-71925
ISBN: 0-87170-143-X
SAN 204-7586

PRINTED IN THE UNITED STATES OF AMERICA

Contributors to This Source Book

G. E. BARGEHR
CIBA-GEIGY (Pty.) Ltd.

C. E. BATES
Southern Research Institute

D. J. BERANT
Giffels Associates

B. A. BETTS
U.S. Bureau of Mines

F. F. BOWNES
Steel Castings Research &
 Trade Assn.

A. J. BROOME
Foseco Pty. Ltd.

J. R. BROWN
Fordath Ltd.

R. K. BUHR
CANMET, Dept. of Energy, Mines &
 Resources
Ottawa, Canada

HARRY E. CHANDLER
Editor, *Metal Progress*

W. CHEEK
Central Foundry Div.
General Motors Corp.

A. J. CLEGG
Loughborough University of
 Technology
England

R. C. CREESE
West Virginia University

R. L. CROSBY
U.S. Bureau of Mines

I. DAVIES
Electricity Council

L. E. ERICKSON
Norton Co.

W. J. EVANS
Ford Motor Co.

A. G. FENNELL
Sterling Metals Ltd.

D. R. FERRELL
Energy Services Engineering, Inc.

P. A. FISHER (ret.)
Magnesium Elektron Ltd.

J. W. FRANCIS
Vulcan Pattern Making Co. Ltd.

A. G. FULLER
BCIRA

P. C. J. GALLAGHER
N L Industries, Inc.

T. E. GARNAR, JR.
E. I. du Pont de Nemours & Co.

S. F. GREENWAY
Ford Motor Co. Ltd.

A. H. GREGORY
Ashland Chemical Co. Ltd.

P. A. HENDERSON
Norton Co.

DONALD C. HERRSCHAFT
International Lead Zinc Research
 Organization, Inc.

F. H. HOULT
W. H. Booth & Co. Ltd.

R. G. HURLEY
Ford Motor Co.

R. H. IMMEL
ARCO Polymers, Inc.

A. JEYARAJAN
Graduate Student
University of Michigan

S. D. KISER
Huntington Alloys, Inc.

M. J. LALICH
Foote Mineral Co.

R. D. LANGMAN
Electricity Council Research Centre

P. H. R. B. LEMON
Borden (UK) Ltd.

N. LUTHER
Pontiac Motor Div.
General Motors Corp.

NORRIS B. LUTHER
Lester B. Knight & Associates

R. F. LYNCH
N L Industries, Inc.

S. MATHER
L. Gardner & Sons Ltd.

NOTE: Affiliations given were applicable at date of contribution.

W. McCORMACK
BCIRA

R. McROBERT
Steelcast Ltd.

S. K. MEHLMAN
Union Carbide Corp.

P. H. MIKKOLA
Central Foundry Div.
General Motors Corp.

R. W. MONROE
Southern Research Institute

J. G. MORLEY
BCIRA

A. E. MURTON
CANMET, Dept. of Energy, Mines &
   Resources
Ottawa, Canada

L. A. NEUMEIER
U.S. Bureau of Mines

R. P. OLLEY
N L Industries, Inc.

J. E. PATCHETT
Norton Co.

J. J. PATERNO
Norton Co.

V. H. PATTERSON
Foote Mineral Co.

R. D. PEHLKE
University of Michigan

SCHRADE F. RADTKE
International Lead Zinc Research
   Organization, Inc.

J. A. RASSENFOSS
Amsted Research Laboratories

G. RICHARDS
Sterling Metals Ltd.

F. W. ROHR
Perkins & Will

R. A. ROWLAND
Electron Corp.

JACK H. SCHAUM
Publisher/Editor
Modern Casting

J. R. SCHOEN
Grede Foundries

W. E. SCHULZE
Caterpillar Tractor Co.

G. F. SERGEANT
BCIRA

M. J. SKUBON
Cober Electronics, Inc.

GLENN W. STAHL
Stahl Specialty Co.

R. P. STANBRIDGE
Foseco, Inc.

R. W. STORY
Norton Co.

W. J. VINCENT
Dartmouth Auto Castings Ltd.

F. W. WALKER
Glow Worm Ltd.

A. WARD
Ward Associates

KEN R. WHALER
Stahl Specialty Co.

T. R. WIGHT
Central Foundry Div.
General Motors Corp.

C. F. WILFORD
Electricity Council Research Centre

R. WILLIAMS
Foseco International Ltd.

THOS. WILLS
Institute of British Foundrymen

R. WRIGHT
CIBA-GEIGY Plastic Additives Co.

PREFACE

The process of metal casting dates back several thousand years. The first casting was performed with nonferrous metals, followed by ferrous metal casting. Then as now, pouring molten metal into a refractory or heat-resistant material containing an internal cavity is the most direct way of producing a desired shape.

Technological advances in the metal casting industry have been dramatic in the last thirty years. Society's demands on the foundry have been as great as those placed on any other industry, if not greater. The need to attain greater productivity, produce castings in thin sections of high integrity, reduce energy in an energy-intensive process, comply with stringent safety and environmental control legislation, and conserve materials has spurred the progress of foundry technology.

This Source Book highlights metal casting process areas that have been subjected to many fresh ideas and presents some recycled ideas for advancing founding from an art to a science. Many new materials and processes are now available to the metal caster, and the choice of one material or process over another has become a difficult decision. A number of overviews are presented here along with detailed articles on individual processes or materials to aid the selection of the most desirable route for a particular operation to take.

Section I is an overview of sand and die casting, giving a state-of-the-art survey and revealing the direction of future technological change.

Section II presents papers on casting techniques, some old and some new, but all fulfilling the demands for castings with better finish, having closer dimensional tolerances and in some instances produced at lower cost than by conventional sand casting. The permanent mold process is not new, but it is being used to a greater extent as aluminum casting production increases. The precision casting processes are old, but techniques have changed greatly to give comparatively close dimensional tolerance control on intricate shapes. Producing expendable patterns from expandable polystyrene is a relatively new casting technique that changes the concept of normal green sand casting regarding material and equipment needs. In combination with green sand molding it reduces some equipment requirements and improves dimensional control. The Japanese vacuum molding process has been available a short ten years, but a number of foundries have been built to exploit this process when medium production is required, because it achieves substantial savings on sand preparation and shakeout equipment needs. The

H process offers an interesting concept in using strings of molds, held rigidly in a horizontal plane and having a continuous runner and feeder system. Squeeze casting combines forging and casting techniques to produce high integrity parts. Although it is in its infancy in this country, the potential looks good.

Technological changes in sand binders and additives have been numerous and frequent in the last decade. Section III presents articles on the base material, sand; on forms of mold materials other than silica aggregate; and on the many binder systems and processes used. "Cold Setting Core Materials" describes sand binding systems that produce cured cores or molds without heat. The cold setting or chemical binder systems have presented new problems in mixing, controlling, and handling bonded sand and have introduced a new dimension in safety control of the binders and gases generated in curing and using cores and molds. "Cold-Box Systems Engineering" addresses itself to these problems and their solution. Energy use savings by curing core binders with microwave ovens is covered in another article. Cores and molds may still need refractory coatings to provide improved casting finish, and the last article in this section reviews coating technology.

Section IV is concerned with tooling. Coreboxes and patterns represent a relatively high expenditure in the over-all cost of producing castings. Current tooling techniques are reviewed, including a study of the use of plastic materials as a means of reducing costs.

Induction furnaces have been adopted by the metal casting industry in great numbers, as melt units particularly for melting the cast irons and to a lesser extent for nonferrous melting. The first three articles in Section V review coreless induction and channel induction furnaces and melting. The article by Harry Chandler, "What Users Should Know About Clean Steel Technology," is an excellent short treatise defining clean steel and describing the many melt systems available for producing it. Although these systems relate to steel producers more than to steel foundries, metal casters tend to adopt these practices in time. A good example of this is the AOD process, which is being used by an increasing number of steel foundries every passing year. The last paper in this section describes a computer process model for cupola melting. The cupola produces great tonnages of cast iron, despite the influx of electric melting, and this process model is a step toward better predicting how to handle the many variables present in cupola melting.

Two metals deserve some reference in this Source Book and we start Section VI with articles on an old material, magnesium, and a relatively new material, compacted graphite cast iron. Because of its light weight, strength, and castability, magnesium has potential future growth as an engineering material. Compacted graphite cast iron was described in the literature 25 years ago, but only in the last five years has it become a production cast metal finding application in a variety of engineering castings. "Fifty Years of Progress in Cast Iron Inoculation" brings us up to date on the various inoculating materials used in cast irons to reduce chill, aid in structural control, and change mechanical properties.

The cleaning of castings has not always received the attention it deserves as an area for technological advancement. In Section VII the handling and cleaning of castings receives that attention. Grinding is a necessary cleaning process for most sand castings, and where silicon carbide abrasives were used extensively in the past, zirconia abrasives are now available, giving improved results in a number of cases. Ultrasonic and sonic nondestructive evaluation of casting integrity is on the threshold of becoming a major inspection tool for cast iron foundrymen. Although ultrasonics have been helpful in linear measurement and detecting internal porosity of castings, graphite form evaluation will be a future viable testing procedure to qualify cast irons. A. G. Fuller's report on this activity as well as his coverage of "Casting Quality Control" should be a valuable reference for foundry inspection personnel.

Section VIII includes articles on a wide variety of technology that will be growing in the future. The need to conserve materials such as sand by reclamation is given an economic look. Computer-aided design (CAD) will be used as standard operating procedure in the near future, to design parts, gating and risering systems, and tooling. The example given of designing a steel wheel casting is only a small insight into what can be done with full-blown CAD/CAM systems as they develop. In an industry where a great amount of heat is used for processing and a great amount of heat is generated by the equipment, thermographic inspection techniques are discussed as a tool for preventing problems due to heat loss or overheating. Ductile cast iron, after thirty years, continues to be produced at an increasing annual rate as compared with other ferrous cast metals. Two papers covering the production welding and the hot rolling and forging of ductile iron could be indicative of further production growth for this material. The last article describes three typical European applications of robotics in metal casting. Although the use of robots is limited at present to tedious or environmentally unpleasant types of work, their full potential has not been realized. Electronic devices coupled with robots offer much help even to small foundry operations.

Casting is and will remain a principal means of forming metal. Utilizing even a small part of the new technology that has been generated in the last decade, plus the increasing amount of technology being brought forth every year, is a distinct challenge to the operating foundrymen. We hope this Source Book will be an aid in meeting that challenge.

*   *   *   *   *

The American Society for Metals and the American Foundrymen's Society extend grateful acknowledgment to the many authors whose work is presented in this book, and to their publishers.

PAUL J. MIKELONIS
Corporate Technical Director
General Casting Corporation

Cover photo: The ability to cast in one piece a cup, saucer and spoon was a test used in bygone days to evaluate the skill of a journeyman molder who applied for a job in a foundry. The cover photo shows the work of Leonard Ivey of Chattanooga, Tennessee, an innovator who went one step further by adding a sphere in the spoon.

CONTENTS

SECTION I
Overview

STATE OF THE TECHNOLOGY IN CASTING PROCESSES

Jack H. Schaum
Publisher/Editor, Modern Casting Magazine

Today metalcasting is a modern technology embracing the disciplines of metallurgy, chemistry, ceramics, sandurgy, thermal dynamics, computerology and a host of others.

Forty years of complete immersion in the World of Metalcasting has convinced me that: "Of all the metalworking processes, metalcasting is the most versatile, the most flexible, the most economical, the shortest and fastest way to get from raw material to finished product." Casting lets you design a shape in such a way as to put the metal where it's most useful instead of being locked into standard commercial shapes. And there is no limit to size or complexity for the metalcasting process.

About 90% of all manufactured products in our country depend somewhere on castings in the process of their creation. No one makes a casting unless someone needs it. Often this need starts on the drafting board of a design engineer. If the shape of the new part deviates from standard wrought shapes and the manufacturer plans to need more than a few, then he should consider casting as the preferred process for making it. You should look on foundries as the "front end" of the manufacturing system.

Although design is normally considered a human responsibility, there is much interest in programming computers to design castings. Engineers at John Deere in Des Moines, Iowa have developed such a program for designing cotton picker cams. Several years ago one of the computer designed cams, cast in ductile iron, won the coveted Annual Design Contest of the Iron Casting Society. It replaced a hardened steel cam and cut total cost by 68% -- an annual saving of $400,000.

The design program represented many years of professional cooperation between computer programmers and design engineers. Once the basic program was developed, the computer could design a new cam to near perfection in 4-6 hours when formerly it required weeks of trial and error, plus strain gauge studies to accomplish such a design - a dramatic breakthrough which is destined to sharpen the leading edge of design technology.

PATTERN TECHNOLOGY
Once the design is firm, the next step is creating a pattern. The pattern may be machined from wood, plastic or metal, depending on the number of mold impressions the customer ultimately want to make from it. Cast-to-size patterns and core boxes are being made with a thousandth of an inch per inch tolerance and an RMS finish approaching 100. In many instances, cast-to-size tooling can reduce costs 50% compared with machined tooling. A recent innovation is the use of flexible rubber patterns to accomodate shapes like the turbine compressor wheel for today's trucks and cars.

Just poking its head up over the horizon is a revolutionary breakthrough in patternmaking -- invented by an Englishman and now being developed at Battelle Memorial Institute.

It uses two intersecting laser beams to harden a liquid plastic and create a complete 3-dimensional model or pattern. A computer is programmed with X, Y and Z coordinates taken from the 3 views of an engineering drawing of the casting desired by the customer. The computer then con-

trols the movement of the 2 laser beams scanning a container of liquid
plastic. Where the light beams intersect, they create a chemical reaction
that will harden the plastic. Afterward the unhardened plastic is washed
away and there is left an exact copy of the image defined in the computerized data. Indeed an exciting concept for making our patterns in the future.

MELTING
Let's next take a look at the state of technology in melting metal for
castings. That old work horse, the cupola, is still around but now it's a
sleek racing thoroughbred with its outer shell water-cooled so it doesn't
need a refractory lining; the combustion air is heated as high as 1000F
and may even have extra 02 fed into it; a second row of tuyeres may be
added; and that belching behometh of pollution is all wrapped up so its
smoke, dust and fume are captured in a bag house or venturi collector.
As a result of this progress, one ton of coke will produce 12-16 tons of
molten iron in a cupola when just a few years ago it could only melt
8 tons!

Another advancement is the practice of duplexing. In duplexing, the
iron flows from the cupola into an electric holding furnace which can be
either electric induction, arc or resistance. These holders act as a
flywheel in the metal system to accumulate large quantities of iron so as
to homogenize the chemical composition and to introduce more heat into
the metal raising its temperature several hundred °F above the cupola
tapping temperature. The iron in the holder is analyzed frequently so
adjustments in composition can be made by adding C, Si, Mn and other alloys
desired to meet the customer specification. Hotter iron opens the door to
casting shapes with thinner walls and less total weight.

The cupola's future is further brightened by the appearance of a new
raw material -- charge pellets made by cold bonding iron ore, coke breeze,
lime and silica. As the pellets are heated in the cupola, the iron oxide
is reduced, fluxed and alloyed with carbon and silicon to produce a high
quality molten iron. Foundries have proven that the cupola will melt pellet
charges ranging from 5% to 100%. And the price will be competitive with
scrap!

Probably the most exciting new technical breakthrough that is beginning
to impact on the steel foundry industry is the AOD Process in which oxygen
and argon are blown through molten steel to produce the high quality, low-
carbon, low sulfur analyses needed for stainless steel. It is predicted
that 75% of all stainless steel will be made this way by the end of 1979. A
number of our leading steel foundries now have units installed and operating.
Some of the advantages are: the ability to use cheaper high carbon chromium
which can save as much as $112 per ton in alloys; the capability for re-
ducing the carbon to as low as 0.01 to 0.02% in spite of using high carbon
materials; and the excellent stirring action of the gas in the converter
produces an extremely clean steel with fewer inclusions and sulfurs as low
as 0.005% max.

Manufacturing engineers stand to benefit in the future from melting
done at central melters equipped with all the expensive pollution and
metallurgical control devices. Smaller foundries will purchase clean

molten metal of known composition delivered by truck and bring it to
temperature in holding furnaces requiring no pollution controls.

Already we see this happening in the aluminum industry. Two ladles,
each holding 18,000 lb of molten aluminum, are being hauled in trucks as far
as 300 miles over our highways from smelters to foundries for casting. The
temperature loss is only 25°F per hour in the well insulated ladles. The
foundry saves cost of 1500-2500 Btu's required to bring each lb of aluminum
from room temperature up to molten superheated condition, plus the time it
would take for the remelting. Just one such smelter is delivering 18
million pounds of molten aluminum a month to customers.

Improved melting practices have given foundries the ability to produce
engineered molten materials at the spout to meet any need. Computers have
been teamed with raw material supplies to calculate least cost charges in
a matter of seconds. And the analysis of this metal is known and guaranteed
before it goes into the mold -- thanks to new high speed spectrographic
equipment. Today's modern iron caster also uses the eutectic arrest of a
cooling curve to determine his carbon and silicon in less than a minute
before pouring.

The old axiom, you cannot control until you can measure, has been
proven with the introduction of the high speed spectrograph which can
determine 6-12 elements to 3 decimal, pinpoint accuracy, and report back to
the melter, by closed circuit TV, his exact composition within three minutes.
Any deviations from the specifications that the customer has defined can
then be adjusted in the furnaces before tapping. As a result, foundries are
holding their chemistry and hardness in a very narrow range, eliminating
hard spots that are the bane of automated machining, and guaranteeing desired tensile properties and microstructures. Rapid microstructure checks
are also made before the metal is poured.

SAND MOLDING
Let's take a look at new developments in sand molding that are giving you
more value for your casting dollar. In the past ten years, hi-pressure molding has raised mold hardness levels to 90 plus. As a result, mold wall
movement has been virtually eliminated, giving you a casting with reliable dimensions, closer tolerances, better finish, uniform hardness, and
less tendency for internal defects -- ideally suited for automated machining operations and often even eliminating much of the need for machining.

Why whittle when you can cast? Our goal in casting accuracy is + or -
nothing and every year we reduce that $50 billion being spent annually to
machine away unwanted metal.

At the same time that molding machines are being designed to improve
casting accuracy, they are being run faster. So today there are highly
mechanized and automated machines producing 300 molds an hour with only one
operator. A high speed pattern shuttle can change patterns during one cycle
of a machine so it is no longer necessary to have long runs between pattern
changes. Coupled with high speed molding is automated pouring that eliminates human error in putting the metal into the mold at controlled rate and
temperature.

The Japanese "V" Process, now being installed in 15 U.S. foundries, uses vacuum to bond the clean, dry, sand grains. The details achieved are fantastic. With no binders or additions in the sand, the process represents an absolute miracle of cleanliness with no smoke or fume generated.

SUCTION CASTING
Super clean castings and thin sectioned shapes are now being produced by the process of suction casting. The mold may be sand, ceramic or graphite. It may be used at room temperature or heated. The mold is held in a fixture that allows suction (vacuum) to be applied to the mold cavity. A small tube (sprue) extends out the bottom of the mold as much as 12 inches. The mold is transported over the melting furnace and lowered so tip of tube is immersed in the molten metal. Vacuum is applied and metal flows up into the evacuated mold cavity. Since metal comes from below the slag layer on the furnace, it is unusually clean. Furnace holds the metal at exact pouring temperature and the casting feeds through the immersed tube so gates and risers are practically unnecessary and yields very high.

Coming up rapidly on our horizon of new technology is the lost-foam pattern process which involves the use of mass produced polystyrene shapes that duplicate the metal part ultimately to be cast. These polystyrene patterns with polystyrene gates and risers attached are buried in dry, unbonded silica sand which is packed by vibration, thus eliminating any need for sand binders or mixers or molding machines. Molten metal is then poured onto the pattern which evaporates instantly under the heat of the metal and the pattern is replaced with an exact duplicate in metal. The process eliminates the need for cope and drag flasks, parting lines and cores. It also reduces the defects attributable to the mold since most defects emanate from the binders added to the sand rather than from the sand itself. Patterns are mass produced like die castings. And the process requires no binders, no mullers, no cores, no mold machines. A miracle of simplicity and economy!

We are very excited about the role this process is going to play in our future. It will give almost unlimited design freedom to you, our customer. Since the pattern does not have to be removed from the mold, it allows designs with undercuts and back-draft and eliminates fins and surface defects. GMC visualizes entire engine blocks and manifolds cast from foam patterns, and Ford is looking closely at this process to cast the hollow pin crankshaft.

CORES
The ability of the metalcasting process to produce shapes with unlimited complex internal configuration is one of its strongest competitive advantages. This is accomplished with sand cores and great progress is being made in coremaking technology.

Today we have shell, CO_2 process, ceramic, hot box, cold box, warm box and chemical set processes for making cores that lend themselves to mass production, as well as single, one-off castings.

What does the future hold for new coremaking methods. Here the cry is -- FASTER, FASTER, FASTER --. The "Go-Go" chemical setting, rigid cores are being made with formulations that set them solid in 3-5 minutes. New self-setting binder systems and mixing equipment are accelerating this rate toward a one minute curing time. Another new technique uses a catalytic gas to harden cores and molds in <u>4 seconds</u>!

Sand casters can now use precision ceramic cores in their molds and produce castings with internal cavities every bit as accurate as an investment caster.

The use of chromite, olivine and zircon sand, instead of silica in selected areas of molds and cores, have eliminated burn-on of sand and give preferential cooling rates in different parts of the same casting.

SAND RECLAMATION

The new technologies of recycling sand through elaborate reclaimers is fast becoming a standard practice as the freight rates on new sand keep rising and regulations against dumping old sand become stricter. Today the technology exists to recycle sand back into the foundry with characteristics as good and often even better than new sand. Sand reclaimers will be as standard in the foundry of the future as sand mullers are today. A number of reclamation systems are available -- wet, thermal, pneumatic, and mechanical -- and often reclaiming can be done for as little as $4/ton.

ELECTROHYDRAULIC CLEANING

A new invention coming from the USSR promises much needed help in the cleaning department where removal of intricate cores from within castings is often very difficult. Called electrohydraulic cleaning, the machine employs the energy generated by a high voltage discharge in water between a special electrode and a casting. The discharge is accompanied by a series of physical reactions including explosive steam shock waves which destroy and remove the sand core from the casting. Pressures up to 225,000 psi occur. When steam bubbles break, there follows an implosion. The rapid alternating explosion/implosion cycles destroy even the strongest ceramic cores without damaging the casting.

More than 200 foundries in USSR are using the process. Mitsubishi in Japan is installing one of these systems for cleaning cores from 130 cylinder heads at a time.

ROBOTS

The age of robots is starting to invade the foundry industry. As labor rates spiral higher and environmental regulations become more restrictive, robots look more and more attractive. In Caterpillar's new foundry, in Mapleton, IL, robots are drying the wash on molds, poking vents in green sand molds, and dipping cores in a wash tank. Investment casting foundries are using robots to apply slurry coatings to wax patterns and to stucco between coatings. Other foundries use robots to pick up hot castings and place them in trimming presses to remove gates and flash.

Robots have a lot of good traits - they don't come to work late, get sick, take vacations or strike. And so far, OSHA has not put restrictions on their working environment!

To pay for the high capital investment in equipment, foundries of the future will operate around the clock, 7 days a week just like the steel chemical, petroleum and glass industries. A new foundry is now being built in Pennsylvania that will have 2 shifts working 10 hours each for 4 days. Then these workers have 4 days off while shifts 3 and 4 each work 10 hours/ day for 4 days. The result is a 20 hr/day, 7 days a week operation. Improved productivity in the future will come from reducing our metalcasting man-hours per ton and working our assets harder!

THE NEW METALLURGY
With the new technology of metallurgy, you are no longer limited to a single property in a casting. Judicious placement of chills and insulators in molds can alter the microstructure and mechanical properties selectively within a single casting. Alloys can be suspended in a wash and applied to specified areas of a core or mold to create localized metallurgical phenomena.

The "In-mold" process for nodulizing ductile iron is smokeless, non-violent and yields high recovery of magnesium. In this process, the treatment alloy is placed in a cavity in the gating system of each mold. It may replace all batch treatments of ductile iron within 5 years. This system is also being used for late additions of silicon to the iron as it flows into the mold, thus eliminating white iron and opening the door to casting thinner sections.

Another way of achieving high alloy recovery involves the use of a pneumatic machine gun for shooting alloys into molten metal in a ladle before pouring.

No longer are you limited to one base metal in a casting. Bimetallic castings are now possible. Lathe beds are being poured with the "ways" cast in hardenable tough alloy iron and the balance of frame in gray iron with its ideal capacity for damping out vibrations. Cylinders, liners and rolls are being centrifugally cast with inner or outer shell in corrosion or wear resistant alloys and backed up with less expensive metals. Cast valves are being lined with teflon to resist corrosive chemicals and cast electric tool parts are being coated on the outside with insulation to guard against electric shock!

The Energy, Mines & Resources Department of Canada has recently developed the CANCOAT Process. Basically, the process permits coating any specified surface of a ferrous casting with either an abrasive-resistant or corrosion-resistant alloy at the same time the casting is poured. Investigations have centered around tungsten carbide and chromium carbide coatings. Laboratory wear tests indicate the wear-resistance of these coatings to be similar to those applied by welding processes. Mild steel, low alloy steel, cast iron and nodular iron have all been successfully coated with this process.

The process uses vacuum to hold the powdered carbide in a graphite mold insert. During pouring the metal diffuses into the powder. The process is particularly suited to applying a super wear resistant surface of tungsten carbide on cast steel digging teeth. To give you an idea of the size of this market, $80 million worth of digger teeth were consumed recently in constructing just one Canadian pipe line.

The Dilex patented process will diffuse alloying elements such as Cr, Al, Co and Ni into low C steel surfaces to produce a thin skin of alloy with great corrosion or wear resistance. The process does not change the dimensions of finish machined parts so it is best done after machining the soft untreated steel. The prlcess can develop a stainless steel surface - 0.003" thick - analyzing 35% Cr on a low carbon (below 0.10%) steel. Steels with carbon above 0.30% form a chromium carbide surface, 1 mil thick, as hard as commercial sintered tungsten carbide. Up to 10% aluminum can be diffused into high carbon iron and malleable iron castings. Combined with chromium the process can produce a wear resistant, oxidation resistant alloy skin.

The process involves sealing the parts in a drum with lead (Pb) plus the Cr/Al and rotating it in a heat treating furnace for 4 hrs at 2000F. Think of the savings in alloys and ease of machining compared with working with solid stainless steel.

Dissimilar steel alloys can be cast in separate shapes and then welded together to form a cast combo with customized properties. Forged or cast turbine blades are now set in dry sand molds and locked into place by pouring the steel inner and outer ring around them.

In Germany, malleable iron castings are being surface decarburized by heat treatment until the outer skin is actually changed to low C steel. Then these castings are welded to compatible steel tubing for automotive components.

Lasers are now providing the alchemy that turns soft iron into hard metal. They are being used to selectively case harden special areas of iron and steel castings such as gear teeth, the inside of cylinder liners, and outside of seal rings. A minimum of energy is consumed since only the skin of the casting is heated by the laser. General Motors is now running a 5 KW CO_2 laser installation three shifts a day, heat treating the inner walls of gray iron cylinder liners for large diesel engines to a depth of 0.010 to 0.025 in. and hardness of Rc 55-60 without distortion. Laser hardening of the casting skin may also improve the fatigue resistance of ferrous and aluminum alloys. Lasers can be used to provide heat for diffusion of powdered alloys like Ni and Cr into selected areas of a casting High melting ceramic coatings can even be applied on lower melting aluminum.

A new baby has recently been born to the cast iron family which may play a role in your future. It's called compacted graphite iron and it has properties that fall between gray iron and ductile iron. It's particularly suited to cast products subject to thermal shock such as exhaust manifolds, brake drums, and ingot molds. Tensile strengths can run as high as 60,000 psi with ductility of 6%. And it's made by treating cast iron with a Mg-Ce-Ti alloy additive.

A high strength, high ductility, austempered ductile iron has recently been developed and made a dramatic appearance in the rear axle hypoid ring and pinion gears for automobiles. Castings are first machined and then heat

treated to a bainitic microstructure which has excellent resistance to
scoring and wear. Other possible uses are track components on agricultural
equipment, heavy truck gears and bar stock for machining small parts to
replace forgings.

We thank our good casting customers for demanding better quality,
high technology castings. Customer pressures have forced great strides in
quality control of every step in the casting process. Today raw materials
don't enter the plant until they run a gauntlet of tests; furnace charges
are calculated by computers; molten metal is analyzed by high-speed spectro-
graphs before it is poured into the mold; mold and core sand is subjected
to a myriad of tests to be sure it is going to behave properly when molten
metal contacts it; casting cooling times are carefully regulated; and the
final quality of the casting is checked for proper microstructure and
properties with hardness, ultrasonic, magnetic and radiographic tests.
Since WW II, metalcasting has moved from an art and craft into a science
and technology; from a labor intensive to a capital intensive industry;
from a polluting monster to an environmentally acceptable neighbor; from
a poorly run, unprofitable operation to a profitable, well-managed business enterprise!

Today alert foundries and progressive users of castings are sitting
down and planning their futures together, since each depends on the other.
Growth industries need growth-oriented foundries to meet future needs.

The Future of Metalcasting will ride the crest of this new Work-A-Day
Technology. The kind that saves energy, conserves raw materials, accelerates
productivity and delivers reliable castings at competitive prices.

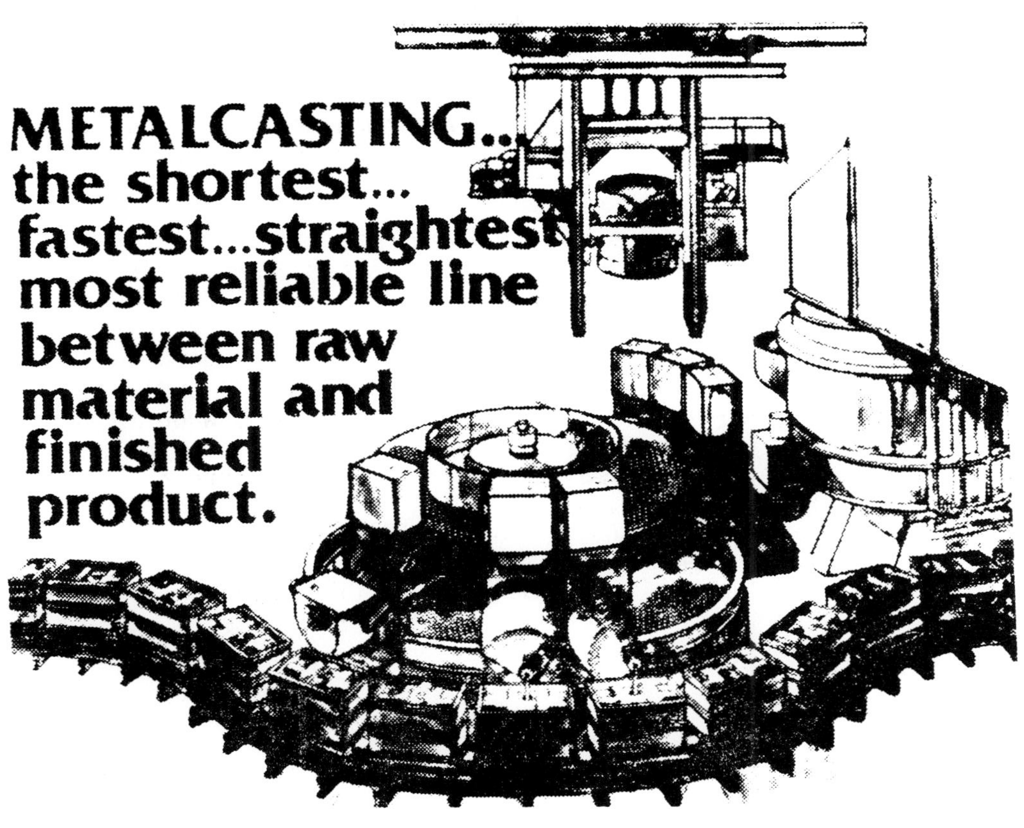

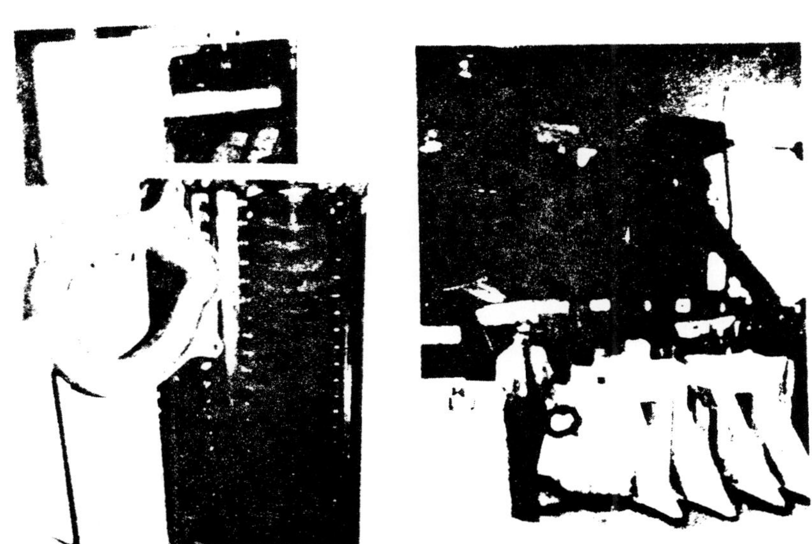

Computer designed ductile iron cam for mechanized cotton picker.

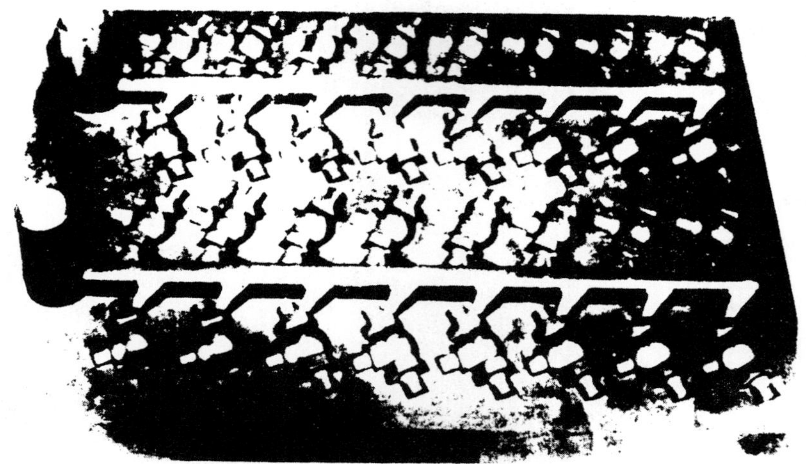

Cast-to-size pattern plate.

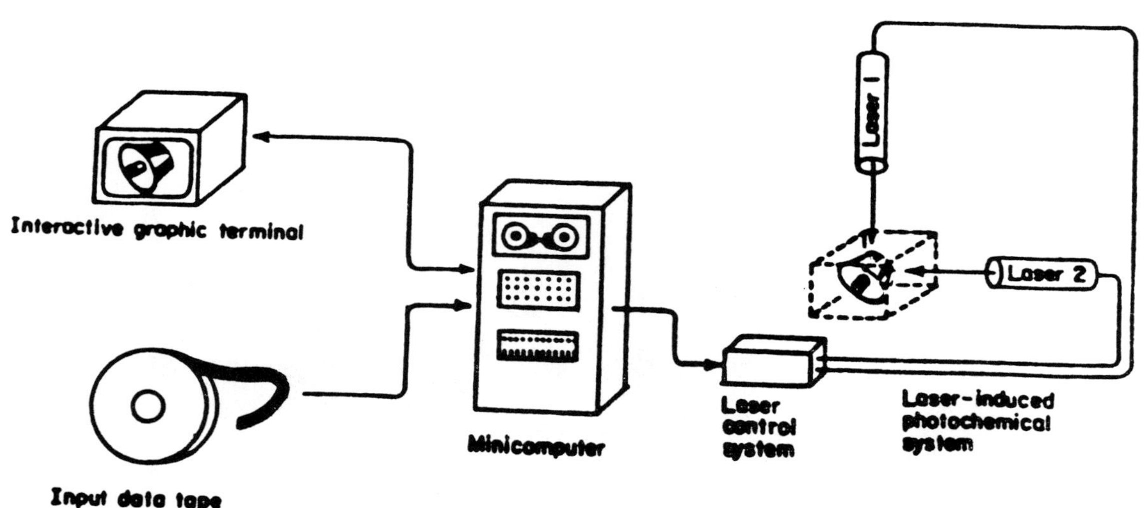

Schematic of computer assisted laser technique for making patterns.

Over-the-road trailer truck hauls 18,000 lb of molten aluminum in two insulated ladles.

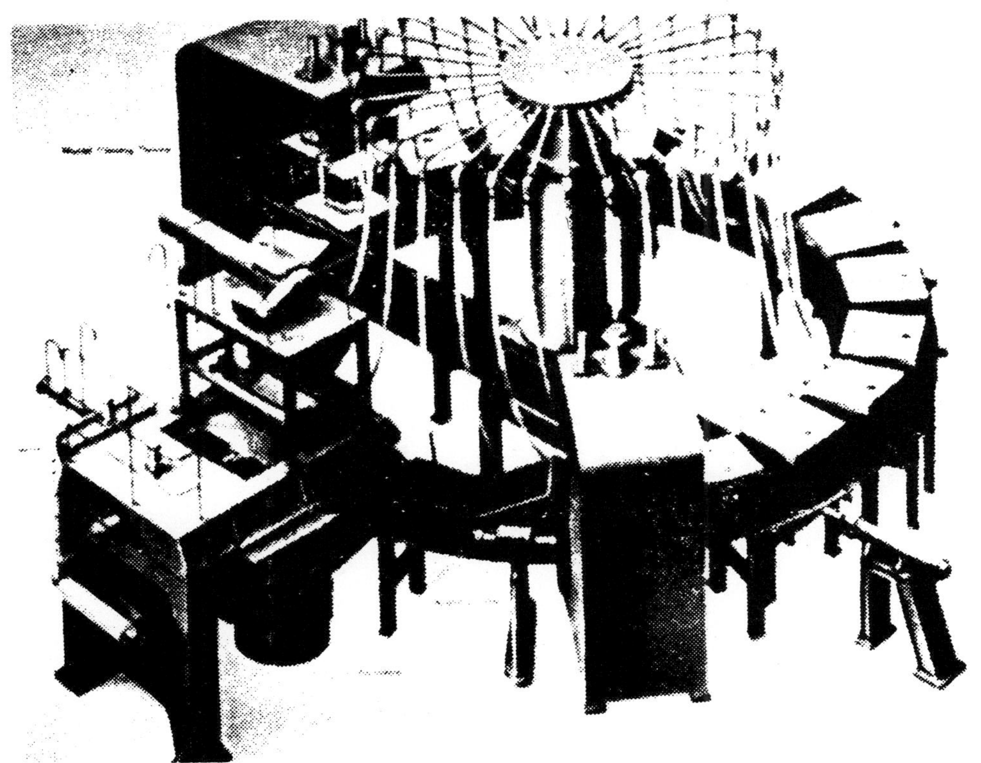

Environmentally clean "V" process uses vacuum to bond sand molds.

Source: SME Technical Paper No. CM79-630, 1979

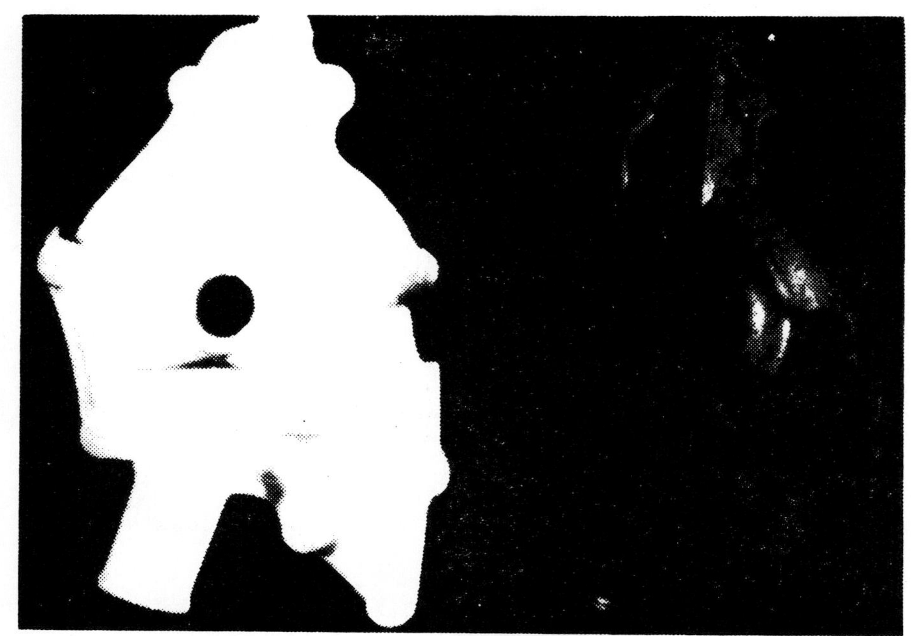

Expendable polystyrene pattern (left) and identical casting produced when evaporated by molten iron.

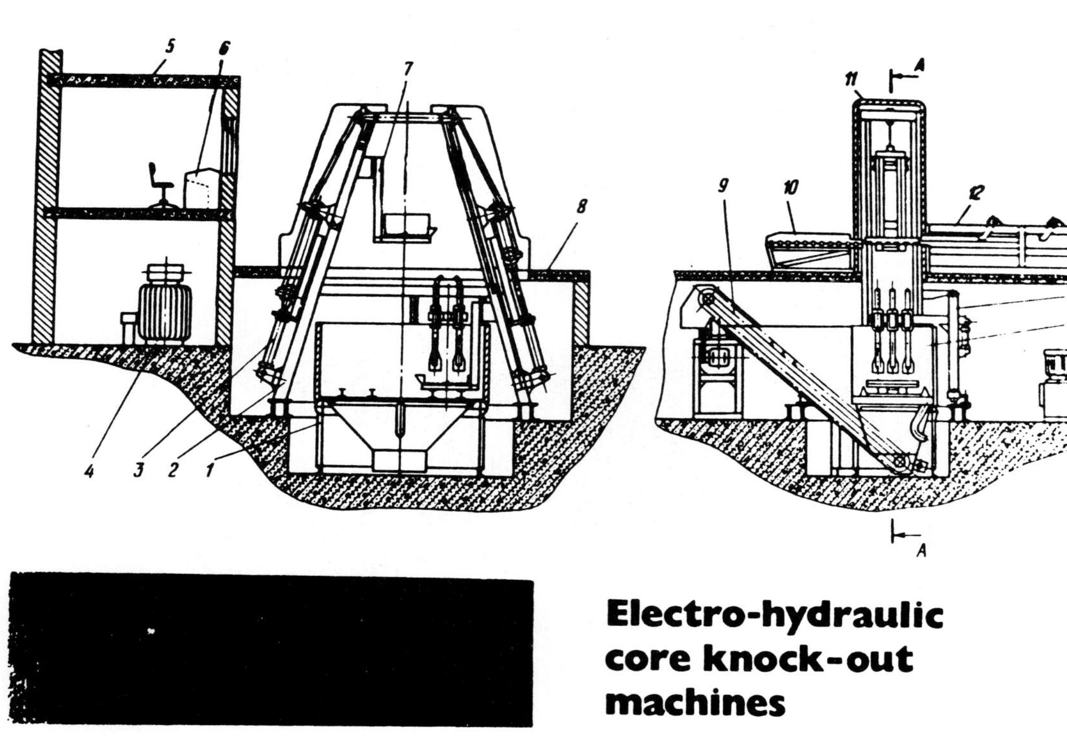

Electro-hydraulic core knock-out machines

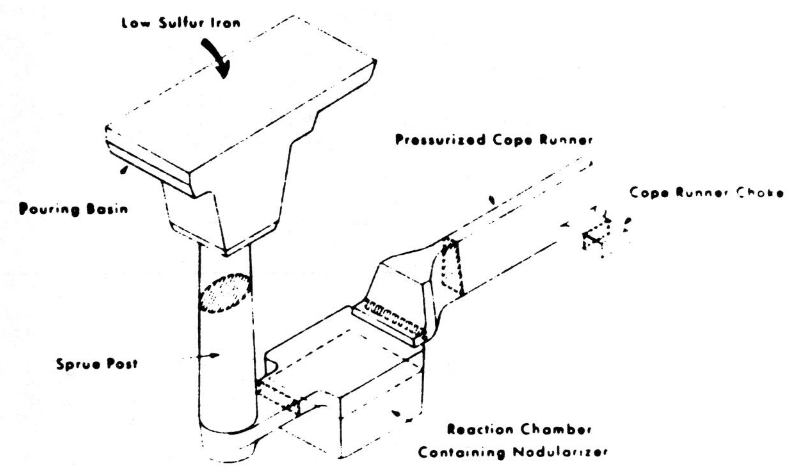

Special gating system designed for alloying iron in the mold during pouring.

Cast-weld combo of three separate steel castings.

EDITOR'S NOTE

Single crystal castings are now in mass production — jet engine turbine
buckets are poured and directionally solidified from a crystal seed.
The entire casting is a single crystal with no weak grain boundaries to
be susceptible to fatigue or creep damage. Ceramic whiskers of B nitride,
Al_2O_3, etc., are being suspended in molten Al and other alloys to triple
their cast physical properties where they create a stitching action in
the grain boundaries of the solidified casting.

   Have you heard of hot isostatic pressing (HIP) of castings to create
premium quality? The HIP process simultaneously applies very high pressures (up to 15,000 ksi) and temperatures (up to 2300 °F) to the casting,
sealing internal voids and other defects. HIP also improves the microstructure of castings so they have better weldability, machinability and
improved dimensional stability during heat treatment. There is even
evidence that stress-induced damages such as creep can be healed by HIP
treatment and the part rejuvenated to its original condition.

   The dream of producing guaranteed sound castings with isotropic
properties to replace forgings is now a reality. So far the applications
have been primarily in aerospace. But the current cost of 25 to 50 cents
a pound may very well prove attractive to more mundane industries such as
automotive, because of the many benefits achieved.

   The new era of metallurgy will conserve expensive nickel and chromium
used to make stainless steel by creating only a surface of stainless
steel. How? By casting plain C steel parts and then treating them in
molten lead saturated with Ni and Cr, causing the alloys to diffuse into
the surface creating an 18-8 stainless steel skin in the casting. Similar
surface effects are being achieved with ion nitriding and ion diffusion
of alloys. Why pay for a solid alloy casting when it's only the skin
that must resist corrosion or wear?

Recent advances in zinc die casting technology

SCHRADE F. RADTKE and

DONALD C. HERRSCHAFT

International Lead Zinc Research Organisation, Inc., 892 Madison Avenue, New York, NY 10017

Abstract

The zinc die casting process is undergoing major changes as researchers uncover new knowledge about the material and the process. International Lead Zinc Research Organisation, Inc. (ILZRO) has been sponsoring such research for years in the U.S., U.K. and Australia. The basic objective of these projects is to improve the zinc die casting process itself so that ultimately all major parameters in the process will be controlled automatically to enable production of zinc die castings at the lowest possible cost and with zero defects. This paper reviews some of the specific developments emerging from all three research areas, including thin-walled die casting, alternative gating systems, die coatings and die materials, pressure and thermal studies, metallurgical properties of castings, and runner design, with emphasis on tapered runners.

Introduction

Perhaps at no time in its history has zinc die casting technology been in a greater state of ferment than the present. The onrush of competitive materials in a world where the pace of change is swiftly accelerating has forced the zinc die caster to discard traditional methods of operation and to adopt new procedures and equipment if he wishes to remain in business. For an industry once marked by its conservatism, zinc die casting is showing strong signs of a rebirth in enthusiasm for the process developments emerging from research laboratories around the world.

International Lead Zinc Research Organisation, Inc. (ILZRO), the cooperative research arm for the world's lead and zinc producers, has been engaged in die casting research since its founding twenty years ago. The level of activity in this area has increased significantly over the past several years with principal projects being sponsored at Battelle Columbus Laboratories (BCL), Columbus, Ohio; BNF Metals Technology Centre, Wantage, England and Commonwealth Scientific and Industrial Research Organisation (CSIRO), Melbourne, Australia. The long term objective of these projects is to improve the market position of zinc die castings.

The emphasis of the research has been to develop techniques to achieve savings in material and energy costs, to reduce the thickness and weight of zinc die castings, to improve their surface quality and to reduce the finishing costs so as to make zinc die castings more competitive with alternative materials. Ultimately it is hoped to control automatically all major parameters of the process.

This paper will describe the major research efforts, some of which are still in progress, of these three groups. The following discussion considers the different philosophical approaches of the various groups and the areas of common agreement amongst them. It should be noted that there are also some differences in the recommendations made by the different groups and further work may help to determine which of the recommendations are most beneficial.

ILZRO/Battelle Research

ILZRO has been conducting die casting research at Battelle Columbus Laboratories for over a decade. The research lead in the early 1970's to development and refinement of the concept of thin-wall zinc die casting. The specific objective of producing large, lightweight thin-wall zinc die castings was achieved through the development of design guidelines for (1) the runner and gating system, (2) overflows and (3) die-temperature control systems. A manual, 'Designing for Thin-Wall Zinc Die Castings', based on the information obtained in this work, was prepared and published.

Beginning in 1973, however, the emphasis of the research in this programme shifted from steps that could be taken to modify existing dies and die casting machines so that they can produce castings with thinner sections to studies of unconventional methods that could lead to improvements in the production of zinc die castings.

This work has involved a study of the use of die casting dies and alternative die materials for heat transfer control and studies of alternative gating systems. These studies include:

1. Design of a constant area sprue
2. Calibration of the shot-system
3. Development of programmes for use in programmable pocket calculators as an aid in die cooling system design
4. Evaluation of alternative die coatings and die materials
5. Development of alternative gating systems

Item 5 is directed towards the development of a heated zinc delivery-and-

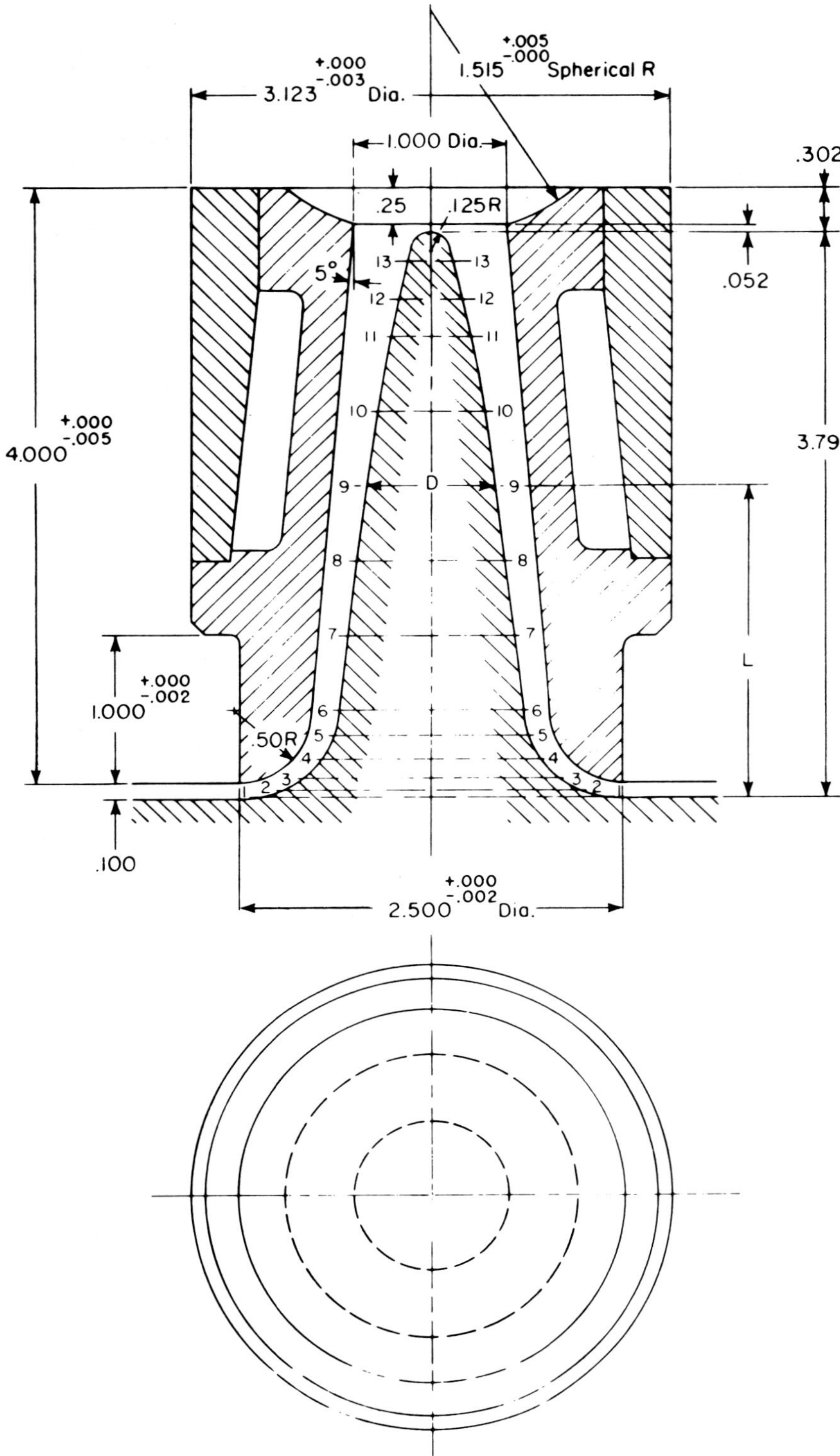

Section	L	D
1-1	0.012	2.257
2-2	0.060	1.949
3-3	0.147	1.666
4-4	0.271	1.432
5-5	0.426	1.272
6-6	0.600	1.210
7-7	1.100	1.095
8-8	1.600	0.973
9-9	2.100	0.843
10-10	2.600	0.700
11-11	3.100	0.533
12 12	3.350	0.432
13-13	3.600	0.306

Fig. 1 (A & B). Constant-area sprue spreader for use with standard sprue bushing (DME large-size water-cooled sprue bushing, long type, DC-200-204). All dimensions are in inches.

return system for introducing zinc directly into the cavity so that the sprue, runner and overflow portions of the castings can be eliminated. These studies will now be briefly discussed.

Design of a Constant Area Sprue

One obvious area for improvement was in the metal feed system for die casting dies. The main aims were to obtain a steady, turbulence free fill of the casting die, to produce a minimum amount of waste material in the form of runners and overflows and to limit die erosion.

Earlier work at Battelle developed the principle that in such a metal-feed system the cross-sectional area of the running and gating system should be maintained constant and should be equal to or less than the minimum cross-sectional area of the sprue, nozzle, or gooseneck.

In a majority of die casting machines used in the U.S. that minimum area occurs at the sprue. In addition, many U.S. die casters prefer to use sprue spreaders and sprue bushings that can be purchased from die casting equipment suppliers.

The more popular standard sprue spreaders and sprue bushings provide a considerable variation in cross-sectional area from the inlet to the outlet. From the inlet, the cross-sectional area through the sprue can vary considerably, depending on how the sprue spreader is set in the die and/or the manner in which the runners are taken off the sprue base. One particular sprue runner design to eliminate flow problems that may occur in the sprue region is one in which the runners are cut into the sprue. That technique reportedly has been used by some die casters for many years and more recently was promoted for more universal application by the Australian Zinc Development Association and by CSIRO. While the use of the sprue runners probably represents the ideal solution, similar results can be achieved by relatively minor changes in the

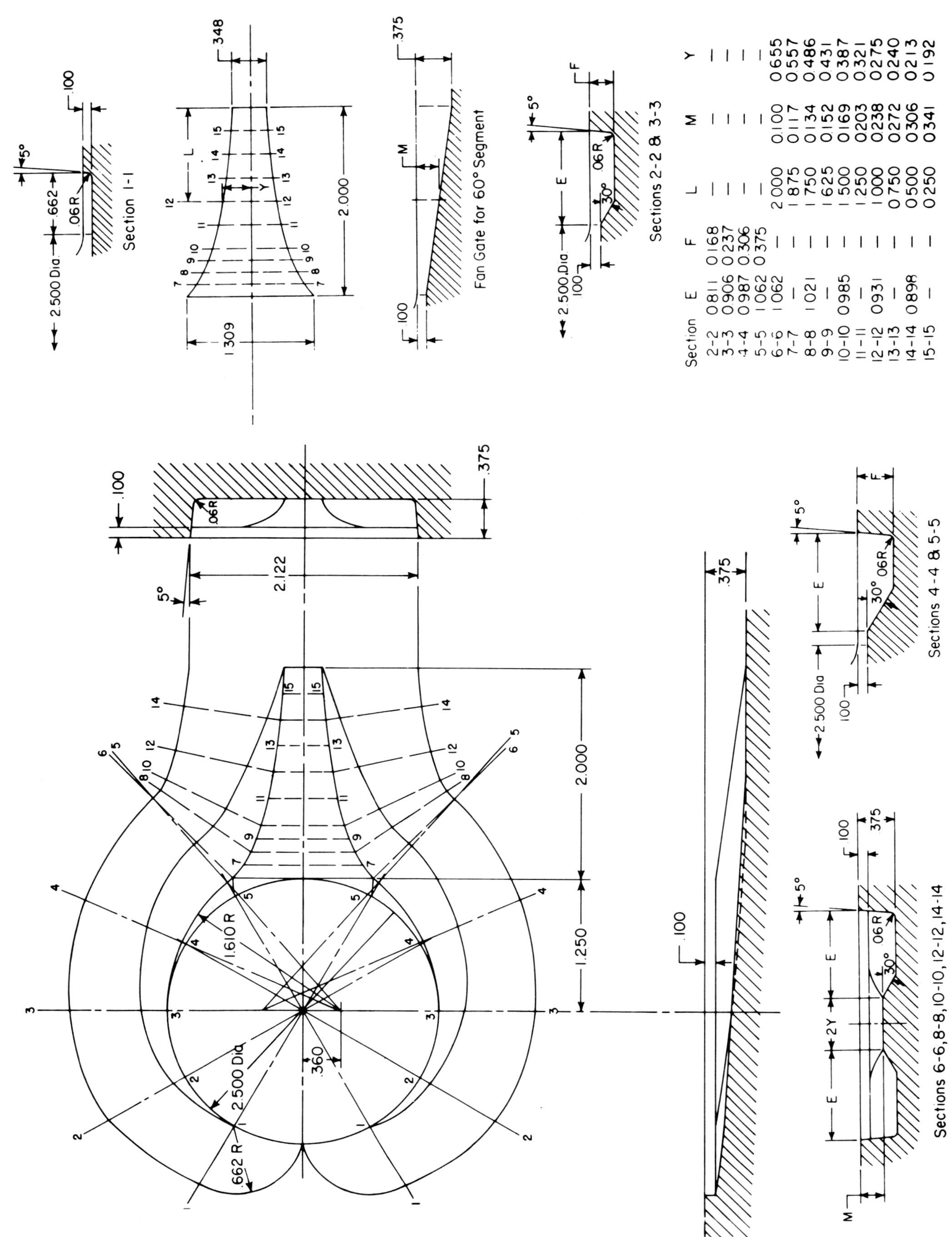

Section	E	F	L	M	Y
2-2	0.811	0.168	—	—	—
3-3	0.906	0.237	—	—	—
4-4	0.987	0.306	—	—	—
5-5	1.062	0.375	—	—	—
6-6	1.062	—	2.000	0.100	0.655
7-7	—	—	1.875	0.117	0.557
8-8	1.021	—	1.750	0.134	0.486
9-9	—	—	1.625	0.152	0.431
10-10	0.985	—	1.500	0.169	0.387
11-11	—	—	1.250	0.203	0.321
12-12	0.931	—	1.000	0.238	0.275
13-13	—	—	0.750	0.272	0.240
14-14	0.898	—	0.500	0.306	0.213
15-15	—	—	0.250	0.341	0.192

Fig. 2 (A & B). Constant-area sprue-base-and-runner system for single-cavity die.

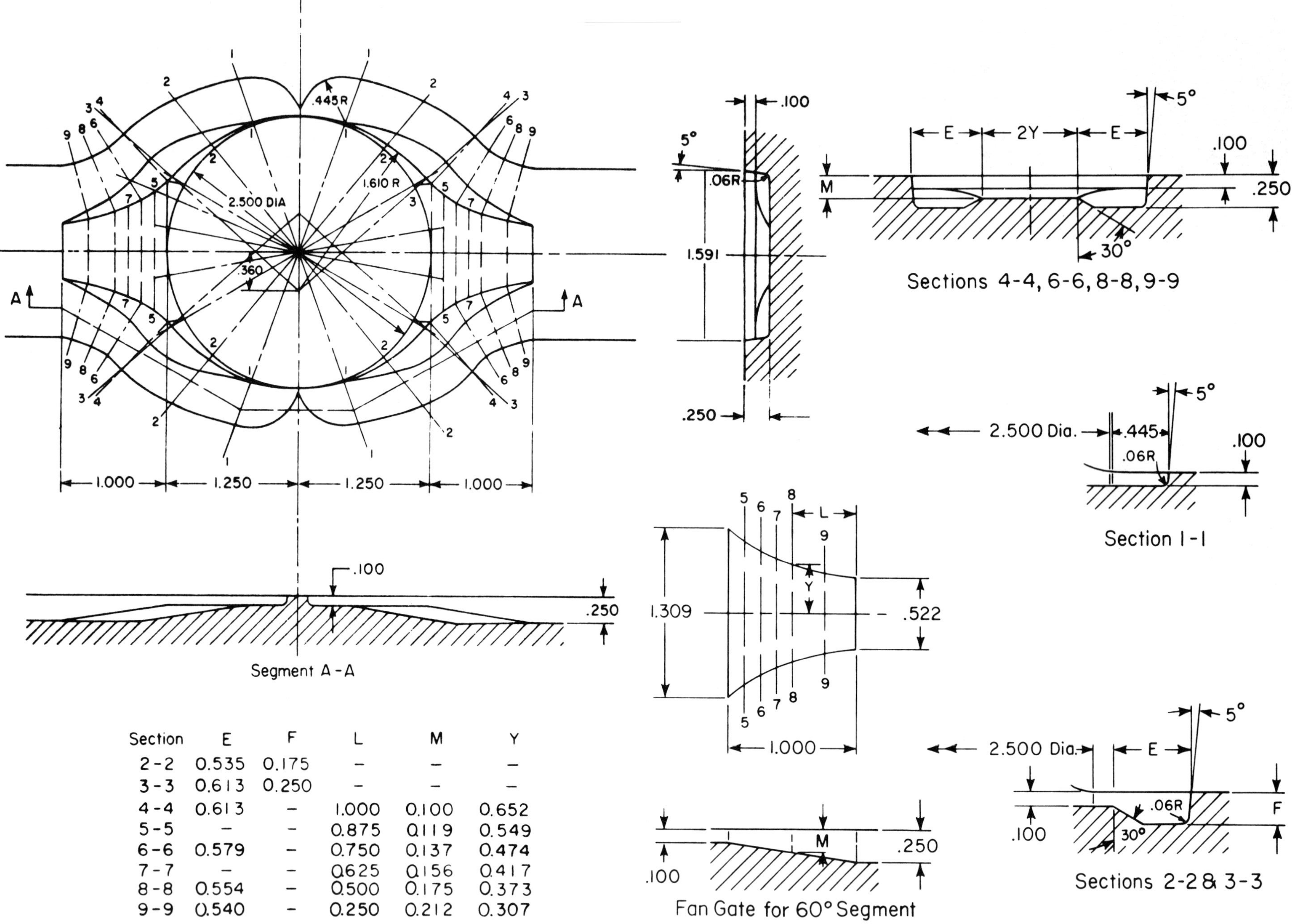

Section	E	F	L	M	Y
2-2	0.535	0.175	–	–	–
3-3	0.613	0.250	–	–	–
4-4	0.613	–	1.000	0.100	0.652
5-5	–	–	0.875	0.119	0.549
6-6	0.579	–	0.750	0.137	0.474
7-7	–	–	0.625	0.156	0.417
8-8	0.554	–	0.500	0.175	0.373
9-9	0.540	–	0.250	0.212	0.307

Fig. 3 (A & B). Constant-area sprue-base-and-runner system for a two-cavity die. (Note: runners are symmetrical).

configuration of the standard sprue spreaders and proper design of the sprue-to-runner transition region. As a result, an analysis of the design of a standard sprue-spreader/sprue-bushing set was conducted and the configuration of the sprue spreader modified so as to have a constant cross-sectional area throughout the sprue. The sprue-to-runner transition regions for a single cavity and a two-cavity die that would result in constant area were also designed. The sprue spreader selected for this analysis was the DME Type 218.

The modified design of the DME Type 218 sprue spreader that will result in a constant cross-sectional area when used with the matching sprue bushings (DME Types DC200-204) is shown in Fig. a(a & b).

The designs of the sprue-to-runner transition regions for a single runner and for a two-runner feed system are shown in Figs. 2(a & b), respectively. For the case of the two-runner system, it was assumed in this analysis that the runners were symmetrical. The techniques used to design the sprue-to-runner transitions are based on the approach that the cross-sectional area of the runner is held constant and equal to the sum of the gate areas fed by those runners throughout the length of the runners. Typical dimensions are shown for a single-runner feed system and a feed system with two symmetrical runners. The same techniques can be used for any number of runners and/or for runners of unequal area. It is believed that use of constant-area sprue configurations and sprue-to-runner transition region designs similar to those shown will minimise flow-control problems in the sprue area during filling of the die.

Shot-system Calibration
During the injection of metal into the die there is normally a two-stage pressure cycle. The first low pressure stage is to ensure smooth filling of the die, the second high pressure stage is to ensure complete filling and prevent excess shrinkage. Therefore, knowledge of the pressures in the system and the resulting metal velocity is important to the die caster.

The original objective in this research was to use analog instrumentation, most of which was available on the research die casting machine, to record pressures and velocities developed during a shot. It was planned that those data would then be reduced manually and pressure-flow curves be plotted for the BCL (Battelle Columbus

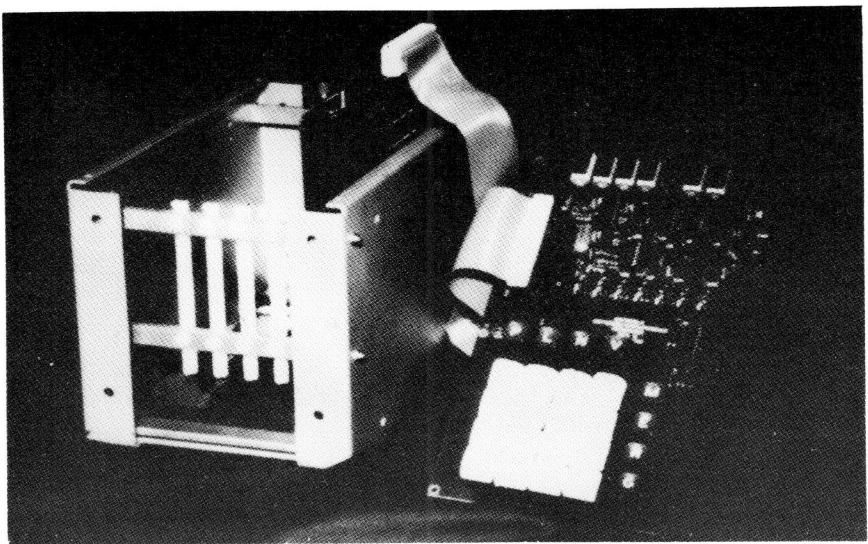

Fig. 4. Components of portable mini-computer system that calibrates die casting machines directly from analysis of operating data.

Laboratories) shot system.

However, it was decided that it would be desirable if the end product of this study was a portable instrumentation package that could be purchased by consultants, machine manufacturers, die casters, or zinc suppliers and used to calibrate and/or monitor machine performance. With this goal in mind, consideration was given to development of a micro- or minicomputer data-processing system instead of the analog data-acquisition system originally planned. Recent developments in computer technology indicated that a micro- or minicomputer system would be less costly, much easier to use, more portable, and probably more rugged than a portable analog recording system. In addition, the minicomputer system can be programmed to make all of the necessary calculations so that manual calculations or manual manipulation of data would not be required, thereby reducing the possibility of error and time required to determine machine performance. Emphasis is placed on the use of the micro-computer as a data-acquisition and recording device and it is the rate at which it can collect this data which determines the value of this tool to measure shot-system performance. A prototype model now under development is shown in Fig. 4.

Pocket Calculators for Die Cooling System Design
In earlier work in the research programme heat transfer analyses had been carried out on the dies to determine the ideal position and size of water-cooling lines relative to the casting to provide the most efficient

cooling and to facilitate the use of an automatic die-temperature-control system.

In those analyses the casting configurations considered were an infinite plate and an infinite strip, each one cooled either by waterlines or fountains. The results of these analyses were correlated into equations in the case of plate castings or strip castings cooled by waterlines or a series of graphs for those castings cooled by fountains. However, the equations are very complex and are difficult to apply. A family of curves for these equations were plotted so that the distance from the casting surface to the waterline could be obtained. However, these curves are limited in use as it is impractical to include a wide range of operating parameters. Thus it was decided to develop a programme for pre-programmable calculators, which calculate the necessary data from the relevant input figures.

This effort has been successful and two commercially available programmable pocket calculators (Texas Instruments TI-59 and Hewlett-Packard HP-67) have been successfully employed, although the evaluation is still in the very earliest stages. The program itself is contained on a small plastic strip that is easily fed into the calculators, as shown in Fig. 5.

Die casters will be able to use the calculators to locate waterlines under portions of a casting that can be approximated by flat strips. Those portions include runners, and overlows, which are usually critical areas for cooling and which can also be approximated as flat strips. It will not be necessary to add or eliminate water-

Fig. 5. Hand-held pocket calculators that have been programmed to predict the location of water lines in dies.

lines on a trial-and-error basis after the die is built. The programs are available for both calculators in either metric or English terms.

Alternative Die Coatings and Die Materials

Research has been carried on since 1973 to evaluate the technical feasibility of using coatings on zinc die casting dies or of using alternative materials for die inserts to control the heat-transfer rate during die filling and during freezing of the casting. It was theorised that the use of proper materials as coatings or as die inserts would provide localised control of heat-transfer rates and also would produce castings with excellent surface finish.

Through 1976, the emphasis of the research in this phase of the program was on the evaluation of candidate die-coating materials. In these studies, the coating materials were applied to sprue spreaders and the coated spreaders then sent to commercial die casters for evaluation. The application methods used were state-of-the-art methods and no preliminary studies were made to optimise the adherence or possible performance of the coatings.

Seven different coating materials were evaluated and none of them withstood the severe die casting conditions experienced in the sprue region. However, based upon the number of shots experienced by some of the oxide-coated sprue spreaders prior to failure or removal from the machine, it appeared that some of the coatings performed better than others. It is possible that further development work on those coatings could result in significantly better performance.

The work showed that the application of an elastomeric material to the die-insert surfaces resulted in a 30% increase in the metal-flow distance compared with that obtained using the uncoated P20 steel inserts. In addition, the solidification characteristics of the zinc were modified in such a way that it appeared that the coating retarded heat transfer.

Based upon those observations, other investigators formulated a boron nitride-containing elastomeric material to be applied to the research die. Evaluation showed that the coating increased the metal-flow distance by about 40% compared with that obtained with the uncoated steel inserts. The first and fiftieth castings produced in the coated cavity and in the uncoated cavity are shown in Figs. 6 and 7. However, the coating as applied is rougher than is desired and a portion of the coating adhered to the casting surface during the first shot. It was concluded, therefore, that if that type of coating is to be applied successfully, more research is needed to improve the adherence and smoothness of the coatings.

One of the later developments included water-base formulations of the boron nitride coating as well as boron

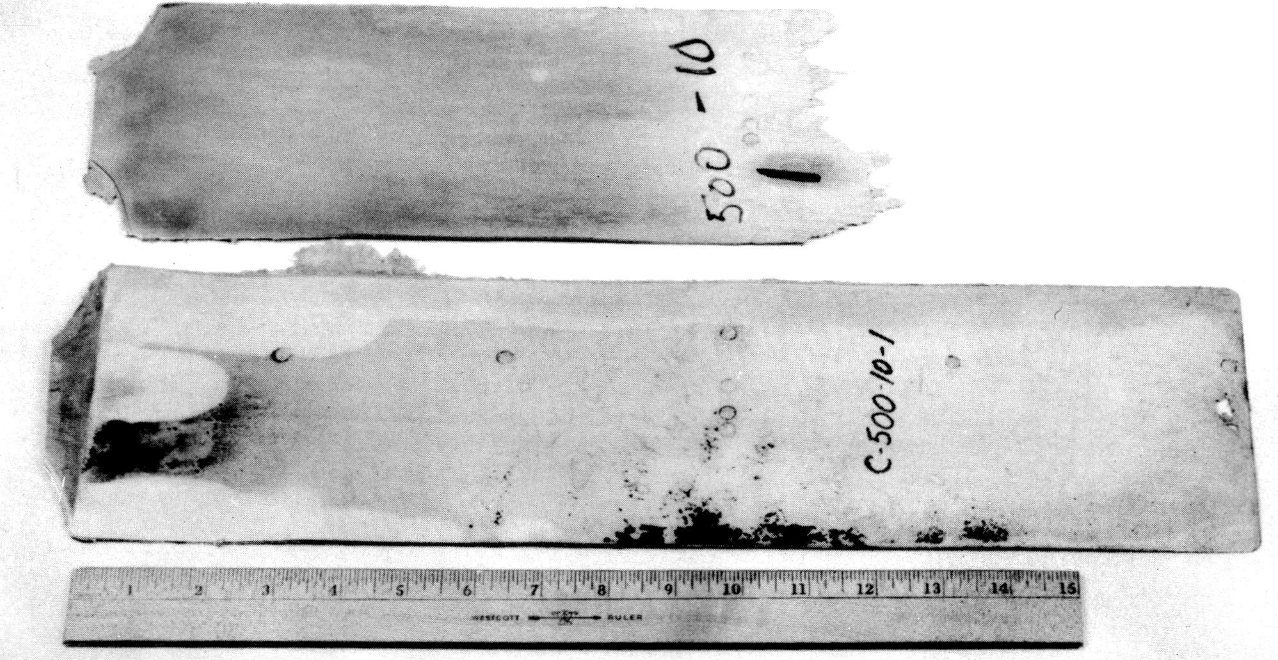

Fig. 6. First-shot castings produced in an uncoated and an elastomeric boron nitride-coated die cavity (castings were about 0.025 inch or 0.635 millimetre thick).

Fig. 7. Fiftieth castings produced in an uncoated and an elastomeric boron nitride-coated die cavity.

nitride coating mixes that contain other compounds. These compounds are now under test to determine the smoothness, adherence and thermal properties of the coatings.

This work lead to evaluation of alternative die materials, with special emphasis on graphite. In 1977, blocks of Grade EDB graphite were obtained and machined into inserts for the research die, shown in Fig. 8(a). Work is now proceeding to evaluate the graphite inserts after they were coated with pyrolitic carbon, Fig. 8(b), to retard heat transfer and provide water resistance.

Development of Alternative Gating Systems

One of the primary aims of die casting research is to produce a casting that can be plated without the need for intermediate finishing, or with only minor finishing, therefore, it would be desirable to produce castings that do not have gates that must be trimmed and polished. For many castings it is also difficult to gate the casting so that a uniform and predictable flow of zinc can be obtained through the die cavity.

One potential technique for eliminating gates is to use a pin-gating system in which zinc is introduced directly into the cavity through one or more small heated nozzles. With such a system, the sprue, runners and gates on conventional castings could

be eliminated. To eliminate overflow gates additional heated nozzles and a heated manifold system might be used to return to the holding furnace that zinc which normally goes into overflows. This, of course, would reduce the trimming and polishing of finished die castings, thereby reducing the intermediate efforts required before plating.

During 1978 two different versions of miniature direct injection nozzles were designed and specifications with required dimensions were prepared and forwarded to a heating-element manufacturer preliminary to supplying a resistive-heated nozzle suitable for this application.

One of the completed designs is a miniature version of a large-nozzle heater with electric cartridge heaters. This nozzle is shown in Fig. 9. It was machined from H13 steel and contains four 6.35 mm x 102 mm cartridge heaters that provide a total heat capacity of 700 watts.

The second type of nozzle being evaluated uses induction heating. It is shown in Fig. 10. That unit was originally designed and constructed with a 200-watt heating capacity. Initial testing of the mininozzle indicated that a heat input of 200 watts would be marginal for the desired application. The nozzle and power supply are now being modified to develop in excess of 500 watts.

Following completion and testing of the modified induction-heated nozzle and testing of the resistance-heated nozzles to be provided by the commercial source, one type of nozzle will be installed in the research die and evaluated under die casting conditions.

Additional Work

Other research tasks being carried on at Battelle at the time of writing include an evaluation of the effects of variation in gate velocity, gate-to-cavity thickness ratio, die temperature on casting quality, and development of improved methods for calculating overflow sizes.

The BNF Work

In the work carried out at BNF, the objective was to establish a simple system of rules to guide die designers. Two aspects of die design which were given priority were the gating design and the thermal design. In this work emphasis was placed on how variations in the gating design and the thermal design affected the soundness of the casting, the pressure tightness and the surface finish.

It should be noted that much of the work was done using aluminium alloys. However, the results obtained using zinc were much in accordance with those for aluminium.

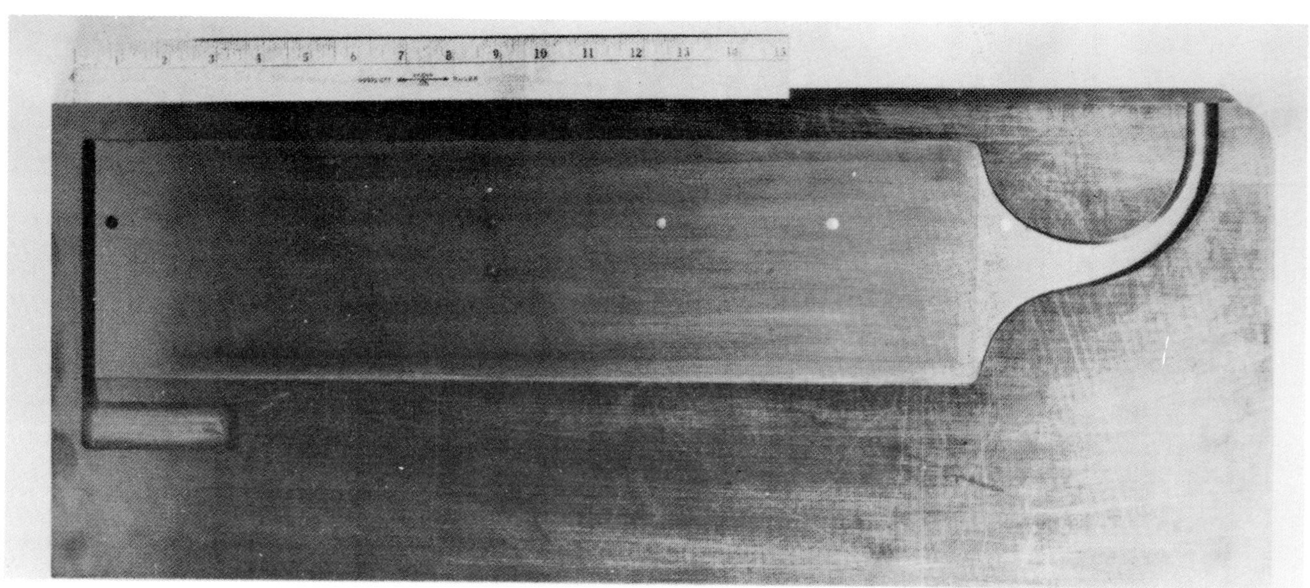

Fig. 8A. Block of graphite machined into research cover die, prior to coating.

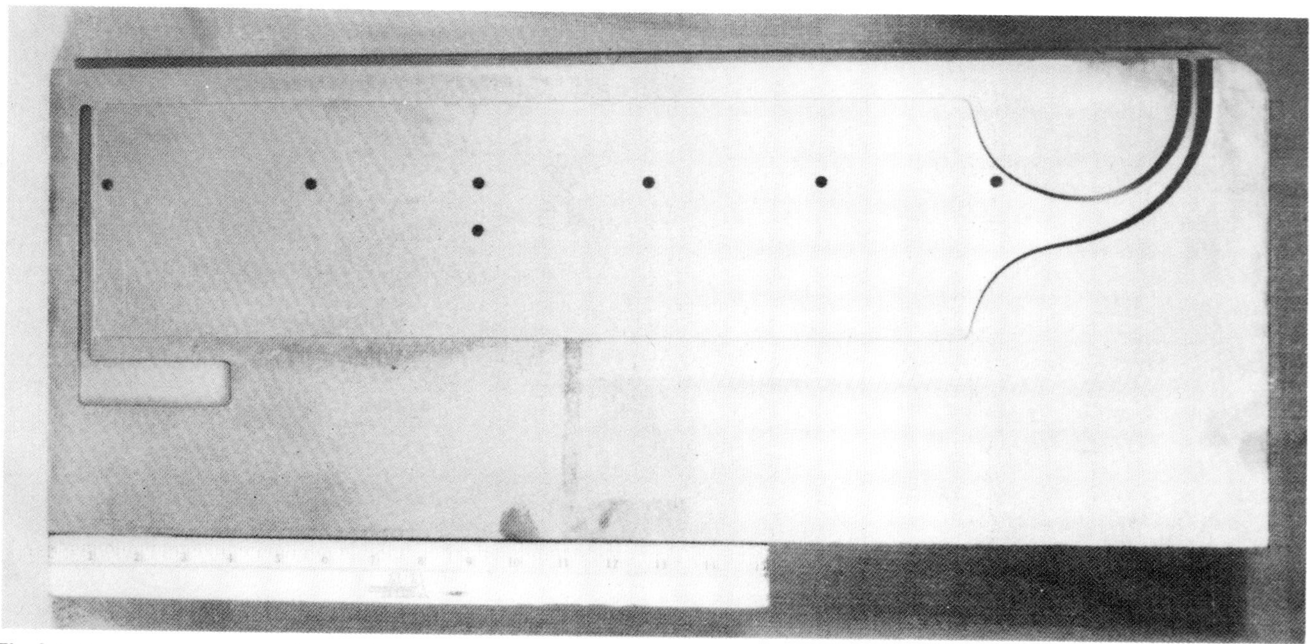

Fig. 8B. The same graphite die after coating with pyrolytic carbon. Costing retards heat transfer on die filling.

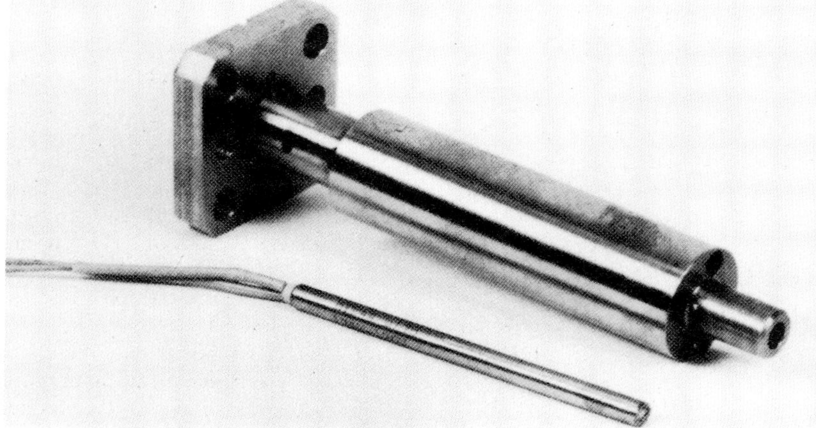

Fig. 9. Experimental mininozzle heated by cartridge heaters.

Gating Design

Gate Thickness. Gate thickness is an important variable. If too thick a gate is used, problems in trimming are encountered and when a good surface is required, porosity at the gate can cause castings to be rejected. Too thin a gate (with the resulting decrease in area) can increase the resistance to fill, lengthening the cavity fill time with a detrimental effect on surface finish. An important consideration is the thickness of the gate in relation to the casting to ensure adequate pressure transmission during solidification. Previous work in this project had indicated that when high speed fill is

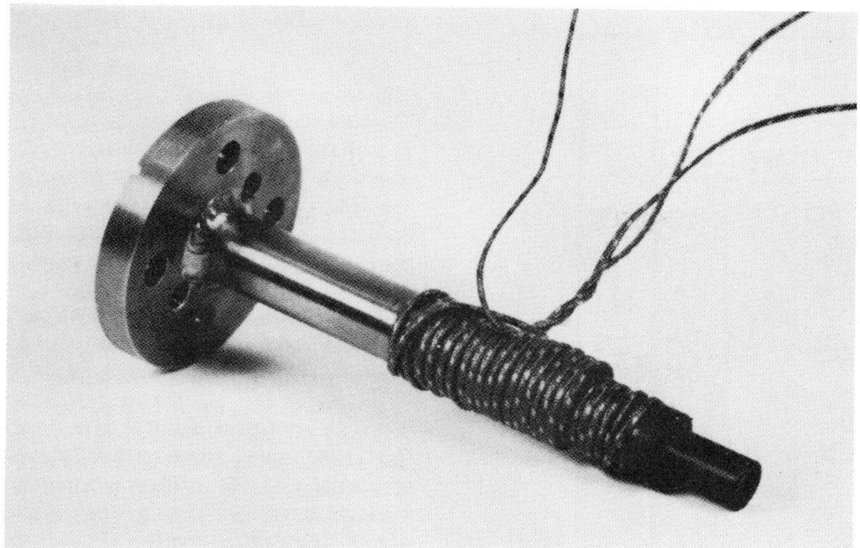

Fig. 10. Induction-heated experimental mininozzle.

used, sounder castings can be produced, in some cases, with a thinner gate.

The research showed that gate thickness has a large effect on the duration of pressure transmission to the casting. Even with the thinnest gate employed, the duration of pressure transmission was much longer than the response of pressure intensification systems used on modern cold chamber die machines, such as the one used in this work. The differences in pressure transmission occurred in the period following intensification, when trapped gases have been compressed and shrinkage due to freezing was the form of unsoundness likely to respond to continuing pressure.

In the case of two aluminium alloys tested (LM24 and LM6) using a hot die, pressure was transmitted for about one-half second with the 0.4 mm thick gate, one second for the 1 mm gate and two seconds for the 4 mm gate. Reducing the die temperature from 395 to 255°C approximately halved the duration of pressure transmission.

In the case of zinc (BS1004A), cast at 500°C into a die of normal temperature, very long pressure transmission times were recorded – over 5.5 seconds with a 4 mm thick gate. Lowering the temperature of zinc to 420°C more than halved the duration of pressure transmission.

The results of the zinc and one aluminium test are given in Figs. 11 and 12. All the tests were carried out using similar injection conditions, the injection speed control valve was adjusted to give the same cavity fill time for comparison purposes.

Although the gate thickness had a large effect on the duration of pressure transmission, it had only a small effect on casting porosity, provided that the final injection pressure was applied rapidly after completion of cavity fill.

During the study of the effect of gate thickness, the effect of several other process variables were studied.

The shot weight affects the volume of gas entrapped in the casting because it controls the percentage shot sleeve fill and the volume of gas in the shot sleeve. It was found that above the necessary minimum shot weight required to ensure complete cavity filling an increase in shot weight has a negligible effect on casting porosity.

Lubrication of the die also has a negligible affect on porosity. However, longer cavity fill times do reduce porosity. This is because gases in the system have longer to escape.

Pressure Tightness

Pressure tightness is an important requirement for some pressure die castings. The gating test casting was drilled and pressure tested, using air at 200 kPa. The results, given in Table I, show zinc castings to be significantly more pressure tight than aluminium castings. The tests involved two gate thicknesses, two die temperatures and two metal pressures. To carry out tests at the lower die temperature, the die was water cooled.

A large reduction in casting porosity, from 4.1% to 1.6% (produced by injecting with and without intensification), produced only a small effect on pressure tightness. As gate thickness had only a small effect on casting soundness, and soundness itself had a small effect on pressure tightness, it is not surprising that gate thickness had little effect on pressure tightness.

The main factor in achieving pressure tightness was die temperature. Lowering the die temperature – insufficiently to cause any deterioration in surface finish – produced a large improvement in the pressure tightness of the castings.

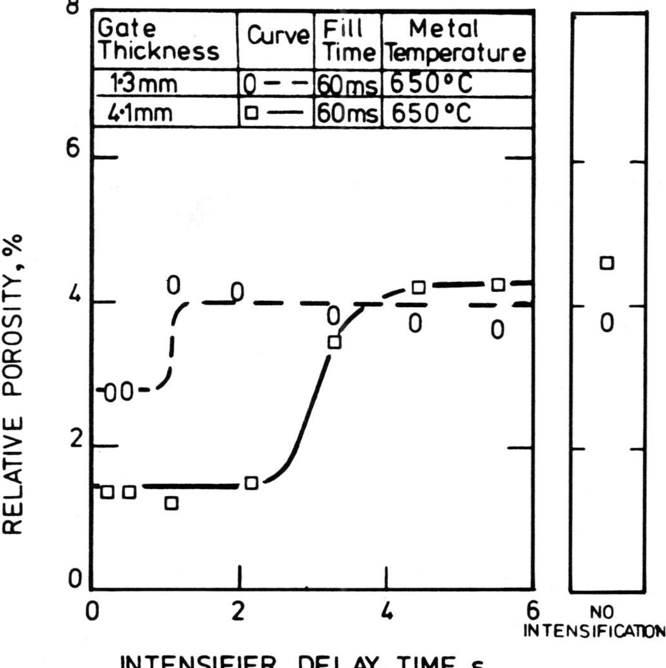

Gate Thickness	Curve	Fill Time	Metal Temperature
1·3 mm	0 – –	60 ms	650°C
4·1 mm	□ —	60 ms	650°C

Fig. 11. Effect of intensification delay on casting soundness for LM6 (A413.2.USA/Al-Si.12. ISO) castings.

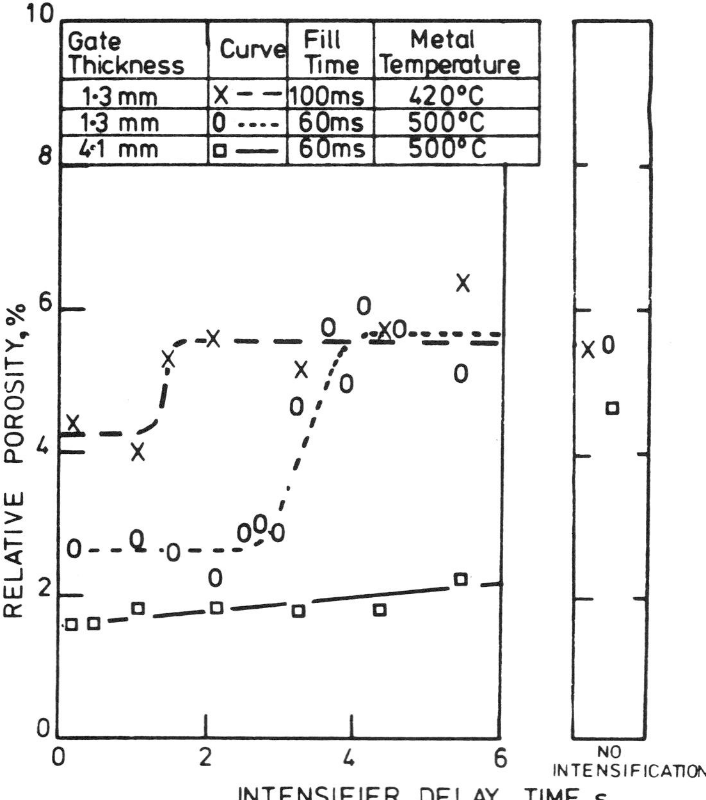

Fig. 12. Effect of intensification delay on casting soundness for the zinc castings.

Use of the gating test die with interchangeable gates gave the opportunity to determine the effects of changes in gate thickness and gate area on overall pressure losses in the metal flow system. The results obtained indicate that losses, probably due to solidification in the shot sleeve and friction between the plunger and sleeve, were significant and variable with injection speed and injection pressure. Increased solidification in the shot sleeve, due to delays between pouring metal into the shot sleeve and commencing the injection stroke and consequent drop in metal temperature, produced a large increase in the pressure loss. Fig. 13 shows a higher efficiency, i.e. reduced pressure losses, with zinc than aluminium, probably because of the increase in internal pressure with zinc owing to its higher density.

Thermal design

The thermal design work has concentrated on how to extract the appropriate amount of heat from the die in order to achieve the appropriate casting rate whilst maintaining the desired mean die temperature.

In this work the gating test die was operated with five different cooling channel configurations to produce large variations in the temperature distribution across the die face. Castings made during these tests were used to investigate the influence of temperature gradients on casting soundness. Large differences in the temperature gradient across the die face altered the porosity distribution but had a small effect on the total porosity of the casting.

Cooling channels passing through a die may be vertical, horizontal, or any angle between. Die casters have posed the question as to the best position to place the cooling water flow valve – on the inlet or the outlet of the cooling channel. If the flow valve were placed on the inlet side of the cooling channel, it was thought that at low flow rates there might be a possibility of the cooling channel running only partly full. This, in turn, could lead to differences in heat extracted by the cooling water from that to be expected. If the valve were placed on the outlet this would ensure the channel being completely filled. Similar problems might also arise with vertical channels, where the inlet and control valve were placed adjacent to the top of the channel.

Tests were therefore carried out to determine whether valve position and channel orientation did affect the heat removed by the cooling water. A simulated heated core was made for this purpose, details of which are shown in Fig. 14. Tests also were carried out to see if the cooling effect of cascade were any different from that of a conventional cooling channel.

Small but definite differences were noted in core temperature for different valve and core positions. Both vertical and horizontal channels, when vented directly to the atmosphere, ran cooler than when a restriction was placed on the outlet. Fig. 15 shows the affect of cooling water flow rate on core temperature for different channel configurations.

Tests also were made using a cas-

Casting alloy	1mm gate					4mm gate				
	Die water cooled	Nominal metal pressure, MPa	Cavity face temperature at injection °C	Relative porosity of casting, %	% of castings failing pressure test	Die water cooled	Nominal metal pressure, MPa	Cavity face temperature at injection °C	Relative porosity of casting, %	% of castings failing pressure test
LM24 (A380.1)	No	18	325	3.6	100	No	19	335	3.6	100
	No	70	335	1.7	100	No	70	325	1.2	90
	Yes	18	230	4.0	80	Yes	19	250	3.5	100
	Yes	70	220	1.6	60	Yes	70	215	1.5	30
						Yes	19	205	3.0	70
						Yes	70	185	1.1	30
LM6 (A413.2)	No	18	340	3.9	50	No	19	345	4.4	60
	No	70	335	2.0	10	No	70	340	1.4	10
	Yes	18	250	3.7	20	Yes	19	240	4.5	30
	Yes	70	245	1.0	0	Yes	70	230	1.0	10
BS1004A (AG.40A)	No	18	215	4.5	20	No	19	230	4.9	20
	No	70	215	2.1	20	No	70	235	1.6	10
	Yes	18	160	3.6	10	Yes	19	160	4.7	0
	Yes	70	150	1.5	0	Yes	70	160	1.5	0

Table 1. Results from pressure tightness tests

cade to cool the core. The effect of flow rate was greater than for a straight-through channel. Core temperatures, however, were similar to those for a straight-through channel.

The CSIRO Work

CSIRO is engaged in two projects devoted to the overall improvement of the die casting process and product. The first is concerned with improvements in the metallurgical properties of castings through control of processing conditions. Success here means that the potential range of applications of die cast compounds can be widened by employing to the maximum the inherent toughness of the zinc-based die casting alloy No. 3 in the production of both lightweight and thick-section castings. The full exploitation of the potential of current casting technology depends on quantifying the relation between the properties of castings and the conditions under which they are produced.

It is now apparent that the design and machine operating criteria used to produce lightweight castings improves the casting integrity and modifies the microstructure of the casting. This raises the possibility of controlling the microstructure of standard zinc alloy 3 to produce castings with a uniformly modified structure and improved mechanical properties through control of metal fluid flow-in of the cavity.

The second project has concentrated on the analysis of metal flow in tapered runners feeding thin gates. In particular the pressure-flow characteristics have been studied to determine their effect on die erosion, turbulence and cavity fill time.

Factors Affecting Microstructure

The microstructure of the die cast components has been widely considered, usually on the basis of the cavity filling with molten metal, followed by turbulence and cessation of fluid flow and solidification proceeding under steady state thermal conditions. However, experimental evidence tends to make this viewpoint untenable when applied to modern thin wall castings produced at high metal velocities through thin gates. The combination of short cavity fill times, thin wall sections and the relative temperature of die and metal in the runner and cavity areas insures that solidification starts to take place under dynamic conditions as the cavity is being filled. These and other data lead to the conclusion that in the

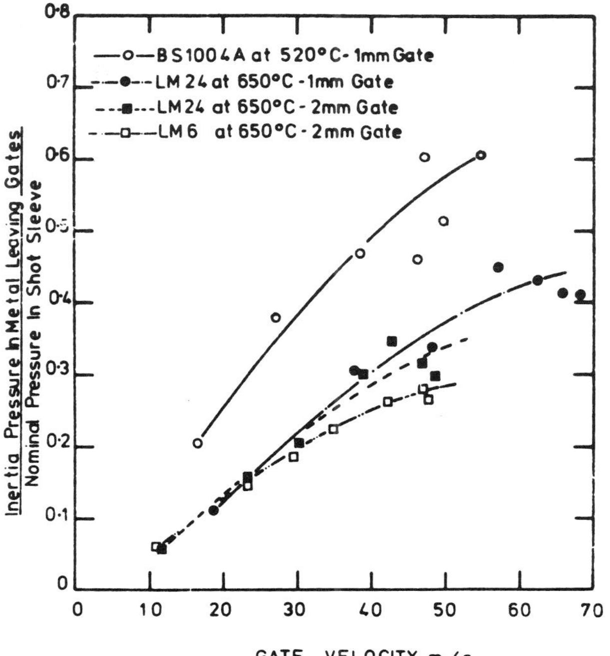

Fig. 13. Effect of casting alloy on the efficiency of the metal flow system.

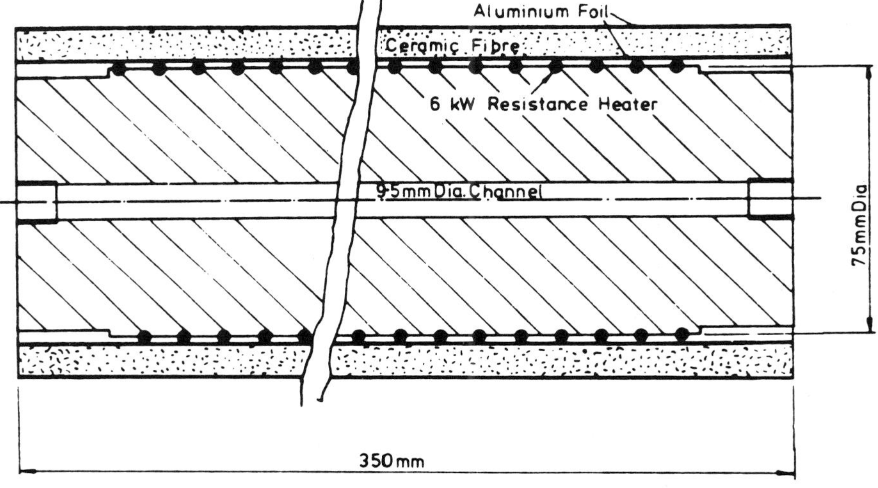

Fig. 14. Electrically heated core used to investigate the effect of cooling channel orientation on core temperature.

production of thin-section castings, the microstructure is dependent on both thermal and fluid flow considerations.

Studies of the hydrodynamics of the metal feed system lead to the findings that the structure of the casting is determined by the rate of shear at the gate and the kinetic energy of the metal stream entering the cavity as well as by thermal considerations. The amount of energy dissipated in bringing the metal to the entry to the gate may also influence directly the solidified microstructure.

These findings were made after it was determined that zinc alloy 3 undergoes a change in morphology, which causes a change in the mechanical properties. This change in morphology from dendrites, through rosettes to spheroids is the result of increasing the fluid flow which causes an increase in the shear rate.

Microstructure and Mechanical Properties

The fracture and tensile properties of zinc alloy 3 castings depend on the microstructure and the casting integrity

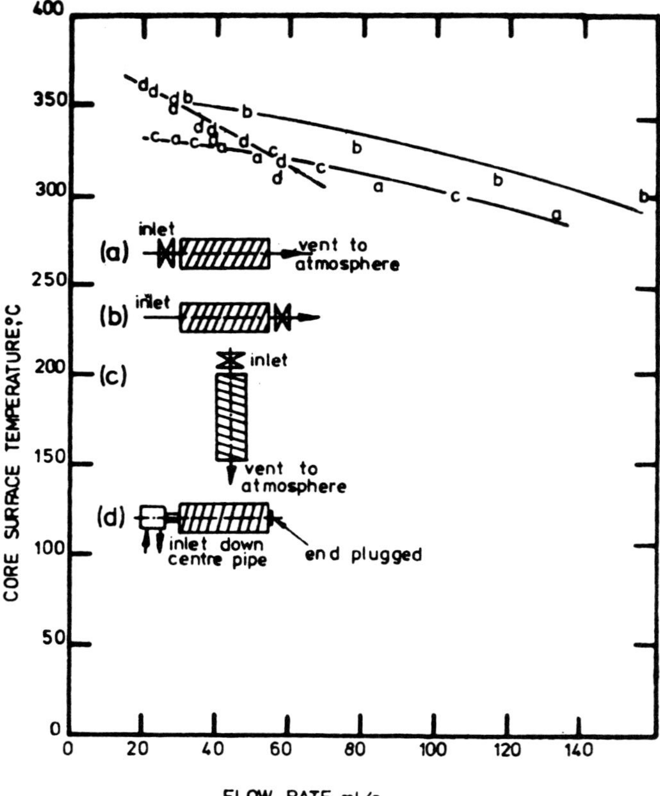

Fig. 15. Effect of cooling water flow rate on core temperature for different channel orientations.

in terms of porosity and surface finish. Both the fracture and tensile properties of alloy 3 die cast in thin sections increased with the degree of shear at the gate, as seen in Figs. 16 and 17. For example, there was an approximate 50% increase in the fracture toughness over the range of shear rates used, the low end of the range being representative of traditional gating and operating practice and the upper end representative of thin wall technology.

The fracture properties were also found to increase with a decrease in die temperature. This is a reflection of the smaller size of the primary phase, whether in dendritic, rosette or spheroid form, at the lower die temperature. Although ultimate tensile strength and proof stress were greater at the higher die temperature, the effect of die temperature on the tensile properties is minimised when the rate of shear at the gate is high, as seen in Fig. 17. In summary, there appear to be advantages from the point of view of mechanical properties in producing castings under conditions which give a high degree of shear at the gate combined with a low die temperature.

The remaining key question, which was studied in the first half of 1978, is whether commercial castings produced under the new design criteria have substantially superior mechanical properties to those which have been attained generally with die cast products. To this end, the mechanical and fracture properties of 15 castings, in current or recent production, were determined and assessed in terms of their feed system design and the energetics of fluid flow in the die cavity.

For convenience, the castings were divided on the basis of feed system design and operating conditions into traditional, modern, and current local production (CLP).

The traditional system used heavy runners and thick gates, the modern system used light runners with a constant area and a medium gate and the CLP used tapered runners with a thin gate.

The changes in the energetics of cavity fill between the three groups of castings were reflected in their morphology. The primary zinc-rich phase in the traditional group of castings was dendritic, that in the modern group of castings was in the form of rosettes, and that in the CLP castings was spheroid.

The mechanical and fracture properties of the three groups of castings are shown in Figs. 18 and 19. The bars represent the spread of values found within each group of castings. In summary, the evolution of die casting technology is now resulting in the production of castings, with a modified microstructure, which have superior tensile properties to the ingot material in the form of test bars.

The fracture properties, such as stress to fracture, energy to fracture and fracture toughness are distinctly superior for the modern and CLP castings, compared to the traditional castings.

It is evident that the mechanical and fracture properties of the castings have been improved by the changes in feed system design and operating conditions which have resulted in increased shear on the metal stream as it passes through the gate and in an increased kinetic energy of the stream as it enters the cavity. This information now has to be quantified to enable these concepts to be used for design purposes.

Sprue and Tapered Runners

In recent years, considerable attention has been given to the design of feed systems for zinc die castings. Gooseneck-nozzle-runner-gate systems have been proportioned for controlled metal velocity at all parts of the system. Long tapered runners and gates have been used to shorten the metal flow path within the cavity. This type of geometry has been very successful in Australian industry, where many new dies have produced thin-walled or other types of castings, with first-shot success and a minimum of die development cost.

The investigations into the flow of molten zinc in goosenecks and nozzles conducted by CSIRO have shown that under normal die casting conditions, high pressures are needed to accelerate the metal and to overcome pressure losses. The results suggest that metal flow is not controlled only by the cross-sectional area available for flow, but rather by the energy losses occurring in sprues, runners and gates, the flow areas being one factor in the calculation.

The current research effort by CSIRO has the aim of providing quantitative design criteria for modern feed systems. The research involves methods of predicting the pressure-flow characteristics of feed systems incorporating sprue runners and tapered runners. The effort included an evaluation of the performance of a range of tapered runners, using an approximate mathematical method

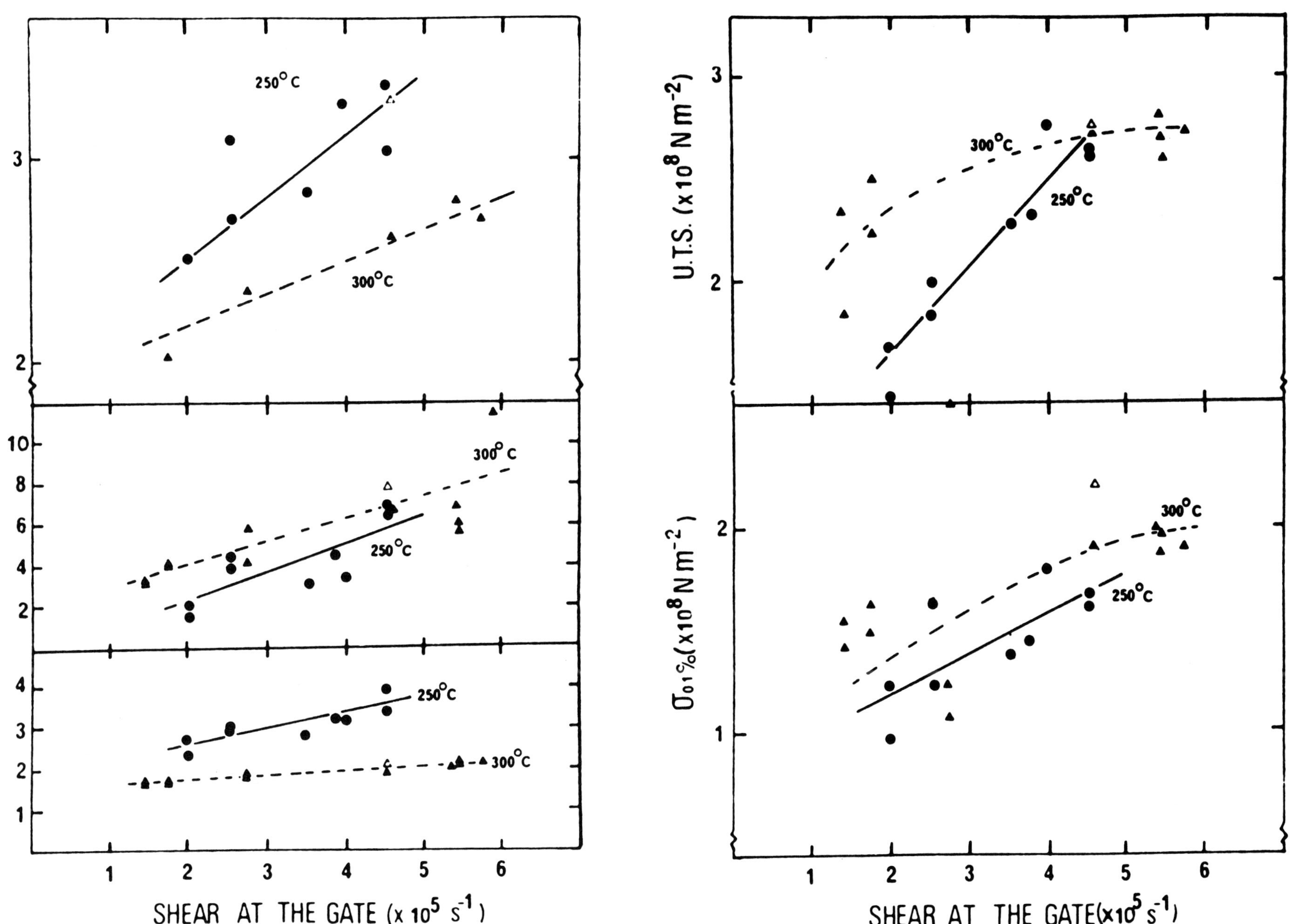

Fig. 16. Variation in fracture properties with increasing shear at the gate.

Fig. 17. Variation in tensile properties with increasing shear at the gate.

Source: *Materials in Engineering Applications*, Vol 1, June 1979, 209-226

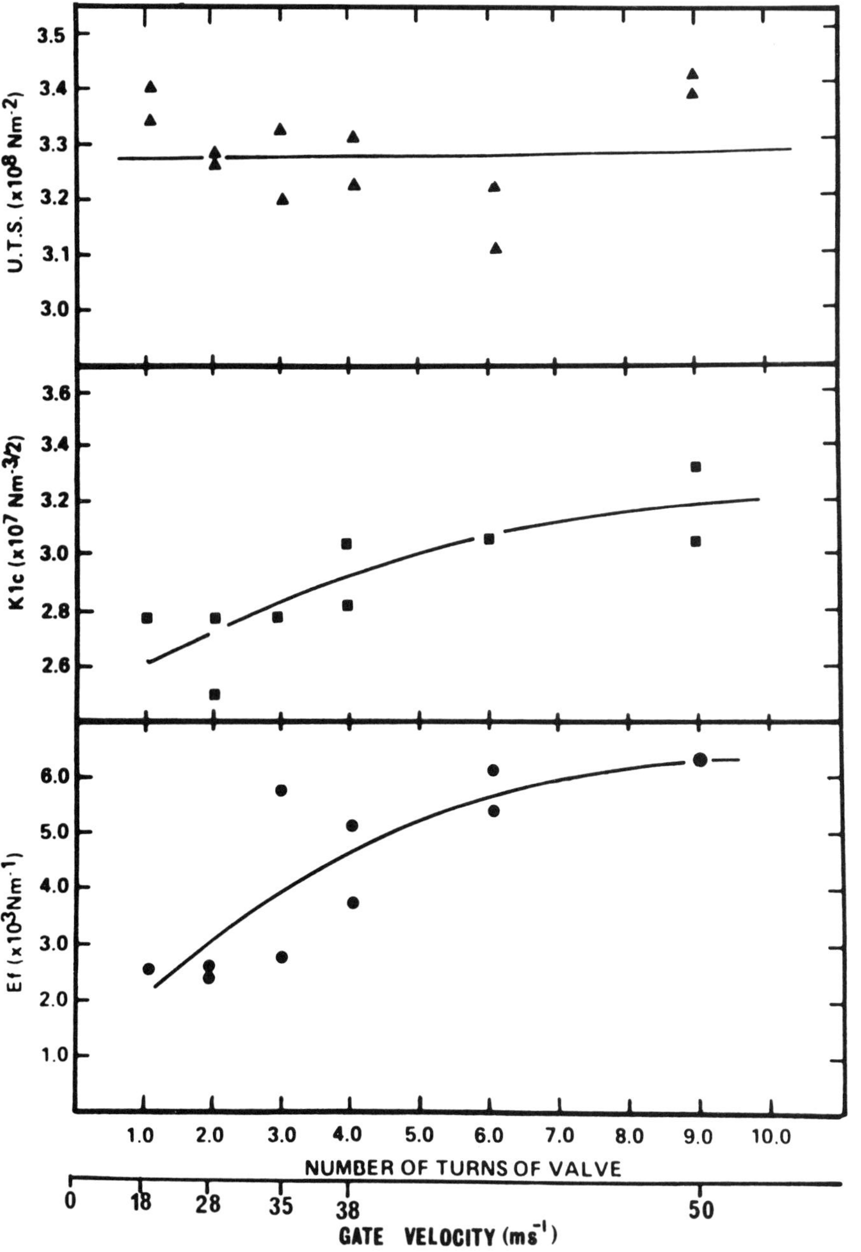

Fig. 18. Range in energy to fracture, fracture toughness and average dendrite size (units of X $10^3 Nm^{-1}$, $\times 10^7 Nm^{-3/2}$ and $\times 10^{-6}$ m respectively) for traditional, modern and current casting feed systems. Dashed line denotes average value.

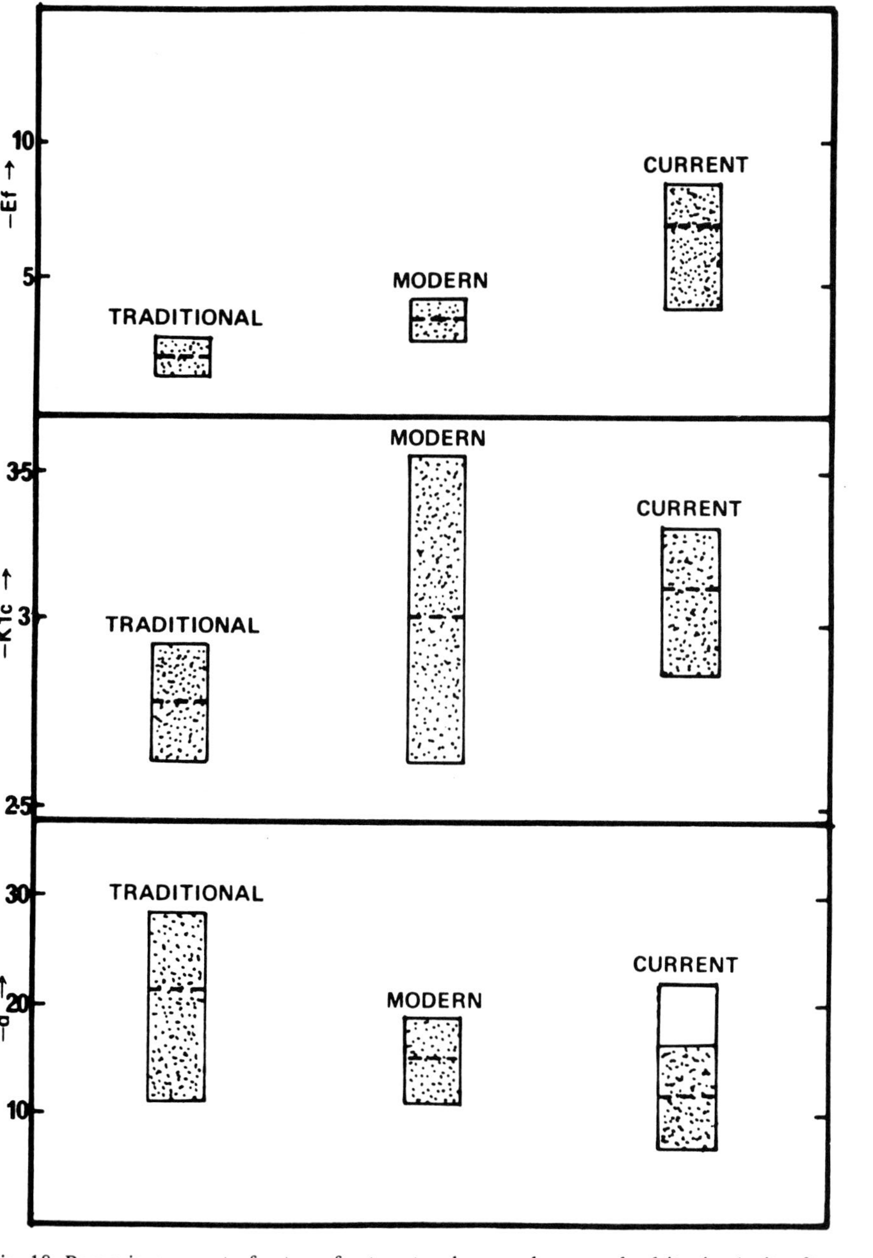

Fig. 19. Variation in U.T.S., fracture toughness (K1c) and energy to fracture (Ef) with increasing number of turns of the shot valve and calculated gate velocity.

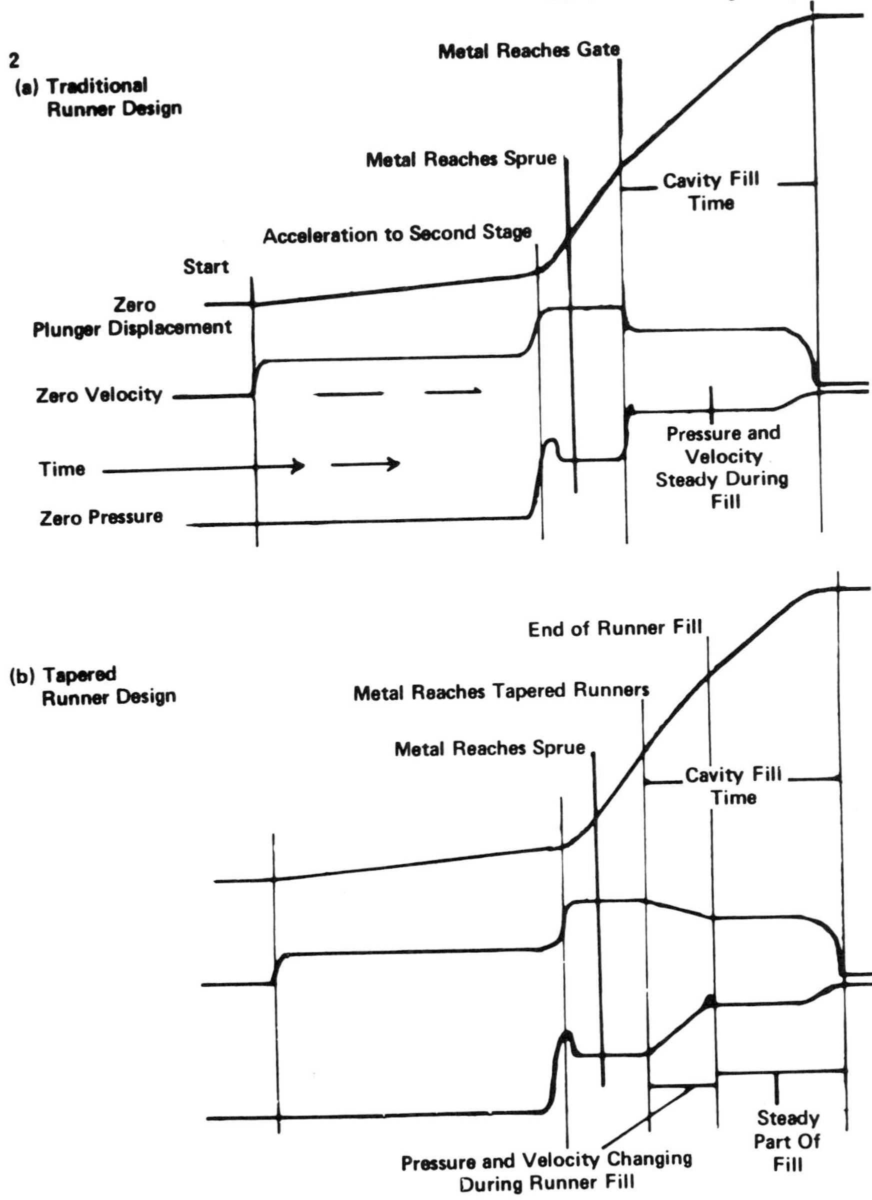

Fig. 20. U.V. traces during cavity fill through traditional runner design.

and applying the CSIRO computer to solve the equation.

The Fill Process and its Evaluation

In the analysis, measurements of ram pressure, velocity and displacement were made for each stage of fill from which the metal pressure and velocity could be calculated. It was found that for a hot chamber machine three stages of fill could be considered.

(a)   Metal arrives at the nozzle exit
(b)   Metal arrives at the gate
(c)   Metal fills cavity and overflows.

This analysis can be used to define the cavity flow time, i.e. the time during which flow occurs through the gate. Alternatively, the cavity fill time could be measured. The studies showed that the tapered runner prevented a sharp rise in pressure when the metal reaches the gate, which is a characteristic of older gate runner designs, as seen in Fig. 20. Changes in pressure and velocity should be smoother. It showed that the resistance to metal flow rises during runner fill, becomes steady then rises again in the later stages of cavity fill. The injection pressure and velocity can be plotted as a $p - Q^2$ diagram, where p is the metal pressure and Q is the metal flow rate. A $p - Q^2$ diagram for a tapered runner system is shown in Fig. 21.

The flow rate is also affected by the design of the casting machine. Older machines with long oil lines and heavy moving parts have a high inertia which results in high metal velocities. This can lead to die erosion and cavitation although it may improve cavity filling. This can be avoided by modern machine design so that the machine decelerates as die resistance is encountered.

Design of Tapered Runners

The design of a runner not only affects metal flow but also affects heat transfer from the metal to the die. It also affects the volume of air required to be vented by the die and the volume of excess metal which has to be clipped from the casting. The design further influences local velocity variations which could cause erosion or cavitation.

A typical tapered runner design is shown in Fig. 22. In such a design there are three key ratios which are expressed relative to the runner inlet area.

1. Runner taper ratio =

$$\frac{\text{Runner Inlet Area}}{\text{Runner End Area}}$$

This taper is important as it affects pressure and gate velocity during fill and is always greater than 1. The higher the ratio, the greater the taper is for the runner. However, it should not taper to zero or this would encourage gate erosion.

2. Runner slenderness ratio =

$$\frac{\text{Runner Inlet Area}}{\text{Runner Length}}$$

In practical use this is limited by the thickness of the gate and for zinc lies in the range of 0.1 to 0.5. Low values would require high pressures and the gate velocity would vary more from end to end, due to increased friction losses.

3. Runner Velocity Ratio =

$$\frac{\text{Runner Inlet Area}}{\text{Gate Area}}$$

For zinc this lies between 1.0 and 2.0. A low value implies high velocities which are required for thin castings, although the high velocities may cause erosion and cavitation problems. A high value implies lower runner velocity.

Seventy-two different combinations of gates and runners were considered, which covered all the size and flow rates most likely to be needed in practical use. For each combination the pressure, runner velocity and gate velocity distributions were calculated for both the steady state and transient flow conditions.

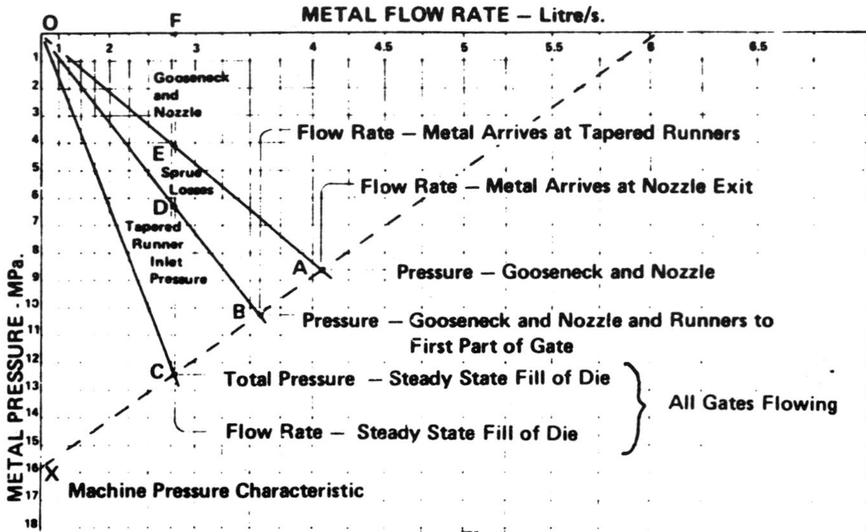

Fig. 21. p – Q² diagram, showing changes in pressure and flow rate filling a cavity through tapered runners.

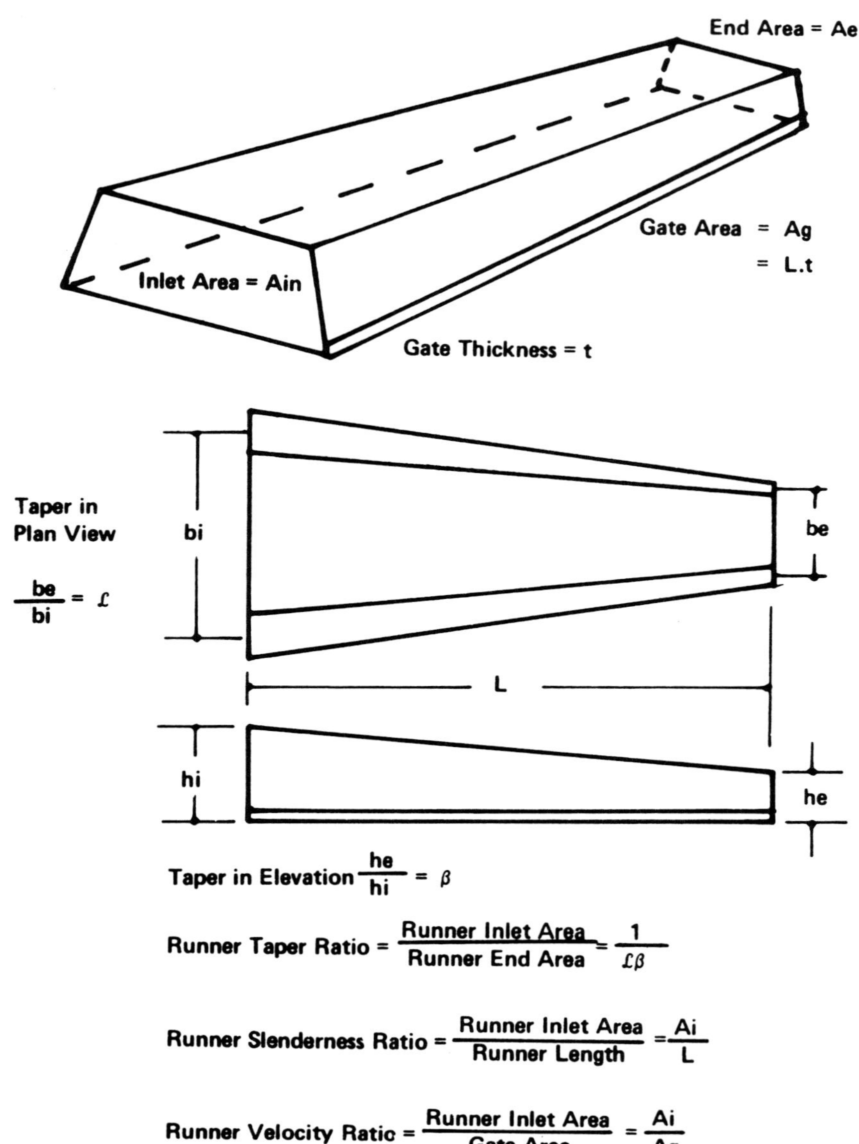

$$\text{Runner Taper Ratio} = \frac{\text{Runner Inlet Area}}{\text{Runner End Area}} = \frac{1}{£\beta}$$

$$\text{Runner Slenderness Ratio} = \frac{\text{Runner Inlet Area}}{\text{Runner Length}} = \frac{Ai}{L}$$

$$\text{Runner Velocity Ratio} = \frac{\text{Runner Inlet Area}}{\text{Gate Area}} = \frac{Ai}{Ag}$$

Fig. 22. The geometry of a straight-sided tapered runner.

It was shown that the optimum cross-section shape for a tapered runner is circular or a close approximation to circular (i.e. a square shape is better than a rectangle). This gives minimum pressure losses and heat transfer during flow. A good finish to the runner is also desirable as this reduces any pressure drops.

Effect of Runner Fill on Cavity Fill
The calculations showed that metal flows through the gate at all stages of runner fill, so that the actual fill time is greatly extended in practice. This is due to the fact that high pressures are generated at the gate during runner fill. The calculations showed that the fraction of the cavity filled during the runner fill time does not vary much with changes in dimensions of the runner and gating system.

The calculations also showed that long, thin walled castings would be almost filled by the time runner fill was complete. Therefore, control of gate velocity during runner fill is of great importance.

Changes of Velocity and Pressure During Runner Fill
Gate velocity is related to pressure in the runner, which pressure is developed due to a combination of pressure losses in the runner and gate plus the need to accelerate metal as it passes along the runner on through the gate.

The pressure rise during runner fill is large enough to slow the injection plunger to some extent, depending on the machine inertia. The machine plunger will slow down further, owing to the step change in runner characteristic at the completion of runner fill. Changes in machine pressure and velocity should be smoother when feeding tapered runners than for a traditional oversized runner design.

In determining affects of gate velocity during runner fill, it was found that, in general, this velocity is most important for long runners which fill a large part of the cavity during runner fill.

Water Analogue Flow Study of Tapered Runners
The calculations carried out so far have predicted flow velocities and pressures. However, they do not provide any information on the flow direction of metal emerging from a gate along the length of a tapered runner or on the pressure shock which occurs at the end of runner fill and which may lead to die erosion.

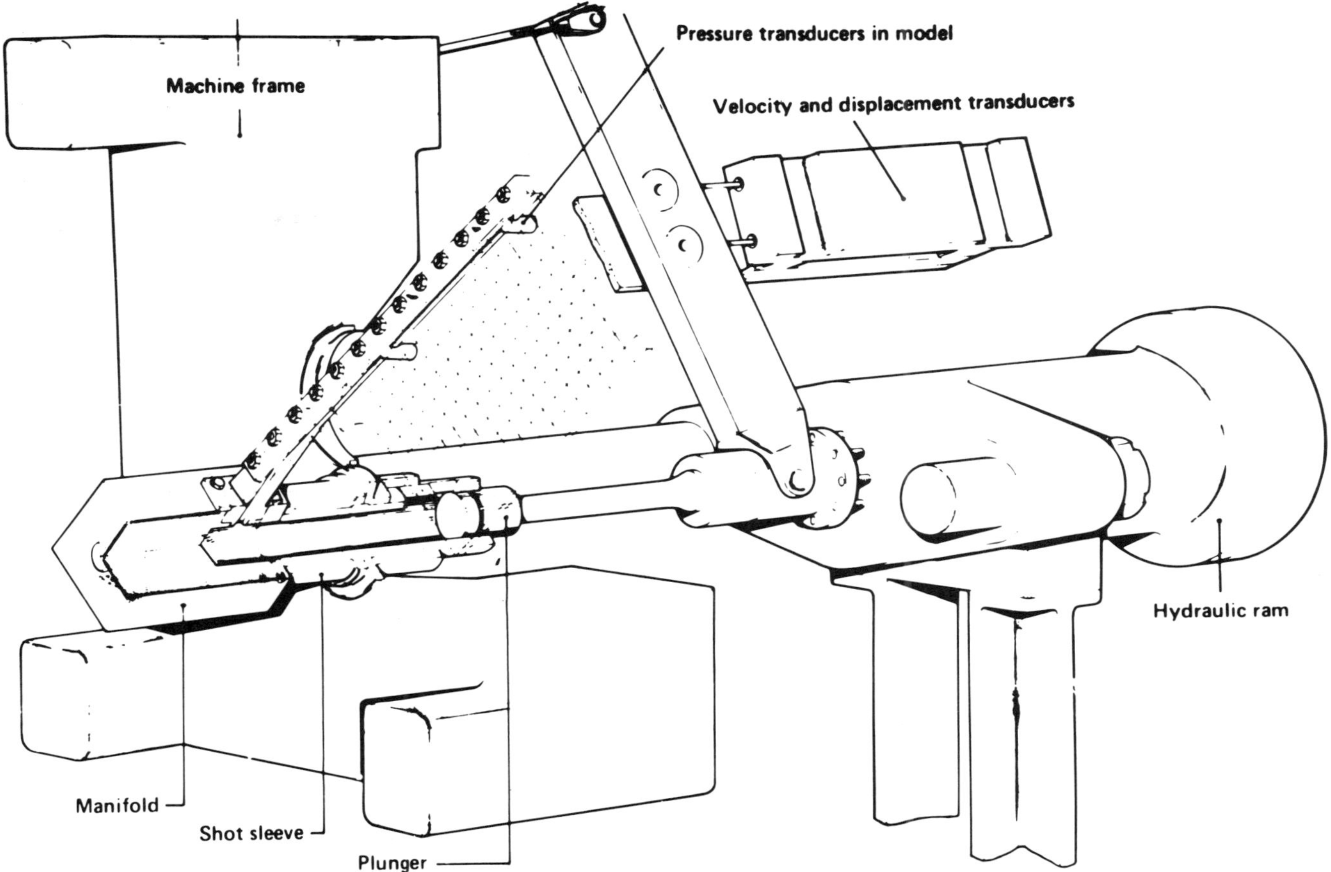

Fig. 23. Water injection system fitted to Lester die casting machine.

An experimental programme was devised wherein transparent models containing water were used to simulate the zinc in the die cavity. The flow velocities were kept similar to those for zinc, however, due to the difference in the physical properties of these fluids, the resulting Reynolds number of water is about half of that for zinc. However, since the flow patterns of both fluids are well into the turbulent region, the use of water was not expected to affect the qualitative results.

The study considered four major areas of interest:

1. How flow performance of a tapered runner is affected by runner and gate geometry and to derive the pressure loss coefficient of the gate from the experimental data, to be used in future theoretical flow calculations.
2. To investigate the performance of a shock absorber and how it is affected by its ingate size so that future shock absorbers can be designed on a sound theoretical basis.
3. To determine the factors which affect the flow angle θ of the 'true' gate velocity. The flow angle θ is the angle of the direction of metal flow from the gate to a line perpendicular to the gate length.
4. To visually demonstrate the phenomenon of the pressure shock which produces a high velocity surge at the end of the runner fill.

Fig. 23 shows the basic arrangement of the water injection system and the transparent runner model.

The specific parameters studied were:

(a) thickness of gate
(b) volume flow rate
(c) runner taper ratio
(d) runner inlet pressure
(e) shock absorber pressure
(f) shock absorber ingate size
(g) shock absorber volume
(h) geometry of gate inlet
(i) width of land at gate
(j) surface finish of runner

Results of Water Analogue Experiments
From this work it was found that the angle θ by which the fluid emerges from the gate does exist and, further, is an important factor to be considered in design of tapered runners. This is so because this angle was found to be very much dependent upon the ratio of the runner inlet area to the gate area. It also is dependent upon the taper and its existence explains why gate speed is much higher than can be calculated simply by dividing the flow rate by the gate area.

A significant improvement in the energy efficiency of the tapered runner can be gained by simply having the gate entry chamfered, as shown in Fig. 24. Further improvements in efficiency can be expected by having a radius at this gate entry.

The existence of the high gate velocity surge, at the 'dead end' of the runner, at the instant when the runner is full, was shown by high-speed photography. It also showed that the magnitude of the gate velocity surge can be reduced or eliminated by having a runner 'shock absorber' at the end of the runner, as shown in Fig. 24. The optimum size of this shock absorber and its ingate size have yet to be established.

These results suggest that the two main proposed modifications of the runner design; that is, (1) incorporating a shock absorber, and (2) making the area at the dead end larger than zero,

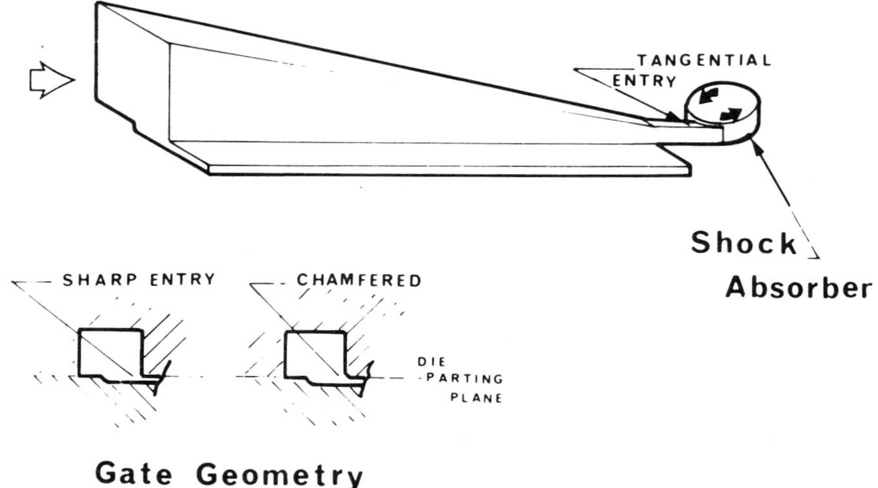

Fig. 24. Shock absorber designed to reduce the velocity surge at the gate and chamfered gate inlet to reduce the pressure losses.

have helped reduce the chances of die erosion.

The computer calculated results were found to be in general agreement with the water analogue results. This suggests that the flow angle, the time to fill the runner, and other flow variables, both during runner fill and steady state, can be accurately predicted.

The ability to predict these variables by computer calculations should provide a useful aid to the die designer in deciding the location of the gates, vents and overflows, and their sizes.

The total amount of fluid which flows out of the gate during runner fill was found to be in the range from 30 to 50% of the runner volume. This is a significant amount which suggests that the traditional belief that cavity fill time commences after the runner is completely full, is not valid when a tapered runner is employed. Therefore, the definition of cavity fill time will have to be critically reviewed.

Discussion

The research studies into the die casting process which are reviewed in this paper have encompassed a wide range of topics, some of which overlap between the various research organisations. However, the philosophical approach adopted by each of these organisations has differed and the conclusions that have been drawn do not always agree. Experience has shown that there is no one absolutely correct way to design die casting dies and that several different designs could each yield successful results.

The main difference of opinion lies in the running and gating system:

In the Battelle approach the zinc die casting is divided into segments. The gate thickness at each segment is proportional to the metal volume required for that segment. The sprue and runner system, upstream from the gate, is designed so that the entire system will maintain constant total cross-sectional area throughout. This is aimed to eliminate flow problems that may occur in the sprue and in the sprue-to-runner transition region. The gate velocity was not considered in this study.

The BNF work included a large number of practical trials, concentrated on the gate thickness. Interchangeable gates were used to determine the effect of gate thickness on pressure transmission to the casting, on the pressure tightness of the casting and on the gate velocity. It was found that increasing the gate thickness increased the duration of pressure transmission, though a reduction in die temperature reduced the duration of pressure transmission. However, a change in gate thickness has little effect on casting porosity and consequently had little effect on pressure tightness. Lowering the die temperature produced a large improvement in pressure tightness.

Decreasing the gate thickness resulted in increased gate velocity.

In the CSIRO approach the gate thickness is small and uniform throughout and is selected in relation to a desired cavity-fill time constant with the maintenance of relatively high gate velocities. In this approach a tapered runner system is used such that the cross-sectional area of the runners decreases from the sprue to the gate, so that the sprue area will be 1.1–1.3 times the gate area.

The choice of design approach to be adopted in any specific case must be the decision of the die caster involved, but it is clear that better results can be obtained by an adherence to a system rather than depending on a trial-and-error approach.

Acknowledgement
We found the condensation of the original reports into a form suitable for the publication a difficult task because we have been so close to the work for many years. We are grateful for the help in this task of Mr. N. Steward of the Fulmer Research Institute.

References

1. Groeneveld, T. P., Kaiser, W. D., 'Die Casting Process Improvement', Progress Report No. 15, ILZRO Project ZM-132A, May 1977.
2. Groeneveld, T. P., Kaiser, W. D., 'Die Casting Process Improvement', Progress Report No. 16, ILZRO Project ZM-132A, November 1977.
3. Groeneveld, T. P., Kaiser, W. D., 'Die Casting Process Improvement', Progress Report No. 17, ILZRO Project ZM-132A, May 1978.
4. Booth, S. E., Allsop, D. F., Kennedy, D., 'Die Design for Pressure Diecasting', 1977 Annual Report, ILZRO Project ZM-204, December 1977.
5. Davis, A., 'Tapered Runners Feeding Thin Gates, 'Die Casting Engineer, 22, 4, July/August 1978.
6. Davis, A., 'The Injection Process in Die Casting', Diecasting Bulletin, The Society of Diecasting Engineers, Australia, October 1978.
7. Murray, M. T., Robinson, P. M., 'Control of Metallurgical Properties of Lightweight Castings', Progress Report No. 2, ILZRO Project ZM-253, November 1977.
8. Murray, M. T., Robinson, P. M., 'Control of Metallurgical Properties of Lightweight Castings', Progress Report No. 3, ILZRO Project ZM-253, May 1978.
9. Davis, A., Siauw, 'Fluid Flow in Gate and Runner Systems', Progress Report No. 2, ILZRO Project ZM-254, November 1977.
10. Davis, A., Siauw, 'Fluid Flow in Gate and Runner Systems', Progress Report No. 3, ILZRO Project ZM-254, May 1978.

Casting Techniques

Reviewing the Permanent Mold Process—Part I

In Part I of this review of the gravity permanent mold casting process for aluminum, the authors cover the basics of mold materials, gating and ejection systems as well as the thermal considerations of the process.

Glenn W. Stahl
Ken R. Whaler
Stahl Specialty Co
Kingsville, MO

Refinements of the gravity permanent mold process, particularly during the past ten years, are rendering it an increasingly attractive process for producing aluminum castings. Improvements in areas such as mold temperature control, automatic ladling, metal filtering and others have not only added to the productive capabilities of the process but also to its ability to produce mechanically sound parts. Following is a review of the gravity permanent mold process as well as some of the process and equipment improvements of the past few years.

Permanent mold casting of aluminum is, of course, not new. Molds of iron were developed about 1650 AD, while molds of limestone and baked clay were known to have been used thousands of years earlier. Since, in permanent mold casting, no force other than gravity is exerted on the molten metal, the process is also known as *gravity diecasting* in many parts of the world. But whether it is called permanent mold casting or gravity diecasting, the process is the same and is capable of producing high quality castings in volume.

Because the metal enters the mold gently and without turbulence and is cooled rapidly, sound dense castings with superior mechanical properties can be obtained even in heavy sections. Since the mold is normally made of metal and is relatively stable, the castings produced are quite uniform assuring a high degree of accuracy. This increased accuracy in many cases helps eliminate all or some of the machining that may be required on the part. The permanent mold process is also adaptable to producing a consistent quality finish on the castings. It also lends itself to the use of expendable cores and makes it possible to produce parts not currently practical through pressure diecasting.

Some possible disadvantages of the permanent mold casting technique include higher tooling costs than the sandcasting process, making it economically impractical on small quan-

Whether it is called permanent mold casting or gravity diecasting, it is a process capable of producing high quality castings in volume.

tity runs; surface finishes, although usually consistent, are not as smooth as can be obtained by the diecasting process; and extremely thin sections are difficult to obtain in permanent molding. Though the process is adaptable to most ferrous and nonferrous metals, many more permanent mold castings are produced from aluminum alloys than from any other metal.

PERMANENT MOLDS
Material Requirements
Before selecting a mold material, it is important to consider some of the requirements that the material should meet.

Heat Transfer—In the permanent mold process the rate of heat transfer in the mold itself is important because the quality of the casting is determined by progressive solidification so that the last metal to freeze is in the gates or risers. This is particularly true in permanent molding of aluminum alloys. Also in casting aluminum alloys it may not be feasible to build a mold out of a material that provides a very fast heat transfer rate since the metal may solidify before it completely fills the cavity. Conversely, a mold material with a very slow heat transfer rate may slow the solidification to where production is adversely affected and does not make the process economically feasible.

Erosion—A mold material's resistance to erosion or alloying is also an important factor. Because of aluminum's high affinity for many other metals, a mold material may actually alloy itself to the piece part, adversely affecting mold life.

Thermal Heat Shock—Mold materials should also possess good resistance to thermal heat shock. Since heat from the molten metal must be absorbed by the mold face before the casting can solidify, it is obvious that the surface of the mold approaches the molten metal temperature and then cools along with the casting. This cycle is repeated for each casting and eventually causes thermal heat checks in the mold face.

Serviceability—Serviceability is another important consideration in selecting a mold material. If the material is soft and easily damaged in normal handling or use, the life of the mold will be shortened. It is also important to consider whether a material is easily repairable by welding or other techniques to restore damaged areas.

Economy—Since the tool cost in permanent mold casting is an important consideration in justifying using the process, the cost of the raw material for the mold as well as the ease with which it may be fabricated are important factors. Some mold materials may be readily cast to shape, whereas others must be machined at considerable increase in cost. Even in cast-to-shape molds, some machining is involved, so machinability of the material is also important.

Considering all of these requirements for mold materials, it is obvious that no one material is ideal for all applications. The actual material selected may be dictated by the design of the casting to be produced, by the metal to be cast or by the facilities available for producing the mold.

Mold Materials
Cast Iron—More permanent molds in use today are made of some grade of cast iron than any other material. Cast iron has a usable heat transfer rate which is applicable for the casting of aluminum. Free carbon, in the form of graphite, in cast iron gives it a resistance to erosion from the molten metal in the casting process. Another strong factor in favor of cast iron is, in many cases, that it can be cast to finished configuration. Or, if this is not feasible, it can be cast with a machining allowance and machined quite readily.

One of the metal's weak points is that it is difficult to repair by welding or other means. It also lacks a high degree of resistance to thermal shock although it is relatively stable as far as thermal warping is concerned. Ductile iron and proprietary engineered gray irons are also used for molds. Many foundrymen feel, however, that the flake graphite structure in gray irons make them superior to ductile's spheroid graphite shapes. For this reason, compacted graphite iron's coral- or wormlike graphite structure and good strength may also make it a viable permanent mold material. Further investigation of CG iron will be required for an accurate determination of its applicability for permanent mold casting.

Carbon Steel—Low carbon steel has also been used to some extent for the building of permanent molds. It has the advantage of being quite low in cost; however, it is not readily cast to shape and it does not machine as easily as cast iron. It has a relatively high degree of resistance to heat shock but very little resistance to erosion from molten aluminum alloys.

A wide variety of casting shapes and sizes can be produced with the gravity permanent mold process.

Alloy Steel—So called high alloy steels containing a higher percentage of carbon and high chrome have been extensively used for permanent molds. They are not as readily cast to shape as the cast irons and are considerably more difficult to machine. These steels do have the advantage of a higher hardness for resisting normal wear and damage from handling. They also exhibit an extremely high resistance to thermal heat shock. Their major disadvantage is a relatively low resistance to erosion or alloying with aluminum.

Carbon—Carbon in the form of graphite is used in selected applications for permanent molds. The principal advantage of the carbon mold is its resistance to erosion by molten aluminum and heat shock. Disadvantages are that it is a soft material and has a tendency to be damaged easily. It can be readily machined into any machinable shape.

Beryllium Copper—Copper with the addition of approximately 2½% beryllium is becoming an important mold material, particularly in the permanent molding of brass and bronze alloys. It has the advantage of a rather high heat transfer rate where this characteristic is desirable. It can be easily cast to shape, but in many cases, it is more expensive than other materials because of the high cost of beryllium.

Other Metals—Other mold materials for consideration are the so-called refractory alloys. These are normally either tungsten or molybdenum or various combinations. Because of the cost and difficulty of fabricating these metals, they are presently being used only for portions of a mold where either resistance to thermal heat shock or high heat transfer rate is particularly desirable.

Physical Shape

Thickness—In considering the physical shape of a mold to produce a particular part, mold thickness becomes important. Because of the heat transfer from the molten metal to the permanent mold and the need to maintain a temperature gradient for progressive solidification of the casting, it is usually advisable to maintain a relatively uniform mold thickness throughout. This dictates a relieving of the back of the mold to roughly conform with the configuration of the cavities. This is very true on larger parts where there is considerable depth of shape to the part. Mold wall thicknesses may vary from 1-2½ in. depending upon the size of the mold as well as the thickness of the sections of the casting being produced. The large ma-jority of permanent molds are built with a wall thickness of 1¼-1¾ in.

Size—Determining how much larger the mold should be beyond the size of the piece part is an important consideration. This is obviously an extension of the parting surfaces. The mold needs to be large enough to allow adequate room for the gating system. Quite often this will be from 2-4 in. on the sides of the mold and somewhat less at the bottom since normally there are no gates into the bottom of the cavity.

In considering the space needed at the top of the mold, allowance for gates and risers should be kept in mind, along with the matter of *head*. Since the only pressure used to get the molten metal into the cavity is gravity, it is important to have molten metal in the gates sufficiently higher than the cavity to create this *head*. This could vary between an absolute minimum of 3 in. and as much as 7-9 in.

Warpage—Another aspect to consider in the overall shape of the mold blocks is that when molten metal is poured into the mold, the inside or the parting line surfaces of both halves of the mold need to start absorbing heat from the molten metal immediately. In the process, the inner or parting surfaces of the mold expand causing a thermal warpage which, in effect, opens the mold all the way around the outer edges.

In the past, some mold designers have attempted to counteract this tendency by building a thick rib or flange into the mold design all the way around the mold. Although this imparts considerable strength to a mold, it only adds to the problem since the back edge of the rib will remain cold and the front edge, near the parting line, will absorb heat and warpage will occur. Careful consideration of the problem will indicate that as the mold itself becomes thicker, it becomes more and more difficult to counteract this thermal warpage. This points then to a very important design concept: *Make the mold weak enough so that the machine is powerful enough to hold it together.* Through the use of intelligent design principles and the available hydraulic mold operating mechanisms, it is possible to overcome this thermal mold warping tendency and hold molds together tightly without external clamping.

Shrinkage Allowances—Since all aluminum alloys suffer volumetric shrinkage on solidification, the mold must be built somewhat larger to allow for this. Most alloys will shrink some 0.007 in./in. from a cold mold to a cold casting if this shrinkage is not retarded by a restriction in the shape of the mold. Cores or other mold configurations may reduce actual casting shrinkage to as little as 0.004 in./in. This amount of shrinkage occurs after the casting is removed from the mold.

PERMANENT MOLD GATING
Gating Requirements

In designing a gating system for a particular part, it is important to consider what is expected of the gating system.

Turbulence—The primary role of a gate is to introduce the molten metal into the mold cavity in such a way as not to create turbulence. Since one of the primary advantages of permanent mold castings is soundness, it is important to take full advantage of this capability. Since turbulence or entrapped air in molten metal is directly related to the speed of flow of that molten metal, it becomes a function of controlling speed to control turbulence. The gating system should not only fill the cavity with molten metal, but should fill it in a *tranquil* manner to minimize air entrapment. Flow from the gating system should be such that no turbulence is created *after* the metal enters the cavity.

Feed—The gating system must also *feed* the casting. Since most alloys used in permanent molding undergo volumetric shrinkage upon solidification, feeding becomes extremely important in producing a sound casting. The gating system, then, must include shrink bobs or risers that will adequately feed all sections of the casting.

Venting—Venting the mold is also a responsibility of the gating system. Since all materials presently being used to make molds are not permeable, the air which is in the cavity before pouring must escape before the molten metal can fill that cavity. In most cases a large percentage of this air must escape through the gating system.

Heat Control—The gating system is also one way of controlling mold temperature. In order to get adequate progressive solidification, it may be necessary to add heat to one area of the mold by using larger runners or risers. If, however, more heat is added than is necessary, the production cycle is adversely affected. Likewise, too much heat added to one spot in the mold can adversely affect mold life. This happens frequently when not enough ingates are used and too much metal flows over one point superheating the mold at that point and causing premature thermal heat checking.

Yield—A gating system must also produce an acceptable casting yield consistent with the economics of the particular part. Also closely related to

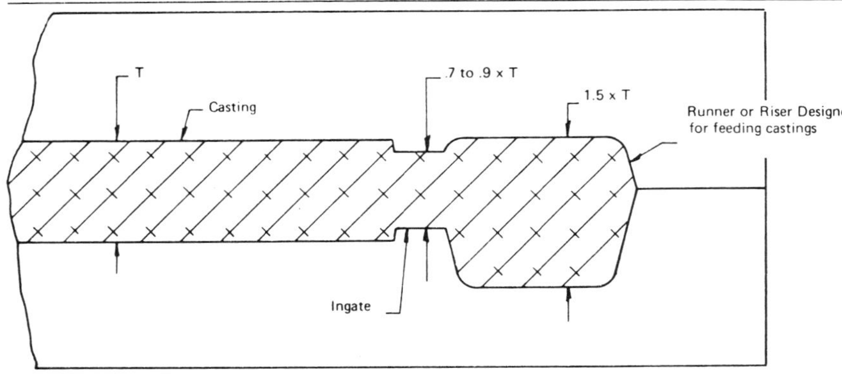

To allow proper feeding of the casting by the riser or shrink bob, each ingate section should be between 70-90% of the section of the casting that is being fed.

the economics of yield is the ease of removing the gate. If the gating system is designed to allow a break off operation or trimming with a punch press, the yield factor takes on less importance. An intelligent gating system design may even eliminate subsequent finishing operations after die trimming or break off. Total cost of gate removal and casting finishing should be considered when designing the gating.

Gating Systems
There are probably more variations among permanent mold foundrymen in their approach to gating than any other aspect of mold design. Most important, the gating system selected should achieve the already mentioned requirements. There are only three basic approaches to gating, although each has many variations. These three systems could be listed as top, bottom or side gating.

Top Gating—Top gating is the simplest and provides the distinct advantage of furnishing more yield or quantity of castings in relation to the amount of metal in the gating system. One problem involved with direct top gating is that it has a tendency to superheat the top of the mold since the metal enters at that point. This can adversely affect mold life since it causes thermal heat checking of the mold in that area. This superheating may also produce a hot spot or shrinkage in the casting near the top.

Top gating is also subject to turbulence problems because the metal cascades over the face of the mold and may entrap air. This simplified gating, in many cases, does not do a satisfactory job of allowing the escape of air in the mold cavity so that it may be replaced by metal. Venting problems may be more severe. A typical example of high production items using top gating in the permanent mold process is automotive pistons.

Bottom Gating—Permanent mold foundrymen have learned that by tak-

ing the metal down below the cavity and bringing it up from the bottom with a bottom or horn gate, much of the turbulence of top gating can be eliminated. With this method, casting yield may suffer because of the amount of metal required to fill the long runner system.

A basic disadvantage of the bottom gate is the fact that the hottest metal enters the cavity last at the bottom of the mold. This makes feeding of the casting from the gating system difficult since feeding is normally much easier to accomplish from above. Bottom gating also tends to superheat the mold at one particular spot which can adversely affect mold life.

Side Gating—The third basic method of gating, that is having the metal enter from the side, is by far the most popular and effective. Metal may enter from one or both sides of the mold cavity depending upon size of the part.

Side gating does not furnish a high yield because of the greater length of the runner system. Labor involved in removing these gates may be minimized through the use of a trim press where production quantities justify the cost of a trim die.

Side gating eliminates superheating of the mold at one spot since the metal enters at a number of areas. In fact, it is advisable in many cases to use a ribbon or continuous gate particularly where trimming will be done with a trim die. Side gating does allow for adequate feeding of the casting at any point along the side or top. Top feeding is accomplished through the use of a riser which is fed hot metal from the gating system. The advent of tilt pouring has allowed the side gating systems to be quite simple since the machine itself takes over the function of controlling the speed of the flow of the stream of molten metal, thereby eliminating turbulence.

Size Relationships
In any gating system, the relationship between the size of the various components and the casting itself is extremely important. The runner or downsprue must be sufficiently large to allow flow of metal to fill the cavity as required and may also be needed to fill the function of allowing air to escape. Any part of this runner system which is expected to serve as a feeder or shrink bob needs to be one and one-half to two times the section of the casting that is being fed.

The ingate actually serves two functions: first, it needs to admit the flow of

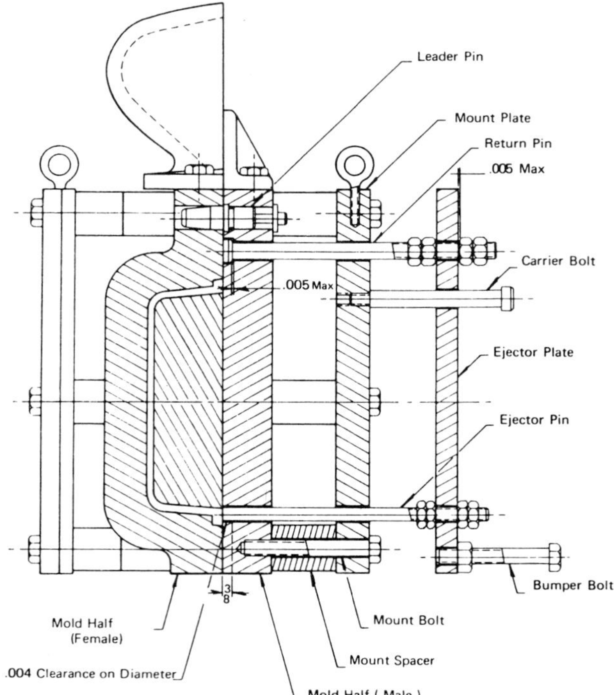

This cross sectional view shows the components of a typical permanent mold assembly with its components.

Removal of a casting from a permanent mold can be made more efficient through thoughtful design of the gating and ejection system.

molten metal gently enough so that cascading does not occur inside the cavity; and second, it must also allow the feeder or shrink bob to feed the casting. For this second function, each section needs to be somewhere between 70-90% of the section of the casting that is being fed. This normally is most important near the top of the die since most sections in the bottom of the casting are fed from hot metal coming in from above the casting itself. The ingate is usually tapered to approach the section of the runner and does not have a total length of more than two to five times its section thickness.

EJECTION SYSTEMS

Early in the development of the permanent mold process it was normal practice to *shake* the casting out of the cavity using some type of hand tool. As castings became more complicated this became more and more difficult and better ejection systems were devised to facilitate casting removal.

Pullback—One early method of making the casting easier to remove was to fabricate the mold in such a way that one section pulled back through a plate. This was the section that normally caused the hanging problem. A typical example of this was in production of cookware where the male plug that produced the inside of a pot was pulled back before mold opening to allow for easy removal. Automotive pistons are also ejected in this way, but here the core removed is a five piece collapsible type.

Ejector Pins—Because of the mold building and maintenance problems with the pullback technique, most foundries now use an ejector pin system. These ejector pins are arranged so that after the mold is opened they may be actuated and literally push the male half off the mold. Although a separate hydraulic cylinder or other device can be used for this ejection movement, most equipment presently in use uses the last part of the stroke of the same hydraulic cylinder that opens the mold.

The accompanying figure shows the components of a typical permanent mold with its components. The return pin has the function of returning the ejector plate to its correct position before the next cycle. The head of the return pin strikes the face of the fixed half of the mold at some point other than in the gating system. The purpose of the head on the return pin is to retain the ejector plate so that it does not move too far.

The bumper bolt strikes a fixed plate on the permanent mold machine and causes the ejector plate to stop its normal movement with the movable half of the mold ejecting the casting. The ejector pin has the function of actually pushing the casting off of the male half of the mold. Since this pin must move back and forth through a hole in the mold at elevated temperatures, clearance is very important. Too much looseness between the ejector pin and the mold will create flash and cause serious problems. Not enough clearance will cause the pin to seize.

Experience at Stahl has shown that to prevent this seizing, allowing the ejector pins to float is preferable to a fixed ejector system used by many diecasters. Clearance should be on the order of 0.004-0.006 in. in the mold. The carrier bolt is provided to simply carry the weight of the ejector plate so it is not hanging on the ejector return pin system causing a binding action.

MOLD MOUNTING

In order to provide for easy changing of molds, most permanent molders are using some attachment to the mold itself to facilitate mold changing. This may be accomplished through the use of mold mount plates which readily adapt to the machine. This may also be accomplished through some type of rail mounted on the molds. A rail system is more feasible in the case of large molds. Lifting eyes are also provided in the mold to facilitate rapid mold changing.

THERMAL CONSIDERATIONS

Warpage—In the permanent mold process, heat from the molten metal is applied only to the inner face of the mold. Most of this heat must be dissipated by the back side of the mold. This means that the inner face of the mold tends to be much hotter than the back. Because of a thermal expansion of the mold material, the mold wants to warp or curve. The function of the mold operating machine is to overcome this tendency.

To do this, mold must be designed in such a way that the machine can perform this function. The mold then should be *weak* enough so that the machine is powerful enough to hold the mold together. Obviously the pressure exerted by the machine needs to be around the outside of the mold and not concentrated in the center.

Expansion—Another serious problem in permanent molding is in the dimensional relationship between the ejector half of the mold and the ejector plate. While the ejector mold half could nominally run at 700F, the ejector plate might normally operate at 200F. Since the coefficient of thermal expansion of both steel and iron is approximately 0.000006 in./in. per degree Fahrenheit, the thermal expansion of the mold and ejector plate may be quite different.

In fact, this difference would amount to 0.003 in./in. of size. So, if two of the ejector pins were both 10 in. apart when the mold was built at room temperature, the ejector pin holes in the mold would be almost 1/32 in. farther apart than the ejector pin holes in the ejector plate. For this reason it is important to have ample clearance for ejector pins in the ejector plate which allows them to *float* and prevents binding.

Reviewing the Permanent Mold Process—Part II

In the concluding section of this review of the gravity permanent mold process, the authors present information on mold types and coatings as well as how the process is being automated and improved.

Glen W. Stahl
Ken R. Whaler
Stahl Specialty Co
Kingsville, MO

MOLD TYPES

A variety of mold types are and have been used for the permanent mold process. These include manual and mechanically–operated molds and range from book molds to sophisticated hydraulic systems.

Book Molds—Originally all permanent molds were operated manually. In order to help the machine operator, hinges were added as well as handles and latching devices creating a simple book-type mold. This type of mold is still used for producing test bars and some short run jobs. It has the advantage of being simple to build, but it is not feasible for deep draw parts because of draft problems in pulling the casting out of the cavity.

Rack and Pinion—The next stage of mold development was a table or base mounted device in which at least one of the mold halves slides on a base or rail for guiding purposes and is moved by a rack and pinion with a capstan to give the operator leverage. As molds became larger it was also necessary to provide some type of clamps to hold the edges of the mold together. These were also operated manually.

Air Cylinders—In order to further ease operator effort, many shops went to one or more air cylinders for furnishing clamping as well as ejection force. A major disadvantage of air cylinders is that to get enough clamping force it is necessary to use extremely large cylinders. Also air does not lend itself to any easy means of speed control and molds can be damaged by the hammering action of opening and closing.

Screw Molds—Molds can also be actuated by the use of mechanical leverage through a screw driven by some power source such as an electric motor. The screw technique has the advantage of being able to develop adequate clamping forces with small horsepower requirements. The disadvantage is in getting necessary speeds where desired.

Hydraulic—More molds are presently being actuated by the use of hydraulic cylinders than by any other means. Hydraulic power offers the advantages of good control and the ability to develop desired speeds as well as adequate clamping forces.

AUTOMATION

Our modern industrial society is based upon the concept of more productivity for the amount of human energy expended. This is more true today than ever before. Through the use of hydraulic forces and the application of modern techniques and electronic controls, it is possible to achieve a relatively high degree of automation in permanent mold casting.

Cycling—One permanent mold function that can be easily controlled is the casting cycle. This is accomplished with the use of industrial timers and allows a high degree of repeatability ensuring a uniform quality level and maintaining desired mold temperatures.

Ejection—Most modern permanent mold equipment also makes use of the cylinder that actuates the movable half of the mold to actuate the ejection system either automatically or semi-automatically at least at the end of the opening stroke.

Mold Changing—Progress has also been made in automating the permanent mold casting industry through various techniques making rapid interchangeability of molds practical. Changing of molds may be dictated by the need to go from one part to another in a jobbing foundry without losing machine time. In high production operations, it may be used where extra molds are provided so that a spare mold may be heated and coated and replace a mold which needs repair of the mold coating or other maintenance. This mold interchangeability should never take more than 15 minutes and, in certain applications, can be done in as little as two minutes.

Pouring—Early in the development of the permanent mold casting technique it was apparent that the skill of the operator in pouring has a definite effect on the quality of the casting. To ensure repeatable accuracy as well as dependable results, the permanent mold machine, itself, has in many ways taken over this function.

One approach to this problem has been to merely tilt the ladle by mechanical or hydraulic forces so that the rate of pour is uniform for each casting. The next step is to attach the ladle to the mold and tilt both ladle and mold. This allows for a simplified gating system because it is easier to control the speed of flow of the stream of molten metal. Since speed creates turbulence and turbulence, or entrapped air, is a direct function of speed, it is apparent that this is a prime advantage in making quality castings.

Programming the Process—The next step in automating the pouring function is to not only control the speed of mold tilt, but also to program this movement. This means the mold may start tilting very slowly and then speed up as the cavity is filled or vice versa. The part design, size, etc being pro-

Book-type permanent molds, such as this one, are still used primarily for producing test bars and in some cases for short run casting work.

duced dictates the programming of the tilt cycle.

Automating the pouring cycle one step further is to automatically transfer the metal from the furnace to the pour cup on the mold. This is currently being done with equipment which has a ladle on one end which dips into the furnace bath then, in synchronization with the permanent mold machine cycle, is lifted with a crank-type mechanism which allows the metal to flow through a pipe-type, refractory launder and into the pour cup of the mold machine.

Accuracy—The accuracy of mold movement of any operating device is extremely important because as the movable mold half moves away from the cavity with a casting in place, there is intimate contact between the hot casting and the walls of the fixed mold half. Any looseness or lost motion in the guiding device on the movable section of the machine will cause a rubbing action between the casting and the mold walls.

This tends to rapidly deteriorate the mold coating in this area and soon results in drag on the casting. The obvious solution to this is a permanent mold machine with preloaded ways to eliminate any lost motion or looseness. If this can be accomplished, higher production will result because mold coatings need not be touched up frequently nor molds removed for recoating. Also it will result in a better casting finish.

Multi-Station Systems—By putting these various elements of automation together in a system using multi-stations, the result can be improved productivity. Currently, thousands of castings are being produced on a daily basis through system automation, making gravity permanent molding in-creasingly attractive for long run casting work.

MOLD COATINGS
Coating Requirements
Since some type of mold coating or dressing is always used in permanent molding of aluminum alloys, perhaps a logical approach in discussing mold coatings is to consider the various purposes to be accomplished.

Insulation—One of the first considerations of mold coatings is insulation. Because molds are normally made of metal with a relatively high degree of thermal conductivity, it is necessary in many cases to use an insulating coating between the molten stream of aluminum and the mold itself. This is necessary to keep the aluminum from freezing before it reaches all areas of the cavity. Variations in thickness of this insulation provide for *progressive solidification* ensuring a sound casting.

Isolation—Another need for a mold coating could be called isolation. Because aluminum in the molten state has a high affinity for almost any other metal, it is necessary to have some barrier between the molten stream and the metal of the mold to prevent the aluminum from picking up metals from the mold. This action is often called erosion or washing. Without some means of preventing this, molds would have very limited life.

Venting—Another important consideration in permanent mold casting aluminum alloys is the need for venting by the mold coating. Since the mold does not have any degree of permeability, that is air cannot pass through it, and since the molten aluminum is actually entering the mold cavity inside a container or sack of aluminum oxide, it is necessary to evacuate the air in the cavity before complete filling can occur. The rough-

Automating the pouring cycle can also add a dimension of control to the gravity permanent mold process. With the system shown here, a ladle is dipped into the furnace and in cycle with mold machine tilts allowing the molten metal flow from the ladle to the pour cup of the mold machine.

ness or textured surface of the mold coating material must then support this film of aluminum oxide in such a way that there are tiny channels for evacuating air between the oxide film and the mold itself.

Lubrication—Still another coating function is lubrication. Since all aluminum alloys shrink as they go through the solidification stage, a casting may grip very tightly to certain areas in the mold where there is a minimum amount of draft. A coating that has lubricating properties will allow for release of the casting in these areas.

Durability—Durability is another important factor to consider in mold coatings. Because it is not economically feasible, in most cases, to use any type of mold dressing for each casting, coatings should be able to adhere to the mold and not flake off making continual repairs necessary.

Coating Materials

Many materials are being used to achieve the above mentioned coating requirements. Some of the materials used for insulation include vermiculite, kaolin or clay, whiting, talc and bentonite. These materials have been listed in their approximate order of effectiveness for insulating purposes. Talc and bentonite are also somewhat effective for mold lubricating.

By far the most effective material for lubrication is a colloidal form of graphite. Any of the above mentioned materials will furnish a satisfactory degree of isolation for preventing erosion when applied in sufficient thicknesses and the coating maintained properly. The venting function of mold coating can best be controlled by the coarseness of basic material and the application technique. As far as durability is concerned, kaolin, talc and bentonite tend to be more durable since they are harder materials than vermiculite.

In any case, these materials must be used with a binder or cement to create adherence to the mold. This is accomplished through the use of sodium silicate (water glass) in various proportions with the addition of water to the proper consistency to make applying it possible. Commercially available coatings using the above mentioned materials are usually more consistent than shop mixed materials. Some newer mold coatings also contain refractory components to provide extended life, helping to reduce machine downtime for recoating.

Coating Application

Usually coatings are applied to the core and mold cavity parts of the mold by spraying. Many foundries used to use a simple type of aspirator spraying device. It has been demonstrated, however, that a high quality commercial gun made for applying paint will accomplish better results in the hands of a less skilled operator. It is even possible to use a brush in applying the mold coating in the gates and runners where finish requirements are not critical.

Dipping removable cores into a graphite-based coating may also be satisfactory. This action will also help to cool the core. Dipping is commonly used for the mold itself when permanent mold casting brass and bronze alloys. Here again, the two functions of cooling and coating are accomplished.

MOLD TEMPERATURE CONTROL

Just as it is important to control the heat content in the aluminum alloy itself during permanent mold casting, equally important is controlling the heat of the mold itself.

Preheating—Because molten aluminum poured into a mold at room temperature would freeze before filling the cavity, it is normal practice to preheat the molds before starting operations. This is generally done by using a gas burner with a configuration that will uniformly heat the entire mold. Although most foundries heat the face of the mold itself, some operators have found they get better mold life by heating the back of the mold. On large molds it is extremely important to heat the mold as evenly as possible because it is possible to break a cast iron mold by heating unevenly.

Mold Temperature Control—Once heated to desired temperature, it is extremely important to maintain that temperature. After the mold has been preheated and coated with the proper materials, mold temperature can be maintained by two factors. One of these is the temperature of the metal being poured into the mold. This needs to be automatically controlled and should not vary more than 50F. The other factor controlling the mold temperature is the cycle or the time interval from one casting till the next. Mold operating temperatures are determined empirically and may range from 550-930F. A thermocouple inserted in the back of the mold attached to a digital readout meter provides an excellent means of monitoring mold temperatures.

Whereas previously it was believed that a thermocouple was not an effective means for monitoring mold temperature, recent experience has shown it can be used for this purpose when located properly in the mold. When determining its placement, design of the mold must be considered because it must allow for rapid disconnect and insertion during mold changing. Ideally, it should be located near the middle of the mold and usually on the female or stationary mold half.

The thermocouple should be inserted so that it comes within ½ in. of the mold cavity. Because the thermocouple expands, it is necessary to drill the hole where it will be inserted slightly larger to allow for removal. In some cases, the thermocouple can be located in a heavy section or a boss and the temperature controlled based on the temperature of that spot. Spring

It has been shown that using a high quality gun made for applying paint will produce better mold coating results even in the hands of a less skilled operator.

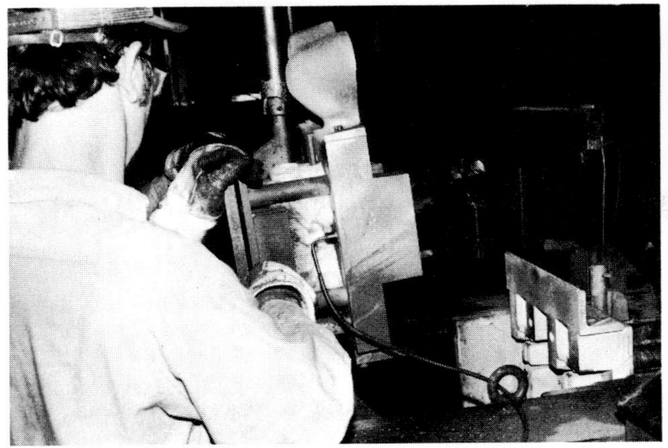

When properly located in the mold (left) a thermocouple attached to a readout meter can be helpful in controlling mold temperature. A spring loaded thermocouple, like the one shown at right, are good for this application.

loaded thermocouples do a good job for this application because they hold the probe tightly against the mold.

Trimming—After the mold has been preheated, coated and placed into operating on a time cycle, it begins to stabilize. This normally is accomplished after 10-15 cycles. Then, and only then, is it practical to begin to trim the mold coating in order to achieve progressive solidification to get a high quality, dense casting by the permanent mold process.

This trimming is accomplished by the removal or addition of mold coating to various areas so the casting starts the solidification process at a point farthest away from the gating and feeding system. It then progresses toward the feeders until the shrinkage takes place in the risers or gating system and not in the casting.

Cooling—If progressive solidification is not possible through trimming the mold coating, it may then be necessary to use exernal cooling in certain spots on the mold. This may be accomplished with the use of an air blast from either compressed air or from a blower system directed toward a particular spot on the mold.

Other techniques involve the use of a compressed air and water spray through specially designed nozzles. This spray can be of such an amount that all of the water actually evaporates on the mold and does not create a drain problem. If more cooling is required, water passages can be built into the mold. A typical application of this type of cooling is in the production of automotive pistons. Virtually all automotive pistons produced in the U.S. are made by the permanent mold process using water–cooled molds.

Chills—Other cooling techniques involve the use of chills. This is achieved by using a plug or insert made of a metal with a higher degree of thermal conductivity than the mold. Some materials that are used include copper which conducts heat rapidly but tends to oxidize quickly, providing itself with an insulating shield. More durable materials for this are tungsten and molybdenum or various alloys of these metals.

Adding Heat—Another technique used to assist in obtaining progressive solidification is the use of external heat on the mold. This can be by means of a gas flame or electric resistance heater to add heat to a specific spot. Because of the inconvenience of such an arrangement, it has largely lost favor.

Replacing this technique is the use of insulating material such as rock wool on the back of the mold to retain heat in a specific spot and prevent the mold from losing its heat. Another source of heat to obtain desired results comes from modifying the gating system. Normally, it is much more convenient to add heat to the mold by making the gating system larger, allowing the molten metal to provide heat where it is needed.

MOLD CLEANING & MAINTENANCE

Although it would be desirable to get as many cycles or castings as possible from one application of mold coating, no presently available material can be considered to be really permanent. Because of the very nature of the insulating materials which are relatively soft, they are subject to wear. This, together with finish problems caused by continual patching or touching up of the coating, has made it normal practice to clean off all the mold coating and replace it periodically. This time interval may be as short as one or two eight hour shifts for complicated molds or as long as 15 shifts on simple molds where finish requirements are not critical.

Cleaning—Mold cleaning can be accomplished in some cases with a wire brush either by hand or with a power tool. This method has the disadvantage of taking considerable time and also has a tendency to damage or round off any sharp edges on the mold. Most permanent mold foundries use some type of air blasting technique. Various materials may be used such as sand, aluminum chips or walnut shells. The prime consideration in cleaning a mold is to remove the old coating in as short a time as possible without removing any of the mold material or the oxide film that has formed on the mold. To do this, it is logical to use a material that is softer than the mold.

Hand Work—After removing most of the material with a blasting or brushing technique, there is careful handwork to be done to remove all chips or drags of aluminum that have soldered to the mold itself. This requires a high degree of skill. In most cases, when there is a buildup of aluminum clinging to the mold after the blasting operations, it indicates a lack of draft in that particular area. The operator should then carefully check to verify this condition and remove not only the aluminum but the source of the problem by assuring that that particular point actually does have sufficient draft.

Checking—While the mold is in this condition, this is the time to check for any maintenance required on the ejector pins, return pins, cores, etc. All ejector pins should float freely in their holes and should be checked for replacement because of rounded ends.

A skilled maintenance person should also check to be sure there are no cores or other areas of the mold that

This method of filtering molten aluminum uses a crucible submerged in the metal bath which has a filter inserted in its bottom. This allows the metal to rise up into the crucible filtering out both dross and oxides.

are preventing the mold from closing fully. This could result in dimensional change on the part or could cause the mold to leak molten metal. Any areas of the mold that are showing erosion or thermal heat checking problems should also be observed at this time.

Repair—If the mold is made of iron and requires repair or alteration, careful consideration should be given to inserting a new piece of iron if it is feasible. If an insert is not practical, welding is normally considered the last resort. Welding is not always used because of the difficulty in getting satisfactory service from a weld in an iron permanent mold. Probably the most successful technique, if welding is required, is the use of spray welding using powdered alloys melted and deposited by an oxyacetylene flame.

INSERTS & CORES

Where casting design indicates the use of inserts, the permanent mold process lends itself well to this application. Inserts can be made of any of the common alloys including aluminum. When using an aluminum insert in aluminum permanent molding, there is normally no metallurgical bonding, only a mechanical bond. Therefore, in the use of aluminum inserts as well as for brass, bronze, steel or other materials, it is necessary to provide some knurling or other means to create a mechanical bond between the casting and insert.

Various techniques are used to hold the inserts in place during casting. For cylindrical inserts such as bearings or threaded inserts, a simple core or pin is all that is needed. If the insert is of a magnetic material, such as steel, an alnico permanent magnet may be embedded in the mold to retain the insert during casting.

When disposable sand cores are used, the process is normally referred to as semi-permanent molding. Nearly any of the coremaking processes in use today are applicable to aluminum permanent mold casting, including shell, coldbox and the nobakes. One process that appears to hold good potential for use in aluminum casting is the furan/peroxide coldbox process (SO_2) because of its good bench life and excellent shakeout characteristics.

TRENDS IN ALUMINUM CASTING

Although many of the trends in process metallurgy and metal filtration are applicable to any aluminum casting process, permanent mold foundrymen can also benefit from the newer techniques. This is particularly true as demands for higher quality castings with predictable mechanical properties increase. Improved process controls are also becoming increasingly important.

Some metallurgical techniques which are gaining prominence include silicon modification with strontium and grain refinement with titanium-boron. Both processes have been proven effective for producing sound, dense castings with superior mechanical properties by improving the feeding of aluminum alloys during casting.

Degassing of molten aluminum is also becoming better understood by operating foundrymen. The secret to hydrogen gas in aluminum is controlling it, not necessarily eliminating it. It has been shown that for nearly every individual casting, there is an optimum level of gas, where the number of casting rejects increases as the gas level either goes above or below that optimum level. It is the responsibility of the foundryman to find that optimum level of gas for the castings being produced and then effectively control and test for it.

The heat treating area of aluminum casting production also may hold good potential for improving casting properties, economically. Work is currently in progress to determine if a shorter heat treat cycle can meet specified mechanical properties produced by the cast, quench and aging process (CQA). For example, this cycle consists of quenching a part cast in A-356 aluminum alloy in water immediately out of the mold and then aging it at 310F for approximately six hours. Preliminary findings indicate that this process can improve both ultimate tensile strengths and percent elongation over that of as-cast or T51 heat treated castings. For this CQA cycle to be viable, though, the casting must be sound and poured with exceptionally clean, modified and grain refined metal.

Possibly the major area of change in aluminum casting is in molten metal filtering. One filtering technique in use at Stahl uses a crucible submerged in the molten aluminum bath of the melting furnace. The crucible has a ceramic open pore filter fitted in its bottom which allows the molten aluminum to rise through it, helping to remove dross and oxides.

Possibly an even more dramatic breakthrough in filtering metal is that some secondary smelters are now offering filtered secondary aluminum. This secondary ingot, though obviously not as pure as primary ingot, is a very clean and dross-free material. It should be noted that even after filtering secondary metal, it in all likelihood, will contain higher levels of tramps elements like copper, iron, manganese, etc. Though these elements may be undesirable in some cases, they can help improve the machinability of aluminum alloys as they offer additional Brinell hardness to the part.

These controls as well as other refinements continue to add to the viability of the gravity permanent mold casting process for producing sound, high quality aluminum components. Any casting requiring good mechanical properties and soundness should be considered a candidate for the permanent mold process.

Precision casting

G. Richards M.I.M. C.Eng.

A. INTRODUCTION

Historically, precision casting dates back to 1600 BC (Shang Dynasty) when the lost wax process was first known to be used. However, it has been in the last 30 years that the greatest developments in the techniques of precision casting have taken place. Naturally, this paper centres around the activities which have been undertaken by Sterling Metals. The following brief resume of the progress made by them is typical of that which has taken place both in this country and in America.

The late 1940s saw the introduction of the Antioch plaster casting process, mainly for the purpose of producing tyre-moulds. At the same time, keen interest was shown by the company in the Croning shell-mould process, and, like many others, the firm investigated its possible application and found outlets for the more simple type of aircraft castings.

More recently, investment techniques have been introduced for producing precision castings in light alloys (including magnesium and aluminium alloys) and steels (including alloy and tool steels, stainless steel, nickel chrome, cobalt chrome, copper base alloys, etc.).

In the intervening years, Sterling Metals have produced about 7,000 different designs of castings, by variations and combinations of these and other processes. At present, about 500 different types of castings are concurrently produced per week, from rubber ball moulds for the toy trade to intricate microwave components for the electronics industry. Castings of all shapes and sizes are made, ranging from a few grams to 1,600 lbs.

Before going any further, perhaps one ought to attempt to define the misleading term "precision casting". It is not claimed to be precise in comparison with machine shop interpretation. It is however, claimed to be precise compared with other casting processes, hence the name. A close tolerance is considered to be eg. $\pm$ 0·003in, whereas an engineer considers it somewhat finer, perhaps a tenth of this, but a sand foundryman somewhat coarser, eg. $\pm$ 0·030in to $+ \frac{1}{8}$in.

Mr. Richards is Manager of the Precision Foundry at Sterling Metals Ltd.   (F 1287)

Scope

Techniques of precision casting have been applied to meet requirements of complexity and accuracy which would not have been possible using conventional methods. No single casting process produces the answer to all the problems which arise in the manufacture of castings: therefore, established foundry production methods such as green—or dry—sand gravity—or pressure—die casting will always be in demand.

The various precision processes have been applied to augment and widen the scope of the foundry industry in general, to enter new fields, and to retain work which could be lost to other new methods of metal forming, such as tape-controlled milling from solid, electro-chemical machining, precision forgings, electron-beam welding, dip brazing, electro-forming, spark-erosion, etc. The processes outlined in this paper probably offer the most suitable methods of producing precision castings.

Pressure die castings are, of course, precision castings, but these cater for a specialised high-volume, commercial quality field.

B. PROCESSES

The processes described below are the Croning shell, plaster and investment; these will be sub-divided into more specialised branches in due course.

Croning shell

As with any other foundry process, the starting point is a pattern. In this case the pattern is metal and in two halves, each half being attached to a heated metal pattern plate. Sand, coated with a thermo-setting binder, is poured on to the plate and allowed to "cook" for a few minutes. The plate is then inverted so as to remove excess sand. This leaves an easily removable shell of baked sand approximately $\frac{3}{8}$in thick on the pattern plates. Cores, where required, are manufactured in a similar manner, the metal corebox being heated in an oven. Shells manufactured from both halves of the pattern in this way are then bolted together and cast.

Possibly the greatest advantages offered by this process are that dimensions remain consistent from casting to casting, the metallurgical characteristics of the castings are

47

good, and very intricate parts can be produced by semi-skilled labour. In cases where dimensional consistency and metallurgical requirements are of greater importance than close specific tolerances, or fine surface finish, then Croning is preferred to plaster or investment, especially in thick sections, due to the more favourable thermal characteristics of the mould and core materials.

Croning shells can be used to make aluminium, magnesium and steel castings though, in the case of magnesium, special inhibitors are necessary in order to prevent mould/metal reaction.

The thermo-setting binder is normally a phenolic resin, although for alloys exhibiting hot-shortness, a urea binder can be used which is less rigid. But, due to this inherent friability, the castings produced tend to be less accurate and possess a rougher surface finish.

Various combinations and permutations of sands and proprietary binders will give high shell strengths, anti-cracking, shock resistance, and limited mould reaction properties.

Plaster

The patterns used in this process can be made of wood, resin, rubber (to facilitate moulding of undercuts and re-entrant angles), and bronze (for close accuracy).

Two main varieties of plaster process are used by the author's foundry, Antioch, and foam-plaster, each having its merits and shortcomings. As a rule, the plaster process is only used in the manufacture of aluminium castings, although thin-walled magnesium castings and some copper castings have been manufactured experimentally.

Antioch process

This was developed in America in the 1930s during work to find a more permeable plaster mould for bronze art-castings. After a while it was realised that it had a strong engineering potential, and it was developed to manufacture aluminium tyre moulds, impellers and various other types of castings.

A slurry consisting of a filled gypsum plaster and water is poured on to a pattern in a moulding box or into a corebox and allowed to set. After setting, the moulds and cores are steam autoclaved for 8–10 hrs. during which time the plaster begins to granulate and become more permeable. They are then removed, dipped in water and allowed to stand for 10–14 hrs. allowing granulation to proceed. This produces a plaster in which the interior of the mould mass is granular, having a grain size similar to that of sand, whereas the surface, which does not undergo the transformation, remains smooth. The interior grains have sufficient inter-granular strength at points of contact for the mass to retain its integrity whilst exhibiting a high permeability (25–80 AFS). In practice the permeability is held at 25–40 AFS.

The moulds and cores are then oven-dried, after which they can be assembled and cast.

Offsetting the advantage of the high permeability of this plaster is the disadvantage of prolonged steam-autoclaving. This is expensive and also tends to erode the surface of the plaster thus detracting from its otherwise excellent surface finish.

Foam-plaster technique

This technique gives high permeability of the Antioch process coupled with the rapidity and low cost of a straight-drying plaster. The procedure is as follows:—

A small quantity of foaming agent is added to water and

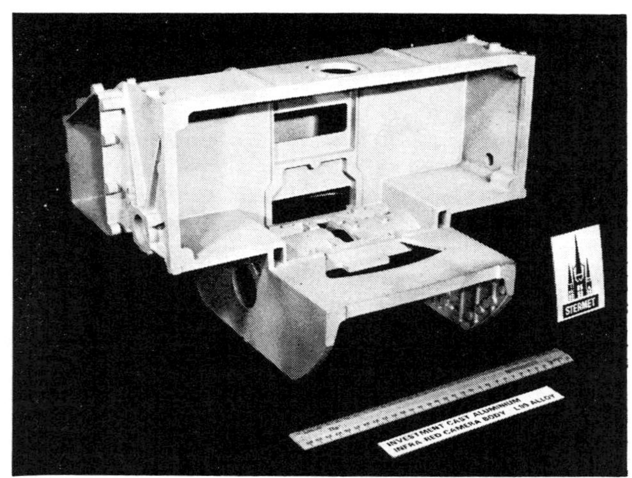

whisked. The plaster is then added and foaming continued until the desired density is achieved, whereupon the slurry is poured into the moulds and coreboxes and allowed to set.

After stripping, the moulds and cores are oven-dried for several hours. This produces a plaster with a permeable cellular structure having a smooth dense skin appoximately 0·020 in thick, which imparts a superb finish to the casting. The moulds and cores are assembled and cast in the normal manner.

The main advantage of foam plaster is that the raw material is comparatively cheaper and, due to the cellular nature, a greater bulk of material is produced for a given weight of plaster. Moulds and cores with high perme-ability can be produced without autoclaving. The moulds and cores are light and rigid, making them easy to handle (a distinct advantage with large cores). Unfortunately, only simple shapes can be produced since the plaster actually expands on setting which causes locking on deep draws, resulting in damage to the very fragile skins during stripping. Coreboxes for this type of plaster are most suitably made from rubber or designed so that the entire corebox can be dismantled.

The merits of the Plaster process include good surface finish and dimensional accuracy. The major disadvantage of the process is that it promotes slow cooling of the castings resulting in poor properties. Also, it is a labour intensive process.

Investment

Investment casting is distinguished from other techniques by virtue of the type of pattern used. Whereas conventional processes employ a repeatedly used permanent pattern, investment casting utilises a disposable pattern, normally wax, although low melting alloys, salts, plastics and even frozen mercury have been used. Wax, however, is the most common pattern material—hence the name "lost wax process".

The pattern is virtually an exact replica of the desired casting shape and is produced by injecting semi-solid wax into a die, after which the die is stripped and the wax removed. If a hollow pattern is required, it may be cored with a soluble wax. Several of these pattern waxes are attached or gated to a sprue so as to form a cluster or tree. In the case of large patterns, wax runners, risers, etc. are attached to the pattern. The assembly is then ready for investing. Investing here implies completely surrounding or "clothing" the patterns with mould material. Two methods of investing are currently commonly used:—

Block mould

The cluster is placed on a plate and surrounded by a can into which is poured the moulding slurry under vacuum. The slurry in this case is similar in consistency and colour to that used in the plaster process, but is somewhat more complex in composition, consisting of water and a blend of plaster, cristobalite, various fillers, stabilising ingredients and additions to aid permeability. Vacuum pouring is used partly to remove air from the slurry and partly to ensure complete filling of complex cores, etc. in the waxes.

The moulds are then baked at up to 650°C for 17–48 hrs. depending on the geometry and size of moulds (an operation which also removes the wax) after which they are cast, whilst still hot, pressure or vacuum assistance being utilised during casting. Aluminium, magnesium and copper alloys can be cast into block moulds.

Ceramic shell

In this case the mould (or ceramic shell) is formed by dipping the cluster into a slurry consisting of a ceramic flour in colloidal silica and then sprinkling it with a fine sand (generally known as stucco). After drying, the cluster is dipped into another ceramic slurry, thence into a fluidised bed containing granular molochite thus stuccoing the surface again. This second sequence, i.e. ceramic slurry dip and molochite stucco is repeated until the desired thickness o shell (usually 5–6 coats) is achieved, drying between each coat. Some mechanisation of this process is frequently employed, so that larger moulds can be produced, and also several moulds produced simultaneously. The invested wax is removed either by shock-firing, or steam autoclaving: the shell is then fired at 900–1000°C for a minimum of 2 hrs. after which it is cast whilst still hot .Most ferrous and non-ferrous alloys can be cast into ceramic shells.

The main advantages of investment casting include accuracy, good surface finish, reproduction of detail, versatility of alloy, suitability for short and long runs, ideal for very intricate components and also where no draft is permissible. The particular advantages of ceramic shell over block mould include better thermal characteristics (and, therefore, better properties from castings), reduced fettling costs, and more economical production. Block mould, on the other hand, is preferred where castings include very thin walls (less than 0·060in) and intricate coring.

Process combinations

It can be readily seen that several of these processes can be integrated so as to take advantage of the beneficial properties of each process. One example is to use a plaster core and CO_2 sand moulds in order to produce waveguides where internal surface finish and accuracy are critical, but the external surface finish is completely unimportant and is, therefore, manufactured as cheaply as possible.

In some cases the CO_2 sand mould can be replaced by a Croning shell mould where greater complexity and/or accuracy of the outer surface is necessary. Similarly, torque convertor castings may be manufactured using plaster vane cores (for high accuracy and surface finish) a shell base core for accurate vane core location, a CO_2 cope for cheapness, and a die ring to promote directional solidification. If production quantities warrant it, the individual plaster vane cores may be replaced by a one-piece core plus a die cope and drag. Shrouded impellers can be made by similar

techniques. In extreme cases up to 5 or 6 processes may be combined where a very complex casting is required.

C. FETTLING

Normally this entails removing risers, gates, runners, flash, etc. from the casting; however, contrary to most expectations, the finishing and checking operations after knock-out often bear the major proportion of the production cost. This may seem surprising since the castings are supposed to be close tolerance castings and as such should need only light fettling. In the case of a precision casting, however, minor surface irregularities—normally masked on a sand casting—are more serious. This means that inspection standards are higher than normal, resulting in a higher standard of work and more careful correction by the fettling shop. In fact, specialised personnel must be trained to work to close tolerances and high degrees of surface finish, usually on castings of a fragile nature.

Closely allied to this is the problem of locations, or pick-up points for machining. As may be expected, the production engineer or machinist does not often have detailed knowledge of foundry techniques and only considers the casting from the point of view of successive machining operations. This is alright for simple shapes, but for cast revolving parts, such as torque convertors which have to be dynamically balanced, this can lead to trouble, because the machinist may ignore consistently cast surfaces.

Hence the foundry is often obliged to machine datum or location surfaces on the castings true to torous surfaces (hubs) or bowl shapes to enable the machinist to finish machining the castings, thus saving scrap. For example, when production of turbo-charger impeller wheels commenced, it was found that, on average, at least 20% would not balance dynamically. After introducing datum locations, machined on by specialists who can split errors in manufacture, the scrap on dynamic balancing was reduced to less than 1%. This, of course, represents a cost saving.

These machining operations can also be used by the founder to fettle and remove casting flash, which might otherwise be an expensive operation.

D. TOOLING

The pattern equipment, as is to be expected with close tolerance work, is of prime importance, not only from the constructional aspect, but also from the point of view of

Source: *The British Foundryman*, July 1979, 162-167

moulding methods, print sizes and shapes, contractions etc. It is the trend in precision casting for the pattern equipment to be made by either specialised patternmakers or tool-makers, since the equipment includes wooden, metal, rubber and to a lesser extent plaster and resin patterns, and light alloy and white metal dies (for wax injection).

It is generally considered necessary when working to fine tolerances to produce pattern equipment to a tolerance of 1/10 of that allowed on the final casting. In designing the pattern, careful allowance must be made for the contraction of the metal during solidification and cooling which in turn is affected by mould material, collapsibility of core, shape of casting, etc. A classic example of the complexities occurring here is when a casting is produced by a combination of processes where a Croning shell core box expands 0·004in per in. during heating, the gravity die expands 0·002in per in. during heating, and Antioch plaster cores expands 0·006in per in. during autoclaving.

Then aluminium contracts 0·012in per in. when un-restricted and no casting is unrestricted; therefore, allowance must be made accordingly.

The machining allowance also has to be made during tooling although this, of course, is not critical, it makes little difference machining 0·090in off to machining 0·010in off. Normally 0·030in machining allowance is preferred and 0·010in grinding allowance on investment castings and 0·090in on shell and plaster castings, though the more the better since any sub-surface defects are then removed during machining.

E. WHY PRECISION CASTING?

Basically because the customer wants to buy a component which is almost immediately usable. Casting metal and con-verting it to swarf is wasteful and costly.

Quite apart from close tolerances, precision casting offers a number of other features; these are mainly:—

> surface finish
> versatility of alloy
> wide size range
> dimensional consistency
> low unit cost
> low tooling cost
> soundness
> intricacy and complexity
> non-directional properties.

It is not claimed that all of these features are available from any one process, but then neither are they usually all required in a single casting. Therefore, by deciding which of these requirements are desirable for a particular application, it is usually possible to fit a process or combination of processes to these requirements.

Newer techniques and alloys are continuously improving each of these features. For example, low pressure techniques when extended to sand or shell mould castings can bring about metal strengths comparable with those of their wrought equivalents. Better mould and die-washers are improving surface finishes. The size of castings attempted becomes increasingly more ambitious.

F. WHICH PROCESS?

During the course of this paper the outstanding features of each process have been indicated but an overall com-parison is obviously useful. The following tables indicate how the processes compare with respect to the most impor-tant requirements of precision castings. The best in each case heads the table; the worst foots it.

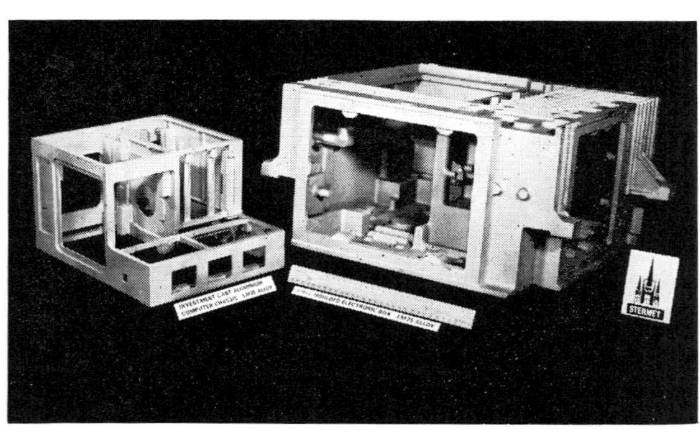

1. Dimensional accuracy

Small (1in)	*Large (> 12in)*
High P. die	High P. die
Investment	Plaster
Plaster	Shell
Low P. die	Investment
Gravity die	Low P. die
Shell	Sand
Sand	Gravity die

Comments:

The graphs represent the best possible tolerances. In al cases except investment casting there should actually be a band rather than a line due to the tolerances across the parting line being wider than those contained in one half of the mould. Also, different grades of sand, plaster, etc. will have an effect on the tolerance achievable. Note that some tolerances deteriorate badly with size, e.g. gravity die com-pared with shell. Large investment castings are used for dimensional accuracy, however, when the accurate dimen-sions are contained within the casting, e.g. hole diameters (2in + 0·012in).

2. Surface finish

Process	Surface roughness (micro-inches)
High P. die	30
Plaster	32
Low P. die	50
Investment	60
Gravity die	70
Shell	125
Sand	500

Comments: Again these are the best achievable; various grades of moulding materials, die-washes etc., will affect these figures. It naturally follows that the ability to reproduce surface detail (engraving etc.) is indicated by this list.

3. Size limitations

Small	**Large**
High P. die	Sand
Investment	Plaster
Plaster	Shell
Shell	Investment
Low P. die	Gravity die
Gravity die	High P. die
Sand	Low P. die

Comments:

These are generally true, but there have been notable exceptions to this order of listing. Note that plaster appears close to the top of both lists, showing its size versatility. Gravity and low-pressure dies tend to cater for middle-range castings. These lists are also approximately true for section thickness.

4. Dimensional consistency

High P. die
Low P. die
Gravity die
Shell
Investment
Plaster
Sand

5. Intricacy

Investment
Plaster
High P. die
Shell
Low P. die
Sand
Gravity die

6. Soundness

Low P. die
Shell
Sand
Investment
Gravity die
Plaster
High P. die

7. Alloy versatility

Sand	any
Shell	,,
Investment	,,
Gravity die	most
Plaster	Low M.P.
Low P. die	,, ,,
High P. die	,, ,,

8. Cost

Since this is the item which is frequently the controlling factor, it is perhaps worthwhile investigating this more deeply.

The customer breaks down casting costs into:—

a. tooling cost.

b. piece per casting.

c. cost of subsequent operations (machining, plating, etc.).

His primary objective, naturally, is to achieve the lowest overall cost of the finished article, compatible with satisfactory performance.

The founder, however, breaks down casting costs into:—

a. tooling cost

b. development cost (of producing satisfactory part)

c. production cost

Obviously, the founder's primary objective is to produce the casting at the lowest price compatible with satisfying the customer's requirements. Naturally, all costs have to be passed on to the customer if the foundry is to stay in business.

Sometimes, development costs are negotiated separately, especially in the case of one off plaster moulding but otherwise an intelligent guess has to be made at these costs at the quotation stage and added on to the piece price. Unfortunately, these are frequently underestimated, since for a one-off, much expensive supervisory time is usually involved.

Piece price		Tooling	
High P. die	lowest	Plaster	lowest
Low P. die		Sand	
Gravity die		Shell	
Sand		Investment	
Shell		Plaster	
Investment		Gravity die	
Plaster		Low P. die	
		High P. die	

Comments

Tooling costs and production costs (piece prices) are fairly

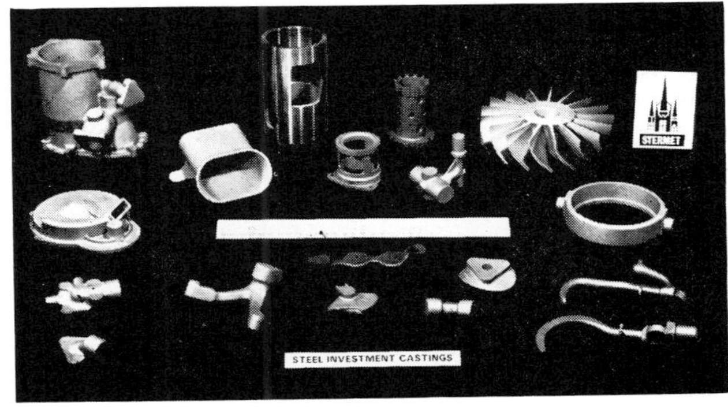

self-explanatory and are frequently inversely proportional to each other since the more sophisticated tooling produces the lower priced castings (e.g.—high-pressure die; and multi-cavity tooling).

Note that 'Plaster' appears in 2 positions in the tooling list depending on whether wood or metal patterns are required.

Frequently, castings are made using e.g. investment or plaster techniques to prove a design before producing expensive tooling for high production quantities to be made by other metal forming techniques. This facility is also used when a design change is envisaged or when only a few more castings are required after the mass-producing tooling has reached its life limit.

G. CUSTOMER LIAISON

As one might expect, the main objection levelled at precision castings is their relatively expensive initial cost. Whilst the production engineer and the designer can often appreciate the ultimate savings to be enjoyed by purchasing a precision product, it is not easy to convince the commercial buyer who frequently looks upon the purchase of castings as so much raw material costing so much per pound; which in any case, once delivered, is no longer of any interest to him. It is therefore advisable, when discussing a new project with a potential customer, that the interest of design, production and purchasing are all represented at the meeting.

The most successful uses of precision castings depend on early discussions between customer and supplier, in order that both parties give vent to their ideas, thus eliminating many potential sources of trouble. Hence, at preliminary discussions the main considerations are:—

a. Design requirements

b. Limitations of processes available

c. Cost and delivery

As discussed earlier, it is often necessary for economic and technical reasons to combine a number of processes (e.g. shell, plaster, and gravity die). It is therefore advisable for designer to leave the choice of process to the founder's discretion.

Frequently, also, the founder can suggest the most suitable alloy knowing the desired properties and ultimate use of the casting. For example, some alloys will run more easily into thin sections, some take a mirror polish, some are more pressure tight, some, whilst good for commercial applications, are difficult to produce to X-ray standards. These considerations are quite separate from strength requirements and typical bad founding characteristics of some alloys such as hot tearing, proneness to gass pick-up, etc.

Another important point is the necessity for limiting close tolerance and surface finish requirements (which cost money) to those places where they are vitally necessary, rather than just putting a note on the drawing asking for e.g. $+0.005$in. and a general surface finish of 60 micro-inches, which if taken literally could raise production costs out of all proportion. Thus, as the standard of perfection increases, so does the cost.

Similarly, the practice of using forging or fabrication drawings when enquiring about castings frequently leads to misunderstanding. It cannot be stressed too strongly that the most successful castings are designed as castings, e.g. unnecessarily thick sections lightened out.

Precision processes encourage the designer to think beyond the limits of normal machining, allowing him in some cases to produce as one integral casting items which would otherwise have consisted of a number of separate components; thus saving weight and space (vital in the aircraft and missile field). Whether or not the final shape he dreams up becomes a reality depends on just how economically it can be produced.

The Shaw Process—a Review

By A. J. Clegg*

The Shaw Process is a precision casting process suitable for the production of castings in a wide range of casting alloys. This article reviews the factors of importance in the production of moulds and castings by the Shaw Process. In particular, the author considers the binder system, refractory aggregates, moulding procedure, and metallurgical and dimensional quality of steel castings produced by the process.

The Shaw Process is a precision casting process capable of the production of accurate castings with excellent surface finish and metallurgical integrity. Moulds are produced using highly refractory aggregates bonded with silica provided by a liquid ethyl silicate binder. A high temperature firing treatment is a feature of the production sequence and this produces an inert mould into which the majority of commercial ferrous and non-ferrous alloys can be cast with confidence.

The process has been used commercially for many years; it was known before the Second World War that silicon esters could be used as refractory aggregate binders[1]. As with most processes there has been a continuous development, in particular with respect to the binder system and the methods of mould production. Many of these developments are the subject of patents and much of the technical information has been restricted to Shaw Process licensees. As a result, the process may not have received the interest and support by the UK foundry industry that its qualities deserve.

At a time when there is an increasing demand for castings with guaranteed quality, dimensional accuracy and metallurgical integrity, it may be appropriate to consider the possibilities that the Shaw Process offers.

Outline of the Process

The mould material is prepared by blending refractory powders, containing a high proportion of fine material, with a liquid ethyl silicate binder and a gelling agent. Careful selection of the refractory material results in two particular advantages:

1 The fine grains of refractory material provide a smooth surface finish on the resultant casting.

2 The selection of a thermally stable refractory material ensures that the mould is not subject to unpredictable dimensional changes in contact with the molten metal during pouring, thus enabling an accurate estimate of casting contraction to be made.

The blended, mobile liquid slurry is poured into the moulding box and around the pattern. Within a short period of time, controlled by the amount of gelling agent, the mould material gels to a rubbery consistency and the pattern can be separated from the mould. The nature of the process permits certain benefits to be gained at this stage:

a Pouring a liquid slurry around the pattern ensures a high degree of contact and therefore accuracy and intricacy in the final casting.

b The mould material sets in contact with the pattern and consequently produces a mould which accurately reproduces the pattern detail.

c The rubbery nature of the mould allows the pattern to be withdrawn without distortion to the mould, thus maintaining the dimensional accuracy of the mould cavity.

On removal of the pattern the moulds are either torched immediately to remove evolved alcohol (Shaw Process) or immersed in a stabilising bath

---

* The author is a lecturer in the Department of Engineering Production at Loughborough University of Technology.

prior to torching (Unicast Process)[2]. Torching pro-duces a very fine crazed surface and interior structure which does not affect the casting surface, ie there is no metal penetration into the fine cracks, but may improve permeability to allow the escape of air/gases during casting. After torching the moulds are fired in a furnace to a temperature of 1,000°C, which ensures that there are no combustible materials in the mould and that a strong, rigid, inert, accurate and stable mould is produced. The stages in the production of a casting by the Shaw Process are outlined in the flow diagram presented in fig. 1.

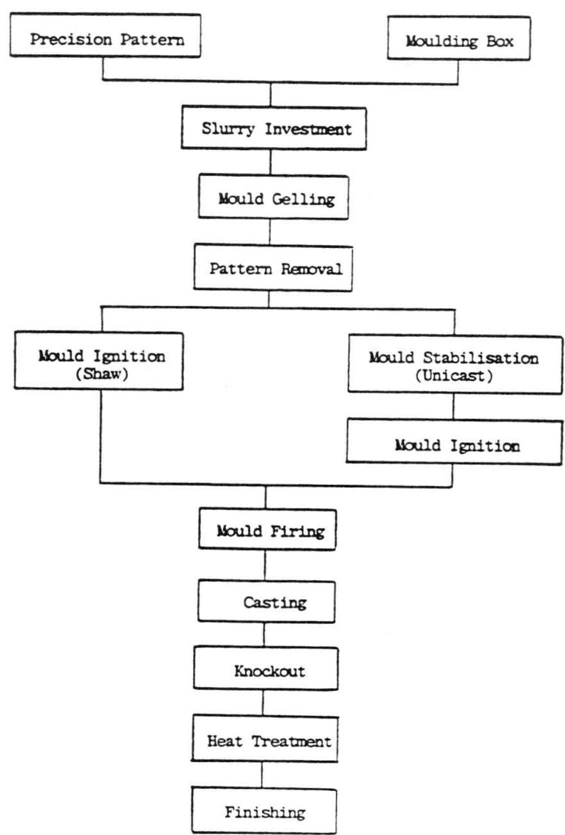

1 Stages in the manufacture of a casting by the Shaw Process.

The advantages claimed for moulds manufactured by the Shaw Process include the following[3]:

1 Good pattern stripping characteristics: the rubbery nature of the gelled mould provides flexibility to the mould, which enables the pattern and mould to be separated without damage to the mould when intricate detail, or even straight draws, are required.

2 Dimensional stability: the excellent reproduction of pattern detail and dimensions is retained by the mould to a great extent after firing and during casting, thus enabling accurate dimensional allowances to be made.

3 Mould strength: this is sufficient to allow moulds to be cast without the need for moulding boxes.

4 Collapsibility and resistance to tears: the characteristic internal structure of the mould material improves its breakdown properties and there is less constraint of the casting during contraction.

5 Resistance to thermal shock: the characteristic internal structure permits expansion of the mould material to occur readily, as a result moulds can safely be poured cold.

6 Resistance to spalling and washing: the nature of the silica bond prevents the generation of inclusions during mould filling.

7 Permeability and inertness: as the mould is inert after firing the only gas to be displaced is that occupying the mould cavity. The characteristic structure provides sufficient permeability to enable the metal to readily displace the gas through the mould.

The Binder

Ethyl silicate is an organic-based silica binder that is free of those alkaline salts which reduce the refractoriness of mould materials. It is produced by reacting silicon tetrachloride ($SiCl_4$) with ethyl alcohol[4]:

$$SiCl_4 + 4C_2H_5OH \rightarrow Si(OC_2H_5)_4 + 4HCl$$

and it is important to remove the hydrochloric acid so that subsequent gelling characteristics may be easily controlled. If industrial ethanol, which invariably contains some water, is used as an alternative to the pure ethyl alcohol, the product obtained is a mixture of tetraethoxysilane and ethoxypolysiloxanes. Tetraethoxysilane contains approximately 28% silica by weight, whereas the ethoxypolysiloxanes which contain an average of up to five silicon atoms per molecule contain 40% silica[5]. This product is known as technical ethyl silicate and because of its higher silica content is generally preferred for use in the foundry industry[6].

Hydrolysis

Ethyl silicate is a stable substance with no binding ability. It is necessary to hydrolise the solution and cause it to react with water in order to produce a solution which will deposit the adhesive form of silica desirable for bonding refractory aggregates[7]. Ethyl silicate and water are immiscible unless a mutual solvent such as ethanol is used[8] and it also serves to dilute the solution to the desired silica content. The hydrolysis may be carried out under either acid or alkaline conditions[4]. However, alkaline conditions usually result in fairly rapid gelation and consequently acid hydrolysis is preferred for foundry requirements. The preferred acid is hydrochloric acid although sulphuric or phosphoric acid are suitable[8]. The acid hydrolysates prepared in this manner have a very good storage life and are marketed as prehydrolysed ethyl silicates which are ready for direct use in the foundry.

Gelation

In order to bind the refractory aggregate the ethyl silicate hydrolysate must be made to gel. It is the gel which provides the bonding action by two methods[8]:

a Air drying bond:

$$3H_4SiO_4 \rightarrow H_2Si_2O_5 + H_2SiO_3 + 4H_2O$$

b Precipitation bond:

$$3H_4SiO_4 \rightarrow 3SiO_2 \text{ gel} + 6H_2O$$

Upon heating, the silicic acid or silica gel binders condense to form a refractory silica cement. It is this silica form which provides the high strength developed by firing.

There are many ways of promoting the gelling of an acid hydrolysed ethyl silicate using the principle of pH control[7]. Hydrolysed ethyl silicate solutions are usually prepared with a pH value of between 1·5 and 3·0, at which they are relatively stable, they are also stable at pH values above 7·0. They are inherently unstable at pH values between 5·0 and 7·0. By adding an alkaline agent to the hydrolysed ethyl silicate solution the pH value of the solution can be increased to a value of 5·0 when the binder becomes unstable and gels. Ammonia, ammonia salts—acetate, carbonate, hydroxide; or organic ammonia salts—piperidine, morpholemar, triethylamine, may be used for this purpose. The actual time for gelation will depend on the particular application and can be varied by adjusting the amount of gelling agent addition.

If accurate control of setting time or a very long setting time is required use may be made of organotin catalysts[5]. The organotin catalysts enable hydrolysis and gelation to proceed readily in the absence of either acid or base. By a special heat treatment process the gelation characteristics can be altered to change the gel time when water is added.

Refractory Aggregates

Many refractory materials may be used in association with ethyl silicate to produce Shaw Process moulds, the list in Table 1 includes many of the possible materials but is not exhaustive. In the selection of a suitable material for a particular application the following factors must be considered:

1 Purity

2 Refractoriness

3 Stability

4 Thermal Expansion

5 Thermal Conductivity

6 Particle Size and Distribution

7 Cost

The mould material must be sufficiently refractory to withstand the pouring temperature of the particular metal being cast, without either melting or softening.

TABLE 1: *Refractory materials suitable for use in the Shaw Process*

Material	Chemical Notation	Melting Point °C
Magnesia	MgO	2,800
Zirconia	ZrO_2	2,677
Calcia	CaO	2,600
Zircon	$ZrO_2.SiO_2$	2,420
Calcium zirconite	$CaO.ZrO_2$	2,345
Magnesia spinel	$MgO.Al_2O_3$	2,135
Alumina	Al_2O_3	2,015
Mullite	$3Al_2O_3.2SiO_2$	1,830
Fused silica	SiO_2	1,723

The melting point value is a good guide to refractoriness providing the material is pure, as very small amounts of alkali metal salts or iron oxide can reduce melting points severely. The refractory selected should exhibit stability, it should not be susceptible to hydration, or reactions with other materials in the mould or the metal. Thermal stability is also important, expansion is inevitable; however, providing that it is predictable, ie constant and reproducible, it can be taken into account when estimating contraction allowance.

Thermal conductivity is in part a property of the material selected but it is also a function of particle size and distribution when that material is used to produce a porous, particulate mould. The property does influence the rate of solidification and heat transfer through a mould which may have an influence on metallurgical integrity.

Cost is a major factor in refractory material selection and is affected by such factors as availability, purity and particle size requirement.

Moulding Mix Specification

The secret of successful mould production and quality castings lies in the mould material mix specification. It is essential to balance the grades of refractory material with the volume of binder and amount of gelling agent in order to produce moulds of consistently high quality. Shaw Process slurry preparation is critical[9], too thin and mould cracking will occur on firing, too thick and detail is lost and air bubbles are trapped at the pattern surface.

When selecting the grades of refractory material to be used the principles which apply to sand moulding can be used as a guide. Surface finish will be improved when finer material is used; however, permeability will decrease and binder requirement increase as a consequence. The strength, measured conventionally by AFS compression and tensile strength, will also be affected by grade selection. Strength will be at a maximum when refractory grades are selected on the basis of using several size gradings which permit infilling of the voids between larger grains by the smaller[7], this is also the most economic way of using the binder. Mix specification is inevitably a compromise between theoretical considerations and practical requirements but a reading of standard ceramic texts[10,11] can be invaluable.

Experimentally determined values[12] of strength

TABLE 2: *Tensile strengths of mullite mixes bonded with silester AR*

Refractory material and size grading	cc. of binder per kg. of refractory	AFS Tensile Strength (lb. per sq. in.)			
		Before firing		After firing at 1,000°C	
		air dried	torched	air dried	torched
1. 100%—300, mesh mullite	456	—	2·0	—	5·6
	422	—	2·7	—	6·0
	400	5·3	2·7	21·0	5·3
	356	—	3·0	—	7·3
	333	9·3	3·2	37·0	8·3
	311	—	3·4	—	11·0
	267	12·0	4·0	58·5	15·8
2. 30%—16/30, 30%—30/60, 40%—300, mesh mullite	267	—	9·0	—	16·7
	222	—	10·7	—	22·7
	178	—	11·8	—	28·3

which emphasise some of these points are shown in Table 2.

Mould Production and Processing

As a principal advantage of the process is its ability to produce castings to consistent and close dimensional tolerances it is essential that patterns and mould locating equipment be produced to a high standard of accuracy.

Pattern Equipment

As a casting cannot contain more detail or be more accurate than the pattern from which it was made, it follows that particular attention must be paid to pattern construction and quality. The Shaw Process utilises permanent pattern equipment and in principle a wide range of materials may be used[13]: wood, plaster, graphite, epoxy resin and metal for examples. However, in practice the requirement for high standards of accuracy reduces the choice. Wood patterns are affected by heat and moisture and their dimensions can vary by as much as 3%, which would severely affect accuracy[13]. Plastic resin patterns may also be subject to distortion when used in conjunction with the Shaw Process[9]. To achieve good dimensional accuracy polished metal patterns are preferred for which aluminium alloys are particularly suitable. Fig. 2 shows an example of a precision aluminium pattern plate assembly and a graphite pattern. Patterns are normally coated with a release agent, for example a 10% solution of paraffin wax in benzene[14], or similar proprietary products and, if sparingly applied and well buffed, pattern release from the mould presents no problem.

Mould Production

Shaw Process moulds may be produced by one of several alternative methods: for small components a ceramic shell is quite acceptable, for larger components a boxless block mould is suitable. If the mould is very large or if demand for a particular casting is great and it is desirable that mould material costs be reduced, a composite mould can be produced. In this case the facing material is Shaw-based but the backing material might be sodium silicate bonded. For mass production applications the process, in each of these forms, lends itself to mechanisation and rapid output of moulds.

In the production of composite moulds, an example of which is shown in fig. 3, additional pattern requirements in the form of a pre-form pattern may be necessary. These may be produced in wood as their purpose is only to produce an oversize cavity in the backing mould. This mould is then positioned over the precision pattern and the gap filled by introducing the mobile Shaw slurry. Permanent pre-form patterns are not essential, the required effect can be obtained by covering the precision pattern with a thickness of material such as oil-sand, CO_2 sand, felt, etc.[14–17] which will permit the oversize backing mould to be produced. To ensure that separation of the backing and facing layers does not become a problem the mould materials selected for the backing and facing layers should have similar thermal expansion

2 Precision aluminium pattern plate assembly and graphite pattern.

characteristics. Permanent metal backing moulds are also suitable[18].

Mould Processing

When the mould has been separated from the pattern it must pass through several stages which can exert a significant influence on the quality of the finished mould. The craze cracking effect in Shaw Process mould material is only produced when ignition follows immediately after pattern stripping. When the mould material is allowed to air-dry for a substantial period the craze cracking does not develop. It is interesting to note that the strengths developed by firing are considerably higher in the air-dried material —but at the expense of permeability, thermal shock resistance and breakdown characteristics[19]. See Table 2.

Coarse cracking may be a problem at internal corners in a mould cavity, particularly if pools of alcohol are allowed to form. Sharp corners are unsuitable in most casting designs and a minimum radius of three millimetres is recommended where practicable. The evolved alcohol usually ignites readily but ignition can be encouraged by the use of a gas/air torch, in either case the moulds should stand on a grid to provide free access of air during ignition.

A further variation in practice exists in the Unicast Process when the separated mould is immersed in a bath of stabilising liquid before torching and subsequent firing[2]. It is claimed that stabilisation for a period of one hour in a bath containing ethyl alcohol, for example, will improve the dimensional stability and strength of the mould. Small moulds can be enclosed within sealed plastic bags to enable stabilisation to occur[20].

After torching moulds are fired for two reasons:

a To burn off any residual organic material and remove any moisture in order to produce an inert mould.

b To improve mould strength through a sintering effect.

A considerable range of baking or firing temperatures, from 300 to 1,300°C, is suggested in the literature. Whereas the first objective should be met by "firing" at 500°C, it is unlikely that any substantial improvement in strength will be obtained with a sintering temperature below 1,000°C. Slightly oxidising conditions are recommended[21] when firing moulds to encourage the burning of residual volatiles. Composite moulds present their own difficulties, with radiant heating in tunnel furnaces being preferred to fire the facing material without excessively heating the backing material[22].

Moulds may subsequently be cast whilst still hot or allowed to cool, depending on section thickness and metallurgical requirements. Mould part location in boxless moulds is most easily obtained by making provision for mould locating pins at the pattern-making stage. This is illustrated in fig. 3 which shows the use of double-ended taper location pins. Mould joints are sealed externally with a luting compound, joint face seals reduce accuracy across the joint, and moulds may be weighted or clamped using metal bands before casting.

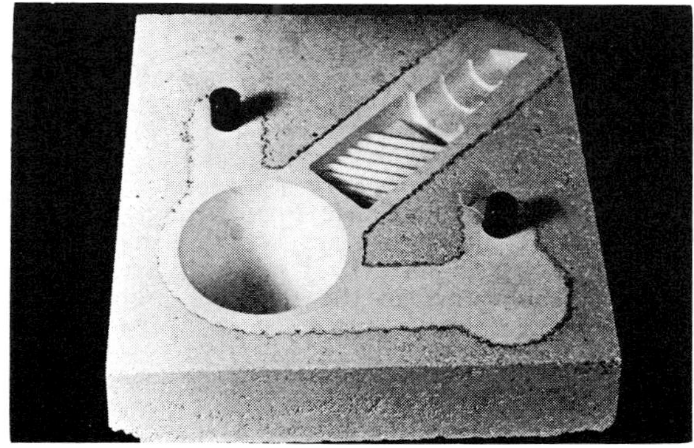

3 A composite mould with Shaw Process facing and sodium silicate bonded backing. Note mould location pins.

Casting Quality

The quality of a precision casting is judged using three major parameters:
1 Dimensional accuracy.
2 Surface finish and appearance.
3 Metallurgical integrity.

Dimensional Accuracy

The dimensional accuracy of any casting is determined by the accuracy of the pattern equipment and by the accuracy with which mould parts are located. It is inevitable that closer dimensional tolerances can be maintained within a single mould half than can be held in dimensions across the mould joint. The mould material also exerts an influence on dimensional accuracy, first through its thermal characteristics, and second through permeability. Neither of these factors can be disassociated from metal temperature which has been shown to exert a major influence on dimensional accuracy[23]. Variations in melt composition may also result in dimensional variations but with careful melting practice this should be a minor influence. As a safety precaution it has been suggested that matching castings, eg die half castings be cast from the same melt[24].

Typical dimensional tolerances suggested for steel castings produced by the Shaw Process are[25]:

For dimensions up to 25 mm. (1 in.)
± 0.08 mm. (± 0.003 in.)
For dimensions between 25 and 75 mm. (1–3 in.)
± 0.13 mm. (± 0.005 in.)
For dimensions between 75 and 200 mm. (3–8 in.)
± 0.38 mm. (± 0.015 in.)
For dimensions between 200 and 375 mm. (8–15 in.)
± 0.76 mm. (± 0.030 in.)
For dimensions over 375 mm. (15 in.)
± 1.14 mm. (± 0.045 in.)

For dimensions across the mould joint line an additional tolerance of between ± 0.25 to 0.50 mm. (0.010 to 0.020 in.) should be allowed.

Surface Finish and Appearance

The surface finish of a casting produced using the Shaw Process is influenced by several factors which include refractory grading, metal and mould temperature, and metal/mould reactions considered in the following section. The finer the grade of refractory filler used in the refractory aggregate the better should be the surface finish. Preheated moulds and high casting temperature are likely to reduce the quality of surface finish[26]. It is possible to obtain surface finish values in the region of two micrometres (80 microinches) cla for steel castings, better for non-ferrous alloys, which compares with five to ten micrometres (200–400 microinches) cla for sand castings.

Heavily oxidised surfaces reduce both dimensional accuracy and the quality of surface finish directly and through the requirement for abrasive shot blasting. The use of inert gas atmospheres to minimise these effects can be worthwhile.

Metallurgical Considerations

The nature of the Shaw Process mould material results in relatively slow cooling of the metal in comparison with other casting processes. This can cause a coarse grain structure which may be unacceptable in certain castings' applications; it can be countered by grain refining additions. The slow cooling is in part responsible for decarburisation and surface oxidation in steel castings[27]. In steels containing chromium the surface may be further affected by pitting[28].

Decarburisation has been shown to be a function of the time that the casting remains at an elevated temperature, that is above 875°C[27]. In practice the severity of the problem can be reduced by one of several methods which include: hexamethylene tetramine placed on the exposed surface of the metal when the mould is enclosed within a container[29], a reducing gas atmosphere[30], or most effectively by a vacuum treatment to remove air from the mould followed by back filling with an inert gas[27]. The occurrence of pitting has been shown to be associated with decarburisation[28]. The two phenomena are caused by carbon monoxide formation. Carbon in the metal reacts with oxygen from the mould atmosphere and the silica which provides the mould bond. This results in decarburisation and the carbon monoxide bubbles deform the casting surface. This implies that it may not be possible to eliminate decarburisation entirely because oxygen will always be present in the silica which provides the bond.

It has been suggested that casting surface characteristics can be altered by passing a gas through the mould material as a steel casting is cooled. By using ammonia[31] a nitrided surface can be developed and a carburised surface produced by using a carbon rich atmosphere[21].

Acknowledgements

The author is working in association with Dr. A. A. Das to whom a Science Research Council grant was awarded to investigate the optimisation of materials and methods in the production of cast dies.

The author would like to thank Professor R. J. Sury, head of the Department of Engineering Production, Loughborough University of Technology, for the provision of facilities and permission to publish this article.

REFERENCES

1 Shaw, C. FOUNDRY TRADE JOURNAL, 1946, **78** (1534), 31
2 Greenwood, R. American Patent 3,172,176.
3 Lubalin, I. and Christensen, R. J. *Modern Casting*, 1960, **38** (4), 83.
4 Emblem, H. F. *Trans. and J. Br. Ceramic Soc.*, 1975, **74** (23), 660.
5 Jones, K. *et al. Tin and its Uses*, 1979, **119**, 10.
6 Emblem, H. G. FOUNDRY TRADE JOURNAL, 1972, **132** (2884), 379.
7 Lawrence, W. G. *Modern Casting*, 1960, **38** (2), 109.
8 Wales, W. F. *Trans. AFS*, 1973, **81**, 249.
9 Paterson, S. *SAE Trans.*, 1967, **75**, paper 660645, 263.
10 Kingery, W. D. *Introduction to Ceramics*. Wiley, London, 1976.
11 Budworth, D. W. *An Introduction to Ceramic Science*. Pergamon, Oxford, 1970.
12 Russell, C. J. MSc Project, Loughborough University of Technology, March 1980.
13 Butler, M. J. *British Plastics*, 1970, **43** (6), 144.
14 Balinskii, V. R. *et al. Russian Castings Production*, 1972, 421.
15 Winter, D. B. FOUNDRY TRADE JOURNAL, 1964, **116** (2474), 427.
16 Dittrich, W. *Giesserei*, 1972, **59** (8), 239.
17 Dunlop, A. American Patent, 2,931,081.
18 Lubalin, I. J. American Patent, 3,242,539.
19 Shaw, N. American Patent, 2,795,022.
20 Scott, R. K. American Patent, 3,213,497.
21 Olsen, C. F. ASTME Technical Paper, CM69-235, 1969.
22 Prasad, J. S. and Watmough, T. *Trans. AFS*, 1969, **77**, 289.
23 Szende, Gy. *Bany. es Koh. Lapok (Ontode)*, 1976, **27** (3), English translation B1S1 14700.
24 Bazzano, E. *Castings*, 1970, **16** (5), 24.
25 *Metals Handbook*, **5B**, 262. American Society for Metals, Ohio.
26 Fominyk, I. P. *et al. Russian Castings Production*, 1976, 467.
27 Doremus, G. E. and Loper, C. R. *Trans. AFS*, 1970, **78**, 338.
28 Yardmici, G., Fischer, F. and Klein, E. *Proc. 4th World Conference on Investment Casting*, June 1976.
29 Shaw, C. American Patent 2,935,722.
30 Lubalin, I. *Cast Forging Dies*. ASIME, Hartford, Connecticut, 1965.
31 Pikman, A. A. *et al. Russian Castings Production*, 1976, 253.

Expandable Polystyrene and Its Processing into Patterns For the Evaporative Casting Process

R. H. Immel

ARCO Polymers, Inc.
Research and Development Department
Monaca, Pennsylvania

ABSTRACT

Expandable polystyrene (EPS) is a thermoplastic material composed of approximately 92% carbon and 8% hydrogen (weight percentages). The basic raw materials used in the manufacture of EPS are crude oil and natural gas. EPS contains a volatile blowing agent which causes the EPS particles to expand when heated. They are capable of expanding to densities as low as 1.0 lb/ft^3 and these low density particles can readily be molded into complex, rigid shapes suitable for use as patterns in the evaporative casting process.

The process for transforming the solid, dense EPS particles into low density, foam patterns involves two basic steps, pre-expansion and molding. The processing steps are relatively clean and the tooling and equipment required to perform these steps is relatively simple and inexpensive. As a result, the economics of producing patterns is generally quite favorable.

An important aspect of foam patterns that should be considered by foundries is shrinkage. Foam patterns shrink with age; however, if tooling is designed with allowance for pattern shrinkage, castings of acceptable dimensions will result. The amount of shrinkage is predictable and castings made from patterns of approximately the same age should be of repeatable dimensions.

EPS has, by definition, a very low order of toxicity. It is, however, combustible and fire is the biggest hazard involved in the processing and storage of patterns. Provision for adequate ventilation in the processing and storage areas along with good housekeeping and safety rules greatly minimizes this hazard.

Introduction

The evaporative casting process, which involves the use of foam polystyrene as expandable patterns for metal castings, was invented by H. F. Shroyer[1] in 1958. The process has been practiced by foundries in the United States since the mid-1960s using patterns fabricated from foam polystyrene billets or board, generally for large, "one-of-a-kind" castings. In recent years numerous foundries have expressed interest in the process for "many-of-a-kind," or production castings, which for production economics preferably utilize foam polystyrene patterns molded to shape.

The technology and nomenclature commonly used for processing expandable polystyrene into molded patterns suitable for the evaporative casting process are, in most cases, foreign to the foundryman. This paper has been written for the foundryman to acquaint him with this technology and nomenclature now in use for the molding of foam patterns.

Expandable Polystyrene, Its Development and Uses

Expandable polystyrene, commonly referred to as EPS or foam polystyrene, is polystyrene which contains a hydrocarbon blowing agent, such as pentane. EPS is transformed from a solid into a low density foam product by the application of heat which causes the thermoplastic polystyrene to soften and the hydrocarbon blowing agent to vaporize. The vaporized hydrocarbon forms a multitude of small, closed cells within each particle of polystyrene. In its foamed or molded state EPS consists essentially of carbon and hydrogen (about 92% and 8% by weight, respectively) and small amounts of oxygen and nitrogen which come from the air that permeates into the EPS after it is foamed.

EPS is manufactured and sold in the form of solid small beads with a bulk density of about 40 lb/ft^3. These beads are available in a number of size classifications varying from a predominance of 0.050 to 0.015 in. diameter.

The basic raw materials for bulk of the EPS manufactured in the United States is derived from crude oil and natural gas. Benzene is obtained from crude oil during the manufacture of gasoline. Benzene is also a byproduct of coke production. The primary source of ethylene is from the thermal cracking of ethane and propane which are constituents of natural gas. Using an alkylation process, benzene and ethylene are combined to form ethylbenzene which undergoes catalytic dehydrogenation

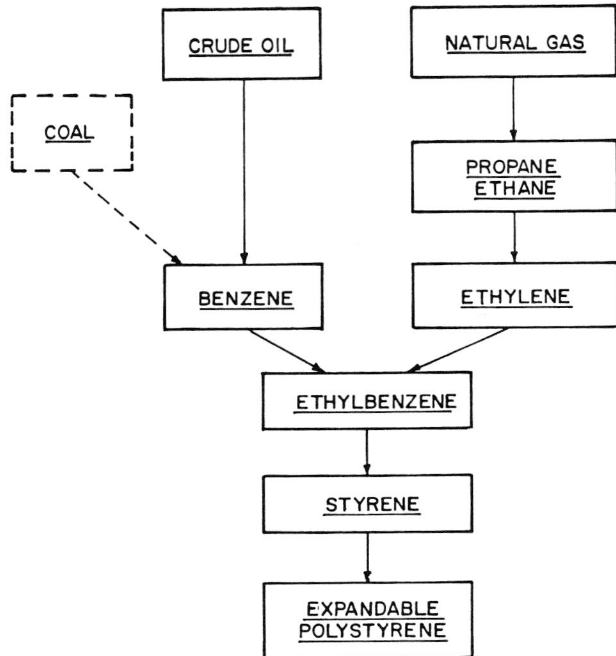

Fig. 1. Simplified flow chart for the manufacture of expandable polystyrene.

Table 1. Expanded Polystyrene Consumption in the United States

	1,000 M.T.	
	1976	1977
EPS BEADS		
CUPS	45	46
PACKAGING	40	54
BOARD AND BILLETS	40	30
OTHER MOLDED PARTS	27	41
TOTAL EPS BEADS	152	171
EXTRUDED FOAM BOARD	20	24
EXTRUDED FOAM SHEET		
DISPOSABLES, NOT PACKAGING	32	37
PACKAGING	78	83
OTHER	21	22
TOTAL EXTRUDED FOAM SHEET	131	142
GRAND TOTAL	303	337

and forms styrene. The polymerization of styrene forms polystyrene which is made expandable by the addition of a blowing agent, such as pentane. Figure 1 presents a simplified flow chart for the manufacture of EPS.

Prior to 1930, styrene and polystyrene were just laboratory curiosities. During the 1930s a number of American companies were actively engaged in the development of processes for the manufacture of styrene. World War II necessitated the acceleration of these development efforts since styrene is a major raw material for the manufacture of synthetic rubber. Nine companies shared their knowledge to accelerate the commercialization of styrene which, along with some government financing, resulted in huge commercial installations for the manufacture of this material vital to the war effort.

Most of the process technology and applications for polystyrene have been developed since the end of World War II. The process commonly in use in the United States is suspension polymerization, a batch process which involves the suspension of drops of styrene in water which contains catalysts. The suspension is heated and the drops of styrene polymerize to form solid, spherical beads of polystyrene. Addition of a blowing agent to the polystyrene results in the product EPS. The EPS beads are dewatered, dried and classified by screening.

EPS was first produced commercially in the United States in 1954. Since then numerous processing techniques and equipment and a variety of applications have developed. The growth of EPS applications has been steady. The 1977 consumption of the material in the United States was estimated to be 171,000 metric t. Hot drink cups, packaging, and insulation constitute the three major applications.

Foam polystyrene is also made by extrusion into board or sheet form. A number of companies produce extruded foam polystyrene sheet which is formed into a variety of items such as egg cartons and trays for packaging fruit, produce and meat. American consumption of extruded board and sheet in 1977 was an estimated 166,000 metric t. Table 1[2] presents a breakdown of expanded polystyrene consumption in the United States for the years 1976 and 1977.

EPS Processing

The transformation of EPS from the raw material beads to the finished molded object, be it a large insulation billet, a hot drink cup or a foundry pattern, involves two major processing steps: pre-expansion and molding (Fig. 2). During the pre-expansion step, the EPS beads are heated, the thermoplastic polystyrene softens and the blowing agent occluded within the polymer vaporizes and expands as the temperature rises and as the polymer softens. The beads are allowed to pre-expand to a bulk density equal to the density desired in the molded object. The pre-expanded beads are then blown into a mold cavity defining the desired shape until the loose, spherical particles completely fill the confines of the cavity. Heat (usually via steam) is again applied to the beads which causes them to soften and expand further to completely fill the void space which had surrounded the spherical particles. The beads pressure themselves against the cavity walls and against each other. Being plasticized by the heat, they weld or fuse together to form a foamed mass the shape of the mold cavity. The molded object is then cooled to reduce the vapor pressure within each cell and to harden the cell walls so the molded object will retain the desired shape upon removal from the cavity.

Pre-Expansion Process

The most common method of pre-expansion utilizes steam in equipment known as a Rodman pre-expander.[3] It is a vertical tank containing an agitator and equipped with means of continuously feeding beads and steam into the bottom of the tank. The steam expands the beads, causing them to rise and overflow from the top of the vessel. Density is controlled by varying the feed rate and/or by mixing air with the steam to reduce its heat transfer efficiency.

Unfortunately, this relatively simple and inexpensive method of pre-expansion is not suitable for the production of foundry

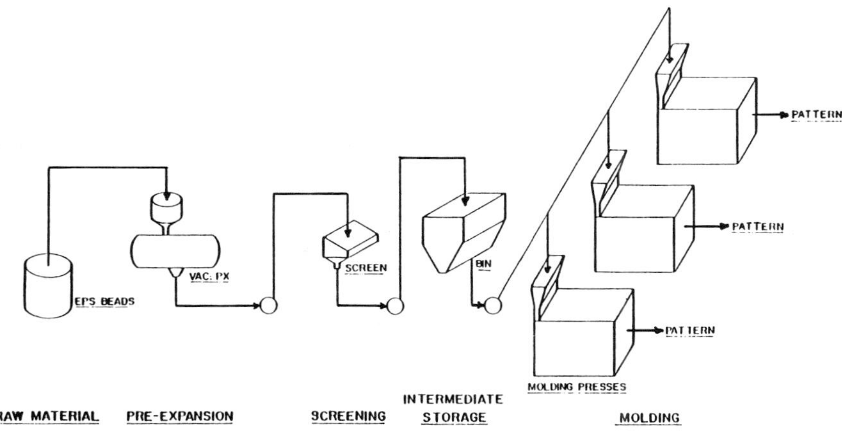

Fig. 2. Schematic diagram of the EPS pattern molding process.

patterns. As has been reported in the past, successful foundry patterns require the use of low density, small beads.[4] Low density is required to minimize the volume of gas formed during vaporization of the EPS by the molten metal and small beads are desirable for surface appearance of the pattern and resultant casting, as well as for ease in filling thin sections often encountered in metal castings. The practical minimum density obtainable with small beads in a Rodman-type pre-expander is about 1.7 lb/ft^3 a density too high for most foundry uses.

Vacuum pre-expansion[5] is at this time the only commercially acceptable means of pre-expanding small beads to densities as low as 1.0-1.25 lb/ft^3 which are required for steel, ductile and gray iron castings. Vacuum pre-expansion is a batch method involving an agitated horizontal vessel jacketed for heat. A volumetric or weighed charge of beads is dropped into the vessel and the beads are heated in the dry atmosphere until they soften and partially expand. The vessel is then subjected to a partial vacuum (15-20 in. Hg) which increases the pressure differential between the interior and exterior of the beads. This added pressure differential causes the beads to expand to considerably lower densities than possible at atmospheric conditions and at the temperature limitations imposed by the beads (temperatures too high result in beads that soften to the point of lumping or agglomerating within the pre-expander). The low density beads must be cooled while under vacuum to prevent them from collapsing during return to atmospheric pressure and subsequent discharge from the vessel. This cooling is accomplished by allowing a small quantity of coolant, usually water, to enter the vessel while it is still under vacuum. The water flashes to steam and, because of the subatmospheric conditions, the steam is colder than the beads. A few seconds of mixing the hot beads with the cooler steam is sufficient to quench-cool the beads and prevent collapse during discharge from the pre-expander. Bead density control is very accurate and reproducible in this method of pre-expansion. The process variables of feed charge, temperature, vacuum quantity and time, water quantity and mixing time, and venting and discharge times are held constant. Control of the only other variable, heating time prior to vacuum, controls the bead density. Figure 3 shows a 100 gal vacuum pre-expander.

There is another commercial method available for pre-expanding small beads to densities as low as 1.25-1.3 lb/ft^3 which is suitable for most Al castings and some gray iron castings. This equipment, known as the Unitex pre-expander,[6] is similar to the vacuum pre-expander except that it utilizes pressurized air instead of vacuum.

After pre-expansion, the beads are screened to remove any small lumps or agglomerates and then are airveyed to immediate storage or directly to the molding press feed hopper.

Fig. 3. 100 gal vacuum pre-expander.

Fig. 4. Molding press.

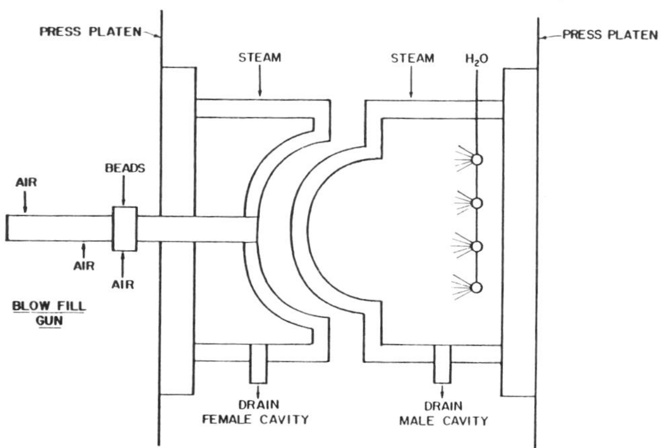

Fig. 5. Schematic diagram of an EPS pattern mold.

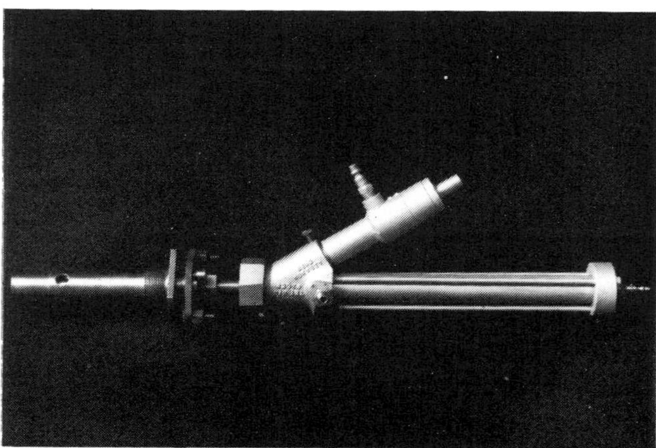

Fig. 6. Vented blow fill gun.

Molding Process

The major items of equipment and utilities required for the molding phase of the EPS conversion process are:

- Molding press
- Mold
- Blow fill gun
- Steam
- Compressed air
- Cooling
- Water
- Electric power.

Molding presses (Fig. 4) for EPS are hydraulic holding presses usually designed to hold pressures of 50 psi. They are generally automatic and equipped with sequencing controls and valving necessary to perform the elements of the molding cycle, namely:

- Close press
- Preheat mold with steam
- Fill mold with EPS
- Fuse
- Cool
- Open press
- Part ejection.

The presses generally are designed to operate horizontally (vertical platens) and are available in a variety of platen sizes ranging from 20 x 20 in. to 52 x 64 in.

Molds

The molds used for EPS pattern molding are usually cast Al with wall thicknesses of about 3/8 in. Depending on the surface quality and dimensional tolerances desired in the foam pattern, the cavity walls of the molds may need to be machined after casting. Each half of the mold cavity is backed by a chamber called a steam chest. During the molding cycle, steam flows into the chest, then into the mold cavity through core box vents or small drilled holes placed in the cavity walls, usually on 1.5-2.0 in. centers. Water manifolds with spray nozzles are located within the chests to provide cooling to the cavity walls. A schematic drawing of a typical mold is shown in Fig. 5.

Multicavity molds are commonly used for EPS molding. Because of the difficulties encountered in filling the molds with small, low density beads for pattern molding, however, it is suggested the number of cavities within a mold be limited to four. The parting lines for pattern molds must be machined to

Table 2. Typical Pattern Molding Cycle

CYCLE PHASE	TIME (SECONDS)
CLOSE PRESS a PREHEAT	15
MOLD FILLING	5
FUSION	10
COOLING	30 - 45
OPEN PRESS a EJECT	10
TOTAL CYCLE	70 - 85

closer tolerances than generally required for EPS molds in order to minimize parting line flash.

Blow Fill Guns

The mold cavity is force fed with pre-expanded beads using blow fill guns. A blow fill gun consists of an air venturi or nozzle connected to an "adaptor tube" which provides a path for the beads from the venturi through the steam chest and into the mold cavity. An air operated piston within the adaptor tube retracts to open the tube for the mold filling operation and remains closed for the other phases of the cycle.

During the filling operation, the venturi action draws the beads from the press feed hopper to the fill gun from which point they are blown or forced into the cavity. The air escapes from the cavity around the parting line of the mold as well as through the core box vents and into the steam chests which are vented to atmosphere during the mold filling phase of the cycle. During the fill cycle, the mold halves are held open a few thousandths of an inch to allow for parting line venting. Since foundry patterns are preferably molded at low density, a more uniform and complete fill is required than when molding higher densities. Any underfilled areas of the part would be of extremely low density after fusion and the foam, upon cooling, would collapse in these underfilled areas. In addition, the use of small beads limits the distance the mold can be held open during fill and hence limits venting of air at the parting line. These difficulties of adequate venting and complete fill were overcome by the development of a vented blow fill gun (Fig. 6) which modifies the common apparatus by using a variable venturi and a vent sleeve. The vent sleeve surrounds the adaptor and provides additional venting of air from the mold cavity into the steam chests. The adjustable venturi allows the use of high pressure air (110-125 psig) to create the velocity (at minimum volume) necessary to move the low density, small beads through the often rigorous path from the gun through thin areas to the extremities of the mold cavity. When the mold cavity is full of beads, no more beads or air can enter and the air feeding the venturi must flow up the suction tube signaling that the mold is full. The mold is then closed completely and the fusion cycle begins.

Fusion Cycle

The vents (drain lines) on the chests close and steam (35-40 psig) enters the chests and flows through the core box vents into the cavity. As explained earlier, the beads are heated, expanded to fill the void spaces and fused together. When the beads expand and fuse, they seal off the core box vents which results in a rapid increase of steam pressure in the chests. This increase in pressure is sensed by pressure switches which terminate the steam flow when activated. At this point the chests are vented through the drain lines and the cooling cycle begins.

Cool Cycle

At the time of fusion, the molded part is soft and the internal cells of each bead making up the molded part are under pressure from the vaporized blowing agent and the expanded air which had permeated into the beads after pre-expansion. It is necessary to cool the molded part to a point where the internal pressure of the beads is low enough that the molded part will not expand upon ejection from the mold. The thinner the molded part, the faster it will cool because of faster heat transfer. Lower densities cool faster because of less blowing agent in the cells and hence less vapor pressure. For low density patterns of about 0.25 in. thickness or less, the time required for cooling of the metal cavity walls becomes the limiting factor for the cooling cycle time. Obviously, cavity walls thinner than 0.375 in. will reduce the cooling cycle times required for patterns of thin section.

Part Ejection

After the cooling phase of the cycle, the press is opened and the molded part is ejected. Low density parts can usually be ejected by blowing the part from the cavity with compressed air. Mechanical ejection can also be used. Careful handling of the molded part is necessary during ejection and for about thirty minutes thereafter since the foam is very soft and easily marred or dented. After part ejection, the press is closed and a new cycle begun. The entire cycle will usually require less than 1.5 min for

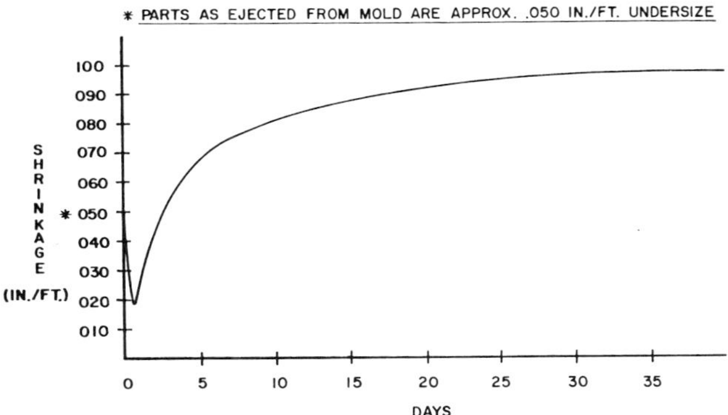

Fig. 7. EPS foam shrinkage by days.

patterns 0.5 in. or less in thickness. Table 2 presents a breakdown of a typical molding cycle.

Pattern Molding Economics

No paper describing the EPS foam molding process would be complete without some discussion on the cost of molding. It is extremely difficult, if not impossible, to generalize on molding costs because of the number of variables involved. In pattern molding the cost per pattern depends to a great extent on the number of component pieces that make up a complete pattern. The size of the component pieces also affets cost because the size of the components often limits the number of cavities in a tool. Another factor affecting pattern cost is the size of the operation, as costs generally are lower for large installations operating the maximum number of hours. Other factors include thickness and density of the pattern (which affect cycle time) as well as labor rates, fringe benefits, overhead, etc.

Despite the inherent dangers of generalization, three typical pattern production cost estimates are summarized in Table 3. The reader is cautioned that these estimates are order of magnitude only and are presented only to point out the effect the size of an operation has on production cost. The numbers presented should not be used for any economic justification studies.

Table 3. Estimated EPS Pattern Molding Costs

SIZE OF OPERATION, NO. OF PRESSES	I	2	4
PRODUCTION, PATTERNS/HR.	60	I20	240
PATTERNS/YR.	360,000	720,000	1,440,000
GRAY IRON (POURED) EQUIVALENT, T/YR.	4,000	9,000	I8,000
INVESTMENT IN PATTERN FACILITIES	$200,000	$280,000	$410,000
PATTERN PRODUCTION OPERATING COSTS $/HR.	37	60	I00
PATTERN COST, ¢/PATTERN	66.2	54.5	46.2

GENERAL ASSUMPTIONS:

1. 3-SHIFT, 5-DAY, 50-WEEK OPERATION.

2. 4-CAVITY MOLDS, 2 COMPONENT PIECES FOR EACH PATTERN.

3. CASTING WEIGHT (WITH GATING), 25 POUNDS.

4. LABOR AT $6.00/HOUR, 25% FRINGE AND 50% OVERHEAD.

5. RAW MATERIAL EPS AT 60¢/POUND (4.5¢/PATTERN) INCLUDED

    IN PATTERN COST ABOVE.

6. COSTS DO NOT INCLUDE PATTERN ASSEMBLY.

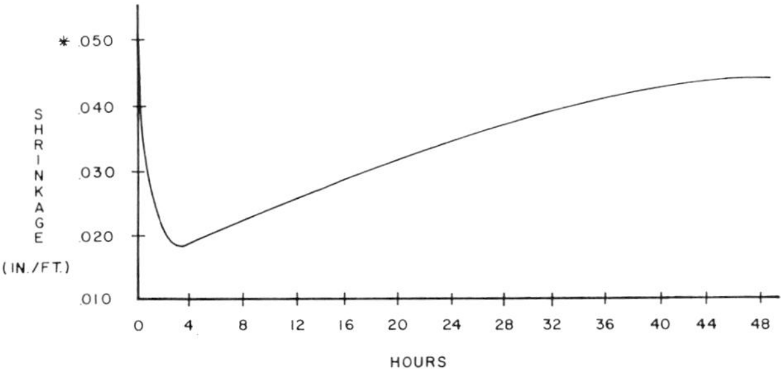

Fig. 8. EPS foam shrinkage by hours.

As indicated in Table 3, the estimated production costs for 1, 2 and 4-press operations are 66.2, 54.5 and 46.2¢/pattern, respectively. This assumes 4-cavity molds operating 3 shifts/day, 5 days/week, 250 days/year. Obviously, operating the molding plant for less than full time will increase the unit costs. These estimates assume the production of patterns for gray iron castings weighing 25 lb (including gating). Neither larger nor smaller patterns would appreciably affect production cost so long as 4-cavity tooling can be used and pattern thickness and density are about the same.

It is to be noted that the cost of drying and assembly of the pattern, tooling depreciation, building investment, and depreciation and profit at the molding plant level are not included in these estimates.

Pattern Shrinkage

Molded EPS shrinks with age and a foundry using the evaporative casting process must consider pattern shrinkage as well as metal shrinkage when designing pattern tooling. The amount and rate of EPS pattern shrinkage will depend to some extent on the size and density of the EPS beads being used. The age of the beads before pre-expansion and molding may also affect the amount and rate of shrinkage; however, to the best of the author's knowledge, these age factors have not yet been quantified for moldings made from the small, low density beads recommended for use in pattern molding.

Extensive determinations of shrinkage have been made for low density moldings made from small, pre-expanded beads which have aged for two hours or less and which were obtained from relatively fresh EPS raw material (nonexpanded) beads, a practice which will undoubtedly be common in most pattern molding shops. Figures 7 and 8 present shrinkage curves for moldings made from small beads at 1.25 lb/ft^3 density, which is recommended for gray iron castings. Figure 7 shows the shrinkage data over 30 days. Molded parts come from the mold undersized (cold mold dimension) about 0.050 in./ft. They then grow about 0.030 in./ft during the first few hours and then start shrinking. After 30 days the patterns have, for practical purposes, stopped shrinking. About 75% of the shrinkage

occurs in the first 7 days of aging. Figure 8 presents the shrinkage data for the same type moldings over the first two days.

Obviously, the dimensions of the casting will depend on the age of the pattern when the metal is poured. If it is practical for the foundry to age the foam patterns for 30 days prior to use, the shrinkage of the foam should not present any problems for casting dimensions.

Safety Considerations[7]

EPS is combustible and the combustion products are carbon monoxide, carbon dioxide, water, carbonaceous residue (smoke) and heat. Carbon dioxide, heat and the depletion of available oxygen are the principal toxic results of an EPS fire.

EPS, both in its raw state and after pre-expansion and molding, has, by definition, very low toxicity. Extended exposure to foamed polystyrene and blowing agent vapors, both in laboratories and commercial molding plants, has not resulted in significant health problems. Skin tests performed with EPS have shown no irritant effect.

The principal hazard associated with EPS is fire and adequate precautionary measures must be exercised during processing of the material and storage and handling of molded parts. The pentane blowing agent normally used is fugitive and evaporates from the beads and parts during processing and storage. Pentane vapor, which is colorless, is combustible when mixed with air in concentrations between 1.4 and 7.8% by volume and these combustible mixtures can be ignited with low intensity ignition sources. Mixtures within these concentrations are explosive only when confined. Adequate ventilation of processing and storage areas is required to keep the processing and storage atmospheres outside the combustible range. In the design of ventilation facilities one must consider that pentane is heavier than air and may accumulate in low spots.

Another important precautionary measure against fire is to eliminate sources of ignition. Smoking and open flames should be prohibited. Equipment should have adequate electrical grounding to prevent static buildup and discharge. Equipment with open flames, such as boilers, should be in a separate building or separated from processing and storage areas by fire walls. Additional safety considerations are available from the raw material manufacturers.

References

1. H. F. Shroyer, U. S. Patent No. 2,830,343 (Apr, 1958).
2. *Modern Plastics*, vol 55, no. 1, p 52 (Jan, 1978).
3. H. Rodman, Jr., U. S. Patent No. 3,023,175 (Feb, 1962).
4. E. J. Sikora, "Evaporative Casting Using Expanded Polystyrene Patterns And Unbonded Sand Techniques," *AFS Transactions*, vol 86, p 65 (1978).
5. R. H. Immel, U. S. Patent No. 3,577,360 (May, 1971) and U. S. Patent No. 3,759,641 (Sep, 1973).
6. S. B. Smith, U. S. Patent No. 4,032,609 (Jun, 1977).
7. "Safety Provisions For Handling, Storing, and Processing DYLITE® Expandable Polystyrene," ARCO Polymers, Inc., Technical Bulletin 87-AP16.

An Evaluation of the Vacuum Molding Process

A. E. Murton, Metallurgical Engineer and
R. K. Buhr, Head, Foundry Section
Physical Metallurgy Research Laboratories
CANMET, Dept. of Energy, Mines & Resources
Ottawa, Canada

ABSTRACT

Information and data are presented on research carried out at the Physical Metallurgy Research Laboratories in Ottawa, Canada, on the Vacuum Molding Process developed recently in Japan. Castings have been produced in all the common metals and alloys and a domestic supply of the plastic film established. In addition, information has been acquired on the rates of solidification in vacuum molds and compared with other molding techniques. The effects of sand blends of different permeabilities and size distribution are also discussed and the degree of carbon pickup in low-carbon stainless steel castings is shown to be related to the thickness of the plastic film employed and the section size of the casting.

Introduction

The vacuum molding process was discovered in 1971, and in the short period of five years more than 50 foundries are using the technique. When the information became available a couple of years ago, we were initially very skeptical, but following a few tests in our experimental foundry and a visit to several Japanese foundries using the process, the commercial feasibility of the technique was verified. Because one of our functions in the Physical Metallurgy Research Laboratories (PMRL) is to be informed on new technology, we undertook a program of investigation into the process and this report deals with the findings of that research.

To date, we have acquired sufficient equipment and expertise to produce small castings using the process in magnesium, aluminum, brass and bronze, cast iron, plain carbon steel and stainless steel. We have carried out work with a Canadian plastic manufacturer to develop a domestic supply of plastic film for the process; we have investigated a number of North American sands and blends to evaluate these for use in vacuum molding; we have determined the rates of solidification of castings when poured in vacuum molds and compared these to the solidification rates for these same castings when they are poured in other types of molds, and we have determined the carbon pickup in low-carbon stainless steel castings produced by this technique.

Equipment and Technique

The method used to produce vacuum molds has been amply described in other publications and will not be repeated here. The equipment and method used in the PMRL is illustrated in Figs. 1-4. The gas-fired radiant heater, pattern and vibratory table are shown in Fig. 1. The plastic film is held under the heater until it is near its melting point of about 100C (212F). This is judged by eye and is characterized by a loss in opacity of the film and the onset of shrinkage of the sheet. When this occurs, the film is lowered down onto the pattern and is sucked around the pattern by the judicious positioning of small holes through which entrapped air can be withdrawn. The flask is then placed over this, clamped in place and filled with the unbonded dry sand. The sand and flask are vibrated to obtain the maximum packing density of the sand, and the excess struck off. A second sheet of plastic film is then placed over this and the air evacuated from the sand. This instantly produces a mold hardness of over 90. The air is allowed to return to atmospheric pressure in the

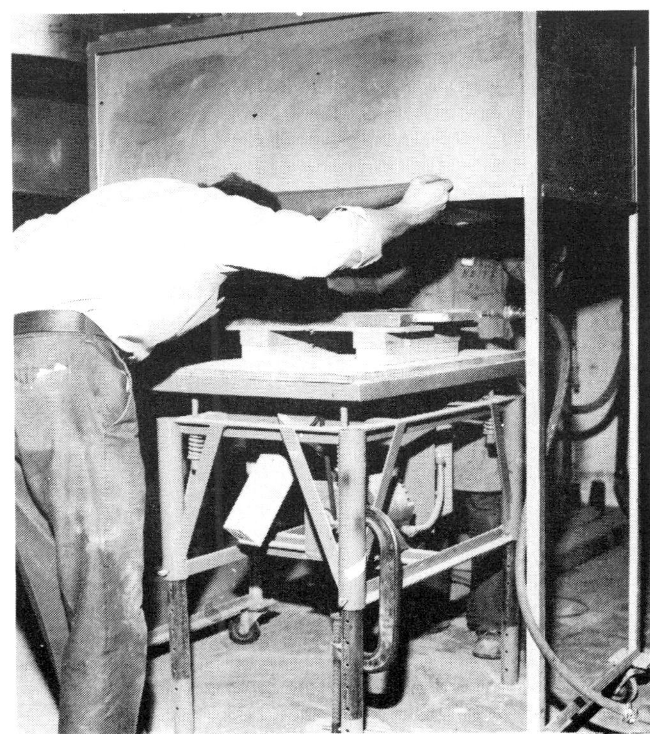

Fig. 1. Photograph showing the plastic film being held under the gas-fired radiant heater. Also shown is the vibration table and the pattern plate.

Fig. 2. This photo shows the completed mold (cope) after it has been removed from the pattern. Because the depth of draw around the sprue is not great, it has been formed directly from the pattern.

be lost, and the sand will drop. This point must be recognized when gating, venting and risering a casting made by the vacuum molding process. The gating design must allow the cavity to be filled as rapidly as possible and also in a manner which permits vents and risers to be located in such a way that the unfilled portion of the casting cavity is always at atmospheric pressure. As a general rule, unpressurized gating systems are preferred to achieve this end, along with gating into the lower portion of the cavity as much as possible. In our work, we have used a relatively thin cope and drag flask, because the test castings poured have, for the most part, been quite thin. This has permitted us to form the sprue and vents directly, as illustrated in Fig. 2. The completed molds, with one ready for pouring, are shown in Fig. 3.

Although the majority of the castings made in our experimental foundry have been quite small, we have made a few which are of a larger size. For this, we had a special flask constructed, measuring 27 x 33 x 6 in. cope and drag. We also devised a pattern manifold box in which we could use either a cope or drag pattern plate. These are shown in Fig. 4. We made some commemorative plaques for use both by the vacuum process and also by the CO_2 process. The differences encountered were: molding time reduced by half, no stripping problems, cast detail was excellent and cleanup was reduced to half the time for the vacuum-cast plaques. However, it should be mentioned that pattern manufacture took longer as considerable time was involved in drilling the numerous fine holes around the letters to ensure the plastic was drawn around the individual letters to reproduce the desired detail. Figure 5 shows the plaque as taken from the mold and wire brushed lightly. One other point to be mentioned when producing castings such as this is that it should be poured uphill to reduce the chances of pockets of unfilled cavity occurring and thus resulting in sand drops.

pattern, and permits the flask to be removed from the pattern, Fig. 2. Sprues, vents and risers can either be formed directly by the plastic film, if the degree of draw required is not excessive, or the sprue, vent or riser pattern can be wrapped in plastic and subsequently fastened to a boss on the pattern plate. After filling the cope with sand, a hole must be cut in the covering plastic film to allow excess plastic on the top of the wrapped pattern to protrude through. The application of a vacuum will partially seal the joint and the use of glue or adhesive tape to hold the excess plastic to the cover film will ensure a vacuum-tight seal and permit the pattern to be removed. It is most important that the plastic covering the joints made by the sprues, vents and/or risers be perforated. This is done to ensure that any gases formed during the initial stages of pouring can escape readily to the atmosphere and subsequently, that the unfilled portion of the mold cavity is always at atmospheric pressure. If this is not done, the pressure difference between the mold cavity and the sand will

Effect of Vacuum Molding on Solidification Rate

The Japanese have reported a slower rate of solidification for castings poured in a mold produced by this technique when compared to conventional methods. They also indicated that the initial rate of solidification was faster but slowed down shortly after pouring ceased. Tests were carried out in which bars having 1/2, 1, 1-1/2 and 2-in. sections were poured in cast iron. U-shaped sheathed thermocouples were located in the center of these sections to record the cooling curves for the bars, which

Fig. 3. Completed vacuum molds; two ready for pouring. Note that the plastic film covering the sprue hole and vents has been perforated. The various vacuum hoses are attached to a small surge tank, which in turn is connected to a primary surge tank and vacuum pump.

Fig. 4. Flask and special manifold for the plaque pattern. The flask walls are hollow and air is evacuated through the sides and the central pipes. The numerous vertical support blocks in the manifold box are necessary to prevent warpage of the pattern plate while subject to the negative pressure.

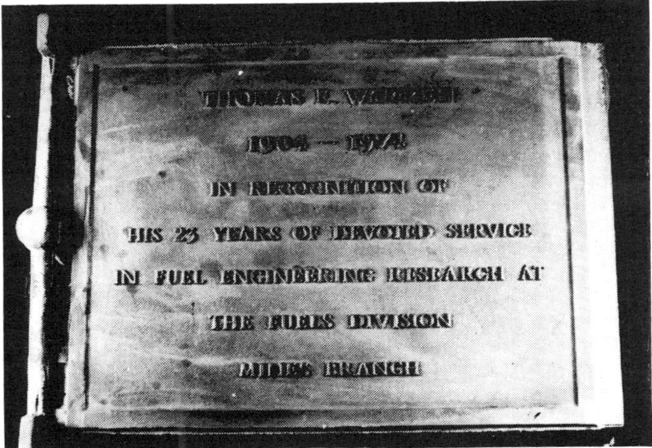

Fig. 5. Photograph of a plaque poured in 85-5-5-5. This casting has been lightly brushed only, and illustrates the surface finish and detail possible with the vacuum molding process.

were sufficiently long to eliminate any end effect. These were poured in green sand, CO_2 sand and vacuum molds made either with silica sand or zircon sand. In addition, to attempt to cover all combinations of bonding, moisture and presence or absence of air, bars produced by the full-mold process in unbonded sand of two different screen sizes were also poured. The results are presented in Table 1, using different criteria for cooling rates. The results for 2-in. section bars are presented graphically in Fig. 6. In this figure, the curves for green sand and CO_2 sand were almost identical, and so only one is shown. The same is true for vacuum molds poured in either silica or zircon sand, and so only the one curve is shown.

These results confirm the Japanese claims for a reduced rate of cooling when the vacuum molding process is employed. In these tests we could not detect a difference in cooling rate in the initial stages. However, our thermocouples were located in the center, whereas the surface of the casting would be the area to check for this feature. Nevertheless, it is most interesting to note this reduction in cooling rate and also the lack of difference in rate of solidification for bars produced in either silica or zircon

sands. We would suggest the following hypothesis to explain these results. In the vacuum mold, there is no moisture, no bond and no air circulation. The removal of heat from the solidifying casting is, consequently, mainly by radiation, either to the air from the exposed surfaces or through the sand in the other locations. In the full-mold castings produced, also with unbonded dry sand, we note the fastest rate of solidification. Here, it is obvious that we have ample opportunity for the removal of heat by air circulation past the casting. This is supported by the fact that when a finer sand is employed, the rate of solidification is slowed down. In the case of green sand or CO_2 sand molds, not only is there removal of heat by the evaporation of the moisture and an intermediate permeability between the two other molding processes, but also a more continuous and direct path for heat removal by radiation is available because of the presence of the bonding agent. Thus, a cooling rate intermediate between these two extremes occurs.

Table 1. Cooling Rates of Gray Cast Iron in Different Sections and Poured in Different Type Molds

Cooling Rate Criteria	Section Size, in.	Cooling Rate – °C per Second					
		Green Sand	CO₂ Sand	"V" Process		Full Mould	
				Silica	Zircon	AFS 25	AFS 45
1190° to 1170°C (Just above eutectic)	½	*	*	*	*	*	*
	1	4.0	4.0	3.33	2.86	4.0	4.0
	1½	2.0	1.82	2.86	1.54	2.85	4.0
	2	0.91	0.91	0.83	0.95	0.44	0.24
1170° to 1140°C (Through eutectic)	½	*	*	*	*	*	*
	1	0.27	0.29	0.20	0.18	0.49	0.35
	1½	0.13	0.14	0.09	0.08	0.18	0.15
	2	0.08	0.08	0.05	0.05	0.16	0.12
1190° to 1090°C (Overall)	½	2.13	2.56	1.82	1.64	3.03	2.85
	1	0.72	0.79	0.51	0.46	1.08	0.92
	1½	0.36	0.40	0.23	0.23	0.49	0.40
	2	0.20	0.21	0.13	0.14	0.41	0.25

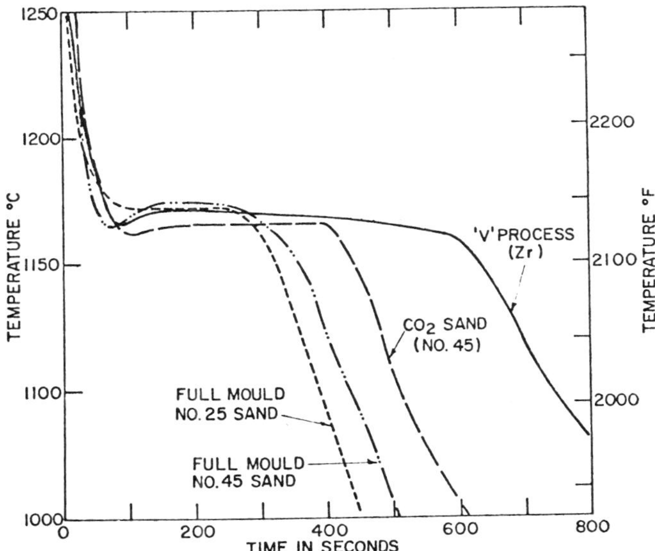

Fig. 6. Cooling rates for gray cast iron, 2-in. section poured in molds produced by different processes. Note: the cooling rate for green sand molds is almost identical to CO_2 sand, and for vacuum molds produced with silica sand is almost identical to that for zircon sand in vacuum molds.

A slower rate of solidification and subsequent cooling may be advantageous or disadvantageous. In cast iron, thinner sections could be poured in vacuum-made molds without carbide formation. On the other hand, in high production, the reduced cooling rate could slow the entire process, especially when one is normally looking at a closed-loop type of situation for an automated vacuum molding line, i.e., a limited number of flasks to form one unit. In steel, the slower rate of cooling may be advantageous in feeding, a longer time being available for the feeding to take place. Increased segregation may also occur — an area where additional research is required.

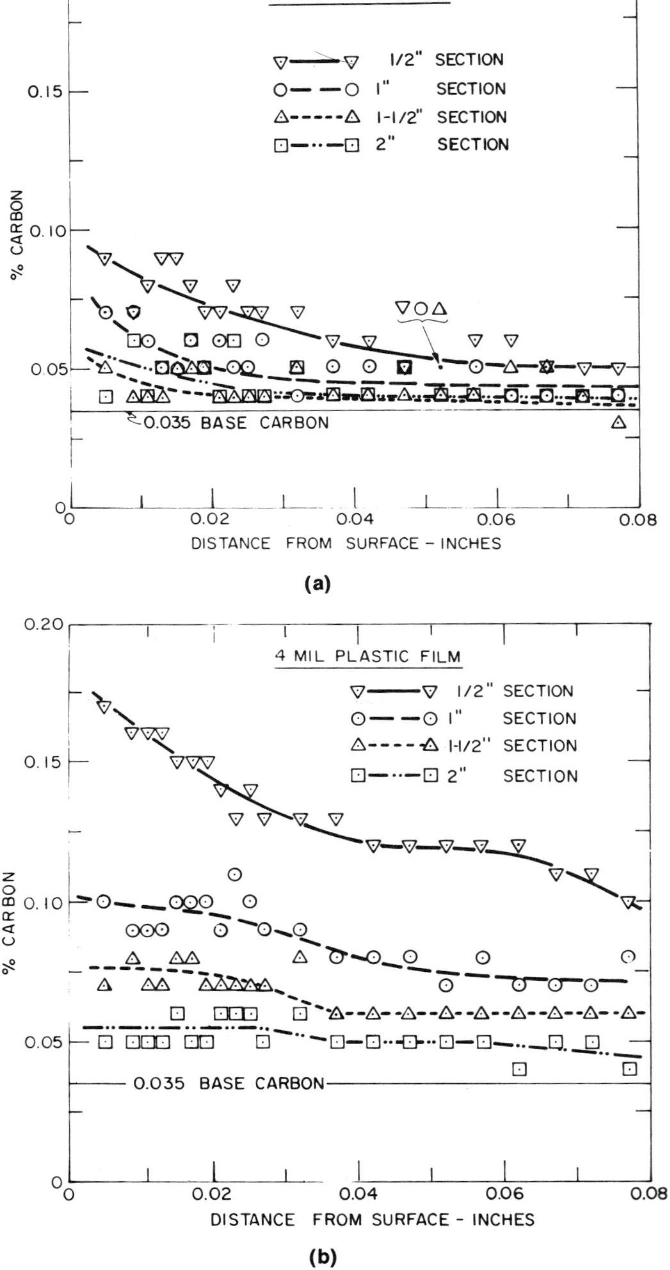

Fig. 7. The effects of plastic film thickness and section size on the carbon pickup in low-carbon stainless steel castings.

Domestic Supply of Plastic Film

A Canadian plastic manufacturing company has carried out relatively extensive work in developing the desired formulation and properties for a film suitable for the vacuum molding process. No detailed data will be presented related to this development project but tests have been carried out to evaluate the various films and compare them with film supplied for the process from Japan. The ethylene vinyl acetate film has been confirmed to be the best film for this process. It would also appear that an 18% vinyl acetate content is the optimum composition. In addition, additives have been made to increase the speed of heating the plastic. This manufacturer can now supply film suitable for the vacuum molding process.

It may be interesting to note that we have tried a large number of different tests to evaluate these films. Tensile properties in both directions have been made both at ambient and higher temperatures, deformation tests in which the ability of the plastic to draw down into different size cups was assessed, and measurements were made on the thickness of the plastic film after various degrees of deformation. However, none of these tests gave results which could be correlated with the production of satisfactory castings. The one test which we have found to give the best results is the ability of the film to deform around the letters on a small plaque. These letters are about 1/4-in. high and are close-spaced. Those plastic films which can conform to the fine pockets on these letters produce the best results. Putting it another way, it would appear that castings can be successfully produced from most of the plastic films, but it is the detail required which is important in the selection of the best film.

Carbon Pickup in Low-Carbon Stainless Steel

It has long been recognized that a low-carbon content is desirable in stainless steels when they are to be used in corrosive environments. Low-carbons are known to improve the general corrosion resistance and weldability and decrease the susceptibility to stress-corrosion cracking, pitting and intergranular corrosion for most grades of stainless steel and nickel-base alloys. Carbon pickup on the surface of wrought products presents no problems; in castings the situation changes. The greater use of organic binders in molding and core sands in recent years can result in significant increases in the carbon content of the surface of such castings, especially when the carbon content is very low in the base metal, and this higher surface carbon can affect the life of the casting markedly. Thus, foundrymen making castings for corrosive environments must take special precautions to ensure that this does not occur. It would appear that the vacuum molding process is ideally suited

Table 2. Carbon Pickup in the Surface of Low-Carbon Stainless Steel Castings Produced by the Vacuum Molding Process

Distance From Surface, in.	% Carbon for Following Conditions							
	0.002-in. Plastic Film				0.004-in. Plastic Film			
	½ in.	1 in.	1½ in.	2 in.	½ in.	1 in.	1½ in.	2 in.
0.005	0.09	0.07	0.05	0.04	0.17	0.10	0.07	0.05
0.009	0.07	0.07	0.04	0.06	0.16	0.09	0.08	0.05
0.011	0.08	0.06	0.04	0.04	0.16	0.09	0.07	0.05
0.013	0.09	0.05	0.04	0.05	0.16	0.09	0.07	0.05
0.015	0.09	0.05	0.05	0.09	0.15	0.10	0.08	0.06
0.017	0.08	0.06	0.05	0.06	0.15	0.10	0.08	0.05
0.019	0.07	0.05	0.05	0.05	0.15	0.10	0.07	0.05
0.021	0.07	0.06	0.04	0.04	0.14	0.09	0.07	0.06
0.023	0.08	0.05	0.04	0.06	0.13	0.11	0.07	0.06
0.025	0.07	0.05	0.04	0.04	0.14	0.10	0.07	0.06
0.027	0.07	0.06	0.04	0.04	0.13	0.09	0.07	0.05
0.032	0.07	0.04	0.05	0.05	0.13	0.09	0.08	0.06
0.037	0.06	0.05	0.04	0.04	0.13	0.08	0.06	0.05
0.042	0.06	0.05	0.04	0.04	0.12	0.08	0.06	0.05
0.047	0.05	0.05	0.04	0.05	0.12	0.08	0.06	0.05
0.052	0.05	0.05	0.05	0.04	0.12	0.07	0.06	0.05
0.057	0.06	0.05	0.04	0.04	0.12	0.08	0.06	0.05
0.062	0.06	0.04	0.05	0.04	0.12	0.07	0.06	0.04
0.067	0.05	0.04	0.05	0.04	0.11	0.07	0.06	0.05
0.072	0.05	0.04	0.04	0.04	0.11	0.07	0.06	0.05
0.077	0.05	0.04	0.03	0.04	0.10	0.08	0.06	0.04

Table 3. Screen Size Distribution of Base Sands Investigated

U.S. Screen No.	Per Cent Retained			
	Ottawa 62	Ottawa 78	Selkirk 77	Ottawa 135
20	0	0	0	0
30	Tr	0.1	Tr	0
40	1.5	0.2	0.1	Tr
50	17.5	2.3	0.7	0.1
70	34.8	22.5	18.3	1.1
100	29.0	40.5	53.2	8.7
140	12.7	27.2	22.4	34.0
200	3.3	7.2	4.2	37.7
270	0.6	0.2	0.5	13.6
Pan	0.1	Tr	0.1	4.8
AFS Fineness No.	62	78	77	135
Permeability	41	54	47	19

to the production of such castings since there is no bond of any description used. The only organic material is the plastic film itself and it is normally only 0.002 to 0.004 in. thick.

To check the possible carbon pickup, two castings were made, each having thicknesses of 1/2, 1, 1-1/2 and 2 in. One casting was made using 0.002-in. plastic film, and the other using 0.004-in. film. The castings were poured in low-carbon stainless steel, the base carbon content of which was 0.035%. The individual sections were separated and millings obtained starting from the as-cast surfaces and at various levels down to 0.077 in. The results of the analysis of these millings are listed in Table 2, and shown graphically in Fig. 7(a) and (b).

These results show that there is a significant carbon pickup in this low-carbon stainless steel. It is also obvious that the amount of the pickup is related to both the thickness of the plastic film employed and the section thickness of the casting involved. The depth to which the effect is significant is also related to the section size and the plastic thickness. If we assume that a carbon content of 0.05% is sufficiently close to the base analysis of 0.035% C to have minimal effect on the corrosion properties of the steel, the following depths of carbon pickup would result.

Section Thickness, in.	Depth of Carbon Pickup, in.	
	0.002-in. plastic film	0.004-in. plastic film
1/2	0.062	<0.077
1	0.027	<0.077
1-1/2	nil	<0.077
2	nil	0.032

A foundry contemplating the use of the vacuum molding for the production of stainless steel castings should be aware of the potential problem of surface carburization in this molding technique. The section size of the casting is very important, and the thickness of the plastic film required to produce the casting should be ascertained. It is important to realize that thinner plastic films can be used when a mold spray is employed, and

Table 4. Effect of Pressure and Film Thickness on the Rated Surface Appearance of Iron Castings in AFS GFN 78 Sand (See Text for Rating Description)

Plastic Thickness, in.	Pressure, inches of mercury		
	10	16	20
0.002	14	7	11
0.003	10	8	11
0.004	11	8	10

this may be the safest method to use if carbon pickup is anticipated.

Effect of Sands and Other Variables

The Japanese have indicated that a blend of two sands having widely separated peaks on the screen distribution curve produce the optimum results. We have carried out a number of tests using a variety of different washed silica sands having AFS gfn of 62, 77, 78 and 135, as well as blends of some of these sands with -270 mesh magnesite fines. The screen distribution of these sands is given in Table 3. The test casting employed is that shown in Fig. 2 and consists of a plate casting with 1/4- and 1/2-in. sections. Additionally, the effects of plastic film thickness, pressure and pouring temperature were also evaluated. Both cast iron and steel were poured in these molds, but not all the variables were investigated for each metal. In general, the cast iron heats were poured at 1330C (2426F) and the mild steel heats poured at 1620C (2948F). The castings were rated from a surface appearance point of view by three different persons using a scale of 1 to 5; 1 being the best. The rating numerals were then added up for each casting and this was used as a guide for assessment of the variable in question. Thus, a casting with a rating of 15 was the worst casting in the opinion of all three evaluators. Tables 4-6 list the ratings and the variables looked at. The pouring temperature effect was tested for iron only, the castings being poured at 1300, 1330 and 1360C (2372, 2426 and 2480F) using sand with an AFS gfn of 78, to which was added 5% fines. These showed the lower pouring temperature to have produced the best casting.

There is still considerable work to be done to more accurately determine the best sands and blends to produce high-quality castings in this process. The work done here indicates that what works for iron does not necessarily work in mild steel. For example, these tests indicate that the best results for iron are obtained when using a sand with an intermediate permeability, whereas with steel a higher permeability is desirable. Also, the 0.003-in. plastic film gave the better results with iron castings, and a 0.002-in. plastic film appeared to be better for the steel castings. One of the factors in this is believed to be associated with the initial amounts of gas formed when the metal first

Table 5. The Effect of Sand Permeability on the Rated Surface Finish of Iron Castings Using Either 0.002-in. or 0.003-in. Plastic Film (See Text for Rating Description)

Sand Identification	Permeability	Plastic Film Thickness	
		0.002 in.	0.003 in.
AFS #78 + 15% fines	11	12	12
" 10% "	16	12	3
" 5% "	26	6.5	5.5
" 0% "	58	7	8

Table 6. The Effect of Different Sands and Plastic Film Thickness on the Rated Surface Appearance of Steel Castings (See Text for Rating Description)

Sand Identification	Permeability	Plastic Film Thickness		
		0.002 in.	0.003 in.	0.004 in.
AFS #78 + 15% fines	11	15	--	11
" 10% "	16	12	--	11
" 5% "	26	8	--	11
" 0% "	58	3	5	7
AFS #62	40	11	8	8
AFS #77	50	5	6	--
" + 5% fines	29	--	3	--
AFS #135	21	5	11	14

enters the mold. When 0.003- or 0.004-in. film was used for steel castings, a considerable amount of "spitting" occurred and some metal was ejected from the vent holes. This did not occur when 0.002-in. film was employed. However, 0.003- and 0.004-in. film could be used with iron castings without this problem. One of the apparent consequences of the spitting was the appearance of rougher surfaces, especially in the drag, which appeared to be associated with penetration. Consequently, iron castings had better surfaces with thicker plastic films, and steel had the best surface when the 0.002-in. plastic film was used. At low permeabilities, evidence of erosion could be detected with iron castings, but in steel castings the surface was irregular and appeared to have been caused by trapped gases. It was also somewhat surprising that the AFS gfn 62 sand, with a lower permeability than the AFS gfn 78 sand, still produced a rougher surface. This may be due to the fact that the AFS gfn 62 sand has a 4-screen distribution as compared to the 3-screen distribution of the AFS gfn 78 sand. Our tests verify the Japanese findings that the 16-in. vacuum was preferred to either a higher or lower vacuum.

Based on our observations, it would appear that the best surface appearance is obtained in iron castings when they are poured at about 1300C (2372F) under a vacuum of 16 in. Hg, using an AFS gfn 78 sand to which about 5% -270 mesh fines have been added and using a 0.003-in. plastic film. For steel castings, an AFS gfn 78 sand should be used with no fines added. A 0.002-in. plastic film is preferred, although there may be a slight advantage in using a 0.003-in. film for the drag only.

Summary

1) The rate of solidification of castings produced in a vacuum mold is significantly slower than that for the same casting poured in either green or CO_2 sand molds, which in turn is slower than the same casting poured in an unbonded sand mold using the full-mold technique.

2) A domestic supply of appropriate plastic film is now available for use on the vacuum molding process.

3) There is a possibility of carbon pickup from the plastic film, especially in low-carbon steels and thin sections. This pickup becomes negligible as the plastic film thickness is reduced and the section size of the casting increases.

4) The tests carried out indicate that gray iron castings are best produced using sand with an intermediate (e.g., 15 to 30), permeability and a plastic film thickness of 0.003 in. A low pouring temperature is also beneficial, and a vacuum of 16 in. was found to give the best results. For steel castings, a higher permeability in the sand was better (over 40), but the best results were obtained when 0.002-in. plastic film was employed.

In general, it is felt that the vacuum molding process has great potential and should be seriously considered as an alternative molding process, especially when a new foundry is being started. The absence of heavy and energy-consuming machinery, such as sand mullers and shakeouts, should reduce the overall capital expenditure and also reduce floor space requirements. To date, the vacuum molding process cannot compete with high-production green sand molding lines. It can, however, reduce cleaning time and should result in fewer defective castings since moisture-related defects are no longer present. Because it is a new process, a training period is essential so that the normal thinking pattern can be altered to cope with the day-to-day problems that arise in any molding process. It would appear, at this stage of development of the process, that it will find its widest application in those shops producing intermediate size castings with relatively light (1 to 1-1/2 in.) sections.

The H Process of Repetition Casting Manufacture

F. H. Hoult, *Managing Director*
W. H. Booth & Co., Ltd.
Rotherham, England

ABSTRACT

The H Process utilizes rigid, double-sided molds produced by coreblowing machines. Each mold carries a half casting impression on each side, complete with runner bar, feeder head and ingate. Molds, when clamped together horizontally, form a pouring system of a top continuous runner and feeder bar. This runner/feeder bar is at one end opened up to form a pouring basin.

Each mold is so designed that, when the string of molds is poured, control of the flow of metal is obtained. Consequential casting is ensured. The first mold is poured and is full before the metal proceeds to fill the second mold. This process repeats itself until the string of castings is poured. This procedure of mold filling ensures a "flow feed" of each casting.

A mold cannot be filled until the preceding mold is full; thus, all metal must pass through the feeder heads of the preceding castings. Each casting has freshly introduced metal passing through its feeder head. Therefore, the maximum yield of metal is obtained in the cast shape and a minimum amount of metal is returned for remelting.

As runner sphere or riser is situated above the top of the casting, all the metal flowing from the riser into the casting heats the surrounding sand, creating a hot spot. This hot spot keeps the runner liquid and facilitates the feed of metal from the riser into the casting through the thin section of runner (ingate).

The H Process achieves the lowest possible ferrostatic casting pressure. Strain on mold joints is reduced to a minimum. Rigidity of molds, increased by horizontal clamping, results in the final castings being precise dimensionally, with minimal mold joints. These factors result in precise castings with little or no cleaning being required.

Introduction

The H or Horizontal controlled pour process of repetition casting manufacture was developed and put into production four yeas ago at the Rotherham England Foundry of W. H. Booth & Co. Ltd., Master Founders in Ductile Iron.

The advantages of controlling variables in casting manufacture have been proved. To explain these advantages and show how they can be applied to the production of castings with resultant efficiency, economy of power and raw materials, is the object of this paper.

Development has been in the production of ductile iron castings, but the application of the H Process to most cast metals has been established. To date, the size, weight and complexity range has been limited.

Description of the H Process

The H Process is a means of incorporating into a rigid sand mold all the factors which contribute to and control the production of a sound, precise casting, resulting in minimum work to be done in finishing or cleaning the casting, and with maximum yield of good castings per unit of metal poured.

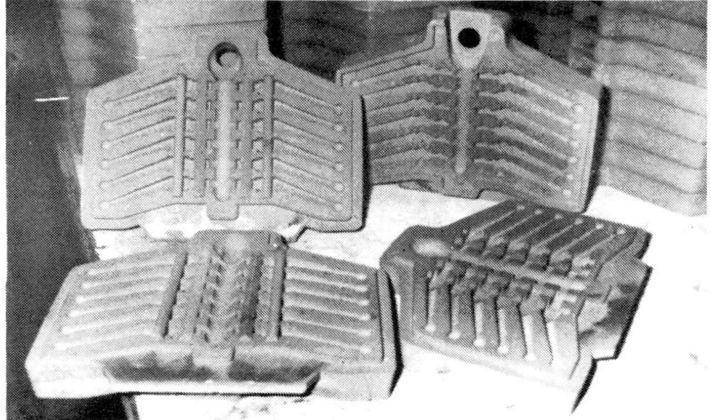

Fig. 1. Double sided H Process resin sand cores showing:
Top right: **Runner Bush.**
Bottom Right: **End core with runner end closed.**
Top Left: **Multiple core.**
Bottom Left: **Shows coreblowing method.**

The factors affecting repetition production of good castings can be listed and classified as follows:

1) a rigid sand mold capable of repetitive, economical production
2) contained within that mold must be a runner system which enables controlled pouring of the metal into that mold, smoothly and without agitation or turbulence
3) during the pouring of liquid metal, the mold must retain its shape, must resist any erosion, breakage or washing away of sand on the face of the mold. The mold, after metal has been poured, must have a means whereby liquid and solid contraction may take place. Liquid metal must be available to feed all parts of the casting that require feeding at the point where it is required and for as long as it is required

Fig. 2. String of H Process castings as cast, and double sided cores complete with feeder/runner.

Fig. 3. Method of vertical assembly of cores. Clamping of molds.

Fig. 4. Mold assembly laid horizontal with runner bush, ready for casting.

Fig. 5. Molds stacked two tier, shortly after pouring of metal.

Fig. 6. Molds being poured, showing control.

Fig. 7. Pouring of molds. Castings exposed as sand burns away.

4) after the metal has set, forming the casting, the mold must collapse or break down, allowing the contraction of the casting

5) it must be possible to remove the gating and risering systems with the minimum of effort and without destroying any of the required casting shape.

The H Process fulfills the above functions in a controllable repetitive way. The following paragraphs will describe the process.

Metal equipment is designed and made which, when fitted to a coreblowing machine, will produce sound sand cores. The cores or molds, whichever name may be given to them, carry a half impression of the required casting on each side of the core. This

Fig. 8. Molds break down 15 minutes after pouring.

Fig. 9. Further mold breakdown showing different castings.

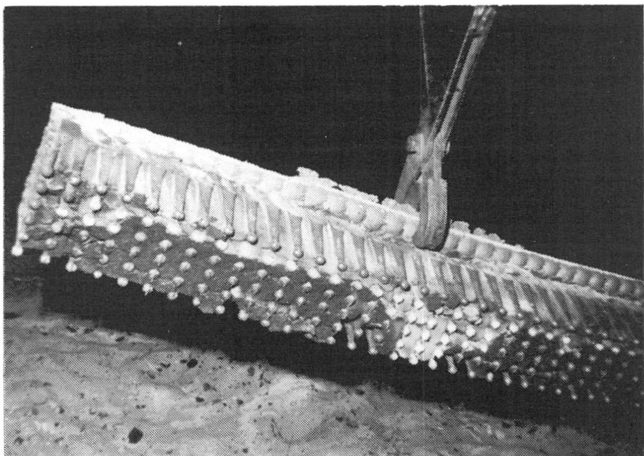

Fig. 10. Castings as removed. Ready for knockoff.

Fig. 11. String of castings as removed from mold.

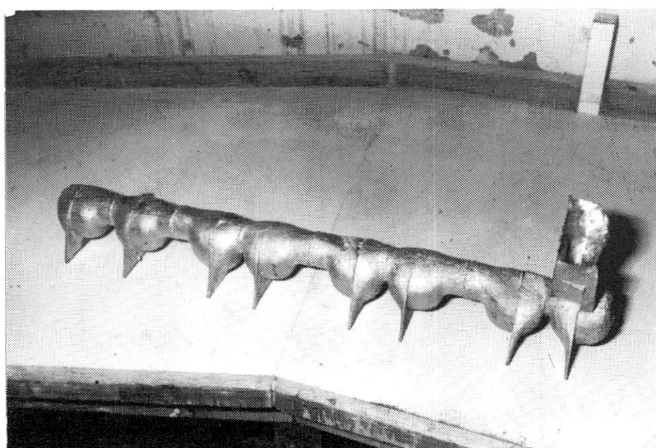

Fig. 12. Runner/feeder bar. Castings broken off runner as returned for remelting.

Fig. 13. H Process strings of castings as cast. Sample castings alongside.

Fig. 14. String of small cylinder castings as cast.

Fig. 15. Knockoff castings shown falling away from feeder/runner.

Fig. 16. Further typical runner/feeder bars as returned for remelting.

core/mold must carry, on one side or both sides, the ingate or runner. Above the runner must be a riser of suitable shape for the particular casting it has to make. This runner/riser must be so designed and made that when the cores are assembled horizontally the holes made by the runner/riser section provide a means of pouring a complete string of castings, each casting with its own runner/riser head.

The cores/molds are assembled vertically, located each on the next core by means of complete location pieces, named for clarity by the author as "Picture Frame Location," and clamped together by means of tie bolts. The assembled string of cores is lowered into a horizontal position ready for pouring. End cores/molds of the string are made so that the feeder/runner is "stopped off" or "filled in" to provide closed ends of the feeder/runner. At one end of the string of cores/molds, the feeder/runner is opened up to form a pouring bush.

Special Considerations in Using the H Process

The design and functioning of this riser/runner is the vital part of the H Process. The riser/runner design ensures that, when metal is poured, its flow is so controlled by shape and size that metal enters only one mold at a time. The size and shape of the riser/runner above each casting is such that the rate at which metal fills the casting is balanced against the rate of metal filling the section of riser/runner above each casting. Consequently, each casting is poured under control and a casting and the riser/runner above each casting must be full or nearly full before the metal passes over the weir into the next casting.

This controlled pour of the string of castings is responsible for the most important secondary function of the H Process. The function of the "flow-feed" of each casting is obtained by the controlled pour. Let us consider the metal entering the riser/runner of casting No. 10. This metal has flowed through the riser/runner above castings 1 to 9. Therefore, castings 1 to 9 have hot metal flowing through the riser/runner head above the casting and feeding of liquid metal will take place as long as metal is required to be fed into the casting.

With the casting being poured in the vertical position, under controlled pouring conditions, directional solidification takes place under good, if not perfect, conditions.

A contributory factor to the success of the flow-feed is that as soon as pouring begins, all metal for the casting flows downward through the ingate and the riser/runner above the ingate gradually begins to fill. Thus, a heat reservoir, or hot spot is created and increasingly functions during the complete filling of the mold. This hot spot, or heat reservoir, made by the heated sand around the ingate, enables this ingate to be of a minimum size, with the consequent advantages of breakoff.

The flow-feed function is responsible for the high yield of liquid metal to finished castings. The string of castings cools evenly and slowly and, when required, the advantages of slow cooling can be obtained.

Thus, the functions of the H Process, controlled pouring and flow-feed, together contribute to give high yield and shrinkage-free castings. The completed strings of castings after cooling can be picked up en bloc, the castings removed by minimum effort, and the single runner returned in one piece for remelting.

Development of the H Process has resulted in the design of the riser/runner being changed to a series of spherical shapes. The name runnersphere has been applied and it is felt that it aptly describes the function of this part of the Process. The runnersphere as a unit from which the complete feeder/runner is made up has proved to have applications to all metals to varying degrees. The size of the runnersphere will vary with different metals and research and development is being carried out on the point.

Squeeze Casting of Brass and Bronze

by R. F. Lynch, Manager, Special Projects,
R. P. Olley, Senior Project Engineer,
Research and Development Department,
Fabricated Products Group,
N L Industries, Inc., Toledo, Ohio,
and P. C. J. Gallagher, Head,
Metals and Fabricated Products Dept.,
Central Research Laboratory,
N L Industries, Inc., Hightstown, N.J.

ABSTRACT

Squeeze casting is a hybrid metalforming technique now emerging as a viable commercial process. Components are produced from molten metal in a one-step operation. These components exhibit the density and property levels normally associated with forgings, combined with the configurational detail and low unit cost characteristic of castings. Process economies derive from the full utilization of molten metal and the direct efficient nature of pressure application achieved in a small sized automated machine.

Compatability with the squeeze casting process has been demonstrated for a broad range of cast and wrought alloys of brass and bronze. Sound, fine-grained, homogeneous metallurgical microstructures have been attained in both as-cast and heat-treated material. Mechanical properties comparable or superior to handbook values have been achieved for forging and casting brass, bearing and wear-resistant bronzes and high-strength Mn and Al bronzes. Comparisons are made between squeeze-cast and conventionally forged or cast microstructures and related to observed similarities and differences in alloy response and resultant properties.

Brass and bronze squeeze-cast components are shown in a variety of solid, hollow and ring configurations, demonstrating process versatility. Prototype components illustrate specific applications well suited to squeeze casting include — valve and regulator bodies and end caps, hydraulic fittings and connectors, motor cups, pump pistons, meter and control housings, asymmetric and variable shape components, flanges and hubs, gear blanks, wear plates, synchronizer rings, straight and flanged bushings and bearings.

Introduction

Squeeze casting is a metalforming process in which a single forging stroke is applied to a metered quantity of initially molten alloy to yield a high-integrity finished part. The material of manufacture is typically dispensed into a die attached to the bed of a hydraulic press. Forming and solidification result from pressure application across the entire surface area of the part as the punch attached to the press platen enters the lower die in which it is a close slip fit.

The process has reportedly been widely applied in the Soviet Union, where it originated, to nonferrous and ferrous materials.[1,2] Interest and research has been evident in Japan, Europe and the U.S.[3-8] Industry sources have indicated limited in-house application of this technology in the U.S. for Cu-base alloys, with some public disclosure.[9,10] However, only in 1974 has a squeeze-casting process become available for the custom manufacture of nonferrous components.[11]

Application of the squeeze-casting process to Cu-base materials of manufacture will be reported upon in this paper.

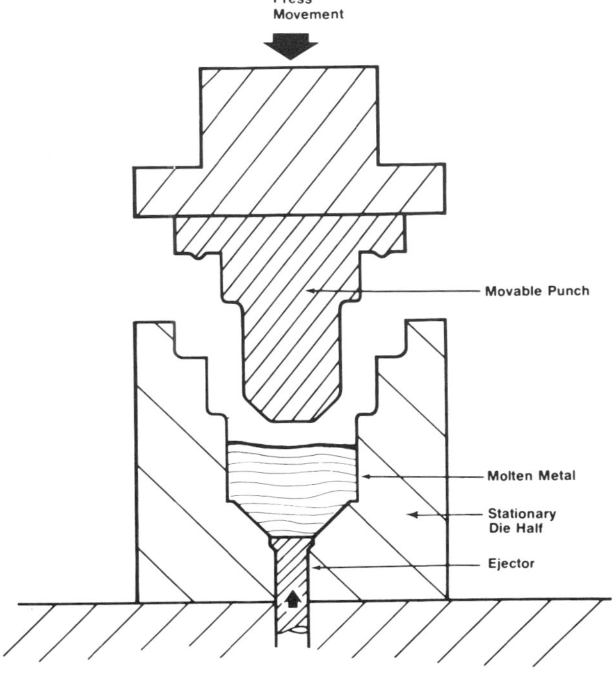

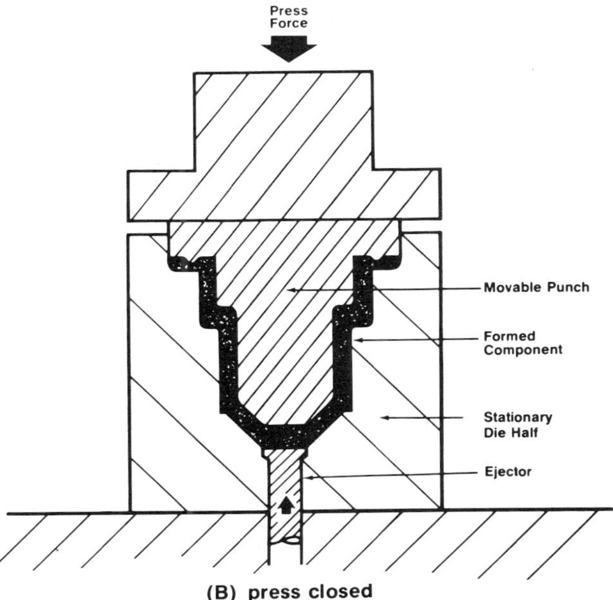

Fig. 1. Schematic of the squeeze-casting process.

Distinguishing process characteristics, component shapes, material properties, microstructures and product selection criteria will be presented for commercial brass and bronze alloys. Parallel results pertaining to the squeeze casting of Al are contained in a companion paper.[12]

Process Characteristics

Features of the squeeze-casting process are apparent in a simplified schematic of process operation, Fig. 1. First, molten metal is precisely metered into the lower die cavity with the punch poised for the forging stroke, Fig. 1A. The punch then forms a molten metal seal as it descends into the die cavity, continues in one stroke to form the part shape, and finally holds the metal under high pressure until solidification is complete,

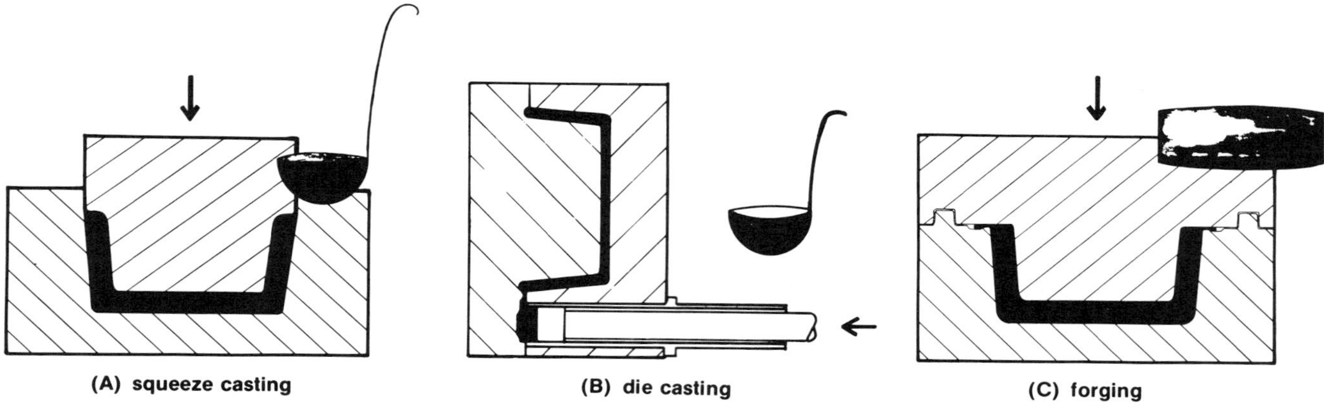

Fig. 2. Schematics of component fabrication by alternate processes.

Figure 1B. The component is then removed from the die using ejector pins or a stripping mechanism, the die is cleaned and recoated, and the cycle is repeated.

This hybrid casting-forging approach has a beneficial effect on material properties, an effect derived from aspects of metallurgical structure on both the macroscopic and microscopic levels. On the macroscopic level, porosity caused by gas entrapment is prevented by the quiescent nature of the process. The metal lies in a stationary pool until pressure is applied, with gases then venting at die seal locations ahead of the moving metal. Shrinkage porosity is prevented by the application of high hydrostatic pressure during formation, which provides liquid to the solidification front from within the part itself, compensating for the decrease in volume upon freezing. This volume change is accommodated by a slight controlled reduction in part height as the transforming material flows under the influence of constant applied pressure.

On the microscopic scale, squeeze castings are characterized by a refined microstructure having fine grains, close dendrite arm spacing and small constituent particles. The fine grain size results from a high level of nucleation and subsequent rate of growth. Nucleation occurs initially in the undercooled region at the die wall and is enhanced as metal movement promotes back melting, dendrite shearing and mixing with rapid cooling. The hydraulic shock encountered at the instant of die closure is felt to contribute further to nucleation.

A high rate of heat removal, second only to that of die casting among commercial processes, occurs because the solidifying metal is forced against the colder die material under pressure, minimizing air gap formation. Close dendrite arm spacing, small constituent particle size and refined grain size all result from this high rate of heat extraction. A separate comparative study of dendrite arm spacing[13] has shown that the heat transfer coefficient, h, during solidification under pressure in squeeze casting can be as high as 0.2-0.3 cal/cm^2 sec C, in contrast to values of 0.02-0.06 cal/cm^2 sec C in permanent mold casting.

The net result of the solidification process as it occurs in squeeze casting is the formation of a porosity-free, homogeneous, refined microstructure. Strength properties are maximized for casting alloys in the squeeze-cast condition, while properties comparable to those in the forged condition are obtained for wrought alloys. Freedom from porosity allows full heat treatment and promotes good toughness and ductility.

Comparison to Alternate Metalforming Techniques

Variably shaped components are produced by a number of metalforming techniques; sand casting, die casting and forging are widely used for brass and bronze. Each alternate method has distinct capabilities and related costs which determine specific areas of application. Squeeze casting provides yet another set of capabilities, unique to this hybrid casting-forging process. Similarities to and differences from alternate processes are noted and discussed. Schematics, Fig. 2, portray the fabrication of a bowl-shaped component by the alternate rigid-die, machine-forming techniques: squeeze casting, die casting and forging.

Squeeze casting is a useful complement to the brass die-casting process, able to produce sound parts with greater section thicknesses than are normally die cast. Complex-shaped, high-tolerance parts up to 0.125 in. thick will continue to favor the high-rate diecasting process. Squeeze casting, however, offers advantages for high-integrity components cast in straight-draw dies with a cycle time which can approach that of die casting. The direct nature of pressure application by the punch to the solidifying component, Fig. 2A, allows a smaller die to be used than is necessary in die casting. For example, the die used to produce a 6.5-in.-diameter, 5-lb annular bronze ring is shown with the component in Fig. 3. A ring-shaped punch, central post and bottom ejector pins are incorporated in the die. The die impressions are contained within an 11-in.-square holding block. Die life is expected to be similar to that in brass die casting; the higher specific applied pressures in squeeze casting replacing the high-velocity impingement of metal spraying into the die cavity through a restricting gate characteristic of die casting, Fig. 2B.

Sand casting provides a low fixed-cost method of fabrication, also allowing complex undercuts and channels to be cast into the part but with a rough surface finish. Many small-sized parts can be cast simultaneously in the same mold, increasing output. In contrast, squeeze casting can provide full-density components without shrinkage or microporosity and having a smoother surface finish and closer tolerances than by sand casting. For example, the shrinkage from the die for squeeze-cast bronze components was found to be only half that experienced in sand casting. Pressures applied during and after solidification forces the metal against the rigid die and prevents shrinkage away from the wall as occurs in gravity-casting methods. Even when full squeeze-cast advantages are not required in a component, the process can prove to be the most cost-effective manufacturing technique by reason of its high production rate and high metal utilization efficiency. Savings result from reduced energy costs for melting, reduced melt loss, elimination of the runner and riser removal step and reduced machining costs.

Several interesting comparisons can be made between squeeze casting and forging. The pressure necessary to consolidate material originally in the molten state is significantly less than

Fig. 3. Squeeze casting die and component.

that which must be applied to form a like configuration in the normal hot working temperature range. Fig. 2C. Thus, smaller capacity presses are required, and the dies themselves tend to be much smaller. The influence of the higher metal temperature relative to that employed for forging tends to affect die life adversely, but is offset by the lower forming pressure required. Part configurations tend to be less restricted for squeeze casting than for traditional forging. Cuplike configurations have successfully been made having a height twelve times greater than the minimum section thickness, with the section itself varying by factors of two or three from place to place. Radii can be considerably smaller than for forgings and fine surface finish is generally obtained. Components currently manufactured as cored forgings are well suited to squeeze casting. The squeeze casting is made with the cores initially extended into the die cavity or incorporated in the punch for the molten metal to flow around, rather than by applying considerable forging force to hot pierce the required cored holes. The use of molten metal and

the high metal utilization level characteristic of squeeze casting combine to provide economic advantages compared to forging where preprocessed forging stock is employed and scrap must likewise be reprocessed.

Configurations, Material Properties

Compatability with the squeeze-casting process has been demonstrated for a wide range of cast and wrought Cu alloys. Satisfactory property levels have been achieved in flat, ring and hollow shapes for forging and die-cast brass, high strength Mn and Al bronze, and bearing and wear-resistant bronze. Sound, homogeneous, fine-grained metallurgical structures have been attained in both as-squeeze-cast and heat-treated conditions.

Over a dozen Cu alloys have been squeeze cast in a variety of solid and hollow configurations, from 0.3-5 lb in weight. Five experimental configurations weighing between 0.50 and 1.25 lb, each having a nominal outside diameter of 2 in., are pictured in Fig. 4. Sectioned castings for the hollow shapes show components with a total height-to-minimum wall thickness ratio up to 12:1. Large hollow and ring-shaped components, also squeeze cast, extend experience to parts having sections from 0.1-1.0 in. thick.

Yellow Brass

Brass is widely fabricated in compositions containing 60-65% Cu, small additions of Si (1%) or Pb (2%) for die casting and forging, respectively, with the balance Zn. Fortunately, standard brass compositions respond well to squeeze casting, and prototype components have been made in a variety of configurations.

Forging brass, CDA 377, in particular, has been squeeze cast with a high degree of success. Tensile properties were determined using flat specimens cut from 0.35-in. wall sections of the high-cup configuration, Fig. 4. Properties comparable to typical handbook values for the wrought (extruded) condition[14] are obtained in squeeze-cast forging brass, Table 1.

Distinct similarities are also noted between micrographs of

Fig. 4. Squeeze-cast brass and bronze components, whole and sectioned.

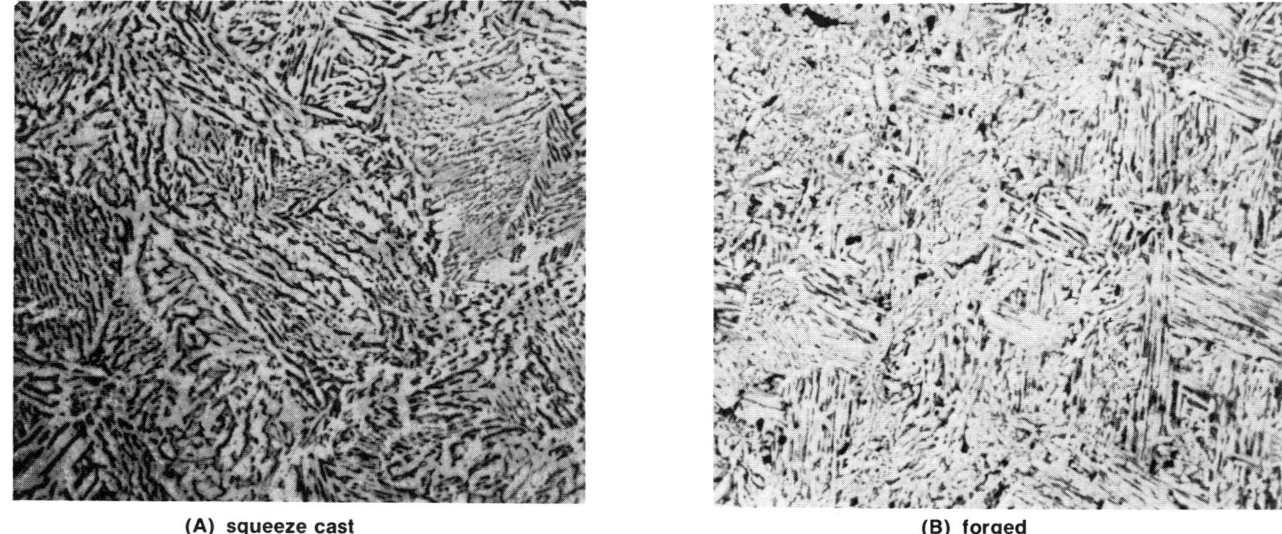

(A) squeeze cast   **(B) forged**

Fig. 5. Comparative microstructures of forging brass, CDA 377. (250 X)

Fig. 6. Macrostructure of squeeze cast manganese bronze ring.

material in the squeeze-cast and forged states, Fig. 5, both showing fine-grained, homogeneous, porosity-free microstructures. A somewhat more isotropic distribution and sometimes finer form of the transformed Cu-rich alpha phase is observed in the squeeze-cast material. Because squeeze-cast properties are isotropic, they compare favorably in all directions with those in the best or flow direction of a comparable forging.

Although Cu-Zn brasses have lower melting ranges than other common Cu alloys, still lower melting alloys would further reduce processing costs by decreasing cycle time and increasing die life. Squeeze-casting process characteristics potentially allow the utilization of new alloys not practical with traditional metalforming methods. Promising initial results have been obtained in Cu alloys containing significant con-

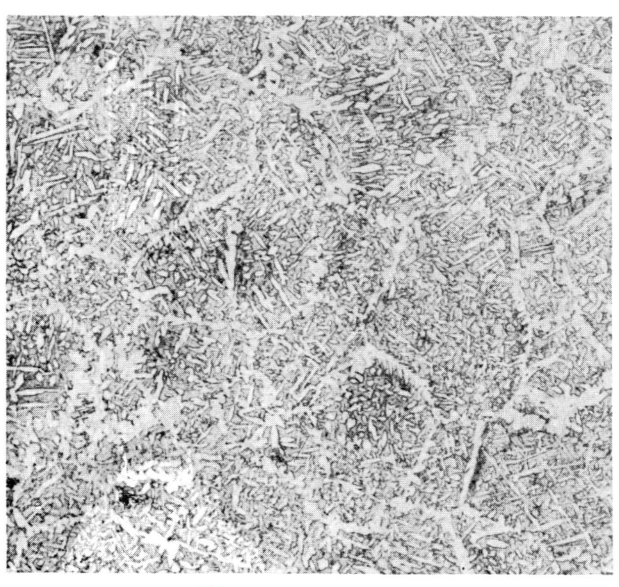

(A) as-squeeze cast   **(B) heat treated**

Fig. 7. Microstructure of squeeze cast aluminum bronze, CDA 624. (250 X).

TABLE 1. — **Comparative Properties of CDA 377 Forging Brass**

	Ultimate Tensile Strength (ksi)	Yield Strength 0.2% Offset (ksi)	Elongation (%)
Squeeze Cast	55.0	28.0	32 (in 1 in.)
Extruded	55.0	21.0	48 (in 2 in.)

centrations of Mn and Si in addition to Zn: ultimate tensile strength — 70 ksi, yield strength — 40 ksi, and elongation — 10%.

High-Strength Manganese and Aluminum Bronze

Manganese bronze alloys designed for forging and casting usage have also responded well to squeeze casting. Mechanical properties for casting, CDA 865, and wrought, CDA 674, Mn bronzes are presented, Table 2. Section soundness and a uniform fine-grain structure are noted, Fig. 6, in the macrostructure of the 5-lb constant 0.75-in.-thick Mn bronze ring viewed whole in Fig. 3.

Aluminum bronzes are noted for their high strength levels, developed through a eutectoid transformation achieved by heat treatment. The corrosion-resistant wrought Al bronze, CDA 624, is well-suited to forming by squeeze casting with properties comparable to typical forged values, Table 3. A uniformity of microstructure is noted in both the as-squeeze-cast and transformed material, Fig. 7.

Bearing and Wear-Resistant Bronze

Copper alloys containing varying amounts of alloy additions, principally Sn and Pb, provide a broad spectrum of balanced bearing and strength properties. Sand casting is commonly used with these alloys, as are continuous and centrifugal casting.

The prototype flanged bearing pictured in Fig. 8 was squeeze cast in the widely used high leaded Sn bronze, CDA 932 (SAE 660). A sound, smooth machined finish was obtained on all surfaces of the component, which weighs 1.75 lb as shown, with a 2.9-in.-diameter flange and 0.20-in.-thick walls.

Promising results have also been obtained on higher Pb alloys having greater lubricity but lower strength. The rapid rate of solidification inherent to squeeze casting leads to a desirably more uniform dispersion of Pb which is subject to gravity settlement when longer solidification times are involved. Such results have been observed in laboratory castings of the 25% Pb, 5% Sn alloy, CDA 943 (AMS 4840A). Large, globular, Pb-rich particles present nonuniformly in a gravity casting contrasted sharply to a far more uniform interdendritic dispersion, in squeeze cast thin sections, Fig. 9A, with only minimal centerline segregation in thick sections, Fig. 9B, of a squeeze-cast cup configuration.

TABLE 2. — **Properties of Squeeze Cast Manganese Bronzes**

	Ultimate Tensile Strength (ksi)	Yield Strength 0.2% Offset (ksi)	Elongation (% in 1 in.)	Hardness (BHN-500 kg)
CDA 865 (casting alloy)	69.0	34.7	13.0	105
CDA 674 (wrought alloy)	75.0	47.0	7.0	——

Fig. 8. Prototype squeeze-cast flanged bearing.

TABLE 3. — **Comparative Properties of CDA 624 Aluminum Bronze**

	Ultimate Tensile Strength (ksi)	Yield Strength 0.2% Offset (ksi)	Elongation (%)
Squeeze Cast	113.5	53.0	13.5
Forged	102.0	50.0	15.0

Wear-resistant applications require alloys characterized by increased hardness in addition to bearing properties. The leaded Sn bronze CDA 925 (SAE 640), containing 11% Sn, is traditionally sand cast and finds application in transmission components and gears. Process versatility is underscored by the breadth of freezing ranges among Cu alloys adaptable to the process, from 20°F (11°C) for yellow brass and high-strength bronze to 300°F (167°C) for this leaded Sn bronze.

A heavy-duty transmission synchronizer ring has been successfully squeeze cast in large quantities in CDA 925. To produce this 6.5-in.-diameter ring weighing 2 lb as-machined, a 3.25-lb squeeze casting was poured which compares to the 7 lb of metal poured for each alternately produced sand-cast ring.

Component configuration, with a height ratio between thick-lug and thin-web sections of 3.3:1, coupled with the solidification characteristics of this extended freezing range alloy made this a difficult squeeze-casting application. Simulated-use testing of squeeze-cast synchronizer rings for 100,000 cycles resulted in full user qualification for this application.

Tensile properties of squeeze-cast CDA 925 leaded Sn bronze reflect the sound, refined microstructure obtained, Table 4, compared to sand-cast material. Results in each case are from flat test bars cut from actual components. Wear testing was conducted for 10,000 cycles under a 120-lb normal load with oil lubrication. Satisfactory results were obtained for squeeze-cast CDA 925, Table 4, correlating with performance during simulated use testing.

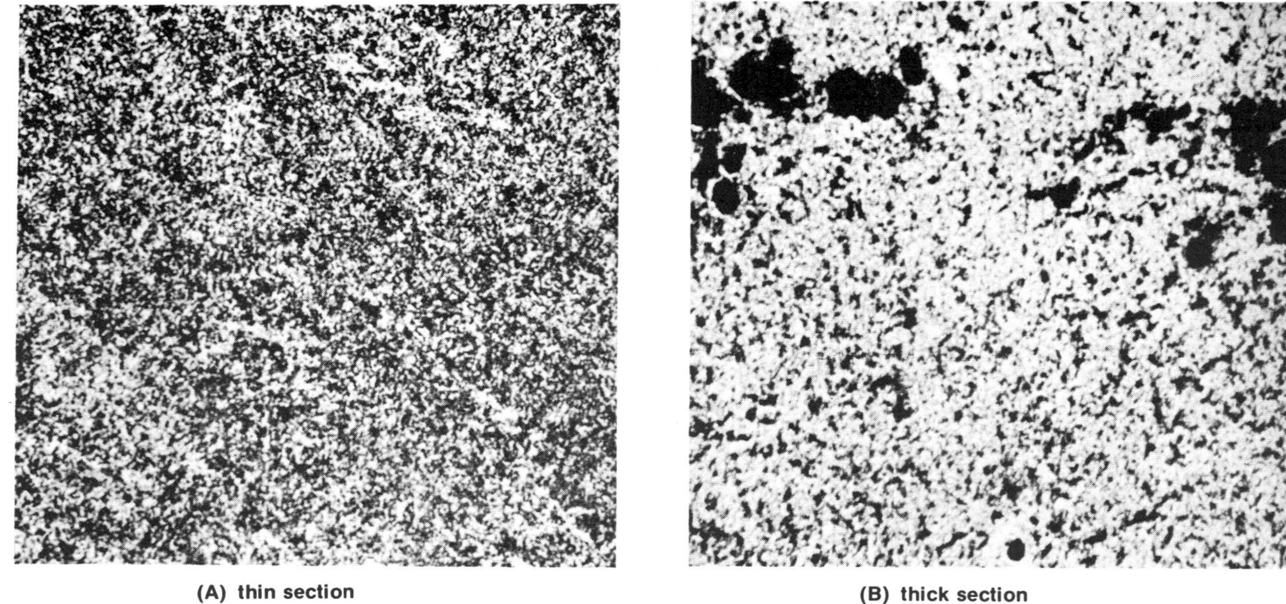

(A) thin section　　　　　**(B) thick section**

Fig. 9. Microstructure of squeeze-cast high leaded bronze, CDA 943. (75 X)

Graphic illustration of the effectiveness of the squeeze-cast process on CDA 925 Sn bronze is revealed by metallographic study. Etched microstructures are compared between sand-cast, permanent mold-cast, continuous-cast and squeeze-cast material, Fig. 10. The structural refinement and homogeneity of the squeeze-cast material is evident, particularly in the reduced particle size of the wear-resistant, Sn-rich delta phase. Scanning electron microscopy further illustrates the difference in constituent particle size and distribution between squeeze-cast and continuous-cast forms, with positive identification of the delta phase particles made by an analysis of Sn distribution, as revealed by electron microprobe scans, Fig. 11.

Product Selection

Alternate process capabilities and related costs ultimately determine products selected for manufacture by a given process. A consideration of squeeze-casting capabilities provides a set of criteria for product selection, to allow maximum utilization of process potential —

1) Broad choice among commercial casting and wrought alloys of brass and bronze.
2) Porosity-free, fine, homogeneous metallurgical structures which are fully gas-tight.
3) High-level mechanical properties as fabricated or after heat treatment.
4) Fabrication of components having thick sections and significant variations in thickness including flat, ring and deep-hollow configurations.
5) Complex surface detail and good surface finish.

Additionally, the high metal utilization efficiency of squeeze casting and the use of molten metal minimize starting material cost. These considerations are particularly important for alloys which are subject to high melt losses, or for which adjustments to alloy chemistry are necessary upon remelting. The high cooling rate attained in this fast machine-casting process allows a high rate of production. Smaller lot sizes are practical compared to some alternate techniques, because a smaller-size die and press are adequate.

A number of brass and bronze applications have been identified as being well suited to squeeze casting. Valve and regulator bodies, end caps, hydraulic fittings and pump pistons, all call upon the pressure tightness, sound mechanical properties, good surface detail and close dimensional tolerance capabilities of the process. Gear blanks can be made in a variety of materials, including the commonly employed casting and forging alloys. The practice of casting large gears in composite fashion, with a bronze ring on a hub of cast iron, is also practical for squeeze casting. Bushings and bearings can also be made well, particularly those types which are at present made by centrifugal casting. Surface features can in some cases be cast in. For instance, in flanged bearings, oil grooves having excellent definition can be formed in the end of the flange. Synchronizer

TABLE 4. — Comparative Properties of CDA 925 Leaded Tin Bronze

	Sand Cast	Squeeze Cast
Tension Test:		
–Ultimate tensile strength (ksi)	44.4	55.4
–Yield strength 0.2% offset (ksi)	26.4	35.6
–Elongation (% in 1 in.)	16.5	19.2
–Hardness, R_E	84	98
Wear Test:		
–Initial static coefficient of friction	0.158	0.192
–Final static coefficient of friction	0.124	0.149
–Final kinetic coefficient of friction	0.121	0.142
–PV	98,100	94,300

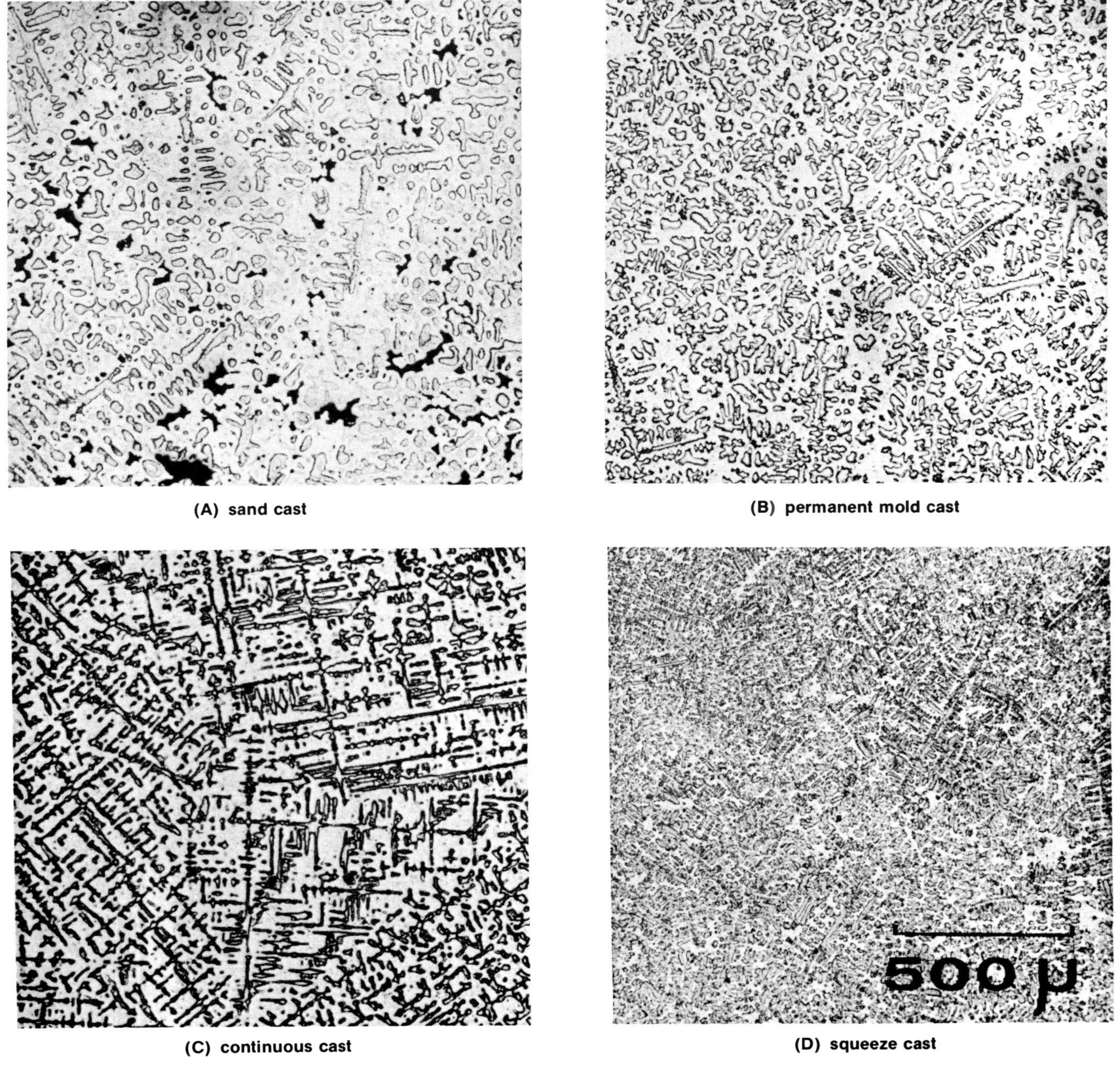

Fig. 10. Comparative microstructures of leaded tin bronze, CDA 925. (55 X)

rings are well suited to squeeze casting by reason of their annular configuration, while in some designs a variability in cross section and a requirement for absence of porosity in the heaviest sections pose problems for casting by traditional technologies. Wear and thrust plates also benefit from the refined microstructures that can be obtained in wear-resistant alloys. Flanges, hubs, meter bodies and control housings are among the asymmetric and variable shaped configurations that are also well suited to squeeze casting.

At this writing, utilization of the process for the custom manufacture of Cu alloy parts is just beginning. Dies for a synchronizer ring and electric motor cup are being completed for production use. However, the overall extent to which squeeze casting will find manufacturing utilization will be determined by the degree to which process capabilities can be matched with component needs. Squeeze casting will prove to be employed for those applications requiring the design freedom and lower cost structure characteristic of casting, combined with the properties and structural integrity associated with forging.

Acknowledgements

The authors wish to acknowledge the support and encouragement of the staff of the Central Research Laboratory, N L Industries, Inc. Their continued assistance in squeeze casting and process evaluation over the extended period during which this work was conducted was invaluable. Assistance provided by members of the Research and Development

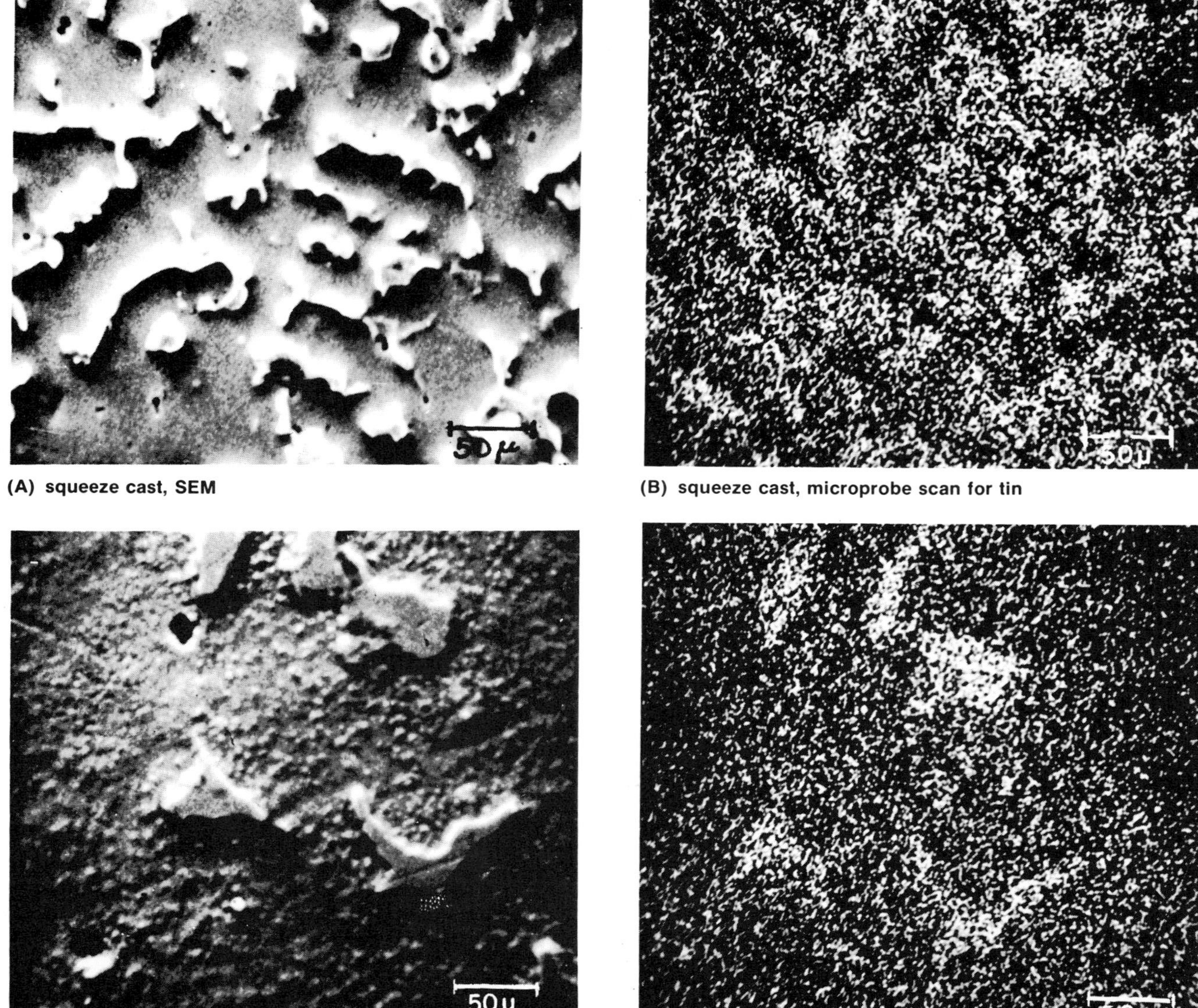

(A) squeeze cast, SEM

(B) squeeze cast, microprobe scan for tin

(C) continuous cast, SEM

(D) continuous cast, microprobe scan for tin

Fig. 11. Comparative size and distribution of delta phase in leaded tin bronze, CDA 925. (250 X)

Department of the Doehler-Jarvis Division, N L Industries, Inc. is likewise gratefully acknowledged. Appreciation is also extended to the personnel of the Bearings Division, N L Industries, Inc. for their strong efforts during manufacturing trials.

References

1. V. M. Plyatskii, Extrusion Casting, Primary Sources, New York, N.Y. (1965).
2. A. I. Batyshev, E. M. Bazilevskii, Yu. A. Evstratov, V. K. Zabaluev, F. A. Martynov and Yu. S. Pugin, "Casting Copper Alloys with Solidification Under Piston Pressure," Russian Castings Production, p. 220-223 (June 1972).
3. W. F. Shaw and T. Watmough, "Squeeze Casting: A Potential Foundry Process," Foundry, Vol. 97, p. 166-169 (Oct. 1969).
4. J. C. Benedyk, "Squeeze Casting," Paper No. 86, 6th SDCE International Die Casting Congress (Nov. 1970).
5. J. C. Benedyk, "Manufacturing Possibilities with Squeeze Casting," Paper No. CM71-840, SME Upper Midwest Tool Exposition and Engineering Conference (Nov. 1971).
6. J. C. Benedyk, "Squeeze Casting: Combining Forging Properties in a Large Casting," Paper No. 72-DE-7, ASME Design Engineering Conference and Show (May 1972).
7. K. M. Kulkarni, "Squeeze Casting Comes of Age," Foundry M & T, Vol. 102, No. 8, p. 76-79 (Aug. 1974).
8. R. F. Lynch, R. P. Olley and P. C. J. Gallagher, "Squeeze Casting," Paper to ASM Delaware Valley Chapter (April 1973).
9. I. Thomas, "How to Hot Hob Beryllium Copper," Plastics Engineering (July 1961).
10. S. T. Brewer, F. R. Dickinson and C. A. Von Roesgen, "Repeaters and Equalizers for the SD Submarine Cable Systems," The Bell System Technical Journal, Vol. 43, No. 4, p. 1243-1273 (July 1964).
11. E. J. Stefanides, "Forging Properties at Casting Costs," Design News, Vol. 29, No. 10, p. 60-61 (May 20, 1974).
12. R. F. Lynch, R. P. Olley and P. C. J. Gallagher, "Squeeze Casting of Aluminum," Paper No. 75-122, 79th AFS Casting Congress (May 1975).
13. T. W. Caldwell, Central Research Laboratory, N L Industries, Inc., Hightstown, N.J., private communication.
14. Metals Handbooks, Properties and Selection of Metals, Vol. 1, American Society for Metals, Metals Park, Ohio (1961).

Fig. 13. Zircon resin shell molds for railway brake beam castings ready for pouring.

Fig. 14. A brake beam casting made in a shell mold such as shown in Fig. 13.

low permeability of such a sand compact. The literature indicates that care must be taken to be certain that zircon sand mixtures which are expected to resist extreme penetration are rammed hard because the shape of the grains resists easy packing. Zircon flour is the base material for the highly refractory mold washes which are used to increase resistance to penetration and to give a good surface finish.

Zircon sand has been used successfully for some time in the production of both iron and steel castings by the resin shell mold method. Mason[12,13] reported in 1968 on the use of zircon shell molding to make iron castings in Britain. Taft[14] reported in 1968 on the use of zircon shell practice in Britain. American Steel Foundries pioneered[15] in the use of resin shell molding with zircon sand for carbon steel castings and for the last 20 years has operated a production foundry which uses only this molding method. Castings produced with the method range in weight from 4 to 250 lb. Figure 13 shows an assembled shell mold for a railway brake beam and Fig. 14 a typical casting made with such a mold. This is a good application for shell molding since the amount of sand used is proportional to both the size and weight of the casting. When made in green sand, the rangy nature of this casting design requires a disproportionately large flask and weight of sand. Zircon resin shell molding has been popular for some time for making stainless steel castings in the US.

Advantages claimed for zircon shell molding for iron and steel castings are that it improves dimensional control, gives good surface finish and, because of its excellent compatibility with phenolic resin, it can be more economically bonded in comparison with other sands. When making steel castings, it is possible to increase casting yield in many cases because of the ability to reduce the size and therefore the weight of the gating system by taking advantage of the ability of the molding material to withstand the erosive effect of molten metal. Because of the high cost of zircon sand, it must be reclaimed for reuse when used as the sole material for making shell molded castings. The sand is economically reclaimed with maximum losses per cycle of 5% and typical values of 3% by heating the sand to the temperature range 1500-1850F (816-1010C) in a rotary kiln or other suitable device in the presence of air so that the resin and other carbonaceous matter are burned. Users of this approach report that sand processed this way can be recoated with resin with equal or higher efficiency than upon first use. Because of zircon's extremely low expansion and chemical inertness, it can be repeatedly reclaimed thermally without grain breakage or size change.

Olivine

Olivine is mineral consisting of a solid solution of two silicates: magnesium ortho silicate which has the mineral name forsterite and iron ortho silicate which has the mineral name fayalite. Small amounts of other minerals are usually present. The quality of the sand depends upon the ratio of forsterite to fayalite; the higher the amount of forsterite the better, since fayalite fuses at a much lower temperature. The tentative SFSA specification for this material stipulates a minimum forsterite content of 80%. The type and the amount of secondary minerals is also important since they can greatly influence the gas content and fusion point of the sand. Olivine occurs as a rock and must be crushed, ground and classified to produce olivine sands having desirable characteristics for making molds. Figure 15 shows typical olivine sand grains. Note that the grains have a very angular shape and a very rough pitted surface character. Large deposits of olivine occur in Norway, the US, USSR and Japan, and smaller quantities are found in Scotland, France and Sweden. Olivine sands have been in commercial use in Norway and Sweden for about 40 years. Beckius[16] has reported that all the steel castings in Norway and 66% of all steel castings in Sweden are produced in olivine sand molds. In addition, many iron castings made in these countries are produced with the use of olivine sands. Four Swedish ingot mold foundries use only olivine sand. The initial reason for the interest in olivine sand was to replace silica sand because of its possibility for causing silicosis in workers who are exposed to silica dust. Usage has

Fig. 15. Olivine sand.

proven that olivine sand is effective in preventing burned-on sand when making austenitic manganese steel castings. It is widely used in Europe for this purpose and it has become a popular molding material in US foundries which produce austenitic manganese castings. Although olivine is the least expensive of the specialty sands, it is considerably more expensive than silica sand. For that reason, it is desirable to reclaim the material. An informal survey showed that there are six manganese foundries in the US having closed systems in which olivine is the only sand used. Pneumatic or mechanical reclamation is used and it is claimed that recoveries on the order of 85% are obtained. Although the sand is readily bonded with bentonite, the angular shape and roughness of the grain and the high acid-demand value for the sand due to its chemically basic nature prevent economic bonding with binder systems using an acid catalyst. Although the sand is sometimes used to make resin shell molds, binder efficiency is relatively low because of the high surface area of the material and its basic nature. It is readily bonded with inorganic binders such as sodium silicates and phosphates.

There have been mixed results with olivine sand when making carbon and low-alloy steel castings. Some have reported that olivine causes cold shuts or wrinkles and that the ability to reduce scabs or other expansion-type defects is not as good as when using zircon and chromite. Sometimes pinholes occur on casting surfaces. Some of the reasons given for the cold shuts and pinholes are that new olivine sand gives off a great deal of steam and carbon dioxide due to the decomposition of impurities in the olivine. This can be overcome to some extent by venting. If the sand is washed and/or calcined before use, the likelihood of such behavior is said to be greatly reduced and it is pointed out that when olivine is used in a closed system casting, quality improves with the age of the system. Ortfeldt[17] reported the results of nine years of operating experience with olivine in a steel foundry. He found that when fresh olivine is used to make carbon steel castings, poor results are obtained. When no more than 15% fresh olivine is mixed with recycled olivine, good results are obtained. Ortfeldt says that large carbon steel castings having excellent surface finish are made with recycled olivine and that a chromite wash gives the best results. One US foundryman is successfully making both carbon and austenitic manganese steel castings in a closed olivine system and found that he follows Ortfeldt's practice and gets good results.

Perhaps the confusion over the ability of olivine for carbon steel castings would be dispelled if some casting experiments were made using olivine which had first been calcined and then washed and graded to remove the fines and correct the grain size distribution which will have been altered by grain splintering occurring during calcining. Experience[16-19] has demonstrated that closed systems using either olivine or zircon sands are free of the hazards of silicosis. This is an important point and can be the deciding factor in selecting the molding material when all costs are considered.

Other Specialty Sands

Although chromite, olivine and zircon are by far the most popular specialty sands in the US at the present time, there are several other promising materials which have now appeared on the market and the industry is determining their best usage. The first of these to appear was aluminum silicate sand which is mined in Florida in the same operation in which zircon sand is a byproduct. The sand grains are clean and rounded, tubular in shape. The sand is chemically neutral, can be efficiently bonded with any binders now on the market, is highly refractory and has a regular, moderate thermal expansion which makes it resistant to expansion-type defects. However, this material has now been withdrawn from the market because of high production costs compared to competitive sands.

Zircore is a relatively new entry in the specialty sand field and is an impure form of zircon sand containing aluminum silicate, iron magnesium silicate and quartz. The sand grains are rounded and somewhat prismatic in character and can be bonded with high efficiency. The material is chemically neutral and can therefore be used with the full range of binders now available. Thermal expansion is moderate and regular and there should be no difficulty with expansion-type defects. This material is offered with the possibility that it can compare favorably in performance to zircon at a reduction in the cost of sand. It should be pointed out that this material contains approximately 10% free silica and therefore does not avoid the silicosis hazard as does the pure zircon sand.

There is mention in the recent papers by Bownes[7] and Middleton[4] of chrome magnesite sand. Kryanin[20] reported in 1954 on the use of crushed chrome magnesite refractories as a molding material for stainless steel casting and indicated that this is a popular sand for improving results compared with silica sand in difficult situations. The chrome magnesite employed is crushed and graded material from used chrome magnesite bricks and is bonded with clay or sodium silicate. It is well known in the steel industry that chrome magnesite refractories are very resistant to attack by molten steel and thus it is not surprising that this material can prevent penetration and chemical attack by the molten steel. Bownes and Middleton in the above papers report that chrome magnesite sands, produced by the fusion of seawater magnesite with chrome are being offered in Great Britain. An informal survey of US refractory manufacturers indicates that this type of material is not now offered in the US.

It is interesting to examine the list of materials which are used in making molds for investment castings. The popular materials are chamotte (aluminum silicate), mullite, fused silica, chromite, zircon, alumina, alpha quartz and cristobalite. The shaw process which is used to make castings with excellent dimensional and surface texture fidelity uses about 70% mullite and 30% chamotte grog and is bonded with ethyl silicate.

Sand Consumption and Cost

Table 3 shows the usage of silica and speciality sands in the US for 1974 and 1975. It is interesting to reflect that if a similar survey had been made in 1950, chromite would not have appeared on the list and zircon and olivine usage would have been almost nonexistent. It must be concluded that the specialty sands are effective since they have grown in usage despite the fact that their prices are all considerably higher than that of silica sand. Their usage can be expected to increase as the combined efforts of foundries and suppliers determine the most effective ways to use them and particularly explore the possibilities for the use of the materials in closed systems where they can be recycled.

Table 3. Ferrous Foundry Sand Consumption (Tons)

	1974	1975
SILICA SAND	11,227,000	9,941,000
CHROMITE SAND	50,000	80,000
ZIRCON SAND	85,000	50,000
OLIVINE SAND		45,000

Molding and Coremaking

Mold Materials for Ferrous Castings

1977 HOYT LECTURE

J. A. Rassenfoss, *President*
Amsted Research Laboratories
Bensenville, Illinois

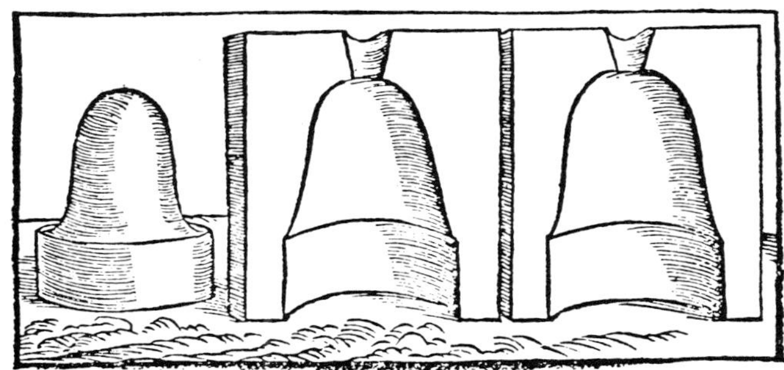

Fig. 1. Green sand mold and core for casting a small bell.

ABSTRACT

Materials which are being used to make one-use, semipermanent and permanent molds for ferrous castings are reviewed. Properties, foundry behavior and characteristic applications for the various materials are described. The economics of sand reclamation are demonstrated and the effect of mold material and process choices on energy usage and the environment are considered. Estimates are made for the consumption of silica and specialty sands and the amounts of ferrous metals cast in sand, metal and graphite molds.

Dedication

Ed Hoyt's career made a great contribution to the growth and stature of the American Foundrymen's Society and I think it very fitting that we honor his memory with this annual lecture series. I'm very pleased too with the honor to have my name added to the list of distinguished foundrymen who have given the Hoyt Lecture. It is a happy coincidence that I have this honor here in Cincinnati because this city has been the scene of two very important beginnings for me. The first was in 1881 when young George Rassenfoss came to Cincinnati from Germany. Here he learned the baker's trade and went forth to central Kentucky to seek his fortune and there met and married an Irish girl and started the family from which my father came. The second beginning occurred when I was a student at the University of Kentucky. Our department had gotten a new arc furnace and some of us wanted to learn how to make steel castings. We arranged for a visit to Cincinnati Steel Castings, to Lunkenheimer and to Sawbrook, and we got a fine reception at each foundry; but at that time, Jack Caine was the chief metallurgist at Sawbrook and Jack not only showed us around but also gave us all sorts of information about sand mixes, risering practices and molding methods and invited us to bring a truck down and take home sand and binders. We took him up on his offer, started to make steel castings and made several visits to learn more about the foundry business and the wonders of Cincinnati. I got hooked on the foundry in this way and it was probably no surprise that I entered the steel casting field when I left the University. I found when I got into the industry that Jack was just as helpful to others as he had been to us and until his death two years ago he was a tireless worker and a technical leader in our industry. In gratitude for that, I dedicate this lecture to Jack's memory here in Cincinnati, his home town.

Introduction

The metalcasting process began when someone poured molten metal into what turned out to be a mold made from suitable materials and found that the solidified material had the shape and the surface texture of the mold that had been made. Biringuccio, in his 16th Century book, *Pirotechnia*,[1] says, "There are many kinds and varieties of earth that are used for the loam compositions for making the molds for casting bronze, brass and other metals." This, I think, was the first reference in a metallurgical text to what we now call naturally bonded molding sand. Biringuccio went on in that text to describe many foundry practices that we use today: green and dry sand molding of brass and bronze; casting of bells with the molding method shown in Fig. 1; and he even described the casting of iron cannon balls in, of all things, permanent molds made of cast iron. Well, they say that you don't have a scholarly paper unless you make a reference to an ancient text; so we've solved that problem now.

Silica Sands

Naturally bonded sand remained the principal mold material for making ferrous castings until the beginning of this century. But in the United States, the abundance of low-clay content, high-purity silica sands and the difficulty in obtaining consistently good steel castings with naturally bonded sand, promoted the development of what we now call synthetic molding sand. Use of the term synthetic in this context does not mean that the sand is artificial. Actually, the ingredients of synthetic sand are, for the most part, natural ingredients and we have synthesized the characteristics of a naturally bonded sand by combining a number of materials. In his Silver Anniversary paper for the molding materials group in 1966, Briggs[2] updated the classic paper of 1939 by Briggs and Morey[3] on synthetic sands for steel castings. The 1966 paper describes the basic philosophy for formulating a steel foundry sand as bonding a clean silica sand of 99% purity with a suitable amount of western bentonite and the addition of enough gelatinized corn flour to impart toughness to the green mixture and act as a cushioning agent to avoid expansion defects. The 1939 paper described the effects of moisture and other ingredients on the characteristics of the sands produced. Sands suitable for making iron castings follow the same general philosophy; however, in iron foundry practice, the western (sodium) bentonite can be replaced in whole or in part by southern (calcium) bentonite and fireclays and wood flour or other inert cushioning materials can be used in place of the corn flour. Seacoal or a suitable petroleum derivative is a common addition to improve surface finish. Suitable silica sands for such mixes occur abundantly in the US. Popular silica sand for steel foundry use in the Midwest is obtained from outcrops of the St. Peter sandstone which occur in Missouri, Minnesota, Iowa, Illinois and Wisconsin. Washed sand from these deposits will characteristically have a silica content on the order of 99.9%. Grains of this sand have a smooth surface and round shape in sieve sizes through 70 mesh and as the sand becomes finer, it tends to be subangular while retaining a smooth surface. The AFS grain fineness number of sands from these deposits ranges from 27 to 120. High-purity silica sand deposits in New Jersey are popular in the Eastern seaboard. In general, the grain shape is subangular but the surface of the grains is smooth. The glass sands of central and western Pennsylvania occur in rock form and require crushing. As a result, grain shape tends to be angular and the surface of the grains is somewhat rough. Figures 2-4, respectively, illustrate the sands which have just been mentioned. As grain shape goes from round to angular and surface character goes from smooth to rough, the total surface area per unit weight increases. For this reason, the rougher, more angular grain sands require greater amounts of organic binders to achieve a given strength level. However, good casting results can be obtained with any of these sands.

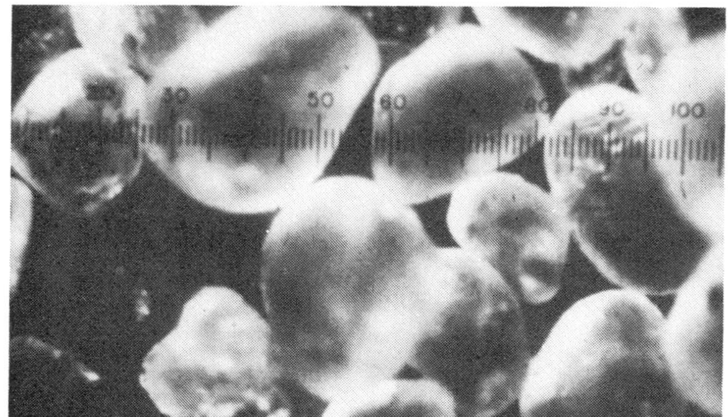

Fig. 2. Ottawa sand — round grain.

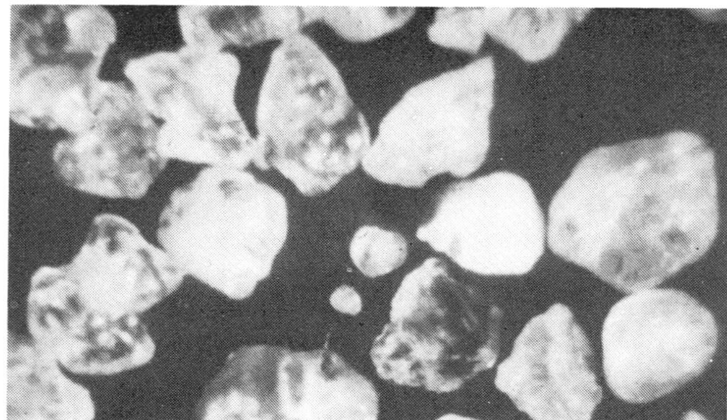

Fig. 4. Pennsylvania glass sand — angular grain.

Most medium- and small-size iron castings do not require silica sands of 99% silica purity. Lake and bank sands of 95% silica content are commonly used in this type of work. Figures 5 and 6 show such sands. Silica sands are readily bonded with clays, bentonites and the wide variety of organic binders which have now been introduced to the foundry industry. With existing technology, iron and steel foundrymen are able to make a wide variety of castings of good quality using silica sand as a base material. Silica sand has the advantages of abundance, ease of bonding with both inorganic and organic binders, ability to be reclaimed for reuse by wet, dry or thermal methods and low cost. Disadvantages of silica sand are its high thermal expansion which requires carefully controlled additions of cushioning materials to prevent and to avoid casting defects caused by deformation and rupture of mold surfaces in contact with molten metal, inability to resist penetration and reaction when in contact with casting surfaces to re-entrant angles or in pockets of large iron and steel castings, tendency to burn-on to the surfaces of steel castings over which relatively large amounts of metal flow or when the steel contains high amounts of manganese (Mn) which attack the silica and the requirement that precautionary means be taken to prevent the occurrence of silicosis by the operators exposed to the silica.

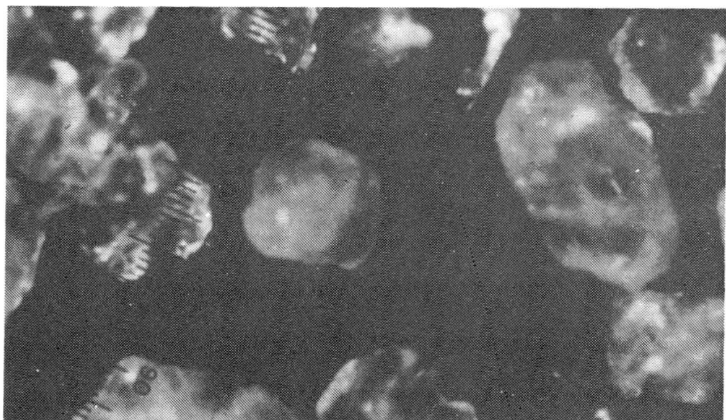

Fig. 5. Lake sand.

Figure 7 illustrates the thermal expansion of various silica materials. Note that alpha quartz — which is the mineral nature of silica sand — expands at a constant, rather high, rate until a temperature of approximately 1100F (593C) is reached. As the temperature increases above this point, a sudden expansion takes place due to the change in crystal form from alpha to beta quartz. It is this expansion behavior which causes scabs, buckles, rattails and other so-called expansion defects and the prevention of which requires addition and careful control of cushioning materials.

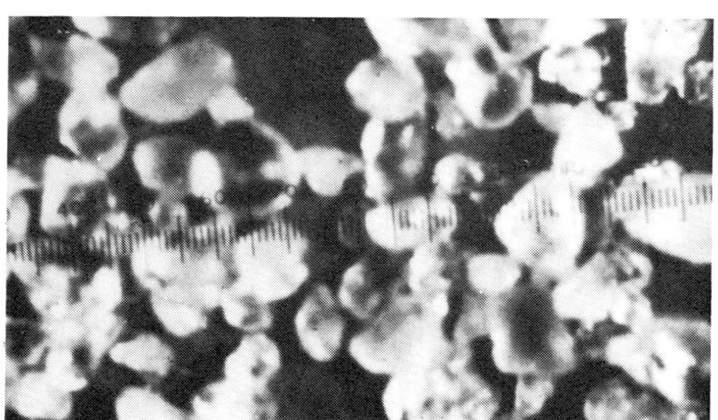

Fig. 6. Juniata bank sand.

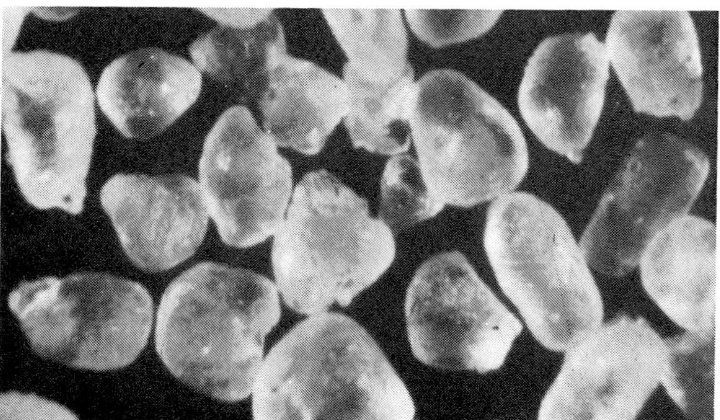

Fig. 3. New Jersey sand — subangular grain.

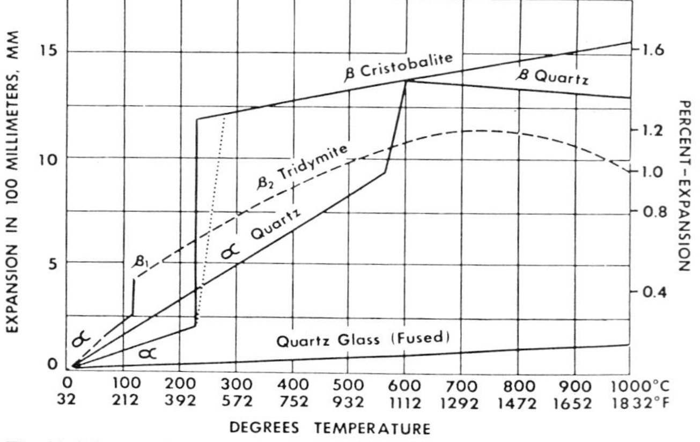

Fig. 7. Thermal expansion of silica minerals.

Table 1. Some Physical Properties of Foundry Sands

	SILICA	ZIRCON	CHROMITE
THERMAL CONDUCTIVITY (K-BTU/hr ft °F)	0.52	0.57	0.63
DENSITY (p-lb/ft^3)	93.2	176.0	150.0
SPECIFIC HEAT (C-BTU/lb °F)	0.280	0.198	0.23
THERMAL DIFFUSIVITY (K/pC)	0.0199	0.0164	0.0183
HEAT DIFFUSIVITY (KpC)$^{\frac{1}{2}}$	3.68	4.46	4.67

Over the past 30 years, a number of speciality sands of materials other than silica have been identified as having properties which make them superior to silica sand for certain applications. Figure 8[4] shows the thermal expansion behavior of most of the candidate materials. the three most popular of these specialty sands at this time in the US are chromite, zircon and olivine. Figure 8 shows that each of these materials has a much lower rate of thermal expansion than silica and none of them undergoes a phase change such as that of silica sand. This data predicts and experience has shown that expansion defects are not a problem with sand mixes made from these materials. Table 1[5] shows some physical properties of these sands in comparison with silica sand. Zircon and chromite have markedly higher thermal conductivity and density than silica and, for that reason, have a higher heat diffusivity. For this reason, zircon and chromite can be expected to chill molten metal at a faster rate than silica. Table 2[6] shows that freezing time of 6-in. spheres made in molds prepared from various materials which might be used to make steel castings. This information ranks silica, olivine, zircon and chromite in the order of increasing ability to chill. As will be shown later, the present pattern of foundry usage agrees with this data since chromite and zircon are used when a mold chill effect is desired.

Chromite

The name chromite sand, as used in the foundry industry, refers to beneficiated and classified chromite ores. Such ores are found in Rhodesia, Turkey, the Philippines, Albania, the USSR and South Africa. At this time, the South African ore mined in the Transvaal has been found satisfactory for steel foundry work in North America. Middleton[4] and Bownes[7] report that Finnish

Table 2. Freezing Time of 6-in. Sphere (min) and Solidification Times Referred to Silica Sand

	FREEZING TIME OF 6-INCH SPHERE (MIN)	SOLIDIFICATION TIMES REFERRED TO SILICA SAND
COPPER CHILL MOLD	4.2	.24
STEEL CHILL MOLD	4.3	.25
GRAPHITE MOLD	5.1	.30
CHROME ORE BONDED	13.4	.79
ZIRCON BONDED	13.8	.82
OLIVINE BONDED	15.8	.93
CHAMOTTE BONDED	17.0	1.00

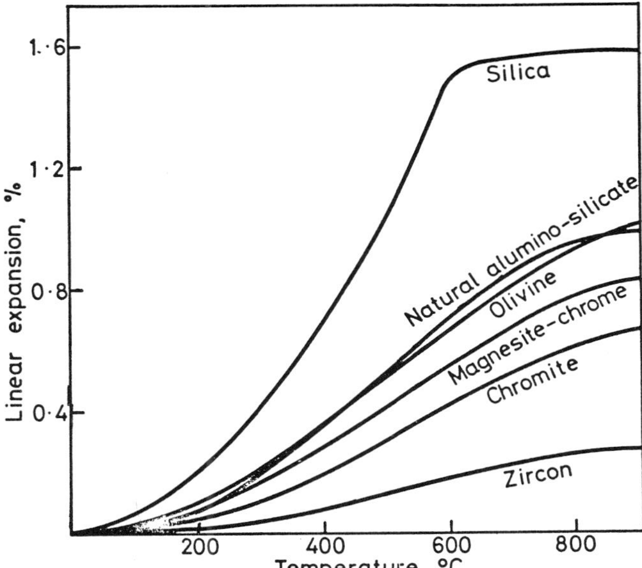

Fig. 8. Thermal expansion of silica compared with other refractory sands.

chromite sand has been found to be just as satisfactory. The foundry grade of chromite sand is a solid solution of six spinels. Oxides of chromium (Cr), iron (Fe), aluminum (Al) and magnesium (Mg) are present and the SFSA tentative acceptance specifications for chromite sand require a minimum of 44% chromium oxide with maximum amounts of Fe_2O_3 (26%), SiO_2 (4%) and CaO (5%). Al_2O_3, MgO and trace elements make up the balance, typical contents of Al_2O_3 and MgO being 14 and 11% respectively. Chromite ore occurs in rocklike form and must be crushed to produce particles of suitable size and distribution. For this reason, chromite sand grains are angular as is shown in Fig. 9. However, it will be noted that the surfaces are relatively smooth and no porosity is present. Chromite sand has been successfully bonded with all the available chemical bonding agents except for acid-catalyzed nobake binders. Cold

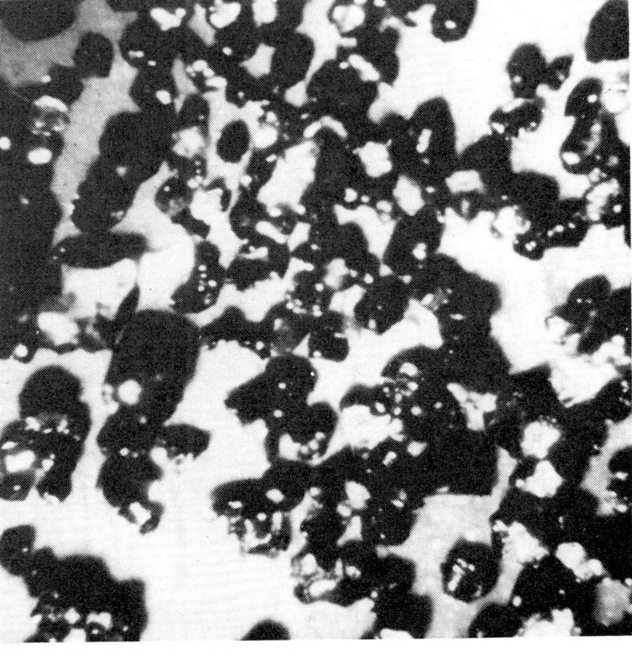

Fig. 9. Typical chromite sand.

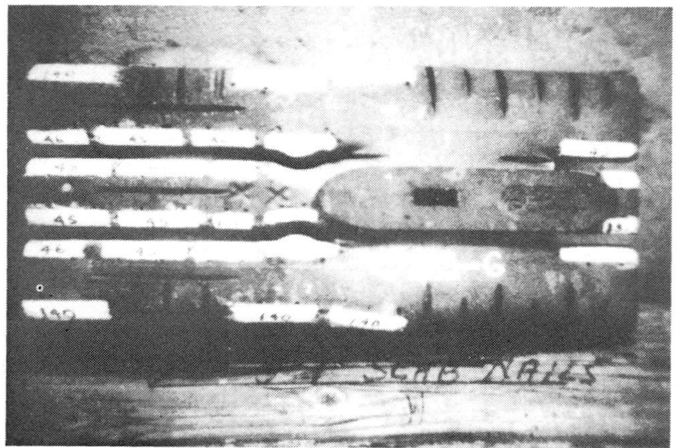

Fig. 10. Railway bolster core with steel chills.

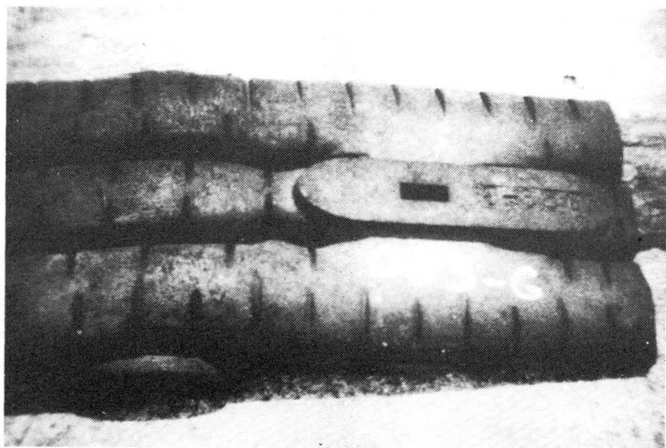

Fig. 11. Railway bolster core with chromite sand facing replacing chills of Fig. 10.

setting difficulties have been experienced with these because of the somewhat higher acid demand of chromite sands in comparison with silica and zircon.

It is generally reported that chromite sand offers higher resistance to penetration by molten steel than zircon, olivine or fine silica sand. Field experience demonstrates that the grading of chromite sand may be coarser than that of zircon sand when used for preventing penetration. Asanti[8] reasons that chromite sands have this exceptional penetration resistance because molten steel and the usual metallic oxides present in molten steel do not wet chromite as readily as comparison materials and the oxides formed upon contact of steel and chromite are of lower fluidity or reactivity than when steel is in contact with other molding materials which might be used. Scheffer[9] reported on the evaluation of chromite, zircon and olivine to prevent penetration and burn-on when making large steel castings. He found that chromite was the best material in this group. He offered the interesting observation that additions of about 5% magnesium oxide were useful in improving refractoriness and penetration resistance in those difficult cases in which the chromite is not quite refractory enough for the job of preventing burn-on and penetration when making large steel castings and states that this is theoretically rigorous.

Chromite is used in steel foundry work as a chill material to replace medium and light steel external chills in the manufacture of complex steel castings such as railway side frames and bolsters. Figure 10 shows a typical core with steel chills in place and Fig. 11 shows a core made with chromite sand, replacing the metal chills. With this change, the cost for removing and reclaiming metal chills for reuse is avoided and the castings have a better appearance. This cost saving is achieved despite the fact that chromite is a relatively expensive material. Sontz[10] and Neff[11] have reported on the development and commercial use of a system for recovering the chromite sand which enters heap sand when chromite is used. This has enabled a further cost reduction and also avoids chromite buildup in the heap sand. The latter effect is important since there have been reports that heap sand containing 15-20% chromite sand showed areas of local fusion on steel castings. Resin shell molds made with chromite sand can be used to make good steel castings. However, the relatively high surface area per unit volume of the angular chromite sand and the relatively low efficiency with which chromite can be bonded with phenolic resin significantly increases the amount of resin required in comparison with zircon or silica sand. Chromite sand avoids the burned-on sand problem which occurs when silica sand is used to make austenitic manganese steel castings but as will be shown later, olivine does a satisfactory job at a lower cost and has become the material of choice for this application.

Zircon

Zircon is the mineral zirconium silicate and occurs in beach deposits from which it is obtained as a byproduct of the winning of some more valuable mineral. Zircon sand is produced in Australia, Florida and Georgia in the US, South Africa, Egypt, Malaya, Russia and China. You will note from Fig. 12 that the grains of zircon sands are rounded and elliptical in shape and the surface is very smooth. The favorable character of the grains, the absence of fines and the chemical nature of the material give it the highest bonding efficiency with organic binders of any sands now available. The average particle size of the available materials ranges from AFS gfn 90 to 130 and the grain distribution is quite narrow, practically all the sand being retained on three adjacent sieves. The first foundry application for zircon sand was for prevention of penetration or burn-on in critical areas when making large steel castings. Because of the narrow grain distribution, penetration can occur when temperature of the sand and metal pressure are high enough. Under these circumstances, zircon flour is added to the sand to reduce porosity and close the openings through which metal can enter. Even though large amounts of flour are added, thermal cracking is not likely because of the extremely low thermal expansion characteristic of zircon. However, if the sand compact must be dried or baked before use, great care must be taken in heating to avoid cracking the core due to gas or steam pressure which can become quite high because of the extremely

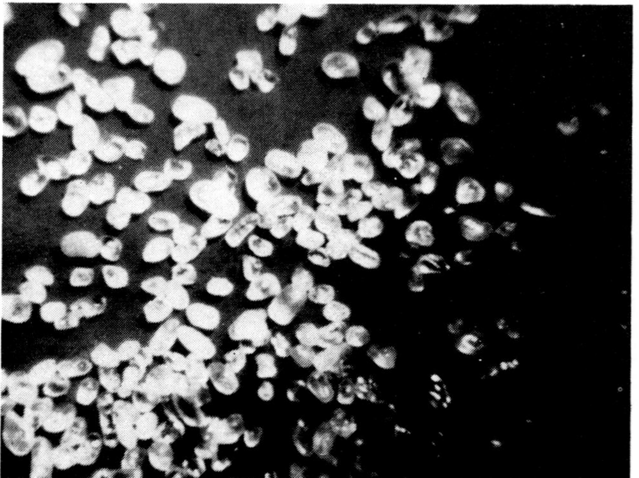

Fig. 12. Zircon sand.

Table 4. Typical Sand Prices

	PRICE/TON F.O.B.	TYPICAL FREIGHT/TON	DELIVERED COST/TON
WASHED & GRADED SILICA SAND	$ 6.30	$13	$ 19.30
CHROMITE SAND	138.00	12	150.00
ZIRCON	150.00	30	180.00
OLIVINE	34.00	22	56.00

It is interesting to see that the amounts of chromite and zircon change positions from 1974 to 1975. The reason for this reversal may be that in 1974, the price of zircon was continuing to rise after the start of an upward price movement in 1972 when there was a worldwide shortage of zircon and foundrymen were working to find a substitute for zircon sand. By 1975, they had determined that they could replace zircon with chromite in some applications and they did so to control expenses. Chromite comes to the US market from South Africa which is under a political cloud and this supply may be interrupted. So it is fortunate that at least for some applications, the foundry industry can use either material. This also suggests that other sources of chromite suitable for foundry use should be developed. Bownes[7] reports that Finnish chromite gives results equal to the South African material.

Table 4 shows typical prices for the various sands along with freight changes for delivery to a mythical foundry in the midwestern part of the US. All the specialty sands are considerably more expensive than washed and graded silica sand. Although olivine is the least expensive of the group, its delivered price is almost three times that of silica sand. The price of olivine has been stable in the US since its introduction to the industry. This is not surprising since there are abundant domestic deposits and supply has always been sufficient to meet the increasing demand. The price of zircon has come back down from a high of $600/ton reached during the abovementioned shortage and should be more stable now because more sources of supply now serve the market.

Table 5 is an attempt to determine how much silica sand is used to make the various classes of castings in the US. When estimating the consumption of new sand for each type of metal, the assumption was made that no reclamation is being done. The only source for this information is the experience of knowledgeable people. This approach yields an estimate of 9.2 million tons consumed in the year 1975. This compares favorably with the 9.9-million-ton total consumption of silica sand estimated by those who are familiar with silica sand usage in the foundry industry and the average of the two estimates should be reasonably accurate.

Table 5. Sand Casting Shipments vs Sand Consumption

	CASTINGS SHIPPED (TONS)		NEW SAND PER TON CASTINGS (TONS)		ESTIMATED SAND USED (TONS)
IRONS	11,067,252	X	.65	=	7,193,714
STEELS	1,713,933	X	1.0	=	1,713,933
ALUMINUM	197,331	X	.25	=	48,333
COPPER	475,860	X	.5	=	237,930
TOTALS	13,454,376	X	.69	=	9,193,910

ACTUAL CONSUMPTION OF SILICA SAND – 1975   9,941,000 TONS

Table 6. Reclamation Methods vs Sand Costs

	NO. RECL. #USED	NO. RECL. COST	DRY RECL. #USED	DRY RECL. COST	WET RECL. #USED	WET RECL. COST
NEW SAND	2,000	$19.30	1,000	$9.65	215	$2.07
WET RECLAIMED	-	-	-	-	2,042	7.71
DRY RECLAIMED	-	-	2,000	3.50	-	-
SAND DISPOSAL		2.00		1.00		.22
TOTALS		$21.30		$14.15		$10.00

Sand Reclamation

It is apparent from the comparison of this rate of consumption with that which can be achieved with reclamation systems that a relatively small fraction of US foundries use sand reclamation. Table 6 shows that both new sand usage and cost can be markedly reduced in a steel foundry when either pneumatic or wet reclamation methods are used with silica sand. Each of these foundries is making the same class of product for the purpose of this illustration. It is typical in making steel castings with a reasonable number of cores to use about a ton of new sand per ton of castings produced. We must also get rid of a ton of sand when a ton of new sand is added. The cost of $2/ton to dispose of spent sand has been used. In an informal survey, figures were found all the way from 25¢/ton (or even less in some cases if they can be believed) up to $10/ton. To some, $2 may seem high, but by next year it will seem a little closer to reality and five or ten years from now, we will look back on $2/ton as fondly as we now regard the legendary nickle beer and the five-cent cigar.

We are all aware of the problems in this case. Dump sites are getting further from the foundry and more work must be done to prepare the material for the dump site so that, hopefully, it will be innocuous; but some do not believe that is going to be the case and the public is being safeguarded by restrictions at dump sites to avoid ill effects on the ground water and streams nearby. Dry reclamation reduces new sand consumption on this kind of work to about 1000/lb per ton but it is necessary to reclaim a ton of sand by pneumatic reclamation in this case to supplement the new sand. If wet reclamation is used, however, it is possible to reclaim almost all the sand; thus, a considerable reduction in cost is obtained as we go from no reclamation to dry reclamation to wet reclamation. Choice of a reclamation system is an individual one. Wet reclamation may be fine in one case, and in another, dry reclamation may be the proper choice.

Perhaps you have wondered how high-priced specialty sands can be used to make castings economically. Table 7 is an estimated cost for sand in a zircon shell molding operation in which the typical sand loss of 3% per cycle is assumed and the ratio of sand to metal is 1.75:1. The zircon costs $180/ton but actual consumption is so low that the cost per ton of casting produced is less than the cost of silica sand when no reclamation

Table 7. Estimated Sand Cost Using Zircon, Resin Shell Molding and Thermal Reclamation

SAND-METAL RATIO = 1.75-1

	USAGE (lbs)	PRICE/TON	COST/TON
ZIRCON SAND	105	$180	$9.45
RECLAIMED SAND	3500	5	8.75
SAND DISPOSAL	105	2	.11
SAND COST/TON CASTINGS			$18.31

Table 8. Estimated Sand Cost Using Olivine — Bentonite Bonded — With Green Sand Molding

	USAGE (lbs)	PRICE/TON	COST/TON
OLIVINE SAND	600#	$56	$16.80
SAND DISPOSAL	600#	2	.60
SAND COST/TON CASTINGS			$17.40

is done. Table 8 is an estimate of the cost for olivine in a closed system in which a bentonite bonded molding green sand approach is used. No reclamation cost is shown. Such foundries have what amounts to a casual reclamation system built into the sand-handling system. For instance, some use pneumatic transport and they make certain to do good dust collection at all the transfer points. When this is done, apparently the system stays in balance and buildup of fines is avoided. Ortfeldt[17] reports running the sand through a edge runner mill to be sure all the lumps are broken before the start of transport. With this approach, consumption is said to vary between 400 and 800 lb/ton. This depends on how many cores are used. Again, the sand cost is somewhat less than buying silica sand and simply discarding it without an attempt at reclamation.

Two of the closed systems making manganese steel castings in the US utilize cold-set binders. One of them uses an alkyd oil binder and the other an inorganic binder based on phosphate. Table 9 is an estimate of the sand cost in such a system when the sand is being reclaimed pneumatically. Sand consumption is 750 lb/ton of castings in this case because sand attrition is higher with a pneumatic reclamation system. This results in a cost at a sand-to-metal ratio of 2.5:1 of about $29. If the sand-to-metal ratio is reduced to 2:1 by economizing on sand usage by closing up all the open spots in the mold, cost goes down to $23. Sand-metal ratio is the key to keeping the cost of sand and binders down. In conventional green sand molding, the average sand-to-metal ratio is about 5:1. In 1976, Howell[21] reported after a study of US foundries that the sand-to-metal ratio which is commonly used in the relatively new airset molding techniques now ranges from 0.4:1 to 4:1. In zircon molding of steel castings mentioned previously, the ratio is 1.75:1. The 1962 AFS Molding Materials and Methods[22] describes a molding method which is called the D-process in which an organically bonded sand is blown between a pattern and a metal dryer having an inside contour which roughly follows the contour of the pattern with a clearance between mold and pattern just enough to allow a suitable sand mix to be blown into place with good packing density. The molds were then baked before pouring. Figure 16 shows molds of a similar nature which were used in making experiments in the author's laboratory. These molds were made of cast iron, were cast to contour and had a nominal section thickness of 3/4 in. Figure 17 shows these molds lined with sand using a core-blowing machine and a sand mix containing 0.35% of a water-dispersible resin and 0.15% cereal binder. With the support provided by the cast iron mold, the amount of binder

Table 9. Estimated Sand Cost Using Olivine Sand, Chemical Binder and Pneumatic Reclamation

SAND-METAL RATIO = 2.5-1

	USAGE (lbs)	PRICE/TON	COST/TON
OLIVINE SAND	750	$56.00	$21.00
RECLAIMED SAND	4250	3.50	7.44
SAND DISPOSAL	750	2.00	.75
SAND COST/TON CASTINGS			$29.19
AT SAND/METAL RATIO OF 2-1			$23.35

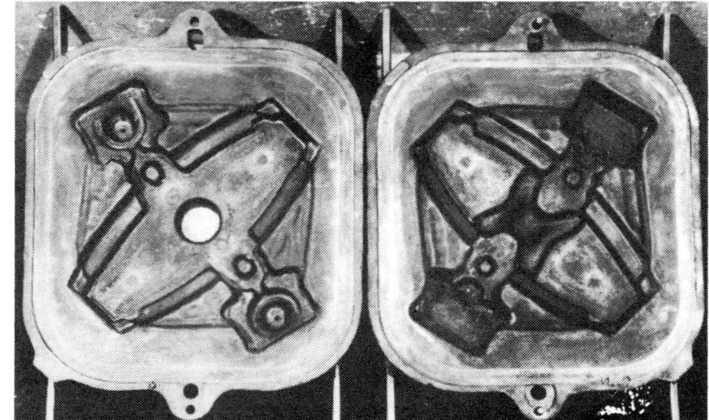

Fig. 16. Cast iron contoured molds ready for sand lining.

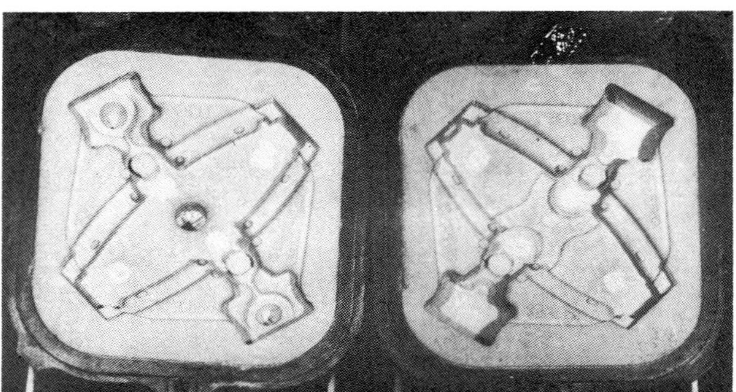

Fig. 17. Cast iron molds of Fig. 16 lined with sand.

need only be enough to resist the washing action of the molten metal. Figure 18 shows an assembly of brake head castings made in this mold. The total weight of this assembly shown is 29 lb, the weight of the two castings totals 24 lb and the sand lining for the molds weighed 8.4 lb. Thus, the ratio of sand lining used to castings made is 0.35:1. The thickness of the sand lining ranges from 1/8 to 1/4 in. in this example. With present day binder systems, lining such molds could be accomplished with greater ease than when these experiments were made.

In a 1976 article[23] there was a description of the Kubota Ltd use of a novel process for making ingot molds. The general approach in this process is to place a flask or cheek around the

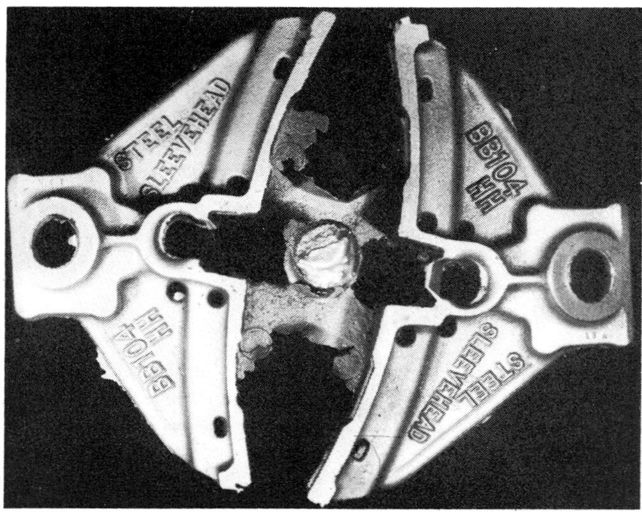

Fig. 18. Railway brake head castings poured in the sand-lined mold of Fig. 17.

Table 10. Typical Energy Usage — Gray Iron Molding and Coremaking

USE	10⁶ BTU PER TON CASTINGS
SAND	.18
BENTONITE	.13
OIL-BASED BINDERS	.55
CARBONACEOUS ADDITIVES	1.34
ELECTRICAL ENERGY	0.17
NATURAL GAS OR OIL	1.90
TOTAL	4.27

Table 11. Typical Energy Usage — Gray Iron Castings

USE	10⁶ BTU PER TON CASTINGS	PER CENT OF TOTAL
MELTING (CUPOLA)	20.74	61.1%
MOLDING & COREMAKING	4.27	12.6%
CASTING & SHAKEOUT	.87	
CLEANING	.85	
HEAT TREATMENT	.32	
IN-PLANT HANDLING	.19	
AIR POLLUTION CONTROL	2.18	
SPACE HEATING	4.54	13.4%
TOTAL	33.96	100.0%

vertically mounted pattern, with the inside of the flask having a contour which gives a clearance of 3-4 in. between flask and pattern. At the same time, an arbor is placed inside the pattern, the arbor having a contour which follows the inside contour of the ingot mold pattern with an approximate clearance of 2 in. Spaces between the flask and pattern and between pattern and core arbor are filled with a urea-furan-bonded silica sand. With this general technique, molds averaging 30 metric tons have been made with a sand-to-metal ratio of 0.32:1. As a result of the high ratio of metal to sand, the organically bonded sand is heated in the presence of air to a temperature which burns a large portion of the binder so that the sand is virtually free flowing at shakeout. The sand used is then reclaimed by mechanical scrubbing which reduces it to grain size, removes fines and cools the sand to usable temperature for rebonding. It is stated that percent recovery of sand varies from 90 to 95% and that the sand has less than 0.5% ignition loss and can thus be rebonded with a high degree of efficiency.

The combination of sand reclamation and reduction of the ratio of sand to metal, can produce a dramatic reduction in sand consumption. It is apparent that this approach is what makes possible the use of the relatively costly speciality sands for making molds in a closed system. All things being equal, the speciality sand will usually cost more than silica sand but the difference becomes so small at the lower sand-to-metal ratios that fairly small increases in yield or reductions in cleaning costs which can be achieved by the specialty sands will more than pay for the additional expense for sand. The use of binders which are more expensive than those which might have been previously used decreases as sand-to-metal ratio decreases and thus binder expense is proportionately decreased. In this way it may be possible in some cases to limit the increase in binder expense to an amount which is less than the savings in other operations.

Energy Consumption

Energy usage in the foundry is a very important subject but it is not possible in this paper to do more than touch the tip of the iceberg. With the rapid rise in the cost of energy and the probability that these costs will continue to rise, it is well to consider the effect of changes in mold materials on energy consumption. Table 10 shows a tabulation of energy consumption in the molding and coremaking area of a typical US gray iron foundry operation using green sand and Table 11 shows energy use by foundry departments. This information is from a report on energy use patterns in metallurgical and nonmetallic mineral processing prepared by Battelle for the US Bureau of Mines in 1975.[24]

Melting is the biggest user and molding and coremaking ranks third and is small by comparison. Certainly there is a good chance to save considerable energy in molding and coremaking but this will not directly affect the biggest usage in the operation. The greatest opportunities for saving energy by changes in molding and coremaking are in changes which will affect energy use in melting and space heating. For instance, if a change is made in mold material which reduces the amount of casting scrap or increases metal yield, it is possible to make a greater energy reduction than by just economizing on energy usage in the molding and coremaking department. If a change can be made from silica dust to some innocuous dust or silica dust can be rendered innocuous and thus the requirements can be reduced for air pollution pickup and air movement in total, not only are the capital and operating expense for air pollution control in molding and cleaning reduced but also energy consumption for space heating is reduced. Tables 12 and 13 show similar energy information for a typical steel foundry. Since the situation is quite similar to that of the typical iron foundry, the same comments apply.

Metal Molds

Cast or ductile iron pipe for the transmission of water and waste products is a large and important class of products for the US foundry industry. At the start of this century, all iron pipe was cast statically in sand molds. The centrifugal casting method was

Table 12. Typical Energy Usage — Steel Molding and Coremaking

USE	10⁶ BTU PER TON CASTINGS
SAND	.25
BENTONITE	.19
CEREALS	.05
OIL BASED BINDERS	.78
ELECTRICAL ENERGY	.22
NATURAL GAS OR OIL	2.50
TOTAL	3.99

USE	10^6 BTU PER TON CASTINGS	PER CENT OF TOTAL
MELTING (ARC FURNACE)	22.89	55.1%
MOLDING & COREMAKING	3.99	9.6%
CASTING & SHAKEOUT	.73	
NORMALIZE	3.00	
CLEANING	1.27	
HEAT TREATMENT	3.04	
FINISH & INSPECT	.34	
IN-PLANT HANDLING	.26	
AIR POLLUTION CONTROL	1.81	
SPACE HEATING	4.24	10.2%
TOTAL	41.57	100.%

	STANDARD MATERIAL	SOME SOIL PIPE MOLDS
CARBON	.26–.34%	.25–.26%
SILICON	.15–.40%	.15–.40%
MANGANESE	.40–.65%	.40–.65%
PHOSPHORUS	.025 MAX %	.025 MAX %
SULPHUR	.025 MAX %	.025 MAX %
CHROMIUM	.8–1.1%	2.50%
NICKEL	.5 MAX %	.5 MAX %
MOLYBDENUM	.15–.30%	.15–.30%

then developed to make iron pressure pipe for transmission of water. When this approach was first introduced, the molds were made of sand. The DeLavaud process developed in France substituted a water-cooled metal mold for the sand mold. When this process was introduced in the US, it gradually replaced the sand molding, centrifugal casting approach so that early in this decade all iron pressure pipe made in the US was being cast in metal molds. Soil pipe was first cast statically in sand molds but at the present time all production in the US is in metal molds. Figure 19 is a simplified cutaway view of a typical DeLavaud casting machine showing the general shape of the mold and its relation to the casting machine. The mold is in the form of a steel cylinder which is rotated at the desired speed in a circulating water reservoir which keeps its temperature down. The mold is rotated as the metal is distributed along the mold and after the casting operation has been finished and the pipe has solidified, the end clamping plate is removed and the pipe is dragged from the mold. The mold also has to endure the alternate heating and cooling caused by making one pipe after another. This severe thermal cycling promotes formation and growth of heat checks in the inside diameter of the mold. Checks are usually first found in the bell end portion of the mold. When they grow to an undesirable size, they are removed by some suitable means and the mold contour is restored by welding. This process is repeated until heat checks grow in the inaccessible barrel area of the mold.

Some producers coat the molds with a ceramic wash prior to the casting operation to control surface finish and prolong mold life while others use no coating but may get a similar effect by spreading the desired innoculant in finely ground form over the surface of the mold prior to the casting operation. Life for molds up to the size needed to make 6-in. pipe ranges from 1 to 50,000 pieces with the average being on the order of 5000 to 6000. For molds of the 24-in. size, life will range from 1000 to 3000 casts with 2000 being typical. Mold life is usually longer when the thinner-wall ductile iron pipe is being cast.

The metal molds are made of steel and Table 14 shows the chemical analysis range for the modified 4130 steel which is the popular choice in the US. This material is a compromise between the need for resistance to wear when the pipe is dragged from the mold and resistance to the formation of heat checks caused by thermal cycling. The usual manufacturing process for molds up to about 30 in. in diameter is to forge an ingot of the desired steel into a solid round of the approximate size desired and create a bulge on the bell end. This blank is then trepanned and machined to finished size. For molds intended to make 30-in.-diameter or larger pipe, a hollow forging is made from an ingot. The usual heat treatment is to normalize the steel for the mold followed by austenitizing the material which is then water-quenched and tempered to 1050F (566C). For certain special cases, the tempering treatment is specified to be 1200F (566C). For certain special cases, the tempering treatment is specified to be 1200F (649C). Table 14 also shows an analysis that is used for some soil pipe molds. It will be noted that the chromium (Cr) content has been sharply increased. This alloy is generally oil-quenched during the heat-treatment operation. Molds of this type are used to make pipe up to 30 in. in diameter. Carter[25] in the US and Tanaka[26] and Yoshimura[27] in Japan have reported the development and use of a practice for sand lining permanent molds to be used in the casting of pipe 16 in. in diameter and up. It has been reported that pipe as large as 94 in. in diameter is being cast with this method. The procedure is to coat a properly preheated rotating steel mold of suitable design with resin-coated sand to a thickness of about 1/8 in. Centrifugal force of the rotating mold distributes and packs the sand and heat in the mold cures the resin. In this way, a smooth, insulating coating is provided on the surface of the mold which makes it easier to cast pipe without wrinkles or cold shuts and greatly prolongs the life of the mold. This incidentally is another application of the idea of lining a metal mold with a relatively small amount of sand.

The Eaton process[28] is a permanent mold method for making iron castings. Figure 20 shows a general view of the typical Eaton casting machine. The iron molds are parted vertically, arranged in a circle on a merry-go-round like device and cooled by an air supply which enters through a central duct and flows radially out to each mold. In completing one revolution, a mold is cleaned, coated with soot, receives sand-bonded cores when the design calls for them, is closed and clamped, filled with molten iron and the castings stripped from the mold as soon as

Fig. 19. Cutaway view of a typical DeLavaud pipe casting machine.

Fig. 20. Typical Eaton casting machine.

they are solid and strong enough to handle. The process is then repeated. Although castings weighing as much as 25 lb each have been made with this process, the usual upper limit of casting size is 10 lb and the average is said to be on the order of 2 lb. Castings made by this method are sound, have a close uniform microstructure and are readily machinable without the occurrence of defects. It should be mentioned that although efforts have been made to avoid chill formation during the casting operation, no dependable pratice has been obtained and for that reason all castings are given an annealing treatment prior to shipment. Castings made by this method primarily go to high-production appliance and automotive applications and must be readily machinable.

Table 15 shows the analysis range for the mold material which has been reported as being best for this service. A number of other materials have been tried but none have given as good results. Table 15 also shows the chemical analysis which Kubota Ltd[23] has found best for ingot molds. Although chemical analyses of the two materials differ in detail, they are both high carbon equivalent (CE) irons.

Kubota has found that ductile iron gives much longer life in ingot molds than gray iron. However, trials of ductile iron in Eaton-type molds has given results which are quite inferior to those for gray iron. No explanation has been offered for this behavior. It may be that the reason for the success of ductile iron in ingot molds at Kubota is due to the overall design of the ingot mold. The mode of failure in the cast iron molds is heat checking and the mold must be discarded when the heat checks become large enough that fins or surface roughness become undesirably large on the castings being made. Apparently, heat checks develop early in the life of a mold but the characteristic of the high-CE gray iron is that the checks grow at a very slow rate.

Molds are cast to size using a master pattern so that duplicate molds can be made at low expense. The cleaned mold is prepared for use by coating with a mixture of pipe clay and sodium silicate whose function is to fill the pores which might be present on the surface of the mold and which could result in surface roughness on the castings. The molds are thoroughly dried after the application of this coating and then attached to the molding machine. The molds are cast with cooling projections on their back to facilitate cooling when forced air is passed over the back of the mold during the time the mold is in use. Mold life is usually on the order of 3000 to 30,000 casts but this might be considerably shortened if a casting design has a severely heated portion such as occurs at a reentrant angle. A method which has been successfully used to combat this problem is to make a mold insert of stellite or some other heat resistant material and insert it in a machined opening in the mold. Design of the insert is such that it will extend past the back of the mold in the same way as the cast-on cooling rods protrude. In this way, a more heat-resistant material is provided at the hotspot and maximum cooling can be maintained to keep the temperature of the hotspot at the lowest possible point. Although other coatings have been tried, carbon soot such as that which can be obtained by the use of a smokey acetylene flame has proven to be the best coating for the mold.

The casting of iron soil pipe fittings in permanent molds has evolved from the general approach of the Eaton process. The casting equipment and general arrangement of the molds are similar but the molds are cooled by water instead of air to be able to cycle rapidly and precisely with much heavier castings than are customarily made with the Eaton approach. The mold material is high-CE iron and mold thickness is about 3/4 in. With this method, soil pipe fittings are made in sizes through 15 in. Section thickness of the castings made ranges from 3/16 in. in the smaller size fitting to 3/8 to 1/2 in. in the larger sizes. Carefully timed removal of castings from the molds and regulated cooling of the castings thereafter by piling them together avoids the need for annealing. Shell cores are customarily used to form inside surfaces of the castings and sand adherence is so little that the castings need not be blasted prior to shipment. Six foundries are known to be using this method and it is the predominant approach for making soil pipe fittings. One

Table 15. Chemical Analyses — Cast Iron Molds for Making Iron Castings

	EATON PROCESS (CUPOLA MELTED)	INGOT MOLDS (BLAST FURNACE IRON)
TOTAL CARBON	3.45 – 3.65%	4.17 – 4.43%
SILICON	2.45 – 2.65%	.57 – 1.01%
MANGANESE	.60 – .90%	.57 – 1.12%
SULPHUR	.12 MAX %	.016 – .033%
PHOSPHORUS	.30 MAX %	.091 – .215%

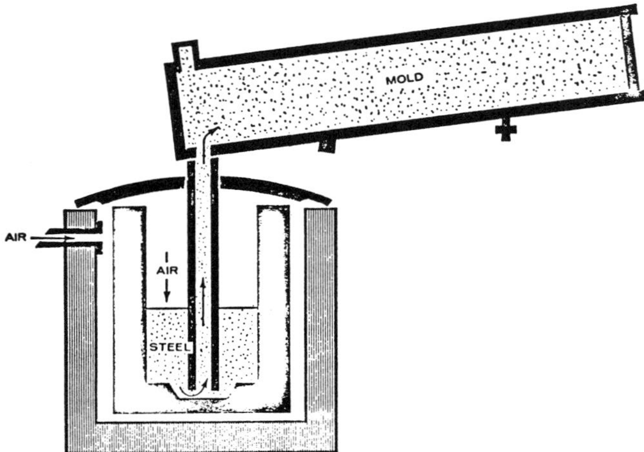

Fig. 21. Schematic cross section illustrating controlled pressure pouring of a slab mold.

Fig. 22. Removing a hot railway wheel from the graphite mold.

Fig. 23. View of the front (at left) and back of typical Griffin wheels ready for shipment. The front of the wheel was formed by the cope of the mold.

of these foundries makes fittings for pressure pipe through the 8-in. size. These castings in general have thicker cross sections and control of dimensions and solidity is more difficult than with the thinner soil pipe fittings. Mold life ranges from 1000 to 25,000 pours per mold with 6000 being a typical figure. Molds fail by heat checking, usually in the gate area or in that part of the mold which forms the casting adjacent to the gates.

Bates[29] reported in 1972 an analysis of the cost of making castings in sand versus permanent molds and concluded that the cost with permanent molding can be less than with sand molding. No one making permanent molded iron castings has either confirmed or denied this prediction. However, it seems obvious from the rapid adoption of permanent molding for fittings that this must be the lower-cost approach. In addition, environmental control in the workplace is simplified and the operation is well mechanized to reduce the number of skilled people required.

Steel Die Casting

Currently, there is considerable activity to develop a process to die cast steel. Cross,[30] Flemings[31] and GKN[32] of Britain have all been active in this field during the past five years. Kulkarni and Miclot[33] recently reported the results of experiments to make a steel casting with the extrusion or squeeze-casting approach. At this time, only limited commercial success has been obtained in die casting steel and it is apparent that cost of the mold is one of the principal obstacles to the growth of this approach. Molybdenum (Mo) and Mo alloys have given the longest mold life but the most economic material to date is H-13 die steel which also is the popular choice for light-alloy die-casting work. When a hotspot shortens mold life, an investment cast insert of the same material has been used to refurbish the mold at a lower cost than complete mold replacement. Draper[34,35] has reported on studies of mold materials and coatings but no solutions were found which offer the order of magnitude improvement which is apparently needed to make this approach a commercial success.

Graphite Molds

Graphite has several characteristics which indicate that it should be an excellent material for permanent molds. It has a very low coefficient of thermal expansion — about the same as that of fused quartz. In addition, it does not crack on either rapid heating or cooling and does not heat check under even the most severe heating cycles. Disadvantages of the material for use in permanent molds are that it is considerably more fragile than any of the metallic materials which might be used and therefore can be easily damaged during casting and handling operations. When heated to temperatures of 800F (427C) or higher in the presence of air, graphite will oxidize. The company which first exploited the possibilities of graphite as a mold material for making steel castings achieved this result by development of a special pouring method now known as controlled pressure pouring and precise mechanized handling of machined graphite molds to make railroad wheels. Figure 21 is a schematic representation of the pressure pouring method. The ladle of molten metal is placed in a pressure vessel. Compressed air is introduced to the pressure vessel in a controlled fashion to push the molten metal up the refractory pouring tube into the bottom of a mold which is clamped to the top of the tube. With this method, metal is introduced into the mold at a high-volume rate yet without turbulence or creation of a jet action which would cause poor surface finish and erode the fragile graphite mold. Figure 22 is a photograph taken at the moment a just solidified hot wheel is being removed from the mold. The mechanization seen here is indicative of that used throughout the process. Figure 23 shows the front and back of a typical wheel after heat treatment, arc washing of riser pads, shot blasting and rough boring of the hub. Surface finish of these wheels is so good and

Fig. 24. View of two graphite slab casting molds which have just been opened to remove the 4-in.-thick slabs which have just been cast.

dimensional control so precise that the wheels are placed in service without any machining of the tread of the wheel. Wheels made by this method have won wide acceptance and more than 10 million have been made in the US. Another company makes wheels in a graphite mold in which the graphite forms only the tread and contiguous features and a silica sand lining in the graphite mold forms the plate and hub contours. Casting of wheels in graphite molds is now the dominant production method in the US and Canada now since more than half of all wheels purchased annually are made by this approach in competition with wrought wheels.

Graphite molds are also used in conjunction with pressure pouring to cast slabs of carbon and stainless steel in the basic steel industry. Figure 24 is a view of such a casting operation at the point where the mold has just been opened for removal of the slab. The mold is constructed of graphite blocks, mechanically clamped in a mold arrangement which accurately aligns them to form continuous planes. To allow rapid cycling of the molds, water cooling is used. Slabs have been cast with thicknesses ranging up to 8 in. and total weight to 40 tons. More than 2 million tons of slabs have been cast with this method.

Usage of Sand and Permanent Molds

Now that the mold materials in current use in the US have been reviewed, it is interesting to put them into perspective. Table 16 shows the amount and percentage of castings made in sand, metal and graphite molds. The total ferrous materials cast and the amount of pipe cast are from widely published government sources. The balance of the information is from the author's informal survey within the industry. Sand molding is still the workhorse material for the industry. But permanent molding is now used to make a significant portion — 15.5% — of ferrous castings in the US. A breakdown of this information reveals that about 16% of all iron castings are made in metal molds and about 12% of all steel castings are made in graphite molds. The absolute quantity and percentage of total output made by permanent molds has risen significantly in the past 25 years. It seems obvious that with the economic and ecological advantages of permanent molding efforts will continue to adapt it to a greater amount of ferrous casting production in the future.

Although no specific information is available which indicates the amount of castings made with the exclusive use of specialty sand, it seems apparent that this is a small percentage of the total. With the rapid evolution of binder and reclamation methods, however, this production segment can be expected to grow. For the same general reasons, continued change and growth is expected in the use of systems which economize on the use of sand by a reduction in the ratio of sand to metal and reclamation of the sand for economic reuse. The next 25 years in the foundry industry should be even more interesting and exciting than the last 25.

Table 16. Ferrous Casting Production by Mold Material Class (1975)

		TONS SHIPPED	PER CENT
IRONS:			
SAND MOLDS		11,067,252	73.2
PERMANENT MOLDS			
PRESSURE PIPE	1,460,000		
SOIL PIPE & FITTINGS	594,435		
INDUSTRIAL & AUTOMOTIVE	55,000		
		2,109.435	14.0
STEELS:			
SAND MOLDS		1,713,933	11.3
PERMANENT MOLDS		224,000	1.5
TOTAL		15,114,620	100%

Summary

We have come a long way since Biringuccio described 16th Century foundry practices. The rate of change in the last 50 years has produced more improvements than those which occurred in the previous 400 years. Foundrymen of this age have a better choice of materials and methods than those of Biringuccio's age but they still must be guided by the same wisdom which he displays. In his admonition and warning to those who want to practice the foundry art, he says in part, "I say that in this, as in every other work, and indeed, in every human action, it is necessary to have good fortune in bringing a work to the perfection of its end. But you yourself can make fortune good. If you always use the necessary care to make the means perfect, the end will never be in doubt. As many times as I have failed, or have seen others fail, it has always seemed to me to have been a result of my own insufficiency." We've been able to improve a lot on the foundry practices which Biringuccio describes but we haven't come up with any more wisdom than he offers for guidance in the development and use of foundry practice.

Acknowledgements

During the preparation of this paper, the author sought and obtained valuable insights and information on current practices from many knowledgeable people in the industry. For this I thank:

Mark Abbot	Jim Kiesler
C. E. Bates	M. Kulkarni
Richard Connors	Harold Lownie
Jim Cregier	William McIntyre
Ray Cross	Don Meves
Stubbs Davis	Alex Morris
Lou Dethloff	P. J. Neff
Frank Dettore	Lou Pettinos
Al Draper	Dee Rose
Robert Fanning	H. W. Ruf
Gene Fisher	James Sale
George Frye	Clyde Sanders
Tom Garnar	Robert Schafer
Herb Gould	James Smith
Richard Green	Pete Smith
Ray Hale	Don Soderlund
Malcom Henley	Tito Tarquinio
L. E. Hopkins	C. P. Voll
R. L. Jones	Richard Wetzel
	Vern Zeller

The AFS staff was helpful throughout. Special thanks go to Steve Erd, AFS Librarian, who performed invaluable service by his thoughtful and thorough literature searchers. Clyde Sanders and George DiSylvestro furnished valuable information and slides. I am indebted to Ed Savich of the Amsted Research Laboratory staff for his skillful work in preparing illustrations and slides and to Mrs. Shirley Brown for preparation of the manuscript. I appreciate the support of Amsted Industries in preparing and publishing this paper.

References

1. Vannochio Biringuccio, "The Pirotechnia," American Institute of Mining and Metallurgical Engineers, p 218 (1942).
2. C. W. Briggs, "Synthetic Bonded Molding Sands — Sand, Clay and Water Systems," AFS Transactions, p 553-572 (1966).
3. C. W. Briggs, R. E. Morey, "Synthetic Bonded Steel Molding Sands," AFS Transactions, p 653-724 (1940).
4. J. M. Middleton, "Alternatives to Silica for Mould Production," BCIRA International Conference — Chemical Binders in Foundries, paper no. 5 (1976).
5. "A Literature Survey of Specialty Sands, Chromite, Zircon, Olivine and Aluminum Silicate," Steel Founders Society of America, Special Report no. 8, 43 pages (May 1972).
6. C. Locke, C. W. Briggs, R. L. Ashbrook, "Heat Transfer of Various Molding Materials for Steel Castings," AFS Transactions, p 589-599 (1954).
7. F. F. Bownes, "Silica and Other Sands," SCRATA Conference on Foundry Practice — Today and Tomorrow, paper no. 1 (Apr 1976).
8. P. Asanti, "On the Interface Reactions of Chromite, Olivine, Quartz Sands with Molten Steel," Cast Metals Research Journal, 4 (1968).
9. K. Scheffer, "Behavior of Chromite in Steel Casting Molds and Cores," AFS Transactions, p 585-592 (1975).
10. A. Sontz, "Recovery of Chromite Sand," AFS Transactions, p 1-12 (1972).
11. P. J. Neff, "The Production Reclamation of Chromite Sand," Steel Foundry Facts, no 312, p 29 (Feb 1975).
12. "Shell Molding in Zircon Sand, Part 1," Iron and Steel, p 274-276 (Jun 1968).
13. R. Mason, "Metallurgical Control in the Production of Zircon-Sand Shell Molded Grey Iron Castings," Iron and Steel, p 517-519 (Nov 1968).
14. D. J. Taft, "Zircon Shell Practice," The British Foundryman, p 69-75 (Feb 1968).
15. R. Hermann, "Makes Large Steel Castings in Zircon Sand Shell Molds," Foundry, p 81-85 (Jul 1960).
16. K. Beckius, "Olivine Sand — Recent Research and Experience," 40th International Foundry Congress, Moscow, USSR, paper no. 11 (1973).
17. J. Ortfeldt, "Olivine Sand — Nine Years' Practical Experience," Foundry Trade Journal, p 17-21 (Jul 2, 1970).
18. G. Isaksson, "A Survey of Silicosis Hazards in Swedish Industries with Special Regard to Foundries," 39th International Foundry Congress, Philadelphia, paper no. 11 (1972).
19. P. J. Neff, "Determination of the Free Silica Content of Respirable Dust in the Presence of Zircon," AFS Transactions, p 701-704 (1976).
20. I. P. Kryanin et al, "Antiscab Molding Mixtures and Mould Washes for Stainless Steel Castings," Liteinoe Proizvodstvo, (in Russian), 2, p 3-7 (Mar-Apr 1954).
21. R. C. Howell, "Can Chemically-Bonded Sand Replace Greensand," BCIRA International Conference — Chemical Binders in Foundries, paper no. 12 (1976).
22. "D-Process," Molding Materials and Methods, 1st ed., American Foundrymen's Society (1962).
23. "Self-Hardening-Sand-Casting Technique Helps Kubota Lead Japanese Mold Makers," 33 Magazine, p 39-41 (Nov 1976).
24. Battelle Columbus Laboratories, "Energy Use Patterns in Metallurgical and Nonmetallic Mineral Processing (Phase 4—Energy Data and Flowsheets, High-Priority Commodities)," for US Bureau of Mines, PB-245-759, p 87-102 (Jun 1975).
25. S. F. Carter, "Controlling Quality of Ductile Iron Pipe," Foundry (Oct 1963).
26. K. Tanaka, "Large Ductile Iron Castings Production," AFS Transactions, p 265-279 (1962).
27. H. Yoshimura, "Centrifugal Casting of Large Diameter Iron Pipe," AFS Transactions, p 117-122 (1968).
28. G. Frye, "Permanent Mold Process as Applied to Production of Gray Iron Castings," Modern Casting, p 52-55 (Oct 1968).
29. C. E. Bates, "Profit Potential in Permanent Mold Iron Castings," Foundry, p 49-51 (Nov 1972).
30. R. E. Cross, "Ferrous Die Casting," Die Casting Engineer, p 14-16, 18, 20 (Nov-Dec 1971).
31. R. G. Riek, K. P. Young, N. Matsumoto, R. Mehrabian, M. C. Flemings, "Rheocasting of Ferrous Alloys," Society of Die Casting Engineers (1975).
32. "New Research Promotes Ferrous Die Casting," Iron Age, p 46-47 (Jun 16, 1975).
33. K. M. Kulkarni, D. Stawarz, R. B. Miclot, "Feasibility of Squeeze Casting Ferrous Components," Proceedings Fourth North American Metalworking Research Conference (Society of Manufacturing Engineers) p 58-65 (1976).
34. F. W. Schmidt, A. B. Draper, "Temperature Distribution and Thermal Stresses in Die Materials Producing Stainless Steel Castings," AFS Transactions, p 285-294 (1974).
35. A. B. Draper, J. M. Samuels, "Ablative Coatings and Die Materials for Ferrous Die Castings," Transactions Society of Die Casting Engineers, paper G-T75-152 (1975).

Recent Advances In The Technology and Application of Silicate Bonded Sand

R. P. Stanbridge
Marketing Director
Sand and Coating Products
Foseco Inc, Cleveland, Ohio
R. Williams
Molding Materials Int. Mgr.,
Foseco International Ltd,
Birmingham, England

ABSTRACT

This paper examines the fundamental technology of sodium silicate (SiO_2:Na_2O) foundry binders. The authors describe efforts to make present silicate practice better and the relevance of new developments to extending the use of inorganic binders.

The history of the use of silicate foundry binders in the United States is summarized with reference to curing with carbon dioxide (CO_2) gas and with liquid esters. Advantages of silicate binders are discussed, emphasizing environmental and safety features. Factors affecting the performance of silicate processes are reviewed, with quantitative illustration of some problem areas and available remedies. Performance features studied include cure characteristics, core storage properties and shakeout.

A typical application exploiting available silicate-CO_2 technology is reviewed. The example describes silicate coremaking practice in a foundry producing steel railroad castings. The techniques developed for obtaining maximum CO_2 process performance in this application are discussed.

The application of ester-silicate molding is illustrated with process development and application background on use of the process for casting steel slag pots.

The nature and direction of continued foundry silicate process development are demonstrated through details of investigations into three different new applications.

Development of a new CO_2 process binder is discussed. Studies of core strength development, core storage stability and shakeout indicate that its performance may permit use in some cold box applications where only organic binders have been satisfactory until now.

Experiments with a new fast-setting, liquid-hardened silicate process are described. The results presented indicate that striptimes below two minutes are realistic and that the process may be an alternative to fast-setting organic resins in some applications. Details include a review of initial foundry performance tests.

A silicate-based binder designed as a core oil replacement is discussed. Data presented indicate that the cure characteristics offer energy savings compared to core oil and it is proposed that this approach offers a useful interim solution to core oven emission and energy problems.

It is concluded that silicate foundry binder use will expand in the United States, both as a result of improvements to existing processes and through introduction of new technology. The range of application of environmentally attractive silicate binders will increase but sand reclamation techniques will require more development.

Historical Introduction

Solutions of sodium silicate have been used industrially as binders and adhesives for at least 120 years.[1] The use of sodium silicate, or "waterglass," as a binder for foundry sand is known to predate the introduction of bentonite as a sand binder. However, until the early 1950s the only practical technique for binding foundry molds and cores with sodium silicate required baking to dehydrate the binder and to rigidify the mold or core.

A detailed review of the physical and chemical complexities of the gelation and dehydration of sodium silicate solutions is not within the scope of this paper. However, it is useful to note that films of silicate solution can be induced to solidify by two mechanisms. Simple evaporative dehydration will produce a rapid increase in viscosity, leading to solidification as a hard glassy film. This sensitivity of physical properties to water content is illustrated in Fig. 1. Gross physical change can also be induced by adjusting the effective silica:soda ratio of a film of sodium silicate solution. Figure 2 shows the effect of the silica: soda ratio on the viscosity of a sodium silicate solution. It can be

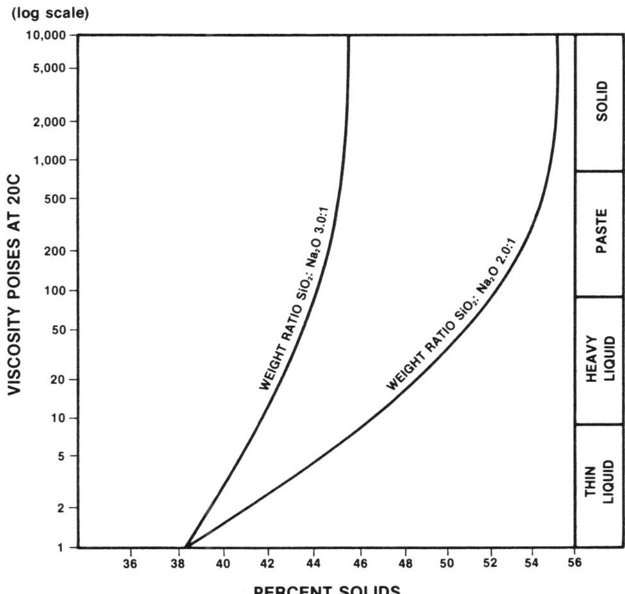

Fig. 1. Viscosity of sodium silicate as a function of solids content. Curves are steepest with higher SiO_2:Na_2O ratio solutions.

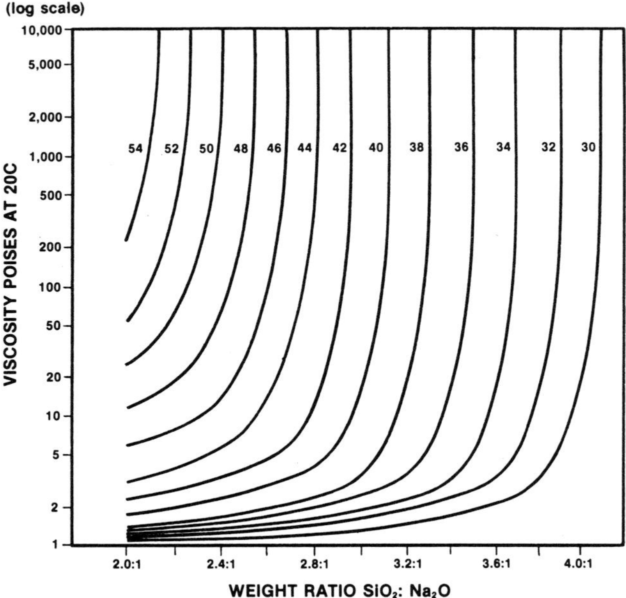

Fig. 2. Viscosity of sodium silicate solutions as a function of silica:soda ratio at various solids contents.

seen that for any given content of solids the viscosity increases dramatically with an increase in the ratio. Thus, by chemically adjusting the effective silica: soda ratio, it is possible to produce a gelation without the need for dehydration. This latter form of gelation is involved in most of the cold-setting sand binder processes which utilize sodium silicate solutions. In practice, the physical and chemical changes are much more complex than these simple mechanisms would indicate. In most cases, both chemical and dehydrative gelation take place concurrently or consecutively and in some cases internal dehydration is part of the chemical reaction. Devising suitable practical methods for cold curing foundry sand coated with sodium silicate solution has offered considerable scope for creative chemistry.

The first practical development to use sodium silicate effectively as a cold-setting binder was the CO_2 process. The CO_2 process achieved acceptance in Europe in the 1950s and its use spread rapidly to the United States. The basics of the CO_2 silicate bond have been well documented. Early work reported by Atterton[2] and later studies by Worthington[3-7] are comprehensive. More recently, practical foundry experience with the CO_2 process has been summarized by Nicholas.[8]

The CO_2 process was the first cold box coremaking and molding system and it introduced the foundry industry to the advantages of fast cold-cured chemical binders as alternatives to heat-cured oil sand and dry sand practice. The simplicity and flexibility of a process for hardening cores rapidly with an inexpensive, odorless and nontoxic gas has led to extensive usage. It is estimated that the United States foundry industry will consume more than 25 million lb of CO_2 process sand binders in 1979.

After the introduction of the CO_2 process, the many advantages of sodium silicate as a foundry binder led to a search for alternative methods of cold curing. Although many users successfully adapted CO_2 process practice to make very large molds and cores, it was felt that the fast cure available was somewhat offset by the skill and expenditures involved in providing venting and manifolding to ensure uniform and consistent CO_2 gassing of large sand masses.

Many non-gassing methods were developed for cold-curing sodium silicate binders and some of these found limited commercial acceptance. Chemical curing agents employed included powdered ferrosilicon (Fe-Si), sodium silicofluoride, anhydrite, dicalcium silicate (2CaO SiO), various types of cement and slag-cement by-products. Some of these curing agents suffered from disadvantages in respect to safety or availability, but all shared the common disadvantage of having to be used in fine powder form. The reaction rate variables introduced by particle size distribution, moisture absorption in storage and mixing efficiency, severely limited the successful foundry application of powdered hardeners. These problems were overcome with the commercial introduction of all-liquid binder hardener systems.[9,10] In these systems the hardeners comprised nontoxic low viscosity liquids of controlled and consistent chemistry. They offered ease of handling, metering and sand mixing, combined with controlled sand-curing characteristics.

The liquid hardeners introduced in the late 1960s and early 1970s were based upon acetate esters of dihydric and polyhydric alcohols. In what has become known as the "ester-silicate" process, hardening of the sand mass takes place progressively as a result of the in-situ hydrolysis of the curing agent. It can be considered that the hardener slowly releases weak acid after dispersion in the silicate binder film, inducing an effective change of the silica:soda ratio with resultant gelation. The development of the "ester-silicate" process greatly extended the use of sodium silicate bonded cores and molds. It is estimated that in the United States the ester-silicate binder system is now used to bond approximately 150,000 tons of foundry sand annually.

Factors Affecting the Application of Silicate Binders

All processes have applications for which they are suitable and which are generally a function as much of the properties of competing processes as of the properties of the principal one in question. The fundamental properties of both the CO_2 and ester setting silicate systems are much the same in each case:

- The binders are odorless, nontoxic, nonflammable and water rinsable.
- The hardeners, gaseous or liquid, are similarly clean and hazard-free.
- The minimal organic content of the binders eliminates smoke and fume at pour and shakeout. All components are formulated without nitrogen (Ni), sulfur (S) or phosphorous (P).
- Little gas is evolved during casting.
- The hot plasticity of the sodium silicate bond is beneficial in minimizing veining and hot tearing problems.
- Large quantities of cores need to be released into green sand systems before molding properties deteriorate.
- Binders can be used with other sands besides silica (chromite, olivine, zircon, chamotte, etc.).
- Binders are relatively insensitive to sand quality.

In examining the specific applications of the various processes, it is appropriate to look at the CO_2 and ester processes separately.

CO_2 Binders

Shakeout

The retained strength of sodium silicate bonded sand which has been heated to the refusion temperature of the binder, 760-871C (1400-1600F), is high. In ferrous castings, particularly those with a high metal to sand ratio, cores made with an unmodified sodium silicate binder can undergo sintering which makes removal difficult and expensive. The extent of the shakeout problem with a given base sand is directly related to the level of the sodium silicate binder which is required to give cores adequate cold strength characteristics, as shown in Fig. 3.

Traditionally, shakeout improvements have been obtained by including sucrose or glucose additions in the sand formulations, either separately or in the form of proprietary sugar-containing sodium silicate binders. In the casting operation, the burn-out of the sugar material weakens the binder film enough to give easier shakeout. This practice has permitted sodium silicate binder use in many coremaking applications where shakeout considerations would have otherwise made its use impractical. However, the use of sugar silicates has not offered the complete solution. These sugar materials do not add to the cold strength of the binder and therefore do not permit any reduction in the total sodium silicate content of the sand. In addition, the hygroscopic nature of sugar severely limits the level of its addition which can be made without creating problems as a result of moisture pickup.

The shakeout available from CO_2 process cores has been greatly improved since the introduction in 1972 of non-hygroscopic polymeric co-binders which interact with the sodium silicate binder and contribute to strength development. Because the co-binder allows core production with a lower sodium silicate content and decomposes thermally to reduce retained strength, there is a major shakeout improvement in comparison with that of systems containing sugar. The co-binders being utilized in the United States to extend and improve CO_2 process application must not be confused with some phenolic resinous additives[8,11] which may compromise the essential environmental advantages of the CO_2 process.

Strength

Relevant mold and core strength comparisons between different chemical binder systems are difficult to make. The tensile or compressive cold strength which produces acceptable performance and casting quality in a given application may be quite inadequate with another. Some applications require more cold strength than is practically available from silicate binders. High bond strength is available from sodium silicate, but at addition levels which generate high mold or core strengths, shakeout performance may then be inadequate. Cold strength has therefore to be compromised for the sake of shakeout performance, with resultant core surface friability or excessive breakage.

Binder addition levels can also be limited by the reduction of sand flowability which is experienced as the total liquid addition level is increased. Co-binder developments have alleviated this situation with CO_2 process cores by increasing the cold strength at which acceptable shakeout is obtained. Gas hardened systems competitive to the CO_2 process, such as amine hardened phenolic urethane resins and sulfur dioxide (SO_2) hardened furanic ones, develop virtually all their available strength in the box. In the CO_2 process, the maximum attainable cold strength and core surface quality is developed by gassing the core to well below its maximum as-gassed strength and obtaining the cold strength through subsequent dehydration. Gassing to develop the maximum strength only serves to reduce that available from the secondary dehydration. In fact, gross over-reaction eventually produces a condition where strength deteriorates below the as-gassed level if subsequent dehydration is encouraged. The characteristics of a typical sugar silicate are illustrated in Fig. 4 which also shows the advantages in this respect of the silicate plus co-binder concept.

In practice, many production CO_2 process coremaking operations have been designed to accommodate the advantages of stripping and handling cores at relatively low as-gassed strengths and promoting subsequent dehydration with a short on-line post drying cycle. This, of course, is most readily applicable where core design does not impose stringent strength requirements on stripping. Many corerooms of production steel railroad foundries have been advantageously converted to the

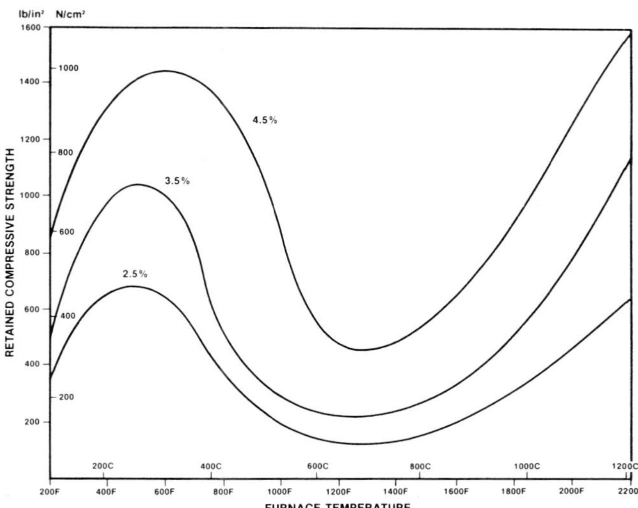

Fig. 3. **Typical retained strength curves obtained for cores bonded with different addition levels of unmodified sodium silicate, cured with CO_2. AFS 55 sand bonded with 47% solids 2.4 SiO_2:Na_2O ratio solution. Cores gassed to 100 lb/in.^{2} compressive strength before heating.**

CO_2 process using the above principles. In large part, success has been made easier by the "chunky" nature of the core designs involved. Experiments aimed at obtaining higher in-the-box CO_2 strength by using warm core boxes or heated CO_2 have not hitherto met with significant success.

CO_2 process binder systems permitting rapid attainment of significantly higher as-gassed strength, without excessive binder levels, will clearly extend the applicability of sodium silicate into production coremaking areas where the organic cold box processes are currently showing growth.

Core Storage

The important aspects of storage stability have been discussed above. The historical storage limitations of CO_2 process cores have been associated with two factors:

1) use of hygroscopic sugar additives for shakeout, with resultant core deterioration due to moisture absorption under humid conditions;
2) lack of control of the gassing cycle, resulting in gassing

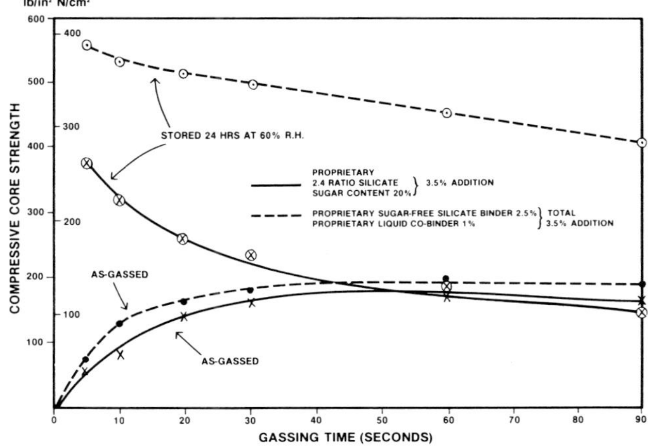

Fig. 4. **Gassed strength development and storage properties of CO_2 process cores. AFS 55 sand was used in these tests. Standard compressive test cores were gassed with CO_2 at 12 lb/in.^{2} pressure and 5 lit/min flow rate.**

Table 1. Humid Storage Properties of CO_2 Process Cores

	3.5% SUGAR-SILICATE*			2.5% SILICATE BINDER 1.0% LIQUID CO-BINDER			2.0% SILICATE BINDER 1.0% LIQUID CO-BINDER		
GAS TIME (SECS.)**	10	30	60	10	30	60	10	30	60
AS-GASSED COMPRESSIVE STRENGTH (LB./IN^2)	102	220	217	143	245	248	168	192	192
COMPRESSIVE STRENGTH AFTER 24 HRS. STORAGE AT 95% R.H. 80°F (LB./IN^2)	218	152	142	328	309	297	268	256	250

* SiO_2: Na_2O, 2.4:1, 20% SUGAR CONTENT

** ALL CORES GASSED AT 12 LB/IN^2 CO_2 PRESSURE, 5 LITRES/MIN. FLOW RATE.

beyond the attainment of maximum as-gassed strength and leading to subsequent deterioration in storage under dry conditions.

Co-binder technology has eliminated the requirement for hygroscopic additives and also substantially reduced the potential for "over-gassing" storage difficulties. Table 1 illustrates the change in available humid core storage characteristics which has been obtained by introduction of the non-hygroscopic co-binders.

Ester Hardened Silicates

Shakeout

Shakeout performance from the ester-silicate binders appears to have been less limiting to growth than with CO_2 binders. In part, this is because the system has been used more in moldmaking rather than coremaking. The situation has been helped by the use of lower addition rates, of higher ratio binders and by the fact that the 10 or 12% of organic material used to harden the sodium silicate is a breakdown agent in its own right. Nevertheless, other breakdown agents have been added to silicates for use with the ester process and these have been almost exclusively sugar. Even with sugar levels as high as 15% in the silicate, ester hardened silicates still are not able to compete with resin binders in the most exacting breakdown applications. However, this is rarely the prime reason for selection of a wholly organic system over an ester setting silicate.

Strength

As with CO_2 silicates, strength needs to be compromised with shakeout performance. Strength is much less of a limiting factor with typical ester setting mold applications than with CO_2 coremaking. Nonetheless, it is strength which governs the wall thickness requirements of a flaskless mold. For flaskless molding, foundries generally have accepted a higher sand to metal ratio when selecting ester setting silicate over a resin process. In flasked molds, ester-silicate strength is not a limitation.

The hot strength of an ester-silicate system is particularly advantageous in comparison with resin binders. The incidence of veining and other problems related to sand expansion are much reduced through the use of ester hardened silicate binders and this has often been a major justification for use.

Mold and Core Storage

Even with the use of sucrose as a breakdown agent, storage is not generally the critical factor with ester hardened molds or cores. It is also less common to store large molds and cores for weeks at a time than to store small CO_2 process cores.

Sand Reclamation

As long as chemical binders were used principally for coremaking or for facing molds backed up with conventional green sand, selection of a binder system was not governed by considerations of sand reclaimability. In fact, sodium silicate bonded sands may have had advantages over other systems because of the potentially beneficial effect of slightly alkaline binder residues on the performance of bentonite bonded sand. However, the major growth being exhibited by nobake binders is now associated with conversion to fully integrated molding and coremaking and with sand reclamation.

Compared to organic binders, sodium silicate systems are at a disadvantage when reclamation characteristics are considered. First, the organic systems benefit because the casting process itself serves to remove binders from the sand. Designing for the minimum sand to metal ratio helps this process. Second, the organic binder residues which adhere to recycled sand after fines removal have a minimal influence on the control and effectiveness of the rebonding process. If the fines are completely removed and total organic content does not become excessive, reclaimed organic nobake sand can be recycled with a new sand demand of 20% or less. In the case of sodium silicate bonded sand, the inorganic nature of the binder precludes thermal removal during the casting process. Binder residue reduction is entirely a function of the efficiency of the reclamation equipment. Practical successes at reclaiming silicate bonded sand in conventional dry reclamation equipment have been reported,[12-16] although some efficiency limitations of conventional techniques have also been examined.[16] The practical problem of reusing dry reclaimed sodium silicate bonded sand without high dilution (about 50%) of new sand, are associated with the harmful effects of binder residues. Silicate binder residues can have adverse effects in two areas:

1) If the sodium oxide (Na_2O) content of the sand is allowed to rise, it leads to reduced refractoriness of the silica sand.
2) Dehydrated binder residues in the sand act physically and chemically to interfere with the hardening reactions and strength development in subsequent rebonding.[16]

Newer binder systems use reduced additions of high ratio silicates with the result that Na_2O buildup potential is minimized. However, it seems unlikely that binder technology will advance to the point that used silicate bonded sand can be processed through dry reclamation units originally intended for organic binder systems at the throughput and reuse levels obtained with the organic binder. It is more likely that dry systems specifically engineered for sodium silicate reclamation will become available.

Wet reclamation of sodium silicate bonded sand remains entirely feasible technically and suffers only from the limitations imposed by sand drying costs and disposal of alkaline wash water effluent. A pilot wet reclamation unit has been operating in an American foundry recently and it is claimed that water recycling techniques have been developed which minimize effluent disposal problems.[17]

Cure Characteristics

Non-gassed organic binder systems are now available offering a wide range of curing behavior. Phenolic urethane systems offer strip times shorter than one minute when utilized with a high intensity mixer and core blower. Alkyd oil systems offer working times in excess of one hour in production of very large pit molds. The acid catalyzed furan and phenolic binders do not approach these extremes, but in the intermediate range they offer excellent work time to strip time relationships and good through-cure properties. Compared to the available organic systems, the conventional ester-silicate process cure characteristics have shown some limitations. In some cases the perceived limitations have been real ones, in other cases they have resulted from misunderstanding of the cure parameters of these systems. Conventional ester-silicate chemistry does not compete with organic binders in respect of work time to strip time characteristics. The organic systems are usually highly exothermic so that after an initial period, the cure reaction becomes self-accelerating. A typical ester-silicate system exhibits a linearity of chemical reaction. Mixing a conventional ester catalyst with its binder serves to demonstrate the gentle nature of the curing reaction. Even with a fast catalyst, there is a progressive increase in viscosity until eventual solidification, with no violence or noticeable temperature rise.

In contrast, an acid catalyst must never be allowed to come into contact with a furan or phenolic resin because an almost explosive reaction will take place. Thus, whereas typically a furan binder system with a working time of 10 min will give a hard, strippable core approximately 30 min after mixing, a typical ester-silicate mix with the same work time would require longer for adequate hardening. Thus, the ester-silicates have proved acceptable in applications where work time requirements in the 20 to 60 min range have been combined with the production requirement for 1 to 3 molds per pattern per shift. The conventional ester-silicate binders have not yet found wide acceptance for faster cycle nobake applications, especially where high strength and rigidity at stripping is essential, as with much coremaking and flaskless mold production.

The nature of the hardening reaction with ester-silicate mixes can be understood by consideration of Fig. 5. This compares the final core strength developed as a function of ester catalyst addition under two conditions:

1) With test cores sealed from the atmosphere to prevent dehydration of the core.
2) With test cores allowed to cure in a controlled atmosphere at 50% relative humidity.

The former condition represents the mode of chemical strength development at the center of a large sand mass or close to the pattern if complete curing is allowed to take place before stripping. The latter represents the strength of a surface exposed to low relative humidity during curing. It can be seen that the so-called "internal" strength is related to ester addition level. With this system, a 15% ester addition level is required to produce maximum internal strength. With less addition, the chemical reaction is incomplete and higher additions appear to weaken the bond by virtue of dilution. In contrast, the strengths recorded when dehydration is possible are relatively insensitive to ester addition level. It seems likely that in the absence of complete chemical gelation, a supplementary glassy bond is developed by dehydration of unreacted binder. This observation has considerable practical significance when supplemented by the additional knowledge that the plasticity of undehydrated silicate gel bond related directly to the completeness of this chemical gelation.

In the acid cured chemical resin binder systems, temperature sensitivity of the reaction's speed is compensated for by adjusting the level of catalyst addition. Within limits, this change is not observed to influence rigidity or through-cure. In the case of the conventional ester-silicates systems, the range

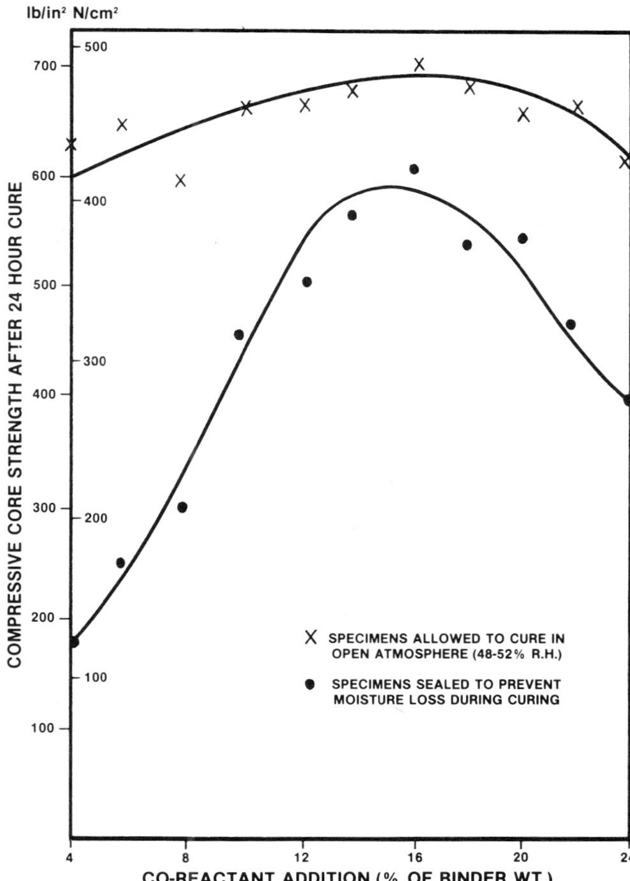

Fig. 5. The influence of curing conditions and co-reactant addition on strength developed by an ester-silicate binder system. Results obtained using a proprietary binder system and AFS 55 silica sand. Binder addition 3%. Co-reactant was slowest available. Sand temperature 22C (72F). Chemical cure of all sand mixtures tested was complete in less than 24 hrs.

within which such adjustment can be made is very limited. Reducing ester addition to compensate for hot sand conditions can lead to loss of through-cure or at least the production of a sand mass which appears cured but becomes subject to deformation and possible collapse after stripping. Increasing ester addition beyond the optimum level has little effect on accelerating the reaction rate with cold sand, but can cut internal strength development. The solution to these difficulties lies in sand temperature control or in recognizing the need to switch to an ester grade with reactivity appropriate to the sand temperature.

Many so-called failures of the ester-silicate system can be ascribed to failure to understand the importance of knowing and maintaining the optimum and rather narrow ester to silicate range. Other problems have arisen when standard commercial grade silicates of unsuitable composition have been substituted for proprietary silicate binders whose chemical composition and viscosity have been designed for compatibility and optimum strength development with ester formulations.

Despite some clear advantages over organic chemical binders, recent growth in the American foundry usage of silicate binders has been modest when compared to the rapid growth in use of cold-setting organic chemical binders. In 1980, American foundry consumption of acid cured and isocyanate co-reacted cold-setting organic binders will exceed 100 million lb, compared to approximately 35 million lb in 1970.

Clearly, despite the many environmental and safety advantages of silicate binders, the newer organic binder systems have been found to out-perform available silicate systems in some important areas. For example, over a decade after its introduction to the United States, the ester silicate system accounts for less than 5% of the total tonnage of foundry sand bonded with non-gassed, cold-set chemical binders. Two decades after the arrival of a binder system cured with odorless, safe CO_2 gas, production corerooms are being converted to the use of cold box core binders cured with gaseous organic amines or sulfur dioxide gas.

In general, one can say that in the United States silicate systems have become identified somewhat as speciality sand binders, used where their advantages are so essential that their disadvantages become of secondary concern. Consumption of sodium silicate in the American foundry industry is equivalent to about 1 kg for every ton of ferrous castings produced. In comparison, other countries consume a much greater quantity. The Germans and French consume approximately three times as much and the British and Japanese just over 10 times as much per ton of castings. Binder selection parameters in Europe generally seem to differ from the United States. Sodium silicate processes are chosen wherever the technical performance of competitive organic binders are not essential.

Outside the gray areas where these opposing views can govern process selection, there lie clear cut applications for resins and silicates where economics and technical performance point to their use. It is possible that in the future strong environmental pressures in the United States will force increasing attention to the use of silicate systems.

Some In-Foundry Applications

With the foregoing background on the advantages and limitations of established and available silicate binder technology, it is appropriate now to review some typical foundry applications where silicates are being used to maximum advantage. These examples will serve to illustrate the current state of the art before considering some new advances in silicate binder technology which will enable even greater advantage to be taken of the positive environmental aspects of sodium silicate binders.

The CO₂ Process at McConway and Torley

In recent years, a major area of growth in the use of CO_2 process technology has been in coremaking for production steel foundries, most notably in the manufacture of railroad car castings. Automated and semi-automated CO^2 process coremaking lines have been designed and installed to take maximum advantage of the production and casting quality benefits available from the process and to utilize continuing advances in CO_2 process binder technology. A new railroad casting CO_2 process coreroom is therefore a most appropriate subject for study.

The Pittsburgh steel foundry of McConway and Torley Corporation is a leading producer of castings for railroad cars and locomotives, with production capacity in excess of 2,000 tons per month. McConway and Torley Corporation has maintained its leading position in this segment of the steel castings market through a program of major capital investment directed towards advances in productivity and in the quality of the finished product.

In the past ten years, melting, molding, coremaking and finishing operations have each been substantially modernized and extensive investment is planned to increase capacity further. CO_2 process silicate cores now account for 90% by weight of all cores used in the McConway and Torley foundry.

Fig. 6. CO_2 process coupler core production. This one-piece core was formerly assembled from four oil sand cores. Gas is introduced through an internal mandrel. Normal production rate on this cure is 150 per shift.

Prior to 1970, CO_2 process coremaking was used at McConway and Torley on a very limited basis for the production of small pin cores. Baked oil sand was the predominant coremaking process, supplemented by some shell core production. The incentive to examine alternative core processes was provided by a desire for improved efficiency in the manufacture of cores for railroad car coupler castings. These cores were being produced in oil sand, the final cores requiring production and assembly of four separate cores. Experiments with hot box and shell as a means to efficient production of a one-piece coupler core introduced soundness problems, apparently associated with the nitrogen content and gas producing characteristics of the resin binders used, and the feasibility of CO_2 process use was therefore considered. Experiments with CO_2 process coremaking showed that sound castings were obtainable from one-piece silicate cores. Some additional advantages were found to be obtainable:

- the incidence of hot-tearing was reduced
- it was possible to produce entirely sound coupler castings with elimination of the extensive chill placement which had become standard practice with oil sand cores.

As a result of these experiments, equipment for automated production of one-piece CO_2 process coupler cores was installed in 1971. The first core shooter installed was equipped with a gas heated core box. It was found that with a heated box at 149-177C (300-350F) CO_2 gassed cores could be produced with exceptional as-stripped surface hardness and handleability. This technique was successfully used for several years but was eventually discontinued in favor of cold core production with a short oven post-dry cycle. This change offered improved control together with reduced maintenance costs.

The successful conversion of coupler cores to CO_2 process manufacture served to demonstrate the general environmental and performance advantages of silicates for this type of castings production and led to additional investment for full conversion to CO_2 process coremaking.

Completion of the conversion to CO_2 process coremaking was implemented in the period 1974-78. Approximately 500 tons per month of CO_2 process cores are now utilized in the manufacture of couplers, coupler yokes and miscellaneous railroad castings. These cores utilize AFS 50 subangular silica sand. Total binder level is 4%, of which 1% is a proprietary liquid co-binder used for its contribution to core strength and shakeout characteristics. Sodium silicate binder is purchased in tank car quantities and stored in a 15,000 gal in-ground tank. Sand is prepared in a batch muller servicing five core shooting stations from a mezzanine level.

Fig. 7. Post drying CO$_2$ process cores. Immediately after strip gassed cores pass through a 400F tunnel. A four minute dry produces rapid strength increase and a hard surface to permit further handling.

Fig. 8. Variety of railroad casting CO$_2$ cores. Core weights range from 2 to 100 lb. Core storage stability is attributed partially to the practice of following a minimum gassing cycle by a short post-dry.

Fig. 9. Cored molds prior to closing. Cores receive a coat of zircon wash in critical areas. Where design permits, cores are mandrelled to reduce costs, assist uniform cure and assist shakeout.

Fig. 10. No smoke from pouring cups. Advantage is taken of CO$_2$ practice to manufacture all pouring cups used at McConway and Torley.

Wherever design permits, core boxes incorporate a retractable internal mandrel. This mandrel reduces sand consumption and shakeout difficulties. In mandrelled cores, CO$_2$ gas is introduced through the vented mandrel, allowing cores to be gassed very uniformly and rapidly from the inside out.

All gassing is automatically timed, with gas times adjusted to provide the minimum as-gassed strength consistent with strippability. All cores receive a short post cure treatment which gives rapid pickup in strength and surface hardness. Cores are transported through a forced draught oven on a powered vented belt. Maximum oven temperature is 204C (400F) and a typical cycle time is four minutes.

Some cores are used uncoated, others receive one coat of a refractory zircon wash. Small cores are dipped prior to entering the drying tunnel. Larger cores, including the one-piece coupler cores, are selectively coated in critical areas by brush application to the warm core on the exit from the drying tunnel.

The combination of controlled gassing and post-cure produces a hard, non-friable core surface. Core storage stability is excellent despite high humidities which arise because of a riverside location.

Shakeout characteristics have not created any production difficulties. Most cores disintegrate completely at the shakeout stage, but because of the contoured and enclosed configuration of the coupler castings, much of the sand is carried over from the shakeout. This retained core sand is typically weak and friable and is removed in the subsequent shotblast operation.

The accompanying illustrations, Fig. 6-10, show the innovative application of a core process which has provided McConway-Torley with productivity and quality advantages, while contributing to environmental improvement and clean working conditions.

The Ester Process at Mackintosh-Hemphill Division Gulf + Western Midland, PA

Use of the ester-silicate process has generally been developed most effectively in the production of large steel castings where erosion resistance and low gas producing potential are extremely important to the economical production of a sound casting. In these applications, the ester-silicate process has provided the ability to convert from dry sand to an improved productivity nobake sand, without the concerns about working conditions which can arise when some organic nobakes are used for large pit molds.

Mackintosh-Hemphill is a major supplier of rolls and other steel castings to the steel making industry. One area of specialization is the casting of slag pots ranging in capacity from 40 to 975 cubic feet. The as-cast weight of the largest slag pot regularly produced is 170,000 lb. These slag pots are cast with both cope (external surface) and drag (internal surface) made in ester-silicate bonded sand.

Fig. 11. Ester-silicate molding for a slag pot. Silicate nobake sand is supplied at 500 lb/min from a mobile continuous mixer. More than 50,000 lb of sand is used to complete this drag.

Prior to 1969, molding practice for Mackintosh-Hemphill slag pot production was entirely with clay bonded dry sand. The use of ester-silicate was first established for producing small cores for general steel castings applications. A desire to improve productivity and increase capacity by reducing the labor content and time required to ram up dry sand molds and by eliminating the extended time required for mold drying, prompted examination of nobake techniques. Favorable prior experience with ester-silicate made the process a leading candidate for evaluation. First production tests were made on small slag pots using

Fig. 12. Slag pot produced from an ester-silicate mold. This 650 cubic foot slag pot is typical of the large steel castings produced in ester silicate by Mackintosh-Hemphill. These pots are cast open end drum and minimum gas evolution is an important binder requirement.

sand prepared on a 200 lb/min continuous mixer, transferred to the molding floor by a temporary conveyor system obtained from a local dealer in used agricultural equipment. Encouraging results led to the casting of progressively larger slag pots in the period 1970 to 1974.

During this time a number of alternative proprietary ester-silicate systems were subjected to careful evaluation before selecting the combination which provided the desired optimum balance of working time, sand flowability and through-cure characteristics. At the same time, tests with organic nobakes led to their rejection for this application. Odor was felt to be a significant objection to the use of the organic systems for these large molds, but the primary reason for discontinuing evaluation was the adverse effect of mold gas evolution.

It was demonstrated that molding man hours could be reduced by approximately 70% by converting entirely from dry sand to ester-silicate practice. Speeding molding and eliminating mold drying doubled molding capacity without any increase in floor area. Equally dramatic was the reduction in cleaning time when a pot was converted from dry sand to ester-silicate practice.

The molding floor is now serviced with ester-silicate sand from a mobile continuous mixer and dry sand molding has been entirely eliminated. It was learned, incidentally, that rated sand output for a continuous mixer based on organic nobake binder use does not necessarily apply to mixing ester-silicate sand. Because of the higher viscosity and addition level of a sodium silicate binder, it was found necessary to reduce sand output from the rated 800 lb/min to 500 lb/min in order to obtain consistent and homogeneous coated sand. Effective and reliable mixing is obtained at this lower output.

With a sand-to-metal ratio of approximately 2:1, up to 5 hr of sand mixer operation may be required to complete the cope or drag for a large pot. A minimum working time of 25 min is desirable to ensure compaction and effective knitting together as sand layers are progressively laid down. With these curing characteristics, stripping is possible within 2 hr of mold completion. Two ester grades are used to provide the desired working time and cure speed under summer and winter conditions. A single grade of proprietary sodium silicate binder is used for all ester-silicate practice. This binder is delivered into a 6,000 gal storage tank.

Mold coating practice has been the subject of exhaustive studies by Mackintosh-Hemphill and the practice which has been developed as providing optimum peel and surface quality uses a heavy-duty, high solids, water-based zircon refractory coating.

Figures 11 and 12 show both the scale of the Mackintosh-Hemphill ester-silicate nobake operation and the high demands placed upon binder system performance. In this application, no other nobake system has been able to offer the combination of performance benefits available from ester-silicate practice.

Mackintosh-Hemphill appreciate the need for an effective low-cost reclamation process for silicate bonded sand. The increased materials cost of replacing dry sand practice with ester-silicate has been overwhelmingly justified by productivity and casting quality improvements, but new sand is a significant element of overall cost and a potentially economic technique for recycling sand would receive close examination.

New Developments in Silicate Based Sand Binders

From the foregoing it was shown that there are a great many advantages to the foundryman in the use of silicate sand binders, but also that application is limited by certain technical characteristics in which competitive binder systems have the edge.

The traditionally attractive features of the CO_2 process (for example) may be listed as low cost, convenience, flexibility and availability. To them may be added the modern benefits of acceptability from the points of view of environment, ecology and energy consumption.

These new factors make it more attractive than ever to put effort into Research and Development (R and D) to overcome the factors which limit applicability. Workers in this field need to ask themselves which advantages and disadvantages are fundamental, that is, which are so characteristic of the systems in question that they cannot be overcome within the definition of the system. Clearly, of the above attractive features, all are quite fundamental. It is not so obvious, but the limitations of strength, flowability and reclaimability which have been referred to in preceding sections are not fundamental. In other words, there is scope for further development of silicate based binders and no doubt most binder suppliers now are investing money in it.

That there is every hope for continuing significant development in silicate binders can be illustrated by three of the latest examples that have appeared on the market. Two of them are fairly straightforward extensions of the existing art and the third is a novel addition to the corebinder scene.

An Improved CO_2 Process Binder

Research into this binder process which resulted in a practical success in 1972 and in substantial foundry use since then, was directed as an alternative to sugar as an effective collapsibility promoter. This research produced the liquid co-binders. These co-binders eliminated the hygroscopic problems associated with sugar and also combined with the silicate binder film to provide additional bond strength and the option of further improved shakeout through reduced silicate binder addition. Continued research has led to further performance improvements. A new co-binder material has been developed which itself offers significantly greater bond strength. In addition, it has been found that the incorporation of certain sodium-based inorganic salts imparts dramatic improvements in the gassing speed and ultimate strengths of CO_2 process cores. By combining the two technologies, commercially applicable binder systems have been developed which appear to offer scope for extending the foundry usage of CO_2 process coremaking.

In Europe, where silicate binders of relatively low SiO_2:Na_2O ratios (2.0:1 - 2.2:1) have traditionally been used for the CO_2 process, the new binder technology has been formulated to provide single addition binders. In the United States, where more reactive binders have typically been utilized (SiO_2:Na_2O ratios 2.4:1 - 2.6:1), very high performance versions of the system have been developed in which the binder is currently supplied in two-component form. Performance details of the system now being used in Europe have been reviewed by Williams.[18] The laboratory studies reported here were conducted with a two-part system developed for the American market.

The new CO_2 binder is supplied as two liquid additions, Part A and Part B. Part A is primarily inorganic and contains the sodium silicate component of the system. Part B is an aquaeous solution of primarily carbohydrate components and contains high molecular weight film-forming materials. The viscosities of the components are lower than typical for conventional CO_2 process binders, offering the advantages of efficient sand coating and flowability. The viscosities of the binders tested are shown in Table 2. The additions are designed for use in the ratio of 3 Parts A to 1 Part B, with total addition levels in the range of 2.0 - 3.5%.

Selected for comparison were two widely used proprietary CO_2 binder systems. Chosen as representative of conventional

Table 2. Viscosity Characteristics of CO_2 Silicate Binders

BINDER SYSTEM	COMPONENT	TYPICAL VISCOSITY (CENTIPOISES)	
		AT 25C (77F)	AT 10C (50F)
COMMERCIAL SODIUM SILICATE SOLUTION (PHILADELPHIA QUARTZ RU GRADE)*	–	1400	4700
SUGAR–SILICATE BINDER**	–	2000	6800
SILICATE PLUS LIQUID CO-BINDER	SILICATE BINDER	350	1950
	LIQUID CO-BINDER	600	1100
NEW TWO-PART BINDER SYSTEM	PART A	95	300
	PART B	200	680

*2.4 RATIO, 47% SOLIDS
**2.4 RATIO, 37% SILICATE SOLIDS, 20% SUGAR

sugar-silicate practice was a formulated binder based upon a sodium silicate of SiO_2:Na_2O ratio 2.4:1 and containing 20% by weight dissolved sucrose. To represent the original co-binder technology, a proprietary two part system was tested comprising a sodium silicate based binder (Part 1) with recommended addition level of 2.0 - 3.0% in combination with its co-binder (Part 2) of recommended addition 1.0%.

Tests were performed using a 55 gfn round-grain Illinois Silica Sand which is widely used for CO_2 process coremaking. Testing was performed at 3.5% total binder addition in each case, again considered fairly typical of conventional binder additions utilized with this sand.

Sand mixes were prepared in a laboratory Hobart mixer. Standard 2 x 2 AFS 3-ram test specimens were prepared and gassed in the specimen tube prior to stripping, with CO_2 gas at 12 lb/in.^{2} and a flow rate of 5 lit/min. Specimens were stripped immediately after gassing and compressive strength determined. Curves of as-gassed strength versus gassing time are shown in Fig. 13. At each gas time the compressive strength recorded is the mean of three tests.

Improvements in strength of the nature cited above automatically lead to improvements in shakeout through the use

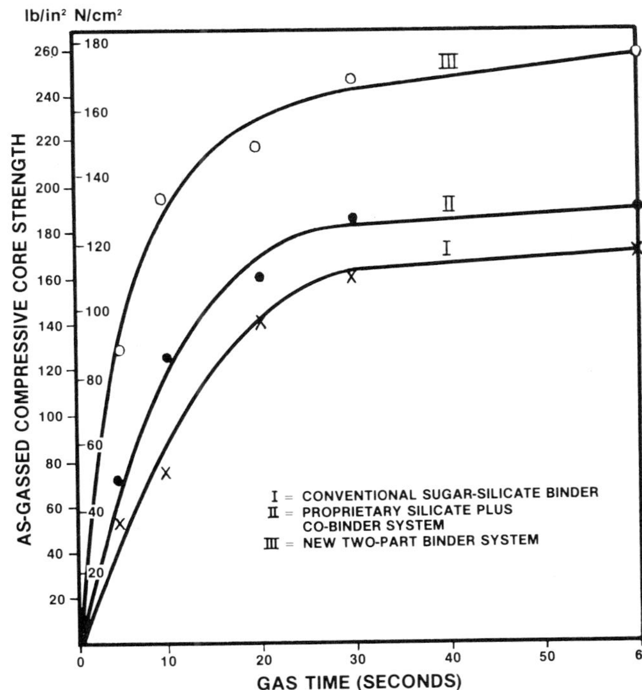

Fig. 13. Comparisons of strength pickup on gassing. Each system was tested at 3.5% total binder addition.

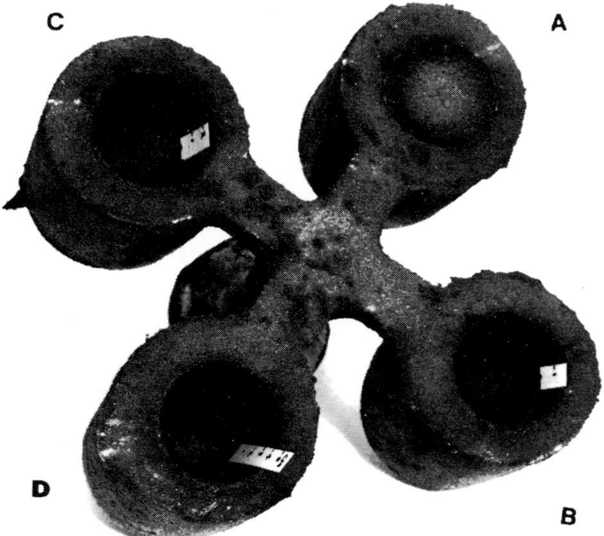

Fig. 14. Shakeout comparison for different CO_2 process binder systems. This test casting was poured in gray iron at 1460C (2660F). Standard AFS cores of equivalent as-gassed cold strength were utilized. Binder additions:
a. 2.4 ratio unmodified silicate (3.5%)
b. 2.4 ratio silicate with 20% sugar (3.7%)
c. silicate plus liquid co-binder (3.2% total addition)
d. new two part binder system (2.8% total addition).

of less binder to achieve a required strength. However, the proportion of soda in the total binder system is also lower than in comparable systems. The improvement in shakeout that results from these two factors may be judged from a simple comparable casting test which has been pioneered by the British Cast Iron Research Association (BCIRA). Four half-closed cylinders are cast off one sprue, the inside diameter of each being formed by a standard AFS test specimen, as shown in Fig. 14.

The casting illustrated compared the three binder systems previously mentioned together with plain sodium silicate of 2.4 ratio, mixed with sand in the approximate amounts to obtain equivalent as-gassed strengths. After cooling, the casting was

inverted and struck 5 times with a hammer. The relative breakdown properties of the four systems are clearly illustrated.

Cores gassed for various times were subjected to 24 hr storage in controlled environments prior to strength determination. Two storage conditions were studied:

1) Normal - 65% Relative Humidity, 24C (75F)
2) Humid - 98-100% Relative Humidity, 29C (85F).

Three replicates were again tested for each gas time.

After plotting the 24 hr strengths versus gas time, Fig. 15 and 16 were developed from this data together with the previously determined relationship of as-gassed strength with duration of gassing. The comparison of stored strength versus as-gassed strength is more practically relevant than a plot versus gas-time, since in production the gas-time of stored cores will be determined by the strength required to strip and handle. Table 3 compares for each system the gas time requirements for different strength levels and the subsequent core strength in storage under the conditions studied.

Consideration of the above comparisons indicates that the performance available from this new system represents a major advance over previously available binder chemistry. The performance benefits of the original co-binder technology when compared to those of sugar-silicate are also clear from these tests. In practice, the wide industrial acceptance of that system attests to the value of that original performance advance and would indicate that the latest advance of similar magnitude will be similarly valuable. The practical significance of these results can therefore be proposed as follows:

- A system is available which reacts 35-50% faster with CO_2 than existing binders, offering faster production and reduced CO_2 consumption.
- The new system offers an improved balance of as-gassed core strength and storage stability versus shakeout, is already considered acceptable, may increase the production rate and may reduce core handling and storage losses. Where strength, storage life or shakeout limitations have hitherto restricted the consideration of CO_2 process coremaking, this new technology may represent an environmentally advantageous alternative to other cold box processes.

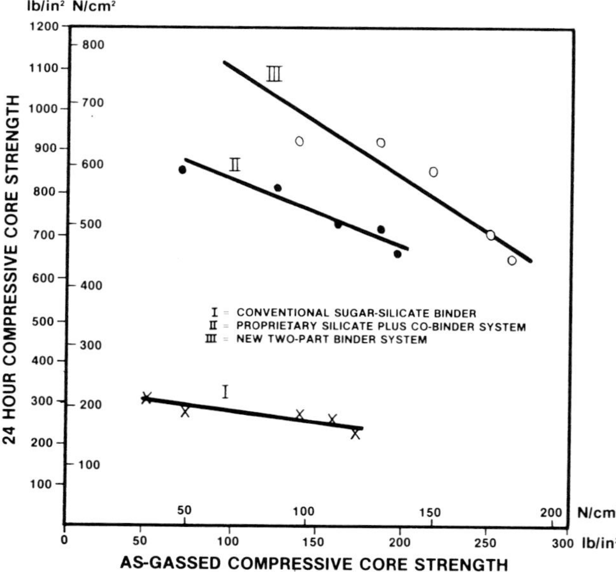

Fig. 15. Core strength after dry storage as a function of initial gassed strength. Cores were stored for 24 hours at 65% relative humidity, 24C (75F).

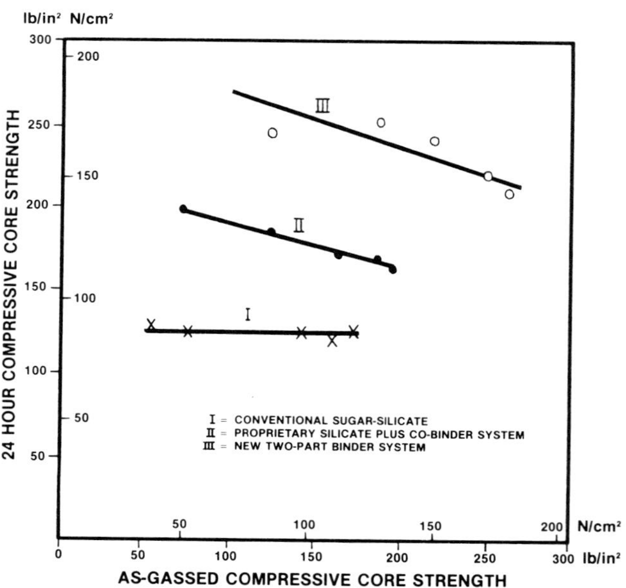

Fig. 16. Core strength after humid storage as a function of initial gassed strength. Cores were stored for 24 hours at 98-100% relative humidity, 29C (85F).

Table 3. Strength Development and Storage Properties For Different CO_2 Process Binders Tested At 3.5% Addition

	GASSED TO 100 LB/IN^2			GASSED TO 175 LB/IN^2			GASSED TO 250 LB/IN^2		
	GAS TIME REQUIRED (SECS.)	STRENGTH AFTER STORAGE		GAS TIME REQUIRED (SECS.)	STRENGTH AFTER STORAGE (LB/IN^2)		GAS TIME REQUIRED (SECS.)	STRENGTH AFTER STORAGE (LB/IN^2)	
		DRY	HUMID		DRY	HUMID		DRY	HUMID
SUGAR SILICATE	14	280	120	60	220	120	N.A.	–	–
SILICATE PLUS LIQUID CO-BINDER	8	810	190	25	705	170	N.A.	–	–
NEW TWO-PART BINDER SYSTEM	4	1080	260	9	900	245	40	710	225

A Fast Curing Liquid Hardened System

Conventional ester-silicate systems exhibit relatively slow cure and do not approach the excellent work to stripping time ratios obtainable with some organic nobakes. As the use of non-gassed chemical nobake molding has been extended from low production rate jobbing applications to the production of smaller castings which hitherto used green sand, the need for faster setting systems has increased. Organic binders are now available which, when utilized via high-intensity mixers or mixer-blowers, provide strip times of less than two minutes. Speed of reaction alone is not enough for practical success, however, as a finite working time is desirable to provide adequate time for the compaction of flowable, unreacted sand mixture prior to rapid hardening. These characteristics have been most readily available up to now from the phenolic-urethane binders. However, a new generation of all-liquid sodium silicate based binders has now been developed which offer considerable promise in fast-curing applications.

To make a valid comparison with the conventional ester-silicate process, examination of working-time characteristics was first undertaken in order to compare cure properties of sand compositions of equivalent working time.

A widely used proprietary ester-silicate was selected for comparison purposes, with a fast-reacting grade of ester and the recommended optimum reactant addition.

Working time was examined by compacting standard AFS compression test pieces at intervals after completion of a 2 min batch mixing cycle (one minute after the addition of ester, a further minute after binder addition). These cores were immediately sealed in individual airtight glass jars, to eliminate dehydration curing, and allowed to harden for 24 hr prior to testing. The results are plotted in Fig. 17. Work time was arbitrarily defined as the time elapsed between mix completion and the loss of 33% of available final strength. The same test procedure was followed using several combinations of the new reactant and silicate binder. A combination of addition levels most closely matching the work time and final strength and conventional "fast" ester-silicate system was selected for further comparisons. The work time characteristics of this system are also plotted in Fig. 17. In both cases the arbitrarily defined work time is 4-5 min, although in the case of the new system the end of the working time is a more sharply defined occurrence.

Cure characteristics were then compared, again under sealed conditions, by breaking duplicate cores at various intervals after mixing. To ensure reproducibility all cores were rammed up within 2 min of mix completion. This comparison, shown in Fig. 18, clearly illustrates the rapid cure characteristics of the new system compared to conventional ester-silicate technology.

In the foregoing tests, a reactant-binder system was selected to match the 4-5 min work time of a conventional "fast" ester silicate. In practice, by varying reactant formulation, it is possible to obtain significantly faster set speeds while still maintaining the desirable high work-time:set-time ratio. Figures

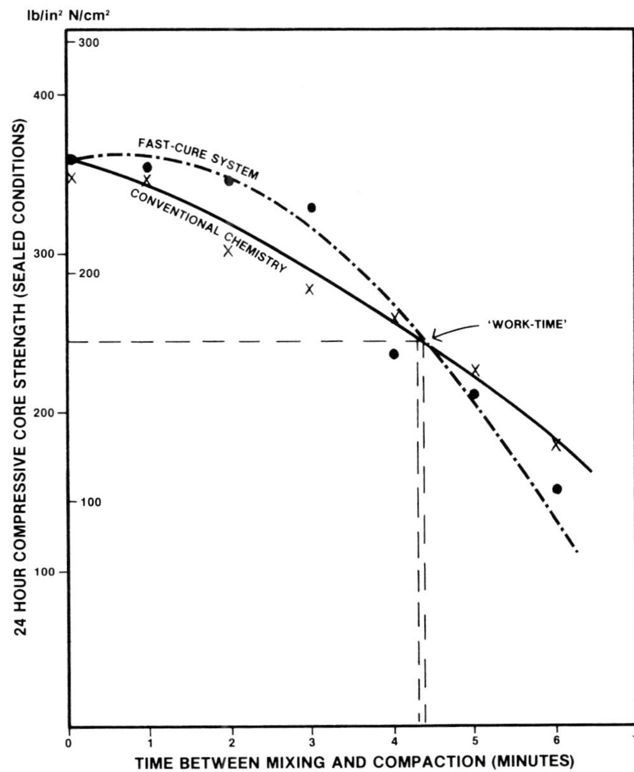

Fig. 17. Work-time comparison of two fast liquid cured silicate binders. Tests were made with 3% silicate binder. Work-time arbitrarily defined as the point at which available 24 hour strength has fallen to 67% of initially available value.

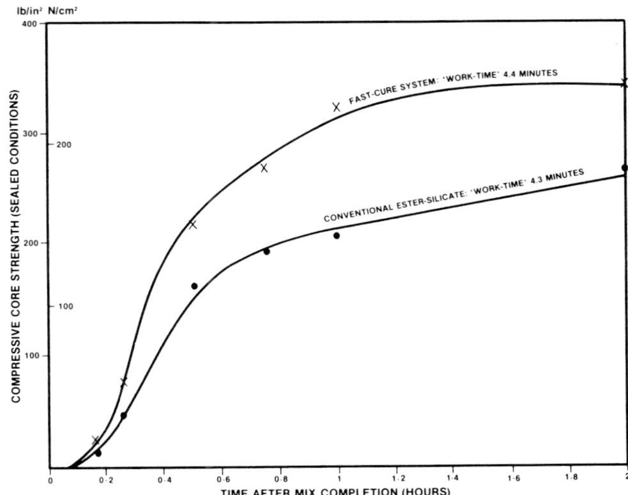

Fig. 18. Cure speeds recorded with liquid-cured mixtures exhibiting similar work-time. All cores were sealed after compaction to prevent dehydration during curing.

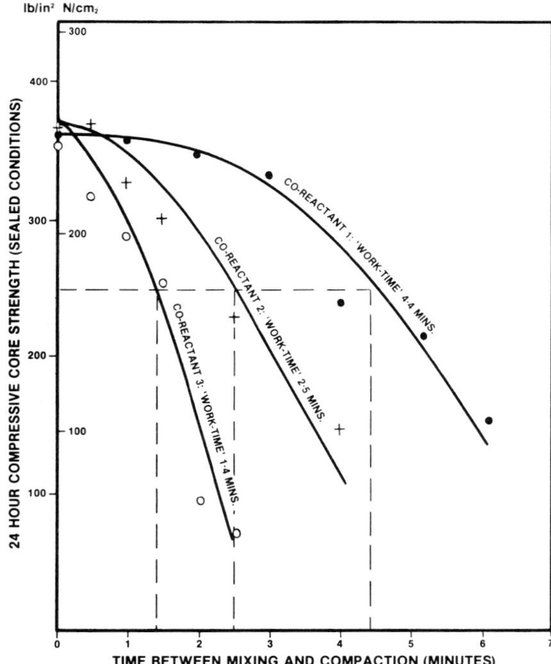

Fig. 19. Work-time characteristics at 22C (72F) using co-reactants of different speeds.

19 and 20 include work time and cure speed plots developed with those reactant compositions designed for faster reaction speeds. The fastest reactant examined in these laboratory tests gave a working time of 1-1/2 min and attained strippable strength 3-4 min after the end of working time. The cure speed attainable in the laboratory is restricted not by the available reactant chemistry, but by the limitations of batch mixing equipment. With commercial high-intensity continuous mixers, reactant chemistry is available which has produced practical set-times of less than 3 minutes.

All of the preceding sand tests were conducted with sand temperatures in the range of 21-22C (70-72F). In order to examine the sensitivity of work-time and cure characteristics to sand temperature variations, a series of tests were performed at

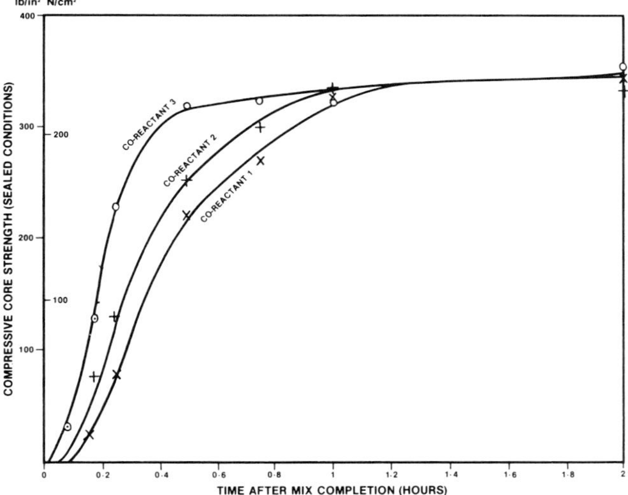

Fig. 20. Cure speeds recorded with batch mixes using different co-reactant grades. All tests were performed at 22C (72F). The fastest system tested cured completely in 30 min under these conditions.

Table 4. Fast-Set Silicate Nobake:Temperature Sensitivity of Cure Properties

SAND TEMPERATURE	WORK TIME* (MINS.)	SET TIME** (MINS.)	RATIO WORK TIME: SET TIME
70F	5.4	10.5	1.9
85F	3.5	5.0	1.4
100F	1.6	1.5	0.9

*TIME ELAPSED BETWEEN MIX DISCHARGE AND LOSS OF 33% OF AVAILABLE CURED STRENGTH.

**CORES COMPACTED WITHIN WORKTIME. TIME ELAPSED BETWEEN EXPIRY OF WORKTIME AND ATTAINMENT OF 50 LB/IN^2 COMPRESSIVE STRENGTH (SEALED CONDITIONS).

sand temperatures up to 38C (100F). Using the previously established criterion for work-time and assuming an arbitrary compressive strength requirement of 50 lb/in.^{2} to define hard set-up, the results can be translated to show the work time: set-time ratio at different temperatures. The results are shown in Table 4. It can be seen that because work-time is relatively insensitive to sand temperature, whereas set-up speed increases markedly, there is a significant improvement in the work-time:set-time ratio as sand temperature is increased.

The availability of sand heaters designed for use with continuous nobake mixers can therefore be fully exploited to obtain maximum curing performance from this new system.

The fast-setting binder system has been applied in several foundries in molding for steel, iron and nonferrous castings.

Some experience of the new system has been gained in the United Kingdom with the Carousel type molding machine, illustrated in Fig. 21, while it was undergoing proving trials in the experimental bay of the manufacturer. The Carousel consists of ten stations each with a tapered molding box. A high intensity mixer fills a box which is then tamped manually. The equipment was designed for use with rapid setting resins, particularly acid hardened furans and urethane types, and required a set-time of 5 min in order to run at its maximum rated capacity of 40 complete molds per hour. Stripping imposes some stringent requirements on the mechanical strength of the mold.

In initial trials the new high speed system proved itself capable of meeting both the strength and speed requirements with total liquid additions of less than 3.8% on an AFS 50 sand. In fact, set times less than 3 min have been obtained with the system, with a work-time of over half this figure.

Early trials in production foundries have confirmed the promise shown in the laboratory, indicating that the new development will extend the use of sodium silicate type binders to more applications.

A Silicate Based Alternative to Oil Sand Coremaking

Although the use of baked oil sand cores has been declining for more than 20 years, at first in favor of hot box and shell processes, and more recently in favor of cold box, the American foundry industry still consumes more than 30 million lb of core oil annually. There is no question that the dimensional, productivity and energy advantages of cold box processes will ensure continued movement from baked oil cores. However, conversion to cold box coremaking requires substantial capital investment. The initial costs for new tooling, coreblowers, sand mixers, etc., are high and although this expenditure can be justified in terms of future cost savings, there may be other immediate demands for available capital, notably in the area of environmental control. The problem becomes even more pressing when the baked oil core process itself is under regulatory pressure due to core oven emissions. Does the foundry spend money to scrub core oven emissions without any net productivity or energy gain, or is the only answer to raise immediate capital and convert completely to a cold core process? A similar expenditure decision may be called for when energy costs and fuel allocations put pressure on core oven operation. There may, however, be an alternative which at least allows the priority of capital for conversion to take its place among other financial considerations instead of becoming a mandatory and immediate need. Replacing core oil with a binder which generated no oven emissions other than steam and which could be cured with up to 50% less energy would remove substantial pressure from the core baking operation. This goal has been pursued through the use of sodium silicate binder technology and it is possible to report that attractive trial results have been obtained with a new system.

A clean, low energy substitute for core oil must fulfill a number of key performance requirements:

1) Sand must be mixed in existing equipment.
2) Cores must be hand rammed or blown on existing equipment.
3) Adequate and consistent green strength must be available.
4) There must be adequate bench life for sand handling through the existing system.
5) Baked cores must combine strength, surface hardness and storage stability with good shakeout properties.

By utilizing binder technology originally developed for the CO_2 process, it was relatively easy to formulate a baking silicate system combining rapid low temperature cure with excellent strength, core surface, storage stability and shakeout properties. Difficulties arise in providing adequate green strength. In extensive testing with various Kaolinitic clays, bentonites and cold gelling cereals, it was found that developing green strength invariably proved detrimental to the final bond strength and core surface. Other difficulties were introduced with shortened bench life and in obtaining consistent green strength development in available foundry mixing equipment. As a result of these limitations, the concept of providing green strength through water activation of conventional powder additives was abandoned in favor of 100% liquid additions. In the practically successful system, green strength was developed during sand preparation by the interaction of two liquid binder components which also provide the chemistry needed for bench life, core properties and shakeout.

The two liquid components of this baking silicate system were mixed in the proportions of 2:1, with total binder additions in the range of 3-4.5%. Sand preparation could be undertaken in virtually any mixer capable of producing oil sand mixes and it should also be possible to use a continuous mixer. Green strength began to develop during mixing, but it was necessary only to produce a homogeneous mix before discharging, as once the components were mixed green strength continued to develop for several minutes. The sand was not used until its green strength development was completed. This was normally less than 15 min, after which sand characteristics remained stable. Green strengths were normally obtained in the range of 1.0 to 2.0 lb/in.^{2}, which was adequate for stripping most cores usually made in conventional oil sand.

The bench life of mixed sand was in the range of 1 to 2 hr, although this was reduced by excessively hot sand or by too much mixing, since both encouraged drying out of the sand mix.

Cure characteristics as a function of time and temperature are illustrated for a typical sand mixture in Fig. 22. Included for comparison is an oil sand mix from an application in which the new system has been evaluated. The potential for energy savings is clearly illustrated by this data.

A variety of cores have been successfully made in oil sand

Fig. 21. **Production molding equipment used for evaluating fast-set silicate nobake.**

production equipment using this system and the resultant casting tests confirmed that satisfactory core properties and shakeout were obtainable. An area of further work concerns the applicability of conventional oil sand wash practice to these baked silicate cores. Typically, where a wash is needed on oil sand cores, it will be water based and applied by dipping or curtain coating. These techniques cannot be used on cold silicate cores without surface damage due to penetration and dehydra-

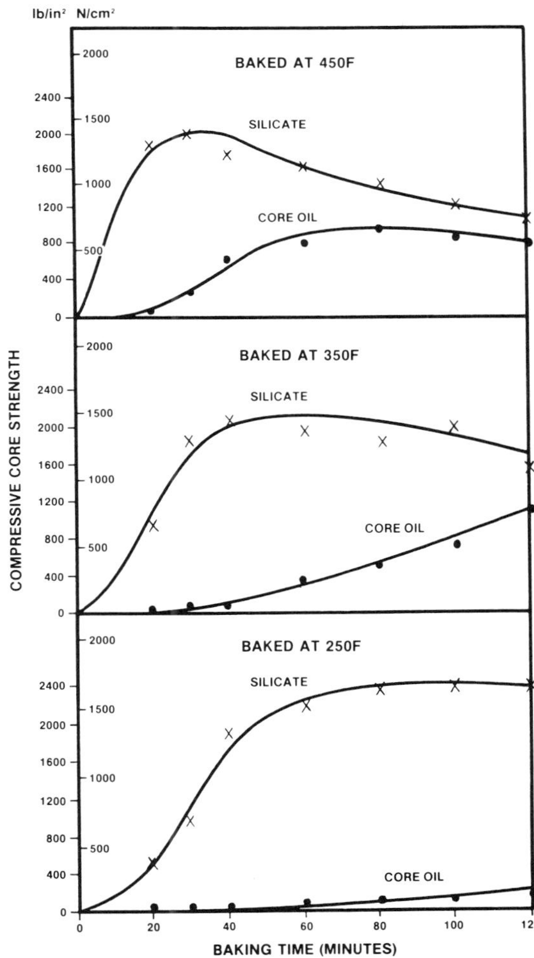

Fig. 22. **Baking properties of green strip silicate system compared to an oil sand mix at various temperatures. The oil sand mix tested used 1.0% core oil, 1.0% cereal and 2.0% water. The silicate system was tested with a total binder addition of 3.5%.**

tion. It has been found that dipping can be accomplished without damage to hot cores and satisfactory results have also been obtained by airless spraying with high-solids, water based core coatings. In some cases satisfactory casting results have been obtained from baked silicate with elimination of the core wash otherwise required on oil sand. Solvent based coatings will of course produce satisfactory performance, but this is a retrograde step from cost and environmental viewpoints. Optimum core coatings practice is receiving further study.

Summary and Conclusions

The use of sodium silicates by the foundry industry really started with the CO_2 process which used certain technical advantages to make mold and core production more economical and convenient.

The development of the ester-silicate process extended these advantages to custom founders of large castings who previously had found the use of CO_2 to be impractical or inconsistent.

In recent years, the fundamental environmental and ecological advantage of the silicate family of binders has become increasingly important and has often been a justification in its own right for conversion either to the CO_2 or ester process.

The need for energy conservation has further highlighted the overall attractiveness of these inorganic, cold setting chemicals.

Suddenly, therefore, the silicate centered processes, having in some countries been out of fashion for some time, are staging a comeback. In the United States, where they have not been used to the same extent as elsewhere, there is more scope for growth than in other countries.

Nevertheless, it has been demonstrated that some American foundries find profitable applications for both CO_2 and ester hardened silicate binders as they exist today.

Many properties of the CO_2 and ester processes which have limited their application have been demonstrated in this paper not to be fundamental ones. It has been shown that through continued development the production speed and overall practical performance of both CO_2 and liquid hardening systems is possible.

Whereas in the past it has been breakdown/shakeout and core storage which have limited the use of these binders, it is now becoming increasingly clear that advances in these respects have shifted the emphasis towards strength, flowability and reclaimability. This shift has not resulted in dramatic increases in the use of the processes only because of the development of new organic coremaking processes with specific technical advantages, but also with fundamental environmental disadvantages.

Binder supply companies have naturally perceived this situation and realized the potential for the growth of silicate based binders, if the disadvantages can be overcome. A great deal of development is known to be in progress around the world and the three examples reviewed in this paper will undoubtedly be joined by announcements of others.

For many applications silicate based chemicals are about to prove themselves to be the sand binders of the future.

Acknowledgements

The authors gratefully acknowledge the permission of the Directors of FOSECO INTERNATIONAL and of FOSECO INC. (Cleveland, Ohio) to publish this paper.

Descriptions of foundry applications are included by kind permission of McConway and Torley Corporation and of Mackintosh-Hemphill Division Gulf + Western. The assistance of D. L. Banks and W. Reuttgers (McConway and Torley) and of R. Anseven and R. Bolton (Mackintosh-Hemphill) is gratefully acknowledged.

References

1. J. G. Vail, "Soluble Silicates," ACS Monograph, no. 116, p 4-10 (1952).
2. D. V. Atterton, "The Carbon Dioxide Process," *AFS Transactions*, vol 64, p 14-40 (1956).
3. R. Worthington, "The Silicate-Carbon Dioxide Bond - Part 1: Some Preliminary Observations," *Iron and Steel*, vol 39, p 176-180 (May 1966).
4. R. Worthington, "The Silicate-Carbon Dioxide Bond - Part 2: The Mechanism With 2.0:1 Sodium Silicate," *Iron and Steel*, vol 39, p 297-300 (Jun 1966).
5. R. Worthington, "The Silicate-Carbon Dioxide Bond - Part 3: A Practical Study of Gassing Techniques," *Iron and Steel*, vol 39, p 331-335 (Jul 1966).
6. R. Worthington, "The Silicate-Carbon Dioxide Bond - Part 4: The Advantages of Higher Ratio Silicates," *Iron and Steel*, vol 39, p 523-528 (Nov 1966).
7. R. Worthington, "The Silicate-Carbon Dioxide Bond - Part 5: The Effect of Storage on The Bond," *Iron and Steel*, vol 40, p 261-264 (Jun 1967).
8. K.E.L. Nicholas, "The CO_2-Silicate Process in Foundries," BCIRA (1972).
9. P. A. Beaney, U.S. Patent No. 3,642,503 (Feb 15, 1972).
10. R. P. Stanbridge, U.S. Patent No. 3,829,320 (Aug 13, 1974).
11. K.E.L. Nicholas and J. G. Morley, U.S. Patent No. 4,020,027 (Apr 26, 1977).
12. G. Horsfeld, "CO_2-Process Sand Reclamation by Means of an Airless Shot Blast Machine," *Foundry Trade Journal*, p 649-665 (May 11, 1972).
13. A. E. Murton, "Using Dry Reclaimed CO_2-Process Sand," *AFS Transactions*, vol 79, p 397-398 (1972).
14. R. M. MacDonald, "The Rebonding of Reclaimed Silicate Sands," *AFS Transactions*, vol 84, p 91-96 (1976).
15. W. D. Cutler and P. Ciciriello, "The San Giacomo Story," Foundry Practice, no. 194, Foseco Foundries International Limited (Sep 1976).
16. R. Williams, et al., "Further Aspects of the Reclamation of Sodium Silicate Bonded Sand," Proceedings 43rd International Foundry Congress: Bucharest (1976).
17. W. Adams (Kenzler Engineering, Milwaukee), Private Communication (Nov 1978).
18. R. Williams, "Carbon Dioxide Binders:Scope for Improvement," Foundry Practice, no. 197, Foseco Foundries International Limited (Jun 1978).

Report from working group P11—
COLD SETTING CORE MATERIALS

The working group was formed in March 1976 to operate under the following terms of reference: "To review the production of cores by cold processes".

Members of the working group have all prepared major contributions to the report, most of which have been based on their own practical experiences.

Many processes described in this report are patented. Care must be exercised not to infringe such rights, and guidance must be sought from the relevant suppliers.

Each process included has been described in relation to its advantages and disadvantages, summarised, and illustrations included. Further general comments have been added as a guide for readers when assessing the processes.

Where little or no experience of the process exists in the UK, the group has drawn on the available information.

CONSTITUTION

The present constitution of the working group is:

A. G. Fennell	Sterling Metals Limited (Chairman of the working group)
F. F. Bownes	Steel Castings Research & Trade Association
J. R. Brown	Fordath Limited
S. F. Greenway	The Ford Motor Company Limited
A. H. Gregory	Ashland Chemical Company Limited
P. H. R. B. Lemon	Borden (UK) Limited
R. McRobert	Steelcast Limited
S. Mather	L. Gardner & Sons Limited
Dr. J. G. Morley	BCIRA
W. J. Vincent	Dartmouth Auto Castings Limited
F. W. Walker	Glow Worm Limited
R. Wright	Ciba-Geigy Plastics Additives Company
Thos. Wills	Institute of British Foundrymen (Secretary to the working group)

(F 1250)

INDEX

Reprinted with permission from *The British Foundryman*, July 1978, 147-170, © 1978 The Institute of British Foundrymen

DEFINITION

Cold setting processes are interpreted as processes where the sand-binder mixture is set to a point where it can be stripped from the box, coated if necessary, and cast without the need to apply heat to the core other than to dry the coating.

INTRODUCTION

Whilst the terms of reference are to review cold core making processes, during the collation and presentation of information, it was found difficult completely to separate core making from moulding, consequently whilst core making is referred to in some detail, moulding practices are also mentioned.

Cold set processes have revolutionised mould and core making techniques particularly in jobbing foundries, they have reduced and in some instances completely eliminated the need for skilled moulders and core makers; many casting defects have been reduced. They have made possible the greater use of boxless moulding techniques, and moulding lines based on cold set processes can often compete with green sand systems, when capital cost and maintenance are taken into consideration.

Over the last two decades a steady progress towards the use of chemical binders has been made for the production of moulds and cores. This change has, generally speaking, been at the expense of oil sand binders for the volume production work of small to medium sized cores, and in place of the loam and dry sand cores for larger work. The reasons for the changes are known well enough. Most of the advantages of chemically bonded systems stem from the single factor that cores are cured whilst still in the core box.

The following are some of the more important advantages to be gained from such processes.

(1) Improved casting accuracy

(a) Accurate cores produced consistently direct from the core box without the problem of inconsistent core sag inherent in the green strip processes.
(b) Intricate casting shapes achieved more easily.
(c) Reduced number of cores to produce the same casting; for example cores previously made in two halves to produce flat surfaces for core drying can be made as one resulting in fewer cores to be placed (or misplaced) in the mould, and larger print areas affording more accurate core laying and improved core support.
(d) Improved core laying techniques can be developed.

(2) Improved casting soundness

(a) Improved core rigidity.
(b) Reduced gas evolution.
(c) Elimination in unwanted variations in metal section which could lead to unsoundness.

(3) Reduced core production costs

(a) Elimination of cost of manufacture, maintenance, and handling of core driers.
(b) Reduction and possible elimination of core support wires, core grids, lifters, sprigs etc.
(c) Reduction of labour previously occupied on stove loading, unloading, handling core driers, wire and core grid recovery and distribution, core checkers, core sizing operations etc.
(d) Replacement of skilled by semi skilled labour.
(e) Increased core production achieved by the development of new machines and adaptation of older machines to produce cores at higher rates.

(4) Savings in floor area

(a) Removal of drying stoves.
(b) Elimination of storage and recovery areas for driers, wires etc.
(c) Elimination of core sizing machines and vacation of areas used during core sizing operations.

(5) Reduced core scrap

(a) Elimination of core sag and drier removal.
(b) Elimination of misplaced reinforcement wires.
(c) Greatly reduced core handling.

(6) Reduced fettling costs

(a) Fewer core joints.
(b) Greater accuracy of core laying.
(c) Reduced core to core and core to mould clearances.
(d) Reduction of wire removal operations and problems of wires fusing on.
(e) Improved surface finish.

CORE BOX EQUIPMENT

It is generally regarded that the costs of core box equipment designed for chemically bonded processes, hot or cold, are considerably higher than equipment for "green strip" processes but when the costs involved in producing and maintaining multiple sets of core driers, box turnover equipment, corebox wear rates, and the shorter life of oil sand equipment are added together the total cost may well come out in favour of the newer process.

If the maximum advantages are to be gained from modern processes it is necessary to pay more attention to detail and a higher quality. A considerable amount of core-box cost results from providing facilities for core ejection, box heating, box mounting, etc., for high speed, high volume machine core making.

Conversion to new processes

When a new process is tried and sampled it is usual practice, where possible, to utilise existing machines and core box equipment. Rarely will old equipment show a new process at anything approaching its best, especially when attempting to convert old oil sand equipment. Core sag and core sizing allowances are no longer required. Core box vents are usually wrongly positioned for processes where gas or vapours are used for setting agents. Stripping rigid cores from worn equipment creates many problems.

Many peculiar results will arise from badly maintained, dirty sand mixing machines, contamination from other processes, haphazard temperature control of sand binders, before, during and after sand mixing, poor shop floor atmosphere, cold damp draughts, roof leaks, etc.

Many conversions to new processes will be condemned if there are initial bad experiences, which may be caused through lack of appreciation of special features appertaining to the particular process under trial.

Machine production of cores is possible with all new processes, indeed, special machines are essential for some and certainly desirable for others. A great deal of machine development has taken place in recent years, and very high rates of production are now being achieved.

All of the foregoing comments are applicable to both hot

and cold chemically bonded processes, and are by no means a complete list, each reader, no doubt, will be able to contribute additional advantages.

No comments so far have been made about disadvantages of the chemically bonded systems. It must be said that none of them are, or are ever likely to be, without their problems. Each suitable process must be examined in detail. The major advantages and disadvantages of each process will be considered in the following sections.

It is not difficult to justify the cost of change from the older methods and processes to the newer ones. The question must be asked, 'which one'? The present cost of energy must give cold processes the "edge".

There are many foundry installations, already deeply committed to hot chemically bonded systems, where it is more difficult to justify the high capital expenditure to change from one chemically bonded to another even though it be from a hot to a cold process.

B SODIUM SILICATE PROCESSES

Sodium silicate processes are widely used throughout many types of foundries, to produce castings from a few grams in weight to many tonnes, across the whole range of metal alloy compositions.

The main types of sodium silicate processes are:
(1) Silicate CO_2 process
(2) Powder Hardeners
(3) Silicate Organic Esters Process

The major difference between them is the method by which the sand binder mix is hardened.

B1 SILICATE CO₂ PROCESS

Range of applications

The silicate CO_2 process is used for the production of moulds and cores for the manufacture of light and heavy castings in a variety of alloys. The process can be adapted to suit small quantity production methods, and high volume techniques.

Binder system

Hardening is achieved by passing CO_2 gas through the sand binder mixture. The chemical reaction is complicated, but in simple terms it is sodium silicate + carbon dioxide gas $\rightarrow$ sodium carbonate + silica gel.

There are several grades of sodium silicate suitable for use as a foundry binder. The difference between the grades is the molecular ratio between the $SiO_2:Na_2O$. Ratios varying from $2\cdot0:1$ to $3\cdot0:1$ can be used. The higher ratios are more sensitive to CO_2 and react more rapidly requiring shorter gassing times.

Mixed sand formulation

The amount of binder required in the sand mix will be influenced by the sand grain size and distribution, and the minimum core strengths required.

Sand

Most commercial sands are usable. The presence of alkalies and many impurities does not greatly affect the cure and, therefore, beach and dune sands can be used satisfactorily. The main requirement of the sand is that it should have a low clay content. The presence of clay seriously reduces ultimate core strength.

Breakdown

Plain binder-sand mixes have relatively poor decoring properties. In general it is necessary to make additions of materials to the sand mix to improve removal of the cores and mould from the castings. Over the years many materials have been used to improve breakdown, for example, coal dust; ground coke; wood flour; pelleted pitch; various forms of sugar. Without exception, all of these materials, whilst giving some improvement to the breakdown properties, cause problems of reduced core strength; shelf life of the cores; and bench life of the mixed sand. By far the best aid to improve breakdown is the use of minimum binder additions. In the more recent years improved materials have been developed which improve breakdown without seriously reducing core strength or shelf life below operating limits. Currently, some excellent formulated silicates are available, which are added to the sand mix as a one shot addition; they permit lower additions of the binder, promoting good breakdown properties and core storage life. With the use of these improved breakdown additives, coal dust additions may not be made specifically to assist in breakdown, but some advantage may be gained with an improved casting surface finish.

Sand mixing

Most types of sand mixers will mix the sand satisfactorily. Good binder distribution is important to the success of the process. The mixed sand is a sticky material, hard crusts and binder rich deposits will build up on the sides of the mills and blades and prove difficult to remove. High efficiency mixers permit lower binder additions. Long milling times in batch mixers will cause rapid temperature increase of the mixed sand, and whilst mixed sand temperature is not critical it is important. High temperatures increase moisture evaporation and cause partial setting of the sand resulting in lower core strength and greater sensitivity to the CO_2 gas. Low temperature sands require extended gassing times, and results in lower core production rates, with a possible danger that some core distortion will occur. A good working temperature ranges between $15°C$ ($60°F$) to $32°C$ ($90°F$). Where continuous mixers are used one shot formulated binders may be preferred.

Mixed sand bench life

Depending upon temperature and ambient shop floor conditions, the mixed sand may have a useful bench life of several hours, but undoubtedly, the best results are obtained from freshly mixed sand.

CO₂

The amount of CO_2 gas required to cure a core will depend upon many factors. Theoretically, the amount required is $0\cdot2\%$ of CO_2 to each 1% of the binder content. In practice it is usually very much higher. Up to 1:1 ratio of gas to binder is normally regarded as acceptable, without the use of special gassing techniques, such as vacuum gassing. The position and number of vents in the core box, sand and CO_2 temperatures, gas pressures and volume, and the method used to introduce the gas into the core, all have considerable effect on the amount of CO_2 used.

CO_2 is supplied commercially in three main forms:
(1) As dry ice
(2) In cylinders
(3) Liquid form for bulk storage

There are considerable cost advantages to be gained from bulk storage, and it is technically more controllable, especially for larger work.

The time needed to cure a core will depend upon core size, sand and CO_2 temperatures, quantity and grade of sodium silicate used, and under normal shop floor conditions should be less than 20 seconds. The best results are obtained by a rapid displacement of the air in the core by the gas.

Core stability

Dimensional stability of cores made by this process can be good. There is little softening of the core during the critical stages of mould filling and metal solidification. Core distortion may occur if the core is insufficiently cured before removal from the core box. Excess binder content and poorly mixed sand considerably extend gassing time required to produce rigid cores free from distortion.

Core storage properties

The core storage properties are considerably influenced by a degree of gas cure. Maximum potential storage life is obtained by minimum gas cure. The ideal gas cure is that which is just sufficient to produce dimensionally stable cores. Correctly gassed cores stored in reasonable shop floor conditions will have an acceptable shelf life of many weeks. Silicate bond is affected by moisture and considerable shortening of shelf life will occur if cores are stored in damp conditions.

Fig. 1—Silicate CO_2 cores for V8 cylinder block

Fig. 2—Silicate CO_2 cores in storage

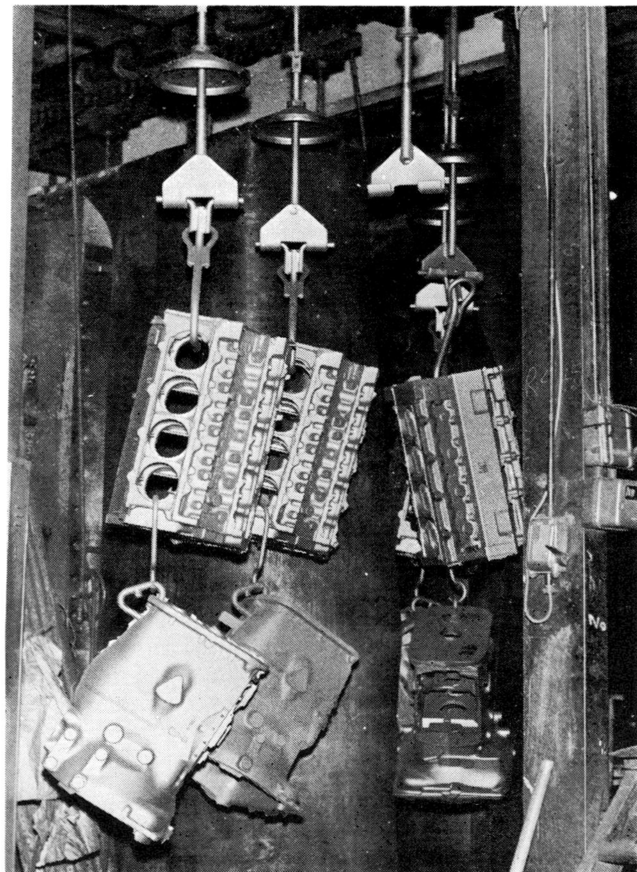

Fig. 3—Castings produced by CO_2 process.

SELF-SETTING SILICATE PROCESSES

These processes involve the addition of hardening agents, which when added to the sand mix react with the sodium silicate binder to form a rigid bond. The most common of these materials are:

Dicalcium Silicates
Ferro Silicon (Nishiyama Processes)
Organic Esters

Slag/cement hardeners

Many calcium compounds including lime, gypsum anhydrite, calcium silicates and calcium aluminates react with sodium silicate causing it to gel. Some of these materials react so fast with sodium silicate that they are difficult to use as foundry binders, since the working time of the mix is too short, but some slags and cements which contain calcium silicates and aluminates give controllable setting reactions.

Dicalcium silicate ($2 CaO.SiO_2$) in particular is a useful gelling agent for sodium silicate, since its initial reaction rate with sodium silicate is slow, giving a good work time to setting time relation.

Dicalcium silicate is a major constituent of a number of metallurgical slags as well as certain types of Portland cement. Unfortunately, while cements are formulated to have constant setting properties with water, their setting time with sodium silicate may vary widely from one batch to another.

The most common source of good quality dicalcium silicate is ferro-chrome slag. The highly basic slags used in ferro-chrome production are known as "falling slags" due to their property of collapsing to a fine powder as they cool owing to a large volume change taking place in the di-calcium silicate crystals as they pass through a phase change. Ferro-chrome slags can be used as made or they can be ground further. A synthetic form of dicalcium silicate is also available for use in foundry binders. This has higher purity and is more active than ferro-chrome slags.

The reactivity of the dicalcium silicate to sodium silicate depends on its crystalline form, purity, and particle size, fine particles being more active. Dicalcium silicate slowly loses activity through storage in moist conditions....

Sodium silicate additions are usually 3–5% (on sand weight) and silica/soda ratios of 2·4–2·9 are commonly used. Dicalcium silicate additions are usually 1–4% (on sand weight) depending on its reactivity and the setting time required.

The setting time varies with:
(1) The reactivity of the powder.
(2) Percentage of powder added.
(3) Ratio of silicate (Higher ratio silicates set faster).
(4) Temperature.

Setting times are usually 10 mins.–2 hours.

The reaction of dicalcium silicate with sodium silicate is not fully understood but no heat or gasses are evolved and the process is very safe.

Ferro-silicon—Nishiyama process

The ferro silicon should contain between 70–80% silicon and be finely ground to 300 BSS mesh.

The reaction between the silicate and ferro-silicon is exothermic. Hydrogen gas is released and fatal explosions have been reported. Because of this danger, the process is little used in the U.K.

Comments

Generally dicalcium silicate and ferro-silicon systems have the advantage of using low cost materials which are added to the sand mix in powder form. Sands which are suitable for the CO_2 silicate process can be used satisfactorily. Usually sodium silicate of a 2:1 ratio of $SiO_2:Na_2O$ is required for use with these powder type hardeners in addition of $3\frac{1}{2}\%$ to $5\frac{1}{2}\%$. Setting agents required will be between 1% and 3% depending on the rate of activity and the bench life required.

Considerable inconsistencies in setting times are often experienced because of the difficult problems of adding powdered materials successfully to sand mixers, with the result that, generally speaking, self-setting powder hardener systems have been superseded by the organic ester processes.

Controlled addition of liquids to sand in continuous mixers is easier than powders. All continuous mixers are supplied with two pumps but a special powder feeder must be fitted to add the powder hardener. Fine powders are prone to bridging in the supply hopper and achieving a constant controlled feed is not easy.

Powder hardeners, being based on slags or cements are variable in activity from batch to batch and their activity decreases with storage time while liquid hardeners have a long shelf life and have closely controllable reactivity.

B3 SILICATE ORGANIC ESTERS

This self-setting system is used for the manufacture of small and heavy castings over a wide range of metal compositions, being especially suited to jobbing and semi-jobbing work of the larger type. The process is widely used for moulding as a facing material subsequently to be backed up by green sand.

Binder system

Hardening is achieved by a two part chemical reaction between sodium silicate and organic esters.

The most significant reaction is a decomposition of the esters by the alkali to form alcohol and an acid which produces gelling of the sodium silicate.

Silicate additions of 2% to 4·5% are commonly used. The higher ratio silicates of 2·5:1 to 2·8:1 perform better with esters. Ester additions of 10% to 12·5% of the silicate are used.

Organic esters

Many organic esters can be used to gel sodium silicate. The speed of reaction depends on the particular ester used and the ratio of the silicate. For example, commonly used esters are:

Glycerol diacetate (diacetin)   — fast
Ethylene glycol diacetate (EGDA) — medium
Glycerol triacetate (triacetin)   — slow

These esters, and others, can be blended to give a wide range of setting times.

Additions above the recommended levels will not improve setting times, but could impair cold core strength properties. Temperature of the mixed sand greatly influences setting times, and control of the mixed sand temperature is vital if consistent production speeds are to be achieved.

Shelf life

The storage life of moulds and cores made in ester-silicate is considerably better than CO_2-silicate. Humid conditions will cause moulds and cores to deteriorate, as with any core binder, but storage lives of about 3 weeks in normal atmospheric conditions are typical.

Suitable sands

This process is used for bonding all the main types of sand.

Olivine, chromite, or zircon sands are widely used in the steel industry for the high pouring temperature alloys. In some cases it has been found possible to replace the most expensive materials with silica sands, because it is possible to obtain glass hard and smooth surfaces on the cores and moulds.

Sand mixing

Continuous trough mixers are particularly suitable for this process, especially for large work. Batch mixers can be used but where more than one mix is required for the same core or mould, there may be some danger of partial setting between the mixes. Problems of inconsistent setting times can result from heat generated during milling of the sand, particularly with batch mills, but during cold conditions some advantage may be gained by purposely milling the sand to raise the temperature to a good working level.

In recent times sand temperature controlling units have been introduced into continuous sand mixing plants, and these successfully control mixed sand temperatures to within a few degrees. Cleanliness of the mill is most important. Contamination by other processes, particularly those which contain acid hardeners, will seriously affect hardening of the sand.

Bench life

The bench life of mixed sand is regarded as half the setting time.

Ramming

Freshly mixed sand has good, free flowing properties, enabling good compaction to be achieved readily. For larger work, hand packing and tucking of sand supplied by continuous mixer is the most popular method used for producing both cores and moulds, although most other means of compaction are also used; vibration techniques in particular give excellent results producing cores and moulds with very good compaction and high surface strength, though problems of "ram-off" may be experienced.

Decoring

Owing to the fact that higher ratio silicates are used core removal is generally better than the CO_2 process. Breakdown additives can be used providing they do not interfere with the setting of the sand.

Fig. 4—Typical self-setting silicate mould being stripped.

Fig. 5—Self-setting silicate mould

GENERAL COMMENTS ON SILICATE PROCESSES

Core box equipment

All conventional materials can be used in the manufacture of pattern and core box equipment. The choice of materials will be determined by the method of production, quantity required, size, etc. Clean strip is possible from all smooth flat surfaces; wooden equipment must be treated with paints which are not affected by the silicate binders; cellulose or epoxy base paints are suitable. Collapsible core boxes may be preferred for hand stripping. Boxes of rigid construction are normally used for machine production, where the cores are removed by mechanical ejector systems, vibration or air ejection.

Core blowers and core shooters are extensively used with the CO_2 silicate process. Trial and error may be necessary to position box vents correctly to satisfy the demands of both good sand compaction and efficient economical gassing. The method of introducing the gas into the cores will greatly influence the optimum vent positions. Normal methods of introducing the gas into the core boxes are:

(1) Probe
(2) Sealed hood
(3) Hollow formers
(4) Vacuum cabinets
(5) Combinations of methods

Ester hardened processes have the added advantage that some slight surface deformation, afforded during the setting time, facilitates easier stripping in jobbing work where the standard of the equipment is often less than good.

Casting quality

Particularly good dimensional accuracy can be achieved by these processes, and dimensional stability of the cured sand during storage and pouring is very good, though some distortion of the cores may occur if they are removed from the core box prematurely. There is little metal/binder reaction to cause unsoundness. It is usually necessary to coat cores and moulds with refractory dressing, to prevent sand burn-on with heavy section castings and with alloys requiring high pouring temperatures.

Gas evolution is low, blow defects are not normally a problem, only minimal core venting is usually necessary.

Finning problems are minimal.

Sand reclamation

Reclamation of sodium silicate bonded sand is possible, but probably more expensive than with other cold set processes. The most widely used method is attrition. Up to 60% of reclaimed sand can be used, the residual soda limits the use of the reclaimed sand for core making and moulding. Na_2O levels over 0·8% may seriously affect refractories of the sand. Incidental reclamation by natural breakdown is not normally harmful to green sand systems. Some foundry green sand systems are wholly maintained by decored material.

Environment and health

Sodium silicate, CO_2, organic esters and most breakdown agents are regarded as relatively safe materials in normal foundry use.

Sodium silicate is caustic; eye and hand protection is essential for operators who handle the binder, and barrier cream should be used by the shop floor operators. Waste material, whilst being non-toxic, is alkaline and the disposal of such material is covered by the Control of Pollution Act 1974, which stipulates that disposal must be carried out under licence of the Local Authority.

No gases or fumes are evolved in the reaction of esters with silicates.

The esters themselves are flammable liquids, having flash points over 60°C. They are powerful solvents and, by dissolving natural greases from the skin can give rise to skin irritation. Direct contact with the skin should therefore be avoided.

Comparative costs

The raw materials from which the binders are manufactured are indigenous, available in abundance, and not subject to considerable world price fluctuations. The mixed sand cost is low; the most expensive ingredient usually is the breakdown additive.

The cost of CO_2 varies greatly depending upon quantity requirements and method of supply. Bulk delivered CO_2 is comparable with the cost of compressed air. Capital cost of plant is modest; most existing foundry equipment can be adapted. Flame off dressings are costly, but it must be appreciated that drying ovens are not required. The cheaper range of foundry sands are generally suitable.

Summary of advantages

(1) Fume produced at core making is minimal.
(2) Gas evolution during pouring and knock-out is low.
(3) Raw materials costs are lower and more stable than for comparable processes.
(4) Relatively short cure times are possible allowing good turn round of equipment.
(5) Capital outlay is modest.
(6) Existing foundry equipment can usually be adapted.
(7) Waste materials can be disposed of safely (under licence).
(8) The materials can be used in relative safety.
(9) Dimensional accuracy; predictability and stability are good.
(10) Adverse effects on casting soundness are minimal.
(11) Sand reclamation is possible.
(12) Residues can be tolerated in green sand systems.
(13) The cheaper base sands may be used.

Limitations and disadvantages

The limitations and disadvantages of these processes are few, but significant.

(1) Core strengths and edge hardness are relatively low.
(2) Poor decoring properties may preclude this process from being used to produce cores inaccessible to direct decoring methods.
(3) Cheaper water base coatings may not be used.
(4) Owing to the relatively poor flowability of the mixed sand, a greater number of sand entry points may be required compared to other cold setting processes, possibly creating greater box wear and more core trimming operations.

ORGANIC RESIN PROCESSES   [C]

Organic binders for cold setting and gassing processes can be divided into two main chemical types:

(1) Phenolic/Furane binders
(2) Urethanes

The applications of each of these binder systems fall into two major areas:

(1) Systems in which the reactive binder components are

mixed with sand to provide a flowable sand mixture which cures in the cold. The curing times range from a few minutes to several hours depending on the binders and the quantity of hardening agent used. Such systems, referred to here as self setting, are also known as cold setting, air setting and no-bake. They are suitable for the production of medium to large size cores and for boxless moulding systems in jobbing foundries.

(2) Systems in which a gas or vapour is passed through an uncured compacted sand/binder mixture in a corebox. The gas induces a rapid reaction between the resin binder components present in the sand. Such systems are characterised by very rapid curing rates, of the order of a few seconds, and are most suitable for high volume core production techniques.

C1 SELF SETTING PHENOLIC/FURANE BINDERS

The cold set organic resin system is used to manufacture moulds and cores for a wide variety of jobbing castings from less than $\frac{1}{2}$ kg to 50 tonne in most grades of ferrous and non-ferrous metals. The process is sometimes used for boxless moulds in competition with green sand where it is often competitive when capital equipment cost is taken into account.

Binder system

There are four main types of organic synthetic resin binders.

(1)	Urea formaldehyde/furfuryl alcohol	UF/FA
(2)	Phenol formaldehyde/furfuryl alcohol	PF/FA
(3)	Furfuryl alcohol/formaldehyde	FA/F
(4)	Phenol formaldehyde	PF

They harden by polymerisation at room temperature under the influence of acid catalysts such as:

(1)	Para-toluene sulphonic acid	(PTSA)
(2)	Phosphoric acid	
(3)	Reactive aromatic sulphonic acids	
(4)	Blends of inorganic acids, e.g. sulphuric and phosphoric acid.	

The most commonly used acids are PTSA and phosphoric acid.

The properties, price and performance of the first three types of resin vary depending upon their furfuryl alcohol and nitrogen contents. The highest price products are those containing the highest FA level and the lowest nitrogen and vice versa.

Phenol formaldehyde cold set resins are the cheapest since they contain no furfuryl alcohol. Early problems of depth of cure, irritant effects, and fume on casting have been largely eliminated and PF resins now offer a useful alternative to furane resins although they require slightly different handling techniques. During the FA shortage many foundries were kept in production by using PF resins. They are normally used with PTSA and reactive organic sulphonic acid catalysts.

Selection of the best binder for any application is based upon price/performance criteria bearing in mind all relevant factors such as:

(a) metal type, weight and temperature and the possibility of metal/mould reactions.

(b) grading and type of sand being used. Is the sand being reclaimed? If so what are the possibilities of mould/metal reaction or reduced moulds rigidity on repeated reclamation?

(c) sand disposal.

(d) ventilation provided at mould or core making and at casting stations.

(e) breakdown at decoring and knock out.

Binder suppliers can offer advice and laboratory tests to assist in the selection of the best grade of resin/catalyst system for any application. Binder levels of 0·7–2·0% on weight of sand are commonly used with 1·5% an average level. Lower levels are possible with high speed mixers (see section on mixers).

Suitable sands

Three or four screen sands of AFS No. from 40–70 of low acid demand (see Appendix) are suitable for the cold set process. The grain size and shape largely determine the binder level required to give adequate strengths. Reclaimed sands from the attrition processes generally require less resin and catalyst than new sand. Normally up to 90% reclaimed sand can be used. (See section on reclaimed sand for tests on resin and acid levels).

Olivine and certain grades of chromite with high acid demand are unsuitable for use with acid catalyst organic binders. Here the urethane type may be suitable.

Chromite and zircon facing sands are used in some steel applications.

Sand mixing

Continuous mixers of the trough and high speed type are ideal for the cold set process. A high speed self cleaning mixer is essential to use binder systems with 3 to 10 mins. setting times.

Batch mixers may be used but it is essential to place the sand as soon as possible after mixing. Overmilling should be avoided.

Both continuous and batch mixers should be cleaned regularly. Some high speed mixers use a neutralising liquid which is sprayed onto the surface of the trough to prevent setting of the mix and facilitate cleaning.

If the same mill is used for silicate, urethane and acid catalysed processes, the mill must be carefully cleaned and, if necessary, treated before changing from one process to another.

Fig. 6—One type of mixer head fitted to a twin screw continuous sand mixer for use with resin catalyst sands

Fig. 7—Cold set sand mix being discharged directly into core box

Bench life

This is normally not a problem with continuous mixers as the sand is discharged directly to the core or moulding box. It should be tucked, tamped and strickled immediately after placing. The practice of "barrowing" cold set mixes from the continuous mixer to other parts of the foundry should only be allowed when setting times are long (about 1 hr). Batch mixers should only be used in conjunction with mixed sand having long setting times.

Ambient and sand temperatures

Constant temperatures and humidities are essential for consistent setting times on cold set mixes of all types, if catalyst levels remain unchanged. It is more usual, however, to alter catalyst additions to compensate for temperature changes. It is important that this is done carefully in a controlled manner, between specific limits, in order to obtain the best performance (binder suppliers will give advice on calibration of catalyst and resin pumps to achieve optimum properties).

It is often an economic proposition to install a sand heater to pre-heat the sand before mixing, as this can save on catalyst consumption during cold weather. In summer months problems can be experienced with hot sand as delivered from suppliers or from the reclaiming plant. Low catalyst levels must be used, but in some instances it is necessary to install a sand cooler after the reclamation unit to enable adequate catalyst levels to be used.

Core/mould stability

Once made satisfactory, resin based cold set cores of the types described can be stored almost indefinitely inside the foundry without deterioration in properties, particularly under dry conditions. No significant change in dimensions is observed.

Shelf life

The shelf life of resin binders and catalysts are stipulated by the supplier and vary from 2 months to 1 year, depending upon the grade. The viscosity of resins increases with age. This factor determines the shelf life, and may affect the behaviour of the core during casting.

Some catalysts crystallise if left exposed to the air without caps on the drums, or if stored at low temperatures.

Core dressing

Isopropanol based flame off dressings are used with most cold set resin binders. However, water based and air drying types can also be used.

Isopropanol based dressings cause slight softening of the surface of the mould or core, so it is important not to apply the dressing too soon after the core is stripped. An interval of half an hour is normally sufficient. Some pinhole porosity defects can be eliminated using iron oxide based dressings.

Core boxes and moulds

Core boxes are often constructed of wood painted with cellulose, epoxide, furane or vinyl paints. Sticking will be experienced with shellac varnishes and they should not be used. Epoxide and metal boxes may also be used, although metal may inhibit setting, due to its acting as a heat sink.

This is particularly true when the boxes are left overnight and have become cold. Preheating metal boxes by flaming has been practised.

Release agents such as wax polish, plumbago and talc are used. The first is applied and polished to a good sheen. The two powders are dusted on.

It is important to remove resin build up once per shift by wiping with detergent in water or industrial methylated spirits.

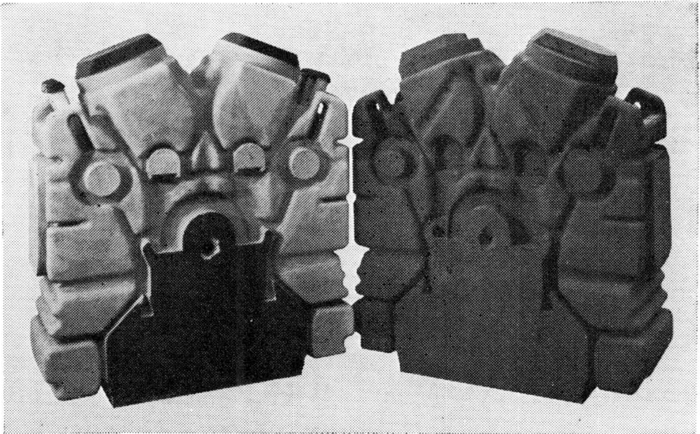

Fig. 8—Cores produced in low nitrogen copolymer resin

PHENOLIC/FURANE GASSING COLD SET RESINS   [C]2

The same chemical types of binder as used in the cold setting process may also be set by the passage of SO_2 gas.

This method of gassing cold set furane and phenolic resins is in early stages of development, and is causing a great deal of interest both in Europe and the U.S.A.

The sand and resin are mixed with a peroxide and blown into the core box. When SO_2 is passed through the core mix it reacts with the peroxide to form SO_3 which dissolves in the water present in the binder to form sulphuric acid and induces rapid exothermic polymerisation of the resin.

Peroxides

Two types of peroxides are used:
(a) organic peroxides such as methyl ethyl ketone (MEK) peroxide where long bench lives, up to 8 hours, are required. From 40 to 50% of MEK peroxide is used relative to resin content.
(b) hydrogen peroxide (50%) where shorter bench lives, up to 3 hours can be tolerated. 20 to 25% of hydrogen peroxide is used relative to resin content.

Resins

Although normal cold set resins of the phenolic and furane types can be used best results are obtained from resins designed specifically for the process.

Sand

Any low acid demand sand of AFS No. 40–80 as used for cold set may be used for this process.

Resin addition

Depends upon the sand surface and grain distribution. As little as 0·7% resin is required for good continental European sands while up to 2% resin is required for some U.K. sand. Average: 1·2–1·5% resin.

Resin types

PF, UF/FA, PF/UF/FA, PF/FA and FA/F polymers can be used.

Mixers

Any type of efficient continuous or batch mixer may be used. The high speed self cleaning continuous mixers are particularly efficient.

Flowability

The mix of resin and peroxide is very easy to blow and can be hand packed into open core boxes easily.

Bench life

Bench life varies depending upon the stability of the peroxide used. An organic peroxide gives a long bench life but hydrogen peroxide a relatively short one. An advantage over other processes such as hot box or urethane cold box is that the mix does not harden until SO_2 gas is passed through it.

EQUIPMENT

Core boxes

The core boxes must be adequately vented, ideally with a sealed gassing system. Painted wood, epoxide or aluminium boxes are necessary when hydrogen peroxide is used. MEK peroxide must be used for cast iron or steel boxes.

Blower

Any normal core blower or shooter may be used operating up to 90 psi.

SO_2 gas generator

For short runs the gas may be obtained directly from a cylinder, but for long runs a vaporiser is required and liquid SO_2 is used from a pressure container. All pipes conveying SO_2 must be in stainless steel including solenoid valves. Gas pressure is normally up to 40 psi.

Ventilation ducting

Because of low TLV (5 p.p.m.) and the pungent nature of SO_2 it is essential to surround the core box with sealed

Fig. 9—Selection of cores in the SO_2 process.

ducting operating under negative pressure to sweep away all fumes. These fumes must not be discharged into the foundry atmosphere.

Scrubber

The fumes are conveyed into a scrubbing tower containing a 10% caustic soda solution. No traces of SO_2 should remain in the air after passing through the scrubber.

Gassing times

From as little as 0·5 secs. to as much as 5 secs. are normally used.

Purging times

5–10 secs. purging with compressed air at 30–90 psi is necessary to remove the residual SO_2 gas.

Total cycle times

From 5 to 15 secs. are possible.

Core strength

Cores are perfectly handleable on ejection from the box and full strength is developed within 5 mins. of stripping.

Strengths equivalent to those obtained from the self setting process are obtained but in a much shorter time. Core strength is not affected by the presence of small proportions of water in the sand or blowing air.

Core odour

Little residual odour of SO_2 remains when the core is stripped from the box and this quickly disappears.

Core storage

No deterioration of cores on storage in normal foundry conditions is observed.

Only 25% reduction in strength is observed after storage at 100% humidity.

Performance on casting

Moulds and cores from the SO_2 process exhibit the same properties as cold set resins of the same composition hardened by liquid acid catalysts.

Environmental considerations

Sulphur dioxide is a commonly used industrial gas and much data is available on its handling and toxicity. It is a self regulating gas in that atmospheres above the TLV are distinctly unpleasant and encourage the work area to be vacated.

Although industrial experience is limited it is felt that the SO_2 gassing system can be made environmentally acceptable provided that the suppliers' recommendations are adhered to.

Peroxides are strong oxidizing agents, which must be handled with care, and in strict accordance with the manufacturers instructions.

GENERAL PROPERTIES OF PHENOLIC/FURANE BINDER SYSTEMS

Casting properties

The following table indicates suitability of resin binders for metals:

Aluminium and alloys:

PF; UF/FA with a nitrogen content of about 17%; specialist resins with high FA and low water contents.

Grey iron:

PF; PF/FA; UF/FA resins up to 9·5% nitrogen content. Some problems may be experienced on certain heavy metal sections with 9·3% nitrogen binders.

Nodular iron; steel:

PF; PF/FA; FA/F; UF/FA resins of nitrogen content below 2% (Preferable less than 1·5% for reclaiming).

Copper base alloys

UF/FA and PF/FA resins are suitable. High quality resins containing 60% FA and over are generally used.

Magnesium alloys

PF, PF/FA and modified UF resins have been used successfully for magnesium alloy castings, but a suitable inhibitor such as potassium fluoroborate must be used.

The nitrogen content of a binder determines its tendency to induce pinhold porosity defects in metal castings and it is for this reason that careful selection must be made.

The breakdown of organic binders is excellent and significantly better than silicate based binders (even hardened with esters and containing breakdown agents).

However, the order of breakdown is approximately as follows:

PF/FA; PF; FA/F: Slowest breakdown
UF/FA of nitrogen up to 7·5%:   Intermediate breakdown
UF/FA of nitrogen above 10%:   Fastest breakdown

Some PF, PF/FA and FA/F resins can cause finning and other expansion defects. The problem of oversized moulds or cores can be experienced in these binders and it may be necessary in rare instances to alter the pattern tolerances.

Environmental considerations

Handling synthetic resins can cause dermatitis and the acids can cause burns so that it is essential to practice good housekeeping and wear gloves and goggles when handling the neat resin and catalysts. All these materials are toxic if ingested.

Fumes can be evolved during mixing, core and mould making and during casting, and ventilation must be installed to bring the concentrations below the threshold limit values (see details in section on Health and Safety).

Dumping of sands and binders

Any surplus resin and catalyst should be sent back to the manufacturer for disposal. Used sand containing PF and PF/FA resins should only be disposed of in sites away from rivers and drinking water sources because of the possibility of phenol leaching. One way of disposing of these sands which is practised both here and in Sweden is to intersperse on the dump with layers of household rubbish that contains bacteria which degrades the phenol.

URETHANE COLD BOX GASSING PROCESS   C3

This process was first demonstrated in the European Foundry Industry in 1968, and is now widely used in both Europe and North America.

Range of application

A wide range of castings made in many alloys are being produced with this process. It is a rapid curing cold resin system, which is particularly suitable for high volume production techniques, but can also be adapted for small volume work. As the curing of the binder is activated by a toxic amine vapour, it is environmentally desirable to produce cores and moulds in well ventilated areas of machines with power extraction.

Binder systems

The binder system consists of three liquid parts:
Part I is phenolic formaldehyde polymer resin, blended with suitable solvents.
Part II resin contain polymeric M.D.I. (Methylene-bis phenyl-isocyanate).
Part III is an amine catalyst.
Part I provides hydroxyl groups which chemically combine with isocyanate groups of Part II and this reaction is accelerated by the presence of amine to form a rigid bond between the sand grains.

Amines

Amines are toxic liquids, which are converted into a vapour before passsing through the sand binder mix.
Two types of tertiary amine are commonly used.
(1) T.E.A. (triethylamine)
(2) D.M.E.A. (dimethylethylamine)
T.E.A. is suitable for most applications, D.M.E.A. is more reactive than T.E.A. It gives off a more pungent fume, has a higher vapour pressure, and is more costly.

The higher cost of D.M.E.A. may be justified by the shorter cure times and greater production rates possible because of the greater reactivity, and in many cases less amine is required.

Amine generators

There are three basic methods of producing amine vapour:
(1) Accurately metered quantities of amine are injected into a stream of air en route to the core box.
(2) A stream of air is bubbled through a bath of amine. The rate of amine absorption into the air may be in the order of 10% to 20%.
The more sophisticated units incorporate heaters to control the amine bath temperature, normally at approximately 25°C to maintain a more constant amine concentration in the air. Amine concentration will vary according to the rate of "draw off".
(3) High pressure cylinders containing a mixture of liquid amine in liquid CO_2 are also available. The mixed liquids are drawn off through vaporising units.

The arguments as to which system is the best can be summarised as follows:

T.E.A. is a cheaper material and suitable for use in the injection type generators, which can be small compact units—one in each machine.

CO_2 is sometimes preferred to air as the carrier; improved permeation of the sand is claimed, thus producing reduced risk of explosion.

Vapour produced from liquid amine in liquid CO_2 high pressure cylinders is claimed to permeate the sand mass with greater ease, consequently, being more suitable for curing complicated core shapes.

Bubble generators are bulky units, and are usually remote from the core making machines, but one bubble generator can service many machnes. A fine vapour is produced, and is considered by many to be superior to injector systems.

The higher vapour pressure of D.M.E.A. leads to better performance of a bubble generator. Reduced cure times will also be possible when used in an injector system. Curing efficiency is improved when the carrier medium (CO_2 or air) is heated.

Owing to the toxic nature of the amine, and the unpleasant fumes given off, amine must be handled with care in sealed containers.

Only minimum quantities should be used to effect the cure. Purging of the core whilst still in the core box is the normal procedure, it serves to force the amine through the core and drive off any excess unused amine from the core before removal from the core box. Total gas and purge times for medium sized simple shape cores may be as little as 5 seconds. Vapour entry points and core box vent positions are all important for quick efficient curing.

Moisture

Contamination of the mixed sand by moisture must be avoided. The M.D.I. in Part II of the resin system has an affinity for moisture. One part of moisture will destroy twelve parts of M.D.I. 0·5% moisture in the sand can result in a reduction of tensile strength by 30% or more, and bench life may be reduced to less than 15 minutes. The source of moisture contamination may be from the sand, compressed air, or the atmosphere. Most successful operations incorporate aid drying units for all air used for blow, gas and purge.

Unnecessary exposure of the sand mix to compressed air should be avoided.

It is the policy to operate with the smallest possible sand cartridge and length of blow time, in order to minimise the possibility of moisture pick-up and loss of solvents.

Core dressing

Casting finish generally is better than from other comparable processes, and often core dressing need not be applied. Where some coating is required most types of core dressing can be used. Flame off coatings may cause blistering problems of the core surface if the coatings are applied too thickly, or if high levels of spirit are absorbed into the core, resulting in high temperatures and long burn off times. This may be a problem associated with dipping techniques.

Core box

High quality equipment is essential. All normal materials can be used. Epoxy lined metal equipment is claimed by some users to be the most successful material, whilst others argue that all-metal equipment, in particular cast iron, is to be preferred.

The mix

The mix sand has exceptionally free-flowing properties, often single sand entry points are adequate for many cores. Amine vapour does not disperse through the blown core as easily as CO_2 will disperse through a silicate bonded core, which makes positioning of the core box vents critical. The positioning of these core box vents correctly for blowing and gassing cannot easily be decided in the pattern shop. It is advisable to blow and gas an unvented core box, and examine the result to help determine the optimum vent positions. Minimum venting is normally recommended for gassing.

Closely fitting core box joint faces are also important. Amine vapour will escape through poorly fitting box face joints, which will result in excessive amine consumption or partially cured cores. Some form of box joint seal is considered necessary by many.

Mixed sand formulation

The amount of binder required in the sand mix will be influenced by the sand grain size and distribution together with minimum core strengths required.

Sand

Most commercial clean sands are suitable, impurities may cause some core production problems and affect ultimate core quality.

Mixing

Part I and Part II resins are mixed into the sand usually in equal quantities. The total resin addition may vary from 1·25% to 2% by weight of sand. Most types of sand mixers can be used satisfactorily. Complete distribution of the binders in the mix is essential, but high temperatures which may be generated during milling must be avoided in order to minimise evaporation of the solvents from the resin.

Good housekeeping for this process is imperative. The sand mixing equipment must be clean, as contamination by other processes may cause serious setting problems.

Bench life

Bench life of the mixed sand may be up to 4 hours, but will depend greatly upon temperature and acid demand of the sand. Owing to the relatively short bench life, sand should only be mixed as required for use.

Core stability

Whilst this process produces cores of good dimensional stability one foundry has reported some slight shrinkage during storage. Shrinkage of 0·001″ per inch has been measured on test cores made from 100% new sand mixes. Test cores made from 100% reclaimed sand have shrunk up to 0·002″ per inch.

Core distortion occurring during pouring and by metallostatic pressures is unlikely to be a problem except with thin section cores.

Core storage properties

Storage properties of the cores are generally satisfactory, but will be seriously reduced by contamination from moisture pick-up during core blowing and gassing stages of core production. Some deterioration of the core may also occur in shop floor conditions by moisture absorption from the atmosphere.

Casting quality

High degrees of casting accuracy and definition can be obtained from this process. The rate of gas evolution during casting presents few problems of blow holes in the castings, thus minimising problems of core venting.

Finning

As with most phenolic resin mixes, finning may cause casting cleaning problems in some iron castings. Iron oxide additions to the sand mix may eliminate minor finning problems and reduce the severity of others, additions of up to 2% can be added without seriously reducing core strengths. Reduction of the core making machine blow pressures, and/or total resin additions may also help in reducing this defect, but will be accompanied by the obvious problems of reduced core strengths.

Cold box resins are to some degree tolerant to sea and dune sands, which are known to have useful qualities when finning problems are encountered.

Lustrous carbon (soot defect)

This defect is sometimes known as resin defect. It is caused by the decomposition of the organic binder by the molten metal. In most circumstances the lustrous carbon formed may be beneficial, producing improved surface finish to the castings. However, excess carbon formed will show itself on the casting surface as shiny wrinkles resembling cold laps. Slow or turbulent mould filling and low pouring temperatures will promote higher levels of this soot defect. The defect may minimise by fast running speeds and higher pouring temperatures. Additions of iron oxide or aluminium sulphate have also been successful in minimising this problem, so too has baking the cores at a temperature of 150°C. The most economical method of control is by minimum binder additions.

Gas defects

Some pinhole defects have been experienced when cold box cores have been used, as they have with hot box and shell processes. Low alloy cast irons and steels seem to be most susceptible to this defect, which is caused by a metal-core reaction. Investigators report that these defects are not related to nitrogen present in the binders, but to hydrogen, probably coming from the solvents used. Additions of 2% to $2\frac{1}{2}$% of iron oxide are claimed to eliminate this problem.

Sand reclamation

Since the breakdown of cold box resins is excellent, sand reclamation presents few problems. It is essential for the reclamation plant to incorporate a re-classifying system because high percentages of binder residues are contained in the fines. If the fines are allowed to build up in the sand system castings defect may be experienced; the removed fines are replaced with new sand. Experience is showing that removal of 50% of the material which would remain on a 140 AFS mesh sieve and all finer material appears to maintain good balance. Contamination from sands used in other processes will cause some problems, one exception being urethane no-bake because it is chemically similar.

Environmental considerations

The environmental problems associated with this process stem from the phenolic resins isocyanate (MDI) and amine materials. Each material presents some health hazards—T.L.V. (Threshold limit values) have been set up by H.M. Factory Inspectorate.

It is necessary that all possible care be taken to protect operators at all stages of handling and use, in the form of skin protection (i.e. gloves and barrier creams) and eye protection. Efficient extraction and ventilation of the machines and working areas, which include pouring stations and decoring areas, are essential.

Instruction to the operators of safe working practices in the handling and use of these materials must not be overlooked.

T.E.A. and D.M.E.A. are both flammable materials, and regard must be given to the requirements of the 'High Flammable Liquids' regulations which deal with this matter.

To reduce the explosion risk, some foundries prefer to use CO_2 or nitrogen as the carrier medium.

Fume cleaning systems

Amine fumes extracted from the core making areas must be cleaned before being released into the atmosphere. There are several systems available for this purpose, some of which are briefly described.

(1) After burning

This method involves collecting the residual gases and passing them through a gas burner, which burns off the amine.

(2) Acid scrubbers, tower scrubber

In a scrubbing or counter current tower the contaminated air is passed upwards through a tower filled with a packing material, which has a large surface area. Sulphuric or phosphuric acid is sprayed onto the packing material and flows downwards through the tower absorbing the contaminant as it does so. The acid may be initially strong, weakening with use until it is discarded or a constant slightly acidic solution may be maintained by constant monitoring.

(3) Impingement scrubber

An alternative system using similar acid absorbers makes use of the Peabody impingement baffle plate to replace the packing material. In operation the contaminated air is drawn through a series of holes in a plate which is flooded with the absorbing liquid. The air carries the liquid with it and strikes a baffle plate above the hole causing intimate mixing of the air and liquid which encourages the absorption of the contaminant.

Both the above units require a secondary section in which moisture particles are removed from the air stream before it is discharged.

(4) Phosphoric acid bubbler

The collected amine contaminated air is passed at high velocity through a bath containing a 7% phosphoric acid solution which neutralises the amine, the turbulence created prevents acidic sludge collecting in the bath. It is important to maintain the correct acid level and acid concentration in the solution.

(5) Ion exchange

The amine containing air is blown through a tray of resin beads, the amine is leached out of the air. The main disadvantage of this system is that at frequent intervals, depending upon the amount of amine being used, the resin beads need to be revitalised by washing in a solution to dilute sulphuric acid. The acidic sludge created by washing must be disposed of in controlled areas.

URETHANE RAPID NO BAKE PROCESS   C4

A wide range of castings made in a variety of metal alloys, particularly cast iron and steel are produced using this binder system. It is a very rapid no bake binder and is normally used together with continuous sand mixers. The curing is activated by the addition of a small amount of a liquid catalyst, usually added to the mixes by accurate metering pumps.

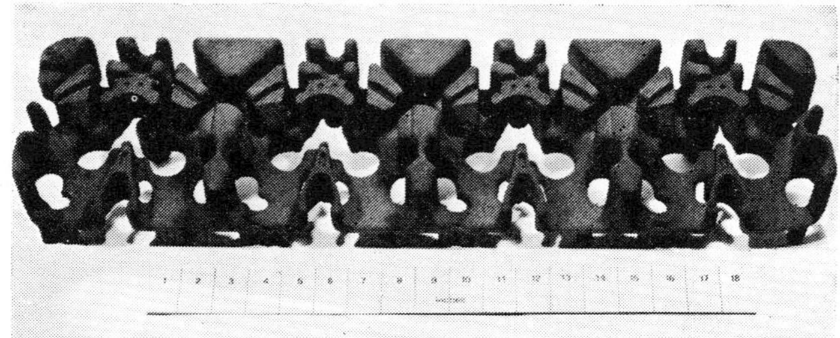

Fig. 10—Urethane cold box core

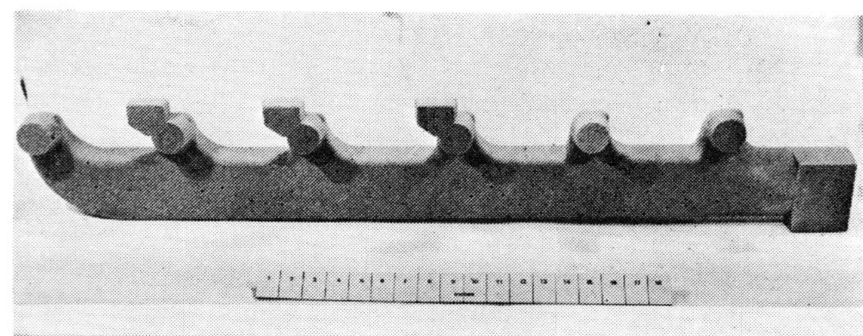

Fig. 11—Urethane cold box core

Binder systems

The binder system consists of three parts:

Part I is a phenol formaldehyde resin blended with suitable solvents

Part II is a polymeric M.D.I. (methylene-bis phenyl isocyanate) blended with suitable solvents.

Part III is the liquid catalyst.

The phenol formaldehyde resin provides hydroxyl groups which chemically combine with isocyanate groups of Part II in the presence of the special liquid catalyst. The resulting reaction provides a tough urethane bond.

Catalyst

The liquid catalyst used to effect the cure in this system is added as a third component to the other materials. It can be pre-blended with Part I (phenolic resin) just prior to use or added as a third component directly into the sand mixer as described earlier.

Curing mechanism

When the three chemical components are blended together in sand, a cross linking reaction will occur. The unique feature of the chemical system is the delay in the initiation of the curing reaction. This means that there is no change in the flowability after mixing with sand until the curing reaction begins. Since the chemical reaction in this system does not evolve side products the cure will proceed at the same rate throughout the system. Once the reaction begins a solid urethane resin bond is formed very quickly. As a result, moulds and cores can be stripped almost immediately after the work time of the sand has passed.

A mould or core produced with a sand mix having a work time of 5 minutes can be stripped from the box after 6·5 minutes, with no fall off in the final cure strength if the sand was still being used up to the end of the 5 minutes work time. This may not be the case with other no bake resin systems, sand used near the end of its work time would produce friable moulds and cores.

Sand

Since urethane no bake resins are sensitive to certain contaminants in the sand, clean, round grain silica sands are recommended. Such sands should have high silica content and be low in fines and alkalies. Special sands such as zircon and chromite are compatible. Certain shore sands have also been found to be suitable and are used successfully. A rounded grain shape provides the best physical properties for a given binder level. Resin demand increases and flowability decreases with greater angularity. Grain fineness of 50–80 AFS sieve size is suitable for most applications.

Alkaline material in the sand can increase the curing rate since the urethane no bake is base catalized. A sand-acidic in nature will reduce cure rate.

Moisture

Moisture in sand will affect the work time strip time ratio, extending the interval between the ending of the work time and the time at which the mould or core can normally be stripped.

Sand temperature

Hot sand will increase the cure rate; conversely cold sand will decrease cure rates. Although the effects of fluctuating temperatures are not as dramatic as with other no bake binders, temperature control is advisable. Optimum cure conditions are achieved with temperatures controlled between 27°–32°C. During cold conditions it is also recommended that the resin portion of the binder system be heated to 21°–27°C to maintain a constant viscosity.

Sand mixers

Most sand mixers are suitable for use with this binder system. Batch mixers are used effectively when the entire batch is used quickly.

Continuous mixers are generally used for this process. Zero-retention mixers provide the best solution as good sand is available immediately the mixer starts and very little or no sand remains in the mixer at the end of the cycle. The mixing capacity of the mixer required will depend on the average size of coreboxes or moulds to be filled and the amount of time available.

Gear pumps are normally used to deliver the liquid components to the sand mixing chamber, the special catalyst can either be pre-blended in Part I resin to produce a two part system, or, metered into Part I using a Tee-junction pipe just prior to entry into the sand. Catalyst levels of 2%–10% based on the weight of the Part I addition is the range used. A typical binder level would be 2·5% total based on a Part I to Part II ratio of 50:50; 45:55 or 60:40 depending on casting conditions and alloys being poured. Dry ingredient additives may be added to the mix by means of a dry powder feeders.

Sand compaction

The excellent flowability of the mixed sand allows for easy distribution and compaction. Vibratory compaction techniques can be used. Hand tucking is adequate around loose pieces; bosses, etc.

Core coatings

Depending on the alloy poured, casting configuration and metal thickness, a refractory coating may or may not be necessary with this binder system. If a coating is required any of the common types can be used. Water based washes are best applied immediately after stripping provided that the wash is dried immediately. Spirit based washes, such as air drying and flame off types are best applied approximately 10 minutes or more after stripping.

Core box materials

All normal core box materials can be used and are used. Some plastic materials can be softened by certain core box cleaning solvents. As the cured core hardens rapidly in the box it should be free from undercuts to facilitate easy stripping.

Core stability and storage

The properties are similar to those of the urethane cold box resins.

Finning

This defect is normally minimal with this process, although some finning may occur as a result of the over compaction of the moulds or cores due to the very free flowing properties of the mixed sand. Iron oxide or fine fibrous materials can be mixed with the sand to prevent or reduce the seriousness of the defect.

Lustrous carbon and gas defects

Both these defects can occur in ferrous castings. The method of control or elimination is the same as described in the section dealing with the urethane cold box process.

Sand reclamation

There are several types of sand reclaimers available and provided that they meet the criteria listed below they will be suitable for this binder system.

(a) Compacted sand reduced to individual grains.
(b) Residual binders reduced to a level below 2·5%.
(c) Metallics removed.
(d) Dust and fines removed.
(e) Grain size distribution controlled.
(f) Any residual chemicals affecting binder cure must be completely removed.

The binder level and the sand to metal ratio greatly affect the ease of reclaiming this material. Nitrogen and carbon are present in quantities sufficient to cause casting defects if fines are allowed to build up in the reclaimed sand due to inefficient screening. If the reclamation is inefficient the following problems can be encountered:

(1) Sand compaction is reduced.
(2) Coating of sand grain surface is impaired.
(3) Binder adhesion (tensile strength) is reduced.
(4) Higher gas evolution occurs resulting in more gas related defects in the casting.
(5) More penetration, burn-on and erosion defects due to lower core and mould density.

Quality control of reclaimed sand

The quality control method for maintaining the reclaimed sand should test for:

(1) Grain size distribution; this monitors the effectiveness of the reclaimer. (Efficiency with which the sand lumps are reduced.)
(2) Loss-on-ignition; this measures the total combustibles and monitors the scrubbing efficiency of the reclaimer. Loss-on-igntition values in excess of 2·5% cause high binder consumption and casting defects from metal penetration and gas.
(3) Carbon; with this binder system the carbon content averages 67·6% of the loss-on-ignition value.
(4) Acid demand value and small changes in the chemical balance of the system which may adversely affect the work time: strip time properties of the mix.

Experience has shown that the reclaimed sand is kept in the correct balance when 50% of the sand on a 140 AFS sieve together with all fines below this removed and replaced with a corresponding amount of new sand.

Environmental consideration

The precautions indicated in the urethane cold box process section should be strictly adhered to.

FAST COLD SETTING PROCESSES   C5

Two fast cold setting resin processes known as the Fascold and Gisag processes, have been developed to produce cores or moulds rapidly by the use of special machines designed for use with selected quick reaction resins and catalysts.

Range of application

Suitable for the production of moulds and cores over a size range up to the limit of the special machinery needed.

FASCOLD SYSTEM
Binder system

Materials used are high quality, highly reactive furane (PF/FA or UF/FA) resins which are hardened by strong acid catalysts. They are similar to those used for the acid catalysed self-setting systems described in Section C1.

Organic acids (sulphonic) are normally used for phenolic furane resin and strong inorganic acid catalysts for urea-furane resins.

Several resin binder and catalyst combinations are available giving setting or strip times from approximately 40 seconds upwards.

For most mould making and for large cores, bonded sand mixtures setting in approximately four minutes are normally used (for special applications slower setting binders can be used). For small moulds and cores requiring no supports, lifting wires, etc. faster setting mixtures are used which are blown into the corebox and stripped in 30–40 seconds. Production rates from one machine can be in excess of 60 cores per hour.

Material requirements

Resin and catalyst are especially produced for use with the process. It is important that the materials are used well within the stated shelf life.

Equipment

The basis of this process is in the mixing machinery. Two types of machines are produced:

(1) a high speed mixer
(2) a dual purpose mixer designed for bench use as a high speed mixer or more particularly for use with a core shooter.

This high speed mixer used in conjunction with a core shooter can operate with the fastest cold setting systems available and strip times of 30 to 50 seconds are easily attained. A high speed mixer used without a core shooter is normally used with resin binder systems giving strip times from 3–5 minutes.

Details of operation

The machines consist basically of two primary screw mixers in parallel, one mixing sand and resin and the other mixing sand and catalyst. These premixed components are then discharged together into a high speed mixer, in predetermined amounts, mixed for about 5–10 seconds and discharged into the corebox. The mixer is then automatically purged with air. Total sand mixing time is about one minute.

Fig. 12—Cores produced in fast cold

The latest machines are fully automatic and self cleaning. They will dispense varying amounts of mixed sand, in sequence, using a variable time switch.

Process requirements

(1) Sand should be of uniform grain size and clean. The process will not tolerate low grade sands with sludge or alkaline materials present.

(2) Mixing equipment must be kept clean and careful attention paid to correct operation of the metering devices for resin and catalyst additions.

(3) The temperature of the sand is also important, cold sands require considerably more resin and catalyst than warmer sands.

The optimum operating temperature is about 20–25°C for maximum throughput and strength with minimum binder requirements.

Typically 1–1·5% binder (on sand weight) and 20–60% catalyst (based on resin weight) would be required for sand at 20°C.

Process properties

Bench life	is very short
Flowability	the mixed sand flows well, and is very easily compacted, provided it is used within its bench life.
Tensile and compression strengths	are good, but lower than with normal furane bonded sands. Final compression strength is about 500 psi.
Gas evolution	is moderate, less than hot box and oil sand but more than silicate bonded sands.
Degree of thermoplasticity	is low
Knockout	is good
Moisture resistance in storage	is good
Smoke evolution during casting	is moderate, and good ventilation is desirable.

Advantages

The fascold process has replaced oil sand, hot box and the CO_2 silicate process in a number of foundries for the following reasons:

(1) No heat or gassing is required for curing.

(2) Owing to high tensile strength of the cured sand a flaskless moulding system can be operated.

(3) The bonded sand has extremely good breakdown and is reclaimable.

(4) The process equipment is easily controlled, giving good quality moulds and cores. A relatively low degree of skill and training is required for operators.

(5) Furane resin binders are used and the resulting castings can be of good dimensional accuracy and surface finish.

Fig. 13—Rapid setting fast cold process

(6) The mixer is purged with air between each cycle and requires relatively limited cleaning during operation.

Disadvantages

(1) In many cases an excess of sand must be mixed during each batch and owing to the very short bench life the excess must be discarded, although the machine can be present to deliver accurately, varying amounts of mixed sand, thereby minimising such wastage.
(2) The process can only be used in conjunction with special high speed sand mixing equipment and/or combined mixing/blowing machines.

GISAG SYSTEM

Binder system

This system developed in East Germany also uses high quality phenolic and furane resins hardened by strong acid catalysts.

Details of operation

Sand and catalyst are premixed (for three to four minutes) in a vertical mixer of conventional type and are fed by a pneumatic conveyor system to the sand hopper of the gisag core shooter.

A predetermined weight of this moulding sand (sand and catalyst) sufficient to fill the corebox is transferred by a feed screw from the storage hopper to the rapid mixer, a mechanical blade type mixer. Whilst the premixture is being metered a suitable quantity of the resin binder component is simultaneously added. The residence time in the fast mixer is approximately ten seconds before the mixed sand is discharged into the shooting cylinder and shot into the corebox.

All operations are controlled automatically. Metering times and mixing times are preset on the control cabinet.

Curing time from coreshooting to stripping from the box is from 15 to 60 seconds. To maximise use of the machine it is recommended that several coreboxes be employed.

[C]6 THE SYNCORE PROCESS

The Syncore process, which has been developed in Poland, relies on solvent evaporation from a mixed sand containing an adhesive resin dissolved in a volatile solvent.

Cores are blown in the normal way and are then hardened using a stream of compressed air or nitrogen at a pressure of 3 to 4 atmospheres for about 15 seconds (time depends on size of core) to remove the solvent. The process requires the development of a physical adhesive bond in the sand, with no chemical reaction taking place.

The resins employed are usually based on polystyrene and the volatile solvent consists of mixtures of chlorinated hydrocarbons and petroleum distillates, which are added to make approximately 50–50 mixtures with the resin.

3–5% of the prepared binder is used; sand mixtures containing 4% of the binder achieve transverse strengths of 1960–2450 kN/m² (285–355 lb/in²) and tensile strengths of 980–1960 kN/m² (140–285 lb/in²).

Moulds and cores obtained using this process are claimed to be suitable for the production of iron, steel and non-ferrous castings, although at present there is no known use of the process in the U.K. for mould or core production.

Cores are said to be non-hygroscopic, can be stored indefinitely without loss of performance and can be used with spirit-based coatings.

Advantages

(1) It is a cold "gassing" process using air, giving good dimensional accuracy of cores and moulds.
(2) The process can be used satisfactorily with silica, zircon, chromite and olivine sands.
(3) The materials used are neutral and do not give rise to corrosion in metering and mixing equipment.

Disadvantages

(1) Chlorinated hydrocarbon solvents are used, which require the use of special extraction of waste solvent vapour and could give rise to fume problems at casting and knock-out stations.
(2) Expensive corebox equipment is required (as for other gassing processes).
(3) Mixed sand has a relatively short bench life unless the sand is stored in well sealed containers before use.
(4) For large moulds and cores the process requires relatively long "gassing" periods compared with other chemically active gassing processes.
(5) The binders used in this process at present are thermoplastic and consequently have little heat resistance, sand erosion from cores and moulds may therefore be a problem during casting.

COMPARISON OF COLD SET RESIN PRACTICE [D] IN THE U.K., U.S.A. AND CONTINENTAL EUROPE

U.K. practice has developed differently from that on the European Continent in the U.S.A.

Normal practice in the U.K. is to use UF/FA resins having 7–9% nitrogen for most grey iron applications where new sand is used. These resins usually contain 50–70% of furfuryl alcohol. Resin additions to the sand are commonly 1·3–1·6%.

Where reclaimed sand is employed, lower nitrogen levels are used and most U.K. iron foundries use nitrogen-free PF/FA resins containing 50–70% FA. Sulphonic acid catalysts are generally used.

Steel foundries in the U.K. normally use nitrogen-free PF/FA-binders with or without sand reclamation.

It would appear that the central research association of the Continent, e.g. France, Germany and Sweden have set the nitrogen contents of binders suitable for steel at 1% and binders suitable for grey iron at 4%. This automatically sets the furfuryl alcohol levels high. Because of this, continental resins are expensive. The sands on the Continent are reputed to be of much higher quality than in the U.K. and thus lower levels of resin can be used, which offsets the higher resin costs.

Cold set resins containing phenol were actively discouraged, particularly in the Nord–Rhein–Westphalia region where most of the German foundry industry lies. It was believed by the environmental authorities that there was a risk of water pollution from used foundry sand containing phenolic resin. UF furanes were therefore used. However, the worries about continuous dumping of phenol may have partly disappeared since it has been found that phenol is rapidly oxidised under normal foundry dump conditions.

Iron foundries using all new sand employ UF/FA resins having about 8% nitrogen and 60–70% FA, but a lower nitrogen content is required when making steel castings and for sand to be reclaimed; this is achieved by using high FA resins (70–90%), and these are very widely used now. Since the FA shortage of a few years ago, there has been some move to PF cold set resins in Germany, although PF/FA resins are still very little used.

Continentals prefer to handle resin of low viscosity, probably because of inadequacy of metering pumps fitted to many of the continuous mixers. In the U.K. metering pumps have been developed capable of handling more realistic viscosities. Resins with the higher viscosities set faster and thus have an advantage over Continental resins.

High FA resins tend to be of lower viscosity than low FA resins, they therefore spread more easily over the sand grains enabling smaller additions to be made, and compensating in part for the higher cost of the resin.

There has also been a tendency for U.K. markets to use lower FA, higher nitrogen products because of their lower prices. Once confidence has been gained in the performance of these higher nitrogen products, it is expected that they will find increasing use on the Continent.

Attempts to introduce into the U.K. very high FA resins of the type used in Germany have met with very little success. In fact, there has been a discernible trend in the last three years towards lower FA contents in an attempt to reduce mixed sand costs.

The use of the urethane based cold box and cold setting processes is more widespread in the U.S.A. than in Europe.

E INORGANIC RESIN NO BAKE PROCESS

Introduction

This system has been developed and recently introduced to the foundry industry in the U.S.A. and Europe and experiences in its use are therefore limited.

Binder systems

The binder is a two part system.

A water soluble resin component is supplied as a clear colourless or yellow liquid with no odour which has low viscosity; it can be pumped and metered easily. Stability during storage of the resin is excellent; resins have been stored for over one year without any change in physical properties or binder performance. The resin is moderately acid and containers must be acid resistant.

The hardener component is a grey or tan material in a powdered form. It is also based on inorganic components and is odourless. In contact with water the hardener provides a mildly alkaline reaction and slowly hydrates, altering its chemical reactivity and physical state. Under normal storage conditions the powder is non-hygroscopic and is unaffected by exposure to ordinary atmosphere, flow properties of the powder are good and there are no difficulties in using standard powder feeders.

Sands

Most common foundry sands can be used with the binder; the one known exception being zircon sand. When used with zircon sand a severe loss of bench life occurs which cannot be adequately compensated for by reductions in hardener level.

Moisture in sand

High water contents of sands can adversely affect performance of the binder.

However, such effects are only observed at moisture levels which would be considered as abnormally high. 0·2% water in sand can be tolerated without serious effect. Moisture in sand, however, dilutes the resin and the degree of tolerance will depend on several factors, i.e. binder level, type of sand, required storage properties of finished moulds or cores.

Sand temperature

The curing time in common with that of other no-bake resins is affected by sand temperatures. Temperatures of 35°C seriously reduce and strip time, whilst temperature of 10°C greatly extend these times. The use of some form of sand temperature control is advised to achieve optimum results.

Sand mixers

Both batch and continuous mixers have been used successfully with this new system. In either case it is preferable to mix the hardening agent in the sand before the resin addition is made. This order of addition reduces premature reaction of the binder.

Sand compaction

Flowability of the mix is comparable to that obtained with silicates, which is somewhat inferior to that of furanes or phenolic urethane resins, and the mix can be readily compacted either by jolting or vibration with appropriate tucking. Maximum flowability is obtained when using a continuous mixer.

Cure time and range

This new binder system cannot be classed as a fast setting no bake material. Work time and strip time can be varied by the amount of hardener added. These times can be controlled in the range of 25 minutes to over one hour. In certain conditions it may be possible to operate at higher speeds, but this is not usual. Development is continuing to try and produce a more rapid reacting system.

A sand mix using 3·3% resin and 25% hardener (based on the weight of resin) could be expected to reach strength in excess of 30 psi within 30 minutes and as high as 100 psi in 24 hours.

Through cure and sag resistance compare favourably with other systems.

Moulds and core coatings

Satisfactorily clean castings have been produced without the application of core wash, for relatively thin metal sections in cast iron. Heavier section work requires a coating to produce a good overall casting finish and minimise burn-on and metal penetration. Graphite and ceramic based coatings are satisfactory for cast iron. Zircon containing coatings cannot be used as they are subject to a strong fluxing reaction between the binder and the zircon powder at metal casting temperatures. However, zircon washes may be used if a primer coating is applied first.

Water based washes can lead to softening of the mould face, since the binder system is water soluble, and are therefore not normally recommended.

Alcohol based washes will not cause softening of the mould face, but mould and core damage has been observed after burning off.

To eliminate the difficulties associated with these two traditional types of coatings a series of air-dry systems have been specially developed.

Core box materials

All conventional core box materials have been used satisfactorily and there is no need for other than normal treatment, i.e. release agents, etc.

Casting performance

Extensive work has already been carried out both in a pilot foundry and in production to assess casting characteristics. Many test castings have been produced in the pilot foundry, the test castings were specially designed to accentuate various well known casting defects such as finning pinholing, erosion and casting strip.

The new binder performed well and exhibited good casting finish without wash, no pinholing, little corrosion and finning. The last two defects were easily controlled with a graphite based core wash. Work has been produced to date in cast iron, steel and non-ferrous castings. Production foundries have confirmed the pilot foundry investigation, and also confirmed the excellent breakdown and shake-out features.

Sand reclamation

Moulds and cores produced with this system are being reclaimed commercially by dry reclaimers. Compared to silicates and organic systems, the new binder seems to 'strip' more easily from the sand grains during the reclaiming operation, resulting in little interference with curing and strength characteristics when using reclaimed sands.

Environmental consideration

The resin system is mildly acid and therefore protective clothing and goggles should be worn. The hardener is a powder, and protection from dust is necessary.

There are no irritating fumes or odours associated with this system at any stage of production from sand mixing to knock-out. During casting, the moulds produced are both odourless and smoke free; at most there will be only a trace of steam.

F FLUID OR CASTABLE SAND PROCESSES

By addition of a special foaming agent, sand mixtures containing 4·1–5% of water can be converted to a fluid slurry which will fill coreboxes without the need for compaction. A controlled amount of air must be entrained into the mix; a special mixer is required.

Cement fluid sand

Cement/sand/water mixtures have been used for ingot mould and other heavy casting production. In some cases special accelerators have been developed to speed up the process, for example, a French process uses cement with molasses or calcium chloride additions. An Italian process uses alumina cement accelerated with lithium chloride solution.

Sodium silicate fluid sand

The Russian fluid sand process uses a high ratio sodium silicate binder with dicalcium silicate as the hardener. A U.K. process uses similar materials but involves a collapsible foam. A french process uses a low ratio silicate together with specially selected blast furnace slag as a hardener. The main application again is for ingot moulds and heavy castings.

Resin fluid sand

In the U.K. process PF or furane resin binders can be used together with an acid catalyst containing a foaming agent. A total of about 4% liquid is used. There is a French fluid process which uses 1–2% furane resin, 0·9% catalyst and 1·0% water. There is a Polish resin fluid sand technique known as the Syn-flo process, but few details are known. The U.K. and French processes differ since, in the French process, self compaction occurs after the foaming bubbles collapse. In this way high strengths are obtained at low resin additions, but at the cost of a core which shrinks on setting.

The U.K. system on the other hand does not shrink. The

Fig. 14—A fluid sand process illustrating fluidity.

Fig. 15—Selection of fluid sand cores

sand is mixed with the acid catalyst containing a foaming agent. A suitable mixer beats air into the liquid, increasing its volume and forming a stable foam in which the sand grains can float. Resin is added and mixed into the catalyst film. The fluid sand catalyst resin mixture is then rapidly used to fill boxes before reaction causes loss of fluidity and foam collapse. A good fluid sand does not shrink on curing.

Sands suitable for cold set may be used for fluid sand.

Fluid sand has the ability to fill intricate boxes with undercuts and fine surface details to produce cores of excellent dimensional tolerances and surface finish.

Setting time is restricted to a minimum of 30 mins. for good sand fluidity.

Casting surface finish is good, although coatings are still required.

Cured strength is lower than with conventional cold set processes and the core density is lower.

Breakdown is excellent.

Fluid sands generally cost up to 50% more than do conventional cold set sands.

The main application is for cores and moulds for particular castings e.g. ingot moulds.

Both batch and continuous mixers may be used, but only specific types will function with fluid sand.

The foam generating acid catalyst and suitable mixers are

essential to the successful use of the fluid sand process.

Once the techniques of operating with batch or continuous mixers have been mastered, very little skill is required to operate the process.

The two main disadvantages are high cost and long setting times. However, continued research may reduce both these factors; if so, fluid sand could have a wider range of applications.

G FROZEN SAND PROCESS

Without doubt the frozen sand process is a truly cold process, cores and moulds are hardened by freezing a mixture of sand and water. The process is still under development by the pioneers.

Range of application

Up to this time a variety of metals including aluminium, stainless steel and bronze have been cast successfully into frozen moulds, most of the development has been with S.G. (ductile) iron.

Castings have been made weighing up to 220 lbs.

Binder system

The binder system involves freezing a mixture of sand, water and clay.

A moisture content of 2% to 6% is used together with a quantity of clay ($1\frac{1}{2}$ to $2\frac{1}{2}$%) to produce a sufficient green strength for conventional green stripping techniques.

Freezing

Freezing is obtained by the use of liquid nitrogen or liquid CO_2.

Moulds and cores are placed on plates inside a purpose built insulated chamber, where liquefied gas is sprayed over the surface of the cores and moulds. With liquid nitrogen temperatures down to —196°C are reached. With liquid CO_2 a mould temperature of —50°C is obtained and particles of dry ice are produced. The sub-zero temperatures of the sand are reached in varying times according to the thickness of the sand mass.

Frozen sand properties

Strength properties of frozen sand are said to be comparable with other cold setting processes, no additional reinforcement is necessary to support the core in its frozen condition.

The moisture content of the sand is similar to that for green sand practice. Permeability is increased by freezing.

Pouring

Molten metal poured into frozen moulds is claimed to be free from agitation or reaction. Fluidity of the metal appears to be improved when cast into frozen moulds.

Core knockout and sand reclamation

Complete breakdown of the moulds and cores occurs a few minutes after pouring. All sand can be returned to the mixer immediately for re-use.

Sand mixing

The sand can be mixed in any conventional type of mill.

Bench life

No special precautions are necessary.

Shelf life

The shelf life of the moulds and cores may be up to one hour in uninsulated conditions. In insulated conditions cores and moulds can be stored for several hours, or as required.

Core and mould production

Conventional oil sand practice is used for making the cores, and green sand moulding techniques are used for producing moulds.

Fig. 16—Frozen sand moulds assembled ready for pouring

Fig. 17—Frozen sand breakdown after casting

Fig. 18—Castings ready for cleaning

Both cores and moulds are turned out on to plate for freezing purposes. Trials and experiments are proceeding to utilise mechanical means to produce cores and moulds.

Cores and moulds which need to be jointed, need only to be placed together in the desired position and the freezing of the joint acts as an effective adhesive.

Casting quality

The dimensional accuracy of the castings and reproduction of the pattern image is good, within the constraint of green strip practice, and the frozen sand process does not produce greater chill defect problems than comparable processes.

H SAND MIXING

Sand mixers suitable for cold processes

The ultimate success of cold setting chemical systems depends greatly upon the quality and homogeneity of the mixed sand. There are several types of mixing units suitable for cold setting sand systems, brief descriptions are given of the more common ones.

1. Batch mixers

Principle: The mixing chamber contains slowly rotating blades (15–40 r.p.m.). A range of sizes if available up to 1 tonne capacity. The mixers are designed to give effective mixing with the minimum generation of heat. Mixing time with a single liquid component is usually 2 minutes. If two components are added, such as with a cold set resin, the catalyst is added first, mixed for $1\frac{1}{2}$ minutes then the binder added and mixed for a further 2 minutes. Mixing longer than this causes heat to be developed and the bench life of the mixture is shortened.

Advantages: Reliable and effective. The mix composition can be changed at will. Powder additions are easily made.

Disadvantages: Not suitable for mixing materials with a short bench life. The relatively long mixing time may allow solvents to evaporate from the sand mixture. The mixed sand is discharged in a batch and must be distributed to the coreblowers or the core-making area. It is usually difficult to guard batch mixers satisfactorily. Cleaning of the mixing chamber and blades may be difficult.

2. Speed muller

Principle: it is a type of batch mixer. The mixing chamber contains two rubber lined wheels, which are mounted on revolving arms, the driven arms rotate at relatively high speeds, the wheels are caused to spin by their contact with the sand being mixed.

Advantage: Reliable, very efficient mixing action.

Disadvantages: Rapid temperature rise of the sand during mixing.

It is not suitable for short bench life processes, and cleaning is difficult.

3. Single trough slow speed continuous mixers

Principle: sand, binder and catalyst are continuously fed into a trough containing a mixing blade consisting of a shaft with a large number of mixing elements. The blade assembly rotates to 80–100 r.p.m. The binders are mixed with the sand as it progresses along the trough. The dwell time of the sand in the mixer is 1–3 minutes. Powders can be added by the use of continuous powder feeder.

Advantages: reliable, good wear characteristics; a variety of configurations are possible, e.g. single arm, articulated, mobile; as well as mixing, sand is conveyed to the point of use; a single mixer with automatic indexing can feed up to 4 coreblowers; a wide range of outputs is possible, from 2–40 tonnes per hour.

Disadvantages: the first sand out of the mixer is not well mixed so there is considerable sand wastage at the start; the trough must be emptied for any stoppage of more than a few minutes; these mixers can be used with self-setting sands having setting times down to 20 minutes, but with shorter setting times, the residual sand in the mixer trough starts to harden during any stoppage so that the wastage of sand can be excessive.

Applications: widely used for self-setting resins and silicates with a setting time over 20 minutes; also used for urethane cold box or CO_2 silicate sand for feeding coreblowers. There is a trend nowadays towards fitting a continuous mixer as an integral part of the core-blower.

4. Twin trough high-speed mixers

Principle: these mixers are designed to avoid the problems mentioned above with single trough continuous mixers when fast self-setting mixtures are used; the sand feed is split into 2 primary troughs and the same construction as the single arm mixers, one trough mixes sand with the binder, the other mixes sand with the catalyst; the troughs discharge into a final blender in which the residence time is only a few seconds, so that fresh mixed sand is discharged into the coreboxes; the final mixer may be a highspeed batch mixer or a continuous vertical mixer, usually in conical form; the final mixer is normally emptied when stopped.

Advantages: can handle mixtures with setting times down to 2 minutes with infrequent cleaning; all the sand from start to finish is well mixed; available in a variety of configurations, so that sand can be mixed and conveyed to the point of use, easy to guard.

Disadvantages: more complex and expensive than single trough mixers and because of the high speed of the final mixer, wear is more serious.

Applications: fast setting resin systems and ester silicates; powder additions can be made if necessary. Can be used to feed coreblowers with rapid setting sand mixtures.

5. Single arm high-speed mixers

Principle: a single mixer trough or tube contains a high-speed (450–1500 r.p.m.) mixing blade, consisting of a shaft containing a number of mixing elements; the tube is fed with sand, powders and liquid additives; the dwell time of the sand in the mixer is only a few seconds; the mixer empties completely when stopped.

Advantages: can handle mixtures with setting times down to 1–2 mins; little cleaning is necessary; the efficient mixing and short dwell time may allow reduced binder additions in some cases; the self-emptying feature minimises sand wastage and it is possible to change quickly from one sand mix to another, for example, to provide facing and backing sand; can be used to feed coreblowers where self-emptying feature avoids any sand wastage.

Disadvantages: It is difficult to ensure that the sand at the end of the mix is well mixed; the high-speed rotation gives higher wear rate than on a conventional single trough mixer.

Applications: all self-setting mixtures including resin and silicates; can be used to feed coreblowers, with rapid setting sand mixtures.

6. Vertical mixer (spinner)

Principle: a very high-speed rotating disc spins binder and catalyst mist on to a curtain of sand.

Advantages: almost instant start and stop; the mixer

empties completely when stopped; the dwell time of sand is 1–2 seconds; no guard is required.

Disadvantages: it is expensive and complex, and only suited to low viscosity binders and catalysts.

Applications: it is particularly suitable for urethane no bake systems.

7. Three stage vertical mixer chamber

Principle: a vertically mounted cylinder divided into three mixing chambers in which mixing blades mounted on a driven shaft rotate to mix the sand; the sand is fed into the top chamber, and the mixing action transfers the partially mixed sand through connecting holes into the second and third chambers, finally discharging the mixed sand from the bottom; powder and liquid additives can be added into each or any of the mixing chambers as required.

Advantages: can handle mixtures with relatively short setting times; little cleaning is necessary; no exterior guarding is required; the units are compact, requiring minimal floor space.

Disadvantages: not suitable for mixing materials with very short bench life; the mixed sand is discharged in a batch,

Applications: used for mixing urethane bonded sands for self-setting and gas setting processes.

☐ SAND RECLAMATION FOR COLD COREMAKING PROCESSES

The reclamation of sand from cold processes does not present any special problems though there maybe some restriction on the use of the sand reclaimed.

Cores break down during casting and cooling. Some of the broken down core sand contaminates the moulding return sand, and is processed into the green sand system. Core butts and partially broken down cores are usually screened out, and discarded as considered not worth recovering for re-use in cores because of contamination with clay and coal dust,

If the amount of de-cored material which gets into the moulding sand is not excessive, it is usually regarded as beneficial, since it behaves like an addition of new sand keeping the sand system free from excessive burnt clay and coal dust. However, the burnt core sand usually contains some chemicals from the bonding process which may be harmful to the sand system if present in large amounts.

Complete reclamation of moulds and cores

Re-use of some or all of the core and mould sand is usually only possible when cores and moulds are made with the same or a similar binder, e.g. furane cores and moulds, CO_2 silicate cores and ester moulds, urethane cold box and urethane no bake systems.

There is danger of contamination problems where sands from several different processes are reclaimed simultaneously.

There are three reclamation methods normally used:—
Wet
Thermal
Attrition

Wet reclamation

In this, the sand is crushed and water washed. This system is effective particularly for silicate sands but is rarely practised because of the high cost of the plant and the plant and the difficulty of treating effluent water so that it can be safely discharged into waterways.

Thermal reclamation

The sand is heated to about 800°C in order to burn off residual organic material. It is a very effective method, providing there are no inorganic additions in the sand, and can restore the sand to its original state. The process is not widely used however because of the high cost of the plant and running cost.

In most of the U.K. the cost of reclaiming sand by thermal methods is usually comparable with that of new silica sand. The process is used, mainly, for special materials such as zircon or in geographical areas where good silica sand is exceptionally expensive. The rising cost of dumping chemical waste sand may make thermal reclamation more attractive in the future.

Attrition reclamation

Attrition is the most widely practised method. The used sand is crushed to grain size, mechanically scrubbed, then screened, air classified and finally cooled. The process does not remove all the residual binder from the sand so that continued re-use of reclaimed sand results in residual binder levels increasing until a steady state is reached which is determined by:

The burn out occurring during casting and cooling.
The effectiveness of the reclamation equipment.
The percentage of new sand added.
The effectiveness of attrition reclamation depends on the type of binder used.

Furane binders

The process is very widely used with furane cores and moulds.

The three main criteria governing the re-use level of the reclaimed sand are loss on ignition, nitrogen content and gas evolution. Experience has shown that in the case of grey cast iron the maximum allowable loss on ignition is around 2·5–3·5%. This can generally be achieved with well-designed attrition plants and regular addition of 10% new sand into the system. To avoid pinhole problems building up with continued re-use of reclaimed sand, it may be advisable to use nitrogen-free or very low nitrogen binders with sulphonic acid catalyst. It is well established that the compression strength achieved with reclaimed sand may be higher than that obtained with new sand so it is usually possible to reduce the resin addition when moulds are made with reclaimed sand. However, the transverse and tensile strengths obtained with reclaimed sand can be lower than with new sand, since these strengths are important cores are often made from new sand. In addition the gas evolution from reclaimed sand is several times greater than from new sand so gas problems may result if reclaimed sand is used for coremaking. A problem recently reported is that of sulphur build-up, where a sulphonic acid catalyst has been used. Sulphur in the reclaimed sand may result in graphite flakes in S.G. iron, or hot tearing in steel castings.

Urethane bonded cores and mould

Attrition is widely practised for these processes, the same conditions apply as for furane resin.

Sodium silicate

The main problem with reclaimed sodium silicate sand is the gradual build-up of soda. If residual soda rises to levels above about 0·8% the refractoriness of the sand is seriously reduced: this generally limits the re-use of reclaimed sand to a maximum of 50%.

Ester-hardened sodium silicate presents a further problem in that a variable amount of ester will remain in the sand. Close to the casting surface virtually all the ester is destroyed by the heat but at the corners of the moulds some residual active ester will remain even after attrition reclamation. Some variability in setting times may result due to this residual ester; at this time no foundry is known to be reclaiming this material by attrition methods.

Sand from mixed processes

Rather limited evidence is available on this subject but it is clear that sodium silicate bonded sand must not be reclaimed together with furane sand because the alkaline silicate interferes with the acid hardening furane processes. Even very small amounts of silicate contamination, as low as 5%, can seriously affect the performance of the reclaimed sand. On the other hand, a small proportion of furane sand, say up to 10% can be tolerated in reclaimed silicate sand.

There is little experience available on the use of urethane bonded sand with furane. It is believed that a substantial amount of urethane bonded sand can be tolerated in a furane reclamation system but the reverse is not true and it is understood that reclaimed urethane bonded sand should be completely free from furane contamination. From limited information, it would seem that mixtures of urethane cold box sand can be intermixed with urethane no-bake sand.

J ENVIRONMENT AND HEALTH

As all cold cure processes come within the general term 'Chemical Binders' and their use often involves the handling of toxic materials with the likelihood of the generation of toxic fumes, consideration must be given to the moral and statutory requirements of environment and health.

The Health and Safety at Work Act 1974 requires work people, in the broadest terms, "to know the chemical value of the materials they handle . . . the hazards which are involved and the precautions which should be taken". Furthermore, as it empowers the Health and Safety Executive to issue prohibition and improvement notices to correct any deficiencies, the foundryman in making a choice of a cold core making process must ask the following quetions.

(1) MATERIALS--What materials are being used and what are their toxic properties in basic and modified form during the foundry processes?
(2) EQUIPMENT—What additional equipment and facilities are required to detect and control the toxic materials?
(3) EDUCATION—What training is to be given in the recognition of the potential hazards and in the safe working and emergency procedures, to fulfil the statutory requirements?

In obtaining answers to these questions, the foundryman should not only consult the suppliers but refer, in close detail, to the manufacturers' 'Health and Safety Data Sheets'.

Potential hazards in the handling and use of resin based binders

Potential hazards are present at all stages in the use of resin based binders for mould or coremaking.

(1) Handling the resin, its hardener and accelerator.
(2) Mixing with sand.
(3) Handling the moulding mixture.
(4) At casting and knock-out.

These hazards can be divided into the effects caused by:

1. Ingestion

It is assumed that provided that drums are labelled "DO NOT INGEST" it is not necessary to cater for this aspect by developing non-poisonous resin systems. (They are not designed to be eaten).

2. Dermal toxicity

Induced by absorption of the material into the skin and its reaction with the skin on contact.

3. Inhalation

The toxic effects of inhaling vapours from the separate components of the resin based binder, any vapours evolved during the curing and decomposition products evolved at the casting or knock-out stage.

Comments are thus confined to effects caused by contact with the skin and by inhalation of vapours or decomposition products.

Contact with the skin

All resins and their hardeners and accelerators are reactive chemicals and thus should be prevented from contact with the skin by wearing stuitable protective clothing.

Gloves and goggles are essential for handling resins— goggles merely to protect the eyes against splashes. For acid hardeners or alkaline accelerators exrta protection in the form of rubber aprons and boots are recommended.

These materials can be rated according to the Primary Irritation Index (Draze Method) (Federal Register 17, September 1964 191:11) P.I.I. indicates the level of effect that the material will have on the skin, should it come into contact with it.

Tests indicate that most PF, UF/FA and PF/FA polymers give responses of 2 or less and thus are "Mildly irritating".

Phosphoric acid hardener is a "severe irritant" with a response of more than 6.

As all resins are used with a hardener or accelerator whose P.I.Is are of the "severe irritant" class, gloves and goggles should be used for handling these materials.

These comments apply to handling the neat resin and hardener or accelerator during dispensing and mixing.

Although it is not necessary to wear rubber aprons and boots to handle the sand mix.

Inhalation of vapours or decomposition products

H.M. Factory Inspectorate produces a list of Threshold Limit Values for substances in workroom air Guidance Note EH15/76 from the Health and Safety Executive.

The following has been extracted from BCIRA Broadsheet 163-1:

The working environment—meaning and use of Threshold Limit Values

Threshold Limit Values (TLVs) are concentrations of airborne substances such as chemicals, gases, fumes and dusts, to which workers may be repeatedly exposed for a working period—either an 8 hour shift or a short time—day after day, without ill effect. A list of Threshold Limit Values is published for use in the UK, with introductory information and is revised annually by the Health and Safety Executive.

TLVs fall into three classes:

(1) Time-weighted average values (TLV-TWA) refer to concentrations averaged over an 8-hour working day (40-hour working week). Nearly all workers not sensitive to the substance can be exposed to This concentration daily without ill effect.
(2) Short-term exposure limits (TLV-STEL) refer to limiting concentration values for 15-minutes exposure. These values are often higher than TLV-TWA

values, but their application is much more stringent. Not more than four such exposures may occur in a day, and they must not recur within 60 minutes. At present, TLV-STEL values are regarded as being only tentative.

(3) Ceiling values (TLV-C) have been allotted to the TLV concentrations for some substances and should not be exceeded, even for a very short time.

How are TLVs used?

TLV-TWA and TLV-STEL values are guides for the control of health hazards, and are not to be used to decide whether the concentration of a substance is 'safe' or 'dangerous'. They allow management to decide after analysing appropriate samples of airborne pollutants, whether action needs to be taken to improve conditions. If a TLV is approached or exceeded there is a clear need to reduce the pollution or the exposure of workers to it.

When a ceiling value has been allotted to a TLV, steps should be taken to ensure that it is not exceeded under any circumstances.

When two or more substances are present as pollutants, the TLVs for each may be modified if the substances have combined biologic.l effects.

At present, if the TLV has been exceeded this fact may be used to support any prosecution brought by the Factory Inspector under Section 63 (1) of the Factories Act 1961, or relevant Regulations.

The main materials relevant to resin binders and their hardeners and accelerators are listed below, together with their respective TLVs.

Substance	TLV-TWA (ppm)
Carbon monoxide	50
Carbon dioxide	5000
Methylene disphenyl isocyanate	0·02
Triethylamine	25
Hydrogen sulphide	10
Sulphur dioxide	5
Phenol	5
Benzene	25 (10) (1)
Toluene	100
Xylene	100
Naphthalene	10
Formaldehyde	2
Acrolein	0·1
Furfuryl alcohol	5
Nitrogen oxides	5
Hydrogen cyanide	10
Ammonia	25
Mek peroxide	0·2
Hydrogen peroxide	1

This list is not exhaustive, and new substances must be added as and when they are used or discovered.

It is essential for the foundry to ensure that ventilation at the work place is sufficient to reduce the level of contaminants in the air to a level as low as is regarded practicable and certainly below the TLV's. (See Guidance Note EH 15/76 for dealing with mixtures of vapour).

Methods for detection of toxic substances in air are described in booklets available from H.M. Stationery Office. There are other quicker methods which may also be used.

Particular attention must be applied to all points where the binders are handled, mixed and transferred to core boxes and to the casting and knock-out stages. Obviously the vapours evolved depend upon a number of factors:

(a) Composition of resin and hardener or accelerator.
(b) Ambient temperature (sand temperature).
(c) Volume of work room.
(d) Amount being mixed (or cast).
(e) Sand to metal ratio.
(f) Temperature of metal (composition of metal).
(g) Ventilation.

Ventilation should be installed to cope with all of these variables at their worst condition.

It must be clearly understood that proper safety precautions and acceptable environmental conditions are the ultimate responsibility of the foundry, and it is incumbent upon the foundries to understand the dangers which may exist when using potentially hazardous materials. All suppliers of these potentially hazardous materials will readily help and advise on all matters relating to safe working practices during the use of such materials.

Cold-Box Systems Engineering

PANEL PRESENTATION

J. R. Schoen, *Grede Foundries, Milwaukee, Wisconsin*
D. J. Berant, *Giffels Assoc, Detroit, Michigan*
R. A. Rowland, *Electron Corp, Littleton, Colorado*
N. Luther, *Pontiac Motor Div, Pontiac, Michigan*
W. E. Schulze,*Caterpillar Tractor Co, Mapleton, Ill*
W. Cheek, *Central Foundry, Saginaw, Michigan*

Introduction

We are in an era characterized by reduced availability of energy. The old cliche of "things are going to get worse before they get better" certainly applies. Natural gas, propane and oil are bound to go up in price and may not be available in sufficient quantity at any price. Consequently, it would behoove the informed foundry manager to consider ways of producing cores and molds other than those requiring heat. The cold-box process, the present subject, requires no heat thereby permitting continuous production and substantial gas savings. Gas saved can be used elsewhere in the plant or the reduced consumption can result in lower operating costs. The process has many other advantages which, at one time, were more important than energy conservation.

The aspect of no heat permits relatively inexpensive coreboxes as opposed to costly cast iron machined tooling. In addition, box growth considerations are eliminated, as are expansion and contraction of boxes which contribute to dimensional problems. Box warpage, which is a function of heat, is also eliminated. Another extremely important benefit is the ease of changing tooling in that no cooldown period is required. Worker comfort is improved and hooding and venting requirements are reduced. Worker safety is enhanced because of the elimination of open flame at the work station.

Advantages of the cold box process in the form of energy conservation and other ancillary benefits make the cold-box method extremely attractive. More and more foundries will certainly be investigating this method as well as other chemically bonded processes that have similar advantages. Cold-box adapts itself readily to high-speed production, although it also has a place in the jobbing foundry. Due to increasing foundry industry orientation in the direction of chemically bonded systems, this paper considers cold-box systems engineering to give the foundry manager a ready source of data for all facility considerations.

Raw Material Handling, Storage and Distribution

Raw materials involved in isocyanate binder systems are classified under six headings: sand, binders, catalyst, carrier, additives and gassing fume neutralizers. The basic ingredient is washed and dried lake or silica sand. Several physical properties of this material are critical and should be kept in mind during storage and handling.

1) Moisture level of the dried silica sand as it is received from the supplier should be below 0.2% by weight and should be kept as low as possible during storage and handling.
2) Sand temperature is very important and ideally should be between 65 and 85F (18 and 30C) at the point of use. A spread from 50 to 105F (10 to 41C) is tolerated by some foundries but uniform temperature is necessary for short-cycle cure time on mechanized coreblowers. This requirement of temperature control indicates that some means of heating and cooling the raw sand as it is conveyed from storage to mixing is imperative.
3) Perhaps the most important physical property to control is sand grain distribution or fines fluctuation. In this discussion, fines are classified as particles 200 mesh and smaller. It is the surface area of these particles which must be considered. Since each particle requires a film coating of binder, regardless of size. For example, a 100-g sample or an AFS gfn 48 sand with 0.4% fines on the 200, 270 mesh screen and pan has a calculated surface area of 11,970 cm^2 and requires 1.5% water to film coat. A similar sample of an AFS

gfn 66 sand with 5.9% fines has a calculated surface area of 17.511 cm^2 and requires 2.2 water to film coat — almost a 32% increase in surface area and consequently binder requirements. A recent screen analysis study showed that over a period of 10 days, sand purchased as an AFS gfn 57 sand and handled without regard for segregation varied at the mixer from an AFS gfn 54 to an AFS gfn 62 and the surface area varied from 10,500 cm^2 to 16,500 cm^2.

In addition to being in the newly purchased sand, fines are also caused by attrition, particularly with pneumatic loading and/or improper dust collection and are concentrated by segregation in the bulk handling process. Since all sand will tend to segregate when conveyed into storage, it is important to design a storage bin or silo which will reblend the sand as it is fed from storage to the mixer.

Standard storage silos and bins can be used for raw sand storage. A wide range of relatively inexpensive metal silos are available with internal pneumatic fill pipes and vent filters. A recent silo installation used an inverted cone to reblend the sand just prior to discharge. A vibrating bin activator was used to feed the sand out of the silo. A vibrating tube conveyor feeds the sand into a spiral elevator, which in addition to elevating the sand, has the ability to heat or cool the sand via a built-in fluid heat exchanger. The sand is fed to a weigh hopper above the mixer for precise batch control. A flat bottom silo or bin with strategically located discharge pipes and a rotary gate would work equally well.

The two-part resin binder system is generally used in approximately equal parts and both parts are available as liquids in either 55-gal drums or in bulk. Part 1 is a phenolic resin in liquid solvent exclusive of water and has a low viscosity index. Outside storage is not desirable. This liquid binder should be stored at temperatures in the 60-90F (16-32C) range. It can be stored colder but becomes too viscous to pump and must be heated first.

Part 2 is a polyisocyanate in liquid solvent and is very susceptible to reaction with moisture. In bulk storage, a blanket of nitrogen or desiccated air should be used. Accurate and repetitive binder measurement into each batch is necessary to control quality and cost.

There are two catalyst systems in common use in isocyanate binder systems: triethylamine (TEA) and dimethylethlamine (DMEA). Both these amines are normally supplied as liquids in 55-gal drums. In the case of TEA, air is generally used as the carrier. It is mixed in a generator at the point of use at each machine. The mixture is flammable and the fumes are noxious and toxic and should be neutralized or incinerated. Most materials are handled in pipes and pressure vessels under pressure.

The DMEA, being more inflammable and reactive than TEA, is carried in CO_2 gas at 70F (21C) (nitrogen can also be used). This gas can be mixed in a remote area and piped to the point of use. However, the mixer house becomes a hazardous area, similar to an acetylene generating house, and must be constructed accordingly. The DMEA can be purchased as a gas in steel cylinders premixed with CO_2. Generally 12% DMEA and 88% CO_2.

In a DMEA gas generating process the CO_2 is required in substantial quantities and generally purchased and stored in bulk at low temperatures (-10F (-23C)) and high pressure (300 psig). An advantage of DMEA carried by CO_2 is that it can be piped and manifolded to a number of machines. At the point of use, each machine will have a surge tank or accumulator and pressure regulator.

Dry compressed air at controlled temperatures is used to cure and purge the cores. Even with dry plant air, it is desirable to use a desiccant or refrigerant air dryer at each machine. DMEA is heavier than air and any residual fumes emitted from the core after removal from the corebox can be collected with a downdraft hood.

Fumes from the gassing process should be treated prior to release to the atmosphere. Several methods are currently available. One uses phosphoric acid, another uses sulfuric acid. Acid is available in carboys in 5-gal to 55-gal containers and in bulk. Ninety-eight pounds of sulfuric acid in a packed tower will treat 146 lb of amine gas.

Another raw material currently used in isocyanate core process is high-purity iron oxide, which is purchased as a powder in 100-lb bags and added at the mixer in quantities of about 2% by weight. It is used to reduce veining and hot tears.

Mixing and Distribution of Sand

Selection of mixing equipment for the cold box process is often predicated on individual choice and preference. All kinds of equipment imaginable have been and are being used successfully by one foundry or another. These units range from batch types such as wheel design mullors, blade mixers and even cement mixers to continuous units including recirculating types, screw designs and zero retention, high-intensity models.

The present research has indicated that all types do a good or at least an adequare job of preparing the sand mixture. The purpose of the equipment, of course, is to blend the sand and binders to coat each grain with a viscous material which is suspended in a solvent. The viscous material consists of two compounds — phenolic resin and isocyanate.

Theroetically, a ratio of 50% of phenolic resin and 50% isocyanate with a total binder level of 1.2% of the sand weight will produce an admixture with good physical properties, good economy and bench life of about 1.5 hr. In practice, however, ratios of 58 to 42, with either constituent in the majority, and binder levels of 1.2-1.75% are not unusual and depend on empirical testing procedures and methods used by individual foundries.

Although most types of mixers will adequately blend the mix, certain types have demonstrated superiority with respect to end results, efficiencies and practicality. For example, batch units seem to offer a more even distribution of binders, better coating characteristics and more control of attrition, thus providing maximum development of the physical properties. The result is that batch mixers can usually be expected to yield consistently good core characteristics with reduced resin requirement.

Continuous mixers, on the other hand, generally assure a closer relationship between sand prepared and actual sand requirements of the core blowers, thus reducing critical bench life considerations. Continuous mixers, although somewhat improved, have often demonstrated an inability to evenly blend sand and binders.

Condition of the sand is of utmost importance and should be free of such impurities as clay, moisture and oily materials. Clays in the sand will absorb the less viscous liquid materials including the isocyanates and catalysts, thus affecting physical characteristics of the mix. Adjustments in the quantity of binders used are possible but must be recognized and compensated made. Presence of clay in the sand will result in either weakened sand or increased use of expensive materials.

Moisture in the sand will also adversely affect the strength of the mixture. As is illustrated in Fig. 1, transverse strength is dramatically reduced as water content is increased. Moisture content should be held to less than 0.1%. Oily materials as may

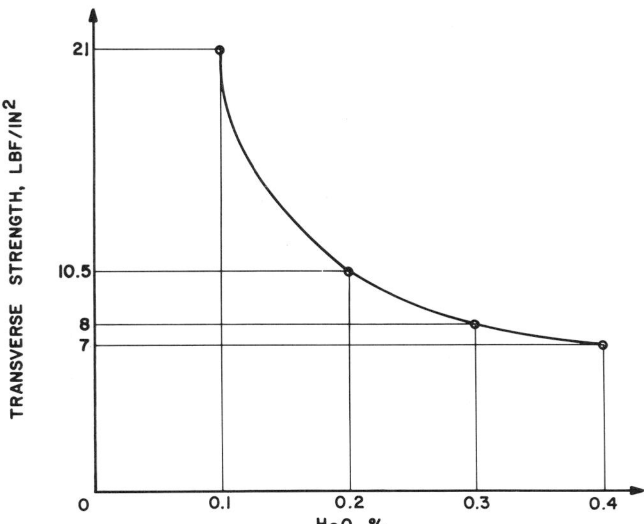

Fig. 1. Sand containing 1.3% binder (testpiece).

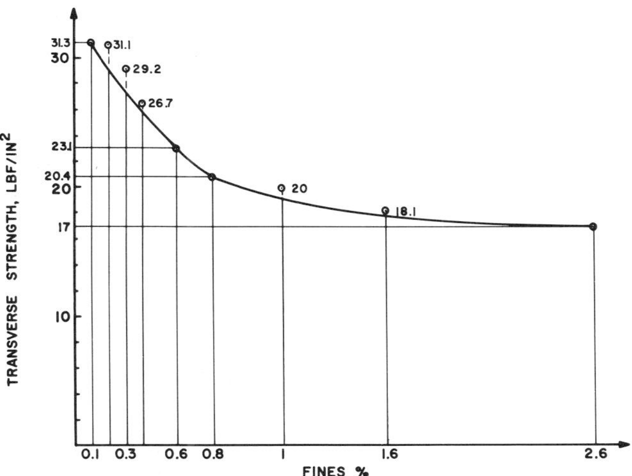

Fig. 2. Resins 1.2% (0.6 + 0.6); blowing pressure: 3 bar.

be introduced through flame drying processes are also detrimental to the sand since they also have an affinity to binders and will cause decreased properties and increased costs. Fines should also be closely controlled since they will increase the surface area to be coated. Impact of excessive fines can readily be seen in Fig. 2.

Since the most common impurity found in sand is fine sand particles, it is important to consider antisegregation devices on storage bins and hoppers to maintain a constant surface area to be coated. These devices will prevent fines from reaching the mixers at irregular rates which will have a significant and unexpected effect on the sand properties.

In addition to the above factors affecting binder usage, Fig. 3 illustrates the effect of configuration of the sand grains. Angular or irregular shaped grains, as can be noted, require more binder than do rounded sand grains. Where sand reclamation is used, angular grains do tend to become rounded after several coatings, the original binder remaining on the grain in the form of a fused-on deposit.

Regardless of the type of mixing equipment selected, several comments are appropriate.

Mechanical parts of the equipment should be designed so as to prevent excessive scrubbing motion against the sand itself which could result in removal of any coating. Movement of sand during the cycle must not damage the coating by sand grains rubbing against each other under pressure. Attrition of the mixture is extremely undesirable since it hinders formation of binder on sand grain as well as causing overheating which tends to decrease bench life and tensile strength of core sand. For batch mixing procedures, maximum blending duration of 3 min is recommended for vertical wheel machines or 45-60 sec for a horizontal wheel machine, as is a relatively low rpm of the mixer — approximately 60 rpm — depending of course on the unit selected.

When selecting the coating machine, it is important to remember that machines are usually designed with specific capacities to maintain good mechanical performance relative to a given amount of sand. This is to assure that no granular damage or undesired heating will occur during the coating operation. It is obvious that any coating machine, working with more than the designed quantity of sand, will develop excessive

heat as sand grains rub against each other. Should the unit not be charged with the designed minimum amount of sand, only partial coating and undesirable aerating will take place. Unnecessary aeration of the sand should be avoided to minimize evaporation of solvents. Excessive aeration will reduce bench life and tensile strength characteristics of the sand as well as increasing binder requirements.

Sand must be fed into the mixing chamber or bowl at a constant rate and temperature — preferably low enough to prevent evaporation of binder additions.

Mixers of all types should be enclosed to prevent solvent vapor loss as well as objectionable odors. Vapor losses result in the same detrimental aspects as evaporation.

Succession of the addition of the binders seems to be somewhat controversial. Some users indicate a firm belief that it is necessary to first add resin to the sand and then isocyanate. Still others insist the two binders can be introduced simultaneously with no deterioration of sand properties or increased binder requirements.

Additive pumps for both resin and isocyanate must be of a variable type equipped with indicators for easy adjustment and recording of volume within very fine tolerance. If continuous operating pumps are used, the metering needle valve located subsequent to the solenoid valve should be equipped with an indicator dial for recording proper setting relative to the amount of sand being prepared. It is important to insure a constant feed rate of the additives and that constant temperatures are maintained, especially for the resin material used. These containers should be stored indoors and the material transferred to tanks with controlled heating elements when required.

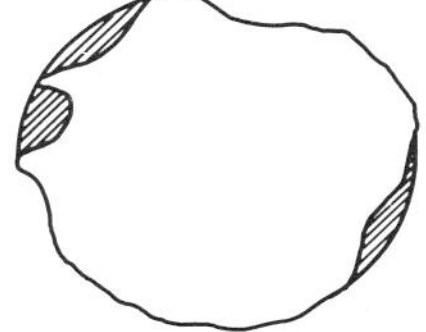

Fig. 3. Coated sand grains.

Fig. 4.

To achieve the best possible sand preparation, all component parts of the coating machine and associated equipment must be maintained and kept clean at all times. Any buildup of hardened material within the unit should be removed as it occurs. Pumps and lines should be flushed regularly to insure constant and accurate feed rates. Self-cleaning cycles can be designed into the mixer hardware to insure that no prepared sand remains in the mixer at conclusion of the cycle.

Distribution of prepared sand between the mixing equipment and the core blowers is dependent on type of mixing equipment selected. If mixing equipment is designed to be an integral part of the coremaking equipment, the distribution system is negligible. If the mixing center is a considerable distance away, the problem becomes more critical.

Fig. 5.

Basically, the same dangers exist in distribution as in mixing. In essence, caution must be exercised to avoid abrasion, evaporation and delays, all of which affect eventual bench life and physical characteristics of the sand. Several system design considerations are worth mentioning:

1) Reduce distance between the mixing center and the core blowers as much as possible.
2) Avoid unnecessary transfers of prepared sand between belt conveyors, through chutes and other components to minimize aeration, evaporation and abrasion.
3) Avoid thin layers of prepared sand on conveyors.
4) Use pneumatic conveying systems cautiously since they can be a source of abrasion and aeration.

Fig. 6.

Fig. 7.

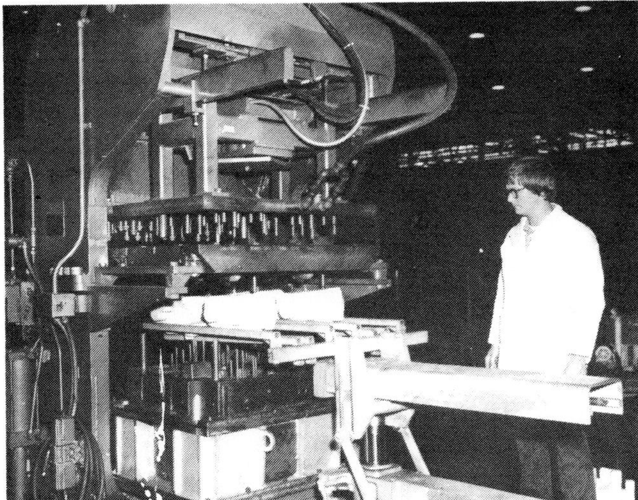

Fig. 8.

5) Control mixing of excessive amount of sand which may have to be delayed along the way because of coreroom stoppages.

Firm recommendations relative to mixer type or systems cannot be offered here since each seems to feature certain advantages or disadvantages which depend on the particular circumstances of the operation. Specific data for acceptable mixing time, temperature and binder content is simply not available at this time, at least not with assurance that such data would be universally meaningful.

Core Machines

The basic machine blows and gasses a core in a box. It can, in the complete units, mix sand, blow and gas the core to be ready to set in the prepared mold. This core can be unloaded by a special sevice, belt conveyor or manually. In some machines the box moves usually automatically from the blower, which fills it, to the gassing unit that cures it.

It is desirable to use sand as dry as possible, because moisture affects physical properties of the finished core. Therefore the machine used must have dry and oil-free compressed air. It also requires only a standard maintenance program which includes preventive activities.

This presentation will show several examples of the basic types of machines, describing specifications and special features.

Pivoting Platen Machines

Figures 4 and 5 show two types of horizontally split pivoting platen machines whereby the drag tooling is rolled 180° for core ejection. The two units are very similar in construction and basically operate the same. The only difference is that the second has two separate sand reservoirs and blow chambers. Both units have built-in rollover devices that deposit cured cores in a runout belt from which they are gathered by the operator. They also can be furnished with a three-stage sand mixer located above the machine. This makes it possible to mix sand at rates required by the core design and machine. Maximum core box size is 48 in. in length, 36 in. wide and a combined height of 23 in. with a blown-sand capacity of 150 lb.

Figures 6 and 7 show 90° pivoting platen machines which use vertically-split boxes. Both machines have pivoting platens moving 90° to a take-away belt conveyor. The conveyor in Fig. 6 is stationary but adjustable. In Fig. 7 the conveyor moves up to the movable half of the corebox which has rotated 90°. This

Fig. 9.

allows the core to be ejected onto the conveyor. The smaller unit has a maximum blown-sand capacity of 30 lb and the corebox size is 15 in. high x 18 in wide and adjustable to 6/6 in. The larger unit has a maximum blown-sand capacity of 130 lb and the box size is 29-1/2 in. wide x 32 in. high and an adjustable depth to 12 in. stationary side and 14 in. on the movable side.

Horizontally Split Parting Machines with Top Blow

Figure 8 shows a horizontally split tooling machine with a moving sand magazine or blow head. This is the first example of the above type. It has a blown-sand capacity of 125 lb. Maximum outside dimensions of the corebox are 28 x 38 in. with a draw of 25 in. The standard unit has 50 in. of daylight space between the table and sand magazine flange. These machines can be equipped with a skip hoist with a capacity of 450 lb as shown in Fig. 9, also a core pickoff device which uses a simple in-up-back and down movement that enables the operator to unload the finished cores while the machine continues in its completely automatic operation, which is 20 sec or less. A programmable controller can also be furnished, which means less maintenance and easier troubleshooting. If a new sand-storage hopper is installed above the machine, a continuous mixer can be fitted to control the prepared sand to the core blower as shown in Fig. 10.

Figure 11 shows a double-station machine and uses horizontally split parting boxes with a top blow. Construction of this machine is tubular steel with four corner supports with a blown-sand capacity of 125 lb per head. Outside dimensions of the coreboxes are 30 x 40 in. with daylight space of 50 in. and a draw of 25 in. It too can use the core pickoff device while the pins are in the eject position for removal. Cycle rate will be from 20 to 25 sec. This machine provides an extremely compact package, less floor space, independently operated stations, floor-level installation, fast corebox change and telltale light monitoring system.

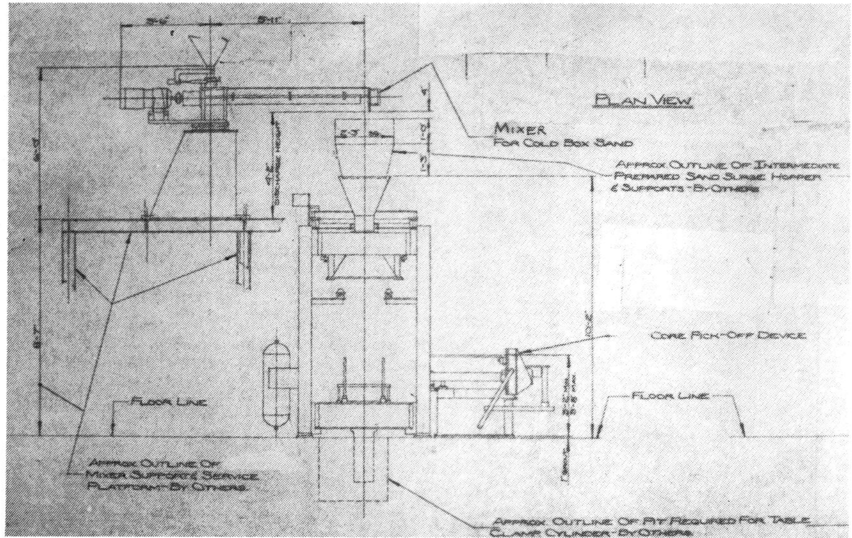

Fig. 10.

Figure 12 shows a horizontally split parting corebox with a top blow, having a stationary blow chamber. Construction of this machine is tubular steel and it is designed to reduce motions to a minimum. The corebox is moved only in a vertical plane, thereby eliminating possibility of sand movement in the corebox prior to gassing. A shuttle-type gassing head is incorporated to reduce the required catalyst. The gassing head is shuttled into the space between the blow plate and corebox after the core is blown and lowered the required distance. At the complete retraction of the clamping cylinder, the core is ejected. The automatic take-away device removes the core and the machine automatically starts the next cycle without the operator's assistance. Maximum corebox outside dimensions are 40 x 40 x 13 in. Total daylight from the top of the table to the blown-sand magazine flange is 50 in. Capacity of blown sand is 200 lb and the magazine is 470 lb. Another variation of this machine type is a coreblower which incorporates a fixed blow head and a double-station unit with one blow head. This machine is 100% air-operated including the lift table. Nominal tooling size is 30 x 40 in. and blown-sand capacity is about 125 lb. The machine shuttles the tooling back and forth in such a manner that one set of tooling is in the blow station while the second is in the gas, purge and strip station. After both cycles are complete, the tooling shifts back and the set that was just blown now goes through the gas, purge and strip cycle and the other enters the blow station.

Horizontally Split Parting Machines with Side Blow

Figure 13 shows a recently developed cold-box machine which was designed to accept shell sand coreboxes. Normally shell boxes can be used with a minimum rework of the box in the area of venting and blowing openings.

Catalytic gas is introduced through the blow opening by an integral gassing manifold with the exhaust-gas discharge through the parting line and cope and drag vents. This is called the open-gassing system. Gasses exhausted must be collected by a positive displacement exhaust-disposal system. Unless a

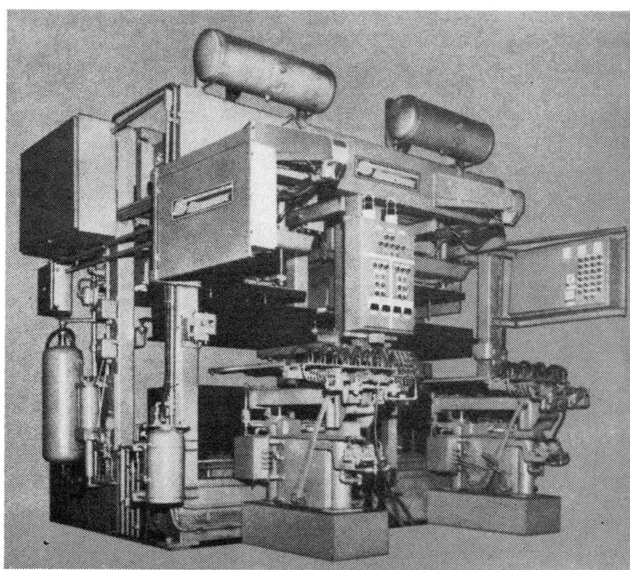

Fig. 11.

Fig. 12.

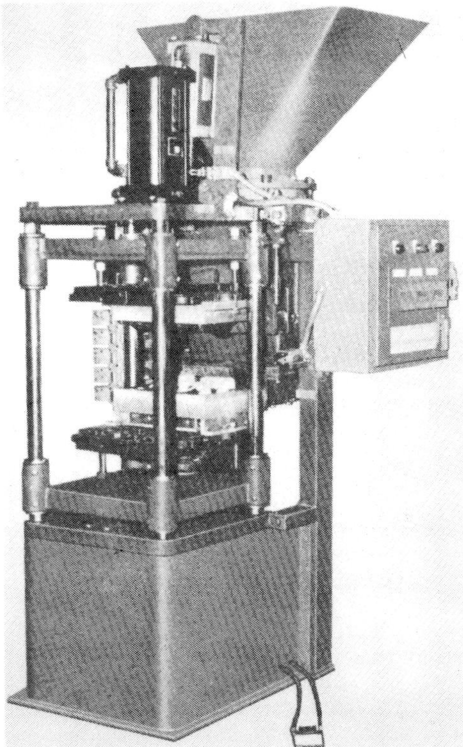

Fig. 13.

Fig. 14.

The lefthand arm positions the sand magazine and head directly over the corebox to begin the automatic cycle. Immediately following the blow operation the righthand arm positions the gassing head directly over the corebox for the cure and purge portions of the cycle.

corebox has an extremely inaccurate seal between cope and drag, sealing is normally not required. This machine provides a practical solution to produce cores for the cold-box process without major corebox modification. Maximum corebox size is 18 x 20 x 3 in. and has a blown-sand capacity of 25 lb.

Vertically Split Parting Machine with Top Blow

Figure 14 shows an automatic core machine and fits the above description. This machine automatically produces cold-box cores using a unique dual-arm design to provide the necessary phases of the cold-box cycle with minimum mechanical motion.

Fig. 15.

Fig. 16.

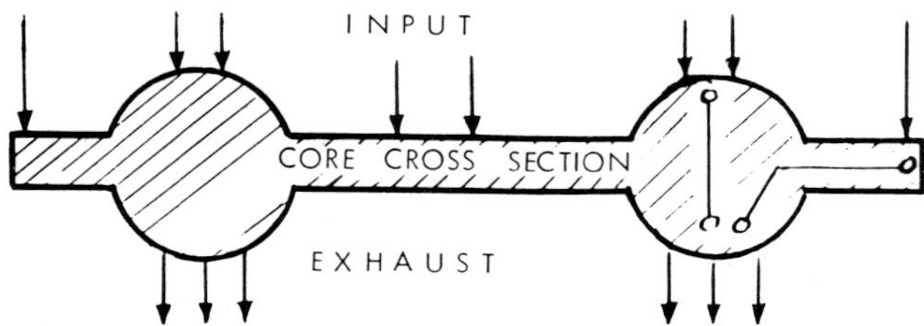

Fig. 17. Balanced gas flow.

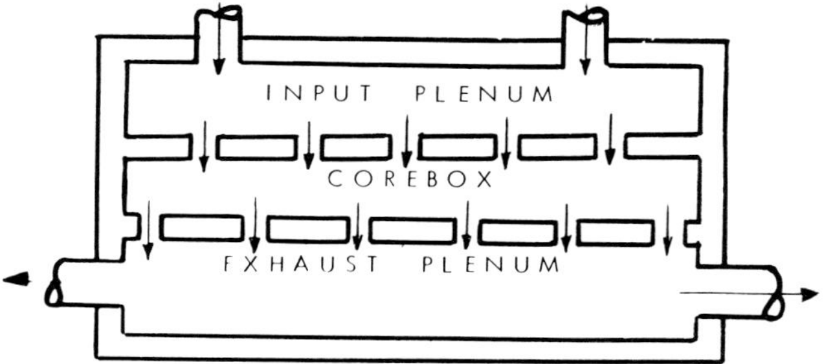

Fig. 18. Simple gas flow.

The center-mounted mandrel assembly incorporates a vented plate which seals against the lower face of the corebox. Gas used to cure the core is collected in a chamber below the vents and is exhausted through a convenient hose connection. Height of the vent plate is widely adjustable for use with a variety of coreboxes. Use of the gassing head and vent plate eliminates or minimizes the need for venting or manifolding of individual corebox cavities. The dual horizontal-clamp cylinders move both halves of the corebox inward and outward to minimize corebox opening and closing time and allow stationary mounting of the mandrel attachment. Maximum core capacity is 30 lb with a corebox size of 17-3/4 x 22 x 12 in. The sandhopper capacity is 200 lb.

Miscellaneous Bench-Type of Machines

Figure 15 shows a small bench-type core blower which is ideal for small cores using the cold-box process. This unit handles a wide range of vertically split and open-face coreboxes. After the core is blown, curing gas is injected at the top of the corebox, passes through the sand, is collected in a chamber beneath the corebox and is exhausted through a scrubbing unit. Blown sand capacity is 4-1/2 lb. It has an adjustable daylight from 0 to 10 in. Side clamps are also adjustable from 0 to 5 in. or 3 to 8 in.

Figure 16 shows a machine which is well suited for tensile specimen cores, using the cold-box process. Tensile-specimen cores are produced efficiently and accurately with this equipment allowing important time savings for laboratory technicians and yielding more useful test results. Precise control of blowing and curing provides important consistency between successive tensile tests. The three-cavity corebox furnished with the machine is completely rigged and factory tested to assure problem-free operation. All necessary vents and seals are provided.

This is only a sample of machines for the cold-box process. Others are also available in these categories.

Cold-Box Tooling Design Considerations

The cold-box process offers many advantages in tooling over certain other processes:

1) Economy in construction as compared to the hot-box process.
2) Flexibility of material selection for overall economy. Aluminum can be used for short-run jobs. Epoxy and urethane tooling has been produced to yield high life with relative economy and lead time requirements as a side benefit. Conventional cast iron tooling with or without special coatings or plating is best suited for long-run jobs.
3) Speed of changing coreboxes is a real productivity bonus as compared to the long times required to change hot-box or shell tooling. In many cases a 2-hr tooling change can be reduced to about 30-40 min.
4) Latitude in core design is greatly enhanced by cold-box due to the fact that loose pieces, vent rods, lighteners and pullbacks are easily designed for cold operation and since the core is fully cured in the box can be retracted without damaging the core. Further improvement is recognized by virtue of the fact that blow tube projections can be restruck by tapper pins in the gas manifold, thus eliminating the need for mudding and patching on cast surfaces.

Certain basic considerations must be part of a cold-box tooling program:

It may be possible to convert existing tooling from such processes as oil sand, hot-box, shell and CO_2 to cold-box with a minimum of rework. Dimensional variations between the two must be checked, allowing for variation of cavity dimensions at operating temperatures as opposed to being cold. Of course, CO_2 should be very compatible because it is a cold tooling process as well.

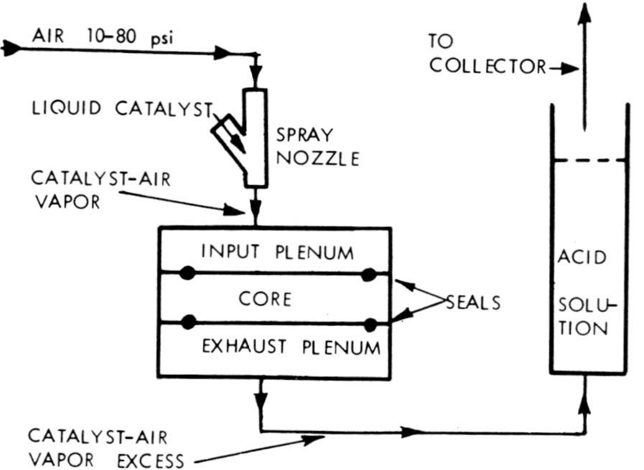

Fig. 19. Catalyst injection and disposal.

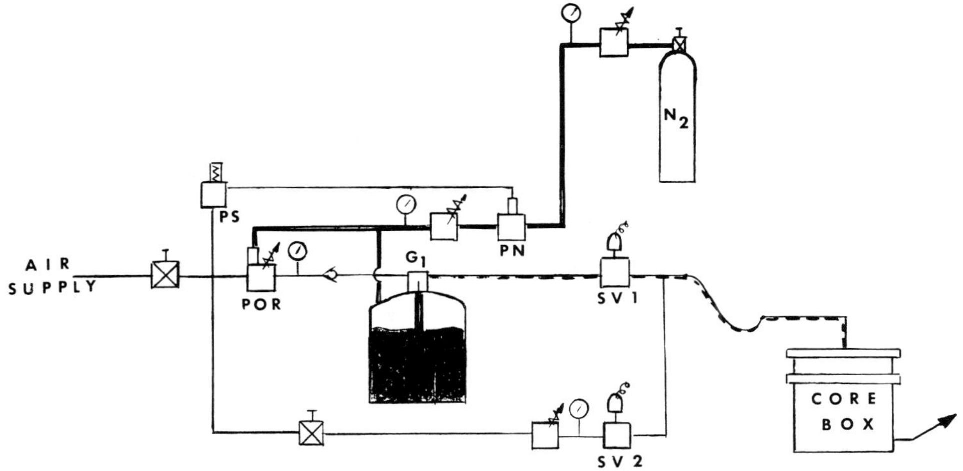

Fig. 20.

The blow, gas and vent area is extremely important to the success of cold-box tooling, even more so than with other processes. In cold-box, in addition to the basic problems of blowing a uniformly dense core, a second problem is uniform permeation of the corebox cavity by the catalytic gas. The task is to blow, gas and purge in the shortest time possible without wasting any expensive catalytic gas. Basic considerations are:

- Allow 0.20-0.35 in.^{2} blow tube area per pound of core (more for rangy lacy cores, less for chunky ones).
- Balance the above area with drag exhaust venting.
- Provide total gassing manifold input area to be 20-30% greater than the drag exhaust venting. This will vary from core to core.
- Stagger drag exhaust vents so as to prevent short-circuiting and promote lateral movement of catalytic gas that will result in uniform cure rate.
- Balance the placement of vents according to geometrical section size, volume, depth and path within the core cavity itself.
- Remember that gas and purge air will follow the path of least resistance, thus both staggering and proportional placement of drag exhaust vents act together to yield the best combination for uniform permeation of the core cavity by catalytic gasses in the shortest possible time with minimum waste of catalyst.
- While generous input and exhaust vents do promote speedy curing of cores, too much venting can result in excessive vent cleaning time and the resultant loss in productivity. Schematic examples of the above are shown in Fig. 17 and 18.

The gas and purge mixtures must be contained within the gassing manifold, the tooling itself and conveyed to whatever scrubber that is used (Fig. 19).

- The underside of the drag box must develop a plenum which is sealed and is large enough to pass the gasses to the outlet ports at minumum back pressure.
- Joints between the gassing manifold and the tooling parting lines can be sealed with either inflatable type or compression seals. Care must be taken when sealing relatively light cope boxes or covers as they have a tendency to lift and separate between the blow and reclamp cycle when the combination of the cope weight and/or swelling of seals is not properly balanced. In some cases it is necessary to clamp the boxes together to defeat this separation and ensure dimensional stability of the finished core.

Last but not least, consideration must be given to compatability of the corebox and its rigging with the process materials such as the resins, parting agents, cleaners and amines.

- The tooling rigging should be fitted with black iron pipe for rigid piping and either teflon TFE or type 2 nylon for flexible connections. If cylinders or valves are mounted to the tooling, they should be so equipped as well.
- Remember that certain corebox cleaning agents cannot be used with corebox materials such as urethane.
- If doubt exists as to the compatability of seal materials, etc with cleaners, amines or resins, check them by soaking together and measuring the time rate of softening, swelling or deterioration. This procedure can assist in developing a program frequency for changing seals, etc before they are troublesome to process and production.

Catalyst Systems

The present discussion of the cold-box process, has included review of methods of sand conveying, mixing equipment for sand and resin and machines to make the core. After sand is blown into the corebox, the core is cured by introduction of a catalyst which causes the binders to rapidly polymerize or harden. Methods and equipment available to the industry to mix, meter and transport the catalyst and purging medium safely and economically are the subject of this part of the paper.

There are two commercially available catalyst materials used today in the industry. These are triethylamine (TEA) and dimethylethylamine (DMEA). TEA is more widely used, as the cost has been lower and the supply is readily available. Until recently, DMEA was all imported into the US resulting in higher cost and very limited usage by foundries. Theoretically, DMEA is a more active catalyst but generalized statements that one material is faster or more economical to use are difficult to make. Due to the more common usage of TEA, this material will be considered in greater detail.

Use of the catalyst TEA to cure the core is done by converting the liquid to an aerosol or gas and then mixing it with a carrier such as air or an inert gas in the correct ratio and introduced into the corebox containing the blown core. This gassing process is stopped after sufficient catalyst has been introduced into the manifold and core.

This normally only takes several seconds, followed by a cure or purge cycle using air to finish curing the amine and sweep the excess vapor out of the core. It is important to maintain the minimum ratio of catalyst to sand and have only the proper

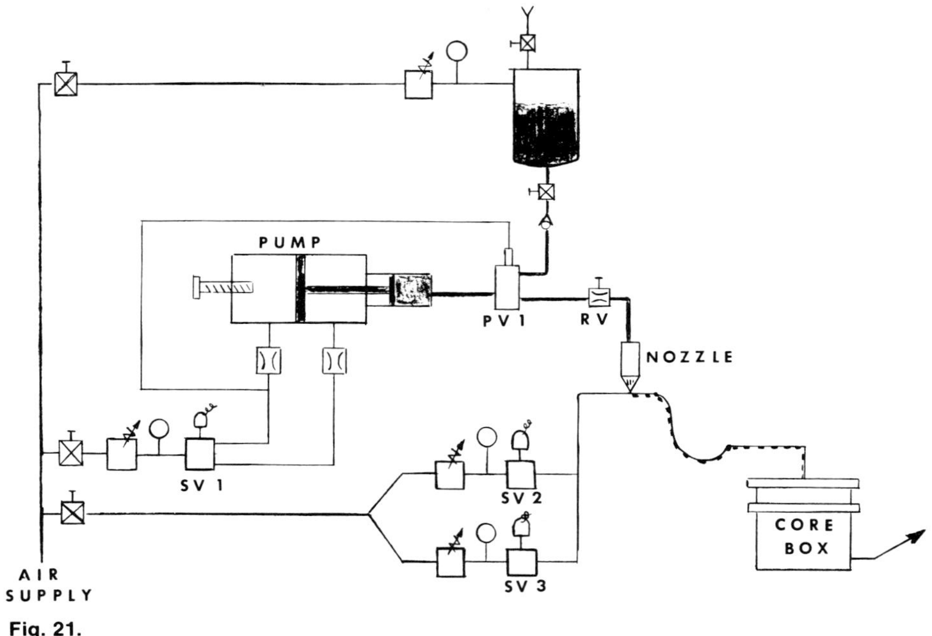

Fig. 21.

amount of amine to catalyze the resin. Too little catalyst will cause undercuring and scrap cores or excess catalyst will result in waste and added cost.

A very important consideration in the gassing and purging operation is the requirement to have no moisture present in the air being used in these processes. It is desirable to have compressed air with a dew point at 40F (4C) or less. Moisture introduced into the core by the air, either in blowing, gassing or purging will result in scrap cores. This demand for dry air may require the addition of refrigerant- or dessicant-type driers on the compressed air line serving the cold cure machines.

Some equipment and methods used for generating the catalyst in gaseous form and dispersing it through the core will now be considered. The simplest method, and possibly one of the earliest used, is for the foundryman to purchase premixed catalyst and inert gas mixture in a commercial gas cylinder. This is now available in TEA and DMEA/CO_2 mixtures. These can be formulated in various concentrations to meet specific requirements. Use of this method is suited for a few small low-production machines or a test specimen blower in a laboratory, as it is more expensive than using the liquid catalyst.

Early in the cold-box equipment development stages, systems were developed to produce a mixture of catalyst-carrier gas at each core machine by using liquid catalyst. These systems operate on the injector principle — forcing a given quantity of the liquid catalyst into a moving stream of carrier gas. One system can be shown as in Fig. 20. After the core is blown, SV-1 is opened allowing air to flow to the corebox. The liquid catalyst, under pressure from the N_2 source, is forced thru the gas injector G-1 and is entrained in the airstream as an aerosal. This unit functions similar to a microfog lubricator — liquid only flows when air movement is present through the unit. When the cure cycle is complete, SV-1 closes and SV-2 opens, allowing air only to purge the core. System pressure is controlled by a pilot-operated regulator (POR) which is controlled and actuated by the nitrogen pressure. Loss of nitrogen pressure will

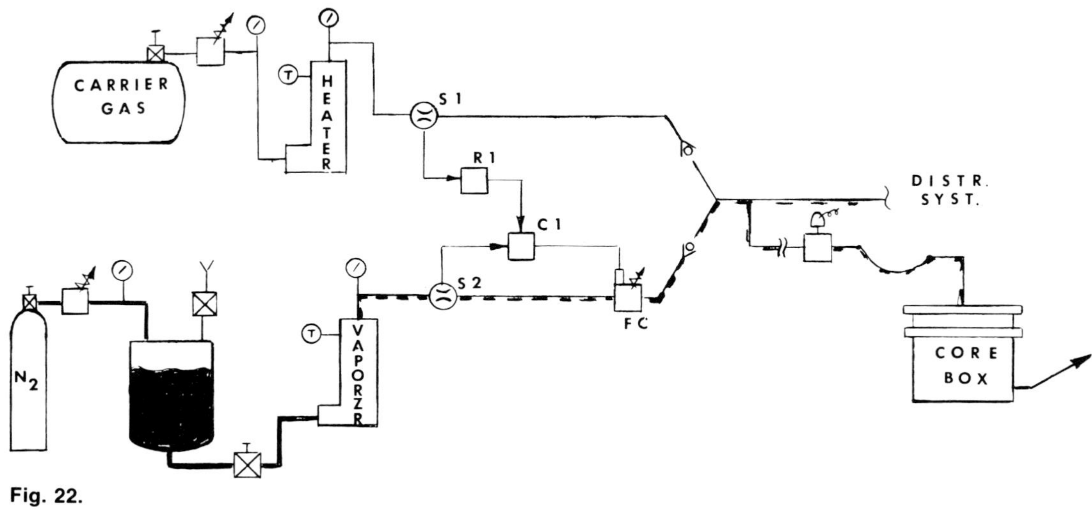

Fig. 22.

prevent the system from operating. Loss of air pressure will trip the pressure switch (PS) and close the nitrogen pilot valve (PN).

A second method of achieving a catalyst-air mixture consists of a pump, nozzle and valving shown in Fig. 21. At start of the cure cycle, SV-1 opens, which allows airflow to the blind end of the pump and to the pilot valve (PV-1) which opens. Pump stroke forces the liquid catalyst through a regulating valve (RV) and through a nozzle which atomizes the liquid. At the same time that SV-1 opens, SV-2 also opens, allowing air to flow and carry catalyst to the corebox. At end of the cure time both SV-1 and SV-2 close and SV-3 opens for the duration of the purge cycle. As SV-1 closes, it shifts PV-1 and the pump stroke, allowing liquid catalyst to flow from the storage tank to the pump chamber. The pump has an adjustable stroke, thus allowing adjustment in the amount of catalyst injected for different sized cores.

As the use of cold-cure process expands and multiple machines are operating in a coreroom, the need for a central bulk generator system producing a true gaseous mixture becomes greater. By assuring the catalyst enters the core as a gas and not as liquid droplets, faster curing takes place and less material is consumed. There are currently two commercially available methods in use to generate a catalyst-carrier gas mixture of constant mass ratio or distribution. These operate on the vaporization principle — gasifying the liquid catalyst and combining with a carrier gas.

The first system operates by converting liquid catalyst to a gaseous state by passing it through a heated chamber. It is then metered through a proportioning control valve and mixed with the carrier gas. During mixing, catalyst gas is then expanded to its partial pressure, effectively superheating the gas and lowering the temperature at which the liquid phase catalyst dropout occurs. Metering of the catalyst gas is continuous as a function of carrier gas flow such that mass concentrations are uniform throughout the process line system.

This proportioning system is shown schematically in Fig. 22. The basic system consists of a liquid catalyst storage tank, inert gas pressurizing system, catalyst vaporizer, catalyst flow control and proportioning valve system and carrier gas pressure regulator and heater. Functionally, as flow of the gas occurs in the process line, due to core machine operation, the carrier gas flow sensor (S-1) detects flow rate and supplies an input signal to a ratio relay or pneumatic amplifier (R-1). Output signals of the ratio relay (R-1) and the catalyst flow sensor (S-2) are input signals to a pressure controller (C-1). The controller output then establishes the setting on the catalyst flow control valve (FC-1) which controls the flow of catalyst gas into the process line. Under normal conditions, the system will maintain a constant ratio of catalyst to gas under all flow conditions. By adjusting controls on the ratio relay (R-1), variation in the catalyst to gas ratio can be achieved. As mentioned previously, the catalyst is vaporized by passing through a heater prior to mixing with the carrier gas. An inert gas, usually nitrogen, is used to provide the pressurizing method to the liquid catalyst and act as a protective atmosphere in the catalyst storage tank.

A second basic method of achieving a controlled gaseous mixture of catalyst and carrier gas involves bubbling the inert carrier gas through a bath of liquid catalyst. It can be shown in Fig. 23 that for any given condition of temperature and pressure, the carrier gas will contain or be saturated with a specific

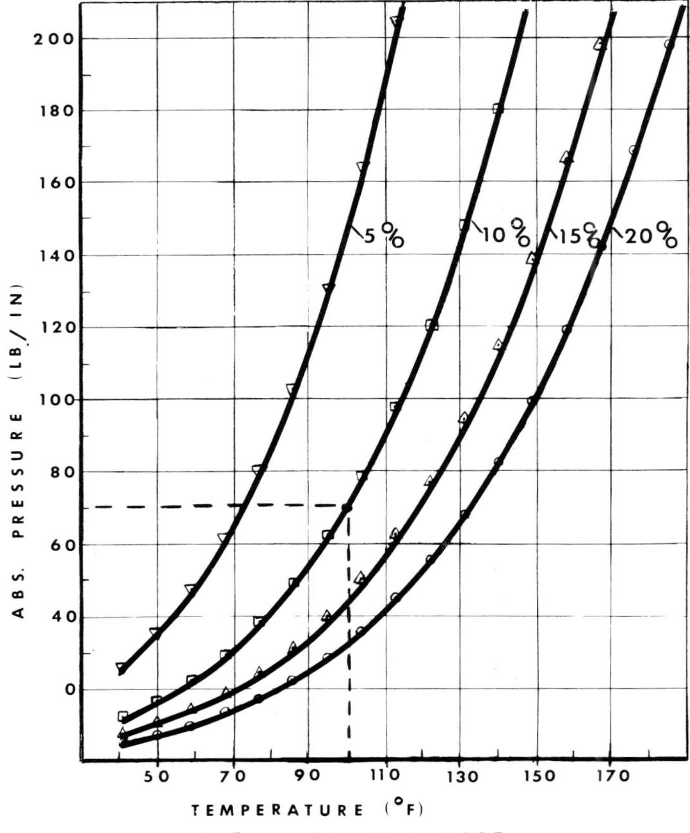

Fig. 23. Pressure vs temperature: 5,10,15,20% TEA-N₂ mixture.

amount of gaseous catalyst. Thus by controlling, for example, the pressure at 70 psi-absolute and the temperature at 100F (38C), the mixture would contain 10% catalyst. Therefore a control system can be operated by heating a quantity of liquid catalyst in a pressure vessel and percolating the carrier gas through this heated liquid and controlling the pressure of the system to the desired level.

This system is shown schematically in Fig. 24. As material is used from the process line, carrier gas flows into the pressure vessel through a bubbler or sparger (B), which is submerged in heated liquid catalyst. As the small bubbles travel upward through the liquid, they become saturated. Prior to exit from the pressure vessel, the mixture passes through a screen or demisting section which eliminates any small droplets of liquid catalyst, allowing only vapor to enter the process line. Liquid catalyst is heated by passing thru a heat exchanger, heated by a circulating hot water system. Pressure is maintained by the pressure regulating valve (V-1) in the incoming carrier gas line. Liquid level is maintained by a liquid level system and catalyst is fed from the storage tank as required. It also is desirable to provide an inert gas blanket over the liquid catalyst in the storage tank.

Several important considerations in design and installation include avoiding the use of any brass valves or fittings in piping systems. The TEA will chemically attack brass but will not corrode steel. It is also imperative to use resiliant sealing seats in any valves or pumps that come in contact with the catalyst. These seats should be teflon or some other amine-resistant material.

A word should be mentioned on capacity requirements, as applies both to ability to deliver adequate catalyst in minimum time and sufficient air for curing and purging quickly. The generator must have the ability to deliver from 0.5 cc/sec for a 1-lb core up to 35 cc/sec for a 350-lb core. In regard to the required carrier gas, the range may be from 50 cfm for 3 sec up to 300 cfm for 10 sec. Since it is not practical to build generators with this wide a range, they are usually made in several capacity ranges. It is essential to rapid and efficient core production to select a unit that has sufficient size to cure the largest core that the core machine will blow. For the central bulk generation system design, an accumulator tank located at each machine is normally recommended. This provides the surge capacity needed for the cycle but allows distribution lines to be kept to a reasonable size. These tanks are usually heated and insulated to

prevent consideration of the liquid amine.

In summary, two catalysts — TEA and DMEA — are available today. Care must be exercised in handling these materials. For proper core curing and economical operation, only as much catalyst should be introduced into the core as is required to advance the resin. Several methods are available for individual machine operation and several designs are available for creating gaseous catalyst and carrier gas mixtures for use in large central systems. Only the user can decide which system is best for his plant and method of operation.

Environmental and Safety

No discussion of the cold box process would be complete without a review of safety and environmental considerations.

Since the introduction of this process, some foundrymen have been reluctant to adopt its use due to the hazardous nature of the materials. The cold-box process should create no safety or environmental problems if the charateristics of the materials are recognized and necessary precautions are taken in their use. These precautions include proper equipment layout, effective material storage and handling, good maintenance and the establishment of safety procedures. Effective planning in these areas will result in a safe working environment. Materials used in the process that require special attention include the binders, catalyst, corebox cleaners and acids that may be used in the treatment of exhaust gases.

The part 1 binder is a phenolic resin dissolved in a petroleum solvent while the part 2 binder is a polyisocyanate which is also carried in a petroleum solvent. Both binders are toxic and act as defatting agents which will irritate skin and eyes on contact. Both resins are also midly flammable with flash points of 182 and 230F (83 and 110C), respectively.

Catalyst used in the process may be triethylamine (TEA) or dimethylethylamine (DMEA). Each of these materials is highly flammable with flash points of 20 and -32F (-7 and -36C), respectively.

In addition to high volatility, these materials are strong alkalis which are very irritating if brought into contact with skin or eyes. Spillage on skin or eyes should be thoroughly rinsed immediately with water.

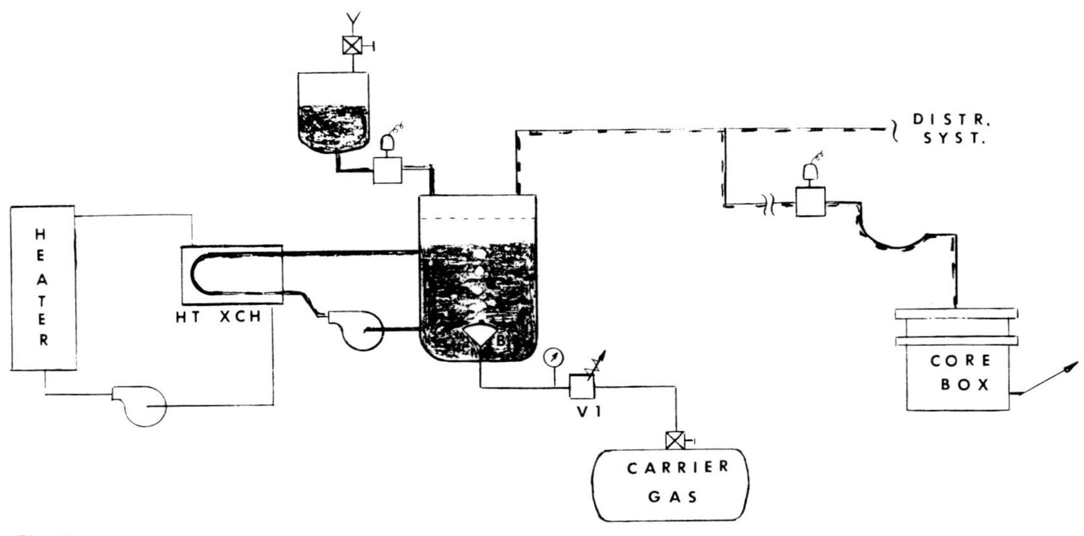

Fig. 24.

Vapor from these liquids can be an irritant to the respiratory system, and is also an explosive mixture when mixed with air at a concentration of 1.2-8% by volume. The normal mixture, if air is used as the gassing medium, will be about 0.2%.

Threshold limit value for TEA is 25 ppm, which compares to a maximum of 5 ppm allowed for phenol or formaldehyde. In addition to this higher allowable working level, the amine vapor will possess a detectable odor at 1 ppm, which becomes very obnoxious at a level of 6-8 ppm. This characteristic acts as a built-in safety system to warn workers before they can be exposed to harmful concentrations of the vapor.

The material used to clean coreboxes and other metal parts is a toxic material which generally requires the same type of care and safety precaution required of the binders.

Sulphuric, phosphoric or other acids that may be used in the treatment of exhaust gas will naturally require special handling, storage and procedures. These materials, however, are only ancillary to this process and will not be discussed further.

Special precautions should be taken by all employees who are exposed to handling of the binders and catalyst. Employees should wear splash-proof goggles and special gloves when handling or transporting the material. In addition, employees handling amine should wear a face shield and a plastic apron when transferring this material to either a transport container or into the generator tank. It is generally good practice to require one man to hold a fire extinguisher during the transfer operation. During transfer of liquid amine, each container must be electrically grounded to preclude an ignition spark and no smoking can be allowed.

This brief overview of the basic process materials indicates that care and planning are required to assure a safe operation. In general, cold-box coremaking should pose no problem to the safe operation of corerooms in the future.

Resin and catalysts are normally shipped in drums, although bulk shipment can be arranged for large users. These should be stored as close as possible to the process to eliminate extra handling.

Storage and handling of the binders and cleaner will require control and safety procedures as with other core materials that are mildly flammable and toxic, plus a spill-control procedure for the isocyanate binder. This material does not, however, require special engineering considerations. From an operational standpoint, it is desirable to store these materials at a temperature above 50F (10C) and in a manner that will not allow accidental contamination with water. The catalyst, on the contrary, will require special consideration. The high flammability of this material makes it necessary to design safeguards into the system.

Storage of liquid amine should be in an outside area or in a building that is isolated from the main plant. If in a building, sprinkling protection should be provided. In one installation, dispensing of amine into 5-gal transport containers is performed in an outside area. Both the drum and receiving container are electrically grounded. Transport containers are then manually carried to the generator station where the generator tank is filled. In some installations the amine is transported in an explosion-proof rubber-tired tank cart. Generator tanks are filled by either gravity or manual pumps. The generator tank must always be carefully depressurized before filling. The generator tank should be able to receive the full contents of the transport vessel.

Transfer of amine from one vessel to another should always be made through pipe or hose, such that the material is never exposed to atmosphere. In this case, hose maintenance is critical. In some operations, transport containers are stored in specially marked cabinets when not in use.

The generator area should be clearly marked as an area requiring special precaution, including the prohibition of smoking, welding or open flames. The area must contain an eyewash, a drench shower and a fire extinguisher. Specific safety procedures and action to be taken in the case of a spill should be prominently displayed.

The generator should be inspected frequently and provided with proper and careful maintenance. An open design of the generator is recommended. If enclosed in a cabinet, a slight negative pressure should be provided through connection with an exhaust system connected to the bottom of the cabinet. Special care should be taken in all piping installations to assure a leakproof system. All piping exposed to amine should be either seamless stainless steel or black iron. The same ventilation technique should be used in tightly enclosed pits that may be subject to collection of amine vapor. It is heavier than air and may settle in the pit if ventilation is not adequate.

Each generator should contain a shutoff valve at the lowest point in the liquid amine line in the generator. The valve seat should be resilient and of amine-resistant material such as teflon. This valve should be closed during periods when the generator is shut down to prevent accidental leakage of liquid amine into the gassing lines or corebox. This is particularly true in a pressurized system. Appropriate valves and electrical equipment should be provided with lockout capability for cleaning and maintenance purposes. Electrical connections to the generator station should be of explosion-proof design.

Added safety in generation of the gassing medium can be achieved through the use of CO_2 rather than air as the carrier. This is particularly advantageous when DMEA is used as the carrier gas. In this type of installation, the gas mixture is nonflammable. Use of CO_2 will increase catalyst cost 20-40% but may result in faster cure. One installation, currently in the planning stage, will use two generator tanks, each capable of receiving a full drum of amine. The generator will use CO_2 as the carrier and will be located externally to the main plant. Gas will be piped under pressure to coremaking operations. Advantages here are that liquid amine will not enter the plant and the gassing medium will be nonflammable.

Gas generated during the core-curing process must be treated before being released to the atmosphere. Gas contaminates include amines, phenol, aldehydes and isocyanates plus other materials. Amines and isocyanates are of major concern at this step in the process.

The gas treatment system from the corebox should be under negative pressure to assure free flow of exhaust gas into the treatment system. The main exhaust pipe should generally be three times as large as the gas input pipe. Current installations for treatment of this gas include pressure acid bubblers, incinerators and packed sulfuric acid towers.

A thorough recent evaluation of these systems indicates that at the present time the packed tower and a refined design of the acid bubbler are preferred. The packed tower has higher initial cost and would be recommended for larger, multiple-machine installations which require treatment of greater volumes of gas. The acid bubbler is lower in cost and is generally applicable to single-machine operations with low volume treatment requirements.

Incinerators are ineffective and use great amounts of energy. Incinerators, if used, should be isolated from the corebox by an effective flame arrestor.

Advantages of the packed tower are ease of maintenance, low acid usage, high capacity, minimum disposal of salts and high efficiency. In addition, treated exhaust can be discharged to the plant atmosphere, rather than outside, which would require makeup air.

Ventilation of the core machine area is an important consideration. One effective method of ventilation is to provide an exhaust air duct on one side of the core machine. Any leakage of gas during the coremaking operation will then be easily drawn into the exhaust. In addition, it is generally good practice to supply fresh air to the coremaker.

Amine gas is heavier than air and may tend to settle or disperse unless effective ventilation is provided. The obnoxious ammonia-like odor will be apparent if ventilation is not adequate. Normal ventilation practices for core storage areas should prove adequate for cold-box cores. This will be dependent on the fact that the cores are properly cured and purged.

Use of cold-box cores in molding operations will result in release of gases containing isocyanates. Tolerance for this material is very low which will require that all pouring, mold cooling and shakeout operations be properly exhausted. Gases from these processes must be carefully evaluated to assure that the working atmosphere is not adversely affected. Ventilation systems that are currently borderline will probably not be adequate to allow the use of cold-box cores.

In conclusion, the cold box process is here to stay. Most foundries are either using it or considering its use for the future. While there are certainly safety considerations, there are no unusual safety hazards that cannot be overcome by good layout of equipment and establishment of adequate and proper procedures.

Mineralogy of Foundry Sands and Its Effect on Performance and Properties

T. E. Garnar Jr., *Research Associate*
E. I. du Pont de Nemours & Co
Starke, Florida

ABSTRACT

Physical properties and performance of foundry sands are directly related to the minerals they contain. Many studies of foundry sands have been made but only a few used the mineralogist's tools and viewpoint. It is the purpose of this paper to relate foundry sand performance to mineralogy and geologic history. Ten silica sands, three olivine samples, four chromite sands, four zircon sands and a sample of aluminum silicate-zircon blended sand were used in this study.

Differential thermal analysis shows silica undergoes a phase inversion to an unstable high temperature form at 563C (1046F) accompanied by expansion of each grain. Chromite undergoes a phase change as it oxidizes beginning at 370C (698F) and peaking at 420C (788F). Olivine, zircon and aluminum silicate-zircon blended sands do not show significant thermal transitions to other phases.

Thermogravimetric analysis shows Michigan bank and dune sands lose significant amounts of volatiles when heated as does olivine. Chromite gains weight as it takes on oxygen when heated in an oxidizing atmosphere. There are indications that thermal expansion of chromite grains is the result of lattice expansion to accommodate the oxygen. Zircon and aluminum silicate-zircon blended sands show very little weight loss on heating.

Friability of olivine grains is probably related to the tectonic origin or perhaps the minerals cleavage. In either case, the lack of grain durability will adversely affect the use of olivine sand in a foundry using sand reclamation techniques.

Geologic origin and mineralogy of Michigan lake and bank sands, olivine and chromite result in basic pH and relatively high acid-demand values. Laboratory tests show that acid-demand values for Michigan lake and bank sands can be improved by removing surface coatings using a caustic scrub. Zircon and aluminum silicate-zircon blended sands are slightly acid and have good acid-demand values.

It appears doubtful that any of the industrial minerals now available or those which might become available in the future will have enough of the physical, chemical and thermal properties required of a superior specialty sand to be substituted successfully for zircon, zircon-aluminum silicate blended sands or chromite sands.

Introduction

Many studies of foundry sands have been made but only a few have used the mineralogist's-geologist's tools and viewpoints. It is the purpose of this paper to review current foundry sands and relate their mineralogy and geology to their performance in the foundry. Ten silica sands, three olivine sands, four chromite sands, two domestic zircon sands, two Australian zircon sands and a sample of domestically produced aluminum silicate-zircon blended sand were selected for use in this study.

Analytical Techniques

Methods used in this study included optical examination, differential thermal analysis, thermogravimetric analysis, x-ray diffraction and bulk density determination. These techniques are described below.

Petrographic Microscope

One of the mineralogist's most powerful tools is the petrographic microscope. It is different from a standard optical microscope in that it has a substage polarizer that restricts light to vibrate in only one plane, a condensing lens to converge light, a rotating stage, a polarizer in the microscope tube to observe interference figures in crystalline minerals and crosshairs in the ocular lens. A typical petrographic microscope is shown in Fig. 1.

This unique arrangement allows minerals to be observed for color, cleavage, opacity-translucence, index of refraction, optical orientation, crystallinity, etc. All these can be used to identify mineral components in rocks, certain inorganic chemical compounds and in this case the mineral components in foundry sands.

Differential Thermal Analysis

Differential thermal analysis (DTA) was developed by H. L. LeChatalier in 1887 to investigate the components of clay. DTA has been widely used in the study of foundry clays. It can also be useful in studying phase transformations and other thermal reactions in nonclay minerals and chemicals. The method uses an instrument designed with a specimen block having two holes (wells), one for the unknown specimen and one for a thermally inert material called the standard (usually calcined aluminum oxide). Thermocouples (chromel-alumel or platinum-rhodium) are placed in the bottom of each well. A general arrangement is

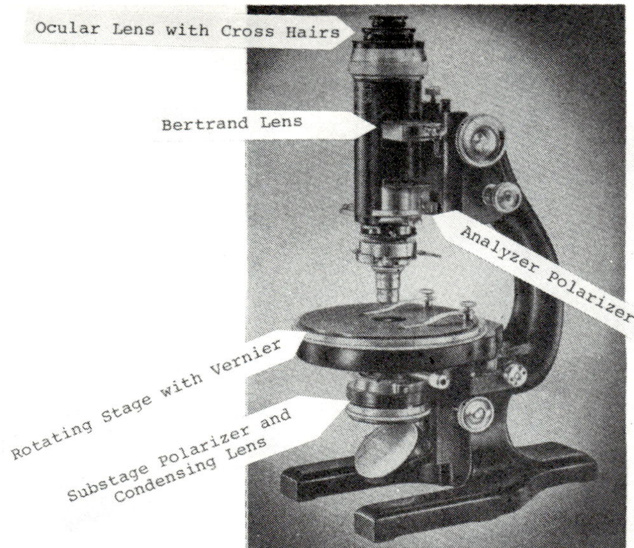

Fig. 1. Petrographic microscope.

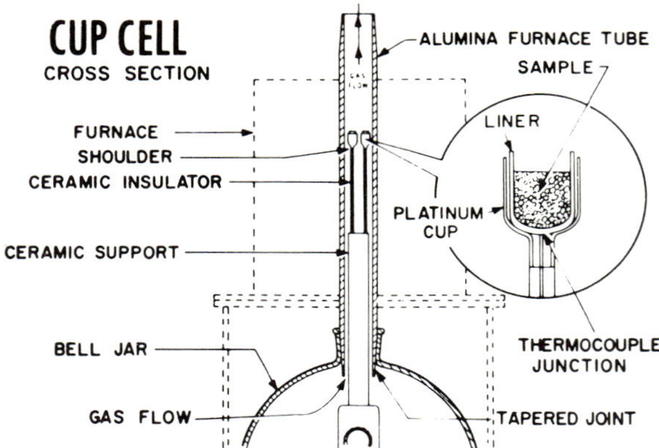

Fig. 2. Differential thermal analysis apparatus. (Courtesy of DuPont Instrument Products Div)

shown in Fig. 2. The holder and thermocouples with the inert material and unknown specimen are placed in a furnace closely controlled to produce a uniform rate of temperature increase. The temperature of the inert material increases uniformly as the temperature of the furnace increases. When a thermal reaction takes place in the unknown sample the temperature of the sample is greater or less than that of the inert material depending upon whether the reaction is exothermic (gives off heat) or endothermic (takes up heat). For the interval of time between beginning and end of the reaction one junction of the difference couple is different from that of the other and an EMF is set up in the differential thermocouple circuit which is recorded in volts as a function of the temperature of the standard. When no thermal reaction is taking place in the specimen the temperatures at both junctions of the different couples are the same and no EMF is set up.

Thermogravimetric Analysis

The method to continuously measure and record the weight loss of a compound was developed in France. It was introduced to this country under the name Chevenard balance and later called thermogravimetric analysis (TGA). Early instruments were very crude and consisted of a tared sample holder which was placed into a furnace. The sample holder was arranged with a mirror so that as the sample lost weight a fine light beam was reflected from the mirror and the weight loss recorded on a sheet of photosensitive paper moving at a rate equal to the heating rate of the furnace. Modern versions of this instrument are much more sophisticated and take very little space. A TGA is shown in Fig. 3. In addition to the studies on pure foundry sands, TGA could provide valuable information to the foundryman on weight loss (or gain) of mold and core mixes as binders burn off and other ingredients yield volatiles.

X-Ray Diffraction

X-rays have been used to study minerals for 65 years. The minerals studied must be crystalline to have parallel planes of atoms. When an x-ray beam strikes a crystal it penetrates it and the resulting reflection is not from a single plane but from an almost indefinite number of parallel planes.

A powder sample is placed between a columinated x-ray beam and a Geiger tube detector. The instrument is so constructed that the sample rotates in the path of a columnated x-ray beam while a Geiger counter, mounted on an arm, rotates about it to pick up the reflected x-ray beams. The slide holding the mineral powder rotates through an angle Θ and the Geiger tube rotates through two Θ. The purpose of this arrangement is to maintain such a relation between x-ray source, sample and Geiger tube that no reflections are cut off by the glass slide on which the mineral powder is mounted. As the Geiger tube scans, a strip chart records the reflections from the specimens. A typical setup is shown in Fig. 4.

Each crystalline mineral or chemical compound has its own x-ray diffraction pattern. Using the pattern it is possible to identify unknown minerals or chemicals by converting the 2Θ to d spacings. Each mineral has its strongest lines at one or more d

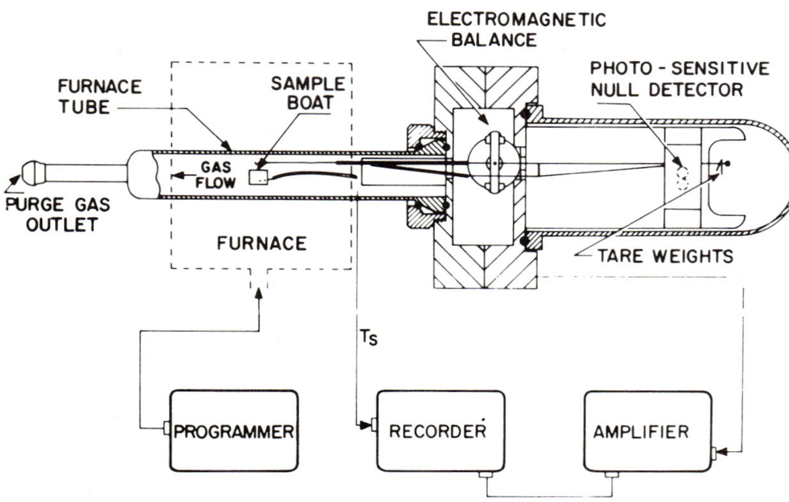

Fig. 3. Thermogravimetric analyzer. (Courtesy of DuPont Instrument Products Div)

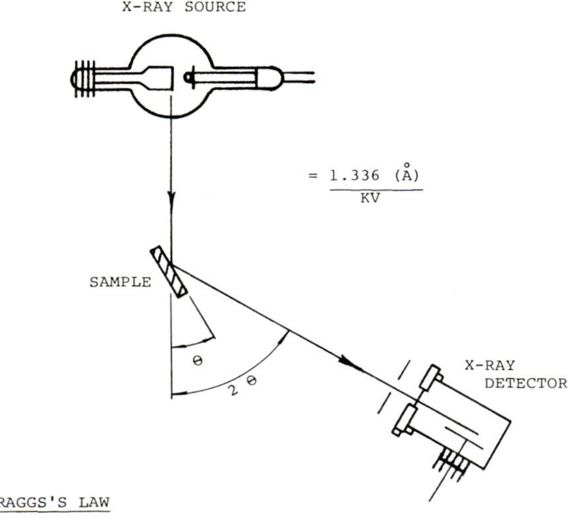

Fig. 4. X-ray diffractometer arrangement.

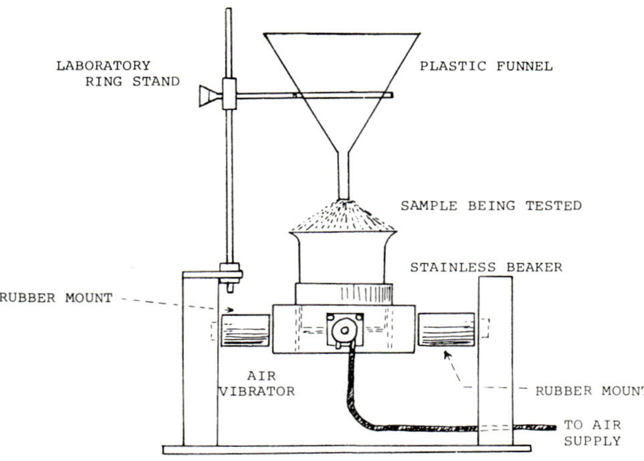

Fig. 5. Bulk density measurement apparatus.

spacings. In the case of silica its strongest line is 3.35 dA. Its next strongest lines are 4.26 dA and 1.37 dA. This method can also be used to identify any unknown crystalline compound whether mineral or chemical. It can be used to estimate the composition of a rock or other mixture of minerals based on intensity of the strongest lines of the constituent minerals. Sharpness of the lines also gives some indication as to the crystallinity of a mineral or compound.

Bulk Density

Determination of bulk density of unbonded sand is important. Two methods are recommended in the literature.[1] The first uses a 100-ml volumetric flask to get a tapped density. The second method recommends using a 100-ml flask and a vibrating mechanism to reduce the amount of time required to compact the sand and if too much sand is accidently added it is necessary to stop, pour out the excess sand and then start again.

Bulk density measurements recorded in this paper were made with a new type of apparatus developed in the author's laboratory. Like the present AFS method, the author's method uses a vibrator powered by air pressure. A container mounted on two rubber supports is used to make the determination. The exact volume of the container is determined by weighing it filled with water then calculating its exact volume. While the unit is vibrating, the sand being tested is poured into a dry container through a funnel suspended above it. When the sand has coned high enough to reach the funnel, vibration is continued for an additional 60 sec; then the container is removed, struck level and weighed. The bulk density is then calculated using the volume previously determined. The reproducability of this method is excellent. The method also has the advantage that smaller or larger containers can be used. It is the author's experience that the larger the sample tested, the more accurate the results. The apparatus is shown in Fig. 5.

Foundry Sand Requirements

Refractory materials to make molds and cores have been very important to foundrymen since metal casting began hundreds of years ago. Early requirements were simple but as sand-mold technology became more sophisticated and better binders were developed, physical and chemical properties of foundry sands became more critical. Basic requirements for good foundry sands are that they must

- have dimensional and thermal stability at elevated temperatures
- have suitable particle size and shape
- be chemically unreactive with molten metals
- not be readily wetted by molten metals
- not contain volatile elements which produce gas on heating
- be available in large quantities at reasonable prices
- have consistent cleanliness and pH.
- be compatible with new chemical binders as they are developed.

Many industrial minerals are available which have some of these properties but only a few have them all.

The most common foundry sand used over the years has been silica sand. This is not surprising because it is the most abundant and one of the most easily mined minerals at the earth's surface. Silica sand and other foundry materials are produced from rocks which formed under a variety of geologic conditions and environments. To better understand the minerals used in the foundry industry, the origin and classification of rocks, minerals and mineral deposits from which they came are discussed.

Geologic History of Rocks and Mineral Deposits

After the earth was formed, an outer shell (crust) formed as molten material solidified and encapsulated the remaining molten mass. The original crust which formed was made up of igneous rock because it formed as a result of crystallization of minerals from molten material. During ensuring geologic time these rocks underwent mineralogical and structural adjustments to physical and chemical conditions which were imposed at depths below surface zones of weathering and cementation. The resulting rocks are called metamorphic. As rains fell and the continental portions of the globe underwent climatic changes, the rocks exposed at the surface were subjected to chemical and physical changes as a result of oxidation and weathering processes. The less stable materials were dissolved or converted to new mineral forms.

The more resistant minerals broke down under rigorous weathering and were transported to rivers and then onward to the seas where they were deposited. These deposits are called sedimentary rocks. Older sediments which had been transported long distances or reworked and redeposited several times often have rounded particles. Younger sediments which have not been transported long distances or which were produced by glaciation tend to be more angular in shape. Particles in a coastal dune or desert environment may have frosted grain surfaces as a result of particle-to-particle contact during

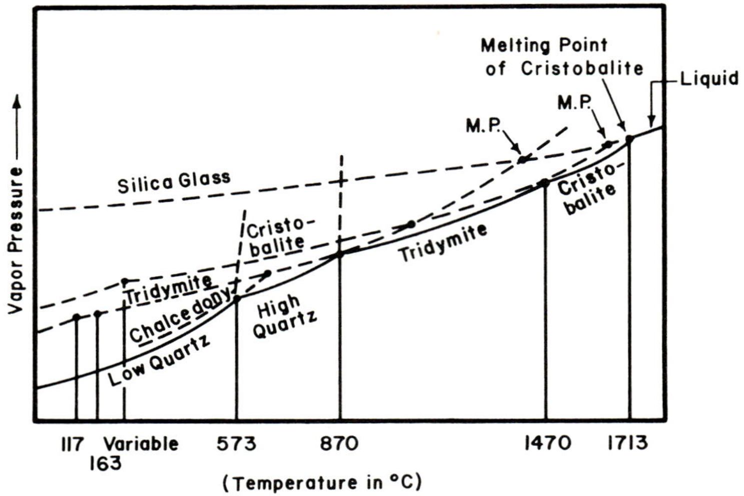

Fig. 6. Fenner's diagram showing stability relations of silica minerals. (After Wahlstrom)

transport by winds. Shape of the particles may also be influenced by the mineralogical character. Some minerals have one or more cleavages along which the particles break forming blocky or needlelike shapes. Other minerals have no cleavage and break in irregular shapes often with shell-like depressions along the fracture surface. These are called concoidal fractures.

A wide variety of minerals are mined and processed for industrial use. Physical and chemical properties of these industrial minerals are a direct result of the geologic environment in which they were formed. The difference in the mode of formation between disintegrates and residual sediments depends upon the ratio between the rate at which mechanical breaking up of the parental rock takes place and the rate at which chemical weathering takes place. In arctic and subarctic regions where mechanical action obtains its maximum, chemical action is minimum. Even finely divided rock powder escapes chemical weathering and is deposited as glacial clay. Conversely, rocks in tropical climates undergo extensive physical and chemical changes due to leaching action and chemical weathering.

The end products of weathering are associated with concentration of those materials, especially quartz and beach sand, heavy minerals that are not notably subject to alteration. All minerals of high chemical and mechanical resistance, as well as most minerals of high specific gravity, can be found in residual sediments. These include quartz and the heavy minerals (s.g. >2.9) — zircon, monazite, magnetite, ilmenite, rutile, garnet, staurolite, tourmaline, kyanite, sillimanite, etc.

Silica Sands

Silica is the common term applied to SiO_2 in the mineral form of quartz. Quartz is used in a number of industrial applications including foundry sands, aggregates, sandblast abrasives, fillers, glass, chemicals, silicon and refractories because it is:

- abundant
- easily mined
- inexpensive
- hard and resists abrasion
- available in a variety of grain sizes
- resistant to metal and acid slag attack.

Quartz is thermally unstable. As shown in Fig. 6 it undergoes an inversion to high quartz at 573C (1064F).[2] This inversion creates an endothermic reaction since it requires heat. The sharpness of the inversion can be seen in the differential thermal analysis curves (Fig. 7 and 8). The inversion is accompanied by expansion of each grain. The expansion which accompanies inversion to high quartz can be seen in Fig. 9. Since the high quartz is an unstable form, on cooling the quartz undergoes a reverse reaction at 573C (1064F) as it inverts back to stable low quartz form. It also shrinks back to its original particle size at the inversion point. If quartz is heated beyond 870C (1598F) it inverts to new forms until the melting point of silica is reached at

Table 1. Summary of Mineralogy and Physical Properties of Foundry Sands

	QUARTZ	OLIVINE	CHROMITE	ZIRCON
COLOR	White to Yellowish Brown	Green	Black	White to Pale Yellowish Brown
CRYSTAL SYSTEM	Hexagonal	Orthorhombic	Cubic	Tetragonal
CLEAVAGE	None	(010)(100)	None	None
MOHS HARDNESS	7	6.5-7.0	5.5	7.5
SPECIFIC GRAVITY	2.65	3.2-4.8	4.4-5.2	4.68
INDEX OF REFRACTION	1.54-1.56	1.63-1.69	2.00-2.12	1.92-2.02
ALTERATION PRODUCTS	None	Hydrous Magnesium Silicates	Hydrous Iron Oxide	Metamict Zircon
MELTING TEMPERATURE (C)	1713°	Variable 1205-1900°	Variable 1760-1980°	2200°

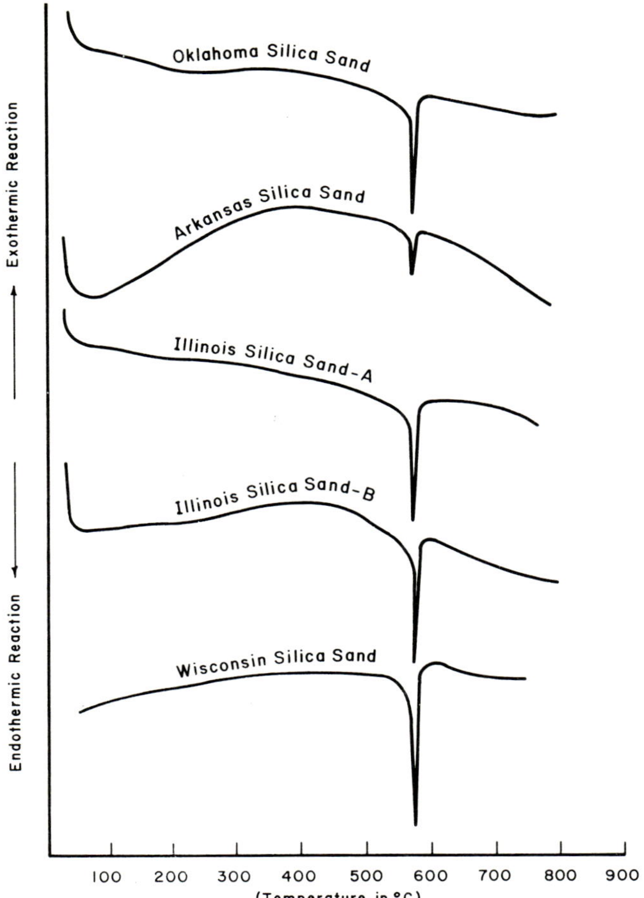

Fig. 7. Differential thermal analysis curves for silica sands.

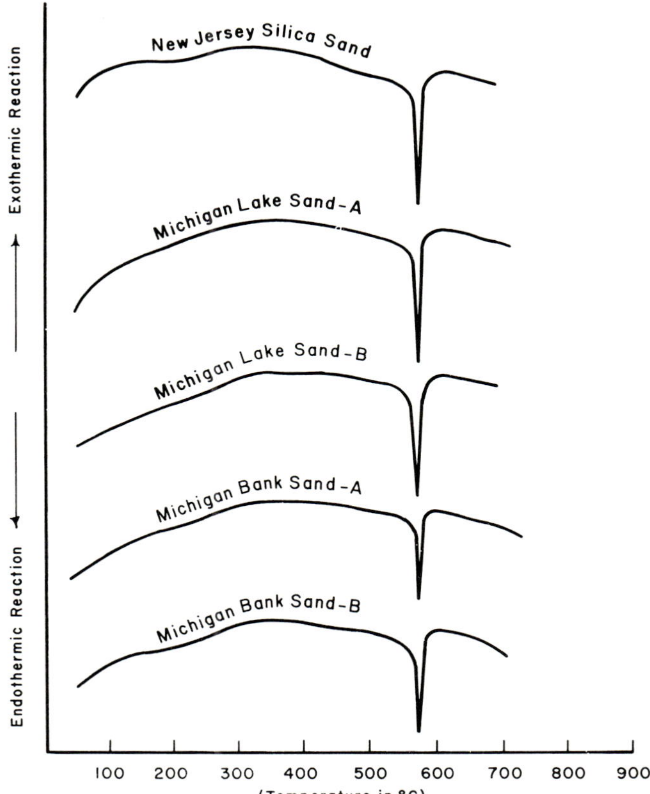

Fig. 8. Differential thermal analysis curves for silica sands.

Commercial Silica Foundry Sands

Ten commonly used silica foundry sands ranging from very pure, white sand with rounded particles to relatively impure bank sands were used in this study. Locations of the samples are shown in Fig. 10. The geology and mineralogy of each of these deposits is discussed below.

1713C (3116F). Silica glass may form on cooling after the melting point is exceeded.

Mineralogy and physical properties of quartz are summarized in Table 1. Quartz is a relatively simple mineral and is unusually constant in composition. If often occurs in massive form, from coarse to fine granular, and sometimes in crypto-crystalline form. It has pyroelectric (develops opposite electric charges at ends of crystals on heating) and piezoelectric (electric potential develops when mechanical strain is applied) properties. Liquid inclusions are very common in quartz. The liquid may be water or CO_2 or both.

Quartz is an essential constituent of many igneous and metamorphic rocks. It is usually the chief and often the only constituent of sandstone, quartzite, gravel, conglomerate and vein rocks and is found almost everywhere.

Quartz sands are used in large quantities because they are cheap but they cannot be used in all foundry applications for the following reasons:

- they are not thermally stable
- they expand on heating and shrink on cooling
- they are wetted by some metals
- their binder requirements, pH, and acid-demand values vary from deposit to deposit
- they are low density and do not provide chill
- they can be hazardous to health and will be more strictly regulated by the Government.

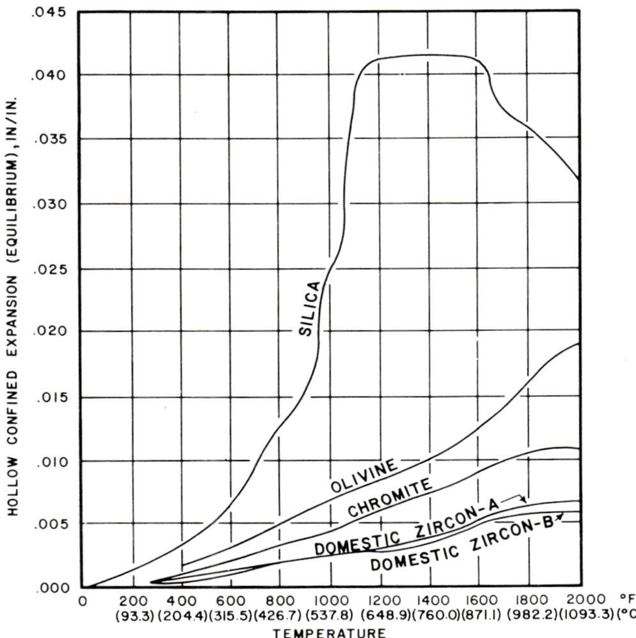

Fig. 9. Thermal expansion curves for mineral sands.

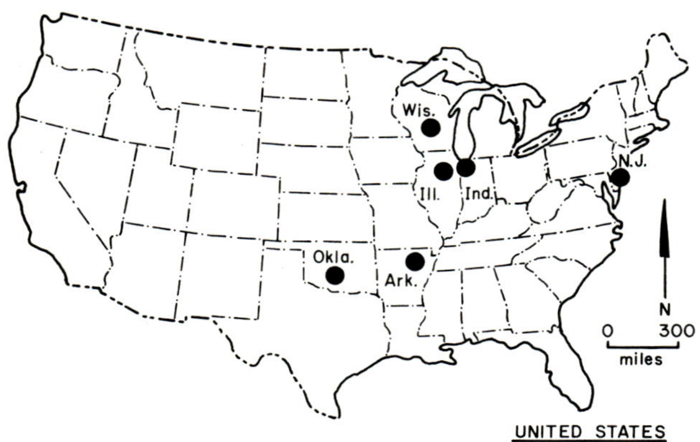

Fig. 10. Location of silica sands studied.

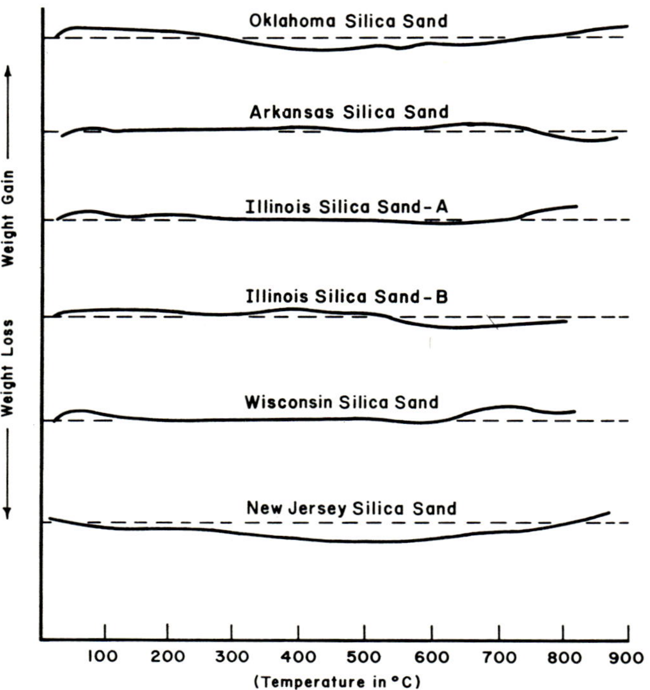

Fig. 11. Thermogravimetric analysis curves for silica minerals.

St. Peters Sandstone

In the early Ordovician period, shallow inland seas covered the central US extending down to Texas and along the present course of the Mississippi. A thick series of limestones ranging from 150-250 ft were deposited in these seas. Following this, the St. Peters sandstone consisting of very pure quartz grains with rounded particle shapes along with feldspars and other minerals were deposited on top of the limestone in Minnesota, Iowa, Illinois, Wisconsin, Missouri, Arkansas and Oklahoma. Later, 600-700 ft of dolomite and shale were deposited on top of the St. Peters sandstone. After 600 million years of leaching, all of the impurities which had been present at the time of deposition have been removed from the St. Peters, leaving one of the most pure silica sand deposits in the US.

The Pleistocene period brought the Ice Age, during which glacier ice advanced southward into north central US several times. After the ice melted, thick layers of rock debris were left covering the Ordovician sediments. As a result much of the St. Peters sandstone is unavailable for mining because it is overlain by glacial drift.

St. Peters reserves are large and for the most part are controlled by the current producers. Even deeply covered areas now unused because of excessive stripping costs or otherwise marginal due to local factors have been explored and acquired by operating companies for possible future use. What little St. Peters sandstone remains under private ownership will soon be, if it has not already been, completely zoned out of the market by urbanization pressures.[3]

Four samples from the St. Peters or equivalent were used in this study. They are from Oklahoma, Arkansas and two from Illinois. Laboratory work is summarized in Table 2. All are white in color, have subrounded to rounded grains and have nearly neutral pH. Microscopic examination of these four samples shows the grain surfaces free from coatings. Particle size can vary from producer to producer with AFS gfn ranging from 63-84.

Thermogravimetric analysis curves in Fig. 11 show essentially no loss or gain of weight on heating. Because of its chemical purity, grain roundness and lack of ultrafines, sands produced from the St. Peters sandstone are among the best silica sands available to the foundryman today.

Wisconsin Silica Sand

The Jordon sandstone crops out in Minnesota and Wisconsin.

Table 2. Physical Properties, pH and Acid-Demand Values for Silica Sand Samples

	OKLAHOMA	ARKANSAS	ILLINOIS - A	ILLINOIS - B	WISCONSIN	NEW JERSEY
COLOR	White	White	White	White	Very Pale Orange	Grayish Yellow
GRAIN SHAPE	Rounded	Rounded	Subrounded to Rounded	Subrounded to Rounded	Rounded	Subangular to Subrounded
SCREEN ANALYSIS (% Retained)						
Mesh / Microns						
30 / 600	-0-	0.1	0.1	0.1	1.7	0.8
40 / 425	0.1	2.4	2.6	1.5	4.1	4.7
50 / 300	0.8	17.3	12.9	13.4	17.2	21.2
70 / 212	9.1	31.8	26.3	32.3	27.4	38.4
100 / 150	52.6	33.4	37.0	35.2	28.5	28.4
140 / 106	28.4	12.1	16.3	12.3	15.0	5.0
200 / 75	7.7	2.4	4.2	4.2	4.7	0.9
270 / 53	1.2	0.4	0.5	0.8	1.2	0.2
Pan / 53	0.2	0.1	0.1	0.1	0.3	0.3
AFS GFN	84	63	68	69	67	57
BULK DENSITY						
Lbs/ft^3	105	105	108	108	108	100
g/cc	1.68	1.68	1.78	1.73	1.73	1.60
pH	6.8	7.4	6.8	6.7	6.9	6.6
ACID DEMAND VALUES						
4.0 pH	1.03	3.72	0.89	1.27	1.60	1.20
7.0 pH	0.19	2.98	0.27	0.46	0.71	-
WT. LOSS ON HEATING	-	-	-	-	-	0.12%

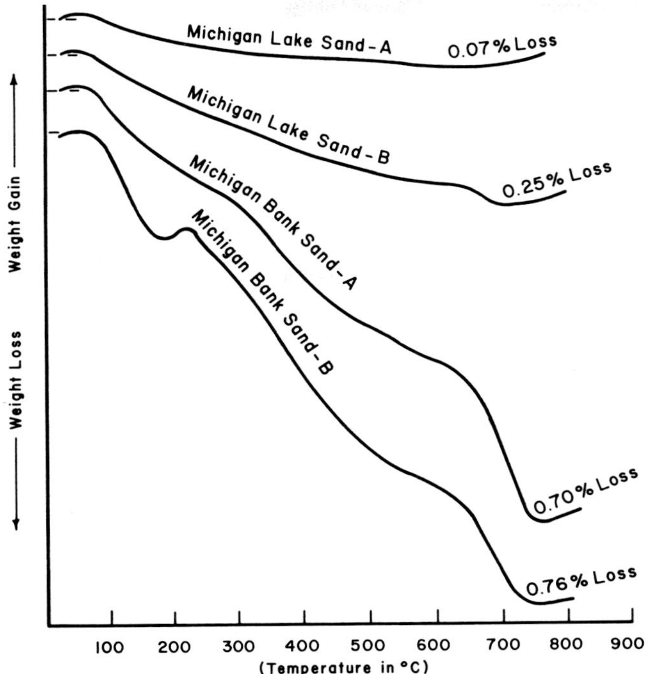

Fig. 12. Thermogravimetric analysis curves for silica sands.

Much of the area underlain by the Jordon is covered by thick glacial drift or dolomite beds. This overburden eliminates large areas from mining because of economic considerations. One sample from this formation was studied. Results are summarized in Table 2.

The sand from this formation is very pale orange in color. The grains are very well rounded; the grain fineness is similar to that of the St. Peters sandstone products; and pH is nearly neutral at 6.9. The thermogravimetric analysis curve in Fig. 11 shows very little gain or loss of weight on heating. The sample contains 0.4% clays ($<$38μm) which may account for higher acid-demand value compared to the samples containing less clays. Scrubbing in caustic removed the clays leaving a very white sand.

New Jersey Silica Sand

Southeastern New Jersey is underlain by the Cohansey sand of Miocene Age. It is a relatively young, immature sand. The sand is unique in that part of it is continental, part is deltaic and part is of marine origin. The Cohansey crops out at the surface in places but gently dips under younger sediments. One sample from this formation was studied. Laboratory work is summarized in Table 2.

The sand is grayish yellow and grains range from subangular to subrounded. The AFS gfn is 57 and the sand has pH of 6.6. The TGA curve (Fig. 11) shows a weight loss of 0.2% on heating to 900C (1652F). The sample contains 0.3% clays ($<$38μm). Despite the presence of clays, the acid-demand value compares favorably with silica sands having less clay. Scrubbing the sample in caustic changed the color slightly from grayish yellow to pale yellowish orange.

Michigan Lake Sands

The origin of Midwest Lake sands is well described in the literature.[4] As the glaciers receded, Lake Chicago, which preceded the present Lake Michigan, passed through several stages of water levels. Beaches were formed at different lake stands. Those were sources of the present dune sands occurring along the eastern and southern shore of Lake Michigan.

Samples from two different producers were used in this study. Laboratory work is summarized in Table 3. Both samples are grayish orange in color with subangular to subrounded grains. Sample A has an AFS gfn of 55 and sample B is 64. Sample A has pH of 6.6 and B has 6.4. Sample A contained 0.4% clays and B contained 1.8%, the difference probably being the amount of treatment the two samples received during processing for the market. After laboratory washing in caustic the color was unchanged in both samples indicating the presence of nonclay coatings on grain surfaces. This may account for the high acid-demand value determined for both samples and the weight loss shown by the TGA curves (Fig. 12).

Michigan Bank Sands

The sands making up this type of deposit had originally been deposited with the coarser lake sands at one time. Later, winds transported the finer particles to form inland dunes. As a result

Table 3. Physical Properties, pH and Acid-Demand Values for Silica Sand Samples

		MICHIGAN LAKE SAND "A"	MICHIGAN LAKE SAND "B"	MICHIGAN BANK SAND "A"	MICHIGAN BANK SAND "B"
COLOR		Grayish Orange	Grayish Orange	Moderate Yellowish Brown	Moderate Yellowish Brown
GRAIN SHAPE		Subangular	Subangular	Angular	Angular
SCREEN ANALYSIS (% Retained)					
Mesh	Microns				
30	600	0.1	0.2	0.1	0.6
40	425	1.3	2.3	0.5	2.3
50	300	20.0	15.2	2.8	7.7
70	212	49.9	33.5	9.9	16.2
100	150	24.1	34.4	33.5	29.8
140	106	3.6	11.7	29.3	23.5
200	75	0.7	2.2	16.3	14.5
270	53	0.2	0.4	5.9	4.4
Pan	53	0.1	0.2	1.8	1.1
AFS GFN		55	64	99	89
BULK DENSITY					
Lbs/ft^3		103	100	102	103
g/cc		1.65	1.60	1.63	1.65
pH		6.6	6.4	6.8	7.1
ACID DEMAND VALUES					
4.0 pH		6.04	11.1	29.8	45.3
7.0 pH		3.20	8.03	13.3	37.9
% CLAYS		0.4	1.8	1.2	1.7
% HEAVY MINERALS		0.8	0.1	0.9	Trace
WT. LOSS ON HEATING		0.7%	0.18%	0.70%	0.76%

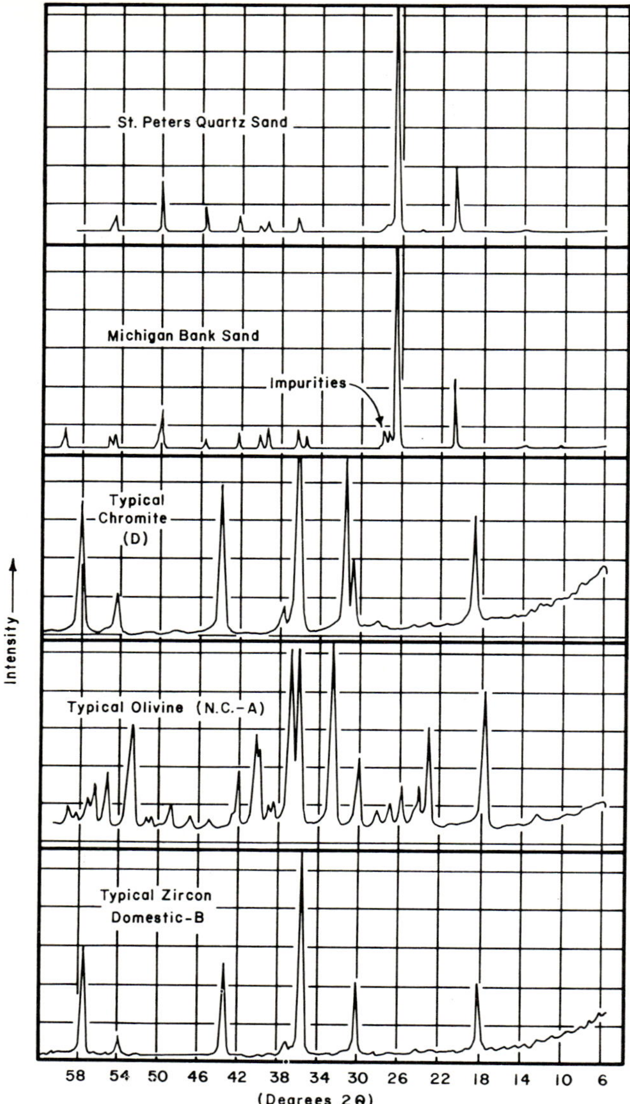

Fig. 13. Typical x-ray diffraction patterns for mineral sands.

of size classification during this process, these sands are generally finer than the lake sands. They normally range from 80-120 AFS gfn.

Two Michigan bank sands were used in this study. Each was produced by a different company. Laboratory work is summarized in Table 3. Both samples are moderate yellowish brown in color and have angular to subangular grain shapes. Sample A is finer with AFS gfn of 99 and B with 89. Both samples have nearly neutral pH. Sample A contained 1.2% clays ($<38\mu$m) and B 1.7%. Dehydration of the clays probably accounts for the weight loss shown by TGA (Fig. 12). The color of both samples remained the same after the clays were removed by caustic scrub. This indicates the presence of nonclay coatings on the grain surfaces. X-ray diffraction of these samples (Fig. 13) shows the presence of impurities.

Acid-Demand Values

Tables 2 and 3 indicate the acid-demand values increase with increasing clay content. To further investigate this observation, all the silica sand samples (except for St. Peters type) were scrubbed in caustic to remove clays. Afterwards acid-demand values were rerun. The results shown in Table 4 confirm significant reduction in acid demand after the clays are removed.

Olivine

Olivine is similar in mineralogy and geologic occurrence to chromite. Large deposits of olivine-bearing dunites crop out in Norway, Sweden, USSR, Austria, Rhodesia, South Africa and the US as shown in Fig. 14. Norway furnishes much of the Olivine consumed in Europe. Deposits in Washington and North Carolina furnish olivine for foundry use in this country. Thermal expansion of olivine is much less than silica sand but higher than the specialty sands as shown in Fig. 9.

Olivine has been moderately successful as a specialty sand. Its grain shape is usualy angular and because of its high pH and acid demand it is unsuitable for certain resin-bonded processes. Impurities lower the sand's resistance to attack from corrosive slags. It is less stable under thermal shock than zircon or chromite sands.

Olivine group minerals forsterite (Mg_2SiO_4) and fayalite (Fe_2SiO_4) form a continuous solid solution series as shown in Fig. 15. Physical properties vary with chemical composition (Table 1 and Fig. 15) which varies from deposit to deposit. Crystal structure of olivine is much more complex than other

Fig. 14. Locations of major olivine deposits.

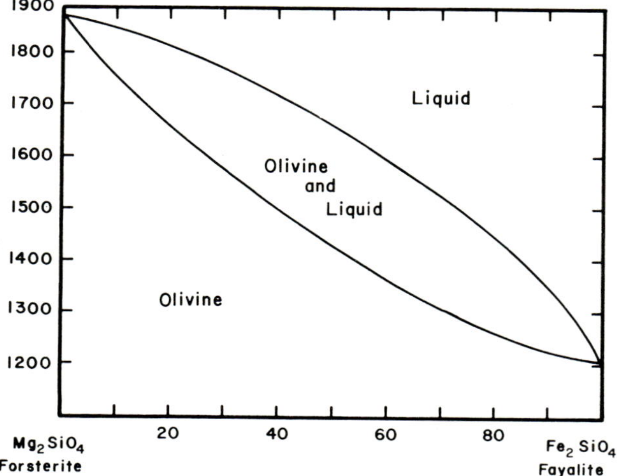

Fig. 15. Phase diagram showing the system Mg_2SiO_4-Fe_2SiO_4, the mineral olivine solid solution series. (After Wahlstrom)

Table 4. Comparison of Acid-Demand Values Before and After Removal of Clays with Caustic Scrub

	Before Scrub		After Scrub	
	pH4	pH7	pH4	pH7
WISCONSIN	1.6	0.7	1.0	0.4
NEW JERSEY	1.2	–	1.3	0.8
LAKE SAND "A"	6.0	3.2	3.9	1.8
LAKE SAND "B"	11.1	8.0	6.0	3.5
BANK SAND "A"	29.8	13.3	19.4	12.5
BANK SAND "B"	45.3	37.9	15.9	9.6

specialty sands as shown in Fig. 13. Inclusions of magnetite (Fe_3O_4), spinel (Mg, Fe, Zn, Mn, aluminates), apatite ($Ca_5(PO_4 \cdot CO_3)_3$), liquids or gases are common. Movable bubbles are found at times. Olivine alters readily; in fact, it is much more commonly found altered than fresh. The most common alteration products are hydrous magnesium silicates (serpentine). The alteration develops in scales and fibers and finally the entire olivine crystal can be transformed. A review of olivine geology may be found in the literature.[4]

Samples Studied

Three samples of commercial olivine were included in this study. Two samples were from Washington and the other form North Carolina. Laboratory work is summarized in Table 5. The Washington samples are identical pale olive color. The North Carolina sample is lighter with a pale yellowish green color. All three samples have similar particle size distribution with AFS gfn between 58-66. The olivine sands have bulk density between 118-120 lb/ft^3, making them the lighter of the specialty sands studied. All three samples are characteristically basic with pH of 8.2-8.7 and acid-demand values are correspondingly high. The differential thermal analysis curves for the samples (Fig. 16) do not show thermal reactions taking place during heating to 900C (1652F). This indicates olivine does not undergo any phase changes on heating. On the other hand, the thermogravimetric analysis curves in Fig. 17 show the olivine samples begin to lose weight immediately upon heating and beginning at 550C (1022F) they begin to dehydrate more rapidly. The samples lost between 0.44-0.73% weight during heating.

Grain Durability

As part of an earlier study of olivine, tests were made to compare the durability of its particles with other specialty sands then under development. The tests were made on a Washington olivine product. The particles were subjected to pressure in two ways. The first test exposed the sand to five successive loadings at 2500 psi (1.72 x 10^4 k Pa). After each test the sample was screened and the surface area calculated. Results of these tests, summarized in Fig. 18 and 19, show that %-200 mesh (<75μm) increased from 1.8-15% and surface area increased from 87-151 cm^2/g. In the second series of tests, the olivine was subjected to four different pressures up to 10,000 lb/in^2 (6.89 k Pa 10^4). The test results summarized in Fig. 20 and 21 show %-200 mesh (<35μm) increased from 1.8-18.6% and the surface area from 87-164 cm^2/g at maximum pressure. The lack of durability of olivine grains tested may be due to the mineral's cleavages or to weakening of the grains during tectonic activity after the deposit was formed.

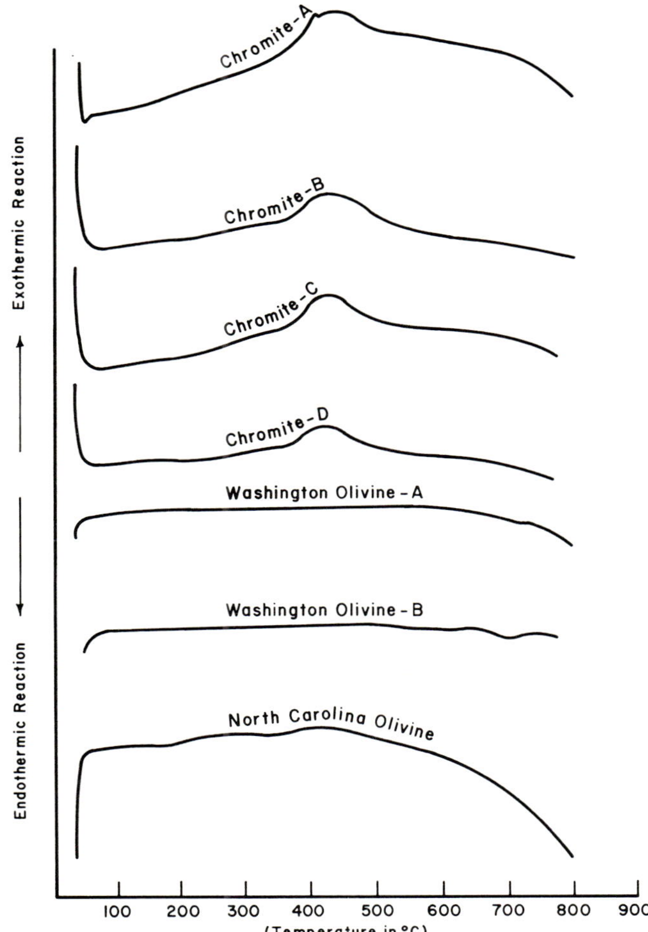

Fig. 16. Differential thermal analysis curves for chromite and olivine.

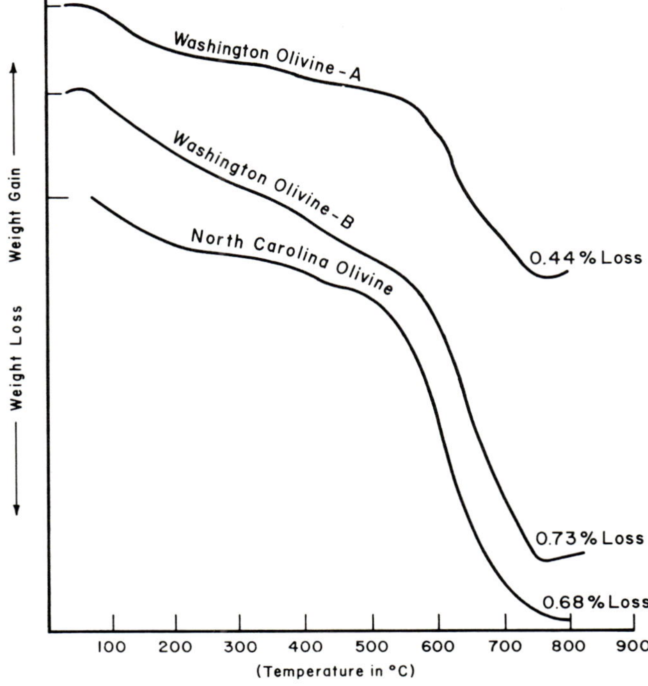

Fig. 17. Thermogravimetric analysis curves for olivine sands. Note the acceleration in dehydration beginning about 550C (1022F).

	WASHINGTON OLIVINE SAND - A	WASHINGTON OLIVINE SAND - B	NORTH CAROLINA OLIVINE
COLOR	Pale Olive	Pale Olive	Pale Yellowish Brown
GRAIN SHAPE	Angular	Angular	Subangular
SCREEN ANALYSIS (% Retained)			
Mesh / Microns			
30 / 600	Trace	Trace	Trace
40 / 425	0.4	0.3	1.7
50 / 300	26.3	24.5	26.1
70 / 212	35.4	40.7	38.3
100 / 150	26.4	19.7	26.3
140 / 106	6.4	6.6	5.3
200 / 75	2.7	4.1	1.5
270 / 53	1.6	2.4	0.5
Pan / 53	0.8	1.7	0.3
AFS GFN	63	66	58
BULK DENSITY			
Lbs/ft^3	120	118	118
g/cc	1.92	1.89	1.89
pH	8.2	8.7	8.6
ACID DEMAND VALUES			
4.0 pH	10.07	11.98	17.79
7.0 pH	8.99	10.35	16.52
WT. LOSS ON HEATING	0.44%	0.68%	0.73%

Chromite

The element chromium (Cr) combines with iron (Fe), magnesium (Mg), calcium (Ca), Lead (Pb), copper (Cu), Phosphorus (P) and sulfur (S) to form a variety of minerals. Of these, chromite ($FeCr_2O_4$) is the only ore mineral of metallic chromium, chromium compounds, chemicals and foundry sands. Most of the world's production comes from South Africa, USSR and Southern Rhodesia. Albania, Turkey and the Philippines also produce chromite. Figure 22 shows locations of world chromite deposits. Although the US has chromite reserves, over 92% have high Fe and low Cr; therefore this country depends entirely on imported chromite. About 60% of world production is consumed in metallurgy, 20% in refractories, 12% in chemicals and 8% in foundry sands.

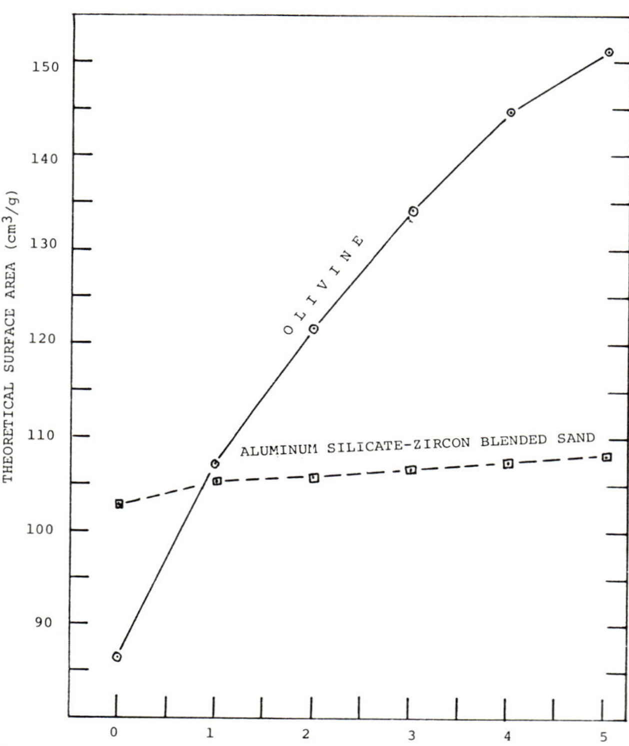

Fig. 18. Percent fines produced by successive loadings at 2500 psi.

Fig. 19. Surface area increase produced by successive loadings at 2500 psi.

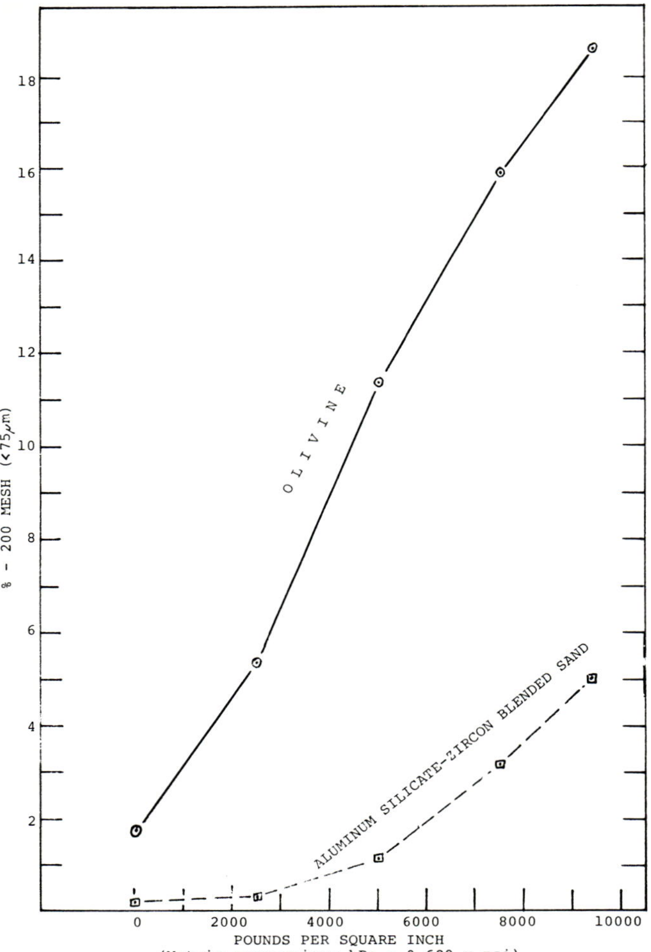

Fig. 20. **Percent fines produced by subjecting mineral sands to increasing pressure.**

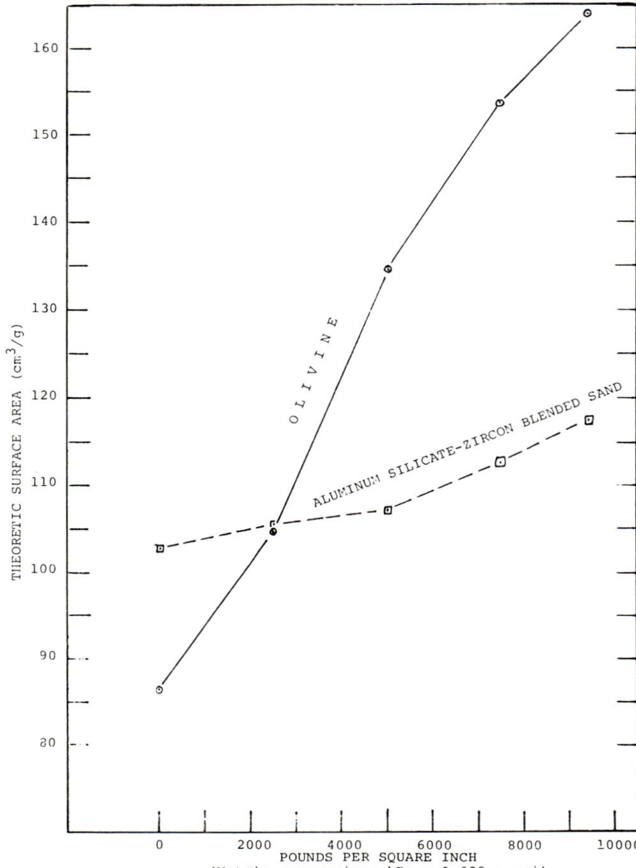

Fig. 21. **Surface area increase produced by subjecting mineral sands to increasing pressure.**

Chromite sand was introduced to the foundry industry in the early 1960s. When zircon was in short supply and later in the face of rising zircon prices, foundrymen adopted chromite as a substitute for zircon. It has been successful as a specialty sand because of the following properties:

- it has good thermal stability
- it has good chill properties
- it is not easily wetted by molten metals
- sizing gives it good resistance to penetration
- it is highly refractory
- it is chemically unreactive.

Some of the drawbacks noted with chromite sands are:

- expansion is nearly twice that of zircon sand
- grains are not rounded as zircon
- presence of hydrous impurities sometimes make chromite sands difficult to use with certain resin binder systems
- being slightly on the basic side increases acid demand over that of zircon
- presence of hydrous impurities may cause pinholing and blowing.

Chromite flour is reported to be a less satisfactory foundry materials than the sand.[5]

Chrome oxide combines with ferric iron (chromite) and magnesium (picrochromite) to form a continuous series. These two minerals combine with $MgAl_2O_4$ and $FeAl_2O_4$ to form a

Fig. 22. **Locations of major chromite deposits.**

complex mineral series called the spinel group as shown in Fig. 23. Chromite mineralogy and physical properties are summarized in Table 1. Crystals are usually octahedral, often massive, granular. Specific gravity and hardness vary with composition (Fig. 23). All chromite products are black with shiny surfaces except for those produced in Turkey, the Philippines and Cyprus. Those have dull lusters. The x-ray diffraction pattern shown in Fig. 13 clearly illustrates the simple crystal structure of this mineral group.

When chromite is heated in an oxygen atmosphere, the ferric oxide is converted to ferrous oxide causing the sample to gain

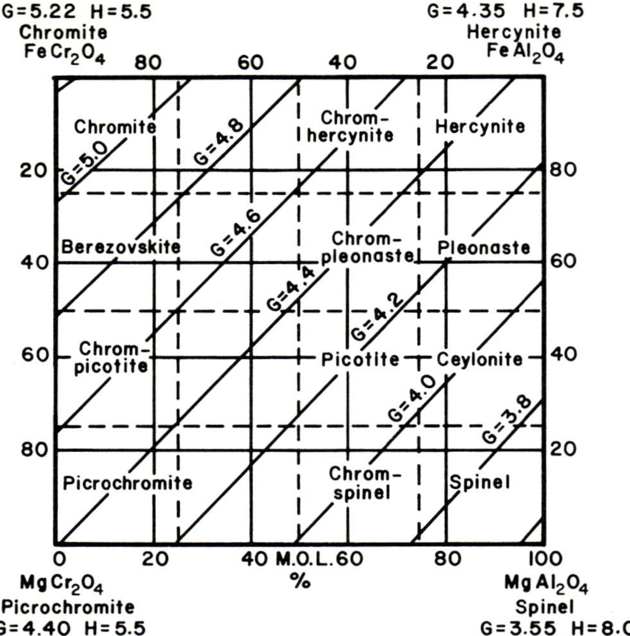

Fig. 23. Variation of physical properties of the spinel group of mineral with composition. (After Winchell)

weight. This comprises any loss that takes place when volatile gases are given off on heating. To determine the loss of ignition, it is necessary to heat chromite in a nonoxidizing atmosphere such as oxygen-free nitrogen or argon. Pure mineral chromite has no water or other volatiles associated with it but volatiles in some commercial chromite sands may be as high as 2.2% because of impurities or alteration products in the product. Middleton and Bownes added serpentine to chromite sands and found that excessive gas associated with hydrated silicate impurities caused pinholing and blowing.[5] Results of their work are shown in Table 6.

Chromite is always associated with ultrabasic rocks such as peridotite, dunite and pyroxenite. These rocks are rich in Mg, Ca and Fe and low in silicon (Si). Minerals with the highest melting points, such as chromite, usually crystallize out first and, if their specific gravity is higher than that of the molten rock, the crystals or grains can settle under the influence of gravity. After the rock has completely cooled, the chromite and other minerals with high melting points and specific gravity will

Table 6. Effect of Hydrated Silicate Impurities in Chromite on Surface Appearance of Casting (After Middleton and Brownes)

Weight of Serpentine Added %	Loss on Ignition of Sand (under N_2) %	Surface Appearance of Casting
0	0.36	Good
7	1.29	Fairly good, pinholing and traces of blowing.
14	2.38	Fair, pinholing and blowing.
21	3.3	Fairly poor, pinholing and blowing

be found as bands or pseudo-strata which can persist for many miles. The resulting rock can undergo extensive alteration through weathering or can go through physical and chemical changes resulting from tectonic folding and faulting or metamorphism.

Commercial chromite is produced from two types of deposits called stratiform and podiform. The Bushveld complex of the Transvaal region of South Africa is an excellent example of stratiform deposit. This deposit is characterized by occurrence of parallel layers of chromite with exceptional lateral continuity.

The podiform deposits might have been formed by processes similar to the stratiform deposit but have since been broken up by tectonic forces such as folding and faulting within the earth's mantle. These ultra-basic complexes are often referred to as alpine type because they are always characterized by intense regional deformation. The Selukwe District of Rhodesia has characteristics of both podiform and stratiform deposits.

Texture and mineralogy of chromite ore vary from deposit to deposit. Chromite grains may range from crystalline to rounded or irregular blocky particles. Another common texture is a nodular ore. These consist of spherical to ovoid aggregates ranging from 5-40 mm in diameter. The nodules are primarily chromite grains 1-5 mm in size in a silicate matrix consisting of olivine or serpentine. In the Transvaal stratiform deposits there are nodules with texture and mineralogy different from that normally seen in podiform deposits.[6]

Samples Studied

Four commercial chromite sands were used in this study.

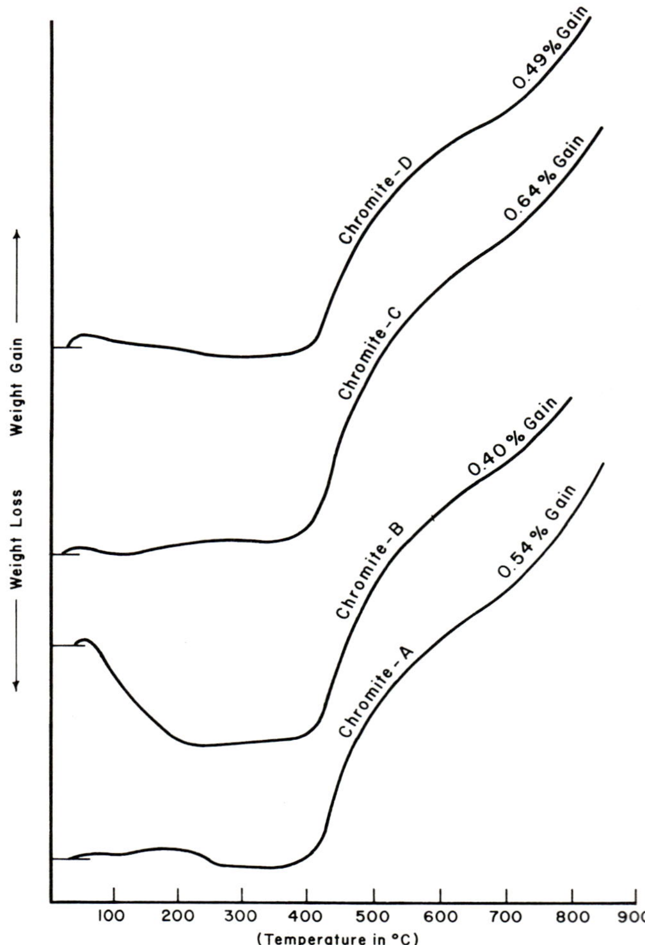

Fig. 24. Thermogravimetric analysis curves for chromite sands.

Table 7. Physical Properties, pH and Acid-Demand Values for Chromite Sands

	CHROMITE SAND A	CHROMITE SAND B	CHROMITE SAND C	CHROMITE SAND D
COLOR	Black	Black	Black	Black
GRAIN SHAPE	Subangular	Angular to Subangular	Subrounded	Angular to Subangular
SCREEN ANALYSIS (% Retained)				
Mesh 30 / Microns 600	8.3	3.1	2.9	5.8
40 / 425	26.9	18.1	14.0	15.4
50 / 300	35.1	35.1	27.7	25.3
70 / 212	19.1	23.9	26.2	21.1
100 / 150	8.5	11.9	21.1	17.8
140 / 106	1.7	4.3	6.8	8.4
200 / 75	0.3	2.2	1.2	3.9
270 / 53	0.1	1.0	0.1	1.4
Pan / 53	Trace	0.5	Trace	0.8
AFS GFN	42	51	53	58
BULK DENSITY				
Lbs/ft^3	190	190	188	190
g/cc	3.04	3.04	3.01	3.04
pH	7.8	7.3	7.8	9.1
ACID DEMAND VALUES				
4.0 pH	4.72	3.20	4.27	4.18
7.0 pH	4.27	2.50	3.72	3.54
WT. GAIN ON HEATING	0.54%	0.40%	0.64%	0.49%

Laboratory work is summarized in Table 7. All are black with shiny luster; angular to subrounded grains with AFS gfn 42-58. Their pH ranges from 7.3-9.1. Acid-demand values ranged from 3.2-4.7. Differential thermal analysis (DTA) curves in Fig. 16 show that all samples underwent a change on heating accompanied by an exothermic reaction beginning at 370C (698F), peaking at 420-430C (788-806F) and ending about 550C (1022F). Likewise, all the samples gained weight on heating as shown by the thermogravimetric curves in Fig. 24. The weight gain is due to oxidation of ferrous iron to ferric iron as discussed earlier. Comparison of the DTA and TGA curves in Fig. 25 shows that as the chromite samples begin to oxidize the mineral undergoes a phase change. It is further reasoned that as more oxygen is taken into the mineral the crystal lattice must expand to accommodate the oxygen.

Zircon

The element zirconium (Zr) combines with sodium (Na), calcium (Ca), manganese (Mn), iron (Fe), titanium (Ti), thorium (Th) and oxygen to form eleven different Zr-bearing minerals. Of these, only baddeleyite (ZrO_2) and zircon ($ZrSiO_4$) are mined on a commercial scale. The other nine Zr-bearing minerals are very rare and do not occur in deposits suitable for mining. Baddeleyite is mined and marketed for refractory purposes; however, its particle size and shape make it unsuitable for foundry sand applications.

Ziron is mined and concentrated for use as a specialty foundry sand, ores of ZrO_2 and zirconium metal, refractories and zirconium chemicals. Zircon is produced in Florida, the east and west coasts of Australia, Malaysia, India and Sri Lanka (Ceylon). Zircon will also be produced in South Africa beginning in mid-1977. Locations of world zircon deposits are shown in Fig. 26.

Zircon mineralogy and physical properties are shown in Table 1. Zircon typically crystallizes in small tetragonal prisms terminated with sharp pointed pyramidal faces. Under certain

Fig. 25. Comparison of chromite thermogravimetric and differential thermal analysis curves showing phase change as mineral begins to take on weight of additional oxygen as ferrous iron converts to ferric iron.

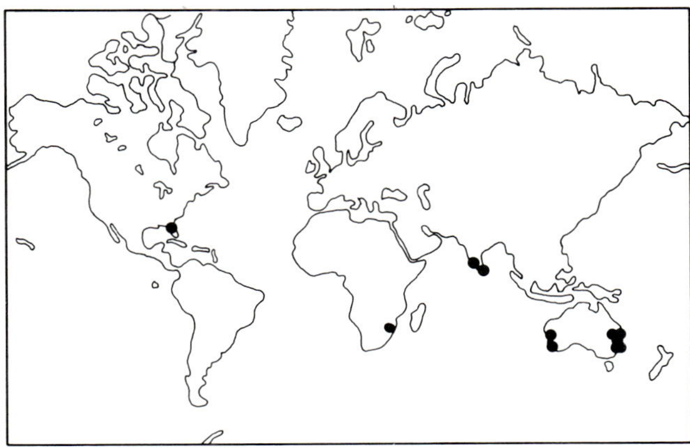

Fig. 26. Locations of major zircon deposits.

geologic conditions, zircon can crystallize with rounded ellipsoidal shapes. Large zircon crystals range in color from yellowish brown to dark brown. Sand-size particles of zircon range from grayish orange to very pale orange to nearly white. Index of refraction of zircon is very high ranging between 1.92-2.0.

Zircon is a common accessory constituent of the acid igneous rock, such as granite, syenite and diorite and is one of the earliest minerals to crystallize from a cooling magma. It is sometimes found in metamorphic rocks such as gneiss and schists. As these rocks are exposed at the surface, they undergo subaerial weathering. The rocks are broken down into smaller fragments and transported downhill by rainwater and gravity. The rocks further decompose to a point that the smaller zircon grains are liberated from the enclosing feldspars and quartz. The mineral particles are transported by streams and ultimately end up along a marine shoreline. Here the action of waves, tidal currents and wind may remove lighter quartz, thus forming a heavy mineral deposit rich in zircon. The more ancient beach sand deposits have consolidated and the grains have cemented together to form sandstones containing concentrations of titanium minerals, zircon, monazite and other heavy minerals. Deposits such as these occur in the four-state area of Utah, Wyoming, New Mexico and Colorado. Since this type of deposit requires more costly mining methods, crushing and grinding, it is less attractive for mining and exploitation than are the more recent, less-consolidated deposits. All the commercial zircon products today are mined and separated from relatively young beach sand deposits occurring on or near active coast lines.

Australian Zircon Deposits

Australia is the major world producer and exporter of zircon sand. Deposits are found on both the east and the west coasts of Australia. In eastern Australia major heavy mineral sand deposits lie between Broken Bay, New South Wales to the south, and Cape Clinton north to Rock Hampton, Queensland. The richest occurrences are concentrated in the area between Yamba and North Stradbroke Island. Although there was some production of zircon from a mine at Ponte Vedra, Florida between 1918 and 1929, the East Australian mines were the first major producers of commercial zircon sands beginning in 1932.

The east Australian heavy mineral beach sands deposits are of both Pleistocene and Holocene age. Zircon and its co-product rutile have been derived from the breakdown and erosion of well-sorted silica sandstone in the late Paleozoic and Mesozoic basins of eastern Australia. Zircon-bearing sands as mined contain less than 1/2% zircon. Zircon from the east coast of Australia ranges in color from dark tan to pale lavender. Because of iron-bearing grain surface coatings, these zircon sands become brownish orange to orange on heating to 1000C (1832F). Grain size and shape vary from deposit to deposit. The eastern Australia zircon is characterized by subangular to rounded particles with many elongated prismatic grains. Particle size distribution and resulting AFS gfn can vary from 82-132 depending upon which mine produced the zircon. There are seven companies now actively mining and producing zircon on the Australian east coast.

Mineral sand deposits of western Australia are found along ancient beaches which lie at different elevations ranging up to 1000 m above present sealevel. The strand line deposits are part of a series of Pleistocene and Holocene sediments which exist as a thin veneer throughout the Perth basin. These overlie much older marine and continental deposits. Shoreline deposits formed during stable Pleistocene sealevel stands. The age of these deposits ranges from fairly young to perhaps over 100,000 years in age. Heavy mineral sand deposits along the southern coast of western Australia near Cable and Bunbury are very rich in the mineral ilmenite (Fe TiO$_3$) but contain relatively small amounts of zircon.

Zircon is fast becoming a major export of western Australia since the discovery of the deposits near the town of Eneabba. These deposits were discovered by local farmers in 1970. The Eneabba heavy minerals occur at different strand lines ranging from 82-120 m above sealevel. The presence of clay, induration and grain surface coatings create difficult mining and mineral separation problems. The mineral grains are extensively coated with ferrugenous aluminum silicates as a result of laterization (surface leaching and deposition at depth by evaporation). The grain surface coatings cause the zircon to take on a light orange color when heated to 600C (1112F).

US Deposits

A small amount of zircon was produced from a beach sand mine located near Ponte Vedra, Florida between 1918-1929. The first commercial production of zircon from this deposit was reported in 1922. The operators of this mine recognized the high refractoriness of this new mineral sand and obtained a US patent covering its use as a refractory. There were no companies producing heavy minerals in Florida from 1929 to 1939. In 1940 a small quantity of heavy minerals including zircon was produced from surface concentrates found in the beach sand near Melbourne, Florida. This operation was later moved to a dune area near Vero Beach, Florida. This deposit operated until about 1963. Between 1943 and 1963 zircon was produced from a deposit midway between Jacksonville and the Atlantic coast. In 1963 production began from a heavy mineral deposit near Folkston, Georgia. Zircon from this deposit is extremely high grade but its very fine-sized particles (AFS gfn 162) does not make it well suited as a foundry specialty sand. Much of the production from this deposit has been ground into flour and sold for use in foundry washes, investment casting slurries, ceramic opacifiers and into the zirconium refractory market. When this deposit was mined out in 1974, mining and concentrating equipment was moved a few miles south to a deposit near Bolougne, Florida where mining for zircon and titanium minerals continues. These two deposits are geologically related and Bolougne zircon particle size is identical to that which had been previously mined at Folkston. Zircon production began from a deposit near Green Cove Springs in 1973. Zircon from this deposit has an AFS gfn of about 140. This is similar in size to the zircon produced from the Jacksonville deposit during World War II.

The largest domestic supply of zircon sand comes from the Trail Ridge deposit in north central Florida. This large deposit was discovered in 1946 and has been in continuous production of zircon since 1950. The deposit was formed at the height of Florida's submergence during Pleistocene times. Sands transported from older sands were redeposited forming the ridge. Waves, currents and wind action removed part of the lighter silica sand leaving an enrichment of heavy minerals in sand dunes on the western flank of this prominent geographic feature. After deposition of the heavy mineral bearing sands the ridge was covered with wind-blown sand which forms an overburden.[7] Unlike the Australian deposits zircon from the Trail Ridge deposit is extremely uniform with respect to particle size and grain morphology as well as chemical purity. The grains are free of surface coatings and have been slightly etched as a result of wind action during formation of the deposit. The grains have no surface coatings; therefore this zircon turns white when heated.[8]

Zircon is chemically and thermally stable; hence, many of the zircon sands mined today have withstood chemical assault in the earth's crust for two or three billion years. Despite its chemical stability zircon undergoes alteration from within the grains themselves. It is one of the rare minerals which is subject to internal alteration resulting from entrapment of trace amounts of Th and uranium (U) within the crystal lattice. Presence of these elements causes the surrounding crystal lattice to break

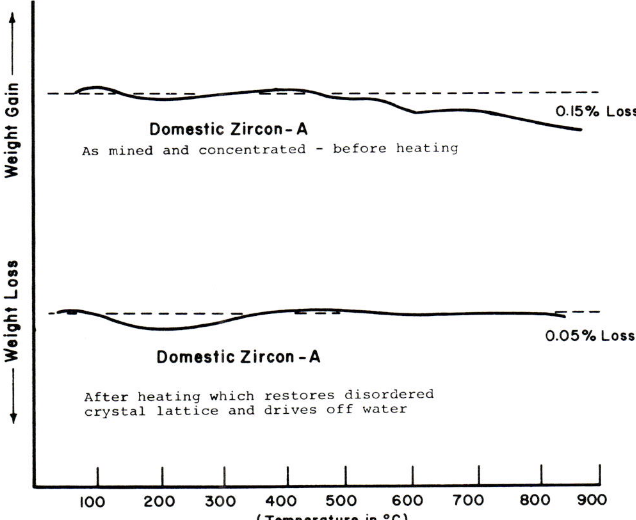

Fig. 27. Comparison of thermogravimetric curves before and after heating zircon to 600C (1112F) during processing for foundry use.

down resulting in a pinkish-purple color. The altered mineral is called hyacinth. As the alteration progresses it produces malacon, a very dark-colored mineral containing up to 10% H_2O. In its final stages the altered zircon reaches an amorphous state called metamict. When zircon in the metamict state is heated, moisture is driven off, the crystal lattice is reordered and the product turns white if there are no coatings on the grain surfaces. The effect of heating zircon can be seen in the upper TGA, Fig. 27. This shows moisture being driven off the zircon beginning at 500C (932F). The lower TGA shows weight loss significantly reduced on reheating the sample since most of the water had been driven off by the first heating. It is also believed that zircon grains are stronger after heating.

Zircon is the only specialty sand having most of the desirable properties for a foundry sand discussed earlier. Its major advantages are:

- lowest thermal expansion of any foundry sand
- has high thermal conductivity and bulk density giving about four times the cooling rate of quartz
- completely unwetted by molten metal
- chemically unreactive
- clean rounded grains which readily accept any binder
- requires less binder than other sands
- volatiles removed by heating during processing
- superior dimensional and thermal stability at elevated temperatures
- pH is neutral or slightly acid.

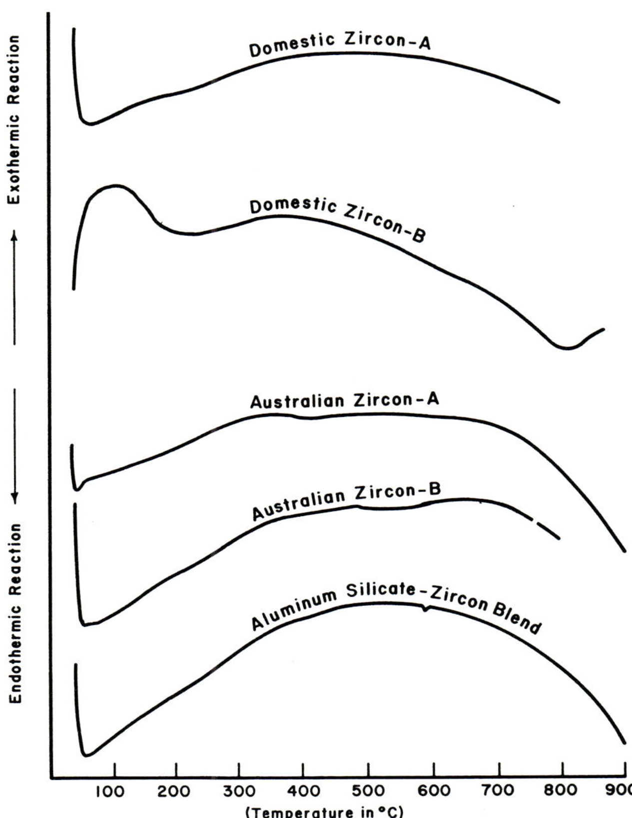

Fig. 28. Differential thermal analysis curves for zircon and aluminum silicate-zircon blended sands.

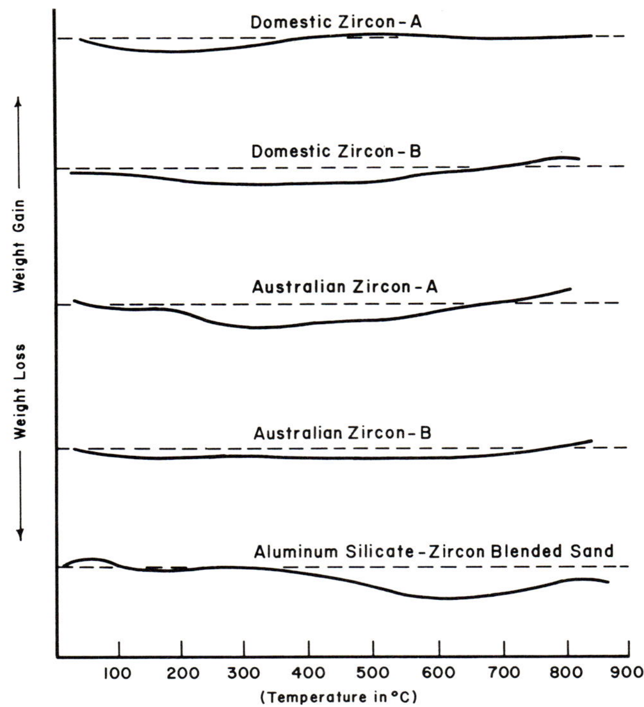

Fig. 29. Thermogravimetric analysis curves for zircon and aluminum silicate-zircon blended sands.

Zircon supply and demand seesaws periodically. In the mid-60s and again in the early 1970s zircon was in short supply. Prices rise during periods of shortages and fall during periods of oversupply. Present zircon prices are low because of a world oversupply which is forecast to last until the mid-1980s, provided beach sand producers mine at current forecast rates. All indications point to zircon supply adequate to meet foundry needs until at least the end of this century.

Samples Studied

Two domestic zircon sands and two Australian zircon sands were chosen for this study. Laboratory work is summarized in Table 8. The two domestic sands are white and the two Australian sands are typically dark-colored. The domestic sands have more rounded particles than the Australian sands. AFS gfn are similar except for the finer domestic B sand. All the samples are near pH 6 except the domestic A sand which is slightly more acid with pH 4.7. Acid-demand values are low except for Australian sand B which is higher, perhaps because of grain surface coatings. The DTA curves (Fig. 28) show all the zircon sands are thermally stable with no phase changes taking place during heating. The TGA curves (Fig. 29) show very little loss of weight due to volatiles for the domestic sands and about twice as much for the Australian sands.

Aluminum Silicates

Kyanite, sillimanite and andalusite all have the same chemical composition (Al_2SiO_5). X-ray study of these minerals shows certain close relations in their atomic structure. The unit cells of all three have nearly the same dimensions. Each crystallize in a different crystal system and form. Kyanite and sillimanite have long been mined and used either raw or calcined to manufacture refractory mortars, cements, castibles and plastic ramming mixes. Imported kyanite and sillimanite occur in coarser forms

and are used in manufacture of mullite brick and for the grog portion of mortar, cement and castible mixes. Crushed kyanite and sillimanite have been tested and found unacceptable for foundry sand use because of their cleavage and irregular shapes produced by crushing.

A mixture of kyanite and sillimanite produced from a beach sand deposit found acceptance as a specialty sand in the late 1960s. This product has been improved by adding zircon and is being marketed as an aluminum silicate-zircon blended sand.

Minerals of the aluminum silicate group are similar in mineralogy, as shown in Table 9. All three minerals are thermally stable up to temperature where they break down to yield 88% mullite and 12% free silica (cristobalite). The mullite is stable to 1810C (3290F). The mullite imparts to the products their desirable properties of high refractoriness, low thermal expansion and resultant resistance to thermal shock, intermediate thermal conductivity, high load-bearing ability even at high temperatures, and resistance to chemical erosion.

All three minerals are similar in origin in that they occur in metamorphic rock. Being chemically stable, all three remain unaffected by normal weathering and erosion processes. Grains of these minerals tend to concentrate as small detrital grains in the heavy minerals fraction of river and beach sands. Many of the heavy mineral beach sands being mined today contain significant quantities of kyanite, sillimanite and andalusite; however, production is restricted to a Florida deposit and to the Eneabba deposit in western Australia.

Aluminum Silicate-Zircon Blended Sands

A Florida aluminum silicate product blended with zircon has rounded grains with AFS gfn ranging between 76-83. Bulk density is slightly lower than zircon and chromite but higher

166

Table 8. Physical Properties, pH and Acid-Demand Values for Zircon Sands and Aluminum Silicate-Zircon Blended Sand

	DOMESTIC ZIRCON SAND A	DOMESTIC ZIRCON SAND B	AUSTRALIAN ZIRCON SAND A	AUSTRALIAN ZIRCON SAND B	ALUMINUM SILICATE-ZIRCON BLENDED SAND
COLOR	White	White	Pale Orange Pink	Pale Yellowish Brown	Pale Brown
GRAIN SHAPE	Rounded	Subrounded to Rounded	Prismatic to Rounded	Prismatic to Rounded	Rounded
SCREEN ANALYSIS (% Retained) Mesh / Microns					
30 / 600	-0-	-0-	-0-	Trace	-0-
40 / 425	-0-	-0-	-0-	Trace	0.1
50 / 300	Trace	0.1	-0-	0.1	3.0
70 / 212	1.9	0.1	0.4	1.0	15.6
100 / 150	20.7	0.2	15.3	10.0	46.4
140 / 106	56.2	8.6	53.8	50.9	25.9
200 / 75	20.4	55.4	28.9	34.5	8.8
270 / 53	0.8	31.7	1.6	3.4	0.3
Pan / 53	-0-	4.0	Trace	0.1	-0-
AFS GFN	102	162	109	114	80
BULK DENSITY					
Lbs/ft^3	187	178	183	176	159
g/cc	3.00	2.85	2.93	2.82	2.55
pH	4.7	5.6	6.7	6.4	6.4
ACID DEMAND VALUES					
4.0 pH	0.63	0.45	0.5	0.91	0.40
7.0 pH	–	–	0.1	0.55	–
WT. LOSS ON HEATING	0.05%	0.05%	0.13%	0.10%	0.05%

than olivine. The product is thermally stable and does not undergo any phase change or heating except for minor amounts of quartz present in the mixture (Fig. 28). Thermogravimetric analysis (Fig. 29) shows only 0.05% weight loss on heating. The mixture has low thermal expansion which takes place evenly throughout heating. Aluminum silicate-zircon blended sands have lower expansion than any of the specialty foundry sands except zircon. The mixture is characterized by relatively coarse, uniform, rounded sand grains with no fines. The grains readily accept binders and can be used with standard coremaking procedures. The mixture has good resistance to shock, high permeability, good chilling action and refractoriness. These properties are believed to result in fast feeding and good directional solidification.[9]

Summary and Conclusions

Each of the foundry sands used today performs differently in each phase of the sand molding process. Some accept binders more readily than others; some perform better in the mold than others and some have better shakeout characteristics than others. Only zircon, aluminum silicate-zircon blended sands and chromite perform equally well in all phases. Use of these three specialty sands will expand in the future. Higher value-in-use and reclamation will offset higher initial cost of zircon,

aluminum silicate-zircon blended sand and chromite because they will produce better castings with fewer rejects and lower finishing room costs.

Use of silica sands in sand molding will decline, not only because they are inferior in molding properties but because they are under attack from hygienists throughout the world due to silicosis hazards associated with them. OSHA and other governmental regulations will certainly restrict future use of silica sands in the foundry.

Thus, with increasing sand disposal problems, government regulation of silica sand and increased use of higher-priced specialty sands, the use of sand reclamation techniques will undoubtedly increase. Zircon, aluminum silicate-zircon blends and chromite will adapt well to most of the reclamation systems in use today. Olivine will not perform well in sand reclamation because grains are fragile and lack the durability required to withstand recycling without producing excessive fines with each use.

Looking at the industrial minerals now available, and those which might become available in the future, none have or will have enough of the physical, chemical and thermal properties required of a superior specialty sand to be substituted successfully for zircon, aluminum silicate-zircon blends or chromite sands.

Table 9. Mineral and Thermal Properties of the Aluminum Silicate Minerals

	KYANITE	SILLIMANITE	ANDALUSITE
COLOR:	Blue to Gray to White	White to Grayish Brown	White to Reddish
CRYSTAL SYSTEM:	Triclinic	Orthorhombic	Orthorhombic
CLEAVAGE:	(100)(010)(001)	(010)	(110)(100)
INDEX OF REFRACTION:	1.71-1.73	1.65-1.68	1.63-1.69
MOHS HARDNESS:	5-7	6-7	7.5
SPECIFIC GRAVITY:	3.6-3.7	3.2	3.2
MULLITE INVERSION:	1200°C	1600°C	1400°C

Acknowledgements

The author is grateful to E. I. Du Pont de Nemours and Co for permission to publish this paper. Thanks are due to many others who assisted in this work: the various mineral producers who furnished the sand samples, A. Graham of Harry Dietert Co who did the mineral sand expansion curves and olivine durability tests, J. Laux and G. Hopkins who critically read the paper and offered suggestions for improvement and the author's many colleagues at De Pont's Experimental Station and Florida R & D laboratories who did much of the work reported in this paper.

References

1. Foundry Sand Hand Book, 7th ed., American Foundrymen's Society, Des Plaines, IL (1973).
2. E. E. Wahlstrom, Theoretical Igneous Petrology, John Wiley & Sons, NY (1950).
3. T. E. Murphy, Industrial Minerals and Rocks, 4th ed., p 1043-1060, AIME, NY (1975).
4. Molding Methods and Materials, 1st ed., American Foundrymen's Society, Des Plaines, IL (1962).
5 J. M. Middleton, F. F. Bownes, "The Properties of Chromite and Its Application in the Foundry," The British Foundryman, p 248 (Aug 1970).
6. H. M. Mikami, Industrial Minerals and Rocks, 4th ed., p 501-517, AIME, NY (1975).
7. E. C. Pirkle, W. H. Yoho, "The Heavy Mineral Ore Body of Trail Ridge, Florida," Econ. Geology, vol 65, p 17-30 (1970).
8. T. E. Garnar Jr., "Economic Geology of Florida Heavy Minerals'" Proceedings of 7th Annual Forum on the Geology of Industrial Minerals, Bur. of Nat. Resources, Tallahassee, FL (1971).
9. W. D. Huskonen, "Spotlight on Specialty Sands," Foundry, (Oct 1970).

Microwave Curing of Core Binders and Coatings

M. J. Skubon
Cober Electronics Inc
Stamford, Connecticut

Introduction

While enormous growth has occurred in fast-reacting sand binder chemistry and core and mold producing equipment, some fundamental problems have been created at the same time. The speed at which these rapidly produced cores and molds are made is not matched downline in the foundry operation so that final molding can be accomplished in the least amount of time from the production of cores and molds.

A major portion of cores and molds need to be coated, post-baked and often glued together. The advantage of microwave processing after initial core production is substantial.

Understanding of microwave equipment capabilities and functional design engineering is necessary for successful applications. While energy savings are significant with microwave heating, its speed, controllability and other unique capabilities must be effectively used to derive effective economical benefits.

The end result can be a very timely tool to answer foundries' growing problems with energy procurement and cost, escalating costs of manufacturing space, shortage of trained technical people and inability to generate sufficient growth capital for economic survival.

Advantages of Microwave Heating For Foundrymen

Advantages of microwave heating in processing of cores and molds are as follows:

- speed of processing - increased throughput
- processes cores and molds at considerably reduced surface temperatures which greatly facilitates handling and promotes stronger labor relations
- low core surface temperature after microwave processing eliminates sand binder thermoplasticity problems
- complete removal of water from cores and molds with associated casting gas defect reduction
- compatibility with low-cost hot-melt adhesives
- potential to eliminate separate core glue curing cycle from coating drying
- ability to use new microwave sand binders now coming on the market which can be water-base and nonpolluting and require lower cost coreblowers and coreboxes
- complete control of processing due to uniform heating and accurate instrumentation which is only possible with microwave electronics
- not dependent on natural gas.

History

The development of microwave technology goes back to World War II and the advent of radar. Initially radar and other military and defense systems were the sole application of microwaves. Today, while these military systems in addition to a gamut of communication applications are developing with a great deal of technical sophistication, the industrial and commercial markets for microwaves are coming of age. Total US industrial sales of home microwave ovens was approximately 2.7 million units in 1977 for a total value of about $980 million. These figures represent only 7% of the total home market. In 1970, by way of contrast, there were roughly 40,000 ovens sold into homes. In industry microwave heating is used in food processing, continuous and batch curing of rubber compounds at fast throughput rates, investment shell dewaxing, curing of plastic foams and, of course, curing of cores and drying of coatings. Microwave heating technology is not simply an interesting new approach to heating and curing — it is a proven tool which the foundry industry needs and is using today.

How Microwave Works

When an electromagnetic wave is propagated in a heatable dielectric material, its energy is converted to heat. To understand this more fully, the molecular properties of dielectric materials must be examined. Water is the major dielectric material in the foundry industry which is heated by microwave energy. The water molecules consist of hydrogen and oxygen ions arranged so that the molecule is electrically neutral. Because of this arrangement, however, the electrical charges within the molecule have a dipole moment and are said to be polar. Different molecules have varying degrees of polarity. An electric field exerts a twisting force on a polar molecule which attempts to align the molecule with the field. When the direction of the electric force is reversed, the molecule attempts to reverse its orientation. However, in so doing, frictional forces created by the molecules rubbing together have to be overcome and energy is thereby dissipated as heat. Friction generates heat and the dielectric becomes hot. Electrical energy which should be stored in the dielectric material is in part lost as heat, a phenomenon called dielectric loss.

Other materials used in foundries which are good heatable dielectrics are phenolic resins of all types, furan hotbox resins, urethane nobakes and all the inorganic systems — sodium silicate and phosphate type. Green sand is also very heatable due to its water content and to carbon and impurities in the reclaimed sand. Reactivity of these binders increases as the temperature from heating increases. This heating rate increase can be rapid; however, with microwave electronics it is controllable. Microwave heating can raise the temperature of the binder to the point at which its exothermic nature carries curing reaction through completion. All the organic binders listed have been cured by microwave to temperatures in excess of 400F (204C).

Oven Selection and Design

Selection of a microwave oven is predicated primarily on throughput of cores or molds to be processed, type of sand binder, size of the cores or molds and amount of water in the binder or coating. Once the quantity, weight and dimensions of the cores or molds required to be processed is known and finally the weight of water borne by the coating and adhesive, then the microwave system can be roughly gaged. The next step is to calculate the power and time required to process the bonded sand and water.

The power is calculated using a value of between 2.0 and 3.0 lb of water removed per microwave kwh and a binder cure rate between 30 and 300 lb of mixed sand per kwh depending on the materials being used. With water the theoretical maximum rate is 3.35 lb removed per microwave kwh. Actual power and time can only accurately be determined by laboratory testing of coated core pieces or better yet entire cores or molds in an appropriate microwave oven, preferably set up in the coreroom. As all sand binders evolve some quantity of solvent vapors when heated, it is important to have sufficient air velocity and flow to carry these vapors out of the oven cavity. This air must be heated so as to maintain air temperature above the dewpoint of water — approximately 130F (54C) — otherwise water and other solvent vapors may recondense in the oven cavity.

These vapors can be made to evolve so quickly that cores or molds crack due to high vapor pressure behind the sand surface. A minimum power-time curing cycle must be achieved so as to maximize the speed capability of microwave and allow for necessary pressure relief within the core or mold.

The oven will be either batch or conveyorized type. Oven cavity length is a function of the residence time for satisfactory heating found in testing. A typical system would have a residence or heating time of 2-8 min. For a conveyorized oven this time is controllable by varying the conveyor speed. This gives full process control capabilities and versatility for processing a wide range of core sizes and materials. Batch ovens require turntables and microwave mode mixers to assure uniform microwave field distribution for even heating. The volume of air and velocity are important in achieving optimum system performance. Air volume and exhaust pressure requirements are calculated from the amount of sand binder in addition to the water contained in coatings and adhesives.

The air flow pattern in a conveyor oven is typically countercurrent to the conveyor flow. Air enters on the exit side of the oven and is exhausted on the input side. This helps for complete water and solvent removal as the last part of the drying process is the most difficult. The air is heated by either flowing it by separate resistance heaters, past the magnetron or microwave power tube or a combination of both. Given the volume of air the size of the heaters can be calculated. This air temperature should be controllable to about 180F (82C).

Electronics involved with the oven selection are the magnetron type and size, the system used to transmit microwave power from the tube to the oven and finally power control and process regulation. Because microwave heating and curing is so rapid, accurate power control is imperative. There are two concepts of microwave power generation. One involves the use of a commercial microwave tube while the other uses a microwave tube of industrial design. Commercial tubes are air-cooled, while industrial tubes generally are water-cooled.

Key equipment features necessary for a modern reliable microwave processing system can be summarized as follows:

1) Reliable continuous operation in the adverse foundry environment via
   - control of electrical environment through cooling
   - steel conveyor meeting most demanding foundry standards
   - use of fully heavy industrial NEMA, JIC and OSHA standards for industrial tubes in sealed environments rather than commercial standards.
2) Ability to operate system without failing, at high microwave power levels into marginal core and coating loads by using the best available microwave tube protection.
3) Versatility to process a wide variety of cores, molds and coatings by selecting adequate microwave power and use of waveguide tuning sections to maximize absorption of power by core or coating.
4) Complete electronic control including full

instrumentation. This is necessary to optimize performance in drying and curing quality and economics and can only be accomplished with a modern electronic system.

Maintenance

As with all foundry equipment microwave ovens require both routine preventive maintenance and repairs. Routine maintenance includes cleaning and monitoring cooling systems, cleaning and lubricating the oven cavity and its conveyor or turntable.

Dirt buildup on the conveyor or turntable and on the walls of the oven cavity will absorb some microwave power and rob the system of efficiency. The latest generation conveyorized systems have shown virtually no dirt buildup in oven cavity walls due to the reliable operation of the support air system. For this reason it is essential that any air filters present be kept clean and blower and fan motors be lubricated.

Magnetron tube life varies between 2000 and 10,000 hr depending on the tube size and type and method of its protection. Commercial-type microwave power tubes have lives of about 2000 hr, while the higher powered industrial tubes are between 6000 and 10,000 hr.

Performance Factors and Operating Costs

Use of microwave to date in the foundry industry has been justified largely on the basis of processing speed. Only recently have accurate operating cost comparisons been made. These comparisons are against conventional resistance-heated and gas-fired ovens, both with supporting high-velocity air. The industry today is beginning to recognize the total-system advantages of microwave.

Whether batch or continuous, microwave ovens are initially more expensive than conventional ovens. They are economically justified through proper use of their speed, smaller space requirement, energy savings and improved core and mold quality.

With microwave heating, water-base coatings can be completely dried with considerably less input of Btu's than is required with a conventional oven. The microwave dried coating is significantly drier also, which means less gas-associated casting scrap such as hydrogen porosity and blows.

With all oven systems there are energy losses involved which are associated with their design and efficiency. In microwave there is roughly a 10% loss in converting from AC to DC power. There is a further 30-40% loss associated with the conversion to microwave energy. The total system operating efficiency of a microwave oven varies between 35 and 45% from AC power into microwave energy coupled into dehydrating water. A conventional oven will operate at less than 10% total system efficiency.

The drying rate of water-base coatings is an excellent example of the speed and efficiency of microwave heating. With conventional ovens about 50% of the total drying time is consumed drying the first 80% of the water. The last 20% of water then consumes about 50% or the balance of the drying time. Because microwave energy penetrates immediately, the entire core or mold mass, all the water or sand binder is energized uniformly and heating starts at the same time everywhere through the mass. No time is wasted while heat convects from the core surface to the interior as occurs in conventional systems.

The drying or curing time cycle required by microwave is only limited by the ability of the core or coating to vent itself without bursting, cracking or blistering from rapid volume expansion due to the binder or coating solvent-forming vapors. A solvent cannot be dried without going from the liquid to the vapor state. This drying time has been as fast as 60-75 sec in actual coated production cores.

With a given microwave oven cavity the amount of microwave power can be increased and drying time can literally be only a few seconds to cure or dry.

Experience to date in drying water-base coatings shows a definite relationship as regards the coating's composition and physical properties. The objective in drying coatings is to couple or absorb the maximum amount of microwave energy into the coating's water. The starting water concentration in the fluid coating is very important. This water concentration can vary from about 30% to 75% by weight. Foundrymen should choose a coating with as little water content as possible. This is easier said than done as the coating is required first to coat the sand surface evenly and penetrate to a critical depth. It also must produce smooth and defect-free casting surfaces. Coating and microwave oven suppliers can help with this selection.

The coating's refractory should preferably be transparent; for example, talcs, micas, pyrophyllite, clays, ground silica, alumina or various alumina silicates are all good refractories that absorb little microwave energy. Carbonaceous refractories on the other hand absorb much microwave power. This can be either through direct absorption or through localized resistance heating between two adjacent particles. Experience has shown coatings based on carbonaceous refractories cure out very quickly initially; however, before all the coating's water is dehydrated the carbon begins to preferentially absorb power and makes complete drying difficult. The surface temperature of fully dried cores coated with carbonaceous refractory coatings will be considerably higher than if an insulating refractory is used.

The coating's depth of penetration is also a factor. This depth should not be excessive; otherwise high water concentrations will be absorbed into the coating and drying time will be increased to remove this deep-seated water. This extra time is required due to high water-vapor pressure deep underneath the core's surface posing a vapor relief problem.

The last factor of consequence is the coating's permeability. High permeable coatings definitely can be dried faster in a microwave oven. This assumes that the foundryman is trying to dry the coating in the least amount of time. Rapid heating rates with low permeability coatings result in core and coating cracking and coating blistering.

Because of the considerably faster drying rates for coatings in microwave ovens, there is a significant reduction in runs, sags and tear drops after drying. Further, the general dried coating film thickness is more uniform. This is due entirely to the water evaporating more quickly and leaving behind a higher solids coating with a higher viscosity which inhibits flow.

Because the sand binder and coating refractories are always somewhat microwave-absorbent, they compete for energy with water when drying coated precured cores or molds. Effective preferential coupling to the water load is achieved by reducing the core's residence time to a minimum in combination with high coupling efficiency to the water load.

The surface temperature of microwave-dried cores or molds

will be a function of the oven's microwave power, residence time, type of sand binder and type of coating refractory. Long residence times — approaching 8 min — combined with carbonaceous coating refractories, would yield coated core surface temperatures of 130-180F (54-82C). This is 100°F (56°C) cooler than is conventional.

Conclusions

The foundry industry's wide expansion into the modern fast-curing chemical sand binder systems, including reclamation, is complemented by microwave heating. While initial microwave equipment costs are high, their wide acceptance appears inevitable. If the fast-reacting core and mold system's processing speeds are to be fully exploited, the speed inherent in microwave is essential.

Current microwave installations and most new interest is for drying water-base coatings. The greatest potential, however, for microwave could well be for rapid curing, pollution-free, novel, sand binder systems.

Mould and core coatings and their application

A. J. Broome

Introduction

From a foundryman's point of view, the purpose of a mould or core coating is to impart a good surface finish to the casting. In order that this may be achieved, the coating must withstand penetration by molten metal thereby preventing the occurrence of such defects as 'burn on' and 'metal penetration'.

Contrary to popular belief, such coatings were not developed only by the modern foundryman. In China during the Han dynasty around 4500 BC it was accepted technology that limestone or clay moulds, baked in the sun to dry, should be coated with a wash of clay prior to the casting of copper. During the reign of King Solomon, the production of bell castings was achieved through the use of a tallow coating over a clay mould. In Europe, according to Biringuccio, typical bronze and brass castings were produced by loam moulding techniques, the moulds being coated with charcoal, crushed cuttle bone, or even tinfoil. Leonardo da Vinci, on the other hand, used river sand moistened with urine or vinegar and mixed with old flour as a moulding medium and coated his moulds with tallow candle and turpentine smoke.

During the twentieth century, however, advances made in coatings technology have been quite dramatic. The earliest of the *modern* coatings appeared in the formulation of clay, molasses and yeast-based washes, whilst charcoal, talc and coke and coal dusts became common around 1905, in the U.S.A.

Heavy cannon were produced around this period by applying a pure graphite and molasses, water-based wash with a camel hair brush. Needless to say, the production rate using these techniques was very low.

In the 1930s coatings utilising shellac varnish, graphite and sodium silicate became common and the first of the *specialist* coatings for steel appeared; this incorporated ferro manganese dust to improve the final casting surface finish and from all reports was fairly successful. Coating fillers comprising charcoal and talc combinations were in common use in the late 1930s and early 1940s but it was not until the 1950s that a variety of proprietary coatings began to appear. Coatings based on zircon became the normal foundry wash during this period and the advent of pastes based on water or alcohol brought about the decline of the home-made blackings in the 1960s. High production rates and the introduction of new coremaking processes in the automobile industry gave rise to the speciality ready-for-use coatings. The total value of coatings used by the western world now exceeds $50 million annually.

The author is Moulding Materials Manager, Foseco Pty. Limited.

Purpose of a coating

Whilst a coating may be used to influence the casting metallurgically, e.g. to prevent microshrinkage in cast iron, its main purpose is to promote a good surface finish. In doing this it may act in several ways:

1) Sand moulds and cores consist of porous masses into which the metal tries to penetrate. If this penetration is superficial, the surface is uneven and rough, but if it is severe a metal-sand crust adheres firmly and the defect is known as metal penetration. The application of a coating which seals the pores may prevent both these forms.
2) The sand may adhere due to melting or sintering. A refractory coating may prevent this.
3) A mould surface which is susceptible to erosion can be protected with a coating which is more erosion-resistant than the surface beneath.
4) Some actions may be either positive or negative. For example a coating may insulate and thus reduce the rate of gas formation. Alternatively a top coating with a high volatile content can help to develop a high positive mould-gas pressure.
5) Coatings may have an effect on other casting defects such as scabbing, sub-surface blowholes and finning.

It would be ideal to have available coatings that exhibit only favourable properties. However, this is not always possible for technical or economical reasons, so one must select a coating with some care.

Constituents of a coating

In any mould or core coating there are four basic constituents that will determine the use, method of application, method of curing and performance when cast:

1) Fillers;
2) Carrier liquids;
3) Suspension agents;
4) Binders.

Fillers

The refractory filler is the active agent in determining the effectiveness of the coating, i.e. in preventing metal penetration, erosion and metal mould reaction. It is selected according to the metal to be cast, the substrate and possible defects.

The refractoriness and the sintering point of this material, together with its thermal expansion characteristics are extremely important. In addition, the filler must be of the correct particle size, must not react

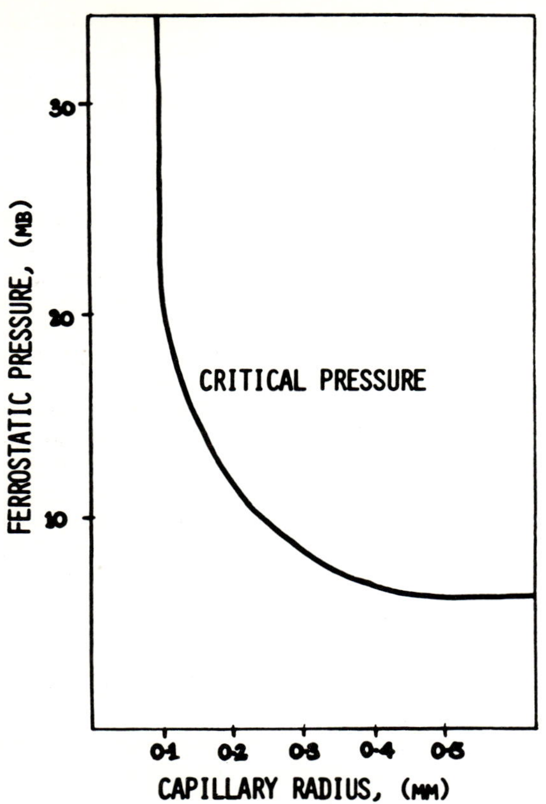

Fig. 1—Critical pressure for commencment of metal penetration as a function of capillary radius

material, the surface tension of the molten metal and the pressure of the metal (Fig. 1).

Refractories which are wet with difficulty or which in doing so form surface slags, act against penetration. If the melting point of the refractory is lowered, the contact angle is decreased and the risk of penetration is increased. In consequence the most important aspects of a refractory are the melting point, reactivity with the cast metal, slag formation, thermal conductivity, thermal expansion and particle size. The properties of the most important fillers are listed in Table I.

Thermal conductivity and thermal expansion of the filler are properties which affect the heat expansion characteristics of the mould substate. Thermal radiation too (white v black coating) plays an important part in the rate of heating of the subtrate (Fig. 2). The fineness is important since it influences the risk of cracking in the fired coating and hence the tendency to penetration and it also partly determines the proportion of binder required; the selected filler consequently should not be too fine. The suitability of various fillers is set out in Table II. Their behaviour may be summarised:

Ground coke – Coke is used mainly in coatings for grey and nodular iron, alone or combined with other refractories. It may vary considerably due to salt contamination. The advantages of coke are high melting point, low reactivity and low wetability for almost all cast metals and its disadvantage is its high porosity which causes slow drying of coating. However coke-based washes have not proved suitable for steel castings even where good insulation of the substrate is required.

Graphite – Natural and electrode graphites are constituents of many coatings. They are highly refractory and are not wetted by many metals, e.g. Wagner and Macherauch reported a high contact angle for pure graphite-based coatings which had reduced the incidence of metal penetration considerably. Pure graphite-based coatings are mainly used in automotive and intricate iron casting production. They have proved unsuitable for steel castings however, owing to solution

with the sand or the metal and should withstand wetting by the molten metal. Certain fillers may contain compounds which evolve gases during casting and these can only be used with discretion.

The penetration of a molten metal into a sand mould is a function of the average pore diameter of the moulding

TABLE I – Properties of some refractory fillers

REFRACTORY	CHEMICAL FORMULA	DENSITY g/cc	MELTING POINT °C
Coke	C	1.6-1.8	—
Graphite	C	2.1-2.3	>3000
Silica	SiO_2	2.65	1800
Zircon	$ZrSiO_4$	4.0-4.8	2200
Mica	$KAl_2(OH, F)_2Si_3AlO_{10}$	2.3	750-1100
Talc	$Mg_3Si_2O_7(OH)_2$	2.7	800-1350
Calcined Magnesite	MgO	3.5-3.7	2800
Alumina Silicate	$Al_2O_3 + SiO_2$	2.6-3.0	—
Chamotte	—	2.6	1700
Alumina	Al_2O_3	3.9	2050

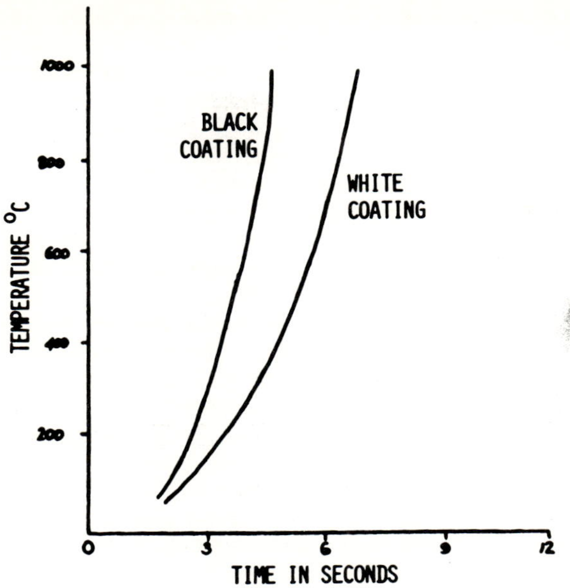

Fig. 2 — Substrate temperature of mould as related to colour of mould coating

in the solidifying metal, but according to more recent Russian investigations the reactivity and thereby the dissolution rate of graphite can be reduced by various high temperature binders. A disadvantage of graphite as a refractory is the high thermal conductivity and heat absorption by radiation, which makes its use unattractive for large castings, particularly when the mould substrate must be insulated.

Silica – Despite the physiological hazards, silica flour is still frequently used outside Australia, mainly for grey iron, especially in conjunction with graphite or coke dust. For instance, research workers such as Middleton, Beeley and Morey found satisfactory behaviour in carbon steels, whilst Kusmenkova reported on good results in castings up to four tonnes weight and wall thicknesses up to 80 mm.

Zircon – In addition to graphite and talc, zircon is the most frequently used refractory. It is used alone or combined with other refractories for steel, grey iron, nodular iron and copper casting. Zircon has many favourable properties. The material can be bonded well and the sintering range begins at temperatures far below the melting point, thereafter stresses are destroyed and spalling prevented. There are scarcely any defects due to erosion, and penetration defects are prevented due to the low reactivity of the zircon and its low wetability.

Mica – Mica comprises a whole series of aluminium silicates which have a flake-like structure. When they are heated water is given off and the structure is changed, with slight expansion. Micas have a relatively low melting point and earlier were used almost exclusively for copper alloys. Recently, mica-containing coatings have been used as insulating coatings in the manufacture of automobile castings.

Talc – Talc is a complex water-containing magnesium silicate, often found in conjunction with magnesium carbonate, which is used as the refractory in coatings for grey and nodular iron and for non-ferrous metals. It has been found that talc additions improve the ability of metal to flow over a casting, whilst dolomitic talc reduces the risk of pinholing in grey iron. The insulating properties have proved useful in controlling the solidification rate in automobile castings.

Calcined magnesite – Due to its high melting point (2800°C) magnesium oxide is used for cast steel, mainly for manganese-containing steels. It must be used in an anhydrous medium, since the formation of hydrated oxides in an aqueous phase leads to a hard packed sediment which is aggravated by ion exchange with the suspension agent, usually a clay.

TABLE II – Suitability of fillers

Alloy	Coke	Graphite	Silica	Zircon	Mica	Talc	Calcined Magnesite	Aluminium Silicate	Chamotte
Non-Alloy Cast Steel	O	O	O	+	–	–	O	–	+
Alloyed Cast Steel	O	O	O	–	–	–	+	–	+
Cast Iron	–	+	–	+	O	+	–	O	–
S.G. Iron	–	+	+	+	O	+	–	O	–
Malleable Iron	+	+	+	+	O	+	–	O	–
Copper Alloys	+	+	–	O	+	+	O	+	–
Aluminium Alloys	–	+	–	+	+	+	–	O	–
Magnesium Alloys	O	O	–	–	–	–	+	–	–

+   Applicable or Common
O   Possible Under Certain Conditions
–   Unsuitable or Uncommon

Carrier Liquid	Formula	Boiling Point °C	Density g/cc	Flash Point °C	Maximum Allow. Conc. p.p.m.	Evaporation Number
Water	H_2O	100	1.00			
Methanol	CH_3OH	64	0.79	6.5	200	6.3
Ethanol (94%)	C_2H_5OH	76	0.80	12	1000	8.3
2 – Proponal	$(CH_3)_2CHOH$	80-82	0.79	12	400	10.5
1 – Butanol	$CH_3(CH_2)_2CH_2OH$	115-117	0.80	28	100	2.5
Light Petroleum	–	80-110	0.70-0.72	20	500	3.3
Methylene Chloride	CH_2Cl_2	39-41	1.33-1.34	–	200	1.8

Aluminium silicates – Harazenk reported that pyrophyllite ($Al_2 Si_4 O_{10} OH$), which has a high melting point, is suitable for the teeming of steel. Sillimanite and kyanite have also been tested successfully for steel castings.

Chamotte – Chamotte is suitable as refractory material for cast steel and grey iron, however pure chamotte-based coatings are not common in many countries.

Other refractory materials – In the last few years various olivine-containing coatings have been tested on the market. However, they have found only limited use.

Chromium ore flour (chromite) has been tried in coatings but many shortcomings, e.g. low fusion point and poor suspendability, limit its use. Rolands has also reported that kieselghur is useful for insulating coatings.

Carrier liquids

The filler is suspended in a carrier liquid, Table III, the choice of which is governed by the moulding material binder system, the operational conditions (drying, exhaust, etc.) and the methods of application.

Water, used with stoved or torched coatings, is without doubt the most important carrier liquid due to its non-toxity, low price and ease of adjustment of the rheological properties. There are few problems provided the formulation is adjusted for water hardness.

For coatings which must be flamed or air dried, e.g. for sodium silicate cores, alcohols (and less commonly petroleum-based solvents) are used. The common alcohols all have suitable burning characteristics. The flame is hot enough to dry and polymerise the binder resin, but not sufficiently fierce to weaken an organic bond, whilst the flash point is not low enough to be dangerous. Isopropanol is the most common: it has medium volatility and low narcotic effect. Ethanol is used alone, whilst methanol and 1-butanol are used only in conjunction with other alcohols. Petroleum-based solvents are seldom employed due to high flammability and high risk of explosion, although the price is low. It is normal to make an addition to adjust the desirable ignition and combustion rates.

Occasionally rapid air-drying solvents are needed, and these are normally of the chlorinated type. They are more hazardous than alcohols and more costly but may be used if adequate ventilation and strong fire precautions are present. Probably 1, 1, 1 trichloroethane is most satisfactory from a safety and toxicity point of view.

Suspension agents

These are used to prevent settling of the refractory materials and to prevent severe penetration of the carrier liquids into the moulding material. Additions of a suspension agent to a mould coating increase the risk of cracking. For instance, it has been stated that there is a connection between risk of cracking, specific weight and bentonite content in zircon-based coatings and the same theory can be applied to all suspension agents and refractory materials.

Bentonites and clays are often used as the suspension agent in aqueous systems. The advantage lies in good thixotropic behaviour and the ability to act as a high temperature binder. The time taken to stabilise coatings which contain bentonite is, however, a drawback.

Celluloses and alginates are also used in aqueous systems. The risk of cracking of the dried coating is lower with alginates than with bentonite, but celluloses produce systems which have a tendency to run and tear. The use of bentonites together with organic thickening agents may lead to breakdown of the thickening system, frequently recognisable by 'thinning out' on the coated mould parts. Knowledge of the colloidal chemical links is therefore important.

Bentones which are bentonites which have become oleophilic due to linkage with long chained amines are used mainly in alcoholic and aliphatic carrier liquids. The amine group is linked to negatively charged bentonite layers and the hydrocarbon chains give the bentonite layers good dispersibility in many organic liquids.

Metal stearates are also used to a limited extent due to their lustrous carbon forming capacity, but a major drawback is the fact that when the coating is ignited they become liquid and tend to cause tears during flaming.

Binders

In aqueous systems, dextrins, sulphite lyes and partly unsaturated carbonic acid esters (linseed oil, etc.) are used as ambient temperature binders. Dextrins and some sulphite lyes are sensitive to bacterial decomposition and the use of anti-fermentation agents is therefore necessary. All ambient temperature binders develop their highest bonding strength only at higher temperatures; however if the drying temperatures are too high, the abrasion resistance of the coatings decreases; there is an optimum temperature which should be specified by the manufacturer.

Linseed oil containing coatings have a strong tendency to cracking which may occur in the intercalated droplets of linseed oil during drying. Drying at room temperature should be avoided due to the hygroscopic nature of some of the binders. Sulphite lyes for example may be hygroscopic due to high alcohol contents.

Bentonites, clays, phosphates and sodium silicate are used as high temperature binders in aqueous systems; however when using the last two compounds, flocculation or peptisation may occur with bentonite. In addition to the conventional water-soluble binders, dispersion binders similar to the systems used in the dye industry are also used. These binders dry irreversibly in air. A residual water content of up to 10 per cent has, however, been measured (depending on the refractory material, suspension agent, air humidity, etc.) and a drying aid (radiator, hot air) is therefore recommended.

Alcohol-soluble synthetic resins such as novalaks, maleinates and ketones, together with natural resins and some alcohol-soluble plastics, are used as binders in alcohol systems. The choice of type of binder is made according to the bonding relationships with the individual refractory, e.g. binders with a good bonding capacity for graphite need not necessarily be suitable for zircon. Migration of the binder towards the surface, skin formation and lustrous carbon formation are also important criteria of choice.

Preparation and application

The object of preparation is to bring a material to such a state that it can be used for a further processing operation. With coatings this is usually done by introducing a carrier liquid and by mixing and stirring. In doing so, chemical and physical processes take place in the colloidal region which may or may not be complete, so that correct preparation has an important influence upon the behaviour of the coating.

Preparation of liquid coatings

Liquid coatings on the market are mainly ready-to-use with either alcohol or water as the carrier liquid. They are usually complicated to make and require special mixers to provide the necessary shear for the liquid to become effective as the carrier. The difficult part of the preparation is undertaken by the supplier.

Preparation by the customer is relatively simple and is a matter mainly of a dispersion process, since separation and sedimentation may occur during storage and transport, particularly in the case of alcohol base coatings.

In some cases a further addition of carrier liquid is then made in order to adjust to the desired consistency and viscosity. The mixing may be carried out manually or better still by means of a paddle stirrer. It is not necessary to apply shear and indeed it is not desirable as it may produce changes in viscosity and flow behaviour.

The modern water base and ready-for-use coatings are designed to be used straight from the container in which they are supplied with no prior mixing or dilution.

This relatively new advance has in fact been known and in use in Australia for a number of years and has spread throughout all forms of casting production. In Europe, the U.K. and the U.S.A. the use of this type of coating is comparatively new and is looked upon as a major technical advance.

Preparation of pastes

As with liquid products, the actual preparation work has already been done by the manufacturer. However, the adjustment to the desired application state by adding carrier liquid is more difficult due to the more viscous consistency of pastes. The paste phase does not absorb the liquid readily and only when the thick mass is broken up mechanically can the liquid phase be absorbed. Therefore it is recommended that the diluent be added step by step. In many cases, the volume of the supply container is not sufficient to obtain a ready-to-use product and it is most advisable to dispense the paste first into a larger container, or in a stationary preparation plant, before dilution.

Application

The common application processes are brushing, dipping, swabbing, and spraying. These methods differ greatly from one another in the technique required. The behaviour of the coating during processing must thus be suited to the application method and the special properties of the process must be taken into account. For every one there are optimum rheological conditions and thus there is therefore no universal coating suitable for all processes. During brushing for instance, brush marks may occur which should flow out, thus there must be flow without the effect of running. Such a flow movement has a very negative effect in dipping, the fluctuations in the coating thickness are too severe and tearing occurs; a dip must therefore have a so-called short flow-character, i.e. it must develop no flow movement or only a short movement in the resting state.

Brushing – Table IV sets out the suitability of coating process for various types of workpiece. Brushing is the simplest and indeed the oldest method of applying the coating. However it takes time and skill if the coating is to have the desired uniformity. This applies particularly when the mould or cores have sharp angles and undercuts.

Nowadays, brushing is used preferably on moulding materials which must absorb little carrier liquid, and on larger moulds. In order to produce as few brush strokes as possible one must work with good, soft squirrel hair brushes and pay close attention to the coating viscosity. The advantage of brushing is that it can be done anywhere on the spot, and that core prints are not coated in the process unnecessarily thereby saving material.

Dipping – Dipping, the most elegant and fastest method, is particularly suitable for mass production, but it can be adapted for 'one off' cores.

The dip tanks can be built into existing production lines without difficulty and in a constant mould part programme the dipping process can be automated; both individual and multiple cores can be lifted by pneumatically controlled equipment and dipped into an appropriate tank. The automatic or mechanised dipping process has the advantage that the dipping speed is usually slower and the process is smoother than hand dipping.

Dipping by hand is completed very quickly and a great

deal of surplus coating is taken up, possibly forming drops which must be removed by shaking. The coating of core prints may be disadvantageous, particularly if the cores later have to be incorporated into larger assemblies. It is not good practice to try to remedy this by reducing the thickness of the coating, since it may become so thin that no useful protection during casting is afforded. For this reason foundries are going over more and more to the dipping of ready-assembled core assemblies.

Swabbing – Swabbing really is an extension of the brushing process but with the emphasis on medium-large jobbing applications. Swabbing can be faster than brushing, provided the correct coating viscosity is achieved, and deep cavities can be coated relatively easily. Application via swabbing can also result in elimination of 'brush' marks and, under certain circumstances, heavy coating thickness can be built up very quickly

Spraying – This coating process has been practised for a long time and specialised equipment has been developed for the application. For very viscous coatings pressure spraying units are preferred. A certain quantity of coating is poured into a container from which it is conveyed under pressure. A second pipe to the gun is fitted with a double nozzle which provides a uniform stream of spray. In order to prevent too severe atomisaton of the carrier liquid, it is necessary to keep the air pressure as low as possible.

The airless process which utilises a high pressure piston pump instead of a pressure vessel is advantageous when using alcohol-based coatings. Unfortunately, the wear on the pistons and nozzles is very high, due to the high spraying pressure of 300-400 bars (4300-5800 p.s.i.).

The airmix process has been introduced recently in some European foundries. This is a combination of the airless and the pressure vessel process. It also operates with a piston pump process but about eight times lower pressure. Nevertheless, wear is still a disadvantage. In

TABLE IV – Suitability of application processes

Workpiece	Coating Method							
	Dipping		Brushing or Swabbing		Flow Coating		Spraying	
Cores – 1 off production:								
Small	A	+	A	+	A	–	A	–
	W	+	W	+	W	–	W	–
Medium	A	+	A	+	A	+	A	+
	W	+	W	+	W	+	W	+
Large	A	O	A	+	A	+	A	+
	W	+	W	+	W	+	W	+
Cores – mass production:								
Small	A	+	A	O	A	–	A	+
	W	+	W	O	W	–	W	+
Medium	A	+	A	O	A	O	A	O
	W	+	W	O	W	+	W	+
Large	A	O	A	O	A	+	A	+
	W	+	W	O	W	+	W	+
Moulds – 1 off production:								
Small	A	–	A	+	A	–	A	+
	W	–	W	+	W	–	W	+
Medium	A	–	A	+	A	O	A	+
	W	–	W	+	W	O	W	+
Large	A	–	A	+	A	O	A	+
	W	–	W	+	W	O	W	+
Moulds – mass production:								
Small	A	–	A	–	A	–	A	–
	W	–	W	–	W	–	W	+
Medium	A	–	A	O	A	O	A	+
	W	–	W	O	W	O	W	+
Large	A	–	A	+	A	O	A	+
	W	–	W	+	W	O	W	+

<table>
<tr><td>+</td><td>=</td><td>Applicable or common</td></tr>
<tr><td>O</td><td>=</td><td>Possible under certain conditions</td></tr>
<tr><td>–</td><td>=</td><td>Unsuitable or uncommon</td></tr>
<tr><td>A</td><td>=</td><td>Alcohol based coating</td></tr>
<tr><td>W</td><td>=</td><td>Water based coating</td></tr>
</table>

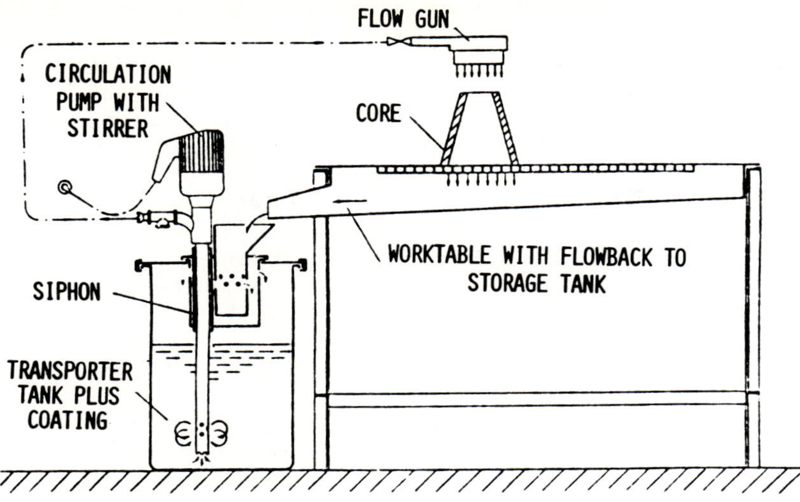

Fig. 3 — Diagrammatic view of flow coating equipment

this process there is a slight atomisation of the carrier liquid.

Flow coating – Flow coating is a new and fast application process which, contrary to dipping, is not restricted to cores, but can also be used for small-medium size moulds. The process is based on a system whereby the coating is conveyed through a fan nozzle from the supply tank by means of a pump (Fig. 3). The stream is fanned out as wide as possible and is conducted over the mould part to be coated. The suspension must be such that the coating will flow as long as coating lasts so that too thick a film build-up is prevented; the coating should thus have a longer 'flowing' nature than a dip coating. Surplus coating liquid is collected in a vat and returned to the supply tank. The shape of the collecting vat should be such that the walls and the vat itself have sufficient slope for the material to flow back to the tank.

The flow coat position of the mould part is critical for a good result. If possible, horizontal surfaces should be inclined so that to thick a coat cannot build up and as few drops as possible will form on the underside. The wear of the pump equipment has been a problem, especially with coatings with hard mineral components such as zircon. More recently pumps have become available which operate according to the rotor/stator principle and whose drive shafts and bearings lie outside the abrasive medium.

The future

In coming years the pattern of foundry technology will favour less labour intensive methods and it is possible that we shall see many new trends. Coatings for increased speed of application will probably become necessary, procedures such as dipping and spraying which require time and in certain cases some operator skill will decline. Methods which will apply the coating in dried form can be envisaged. More complex systems such as electrostatic and hot spraying may be perfected. The use of 'total solids' type coatings wherein a liquid carrier is chemically changed to the solid form which then binds the filler to the substrate is a distinct possibility. These 'catalysed coatings' no doubt will find the same application and acceptance as did the chemically bonded moulding sands not too many years ago.

Along a similar vein, work is already underway on coatings that utilise carrier liquids which can be hardened via the use of a photoinitiator source, e.g. special light radiation. Coating with low liquid 'foamed' carrier systems are also envisaged, these super mobile materials requiring very little drying energy due to the very minute liquid component.

The foundries of tomorrow will certainly have much more rigorous environmental controls. Coatings which are cleaner and less hazardous will certainly be to the fore and it is possible that in 10 years time we may see almost no flammable coatings in use. Thermosetting coatings, requiring only a very fast application of heat to promote a thermo-chemical set, will no doubt come into their own.

It is difficult to predict just when these developments may take place but they will evolve along with more stringent requirements for casting quality, energy conservation and environment improvements.

Economics of coatings

Whether a coating is used is dependent on the 'saleability' of castings in the coated and uncoated state.

The cost of materials is only one per cent of the total manufacturing cost and the cost of the coating process is usually less than 10 per cent of the cost of moulds and cores. On the other hand a suitable coating can substantially aid surface finish while reducing fettling costs and scrap losses. Thus the choice of coating is dictated not by its cost but by its favourable influence on quality and economy of production.

The benefits have been realised and the use of coatings is rapidly increasing. They have certainly come of age and through ever increasing technical improvement, form an increasingly important part of successful and profitable casting production.

Tooling

Practical patternmaking techniques for the foundry industry

J. W. Francis (Associate)

Introduction

The increased overall competition which the foundry industry faces from alternative methods of manufacture, and even initial material specification, make it essential that modern patternmaking equipment is of a standard only recently being appreciated by many users and their customers.

Current developments in various foundry manufacturing processes also make it necessary to ensure that sufficient attention to detail is given when planning the specification for pattern equipment.

Casting users are becoming increasingly more demanding in their requirements for sound, accurate and consistent castings and this can only be achieved with high quality pattern equipment correctly designed to suit the particular requirements of the foundry's manufacturing process and the customer's specifications.

Due to the demand for castings of consistent dimensional accuracy, the use of core loading fixtures, mould and core checking gauges and casting machining/checking fixtures is growing and patternmakers must be able to offer a wide range of these items to suit individual requirements.

Availability

Historically, patternmaking has been divided between wood and metal. Since the second world war, resin, and in particular epoxy resin, has added a third material from which to manufacture the equipment. Therefore, the first decision to be made is between:

(1) Wood
(2) Resin
(3) Metal

In order to assess which to use, an accurate estimate of the required number of castings to be produced is essential. Far too often customers are guilty of under-stating production requirements and of pressuring the foundry to use equipment which is unsuitable. This short-sighted approach is, at last, beginning to be replaced by more realistic attitues, but foundries must make customers aware that correct initial planning will eventually save them money by longer pattern life and also in the higher quality castings which they will receive.

Once an accurate estimate of production is made, the remaining factors which govern equipment specification are those production methods at use in the foundry and any casting finishing the foundry may have to do in excess of normal.

The author is a director of The Vulcan Pattern Making Co. Ltd. (F. 1340)

Wood

The traditional material for general patternmaking is available in a wide selection of grades e.g. mahogany (African and South American), oak, pine, jellutong, English lime. Also there are a variety of plywoods available with mahogany or birch faces, together with a few specialised heavy boards. Most of these materials are available in differing grades to suit particular requirements.

Of these, mahogany is principally used where either regular production is expected or very accurate castings are needed. The two main types available, African and South American, are both of high quality, although the African is becoming more difficult to obtain in the top grades. However, to balance this, British Honduras is available once again and generally offers a better quality than the Brazilian, which has become fairly popular with the recent supply difficulties which the other countries have experienced.

Canadian yellow pine is an excellent general-purpose timber and is readily available, kiln dried, in varying grades. With the additional use of English lime recently, these two timbers provide a good material for the manufacture of short run equipment and large one-off work.

Jellutong is a relatively 'soft' wood and, due to this, is mainly used only on one-off work. It is also readily available in a variety of grades and, together with polystyrene, where possible, can be a fairly inexpensive material for one-off work both from its initial cost and the speed with which it is possible to work.

Of the plywoods available, the one best suited to patternmaking is the mahogany faced with, preferably $\frac{1}{8}$ in. thick facing veneers each side. This allows the faces to be worked flat, if necessary, without breaking into the glue lines and, when painted, displays excellent wear characteristics on *drag faces* even where airset sands are used. Also, as a very stable material, it keeps well in store.

There are available certain types of *heavy* hard-wearing wood boards and, in general, the pattern shape is machined from solid blocks or laminates. Stability and wear resistance are claimed to be very good. However, it is a very heavy material in relation to other woods or resin and also it is fairly expensive to work, and applications are generally limited.

Providing the equipment is correctly manufactured and carefully handled and stored once in the foundry, there is no reason why a mahogany set of pattern equipment should not be capable of producing good quality accurate castings consistently.

Even when used with demanding cold set mould or

coremaking sands, either CO_2 or chemically hardened, providing thorough care and planning is taken with the construction of pattern and coreboxes, regular production can be maintained.

The main point to consider, particularly with coreboxes, is that wherever possible, the mould or core should not be allowed to *drag* along the faces of the timber.

In Figure 1 a corebox is illustrated which has been constructed to virtually eliminate wear due to drag. The corebox in this case consists of a mahogany frame, generously tapered, which has the main print in the bottom, and sets into which the four sides are located. After filling with sand, the box is rolled over by means of trunions either end to ensure minimum damage during handling, and the frame lifted away. The four sides can then be easily removed from the core with virtually no wear taking place.

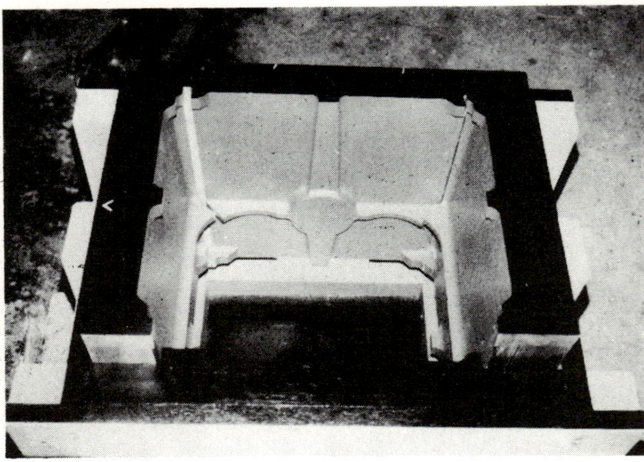

Figure 1

This principle is essential if patterns and coreboxes are to withstand the abrasive action of such systems now in general use. An added bonus of this method is that, when employed on both pattern and corebox construction, taper can be virtually eliminated from most areas with resulting benefits to metal sections and customers design requirements.

When constructing large equipment which cannot be constructed using the previous method, great care must be taken to ensure sufficient strength is achieved to withstand the strain of constant stripping from airset moulds.

Figure 2

Figure 2 illustrates the pine skeleton frame which has been constructed to provide the strength and rigidity necessary. The frame has been built up with 4 in. $\times$ 3 in. material half lapped together and adequately glued and screwed. Then 6 in. $\times$ 3 in. runners have been used to tie the frames together and steel angle iron fixed securely to give additional strength.

Figure 3

The plywood sheets, Figure 3, are then fastened to the frame and form a strong, and when painted, smooth surface which will support continuous production for a number of years. Figure 4 also illustrates the holes which are in each end of the pattern to facilitate easy *rollover* after stripping. It is essential, particularly with large equipment, that the patternmaker builds into his equipment the means by which patterns and coreboxes can be easily handled in the foundry. It is often found that as much damage to the equipment is done by rough handling as by actual moulding and coremaking. Rapping and lifting plates, carefully positioned, are essential to prevent damage to patterns, and trunions and lifting straps on the side of coreboxes can not only protect the equipment but also help to improve productivity in the foundry.

Figure 4

Another area where the patternmaker can help to give not only a better casting but also improved productivity in the foundry is fettling. Fettling is a continual problem to the industry and, in particular, where worn wood equipment is in use, additional work is created by virtue of poorly-fitting loose pieces and badly located cores. The excess flash created is often removable only by handwork which adds further to the problem.

A great deal can be achieved in reducing the fettling content of a casting, therefore, if the patternmaker takes special care over the location of loose pieces and the accurate setting of cores. This is particularly noticeable where cores are forming external undercuts on the outside of a casting, as shown in Figure 5.

Figure 5

One method of ensuring accurate cores is to cut the pattern to the final shape regardless of undercuts and to produce a resin corebox by the following method.
(a) Area to be cored *shuttered* off, carefully producing the print shape required.
(b) Fill-in with resin using the cast or laminate method.
(c) Strip resin from pattern. This is then a model of the core required.
(d) Produce a resin corebox from this model in the normal way.
(e) Reposition the resin core model back on to the pattern to act as the print, Figure 6.

Figure 6

It is apparent that, by using this method, the corebox will produce a core that is an exact match of the moulded pattern profile and also a true fit to print and sets. This will result in cleaner castings and much reduced fettling times.

Where it is necessary to have loose pieces on wood equipment the method of location must be as positive as possible, Figures 7 and 8. Where the loose piece to be

formed is small, a dovetail location provides an excellent method of ensuring a true fit with minimum movement during moulding and if a metal dovetail can be incorporated then little wear should take place.

Figure 7

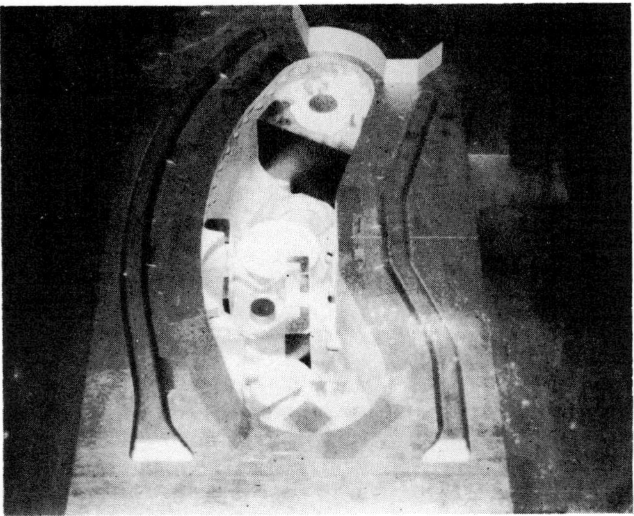

Figure 8

However, on larger pieces it is preferable for the construction of the pattern to be such that bolts can be accommodated. This is achieved by constructing the pattern with a hollow interior and a moulder can then bolt the loose pieces in position before moulding and, after rollover, remove the bolts from the back of the pattern before stripping.

It is also important to provide the moulder with a method of removing the loose pieces from the sand without damage. This can be achieved by using either hand-holds for large work or wrapping plates with a threaded hole for smaller pieces.

By use of this method, together with dovetails, it is possible to ensure that the loose pieces are not only correctly fitted when new but will continue to give a minimum of flash or mark on the casting after regular use.

Where production requirements are very limited it is possible to use pins to set the loose pieces in place. This method, which consists of simply drilling two holes through the loose piece while it is held in position and then using steel rod to dowel the piece in position, is not very accurate as it relies on the moulder *holding* the loose piece in position with sand while the pins are withdrawn and then continuing ramming up with the resultant risks of movement. However, it is considerably cheaper than the other methods described and is therefore used where tolerances allow and very limited quantities are required.

Generally it can be taken that, with the care and attention to detail which the patternmakers must have regard to in the construction and planning of patterns, most of the customer's requirements can be met from one-off to steady continuous production for a limited period. However it is very important for the foundry to realise that wood is a *living* material and, even when painted, great care should be taken to ensure it is not subjected to variations of temperature and humidity if the accuracy is to be maintained. Where a job is taken out of production and put into store, it is necessary to ensure that, wherever possible, conditions in the store should be similar to those in the foundry it is *most* important that the equipment must be kept dry. When consideration is taken of the cost involved in obtaining modern equipment, it is only logical to look after it. The cost of a dry clean store is well worth it in the long term.

Resin-based paints have helped to protect the timber from modern chemicals used in the foundry and, if airset sands are to be used, it is necessary to use this type of paint instead of the spirit-based traditional varnishes which have not got sufficient resistance to attack from chemicals, and soon lead to poor mould and core quality due to stripping problems.

Resin

Where production warrants the additional costs, resin pattern equipment provides an excellent tool for foundry use. Not only will it stand up to the abrasive action of the sand for longer than wood but it is also not subject to the temperature and humidity problems which affect wood, particularly in storage. Also, an excellent surface finish can be obtained together with very positive loose piece and core location. Figure 9 illustrates the use of aluminium loose pieces where thin rib sections are to be formed, and the location of these pieces obtained by moulding them into the box is excellent.

Figure 9

There are many different types of resin systems available to the industry and each one has its own particular applications. However, in general, a basic system suitable for a wide range of work can be based on the conventional three part mix:

(1) Resin
(2) Hardener
(3) Accelerator

To these, a suitable filler can be added to give the required properties for either gel-coat or cast applications. Additionally, further additions of suitable filler will provide a very useful backing-up mixture for giving extra reinforcement to laminated structures. The type of filler chosen should be easily mixed with the resin, give good stability after curing and have sufficient hardness to withstand the wear of constant use in the foundry.

Many different types of filler are available, and of these slate powder and marble flour have proved excellent for gel-coats and casting applications in general. (Slate powder can be used for both methods, but experience and experimentation will prove which are the most suitable for specific applications.)

When deciding on a resin system of this type, it is very important to ensure that all the parts are available in suitable containers which allow easy measuring out by weight and can be resealed with a minimum of spillage. The reason for this will soon be discovered as the *treacle-like* consistence of the resin will soon create a mess if not carefully handled. The best system is to have large drums with taps which allow the material to be poured directly into the mixing bucket close to the scales.

Care with mixing proportions is especially important as the ratio between the three parts is the determining factor controlling cure time and heat generation. With large work too much accelerator can result in temperatures that will damage the master model, particularly if in wood, and result not only in the loss of the resin part but costly damage to the model.

Many systems are now available in two part mixes, some with filler already added, and pre-measured, so that it is only necessary to add one can to the other and apply.

However, these systems can be very wasteful and are best used on specific applications.

Gel-coat application does provide an excellent field for using two-part systems, but the constituents must be in easily resealable containers.

There are a variety of *special* resins available which are pre-mixed and contain various special purpose fillers for which superior wear resistance is claimed. Silicon carbide is one common example of this type of material and, to a certain extent, they do provide better wear resistance than general resins. However, despite many attempts to find a suitable substitute for metal, either aluminium or cast iron, none has yet been found. This is particularly the case where modern *Cold Set* chemical systems are in use for machine coremaking.

A recent addition to the resin field is the *Quick Set* material which, again, comes as a two-part sytem which, when mixed, will set in approximately thirty minutes and be stabilised for stripping after ninety minutes. These systems are almost entirely used for cast applications and the main limiting factor is the pot life which restricts the size of job it is possible to cast. However, developments are continually taking place with these systems balancing pot life with curing time.

Having decided which type of material to use, the next decision is how to produce the resin part. This is decided in most cases by the complexity of the casting. The most commonly used technique is to manufacture the pattern or corebox in wood, take a resin mould or model from it, and then finally produce the part from this. (Figure 10.)

Figure 10

In some instances, particularly with cores, it is possible to cut a wooden model straight from the drawing and, off this, produce the resin corebox. This method is extremely useful when making coreboxes for patterns which have been made by the conventional approach as it is possible to try the wood core models in the resin mould to check fit and metal thickness before producing the resin corebox, (Figure 11). A further benefit is that, in the case of complex core assemblies, it is possible to supply the moulder with a permanent mould complete with model cores so that the method of core assembly and support can be thoroughly finalised before the sand cores are made.

Figure 11

As stated previously, resin pattern equipment is very stable after final curing. It is necessary, however, to ensure that, where lamination techniques have been used, sufficient care is taken with the thorough application of glasscloth and matting with a backing up of suitable thickness particularly where the back of the pattern or corebox is to be used for wrapping.

The question of supporting the resin corebox when in use is one which must be given a great deal of consideration especially where it is necessary to keep large areas flat. The best method of support in these cases is a fabricated steel frame which can be tailor-made to suit the exact requirements of each job. These frames should be designed not only to give rigidity and protection to the corebox during coremaking and stripping, but should also incorporate aids to foundry handling such as lifting straps and rollover trunions, (Figure 12). A great deal of damage to the equipment can be prevented by these simple fittings and also the speed and ease with which large cores can be produced will soon repay the initial cost.

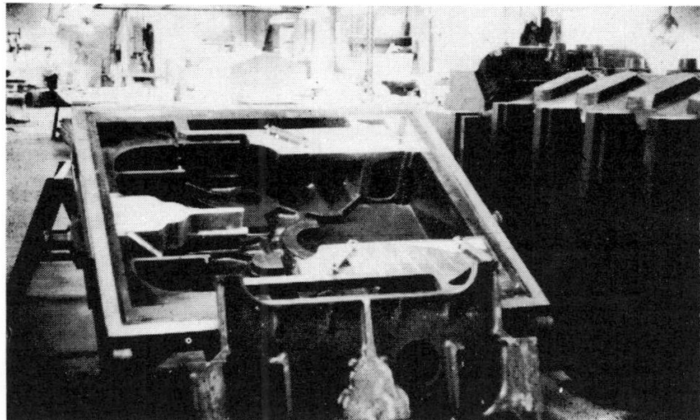

Figure 12

Where the size or shape does not justify a steel frame, then aluminium castings provide an ideal alternative. These can be used either to construct a frame into which the resin can be keyed, or in the case of patterns, a rigid shape on to which the resin can be moulded or cast, and which subsequently provides an excellent area for bolting down on to the moulding plate where required.

In some instances where low production restricts the use of metal frames due to cost, it is possible to use hardwood providing the timber has been carefully seasoned to eliminate warping and distortion as far as possible. However, for the majority of small and medium sized work, the resin will not need any additional support for low production if correctly constructed.

Loose pieces can be used with excellent effect on resin equipment, as it is usually possible to *cast them on* the core or pattern model and ensure a good location. Dovetails provide the most common form of fixing the loose pieces on to the main part and very little wear takes place when in use.

Figure 13

Figure 13 illustrates a complex large core produced from a resin corebox which consists of three main parts fitted to slide easily into a tapered steel frame. The many loose pieces are in this instance bolted into position through the corebox sides before the sides are slid into position. The box is then filled with sand, rolled over, (with trunions), bolts withdrawn from loose pieces, steel frame lifted away (lifting straps), and the main parts of the corebox drawn off leaving the loose pieces to be finally removed.

Where regular production is required for cold set cores, especially CO_2, it is sometimes possible to achieve this using resin coreboxes which have been cast on to metal *backings*. The backings can be of either aluminium or cast iron and, when an irregular shape is to be followed, castings can be produced to enable the cast resin thickness to remain reasonably constant. The method of manufacturing these coreboxes is to produce a master model of the core in resin or metal and, after setting, the metal backing-up pieces in position, cast the resin into the resulting mould. The resin section thickness may vary, although the more constant it is kept, the better cure will result. $\frac{3}{8}$ in. (10 mm) is a good section to achieve if possible, (Figure 14).

Figure 14

An advantage of this system is that, although the corebox will not have a life equal to aluminium or cast iron, it is relatively cheap and easy to manufacture replacement coreboxes by *burning off* the old resin and recasting the box from its permanent model. This ensures that, even with complex blends and curves every replacement box will be identical. Furthermore, the resin will adequately hold corebox vents and any steel inserts necessary.

The area where most care is needed is the surface preparation of the metal backing to ensure good adhesion by the resin. Shot blast and degrease is essential, even when using new castings which have not been previously burnt off.

Metal
Where sustained production is required with corresponding consistency of dimensional accuracy, metal pattern equipment will be required, especially if the foundry methods produce heat, as in the shell mould process for example, to cure the mould or core.

Cast iron and aluminium have proved themselves to be excellent materials for the production of pattern equipment and, although occasionally other metals are used, these two offer satisfactory properties to meet most foundry production requirements.

Grade 17 grey cast iron is the most popular type of iron used, as it combines excellent wear resistance with relatively free machining and working properties for the pattern shop. It is always desirable to ensure the castings have been stress-relieved prior to final machining and fitting and, in the case of equipment to be used for hot processes, this becomes essential to avoid distortion in use.

SG iron has become increasingly popular for use in equipment, particularly from the patternmaker's point of view, because of its easier machining properties. These become even more noticeable when copy milling is to be carried out.,

Aluminium, while not providing the same degree of wear resistance for the foundry industry as cast iron, does possess a reasonable degree of hardness and this, coupled with the much easier machining properties, enables equipment to be manufactured for less cost than cast iron.

Aluminium is available either as castings or forged billets. LM25 and LM4 are the two most popular grades of cast aluminium used and, dependent upon job configuration, it is a choice of which will give the *sounder* product. LM25 has advantages over LM4 and, if the casting is to be machined all over, then a better result is usually obtained than with LM4, although at a slightly higher casting cost, and the need to heat treat to LM25TF if machining properties are to be fully realised.

Due to the need to eliminate any surface defects on the faces of patterns and coreboxes to be used for cold set or shell mould techniques, forged billet is being used more frequently wherever possible. This is available in a variety of grades and the two most commonly in use are NP8 or HP30. Both are produced in a wide range of plate thicknesses from 0·030 in. to 6 in. – if required. The main disadvantage in using billet is the amount of additional machining created as no part of the profile can be followed, as in the case of a casting. However, for equipment to be used in the processes previously mentioned, this initial cost can be offset, not only against the saving from not having to produce a *master pattern* in order to obtain a casting, but also against the additional work created when having to fill areas of porosity, found after machining, occurring on draw faces. If this is not done, the working life of the equipment will be greatly reduced.

The original method of producing metal equipment was for the patternmaker to manufacture a wood or resin *master* as accurately as possible and then ask the foundry to make a casting from it which after a minimum of hand cleaning up could be used by the foundry.

However, with accuracy becoming increasingly important, it has now become generally accepted that, in order to meet customer's requirements, and also to offer further improvements in casting quality, fully machined patterns and coreboxes are needed. This is becoming more the case as developments in foundry plant require equipment to be complete with ejector systems, blow plates, and many sundry parts to enable cores or moulds to be automatically produced.

However, certain patterns do lend themselves to the original *cast to shape* method and many advances have been made by specialised foundries to enable better castings to be produced. Yet, no matter what the foundry achieves in maintaining accuracy it will only be as good as the original wood or resin master pattern and it cannot compete with a fully machined product. This applies equally to cast iron and aluminium.

Generally the principles employed in the construction and layout of metal pattern equipment are similar to those used in wood or resin. However, due to the high cost of modifying metal equipment, it is necessary to ensure that sufficient care is taken in the planning stage to eliminate production problems. It is often beneficial to draw out the equipment in full so that various allowances can be made at the planning stage for ease of use in the foundry. Many of these allowances will have been established by trial and error to suit the particular mould or coremaking machine that is being used. They include vents, heaters, ejector distribution and clearance, blow tubes, clearance for moving parts and many other factors which aid in the easy maintenance of equipment in the foundry.

Ejector replacement is an area where many foundries have problems with coreboxes and it is possible to save not only hours of down time but also patternmaker's time by developing a simple system to enable ejectors to be changed with the corebox on the machine. Also, the ease with which patterns and coreboxes can be taken on and off the plant in the foundry will determine the downtime required for changeover and simplicity will pay dividends.

Conclusion

The paper has tried to illustrate a variety of factors which need to be taken into account when deciding on the material to be used and the method of construction suitable for producing various types of pattern equipment.

It is possible for the foundry not only to save a great deal of production time but also, in many cases, equipment costs by consulting fully with the patternmakers before ordering. This is particularly the case with metal equipment which must be tailor-made to the foundries requirements.

If full advantage of the many new developments in the industry is to be taken then it all starts with good quality patterns. In order to meet its customers' existing requirements and offer further advances in dimensional tolerances achieved, the equipment needs to be of the highest standard.

Much has already been lost to other manufacturing industries and if the trends are to be reversed and casting is to expand its share of the markets and develop into others, then all those involved in the production of castings from the patternmaker to the fettler must work together. This will become increasingly important as the machine tool and automotive industries contract. New industries are developing and it is only by offering high quality competitive castings that the industry will achieve growth and expansion instead of the current contraction which is so apparent.

Acknowledgements
The author wishes to thank Sterling Metals Ltd., Midland Motor Cyl. Ltd., and A & N Foundry Ltd., for their help and assistance with the illustrations.

SPECIALITY EPOXY AND POLYURETHANE RESIN SYSTEMS FOR THE MANUFACTURE

OF FOUNDRY PATTERN EQUIPMENT

by

G.E. BARGEHR

Technical Sales Representative
CIBA-GEIGY (Pty) Ltd, Spartan, Kempton Park, R.S.A.

<u>SYNOPSIS</u>

Growing competition from other production processes has forced foundries
to devote more attention to the accuracy and consistency of their products. The situation today is typified by the widespread adoption of
semi-automatic or fully automatic moulding techniques. At the same time
epoxy and polyurethane resin pattern plates have widely replaced both
the unsuitable wooden plate and the uneconomical metal plate. The advantages of resin pattern equipment and the primarily used speciality
tooling resin systems are discussed.

<u>INTRODUCTION</u>

Two of the main contributors to the cost effectiveness of a modern foundry are the reduction of production times and the raising of standards of
accuracy and finish. Pattern-making materials such as epoxy and polyurethane resins, have played an essential part in furthering these ends.

With conventional pattern-making methods employing wood, greater accuracy
can only be achieved by the use of better and more expensive timber and
by giving more time to fabrication. This additional expenditure and
time spent is justified in the making of master-patterns, but when
production runs are to be long, requiring many patterns of the same
design, substantial savings can be accomplished by making them in resin
via a master mould. At the same time accuracy and finish are fully preserved. Another advantage of resin patterns is that they stand up very
well to mechanised moulding procedures. When properly designed in epoxy
or polyurethane they will perform comparably with metal, but will however show considerable cost reduction.

<u>PATTERN MAKING MATERIALS</u>

As hand moulding has been gradually replaced by mechanised procedures,
so pattern making techniques have been progressively modernised. Wood
patterns, occasionally metal ones, mounted on either wood or steel plates
were at one time used with all types of foundry equipment. Wood patterns
although fundamental as originals or masters were not very satisfactory
from the point of view of dimensional accuracy, especially when they had
to be used over and over again. Metal patterns were found to be not
always economically acceptable, particularly for short and medium production runs.
The removal of patterns by hand has been made obsolete by tremendous
improvements in mechanical moulding techniques. Patterns are now demoul-
ded by a variety of methods e.g. jolt squeeze or high pressure moulding
machines. Whilst wood was satisfactory in conjunction with hand demoul-
ding, it is not adequate when used with modern machines, particularly
when the high pressure process is used. The higher the quality requirements for the finished casting, the greater the disadvantages of wooden
patterns. The following characteristics are particularly unfavourable:-

- warping of unseasoned or poor quality wood.
- surface roughness at gluelines in plywood.
- dimensional changes resulting from the hygroscopic properties of wood.
- some loss of accuracy with each release.
- release properties can be checked only in the foundry instead of in the patternshop.

Careless removal of the pattern from the mould or, of the core from the corebox, causes dimensional variations from casting to casting. In future, as quality requirements and the pressure from competitors continues to increase, wooden patterns will be suitable for production work only to a limited extent. The relatively complicated manufacturing methods and high costs of metal patterns are major factors in bringing synthetic resins to the fore.

Epoxy pattern plates (upper and lower mould halves) for use on an automatic machine.

EPOXY RESINS

Due to the disadvantages of both wood and metal patterns, synthetic
resins were considered at an early stage. They were found to offer a
compromise between the softness of wood and the hardness of metal. An
extensive test programme showed that epoxy resins were by far the most
suitable plastic materials for pattern making. They are superior to
other synthetic resins in many respects, the main ones being :-

- No volatiles are released in curing.
- There is very little curing skrinkage, and what there is can be
  reduced still further by adding fillers such as silica flour or
  silica sand, slate flour, chalk flour, iron powder, aluminium
  granules, glass cloth or cotton flocks. These fillers do not participate in the curing process but they lessen the amount of exothermic heat produced and this minimises shrinkage. Furthermore,
  the addition of filler permits the casting of larger batches and, at
  the same time, offers a considerable reduction in the cost of the
  casting material. The addition of fillers also allows the mechanical
  properties of the castings to be adjusted within certain limits to
  meet the requirements of specific production situations.
- Pattern equipment made from epoxy has high dimensional accuracy and
  mechanical strength, making possible the production of castings of
  exceptional quality and precision.
- Epoxy pattern equipment exhibits good thermal properties, excellent
  ageing characteristics, and high resistance to moisture plus almost
  all of the chemicals likely to be encountered in a foundry.

The production of epoxy patterns starts with an accurate master pattern,
usually made of wood and/or epoxy modelling paste. The pattern is then
coated with a sealer and treated with release agent. An epoxy negative
is built up on this master pattern - that is to say an epoxy gelcoat is
brushed on, backed with a glass cloth laminate and reinforced with a
heavy-filled backing mix. A negative made this way provides a mould
from which any number of accurate, identical patterns can be made. It
can be kept as a master mould in case the need should arise to make
replacement patterns.

An epoxy master model paste is being finished down to the rib contours
with woodworking tools.

Epoxy patterns and fittings offer the following main advantages over
traditional wooden equipment :-

- greater hardness
- excellent dimensional stability
- can be removed from the mould without vibration.
- when stored for long periods and used for repeat orders, suffer no
  dimensional changes.

Other advantages are :-

- minimum flash is produced at parting lines on the casting.
- surfaces perpendicular to the plane of the mould have good sliding
  properties.
- checks on wall thickness can be carried out in the patternshop so
  that trial castings are unnecessary; this saves a considerable
  amount of time.
- the release properties of the casting can also be checked in the
  patternshop.

- Checking wall thickness and core clearance -

Laying the core patterns into the finished negative moulds affords
a simple and effective check of wall thickness and core clearance.

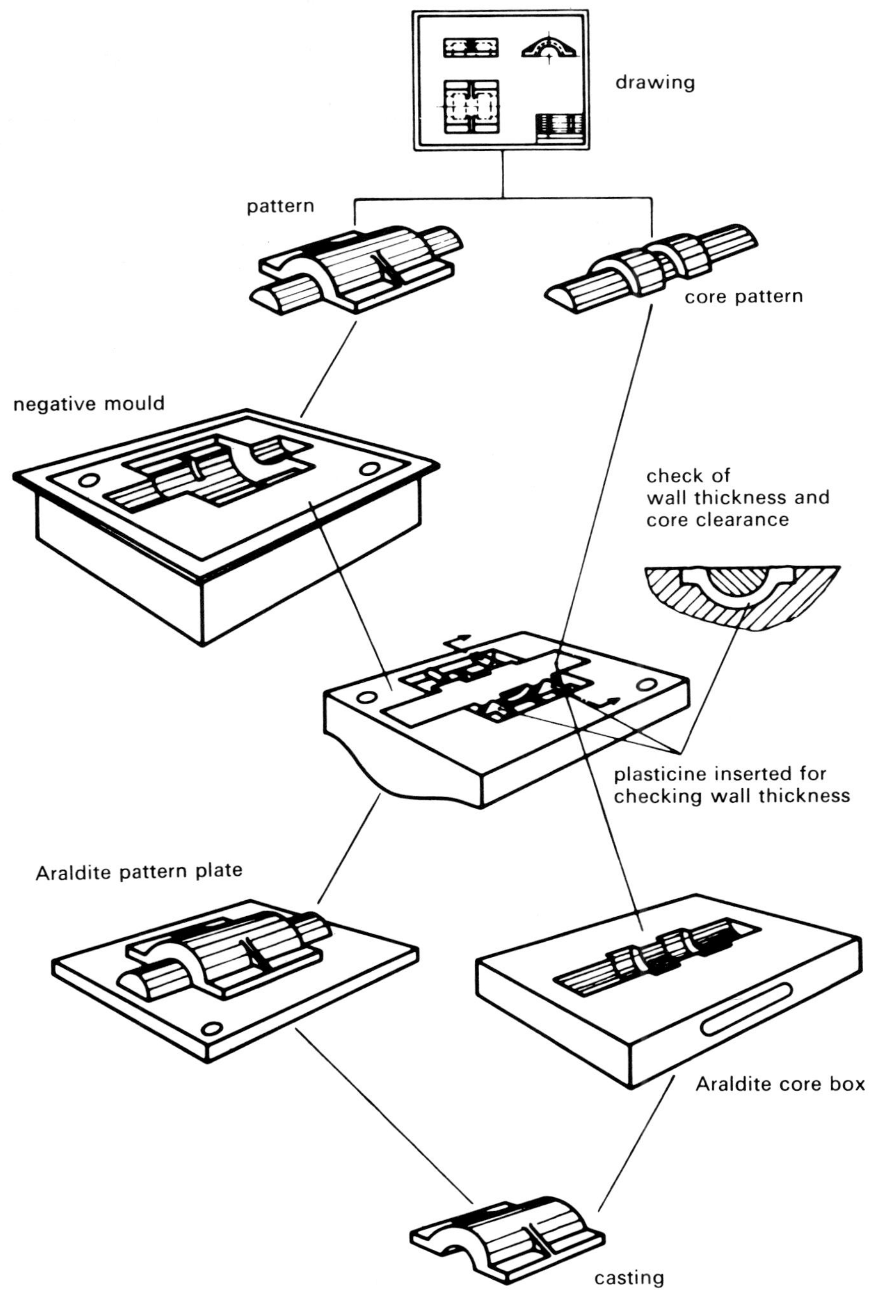

Source: International Foundry Conference 79

POLYURETHANE RESINS

With the development of Polyurethane elastomers the replacement of metal
patterns/coreboxes is within reach. In the initial phase, with soft,
very elastic systems, considerable problems were experienced, but extensive research has led to the development of specialised polyurethane
systems which in many respects are very suitable for high performance
foundry equipment.

Full scale production tests on sand slinger machinery have shown the
abrasive resistance of polyurethane resins to be generally superior to
that of metal. Tests with polyurethane faced pattern plates on normal

and high-pressure moulding machines have shown very satisfactory results
also. The limits in terms of applications are not yet completely known
but probably not all problems can be solved completely with these
materials. Polyurethane elastomers can at this stage not be recommended
for the manufacture of heated pattern plates used on disamatic machines
and for moulding processes where the pattern equipment comes in contact
with warm sand.

Polyurethane elastomers are extremely tough, resistant to wear and
abrasion, thus making them highly suitable for patterns and coreboxes in
the foundry industry. Having selected a suitable system, the chemical
compatibility of the various cleaners, parting agents, sand binders and
their curing agents should be checked prior to production. In foundries,
where the so called 'cold-box-core-technique' is used, the different
types of sand result in a wide variation in the quantity of binders
required. If, due to incompatibility, a chemical attack occurs, this
could result in light swelling on the surface of the pattern. Petrol or
petrol based products have only very limited effect, especially on very
hard polyurethane casting resins. Isocyanates and triethylamine can
affect the surface, in which case, contact time and concentration
will be the major factors influencing the degree of damage sustained.
Another important factor is the hardness of the polyurethane system
chosen.

In some foundries, polyurethane faced patterns are used with great
success, yet in other foundries epoxy resin pattern equipment performs
better. Practical comparison tests must be carried out to determine
which material is more suited to the particular foundry and application
required.

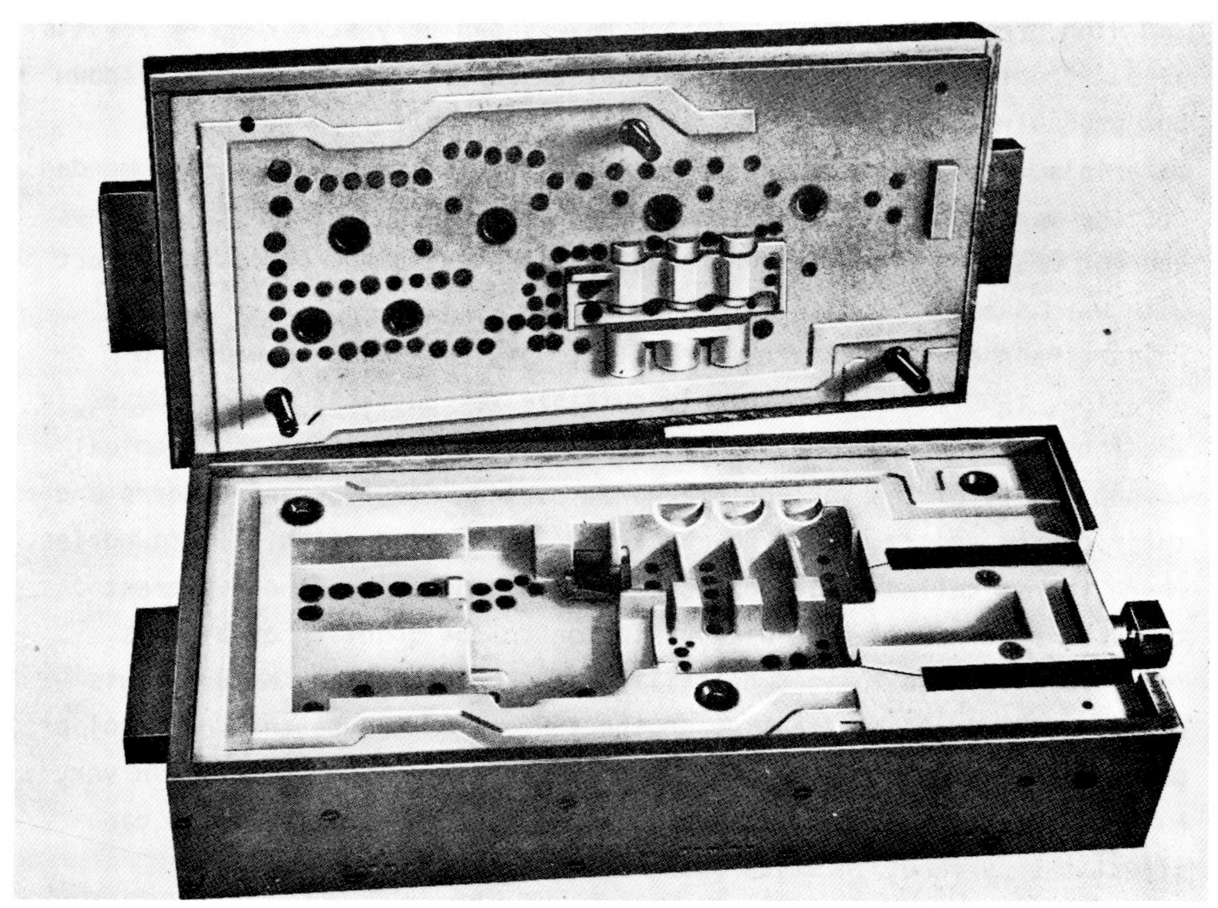

Polyurethane core box for use on a core shooter

1. Cold Curing Epoxy Resin Systems

Gelcoats, casting and laminating resins normally used for the manufacture of pattern equipment are cold curing systems and post curing,
although desirable, is not always essential. Provided the workshop
temperature is above $18^{o}C$ and the equipment is left at this temperature
for a minimum of 48 hours, acceptable properties will result for most
applications. The optimum physical properties in abrasive resistance,
compressive strength, heat distortion temperature, chemical resistance
a.s.o. are only obtained after curing for 24 hours at room temperature
plus post curing for a minimum of 14 hours at $40^{o}C$.
Irrespective of the manufacturer of the cold curing resin system, they
are all similar in that post curing is essential to obtain the best

possible results.

2. Semi Heat Curing Epoxy Resin Systems

Resin systems with a high "heat distortion temperature" must be used if
the pattern equipment is subject to more than $50^{o}C$ as with in most cases
on Disamatic moulding machines. Practical experience has shown that
the temperature resistance of the epoxy system used is an extremely
important factor in respect of the abrasive resistance properties. The
results obtained with semi heat curing systems are excellent. The only
significant difference in the processing, is the different curing
schedule. After precuring for a minimum of 14 hours at room temperature,
post curing of at least 12 - 15 hours at $80^{o}C$ is required to obtain
optimum results.

3. Cold Curing Polyurethane Elastomers

A typical polyurethane elastomer used for the manufacture of precision
pattern equipment has a potlife of 20 - 40 minutes and can be processed
at room temperature. Good mechanical properties are developed after
curing for about 10 days at room temperature. However, in order to
speed up production and also to obtain the maximum physical properties,
a post curing for 8 hours at $40 - 50^{o}C$ after the polyurethane has gelled
is generally advisable.

FIBREGLASS (Finish = Sizing = Binder)

Glass is made from a large number of materials such as, inter alia, sand,
limestone, boric acid and Kaolin and its composition is precisely
defined. After heating the mixture to $1600^{o}C$ it becomes fluid and slowly flows onto the spinning plates and then through the spinning holes
which have a diameter of approximately 2 mm. The filaments so produced
are wound over a rapidly rotating collect and then wound onto a card-
board tube, known as a cake. The diameter of the endless glass filaments depends mainly on the drawing speed.
The filament diameters are generally in the region of 3 - 16 microns.
Between 50 and 400 such filaments are assembled to a bundle, which is
called a strand.

<u>SIZING:</u>

As soon as the filament leaves the spinning plate, it is covered with a thin film of "size" also called the "finish".

The very nature of the size influences the processing qualities of the glass fibre cloth and also to a large extent the mechanical properties of the finished article for which it is used. The size of finish has to ensure rapid and complete impregnation of the glass-cloth and has also to act as a primer or bonding agent for the epoxy resin to the single glass filaments.

Glass fibre cloth with sizings such as Dexol, Silane or Volan-A must be used to ensure good 'wetting' of the fibreglass and optimum mechanical properties of the laminate made with it.

<u>LAMINATING:</u>

Glass cloth-reinforced laminated moulds, patterns and core boxes combine rigidity with very light weight, exceptional dimensional stability and high impact strength. The most suitable technique for producing large sized pattern equipment is the hand lay-up method, whereby a number of laminations of fibre reinforcements are laid up in a mould and are pro-gressively impregnated with laminating resin. Pattern equipment is usually of complex shape and therefore a cloth type which drapes and forms easy over sharp edges and into corners should be used e.g. staple fibre glasscloth.

<u>HYGIENIC PRECAUTIONS FOR HANDLING PLASTIC PRODUCTS</u>

EVERYTHING IS POISONOUS -
NOTHING IS NON-POISONOUS
ONLY THE DOSE CAN MAKE SOMETHING NON-POISONOUS

Theophrastus Paracelsus
(1493 - 1541)

Occupational safety has become a major concern that will be a dominant factor in industrial relations in the years to come. Every operation

with chemicals whether in the household or in business and industry is
recognised to involve some degree of risk. However, once this risk is
understood it is possible to counteract it by suitable measures.

A. Toxic Properties of Epoxy Resins and Hardeners

Various types of epoxy resin systems are offered to the South African
Foundry Industry. All of them are toxic to a certain degree. It is
or should be the responsibility of the manufacturer or supplier to offer
resin systems with a relatively low hazard level.
Fortunately, the most commonly used system in the South African Foundry
Industry has a very low toxicity level. It can be regarded as 'safe'
and processed without endangering the health of those who use it, provided the few simple rules and precautions are observed when handling
epoxy resins and hardeners.

B. Toxic Properties of Polyurethane Elastomers

Also here it should be the responsibility of the manufacturer or
supplier to offer custom designed polyurethane systems with a low
hazard level. Constant research has led to the development of specialised polyurethane systems which can be regarded as 'safe' provided all
reasonable handling and working procedures are observed.
In general, hazard-free handling of chemical products can only be
ensured if the personnel involved are extensively instructed on the
risks in question and given all details of the necessary protective
measures. Experience shows that the observance of safety measures must
be continually monitored. Supervision of the effectiveness of safety
measures e.g. the functioning of local extraction facilities must be
ensured to reduce the risk of contamination to its absolute minimum.

SUMMARY

The use of specialised epoxy and polyurethane resin systems in the production of patterns, pattern plates and core boxes has been part of
everyday foundry practice for years. With the increased pace of foundry
modernisation, the demands made on pattern equipment have become more
and more stringent. If the close tolerances being specified today had
to be met with traditional pattern materials, very expensive machines

and highly skilled personnel would be needed. Fortunately the required
close tolerances can easily be achieved at much lower cost by using
specialised epoxy and polyurethane resin systems.

Melting

Effectiveness of Coreless Induction Melting

by P. H. Mikkola and T. R. Wight,
Central Foundry,
Division of General Motors,
Saginaw, Michigan

Introduction

The success of a melting operation depends on the combined result of equipment utilization and operational efficiency. This combined result is the effectiveness of the overall operation.

$$\text{Effectiveness (E)} = (\text{Utilization})\,(\text{Efficiency})$$

Utilization is defined as putting to use and is normally measured as the percent "on time" of equipment over the available or scheduled hours of operation. Efficiency is a measure of producing desired results with a minimum of waste. It is expressed as a percent of the minimum effort or work required to accomplish a given task, divided by the total amount of work or effort used. When efficiency and utilization are combined, the result is a percentage measurement of the achievement of desired results from the equipment and operation. This measures the effectiveness of the operation.

For example, a molding line operates 90% of the scheduled time and produces 92% good castings (8% scrap). The effectiveness of the operation is (0.90) (0.92) = 0.83 or 83%.

This paper deals with the effectiveness of induction melting by explaining the limiting factors in operational melt rates of coreless furnaces. Some of the factors covered influence equipment utilization while others influence equipment efficiencies. All are extremely beneficial to the melting operation since they point out methods of maximizing capacity or decreasing processing cost.

Basic Concepts

A coreless induction furnace is basically a refractory crucible surrounded by a copper coil connected to an alternating current. The current passing through the coil induces a secondary current in the metallic charge within the crucible. The electrical resistance of the metallic bath generates the heat energy necessary to heat the bath to its desired temperature. The thermal energy generated by the power (I^2R) loss within the metallic bath makes the coreless induction furnace a most efficient method of melting iron. Seventy-five percent of the energy delivered is used for increasing the temperature of the iron. The majority of the remaining losses are carried away by a water cooling system which cools the power coil preventing overheating of the copper.[1]

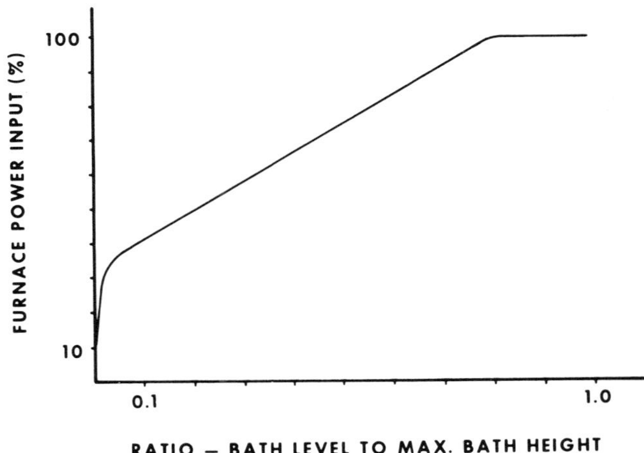

Fig. 1. Power input as a function of furnace bath level.[2]

A coreless induction furnace normally operates on a continuous batch-type cycle. The cycle begins with a full furnace at the tap temperature. A load of a given size is tapped out and an equivalent size charge dropped into the furnace. The new charge melts, actually dissolves, in the molten bath and the total bath is reheated to the tap temperature. The tap and charge size are chosen such that the level of the liquid bath is maintained above the upper edge of the coil. This enables use of full power capabilities of the coil. Figure 1 shows the relationship between furnace bath level and the percent of nominal power available. To maintain the maximum power availability, the bath level must be maintained within the top 20% of the maximum bath height.

Figure 2 shows the relationship between the size of the liquid heel and the melting rate. Note that the melting rate can continue at 100% capacity when the liquid heel is greater than 70%.

Following is a typical example of this concept with 65 ton units. The bath is heated to 2800F (1538C) and 7 tons is tapped into a transfer crane. A 7-ton charge is then dropped into the remaining 58-ton bath. The charge melts and the bath temperature is again brought up to tap temperature. During this entire cycle power remains on the furnace. It takes 430 kwh of energy to heat one ton of iron from room temperature to 2800F (1538C).[3] Considering the efficiency of the induction furnace the total energy consumption required to heat one ton of iron from room temperature to 2800F (1538C) is about 530 kwh. This is shown pictorially in Fig. 3.

Preheating

Preheating of induction furnace charges is accomplished through the use of gas-fired burners. It can be used to simply dry

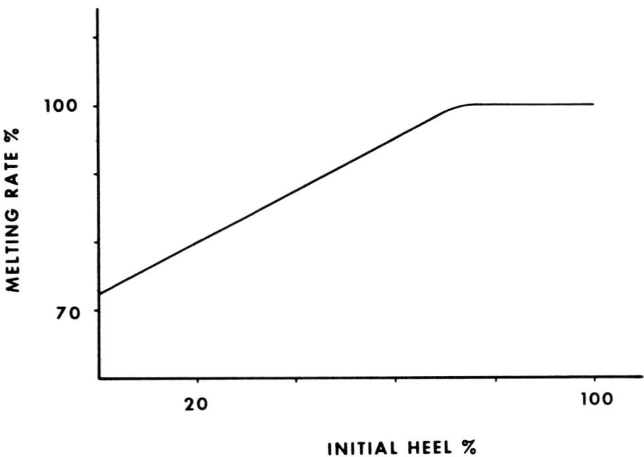

Fig. 2. Melting rate as a function of initial heel.[2]

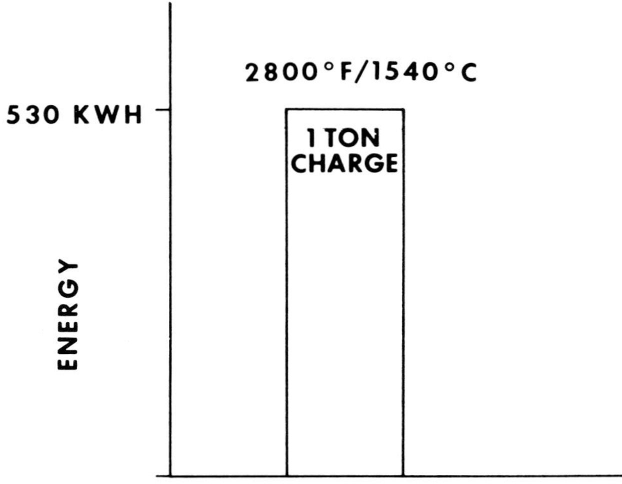

Fig. 3. Total heat energy/ton required.

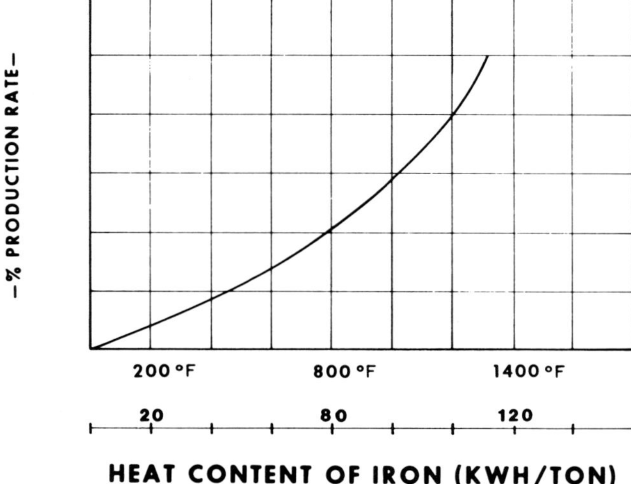

Fig. 4. Graph shows that increasing the average preheat temperature reduces the energy required from the furnace and therefore increases the melt rate of the furnace.[4]

and degrease the charge materials for safety reasons, or it can be used to heat the charge to relatively high temperatures thus reducing the total energy requirements of the induction furnace. Large preheaters are normally used in conjunction with afterburners. The afterburners complete the combustion of the exhaust gases so the system complies with environmental standards.

A charge preheating system can be designed to accomplish three objectives in the operation of induction furnaces —

1) Preheating removes moisture and oil from the charge enabling charges to be placed into the molten bath without danger of explosion.
2) Preheating the charge reduces the total energy input required of the furnace and increases the melting capacity of the furnace.
3) Preheaters are gas-fired and gas is less expensive than electricity, in heating scrap from room temperature to 1000F (538C). The total energy cost of the melting operating is reduced through the effective utilization of preheaters.

To accomplish the above, a preheater system must contain the following operating characteristics —

1) The charge must be heated from room temperature to a preselected average temperature (usually 800-1200F [427-649C]) in a relatively short period of time. This is necessary to prevent oxidation of the charge and to make the preheat cycle time compatible with the induction furnace cycle time.
2) Once the maximum preheat temperature is reached, the charge must be transfered quickly to the furnace. This prevents excessive heat loss prior to charging.

The following example illustrates the role preheating plays in induction furnace effectiveness.

For the purpose of calculating a melting rate and then maximizing that rate to improve furnace effectiveness, assume the following set of parameters. An induction furnace has a holding capacity of 20 tons and it is used to melt 2-ton white iron charges. The maximum available power is 5 Mw (or 5,000 kw). The melting rate for this situation would be calculated as follows:

1) Total energy required per charge;   (Fig. 3)
   (530 kwh) (2) = 1060 kwh

2) Time required to heat charge from room temperature to 2800F (1538C);
   1060 kwh/5000 kw = 0.212 hr/chg. or 12.7 min/chg.

3) (60 min/hr) x (1 chg./12.7 min) = 4.7 chg./hr

4) (4.7 chg./hr) x (2 tons/chg.) = 9.4 tons/hr = melt rate

Figure 4 shows the effect preheating the charge has on the melting rate of the furnace. A charge preheated to an average temperature of 1000F (538C) increases the melting rate by about 30%. This would increase the melt rate of the above example to 12.2 tons/hr. This is shown pictorially in Fig. 5. The total energy required is still the same, 530 kwh/ton. However when a gas-fired preheater is used to obtain 1000F (538C) or the first 100 kwh energy, the furnace is required to supply only an additional 430 kwh, thus decreasing the melt cycle.

A gas-fired preheater is considerably less efficient at transferring heat energy to the charge than is an induction furnace. Normal efficiencies are approximately 50%. The price difference between gas and electricity offsets the reduction in efficiency and the net result is a reduction in operational cost. If gas prices were increased to offset the operational cost

206

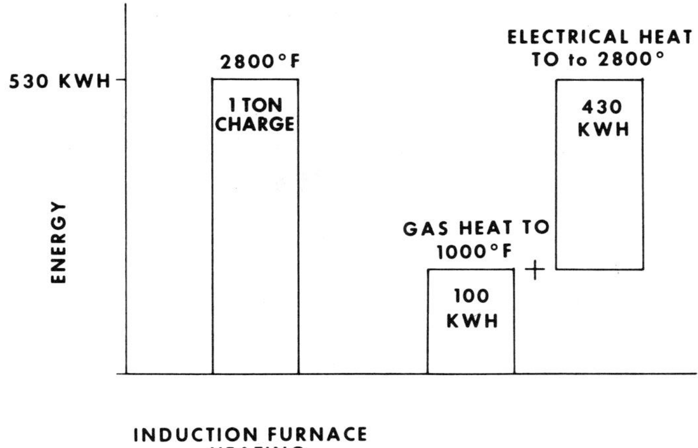

Fig. 5. Preheater and furnace energy — gas and electrical.

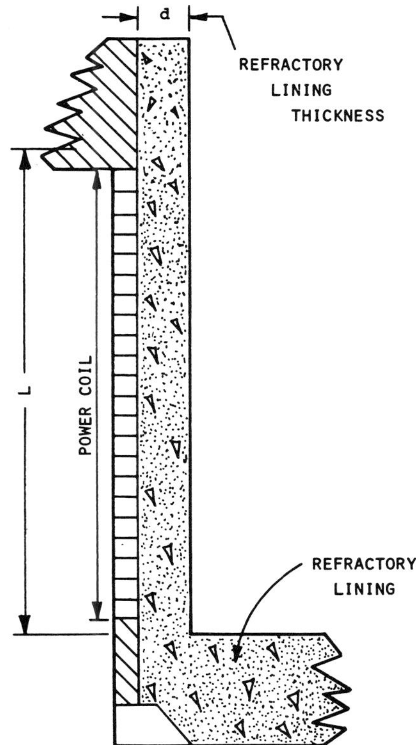

Fig. 6. Refractory cross section of an induction furnace showing L (power coil length + 10%) and d (refractory lining thickness). The cross-sectional area is equal to L x d.

advantage, the economics of charge preheating must be justified on the net increase in melting capacity. In this calculation the capital investment of the total induction furnace facility determines the advantage of preheating.

Refractories

Induction furnaces are generally lined with dry vibrated silica refractory between the molten iron and the power coil. The amount and design of this refractory determines the relative location of the iron with respect to the coil. The closer the iron is to the coil, the better the magnetic coupling and the more power can be obtained with the same furnace voltage. Conversely, the thinner the refractory, the shorter its useful life is as a furnace lining. The balance between refractory thickness and power input can determine the success or failure of an induction melting operation.

The distance between coil and iron is referred to as the average lining thickness. This thickness is important over the entire length of the power coil plus approximately 10%. Figure 6 shows the average thickness (d) and the length of the power plus 10%, (L). This area, (d) x (L) is the effective area acting on the magnetic coupling. Since the length of the coil is constant, the magnetic coupling is inversely proportional to the average lining thickness. Another method of measuring this average lining thickness is the equivalent of the electrical impedance as shown below.

$$\text{power} = \frac{(\text{voltage on coil})^2}{(\text{impedance})}$$

$$(\text{impedance}) = (\text{avg. lining thickness}) = V^2/P$$

Typical curves of V^2/P over a campaign for two different lining designs are shown in Fig. 7. Slopes tend to be steep during the first weeks of a campaign.

It must be understood that the V^2/P values are the average lining thickness values. They cannot detect isolated extreme wear areas and must be combined with meaningful lining measurements to determine refractory life. Meaningful data can be collected on an induction operation by combining the actual V^2/P values with true average refractory thickness to develop a graph as shown in Fig. 8. From this type of graph, specific values of V^2/P can be determined to indicate when to reline furnaces. In addition a graph developed for a specific furnace can determine the ideal average lining thickness that would maximize power input into the furnace. Some idealized V^2/P values are developed for a number of different furnaces in Table 1.

TABLE 1. — V^2/P Using Maximum Voltage and Power For Typical Power Ratings of AFS Furnace users

Power limit	Voltage limit (V)	Ideal V^2/P
2400	1450	876
5400	2950	1612
6250	2650	1123
9600	2650	732
21500	2850	378

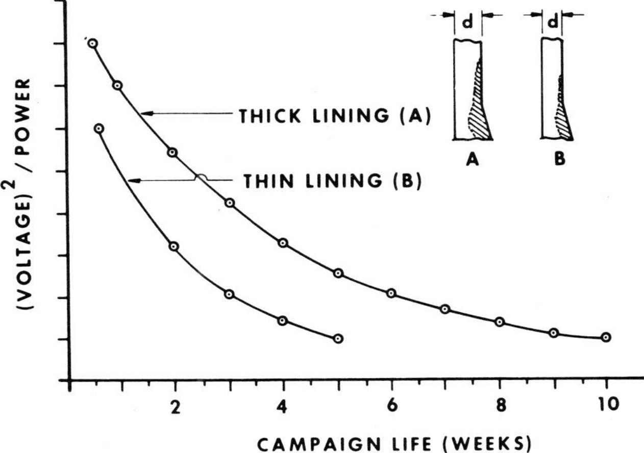

Fig. 7. Two plots of V^2/P over campaign life for two different lining thicknesses and designs. The cross hatched areas show wear experience with both designs.

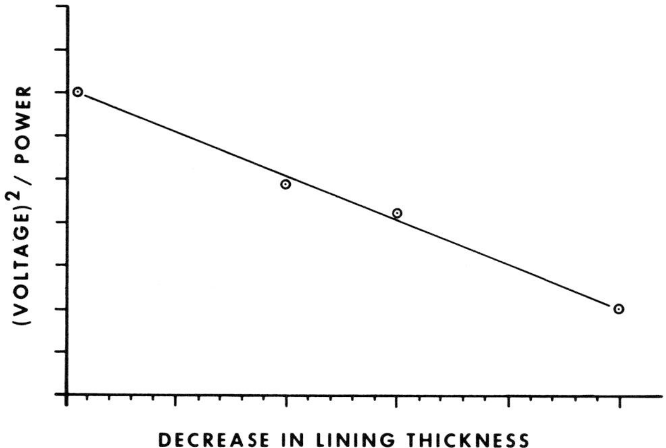

Fig. 8. V^2/P versus decrease average lining thickness as measured during campaign.

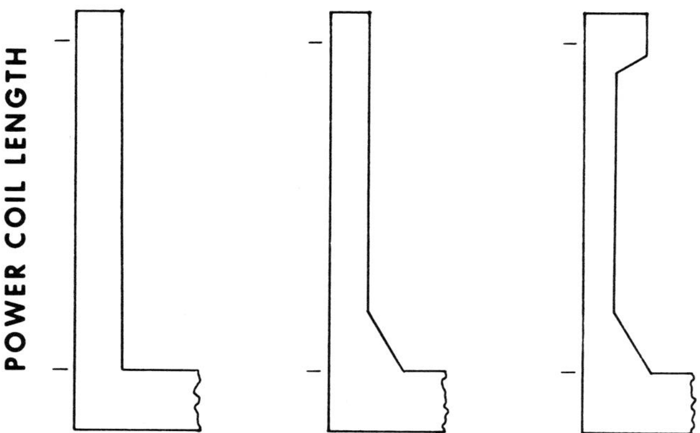

Fig. 9. Three typical refractory designs with equal total refractory area between power coil and iron bath. Design changes are greatly exaggerated to show concept.

The average thickness over the length of the coil can be maintained if the area, (d)(L), remains constant. Then, variation to the refractory design can be made to maximize refractory performance. Figure 9 shows three different designs maintaining the same average lining thickness. Design 1 would be used in a furnace which shows uniform wear over its length, design 2 for a furnace showing maximum wear in the bottom area and design 3 for a furnace with wear on the upper and lower section of the crucible.

The refractory design balance for the induction furnace should maximize input power with the thickness of refractory sufficient to maintain a scheduled refractory repair without causing loss of foundry utilization. By employing too thick a lining, 20% of the furnace melting capacity is lost. The difference between a properly designed lining and an inefficient design may be only 1/2 in. in average thickness. This important principle cannot be ignored in the operation of induction furnaces.

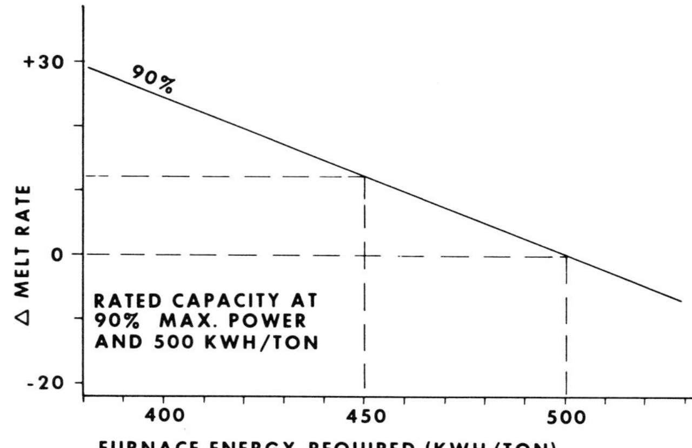

Fig. 10. Graph shows that by reducing the energy required through charge preheating, (from 500 to 450 kwh/ton) and by increasing the available power by 10% through efficient refractory design, the melting rate of a furnace can be substantially increased (by 24% in this example).

Conclusions

Induction melting capacity can be maximized through efforts in the following areas —

1) Maintain maximum power for maximum amount of cycle time.
   - Tap and charge furnace with power on.
   - Slag furnace with power on at reduced level.
   - Do majority of slagging external to furnace. (Slag off hot metal carrier rather than furnace).
   - Charge only clean metallics to reduce amount of slag.

2) Minimize total energy required of furnace.
   - Preheat charges to 1000F (538C).
   - Select charge materials such that charge density is conducive to effective preheating.
   - Maintain tap temperature at as low a level as possible.

3) Maintain available power at highest possible level.
   - Maintain bath height above the top of the power coil.
   - Design refractory lining to maximize available power.

Figure 10 shows the combined effect on melting capacity of minimizing energy required through preheating and of maximizing available power through effective refractory design. The graph shows that a 10% reduction in energy required (from 500 kwh to 450 kwh), and a 10% increase in power available results in a 24% increase in melting capacity. A graph of this type can be developed for any size furnace and should be used to measure the present and maximum capacity of that melting operation.

References

1. D. A. Bertram, verbal communication.
2. H. G. Heine, "Coreless Induction Furnaces," Proceedings First AFS Electrical Iron Melting Conference, p. 5-17 and 5-18 (Nov. 1969).
3. H. & G. Etherington, Modern Furnace Technology, 3rd Edition, p. 120, Charles Griffin and Company Limited, London, England.
4. I. L. Spencer "What Can You Really Expect From Preheating," Modern Castings, Vol. 64, No. 11, p. 72-73 (Nov. 1974).

The channel furnace and its application in the iron foundry

by C.F. Wilford and
R.D. Langman
of The Electriclty
Council Research
Centre

1. INTRODUCTION

The channel induction furnace provides an ironfounder with the opportunity of storing molten metal and it is now increasingly recognised that such a facility can prove a very sound investment. The use of a storing vessel implies the continuous use of energy. Even in the cases where metal is required for only short periods, energy is required continuously by the unit. In an ambience of increasing energy costs, it therefore becomes necessary to examine closely the need for metal storage. It is unfortunate that the overall economic savings obtainable from a channel furnace installation cannot easily be quantified, at least not until such a unit has been employed in a foundry for some time, but some accepted advantages of the investment are worthy of consideration:—

(a) Overall metal cost: By removing the necessity for a melting furnace to match the vagaries of a moulding track, the storage capacity of a channel furnace will assist the melting unit to operate more efficiently at a steadier metal supply rate. A cupola-channel user has claimed[1] that an original 18·3% coke rate was reduced to 15% after the installation of the channel furnace storage facility (for the same cupola metal temperature). Additionally, an increased proportion of steel in the charge was possible. Out-of-specification liquid metal can be returned to the holding furnace, as can any excess liquid metal.

(b) Quality assurance: The facility of having a stored volume of metal whose quality can be monitored at, for example, hourly intervals is recognised as providing a valuable framework to the assessment of casting quality throughout the production time. The concept, for instance, of storing iron which has already been de-sulphurised and magnesium treated, offers considerable advantage to foundries making repetitive SG castings.

(c) Production reliability: One easily assessed area of non-productive time in a foundry is 'time waiting for metal' and this can be reduced by selecting a suitable combination of melting and holding furnaces. A stored volume of metal offers obvious advantages but this itself becomes a (continuous) component of the cost of a casting and, therefore, careful optimisation of this is important.

Further advantages to be gained from a channel furnace installation are discussed in more detail later in this paper.

2. The engineering status of the channel furnace

The early development of the furnace (Fig. 1) has been described elsewhere[2]. The period since the early '60s has been full of activities. Several manufacturers concentrated on developing a horizontal drum furnace with a quick change inductor. An earlier paper[3] reviewed the technique for changing the inductor and by the early '70s all leading international furnace manufacturers were able to offer such inductors. By comparison with both the arc furnace and the coreless induction furnace, the channel furnace remained under powered, inductor rating seldom exceeding 600 kW. The multi-inductor furnace became an accepted method of increasing the kW/t ratio and furnaces involving up to four inductors were constructed. There were, of course, two primary constraints for high powered inductors: that of removing I_2R losses in the furnace coil and the control of inductor/vat temperature differential with its consequent impact on the refractory in the loop zone.

Gosling and Cubitt[4] reported upon early results of shaping the inductor and its throat, which it is claimed adjusts the pinch effect inside the inductor and emphasises the effective lift in such a way that metal can be caused to migrate from the inductor to the vat. This effect has been the cause of some controversy but the use of changing inductor geometry does seem confirmed with ensuing reductions in inductor/vat temperature differential, or, alternatively, for a given differential, increase of inductor rating.

The concentration on choosing the right refractory materials, better ability of absorbing internally generated losses and improved detailed design, has seen a return to the single inductor horizontal drum furnace rated up to 1·5 MVA. Figure 2 shows a typical design. Of special interest is acceptable throat/vat mating surface to permit inductor replacement and the water cooled bush which controls the heat flow within the inductor coil refractory zone.

With such developments the possibility of a rapid increase in the kW/t ratio becomes alive and within the last decade the vertical drum channel furnace has been widely accepted with inductor ratings as high as 2 MVA occurring. A cross section of a typical 60 tonne melting and holding unit is shown in Fig. 3. It will be appreciated that providing the foundry's operating cycle permits, this furnace is acceptable by encompassing the duties of both melting and holding furnaces within one body. The furnace causes little complication in providing electrical supply, and it is capable of using off-peak tariff opportunities.

More recently the vertical channel furnace has been adopted for the production of ingot moulds. In this interesting application, returned spent moulds need to be re-cycled and melted. It will be appreciated that the size and mechanical strength of the moulds are such that it is a costly excercise to break them into smaller pieces. The vertical channel furnace with a carefully constructed throat and vat geometry can melt the moulds in situ.

The Authors have suggested the concept of gas injection for promoting metal circulation from the inductor by creating pneumatic lifting and to extend this concept to the use of the furnace as a gas/metal reaction vessel. The present position is that the interaction between such injected gas, the resulting metal movement and the super-imposed electro-magnetically derived motion is complex. This concept appears to be receiving more concentrated effort in the steel refining and aluminium alloy control areas than in iron foundries.

For the near future, the Authors see no real general need for furnaces in excess of 100 tonne iron capacity with the opportunity of using inductors up to a rating of 3 MVA; perhaps greater strides will be made in the application of the

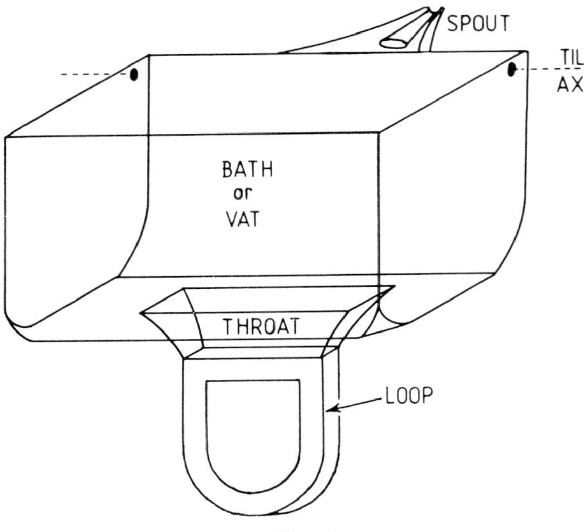

Fig. 1

Fig. 2

Fig. 3

furnaces for tonnage non-ferrous melting and treatment, but work in these sectors is still very much in the development stage.

3. Some applications of channel furnaces in ironfoundries

3.1 Maintaining metal temperature

The simplest application of a channel furnace is to hold a volume of metal at a constant temperature, the energy input being chosen to equal the energy losses from the system. In this situation, it is unwise to select a furnace of excessive power rating since this will be associated with an increased conducted heat loss from an unnecessarily large metal loop and inductor box. An obvious design feature in such an application is the use of heavily insulated refractories, both in the inductor box and furnace bath but it is now commonly accepted that relatively steep temperature gradients through the hot face material is important in reducing both refractory wear and metal penetration which, should they occur, increase the heat losses from the metal anyway. Hence, modern practice is not to maximise insulation behind the hot face refractory.

The effect of furnace capacity on the holding power required per tonne of metal stored is shown in Fig. 4, which has been prepared from data provided by several UK ironfounders. These results, covering several furnace manufacturers, indicate fairly similar heat losses per tonne with furnace capacities exceeding 25 tonnes. Thus the inductor power may be considered only in relation to overcoming circuit inefficiencies and standing heat losses.

In order to decide on the optimum storage capacity of a channel furnace, each foundry has to assess its own requirements carefully. This decision making necessitates an accurate knowledge of the rate at which metal will be supplied to the furnace, taking account of the frequency of melting plant downtime and its duration, together with the expected demand from the moulding tracks and their possible downtime frequency and durations. Simulation of these foundry operations will then allow the optimisation of channel furnace storage capacity, noting that increased capital and running costs are associated with increased channel furnace capacity. This approach is to be recommended, being much sounder than the often used empirical solution of choosing the channel furnace to store one hour's production from the melting units.

Because of the extra heat losses incurred with worn refractory linings, badly fitting lids or incompletely dried refractories found immediately after priming with liquid metal, the values of power quoted by manufacturers to simply retain metal temperature are wisely treated as minimum values when choosing furnace rating, and should be exceeded rather than adhered to precisely. However, as already noted, there are penalties in selecting an excessive high power rating.

With a metal surface which is reasonably slag-free and contained in a well-sealed body, the power input required to hold temperature in a given channel furnace is fairly insensitive to the quantity of metal in storage. It is, therefore, sensible to retain a fairly full furnace during periods of

---

Fig. 1 (top, left)—Schematic view of a channel furnace

Fig. 2 (middle)—Cutaway view of a horizontal drum furnace (courtesy of Inductotherm Ltd)

Fig. 3 (bottom)—Cutaway view of a vertical channel furnace (courtesy of Inductotherm Ltd.)

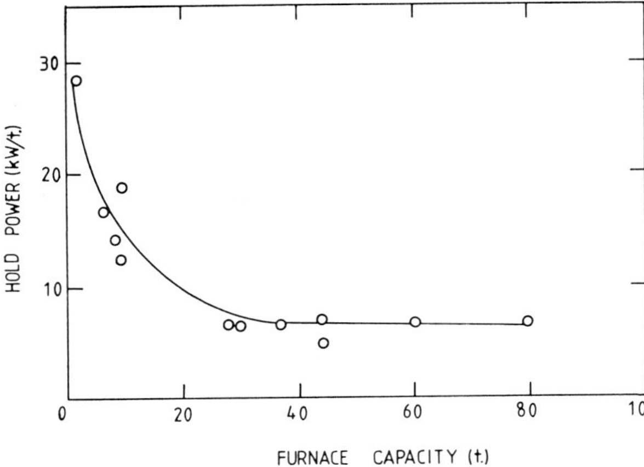

Fig. 4—Holding power required for various industrial channel furnaces—melting furnaces not included

minimum supervision, since an incorrect power input (compared to the exact value required to maintain temperature) then has minimal effect on undesirable bath heating or cooling rate. Heat losses are largely governed by body case temperature and, because the body refractories have a high inertia to temperature change, it follows that much of the energy saved by reducing power input over a weekend is largely taken from the hotter end of the refractory wall. This energy then has to be returned during the following working week and the overall energy saving is often negligible. The conclusion that the furnace is best left full of metal at its normal tapping temperature, however, may be modified in practice by metallurgical considerations (to be discussed later).

3.2 Superheating liquid metal

The specific heat of liquid iron is relatively low, being around 0·22 kWh/t per °C. Once sufficient power is applied to a channel furnace to balance heat losses, further power input is converted to superheat metal at an efficiency of some 95%. This energy transfer efficiency is virtually independent of the base temperature which accounts for the economic strength of the unit in comparison with, say, a cupola for superheating iron.

Power requirements, above that to maintain metal temperature, are shown in Fig. 5 for two selected superheat levels and a range of metal throughput. More generally for a superheating range θ°C, specific heat C, kWh/t per °C and a metal delivery (and acceptance) rate of T t/h, the inductor rating is given by:

$$P = \frac{\theta \times C \times T \times H}{\eta}$$

where H is the holding power, kW
and η is the circuit efficiency.

Because the energy transferred to the inductor loop is independent of the quantity of metal contained in the furnace (assuming the minimum heel is exceeded), it follows that the temperature rise observed in the bath is dependent not only upon the power applied but also upon the quantity of metal contained in the furnace. It must be noted that the batch charging of metal to be superheated also will effect the temperature of metal tapped from the furnace and the extent of this temperature fluctuation depends upon the temperatures of the incoming metal and that already in the furnace, together with their relative quantities. It is interesting to quantify these effects and some examples are shown in Fig. 6 to illustrate the calculated fluctuations in bath temperature) which are obtained with various charging and

tapping practices. The thermal inertia of the refractories has been ignored in the calculation.

Clearly, the only situation which provides a constant metal temperature at the spout of the furnace is that which involves a continuous and steady stream of metal (at a uniform temperature) being fed at the same rate as a continuous metal off-take, provided that the energy required for superheating is precisely applied. As soon as these criteria are not adhered to, then metal bath temperature fluctuations occur. With the batch charging of liquid metal to be superheated, fall in bath temperature is inevitable and, for a given superheat requirement, the temperature fall which is acceptable can only be obtained by the correct selection of the weight of metal added relative to the quantity of metal already contained within the furnace. These effects are demonstrated in Figs. 6(a) and (b) which show the results of two charged metal weights (10 t and 2·5 t) added to two levels of residual metal in the furnace (15 t and 5 t).

Two situations causing excessive bath temperature are demonstrated in Fig. 6(c). Both a delay in the supply of metal to be heated and a higher rate of metal off-take will result in excessive metal superheat, assuming that the power input is not deliberately reduced. As before, a relatively large stored volume will reduce the extent of this effect but the more sensible solution is to monitor metal temperature and to reduce power input accordingly. Several foundries now use a total radiation pyrometer to sight onto the (clean) metal issuing from the pour spout and the information from this, after suitable calibration, is fed back to the timers which control energy input.

3.3 Hmogenisation of liquid metal

Ever increasing standards of casting quality are being imposed upon the iron foundry trade who recognise that consistent molten metal is a prime requisite, involving chemical composition, nucleation characteristics and metal temperature. Both chemical composition and the metallurgical

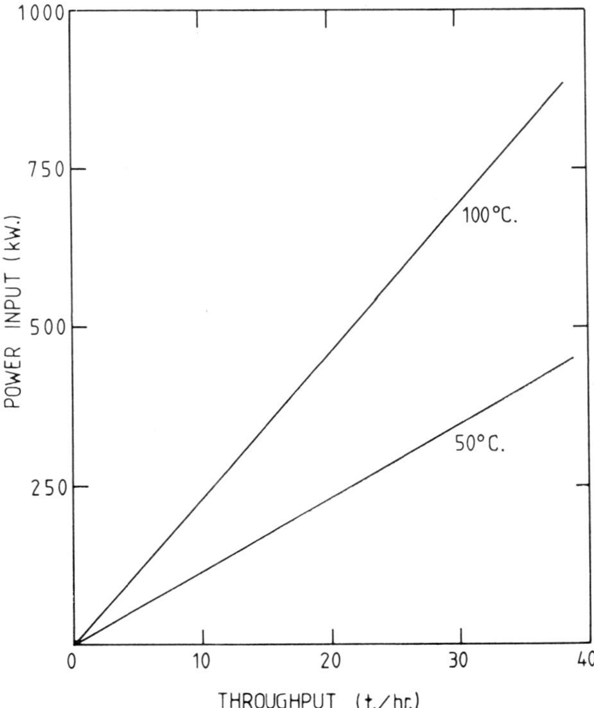

Fig. 5—Calculated superheating rates obtained with additional power inputs above that to hold temperature

Source: *The British Foundryman* (supplement), June 1979, 196-205

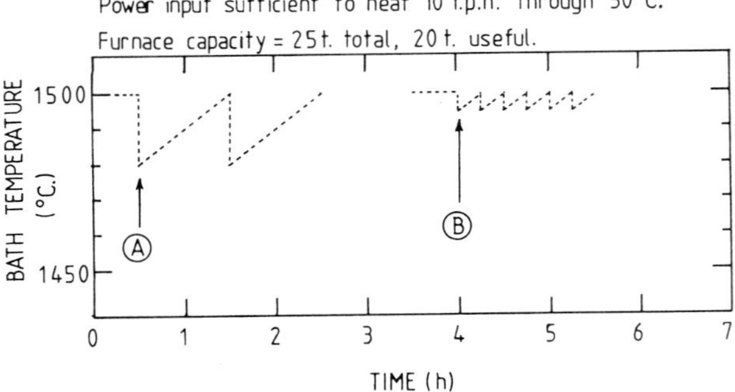

(A) 10 t. metal at 1450°C. added. Then 10 t. tapped when 1500°C. attained.

(B) 2·5 t. metal at 1450°C. added. Then 2·5 t. tapped when 1500°C. attained.

(a) Furnace capacity of 25 tonne total, 20 tonne useful

(A) 10 t. metal at 1450°C. added. Then 10 t. tapped when 1500°C. attained.

(B) 2·5 t. metal at 1450°C. added. Then 2·5 t. tapped when 1500°C. attained.

(b) Furnace capacity of 15 tonne total, 10 tonne useful

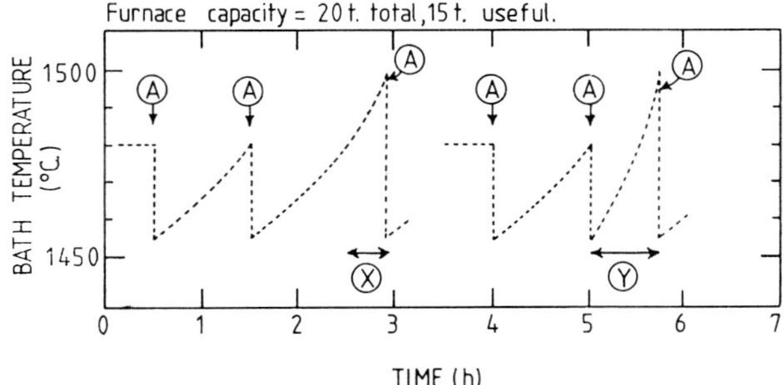

(A) 10 t. metal at 1430°C. added.

(X) Delay in metal supply.

(Y) Doubled rate of metal off-take.

(c) Metal charged hourly and tapped at 6 minute intervals

Fig. 6—Influence of selected variables on calculated channel furnace bath temperatures

nature of the melt have become more difficult to control as charge mixes are cheapened by replacement of pig iron by more variable materials, such as steel and cast iron scrap. The problem of having to make a consistent melt using increasingly inhomogeneous charge materials can be approached in a number of ways:

(i) Selection of charge storage and handling systems designed to encourage charge homogeniety;

(ii) Application of on-line analytical routines combined with rapidly imposed corrective melt adjustments;

(iii) Improve homogenisation of metal in the liquid state.

The use of a channel furnace to provide the third alternative will be discussed in this section, the two related aspects of chemical composition and metal nucleation needing to be considered.

3.3.1. Chemical composition

The modern design of channel furnace employed to hold metal from another prime melting unit is such that the efficient sealing of metal from the atmosphere provides an extremely low rate of carbon and silicon losses from the molten bath. For example, a loss of only 0·03% carbon has been measured during a 14 day period on a 90 tonne furnace. The channel furnace, therefore, acts virtually totally by dilution of the incoming metal and the extent of homogenisation is dependent upon metal throughput and furnace capacity.

The actual homogenisation of chemical analysis quoted by a ductile iron foundry is demonstrated in Fig. 7, the results being obtained during a particular day when some 80 tonnes of cupola metal was passed through a 13·5 tonne capacity channel furnace in a period of $16\frac{1}{2}$ hours. Although the improvement in consistency of analysis is remarkable, it is worth noting that this particular furnace capacity is probably near the acceptable minimum for such throughput and many founders would operate with a furnace of at least 50% greater capacity, thus producing a further improvement in consistency of analysis.

As well as the useful homogenisation of the major alloying elements obtained, the stored volume also produces homogenisation of tramp elements. Typical tramp element levels in cupola charge materials have been usefully tabulated by Else[10] and the results for various types of scrap steel are shown in Table I. Else also shows that an iron of high trace element level is around 40 N/mm$_2$ stronger in tension than one of low trace element level, but at an identical carbon equivalent. This strength variation will be associated with a variation in hardness and it is this which can provide grounds for a machine shop to criticise the quality of the castings produced from such variable metal.

Together with a general homogenisation of trace element levels as described above, the dilution of more damaging contaminants associated with individual pieces of charge (e.g. lead) will also occur and, again, this effect can allow the recovery of a particular batch of melt which otherwise would cause unacceptably high levels of casting rejects.

3.3.2. Metal nucleation

There are many factors which effect nucleation of molten iron (see for example ref. 11). Batches of molten iron tapped from a prime melter can have an excessive variation in nucleation level. The problems resulting range between the occurrence of graphite eutectic in malleable iron castings or excessive shrinkage in grey iron castings, caused by the presence of over-nucleation, through to chill formation in thin sections of grey iron castings, or excessive softness associated with ferrite formation in such castings, caused by

TABLE I—Typical tramp element levels in various grades of scrap steel used as cupola charge (10)

Material	Typical Percentage of Elements Present					
	Sn	*As*	*Cu*	*Cr*	*Ti*	*Mn*
Short Heavy Steel Scrap, Constructional	0·01/0·03	0·01/0·04	0·2/0·4	<0·15	0·02/0·04	0·4/0·9
New Production Bales (No. 4 Bales)	<0·01	<0·01	0·01/0·1	<0·05	0·01/0·25	0·1/0·3
Black Bales (No. 5 Bales)	0·02/0·1	0·01/0·025	0·2/0·5	<0·15	0·02/0·03	0·2/0·4
Fragmentised Scrap	0·02/0·1	0·01/0·04	0·2/0·7	0·5/0·4	0·02/0·03	0·2/0·4
Destructor Scrap	0·2/1·5	0·01/0·015	0·05/0·2	<0·015	0·01/0·02	0·02/0·2
Detinned Scrap	0·05/0·1	0·01	0·02/0·2	<0·05	0·01/0·02	0·05/0·3

under-nucleation.

In order to understand the effect of passing metal through a channel furnace, it is necessary to consider a hypothetical model of the nuclei in cast iron melts and much of the following reasoning is based on studies carried out by Swinden and Wilford at the Electricity Council Research Centre.

The metal produced in a prime melter may have a nuclei population which can be indicated as in Fig. 8[12]. One component of the nuclei population is caused by graphitic material or ferro-silicon added to a steel charge in an electric melting furnace or by graphitic material added (or formed) in a cupola. It is known that this family of nuclei is relatively unstable and 'fading' occurs with time. This fading process leaves a residual population of nuclei (termed a 'base level'), the extent of which depends primarily on the carbon equivalent of the iron and also upon the sulphur content, found by Moore to affect nucleation level[13]. It is worth noting that the manner in which these nuclei operate during solidification will be influenced by factors such as cooling rate, carbon equivalent, silicon content together with trace levels of other graphitising or carbide forming elements, etc. As with any nucleation and growth mechanism, the larger nuclei will begin to operate during the early stages of graphite eutectic solidification and, if they are present in sufficient numbers (for a given cooling rate), then cooling will be arrested and recalescence may occur. However, if insufficient coarse nuclei are present solidification will begin at a lower temperature when smaller nuclei become active, resulting in an undercooled graphite structure. If solidification is not complete before the temperature reaches the equilibrium carbide eutectic temperature for the particular metal composition involved, then carbides will nucleate causing a rapid increase in as-cast hardness.

The channel furnace can now be considered as a metal reservoir causing fade of one component of the nuclei present in the metal taken from a prime melter. Thus, metal poured from the furnace will generally be less well nucleated than that supplied to the furnace, unless this metal supplied is already 'faded' and hence at a 'base level' of nucleation. However, there is now evidence that metal stored even for many hours at the comparatively high temperature of 1500°C continues to retain its natural, base level of nucleation as long as its chemical composition is held constant. This is demonstrated in Figs. 9 and 10, the former showing wedge test samples taken from metal whose carbon content was falling at an average rate of 0·017% per hour, whilst figure 10 shows the wedges poured with metal held at a constant chemical analysis.

If the nucleation level of iron taken from a channel furnace is assessed (for instance, by wedge test) then the following pattern is to be expected. The metal in the furnace before start-up of the foundry will be at a base level of nucleation, whose level is governed by the chemical analysis of the metal and, in particular, its carbon equivalent (which controls the frequency distribution of nuclei sizes) and its silicon content (which governs the on-set of chill formation during solidification). As fresh metal is admitted to the furnace, the

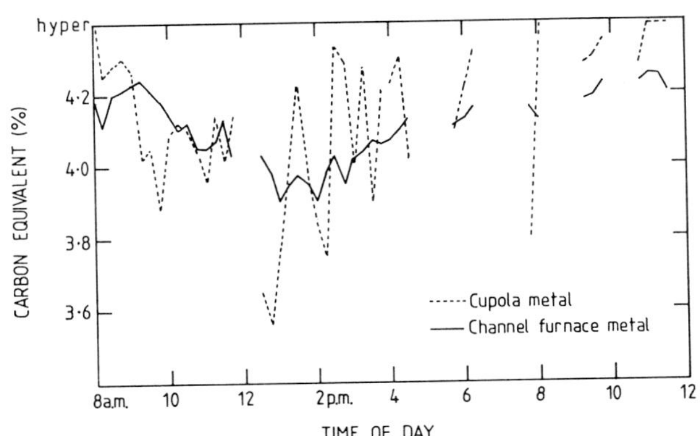

Fig. 7—*Variation of cupola and channel furnace metal analyses throughout a day (9)*

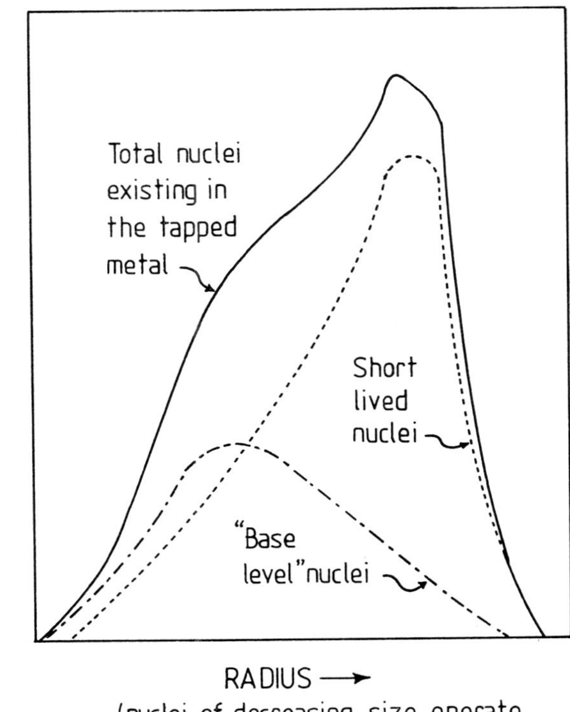

Fig. 8—*Frequency of size distribution of graphite nuclei in metal tapped from a melting furnace*

Source: *The British Foundryman* (supplement), June 1979, 196-205

nucleation level of the metal will change either towards reduced chilling tendency (if the new metal is well nucleated or at a higher carbon content) or, alternately towards an increased chilling tendency (if the new metal is poorly nucleated and at a lower carbon equivalent than the existing metal in the channel furnace). During the working day, the metal taken from the furnace will have a chilling tendency reflecting that of the incoming metal (although, of course, this will be diluted by the stored volume), together with any variations in carbon equivalent and silicon content of the tapped metal. Finally, as the metal is stored in the furnace until the next pouring shift, it will again approach a base level of nucleation which will be controlled by the chemical analysis.

Recent experience with production units tends to confirm the above sequence and it is clear that excessive loss of metal

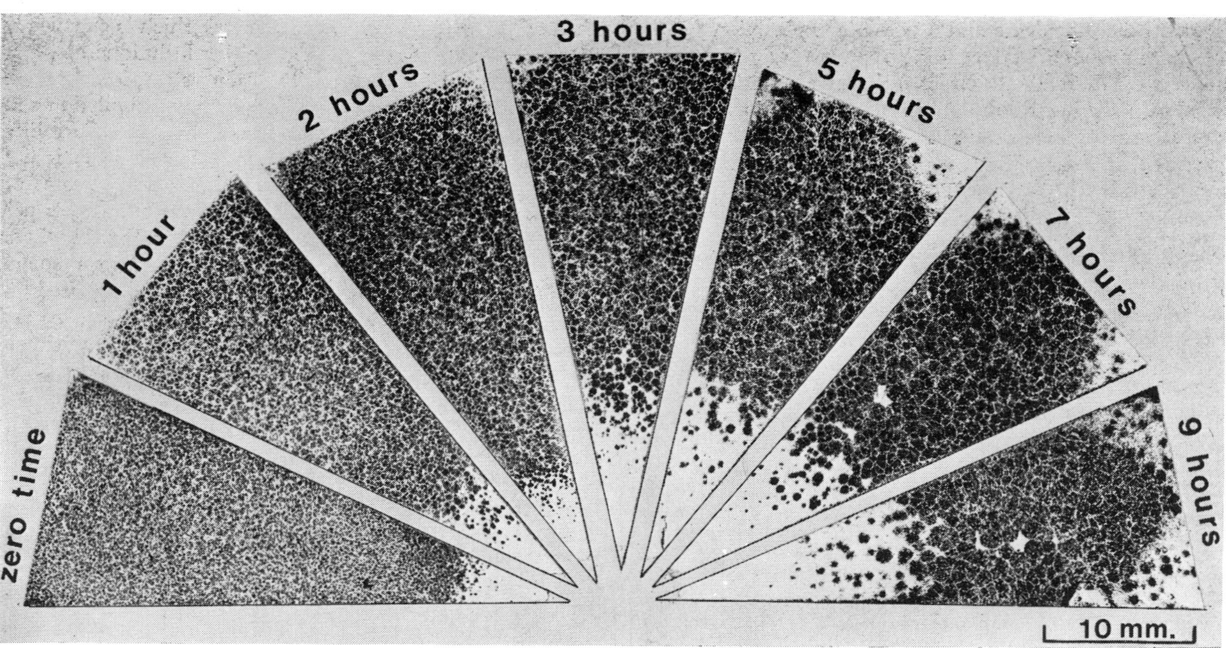

Fig. 9—Appearance of chill test wedges poured with iron of falling carbon equivalent

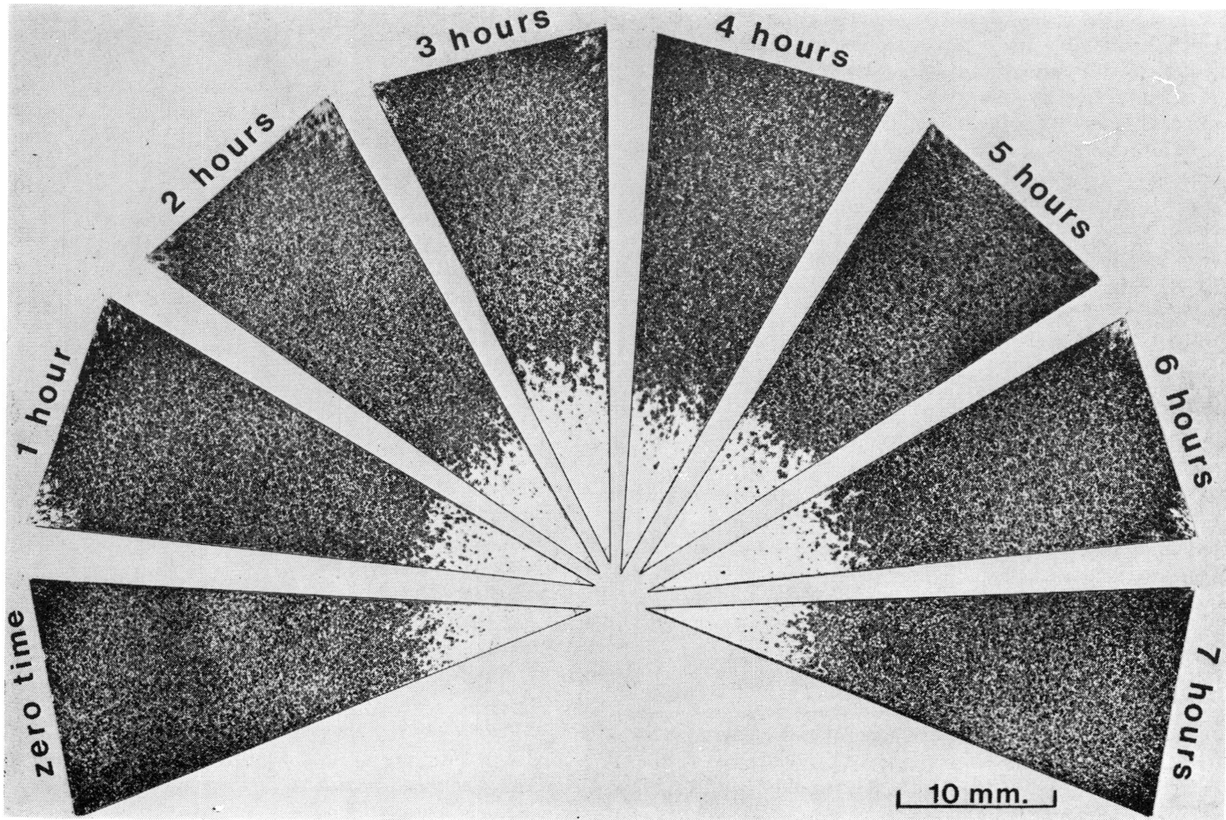

Fig. 10—Appearance of chill test wedges poured with iron of constant carbon equivalent

214

nucleation is not a problem during metal storage, at least if a sufficiently constant carbon equivalent is retained as occurs with a well sealed furnace. However, there is a further aspect which must be noted—that of response to inoculation

Many founders making thin section castings, requiring a high level of machinability, use an inoculation technique to eliminate the formation of carbide eutectic during olidification. It is now recognised that molten iron requires the presence of silica inclusions for it to respond well to ferrosilicon base inoculants and a possible explanation for this has been proposed[12], depending upon the reaction between silica and a de-oxidant to release a local concentration spike of silicon during solidification. It follows from this that excessive superheating of liquid iron to temperatures in excess of the 'inversion temperature' causes loss of silica inclusions by the reaction:

$$SiO_2 + 2(C) = (Si) + 2(CO)$$

and this will reduce the melt's response to inoculation. The precise value of the inversion temperature is dependent upon melt analysis but generally it can be stated that metal superheated increasingly above 1450°C will tend towards a reduced response to a given inoculating addition, and it should also be noted that excessively high temperatures in the furnace will tend to accelerate carbon loss from the melt by oxidation which can cause problems by reducing the base level of nucleation.

3.4 Metal melting

As the power levels of inductors have been increased over the last decade, it has become possible to provide a useful metal melting rate from a channel furnace. The design of such a furnace normally consists of a vertical cylindrical bath having the inductor attached to the base. Since the walls of the furnace body are associated with a relatively low conducted heat loss, there is no excessive penalty in storing a large volume of metal and the unit is therefore especially useful for melting during off-peak electricity tariff periods combined with pouring during the day-time casting shift.

It is useful to consider the basic design and operational features of such vertical channel furnaces, used in a melting role, in order that their advantages and limitations can be understood.

In order to optimise the cost of molten metal produced by the installation, it is necessary to select the power rating carefully since a compromise situation is involved. On the one hand, melting only during off-peak hours (e.g. midnight to 6 a.m.) is advantageous regarding both electrical unit costs and reduced maximum demand charges. However, an attempt to contain all the melting operation to the off-peak period may well generate an extra maximum demand level which is no longer economical in that the saving in cost of electrical units is insufficient to balance the increased maximum demand charges. If the furnace power rating is increased to accommodate melting only during a limited period of time, the capital cost of the installation rises and the standing heat losses usually also rise, this causing an extra increment in metal cost. Manning of the furnace during the melting period has also to be taken account of and this itself can encourage a foundry to melt at a particular rate governed largely by labour availability. Each individual foundry has to weigh such factors carefully during the selection of a new installation.

Apart from metal costs, the operational advantages of vertical channel furnaces can prove very attractive. The melting power will be continuously applied during the whole of the charging period and this produces a very consistent and repeatable melting rate. Modern practice is to add weighed batches of charge materials in conjunction with a kWh meter which inputs a preset value of energy which the operator correlates with charge weight. Excessive superheating of the melt can thus be avoided. Metal of sufficient quantity and at the correct temperature and with the required metallurgical characteristics is reliably available at the start of the casting shift. This metal then continues to be available until casting finishes. Evaluation of grey iron melt systems operated in American foundries[14] indicates that the channel furnace melter produces metal which is chemically more consistent than any other melting technique.

Although the furnace readily produces consistent metal, its lack of surface stir prevents rapid alloy assimilation and the necessity to retain a molten heel also reduces flexibility in changing between grades of iron. This lack of flexibility also applies to metal throughput in that reduced rates of daily metal demand cause rapidly rising energy levels per tonne of metal cast. This is demonstrated in Fig. 11 where the energy consumption calculated for various metal throughput rates is shown, the calculation being based on results measured on a production furnace. However, it must be noted that this trend is common to any furnace using a molten heel, although a melting channel furnace is more sensitive than one used simply for holding metal because of its greater heat loss from the more highly powered inductor.

4. Some recent studies involving channel furnaces

This section describes three projects carried out by Electricity Council Research Centre staff as part of a continuing work programme involving the study of electric furnaces used by the foundry industry.

4.1 Energy Balance

The results of a heat balance exercise[15] conducted on a 40 tonne capacity vertical channel furnace are shown in Fig. 12. A detailed description of the furnace installation has also been given[16]. The total calculated convective and radiative heat losses from the furnace is 147 kW and the water cooling circuits collect 112 kW, giving a total loss of some 260 kW for a full furnace at 1450°C. Allowing for the expected electrical supply losses, estimated as in Table 2, the holding power requirement for this furnace is 265 kW. This value applies to a fairly new cast magnesia lining in the inductor box and it is expected that this source of heat loss will increase significantly throughout the life of the lining.

When the furnace was used in its melting role, increased losses were found to occur due to opening of the lid for admittance of fresh charge. Each lift of the electromagnet transferred an average of 250 kg of charge, requiring the furnace lid to be open for a lengthy period during filling of the furnace. In practice, the charging of 19 tonnes over one night caused a loss of 36 kWh/t charged, associated with lid off time. De-slagging time involved a further loss of 5 kWh/t charged.

An average melting rate of 1·75 tonnes/hour at 1500°C was achieved from an average power input of 1060 kW. The overall melting efficiency of around 65% for the furnace is thus rather poorer than that expected from an equivalent coreless furnace (70–75%). The device does provide a metal holding capability in the same unit and a significant capital cost saving occurs in comparison with a coreless-channel furnace combination.

4.2 Continuous measurement of refractory temperature

Originally attempting to derive the temperature of molten iron within the loop of a channel furnace, a technique of continuously measuring refractory temperatures in an inductor box has been developed[17].

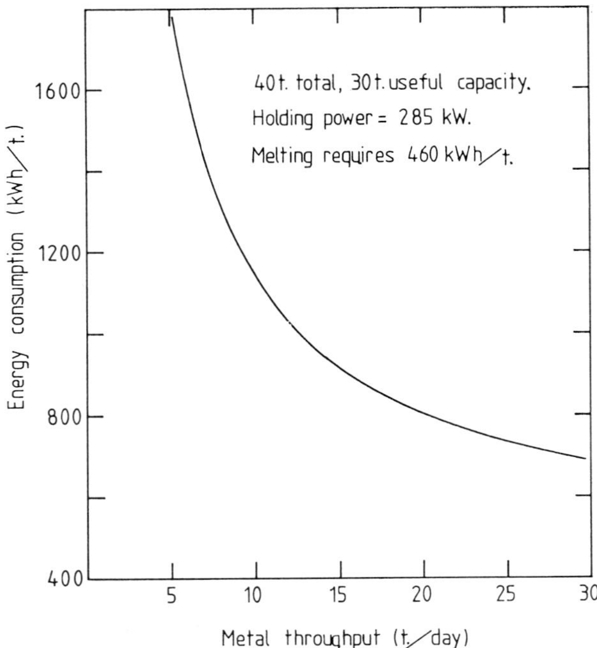

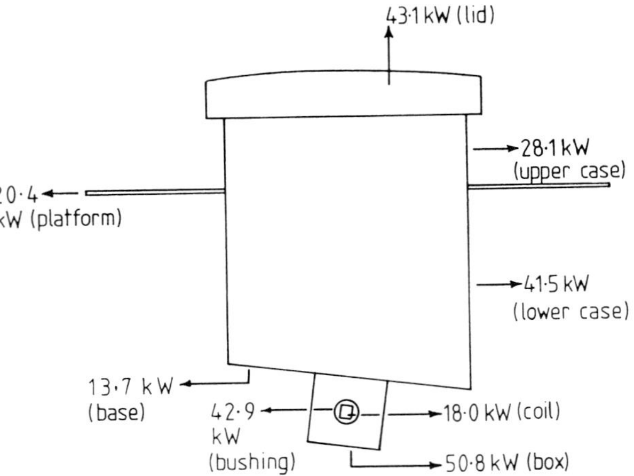

Fig. 11 (left)—Calculated energy required to produce various daily throughputs of molten iron from a vertical channel furnace

Fig. 12 (above)—Measured heat losses from a 40 tonne vertical channel furnace used for melting grey iron (15)

The technique is based on installing metal sheathed thermocouples (1·55 mm dia.) during the installation of the rammed inductor box refractory, the chromel-alumel hot junctions being positioned further from the metal than the 1100°C isotherm (this being the recommended continuous operating temperature of the thermocouple). Three suitable positions in the inductor box are shown in Fig. 13, the hot junctions being some 3–4 cm from the loop former.

The main findings from the work were as follows:

(a) The thermocouples could reliably be installed at a chosen distance from the metal loop.

(b) On no occasion was metal penetration into the inductor refractory found to be associated with the installed thermocouples. It may be noted here that earth leakage wire probes are normally installed in the case of coreless furnaces without giving trouble (unless excessively large diameter wires are selected).

(c) A thermocouple installed near the bottom of the loop responds to change in inductor energy input after 50 to 80 seconds.

(d) A thermocouple near the top of the loop responds to bath temperature fluctuation (caused by solid charging) after 120–180 seconds.

(e) The thermocouples continued to give meaningful readings for the several weeks of the trial campaigns.

It is concluded that suitable metal sheathed thermocouples correctly positioned in an inductor box refractory can be used to provide useful information. The incidence of excessively hot or cold metal in the inductor loop will be indicated and this can be used to prevent catastrophic failure of refractory material. The thermocouple will respond to erosion of the adjacent refractory and this provides a more sensitive technique than the normal loop resistance or reactance measurement. However, the thermocouple is not sensitive to build-up on the refractory since, presumably, this is too thermally conductive and the temperature at the hot junction is hardly affected. Finally, by correct calibration of the thermocouple output, the power input can be controlled during a pouring shift to produce a more consistent tapped metal temperature.

4.3 Holding magnesium-treated iron for SG castings

The usual sequence of operations for producing spheroidal graphite iron castings is to treat batches of liquid iron, of the

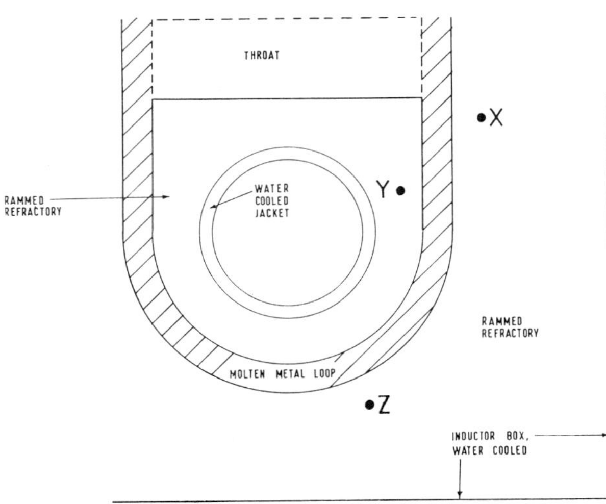

Fig. 13—The manner in which thermocouple hot junctions can be positioned adjacent to the metal loop in a channel furnace inductor box

TABLE II—Estimated power supply losses for a 40 tonne vertical channel furnace (15)

Component	Power Input Level	
	1100 *kW*	270 *kW*
Transformer iron	2	2
Transformer copper	28	1·7
Balance reactor	3	0·2
Balance capacitor	1·5	0·1
PF correction capacitor	1·5	0·1
Total loss	36 kW	4 kW

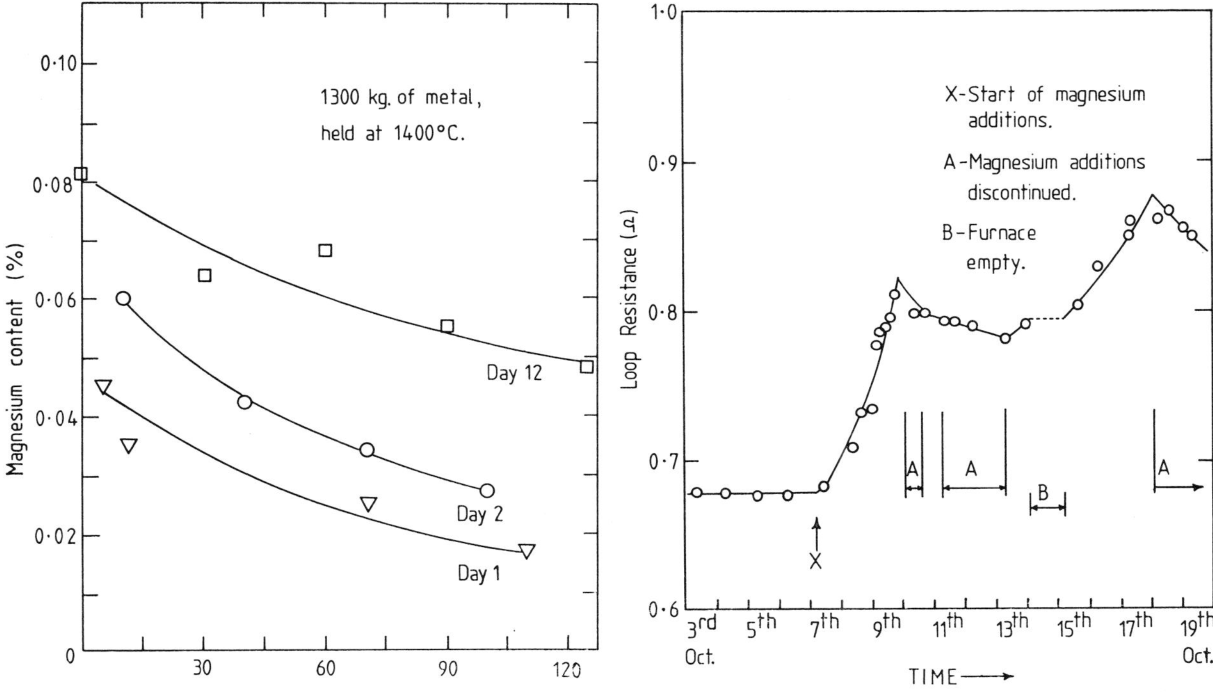

Fig. 14—The rate at which magnesium is lost from molten iron held in a channel furnace (20)

Fig. 15—The effect of holding magnesium treated metal on the electrical resistance of the channel furnace loop (20)

correct composition, with magnesium in a suitable form and then to inoculate the treated metal to promote graphitisation during solidification. In practice, numerous difficulties can arise when operating these techniques. One particular problem is that the dissolved magnesium content can be variables because of the metal's volatility and also its loss by chemical combination with sulphur in the iron. Taken together with the tendency to reduce the target magnesium content because of cost considerations, excessively low levels of magnesium in the castings poured from a given treated ladle can occur, resulting in poor nodularity of the graphite. The ability to store several magnesium treated batches within a single furnace and then to draw off metal with a controlled magnesium analysis at an accurate temperature obviously offers significant advantages.

Two approaches have been made. The first, which has been described already on several occasions (e.g. [18,19]) uses a pressure-tight channel furnace together with argon gas to force metal up the teapot outlet spout for tapping. The second approach has been to use a channel furnace of the bath tub type, freely open to the atmosphere and employing the usual wet rammed high alumina refractory in the inductor and furnace body, rather than the magnesite lining used in the first case. Detailed results of this work have also been published[2].

The results obtained with the two approaches are similar and can be summarised as follows:

(a) The rate at which magnesium is lost from an iron melt held at 1400°C depends upon the magnesium content of the iron and also the time during which the furnace has held magnesium treated iron. This effect is demonstrated in Figure 14.

(b) The holding of magnesium containing iron in a channel furnace causes the loop's electrical resistance to increase due to build-up of magnesium compounds on the inductor box refractory. Some reduction of resistance occurs in periods when the metal does not contain magnesium (see Figure 15) but complete removal of the build-up is not expected.

By holding magnesium treated metal in bulk prior to casting, acceptable nodularity is obtained at lower magnesium levels than normal. This is presumably due to the improved separation of magnesium oxide and sulphide inclusions, leaving dissolved magnesium only to be measured by analysis. However, the loop resistance increase occurring will eventually limit power input and manual rodding of the deposit or inductor box change will eventually be necessary.

5. The future perspective

The position of the channel furnace as an efficient metal holding and dispensing unit in the iron foundry has been confirmed in the past decade. As a melting unit the vertical channel furnace offers major operational advantages and the device is expected to find increasing application in foundries where consistency of molten metal is vital.

SUMMARY

The channel furnace has received considerable attention from ironfounders and plant manufacturers in the decade or so since the mid-sixties. This paper reviews the key issues which have led to this attention.

Advantages to be gained from the installation of a channel furnace are described in some detail and the considerations involved in selecting the optimum size and power rating are indicated. The results of some recent experimental studies involving channel furnaces are described.

The future of the channel furnace within the ironfoundry seems assured because of its ability to improve metal quality, environmental conditions and conserve energy in an industry which is so highly energy dependent.

REFERENCES

1. P. WAKEMAN: IBF Seminar—Electric Melting and Superheating Practice Up-dated, Europa Lodge, 8 November 1978.

2. V. Pachkis and J. Persson: Industrial Electric Furnaces and Appliances, Industrial Science Publications Incorp., New York, 1960.

3. B. Olausson: Quick Change Inductor for Channel Furnaces, ASEA Journal, 41, 5, pp. 65–70, 1968.

4. E. B. Gosling and M. C. Cubitt: Channel Furnace Lining Practice in Europe and the USA, BCIRA Conference, Loughborough, 1967.

5. R. D. Langman: The Electricity Supply Network in England and Wales and Electric Metal Melting, British International Committee for Electroheat Conference, September 1978.

6. Private communication with E. Calton, BSC.

7. C. F. Wilford and R. D. Langman: Channel Injection Furnaces VII, UIE Warsaw, 1972, Paper N143.

8. See for example:
Smith, L., The Development of the Channel Injection Furnace for Demagging in the Secondary Aluminium Industry, Journal of Metals, September 1978.

9 P. J. Mikelonis: Cupola-Channel Induction Furnace Duplexing for Ductile Iron Production, Proceedings of 1st AFS Electric Ironmelting Conference, November 1970, pp. 4–5 to 4–8.

10. G E Else: Raw Materials in Cupola Practice, British Foundryman, 71 (2) February 1978, pp. 35–42.

11. D. J. Swinden and C. F. Wilford: Review of Factors influencing Formation of Graphite Eutectic Cells in Grey Cast Iron, Foundry Trade Journal, 1 April 1976, pp. 491–510.

12. D. J. Swinden and C. F. Wilford: The Nucleation of Graphite from Liquid Iron: a Phenomenological Approach, British Foundryman, May 1976, pp. 118–127.

13. A. Moore: Some Recent Advances in the Practice and Understanding of Inoculation, British Foundryman, 67, March 1974, pp. 59–69.

14. D. D. Day: Evaluation of Grey Iron Melt Systems, Trans. AFS, 1977, pp. 417–420.

15. C. J. Edgerley: The Performance of a 40-tonne Vertical Channel Furnace Melting Grey Iron, ECRC/N1036, March 1977.

16. F. W. Walker: Some Experiences in the Use of Vertical Channel Induction Furnaces for the melting of Grey Cast Iron, British Foundryman, August 1975.

17. C. F. Wilford: Continuous Indication of Molten Cast Iron Temperature in a Channel Furnace, ECRC/N553, November 1972.

18. G. Bylund, et al.: Holding Nodular Iron in a Channel Induction Furnace, Trans AFS, 1975, pp. 385–392.

19. Anon., Holding Nodular Iron in Channel Induction Furnaces, Metals & Materials, October 1977, Foundry Technology.

20. C. F. Wilford: An Investigation into the Use of Electric Furnaces to hold Magnesium-treated Iron for the Production of SG Iron Castings, ECRC/R963, July 1976.

Recent Electric-furnace Developments for Non-ferrous Foundries

By I. Davies*

In this article, the author describes the ever increasing use of electric melting and holding in the non-ferrous sector of the foundry industry. He discusses the use of resistance furnaces and gives detailed cost breakdowns compared with fuel-fired units, and indicates the relative efficiencies of the different types. He then looks at the advantages which the coreless induction furnace offers to the aluminium and the copper-rich alloy founder. He finally looks at other types of electric melting units, and looks specifically at reverberatory furnaces, channel induction furnaces and low-energy furnaces.

The last few years have seen marked increases both in the variety of furnace designs available to, and in the number of electric furnaces installed in, British non-ferrous foundries.

Most of these furnaces have been installed in the smaller foundries, employing 100 people or less, which represent approximately 80% of the total number of aluminium and copper foundries. This expanding use of modern equipment by firms, which often employ no specialist electrical staff, is indicative both of the reliability of the equipment and the vigour and enterprise of the companies installing it.

It would be foolish to pretend that foundries using electric furnaces for the first time have encountered no problems. Several have experienced problems, usually as a result of continuing to use metal treatment processes which are no longer needed with the

new melting system. Such initial difficulties are quickly overcome. Many foundries have installed a single furnace and, having proved its viability, have commissioned further furnaces.

In all, up to 80 coreless induction furnaces have been installed by British copper alloy founders in the last five years while approaching 300 resistance crucible furnaces were installed here by aluminium founders in the same period (fig. 1). There have also been crucible furnaces installed for copper alloys and induction furnaces for aluminium, while more novel furnace designs have also appeared recently.

Resistance Crucible Furnaces

While this type of furnace has been available for over 40 years, fuel costs tended to be very low in the UK in the past and consequently most foundries preferred to use gas or oil-fired crucible furnaces, despite their relatively low operating efficiencies.

Large numbers of bale-out furnaces, typically of 136-182 kg. capacity, are used in aluminium foundries —possibly as many as 4,000 units in total, although numbers are difficult to estimate since many companies built their own furnaces as and when needed.

While fuel was cheap, these small, simply constructed, furnaces used to be economic to operate despite their high heat losses, but over the last few years fuel costs have risen dramatically while supplies have become less secure. About five years ago, fuel prices had risen to the point where electric and non-recuperative fuel-fired crucible furnaces were competitive in both capital and running cost.

Since then the running costs of such fuel-fired units have risen faster than those of comparable electric furnaces, while environmental and other

---

*Mr. Davies is a leading non-ferrous specialist with the Electricity Council.

requirements have proved to be more easily met with the electric units.

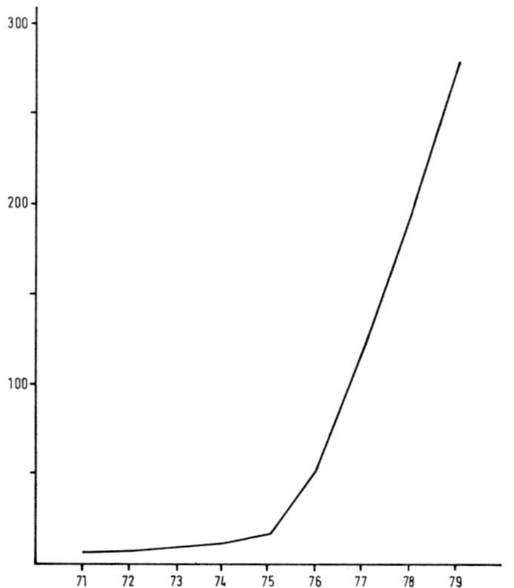

1 Numbers of resistance crucible furnaces installed in UK aluminium foundries since 1971.

Fortunately, although UK foundrymen had almost exclusively used fuel-fired furnaces up to five years ago, there has always been a thriving market for electric furnaces in Europe. So UK and continental furnace makers had developed their early designs to incorporate improved refractory and element materials and more advanced control systems. The more advanced technology has been established because foundrymen abroad have borne the brunt of development problems.

From small beginnings sales have developed very rapidly and there is every expectation that this growth will continue and assist UK aluminium foundries to compete with their continental counterparts.

The resistance crucible furnaces now being installed appear to be very simple in construction—the basic features are illustrated in fig. 2. They consist of a crucible on a stand, surrounded by heating elements, which in turn are backed by refractories and insulation.

This apparent simplicity is deceptive. A great deal of research and development work has gone into the selection of suitable materials for heating elements, element supports, refractories and insulation. It is doubtful whether a 'self-built' design could match the commercially available equipment in performance, reliability, efficiency or safety. Similarly, users need to resist the temptation to install slightly cheaper materials when carrying out routine maintenance and replacement of components or linings. Such modifications could be false economies.

Cost Comparisons

When comparing the performance of electric bale-out furnaces with conventional fuel-fired bale-out units it is easy to show that, for a comparable melting/holding duty and comparable capital cost, the electric equipment is cheaper to run in terms of energy cost. Some figures issued recently are shown in Table 1. These are conservative, showing much higher crucible, lining and element costs than are experienced by many electric furnace users.

The fuel consumption figures shown depend on furnaces working at their maximum efficiency of around $22\frac{1}{2}\%$. As most users are aware, in an industrial environment it is very difficult to maintain such efficiencies for long, and many such furnaces operate at about half this efficiency in practice. Despite this fact, the figures still show a balance in favour of the electric route.

For foundries considering electric melting and holding, it will be necessary to substitute their own costs for fuel/energy, refractories, maintenance and other inputs and draw up their own equivalent to Table 1.

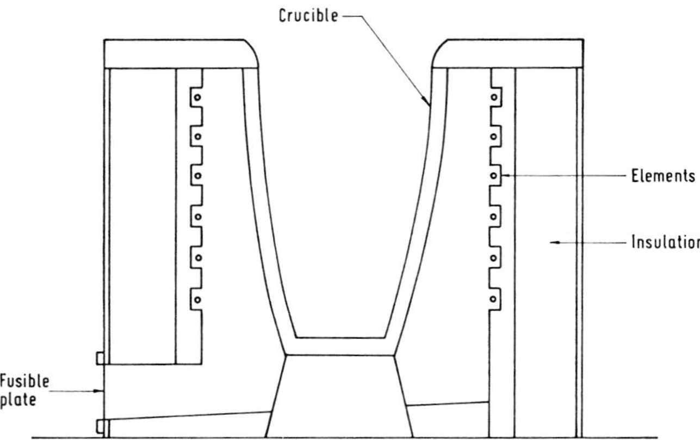

2 Typical resistance crucible furnace construction.

Electric and non-recuperative fuel-fired furnaces are very similar in capital cost. However, if more efficient recuperative units are used in such a comparison, capital cost will need to be allowed for. There may also be a need to include an additional allowance for maintenance for recuperative furnaces.

Ultimately energy costs depend on the efficiency with which the various furnaces operate. Electric furnace efficiency is typically 80-85% and this efficiency is fixed by furnace design. It can only be changed by rebuilding the furnaces with different refractories. The operator cannot change it, and it does not alter in operation.

This is not usually true of other melting systems. To assist users of such systems the BNF Metals Technology Centre has developed a diagnostic trolley to assist in burner adjustment. Given a suitably qualified operator and a disciplined workforce, this is a valuable tool which enables fuel-fired furnaces to be operated at close to their theoretical efficiencies (allowed for in Table 1).

Even so, electric resistance crucible furnaces can compete against efficient fuel-fired units. It may be a closer comparison against the best recuperative designs, but the minimal maintenance, immutable efficiencies, environmental advantages, and security of energy supply, all weigh in their favour.

Tilting crucible furnaces are now available and in use for bulk melting in several aluminium foundries. Some designs have melting rates comparable with those of tilting fuel-fired crucible furnaces.

Typical uses include feeding low-pressure diecasting machines, providing molten metal for sand casting operations and feeding holding furnaces for die-casting operations. Only about 18 such units have been installed in the UK so far.

While this figure appears low, it reflects the traditional balance of tilting and bale-out furnaces in the UK aluminium foundry industry and is in proportion with current bale-out furnace sales. Designs have been well proved overseas and those installed in the UK to date have been as reliable and economic in operation as their bale-out counterparts.

For melting and holding aluminium, resistance fur-

TABLE 1: *Cost comparisons of electric and fuel-fired bale-out furnaces.*

	Gas £	Oil £	Electric £
Fuel/energy cost per year	4,700	5,708	3,553
Lining cost per year	400	400	580
Crucible cost per year	822	822	548
Element cost per year	—	—	360
Melting loss per year	735	735	735
TOTAL Costs per Year	6,657	7,665	5,881
Cost per tonne of metal melted	27.20	31.40	24.00

(Two shifts per day operation, five days per week, melting at a rate of 65kg. per hour. Gas: 24p per therm, Oil: 10.5p per l, electricity: 2.8p per kWh.).

naces are normally equipped with metallic heating elements. These will also operate at a temperature high enough for melting brasses. More recently, a growing interest by copper alloy founders in the use of resistance furnaces for higher temperature alloys, bronzes and gun metals, has led to the use of non-metallic heating elements.

A number of furnace suppliers now offer units with silicon carbide elements which normally operate at temperatures between 1,400 and 1,600°C. Designs for bale-out, tilting, and push-up furnaces now exist

3 Bale-out furnaces of 250 kg. capacity, melting and holding LM4 and LM6.

4 Typical medium-frequency coreless-induction furnace installation melting copper alloys.

which incorporate this type of heating element. About 20 furnaces of this kind have been installed here so far for melting copper alloys, but they show considerable promise, being rather cheaper to install than induction melting systems, although not providing quite the same degree of production flexibility.

Coreless Induction Furnaces

Most of these have been installed in copper alloy foundries. The most popular seems to be the 'push-up' type shown in fig. 4. It is difficult to compare with fuel-fired furnaces since a typical copper alloy foundry is often able to replace between three and ten conventional fuel-fired furnaces with the combination of a push-up furnace, power supply and tilting furnace body to provide the occasional large melt.

It is also possible to make considerable changes in working practice. Since the induction melting system can usually produce molten metal in less than 20 minutes, moulding, melting, pouring and fettling can be carried out throughout the working day.

Rapid melting and the ability to change alloys quickly give the foundry the ability to adjust molten metal supply to match the output of moulds and overcome delays and temporary stoppages in parts of the moulding process. If one machine stops, a different alloy can be quickly supplied to another moulding line. In some foundries, using a wide range of alloys, this flexibility is extremely valuable and makes a big contribution to profitability.

This type of furnace melts quickly because it operates at 'medium frequency', usually either 1,000 or 3,000 Hz, and this enables large amounts of energy to be applied to relatively small furnaces. Push-out furnaces are normally not much larger than 100 kg. capacity because of the difficulty of handling large crucibles full of molten metal.

Apart from the various push-out furnace designs there are several others available to the foundry. Other designs of removable crucible furnaces such as the lift coil and lift-and-swing units also provide the ability to change alloys rapidly. Where a foundry operates in an area with a high water-table, such designs may be preferable to the push-out types, since they minimise the need for excavation. There are also roll-over designs available for investment foundries.

The medium frequency power supplies are not normally rated at less than 100 kW., enough to melt 250 kg. of copper an hour. Since many units are of less than 100 kg. capacity it is hardly surprising that melt times of 15 or 20 minutes are common.

Most power supplies are solid-state inverters or converters and while there are technical differences between the various designs these are of little interest to the foundryman provided price and performance are right.

While the interior of one of these units might appear dauntingly complicated, they are extremely simple to operate. It is usually only necessary to set the power required—anything between zero and 110% of the unit's rated output—and the equipment does the rest. Operating one is normally no more difficult than controlling the volume on a radio set.

Maintenance is minimal. The power supply itself should need none apart from routine cleaning to remove metallic dust and dirt. The furnace and its water cooling system may need maintenance but foundry staff or local electricians and plumbers ought to find this relatively easy.

are dissimilar, it is usually possible to install extra furnace bodies to cater for these, the whole melting installation being energised from the same power supply.

As a rule the cost of installing induction melting, while not small, is usually far less than that of installing fuel-fired equipment to produce the same tonnage of the same alloys.

In addition, many users of the electric equipment

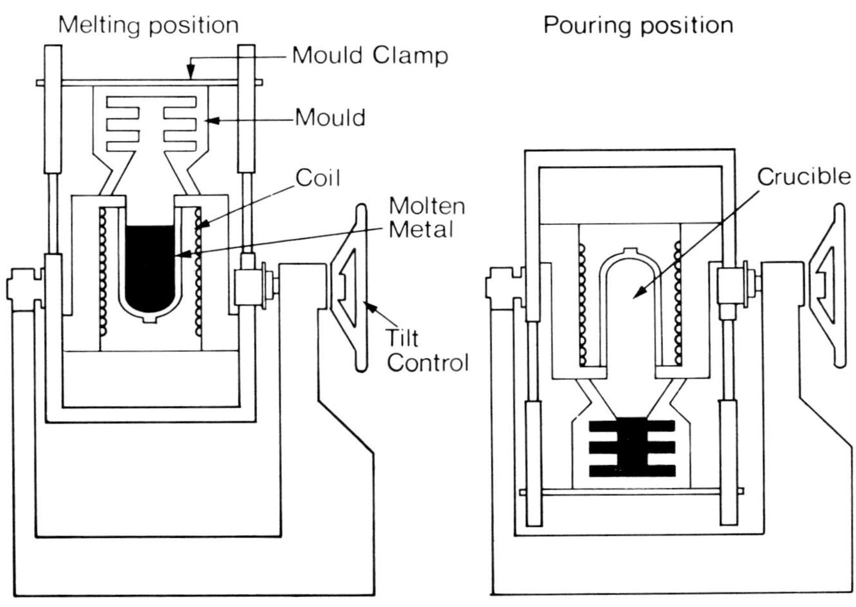

5 Alternative induction furnace designs.

Tilting Furnaces

Although the push-out furnace, with its high melting rate but relatively small capacity, meets most of a foundry's metal needs, it is often the practice to also install a larger tilting furnace body as well to cater for the need to produce the occasional large casting.

Foundries which habitually need to produce large castings or which have mechanised mould production to a fairly high degree would use tilting furnaces in order to achieve the necessary metal availability.

Tilting furnaces are not quite as flexible as push-out designs in terms of alloy changes. Powered at medium frequency, high melting rates are achieved and it is easy to melt out a fresh charge in a cold, empty, furnace. However, a tilting furnace requires a fixed lining and more care is needed when making alloy changes to avoid cross-contamination between melts.

Fortunately, most foundries needing to use larger furnaces melt only a few alloys, and, even when these

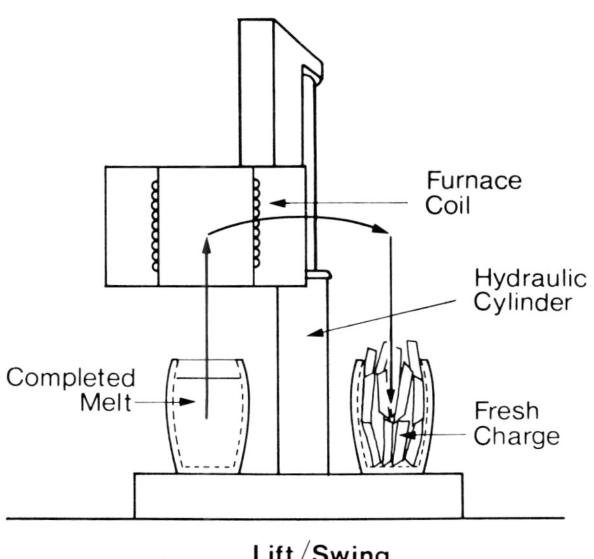

report improved metal quality, fewer rejects and much better working conditions, so that operators are no longer reluctant to approach the furnaces to charge foundry returns or check metal quality and temperature. Reduced metal inventories and ease of recruitment are real benefits.

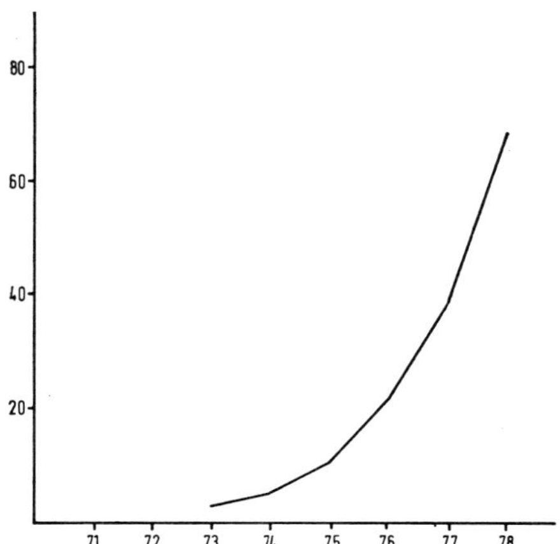

6 Numbers of induction furnaces installed in UK copper alloy foundries since 1971.

There have been a few difficulties when foundries have first adopted induction melting. Operators have not always realised that, having completed a melt in 20 minutes or less, if power is not reduced, these furnaces can superheat the metal at the rate of 200°C per minute. Another is the continued use of phosphor copper to deoxidise the melt, even though the new furnace may not have oxidised it. Excess phosphorus then produces metal/mould reactions and porous castings.

While most induction furnaces of this type have been installed for melting copper alloys, they are now also being used for melting aluminium. Mains frequency coreless furnaces have been used by primary aluminium producers, rolling mills and large foundries to reclaim fine scrap and returns for many years—it is a job they do well.

The average foundry cannot use mains frequency furnaces, however, because of the time needed to start them up. They do not suit single shift operation. Medium frequency units overcome this problem and, because for a given power density they produce less agitation in the melt, they show significantly higher metal yields when melting ingot and normal foundry returns than a mains unit could.

While a few of these furnaces are dual-purpose, where foundries originally installed them for copper alloy melting and now use them for aluminium as well, there are also installations where they are purely used for bulk melting aluminium in foundries. Interest in this latter application is growing.

Other Electric Furnaces

There are several other furnace designs that have emerged in the last few years. One is the "electric reverberatory furnace". This is an American development, brought about by natural gas shortages. There are several US manufacturers but all their designs are similar in appearance and performance. A typical unit is illustrated in fig. 7. They are available with

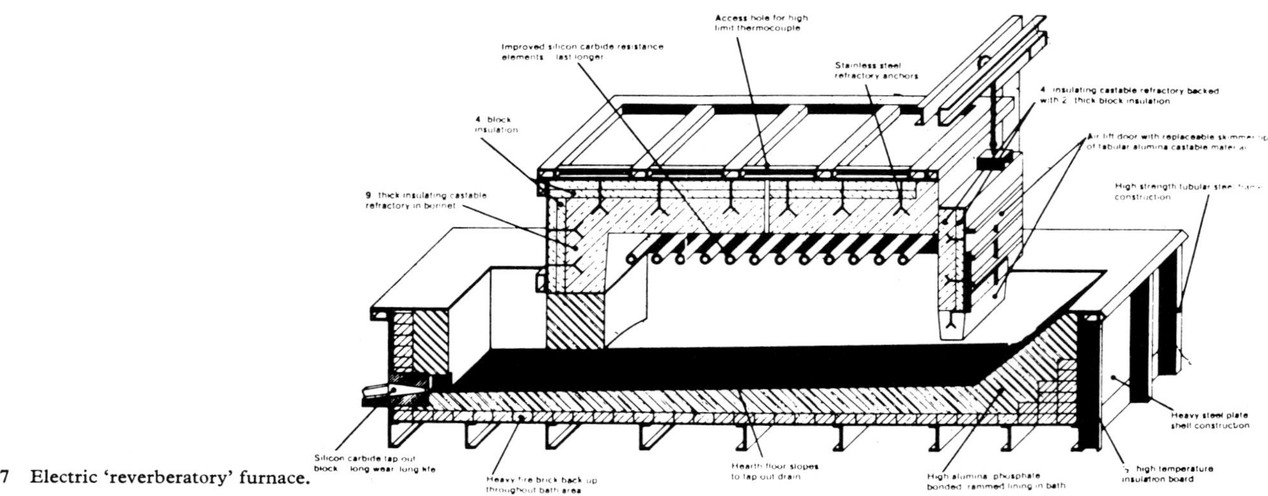

7 Electric 'reverberatory' furnace.

Such difficulties are usually quickly diagnosed and remedied. In one sense the change to electric melting could be likened to swapping a family saloon for a formula one racing car. You need to be careful until you understand what the equipment can do.

melting rates up to 1.75 tonnes per hr of aluminium, and although some UK foundries may find them suitable for particular melting operations it is unlikely that they will achieve very widespread use here for two reasons.

First, hourly melting rates are typically one twelfth to one seventh of the total furnace capacity, which means that their suitability for use in smaller foundries, those needing to melt a variety of aluminium alloys or those working only one shift, may be doubtful simply because of the time needed to melt out from cold.

Second, some alloys produce relatively large amounts of dross when melted and necessitate the frequent use of salt fluxes. Fumes from these fluxes may attack the heating elements and shorten their life. Careful furnace operation and confining the use of fluxes to charge and bale-out wells should give satisfactory service. Where the alloys used are particularly prone to oxidation, it may be desirable to choose an alternative for melting these alloys.

Another development is the improvement in channel induction furnace designs for aluminium melting. Channel furnaces have been used as holding furnaces in aluminium foundries for many years, although few new units have been installed for some time. A number are also in use as melting furnaces.

Historically, when used primarily for melting aluminium, these furnaces had to be limited in power, to prevent the channels from becoming rapidly blocked with oxide deposits. Practical upper power limits tended to be no more than 250 kW. per inductor while even the small low-powered bale-out units required frequent "rodding" to prevent oxide build-up.

Recent developments have overcome this problem. For the small bale-out units, a gas-injection system developed jointly by the Electricity Council Research Centre and the Midlands Electricity Board eliminates the need for periodic channel clearance.

Variations of this technique, using different kinds of gas injectors, have also increased the power that can be applied to channel bulk melters. An extension of this technique has also been developed for the "demagging" process used by secondary aluminium producers.

In the last two years or so a new design of high power channel inductor has also appeared. This does not use gas injection, but relies on improved electrical and mechanical design to boost power input to over 1,000 kW. per inductor—a melting power of over 2.5 tonnes per hour per inductor. One furnace in Europe is now fitted with four of these units. So far, this development has been aimed at the primary producers and the larger wrought aluminium producers, but there is no reason why furnaces of this kind should not be used by foundries as well.

Low Energy

A further development in aluminium furnaces comes from Sweden. This is the "Holimesy" system which relies on very thick, thermally efficient, refractories, and limited areas of accessible melt surface to achieve extremely low energy requirements. Ideally these furnaces require feeding with molten metal for their melting ability is rather limited. They are best used with autoladles or small hand held ladles. A 250 kg. furnace normally has a bale-out area only 250 mm. sq. Such a furnace would have an energy consumption equivalent to about half the electrical requirements of a fuel-fired furnace of comparable capacity.

However, if a foundry needs to use large ladles or if bulk melting is not possible, despite the undoubted energy savings achievable, use of this kind of furnace may not always be easy. Nevertheless, even if there are difficulties in fitting this equipment into existing foundries it is worth consideration for new pressure diecasting foundries, since there is a complete system available: Holding furnaces, launders, bulk supply furnaces and remelting furnaces.

In addition to these developments for aluminium melters, there are a few for copper alloy users. The most recent involve the use of furnaces employing graphite resistance elements for heating. Because of the nature of graphite, the elements need protecting from the atmosphere. Consequently, the element chamber is filled with an inert gas such as nitrogen and this is sometimes fed through the melt as well.

The earliest use was for furnaces producing continuously-cast copper alloy rod or strip. This kind of furnace has now been developed for wider application. It provides good melt homogeneity while effectively deoxidising the metal. Powder production is one of a number of further possible applications.

Conclusions

Electric furnaces installed to date have been installed because they are profitable. The reasons for this are many; lower energy costs, less maintenance, improved environment, lower metal costs, and so on. The newer furnaces, briefly mentioned in this article, should enhance the range of equipment available to the foundryman and offer opportunities for increased profits in the future.

What Users Should Know About Clean Steel Technology

By Harry E. Chandler

·"Customers are becoming more demanding in their requirements."

"User properties of steel are strongly influenced by their cleanness."

What is perhaps the dominant theme in steel selection and specification today is pinpointed by the first quotation, which is from a specialty steel company's 1980 annual report.

The second quotation, which is from a brochure for the Second International Conference on Clean Steel, held in Balatonfüred, Hungary on 1-3 June, pinpoints the area of steel technology many users are counting on to meet these more demanding requirements.

To bring *Metal Progress* readers up to speed on this fast-growing technology, this article:

1. Presents case histories that illustrate some ways in which steel cleanness is being utilized to improve the performance of parts in service.

2. Gives some answers to fundamental questions like: What is clean steel? How clean can it be, or should be? How is cleanness measured? What standards are used? What clean steel technology is in use?

3. Provides an inventory of clean steel technology available from selected steel producers in the United States.

Examples: Why Clean Steel Is Being Specified

A case history illustrating why steel users are becoming more demanding involves what looks like the world's biggest windmill atop a 4000 ft (1200 m) mountain in North Carolina (Fig. 1). The huge wind turbine generator is ten stories tall (140 ft [45 m]), has a housing roughly the size of a boxcar, and a 200 ft (60 m) in diameter propeller. Spars for its blade are made from a specially processed clean steel, ASTM A533-B Cl 2 plate. In this instance, enhanced fatigue resistance is the incentive.

About 81 tons (75 t) of this steel were used to form the blade spars. Blades must stand up to the stress of almost continuous rotation and ever changing wind speeds. A wind of 25 mi/h (40 km/h) will turn the blade at 35 r/min, producing about 2000 kW of electricity.

A calcium-argon ladle treatment process was used to produce this steel.

Sulfur is reduced to 0.010% or less. The sulfide inclusion count is reduced, and inclusions that survive are shape controlled to resist flattening into potential starting points for fatigue cracks. Other important properties include through thickness ductility and good notch toughness.

Another Example — ESR (electroslag refining), another clean steel process, reduces both sulfur and oxygen significantly, resulting in a material that is extremely clean and virtually free from nonmetallic inclusions. Inclusions that remain are widely dispersed due to progressive solidification.

Grades processed by ESR available from one supplier include: AISI carbon steels, 1006 to 1060; AISI alloy steels, 4100, 4300, and 8600 series; mold steels (hot work die steels); ASTM-ASME grades A203, A204, A302, A387 — grades 11, 12, 21, 22 — A514, A517, A533, A537, A542, and A543; AMS grades, including 6358 (AISI 4340); Mil-S-grades 16216 (HY-80/100), 18729C (AISI 4130), 24238, and 24371 (HY-130); ABS hull steels; and ASME code approved grades per section VIII and Section III.

User benefits of these steels are reputed to include more uniform and improved mechanical properties in all three directions, generally high ductility, plus fatigue resistance. Fracture toughness is said to be "better than that of electric furnace steel."

Potential benefits in fabricating operations include: more uniform properties promoted by even response to heat treatment; reduced hazard of cracking during severe heat treating operations; a decrease in the likelihood of hard spots in machining; improved ductility which translates into improved forming properties in both hot and cold working; and better through thickness properties, which reduce the potential for lamellar tearing of welded parts and increase the reliability of parts stressed in the through thickness direction.

Case history: An ESR processed ASTM A302 Gr B plate steel was selected for the drive gear on a mass cement kiln. Homogeneity of the alloying elements resulting from ESR increases hardenability, allowing a reduction in gear face width.

Some Basics, Including 'What is Clean Steel?'

As indicated by the case histories in the previous section, oxygen and sulfur are among the culprits in the clean steel concept. They are present in the form of oxides and sulfides and form nonmetallic inclusions, such as Al_2O_3. The size, shape, and

226

Source: *Metal Progress*, October 1981, 52-59, © 1981 American Society for Metals

In vacuum induction melting, impurities are removed as gases. There is no contamination from the air or slag. Charges may be solid or liquid. (*Carpenter Technology Corp.*)

Fig. 1 — Blade spars for wind turbine generator were made from clean steel (low sulfur, and the shape of sulfide inclusions was controlled) to improve properties. The ladle treatment used is based on calcium-argon. (*Lukens Steel Co.*)

frequency of inclusions have powerful influences on cleanness. Other bad actors are phosphorus, hydrogen, and trace elements, such as lead, bismuth, antimony, and tin.[1]

The key point to remember is that a steel 100% free of oxygen, sulfur, phosphorus, hydrogen, and the trace elements is not possible, nor would it always be desirable — pure iron, for example, has low tensile strength. In fact, the exact relationship of cleanness to quality has yet to be fully established in all instances.

Commercial steels always contain certain amounts of oxygen and sulfur, for example. The exact amount present depends largely on the process. For instance, stainless steel made by the electric induction process has these approximate contents: O_2, 100 to 150 ppm; and S, 50 to 200 ppm, without sulfur cleaning. By comparison, stainless steel processed by ESR contains about 20 to 30 ppm O_2 and 10 to 30 ppm S.

The latter example illustrates that although inclusions cannot be eliminated, they can be reduced in number. Also, their size and shape can be changed. In addition, it must be kept in mind that all impurities do not originate from the steelmaking process per se. There are also outside sources like slags and ceramics to contend with. Finally, it has been estimated that iron with an impurity content of only 1 ppm is still quite "dirty" in the fullest sense of the term. Each gram of this material would contain about 10^{16} atoms of other elements.[1]

By usual standards, a steel low in oxygen and sulfur is considered clean; and it is felt that there is something of a direct relationship between inclusion size and steel quality. This translates to a practice which reads, "the smaller the inclusion, the better the quality." There is a hitch, however. The exact point at which

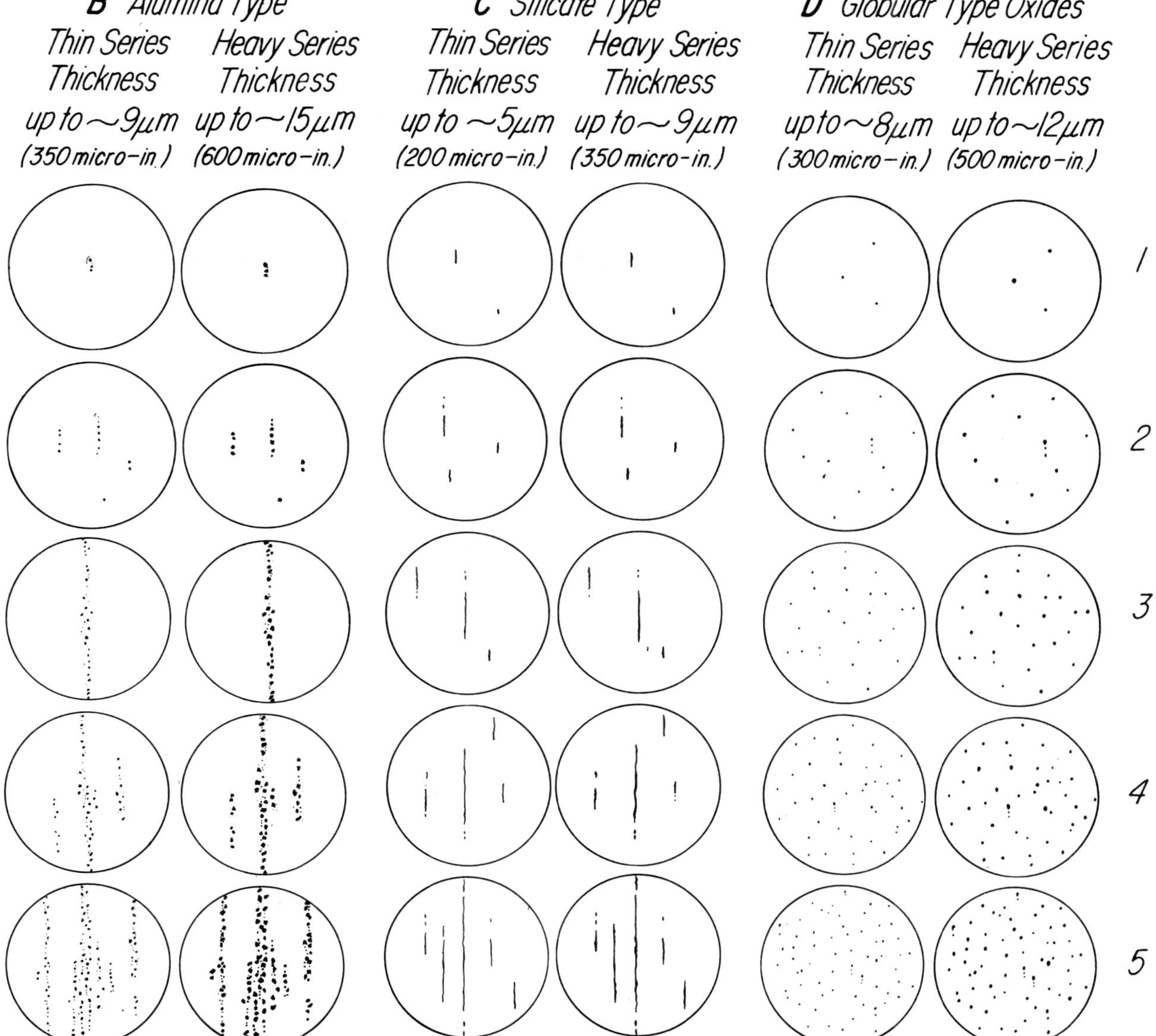

Fig. 2 — JK chart is among standards used to classify inclusions by size, number, and distribution within a given field. This information is used by specifiers. However, it does not amount to a rating system which indicates, for example, that a given size inclusion is detrimental to steel properties. JK standards shown do not include those for vacuum melted steels, for which there are seven more categories. Also, round fields shown duplicate shape of those seen in microscope. In other standards for newer inclusion counting systems, the fields may be square to show the areas covered.

inclusion size becomes acceptable is not known. In general, the aim is to avoid large inclusions. How large is another question.

There is another complication: inclusions are not evenly distributed throughout a steel, meaning that a given sample may not be representative of the whole.

Standards for phosphorus and hydrogen are more precise than those for oxygen and sulfur. A phosphorus level of 0.035% or lower is generally considered clean. Likewise a hydrogen level of several ppm. Phosphorus content is lowered by such means as oxidation and selection of raw materials. Hydrogen content is reduced via vacuum metallurgy or flushing with inert gas.

Trace elements are another matter. As is the case with oxygen and sulfur, it is difficult to quantify their exact effects, even though it is known that they can have adverse impacts on such properties as hot ductility, creep strength, weldability, corrosion, and embrittlement.

Mean content of trace elements ranges from 0.1 to 50 ppm. A German specification for pure nickel, for example, calls for 5 ppm As; 1 ppm Pb, Sn, and Sb; and 0.2 ppm Bi.

Table I — Partial Inventory of Clean Steel Technology in the United States

Process[1,2]

Company	AOD			EBR			ESR			LR			Q-BOP			VAR			VD			VIM			VOD		
	a	b	c	a	b	c	a	b	c	a	b	c	a	b	c	a	b	c	a	b	c	a	b	c	a	b	c
AL Tech Specialty Steel Corp.	2	-	-	-	-	-	1	-	-	-	-	-	-	-	-	6	-	-	-	-	-	-	-	-	-	-	-
Armco Inc.	2	-	-	-	-	-	-	-	-	-	-	1	-	-	-	2	-	1	3	-	-	2	-	-	-	-	-
Babcock & Wilcox Tubular Products Group	1	-	-	-	-	-	1	-	-	-	-	1	-	-	-	-	-	-	2	-	-	-	-	-	-	-	-
Bethlehem Steel Corp.	-	-	-	-	-	-	2	-	-	5	-	3	-	-	-	-	-	-	7	-	1	-	-	-	-	-	-
Cameron Iron Works	-	-	-	-	-	-	-	-	-	-	1	-	-	-	-	14	2	-	1	-	-	1	-	-	-	1	-
Cannon-Muskegon Corp.	1	-	-	-	-	-	-	-	-	-	-	-	-	-	-	-	-	-	-	-	-	3	-	-	-	-	-
Carpenter Technology Corp.	3	-	-	-	-	-	4	-	-	-	-	-	-	-	-	10	-	-	-	-	-	3	-	-	-	-	-
Crucible Specialty Metals Div., Colt Industries	1	-	-	-	-	-	-	-	-	-	-	-	-	-	-	1	-	-	-	-	-	-	-	-	-	-	-
Crucible Stainless & Alloy Div., Colt Industries	1	-	-	-	-	-	-	-	-	1	-	-	-	-	-	6	-	-	1	-	-	-	-	-	-	-	-
Cyclops Corp.	2	-	-	-	-	-	3	-	-	-	-	-	-	-	-	7	-	1	-	-	-	3	-	1	-	-	-
Guterl Special Steel Corp.	-	-	-	-	-	-	2	1	-	-	-	-	-	-	-	1	-	-	-	-	-	-	1	-	-	-	-
Ingersoll-Rand Oilfield Products Co.	-	-	-	-	-	-	2	-	-	-	-	-	-	-	-	-	-	-	1	-	-	-	-	-	-	-	-
Jessop Steel Co.	1	-	-	-	-	-	-	1	-	-	-	-	-	-	-	-	-	-	-	-	-	-	-	-	-	-	-
Jones & Laughlin Steel Corp.	1	-	-	-	-	-	-	-	-	2	3	-	-	-	-	-	-	-	-	-	-	-	-	-	-	-	-
Joslyn Stainless Steel Div.	1	-	1	-	-	-	-	1	1	-	-	-	-	-	-	-	-	-	-	-	-	-	-	-	-	-	-
Latrobe Steel Co.	-	-	1	-	-	-	1	-	1	-	-	-	-	-	-	8	-	-	1	-	-	1	-	-	-	-	-
Lukens Steel Co.	-	-	-	-	-	-	-	-	-	1	-	-	-	-	-	-	-	-	1	-	-	-	-	-	-	-	-
National Steel Corp.	-	-	-	-	-	-	-	-	-	-	-	1	2*	2	1	-	-	-	2	-	-	-	-	-	-	-	-
Phoenix Steel Corp.	-	-	-	-	-	-	-	-	-	1	-	-	-	-	-	-	-	-	1	-	-	-	-	-	-	-	-
Republic Steel Corp.	1	-	-	-	-	-	1	-	-	1†	-	-	2	-	-	6	-	-	2	-	-	-	-	-	-	-	-
Teledyne Vasco	-	-	-	-	-	-	-	-	-	-	-	-	-	-	-	2	-	-	-	-	-	-	-	-	-	-	-

1. AOD, argon oxygen decarburization; EBR, electron beam refining; ESR, electroslag remelting; LR, ladle refining; Q-BOP, a modification of the basic oxygen process; VAR, vacuum arc remelting; VD, vacuum degassing; VIM, vacuum induction melting; VOD, vacuum oxygen decarburization.
2. a, on stream; b, under construction; c, on drawing boards.
*LBE process, top and bottom blown. † Experimental only.

It has been reported that such elements as lead, zinc, antimony, tin, bismuth, selenium, and tellurium can be reduced or eliminated with argon-oxygen decarburization (AOD) and vacuum metallurgy processes.[2,3]

Two quantitative metallographic techniques — one old and one new — are used to count inclusions and to determine their sizes and shapes.

The old way is a manual, microscopic method based on linear analysis and point counting. What is seen in the microscope is compared with such standards as the JK chart (Fig. 2).

The newer automatic image analysis systems are based on video scanning, and minicomputers may be incorporated into the facilities. This is obviously the much faster technique and more parameters can be studied at a given time.

With the manual method it takes about 15 min to check out 100 fields, each with an area of 0.5 mm^2. The total surface area represented is about 50 mm^2.

With the automatic method, by comparison, it takes 50 s to calculate and print out the oxide and sulfide inclusion counts in 500 fields, each measuring 0.32 mm^2. The total area in this instance is 160 mm^2.

To put it another way, you can manually count ten inclusions per minute. The rate with the automatic image analyzer is several thousand per minute.

Profiles of Nine Clean Steel Technologies

In this section, argon-oxygen decarburization, electron beam refining, electroslag refining, the Q-BOP, vacuum arc remelting, vacuum degassing, vacuum induction melting, and vacuum oxygen decarburization are described briefly. A ninth technology called ladle refining (also injection steelmaking or metallurgy) is given separate attention because it is the most talked about process today.

For example, a steel company research executive cites the trend to what he terms injection steelmaking and predicts it will "blossom more in the 1980's in both electric furnace and basic oxygen steelmaking. At (our) Ashland, Ky. and Middletown, Ohio (plants) we are now removing sulfur from hot metal in torpedo ladles by injecting reactants. In Houston we are desulfurizing finished steel in the teeming ladle."[4]

This official next singles out what could be termed a byproduct of this technology that is expected to bring about a minor revolution in steelmaking.

He explains, "This trend is going to grow with the effect that both our basic oxygen vessels and our electric furnaces will become more and more melting vessels and the final refining and alloying will be done in ladles and degassers.

"By 1985, a typical electric furnace will be a fabricated, water cooled structure with no refractories above the bath, with water cooled electrodes, hogging down its metallic charge in a minimum melt time with the aid of scrap preheating, oxy-fuel burners, bath oxygen injection, and super high power inputs.

"Specialty steels will be duplexed in AOD vessels, or remelted in vacuum arc remelting, electroslag remelting, or similar processes. Carbon steel will be cast as crude hot melts with final refining, sulfur and phosphorus removal, alloy additions, and temperature homogenizing in the ladle or some special vessels."

He concludes with the following, "The Japanese are already producing low silicon hot metal to feed their basic oxygen vessels, and they are planning to further remove silicon by the injection of mill scale and soda ash.

"After desulfurizing and phosphorus removal, this relatively pure hot metal will be charged into a basic oxygen vessel, brought up to a manageable temperature, and finished by injection methods and temperature balance. This 'pure iron' finishing has great advantages over conventional practice in yield, process control, and probably in refractory life."

You can get an idea of the results that can be obtained from a Japanese paper on the LF (ladle-furnace) process.[5]

In a commercial test combining the BOF and LF process, it is reported that the oxygen content of molten steel was less than 0.002%; sulfur content was 0.002%; and phosphorus was less than 0.010%. Only fine inclusions of less than 10 μm (400 micro-in.) survived the treatment.

Benefits for heavy plates and hot strip products are said to include improved notch toughness, lamellar tearing resistance, and bending properties. In addition, there was a good gain in reduction in area in the short transverse direction.

Definition — The ladle refining process is relatively simple. By one definition,[6] "Injection metallurgy is a refining and alloying process in which refining (deoxidation and desulfurization) is achieved by injecting solid materials, such as calcium, into molten steel. The solids are introduced in powder form, while a carrier gas, usually argon or nitrogen, is used as a transporting medium.

"In a typical application, the injection station consists of the lance, a lance carrying device, a powder dispenser unit outfitted with related control elements, and the cover (which sits on the ladle to control the atmosphere and splash during the injection process."

Summary descriptions of other clean steel processes follow.

Q-BOP — This is essentially a modification of the basic oxygen process. Oxygen and other gases are blown directly into the melt through the bottom or both the bottom and top of the vessel. An advantage of the Q-BOP over the BOF is that it provides an easier way to produce low sulfur and low carbon steels. In terms of cleanness, the steel is close to that obtained with electric arc furnace, double slag practice.

Vacuum Induction Melting (VIM) — A charge of solid scrap (in some plants, liquid charges), alloys, ferroalloys, and other materials is melted in a refractory lined crucible under vacuum by heat generated from electric currents induced in the charge. Impurities are removed as gases, and there is no contamination from air or slag. The molten metal is cast under vacuum into ingots for processing to mill products or into electrodes for remelting to obtain further refinement and improvement in ingot structure.

Vacuum Degassing — Molten metal from a melting furnace can be degassed in a separate vessel by exposure to pressures substantially below atmospheric

Fig. 3 — In vacuum degassing, dissolved gases such as oxygen, carbon monoxide, hydrogen, and nitrogen are sucked out. Molten metal from a furnace is degassed in a separate vessel. (*Armco Inc.*)

(Fig. 3). Under these conditions, dissolved gases such as oxygen, carbon monoxide, hydrogen, and nitrogen are sucked out. Efficiency is increased by stirring the molten metal to maximize exposure of the entire bath to the vacuum-metal interface. Some facilities are equipped with graphite electrodes or induction coils to reheat the molten metal during or following degassing. In some instances, vacuum degassing is used as an alternative to argon stirring in continuous casting. The purpose is to avoid stratified temperature layers in the molten metal. Vacuum degassing acts as a shroud around the stream and makes it possible to get cleaner and more uniform steels than you can in conventional electric furnace practice.

Electron Beam Refining — Molten metal from the vacuum induction furnace is further refined by passing it through a very high vacuum in a water cooled copper trough, during which time it is heated locally to very high temperatures by beams from electron guns. Refining efficiency is extremely high due to the combination of a high vacuum and intense localized heating. During this part of the process, there is no contamination from air, slag, or crucible. Molten metal is continuously cast under vacuum for processing to mill products or for remelting to obtain an improved ingot structure.

Vacuum Arc Remelting (VAR) — As-cast electric furnace or VIM electrodes are progressively remelted and solidified in a water cooled copper mold under vacuum by an electric arc generated between the electrode and the molten metal above the solidifying ingot (Fig. 4). Some refining occurs (impurities are removed in gaseous form), and there is no contamination from the air, slag, or crucible. The progressively

Fig. 4 — In the VAR process, impurities are removed in gaseous form. There is no contamination from the air, slag, or the crucible. (*Cyclops Corp.*)

solidified ingot is completely homogeneous and free of shrinkage defects and is processed directly to mill products.

Electroslag Refining (ESR) — As-cast electric furnace melted or vacuum induction melted electrodes are progressively remelted and solidified in a water cooled copper mold under a blanket of molten flux. Melting is due to heat generated by the resistance of the molten flux to electric current passing between the electrode and the solidifying ingot. Refining occurs as molten metal passes through the flux, and impurities are removed as gas or by reaction with the flux to form slag. The progressively solidified ingot is completely homogeneous and free of shrinkage defects. It is directly processed to mill products.

Vacuum Oxygen Decarburization (VOD) — Molten metal from the electric arc furnace which contains relatively high levels of carbon and chromium is decarburized under vacuum by lancing with oxygen from the top, much as you do with the BOF. Under these conditions, it is possible to remove carbon without excessive oxidation of chromium. This permits the use of lower cost, high carbon raw materials in the production of stainless steel. Efficiency is greatly increased by stirring the molten metal (with argon or by induction) during decarburization. Dissolved gases, such as hydrogen, carbon monoxide, and nitrogen, are sucked out during the exposure to subatmospheric pressures. Very low carbon can be obtained. VOD is one of two duplex melting or ladle refining processes (AOD is the other) used primarily for stainless steels and other specialty alloys including tool steels.

Argon-Oxygen Decarburization (AOD) — An argon-oxygen mixture is injected below the surface of the metal near the bottom of the vessel through tuyeres. There is a great deal of turbulence, causing intimate contact of the oxygen and carbon. Argon gas reduces the partial pressure of carbon monoxide in the system and allows the carbon-oxygen reaction to proceed until very low levels of carbon and oxygen are reached.

The ratio of argon to oxygen is increased as decarburization occurs. Slag-metal reactions, such as the reduction of chromium oxide and desulfurization, are enhanced by stirring with pure argon after the desired carbon level has been reached. Argon injection also promotes removal of dissolved gases. All types of stainless steel, many nickel and cobalt base alloys, many tool steels, and many low alloy steels have been made by the AOD process.

Who Has What Clean Steel Technology

Metal Progress invited 33 steel producers in the United States to participate in a survey of clean steel technology; 21 responded; results are summarized in Table I. Note that the study covers units under construction and on the drawing board in addition to units on stream. ⬥

For More Information: Mr. Chandler is editor, *Metal Progress*, American Society for Metals, Metals Park, Ohio 44073; tel: 216/338-5151.

References
1. "Clean Steel — A Debatable Concept," by R. Kiessling: *Metal Science*, May 1980, p 161.
2. "Nickel From a Swedish Point of View," by E. Wallen: Amax Nickel Seminar, Stockholm, December 1977.
3. "Fighting the Impurity Problem": Inco Ltd., July 1973.
4. "Steelmaking in the Future," a talk by Lawrence C. Long, director-process research, Corporate Research & Technology, Armco Inc.
5. "Refining of Quality Steel Using BOF-Ladle Furnace Process," by H. Kajioko et al: Nippon Steel Technical Report No. 11, March 1978.
6. "Ladle Injection Metallurgy: Where It's At, Where It's Going, and Why": *33 Metal Producing*, April 1981, p 53-57.

Theory of AOD Steel Foundry Applications

S. K. Mehlman, *Project Engineer*
Union Carbide Corporation, Linde Div.
Tarrytown Technical Center
Tarrytown, New York

ABSTRACT

The history of argon oxygen decarburization (AOD) and the theoretical basis of the process are reviewed. The growth of AOD in the stainless steel industry and into new fields is discussed with emphasis on the steel foundry industry. Refining practices for a stainless heat and a low alloy heat are compared. Quality benefits inherent in AOD processing are discussed. The key processing advantages of AOD are summarized in order to explain its wide acceptance as a multifaceted refining tool.

Introduction

Argon oxygen decarburization (AOD) is a patented Union Carbide process that has gained worldwide acceptance for the production of stainless and other high alloy steels. The AOD process differs in several ways from other pneumatic steelmaking processes in that AOD utilizes commercial high-purity Ar (and nitrogen) as well as oxygen and that the gases are injected into the metal through submerged side-mounted tuyeres. Manufacture of steel by AOD is a duplex process: meltdown in an electric arc furnace followed by refining of the metal in a separate AOD vessel.

The AOD process was conceived in 1955 and after considerable development work the first commercial vessel went into operation in 1968 at Joslyn Stainless Steel. In 1977 AOD produced the majority of the world's stainless with vessel sizes ranging from 5 to 175 t in capacity.

While AOD was developed for the production of stainless steels, specialty grades such as tool, valve and Si steels are produced as well. Superalloys and military grades such as Hy-80 and Hy-130 are also being made. In steel foundries, carbon and low alloy steel castings as well as stainless steel castings are now being processed through AOD vessels.

Historical Development

As a result of laboratory research and development programs in the 1950s at Union Carbide Corporation's Tonawanda Laboratories studying the effect of Ar on oxygen assisted decarburization, the potential for a valuable commercial process appeared. Consequently, in the early 1960s the Linde Division of Union Carbide Corporation entered into a joint development program with Joslyn Stainless Steel to explore a practical steelmaking method to utilize the advantage of Ar dilution.

The original work at Joslyn was performed in an arc furnace and the overall results were less than encouraging. The experiments showed that the relatively shallow bath in an arc furnace precluded attainment of equilibrium even when using multiple submerged lances. The difficulties encountered forced abandonment of further arc furnace development work and focused future experiments on developing a successful separate refining vessel.

Subsequent work in a separate refining vessel was faced with the problem of determining how to inject the gases successfully. Two approaches were taken: separate injection of oxygen and Ar and combined injection.

The combined injection program had to develop a successful submerged injector for oxygen-containing gas mixtures. In separate injection experiments, on the other hand, only Ar was injected below the bath surface; BOF-type lances were utilized for the oxygen. Various methods were investigated including sonic and supersonic tuyeres and porous refractories of numerous designs. Eventually a simple tuyere of similar construction to those used today was developed.

Comparison of the separate injection and combined injection indicated similar refining efficiencies measured in terms of Cr oxidation, except at very low carbon levels less than (0.03%C). That evidence, coupled with the development of a successful tuyere for combined Ar and oxygen injection, gave birth to the AOD vessel as we know it today.

The successful experimental heats at Joslyn led to a commercial 15 t installation in April, 1968. By December, 1969 over 1,000 heats had been refined in Joslyn's AOD vessel and the results were reported at the Electric Furnace Conference.[1] By the end of 1970, AOD vessels spanned two continents refining stainless steels and superalloys in heat sizes from 5 to 55 t. In 1972, Cartech's Reading, Pennsylvania shop began routine refining of low alloy steel.

In October, 1973 the first foundry AOD system went into experimental operation at the Portland, Oregon foundry of Esco Corporation. The results of operating the initial 4 t bare-bones experimental unit resulted in Esco installing a 20 t AOD vessel in September, 1976, which was the first commercial foundry AOD installation. In August, 1977 Quaker Alloy Casting Company became the second foundry AOD licensee. In November, 1978 Esco's Newton, Mississippi foundry became the first AOD vessel to refine exclusively low alloy steel.

Today AOD vessels are refining all types of steels in heat sizes from 4 to 175 t.

Stainless Refining

Stainless steel is nominally an Fe-Cr alloy with minimal amounts of carbon. In refining stainless, it is generally necessary to decarburize the molten metal bath to less than 0.05%C. The problem is that Cr is quite susceptible to oxidation. Consequently, the steelmaker must find a way to oxidize the carbon without oxidizing large quantities of Cr. Prior to the introduction of AOD, this was accomplished by withholding most of the Cr until the bath had been decarburized by oxygen lancing. After the bath was fully decarburized, low carbon Fe-Cr and other low carbon ferroalloys were added to the melt to raise the Cr and other alloying elements into specification.

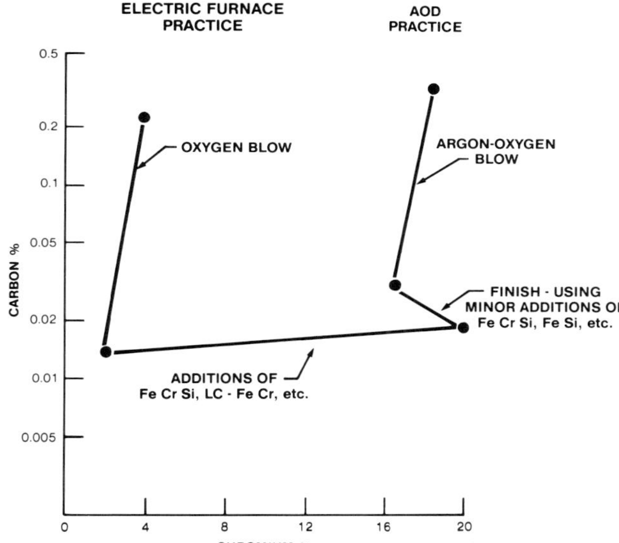

Fig. 1. Composition changes in refining 304-L stainless steel.

In AOD processing, on the other hand, decarburization can proceed readily in the presence of specification levels of Cr without its excessive oxidation. This practice enables the exclusive use of high carbon ferroalloys in the charge mix, avoiding the substantially more expensive low carbon ferroalloys. Fig. 1 compares schematically the refining steps in the two processes.

Physical Chemistry Considerations

The principal reactions in the refining of stainless steel are:

$$\underline{C} + \underline{O} = CO_{(g)} \tag{1}$$

$$3/4\ \underline{Cr} + \underline{O} = 1/4\ Cr_3O_{4(s)} \tag{2}$$

(In these reactions, $\underline{C}$, $\underline{Cr}$, $\underline{O}$ represent carbon, Cr and oxygen, respectively, dissolved in the metal bath). The C-Cr-O equilibrium in stainless steelmaking may be represented by the difference of equations (1) and (2):

$$1/4\ Cr_3O_{4(s)} + \underline{C} = 3/4\ \underline{Cr} + CO_{(g)}$$

Applying the law of mass action, we may write:

$$K = \frac{(\%Cr)^{.75} \times P_{CO}}{\%C}$$

A chromium oxide term does not appear in this expression because chromium oxide forms a separate solid phase and as such has a constant chemical activity of unity. Equilibrium thermodynamics determines a constant value of K.

As decarburization proceeds, the carbon content of the steel decreases. In order to maintain a constant value of K, either the Cr content must decrease or the partial pressure of CO must be reduced. In electric furnace practice, it is impossible to decrease the CO partial pressure effectively and hence decarburization cannot proceed without excessive Cr oxidation. In the AOD process the partial pressure of CO is reduced by injecting Ar with the oxygen. By increasing the quantity of Ar (decreasing the percentage of oxygen in the feed gas) the partial pressure of CO is reduced. As the C level drops, the CO partial pressure must be reduced in order to minimize Cr oxidation.

Thus, decarburization begins at high C levels with a high percentage of Ar in the gas. The oxygen/Ar ratio is decreased in two or more stepwise changes as decarburization proceeds. In

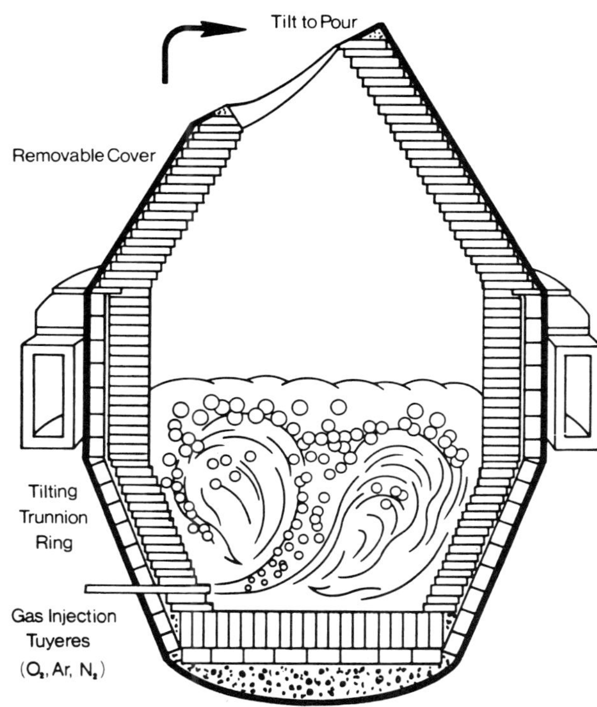

Fig. 2. Gases are injected into typical AOD vessel through multiple tuyeres to refine the metal. Vessel is mounted on tilting trunnion ring to facilitate charging, sampling and tapping.

this manner, decarburization proceeds with a minimal amount of Cr oxidation.

Fluid Flow Considerations

The importance of good mixing for effective pyrometallurgical reactions has long been recognized. A high degree of contact between slag, metal and gas is critical. Herein lies one of the major advantages of the AOD process: the vessel is a side-blown reactor which provides excellent mixing. This phenomenon is illustrated in Fig. 2. Furthermore, because the gas is injected via submerged side-wall tuyeres, the gas bubbles have a relatively long residence time as they circulate through the rolling bath.

The overall effect of the intimate gas-metal-slag contact is the attainment of near equilibrium conditions as evidenced by:

1) Complete utilization of the injected oxygen
2) Excellent recovery of alloy additions
3) Excellent and rapid desulfurization
4) Efficient degassing
5) Temperature and composition homogeneity.

The intimate metal-slag contact also reduces the number of inclusions in the molten steel.

Producing a Heat of Steel in an AOD Vessel

A heat of CF8M stainless steel (nominal composition: 0.06% C, 0.45% Mn, 1.25% Si, 19.3% Cr, 9.5% Ni, 2.2% Mo) was chosen for consideration. Before refining in an AOD vessel, the charge materials (ferroalloys, scrap) are melted in an electric arc furnace. Generally there is nominal control of the Cr, Mn and Ni concentrations to midrange specification. The C content at meltdown can vary from 0.5-3.0% depending on the stainless scrap content of the charge. Once the charge is melted down, the arc furnace is tapped and the slag is removed. The metal is weighed and charged into the AOD vessel.

Table 1. Analytical Results of Heat of CF8M in 5 t AOD Vessel

Conditions	Weight lbs.	Temp. °F	C	Mn	Si	P	S	Cr	Ni	Mo
Transferred Metal	11,700	2910	0.68	0.36	0.67	0.020	0.035	18.36	9.54	2.23
Blow-1	–	3050	0.33	0.25	0.08	0.020	0.035	17.88	9.55	2.29
Adds[1]	480	–	–	–	–	–	–	–	–	–
Blow-II	–	3020	0.081	0.20	–	0.021	0.035	17.58	9.58	2.38
Blow-III	–	3120	0.062	0.20	–	0.021	0.035	17.52	9.58	–
Add. Red. Mix[2]	600	–	–	–	–	–	–	–	–	–
After Reduction	–	3020	0.054	0.39	0.99	0.020	0.015	19.31	9.57	2.26
Tap	12,000	2930	0.054	0.46	1.01	0.021	0.013	19.33	9.59	2.26

NOTES:
1. After Step 1 blow, 200 lbs of charge Cr, 20 lbs of Std. Mn, 20 lbs of Ni Briquettes and 160 lbs of lime were added. Prior to this addition, the temperature was measured and vessel sampled.

2. Reduction mix consisted of 310 lbs FeSi (75%), 240 lbs lime, and 25 lbs spar.

Total Gas Usage: 3210 ncf of oxygen; 2540 ncf of inert.

After charging the metal into the AOD vessel, a complete chemistry sample and a temperature measurement are taken. Relevant heat data are listed in Table 1. At this point the vessel is turned up into the vertical blowing position and a mixture of oxygen and nitrogen at a ratio of 3/1 oxygen/nitrogen is blown at a total flow rate of 2000 ncfh/t. A specific quantity of oxygen is blown, calculated from the simple stoichiometric reactions: oxidation of C, Si and a small percentage of metallics (Fe, Cr, Mn, etc.).

The behavior of C, Cr and temperature during the heat is illustrated in Fig. 3. After completion of the 3/1 blow, the vessel is sampled and scrap and/or ferroalloys are added prior to resumption of the oxygen/Ar blow at a ratio of 1/1. The calculated amount of oxygen is blown to attain 0.10% C in the bath, the vessel is sampled and blowing is resumed at a 1/3 oxygen/Ar ratio to attain the desired C level.

At the conclusion of the oxygen/Ar blowing period, the reduction mix consisting of stoichiometric quantities of Si alloy and lime is added. If low S levels are desired, a second desulfurization slag may be used. The addition of both the reduction and the desulfurization slags is followed by a pure Ar stir. After these steps are completed, any necessary final additions are made to precisely meet the specifications just prior to tapping.

Non-Stainless Refining

The production of low alloy and carbon steels in AOD vessels differs from the production of stainless steels in that temperature considerations override C control considerations. In refining stainless steels, the blow program is designed to achieve the low C levels required without excessive metallic oxidation and resultant high temperatures. In contrast, the efficient decarburization of low alloy steels will occur at low temperatures. Hence consideration must be given to generating heat in the AOD vessel to allow the steel to be tapped at the required temperatures. These considerations are particularly important in foundry AOD systems, which tend to be small vessels requiring relatively high tap temperatures.

The oxygen calculation for a low alloy heat is similar to that for stainless except oxygen must also be blown for the "fuel" element (e.g., Al) which is added to generate heat. The operating steps in refining low alloy steel are even simpler than those for refining stainless steel. The 3/1 oxygen/inert ratio is used exclusively for decarburization and fuel oxidation. After the

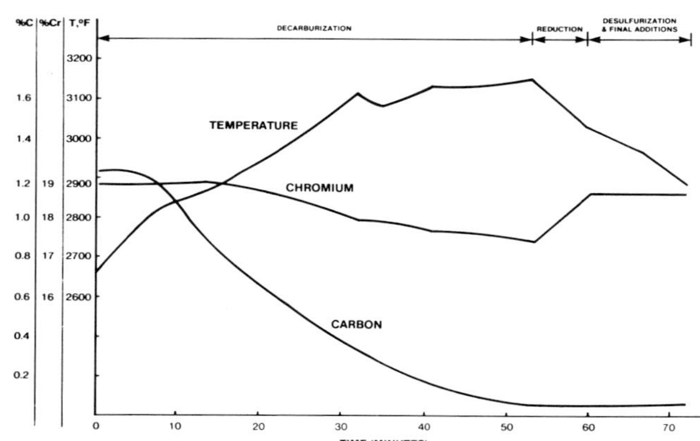

Fig. 3. Behavior of temperature, C and Cr during AOD processing.

completion of this blow, Fe-Si with required lime is stirred into the bath to deoxidize and desulfurize it. After completion of this Ar stir and a final chemistry check, the steel is ready for tapping. Shops refining both stainless and low alloy steel find that heat times for low alloy are approximately half of those for stainless.

Quality Considerations

The ability of the AOD vessel to degas and desulfurize the steel efficiently enables the steelmaker to produce an exceptionally clean steel in the decarburization vessel. Complete deoxidation may be accomplished in the AOD vessel. In low alloy steels, oxygen levels of 20 to 40 ppm are common, as compared to 60 to 110 ppm for arc furnace product. Deoxidation, coupled with the excellent mixing of the steel with a fluid basic slag, enables the steelmaker to rapidly obtain exceptionally low S levels. In low alloy steels, S levels of 0.005% and less are routine with a single slag practice. The Ar stir, which concludes every heat, assists in degassing of nitrogen and hydrogen. Nitrogen levels are normally 1 to 3 ppm.

The lower levels of gas and impurities in AOD refined steel result in higher quality molten steel and castings. The benefits of AOD processing include: improved microcleanliness, improved fluidity, improved cleaning room performance and improved mechanical properties (toughness and/or ductility).

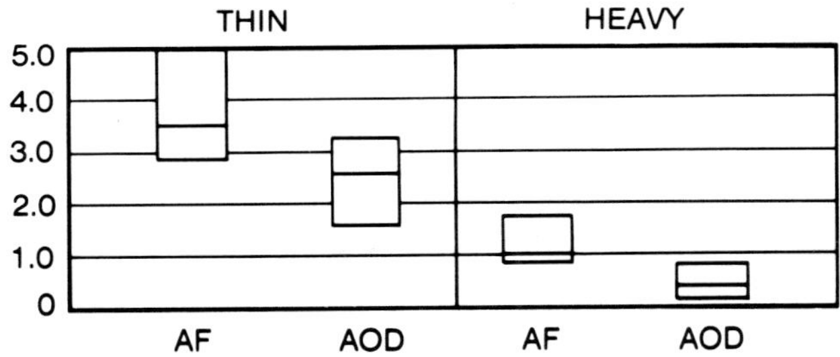

Fig. 4. J-K cleanliness comparison for globular oxide inclusions of arc furnace and AOD refined centrifugally cast CF-8. The horizontal line indicates the average value for all heats evaluated.

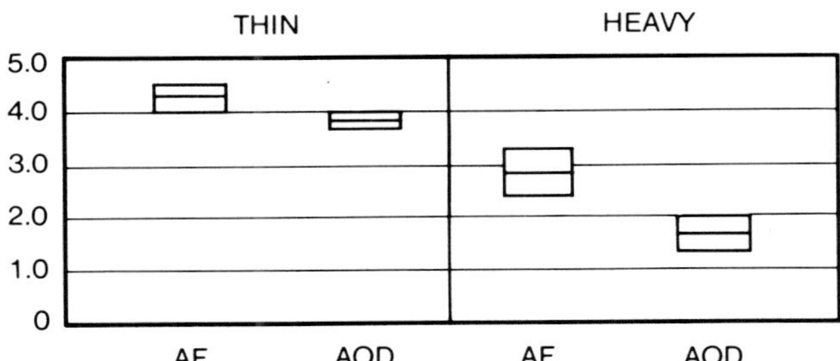

Fig. 5. J-K cleanliness comparison for globular oxide inclusions of arc furnace and AOD refined centrifugally cast CA-15. The horizontal line indicates the average value for all heats evaluated.

The improvement in microcleanliness is the most fundamental of these improvements. One interesting example of the microcleanliness improvement is centrifugally cast high alloy steels. Standard J-K cleanliness evaluations were made on arc furnace refined and AOD refined heats of CF 8 and CA-15. The excellent microcleanliness of the standard centrifugally cast product was evidenced by the presence of only Type D (globular oxide) inclusion. Likewise the AOD refined material revealed evidence of only Type D inclusion. The outstanding improvement is illustrated in Fig. 4 and 5.

The reductions in indigenous inclusions, S levels and dissolved gas contents result in a steel with increased fluidity. The fluidity increase has allowed most foundries to decrease their pouring temperatures by approximately 56°C (100°F) for stainless castings and 28°C (50°F) for low alloy castings. The lower pouring temperatures reduce the severity of mold-metal reactions and related surface defects.

These improvements in molten steel quality (S level, gas content and microcleanliness) result in fewer post shakeout repairs. Savings in the cleaning room result in decreased repair and rejection rates. The defects found in castings subject to repair are generally less severe than those found in arc furnace (or induction furnace) product. An extensive study at Esco's Portland, Oregon foundry indicates that the AOD refined metal produced castings which on average require 25% less grinding and welding repair.[2] The potential savings available are particularly significant for castings which are subject to an involved repair/x-ray cycle. The introduction of AOD to such a foundry will result not only in lower grinding and welding costs but also in reduced x-ray and heat treatment costs as the number of cycles is reduced. Examples of defects typically reduced or eliminated by AOD processing are aluminum nitride, hydrogen fish eyes, and hot tears.

Many of these same factors result in the improved fracture toughness of AOD processed materials, making these steels particularly attractive for performance in harsh environments, such as on the arctic pipeline. Improvements in low temperature toughness (as measured by the Charpy test) of well over 50% have been reported for HY-80[3] and WCB and LCB.[4]

The quality improvements of AOD refined steel arise principally from the fact that complete decarburization, deoxidation, desulfurization and degassing all take place effectively in the same refining vessel. Consequently, the three steps of steelmaking — melting, refining and casting — are allowed to take place in their own special vessel: electric arc furnace, AOD vessel, and ladle, respectively. Each is allowed to perform its own specialized task without being required to do part of another's.

Summary

The most important aspect of AOD is the reduction of the partial pressure of CO by Ar enabling the removal of C in the presence of large concentrations of Cr. That fundamental advantage allows the use of low cost raw materials and results in increased yield in the manufacture of stainless. As a result of the inherent efficiencies of AOD, the process allows precise chemistry control and is quite flexible in its ability to handle a wide range of starting conditions. The excellent mixing in the AOD vessel allows production of a fully deoxidized, desulfurized and degassed product in one refining vessel. This higher quality steel yields added benefits to the foundry in terms of reduced post shakeout repairs. In addition, there is considerable evidence that the improved quality of AOD metal results in improvements in physical properties as compared to arc furnace practice.

References

1. J. D. Ellis and J. M. Saccomano, *Electric Furnace Proceedings*, vol 27, p 76-79 (1969).
2. L. J. Venne and C. Z. Oldfather, "Enhancement of Physical Properties by Argon Oxygen Decarburization," *AFS Transactions*, vol 86, p 139 (1978).
3. W. A. Palko, J. P. Byrne and C. A. Zanis, "Processing of HY-130 Steel Castings," *AFS Transactions*, vol 84, p 193 (1976).
4. J. W. Juppenlatz and S. W. Gearhart, *Steel Foundry Facts* (1979).

A Process Model of Cupola Melting

W. J. Evans
R. G. Hurley
Ford Motor Company
Detroit, Michigan
R. C. Creese
West Virginia University
Morgantown, West Virginia

ABSTRACT

A mathematical model of the cupola iron melting process has been developed as a tool for analysis of fuel and materials usage. The model allows investigation of the effects of changes in operating variables on coke usage and charge material cost. The operating variables which can be studied are blast rate, hot blast temperature, coke quality, oxygen enrichment, furnace dimensions, ambient conditions, and product chemistry and temperature specifications.

This paper describes the method used to construct and verify the model, including the architecture, key mathematical expressions used in combustion calculations, and a summary of cupola monitoring procedures used to obtain validation data.

Several examples of model applications are presented, consisting of studies of the effects of both blast temperature and oxygen enrichment on coke rate. Advantages of the modeling approach are that conditions normally not encountered in a production cupola can be studied and that operating variables can be isolated and examined from a common baseline. In addition, useful projections of coke and melt rate effects can be made with no risk to production facilities.

Introduction

The cupola has been the primary iron melting furnace of the foundry industry for many years. A great deal of operating experience has been accumulated but there is still relatively little understanding of the fundamental chemical reactions occurring in the process. Nevertheless, cupolas are operated successfully in both large and small foundries and are preferred over other melting facilities by many foundrymen.

There have been many technical improvements in the process and the modern cupola is very different from early units. Developments such as protruding tuyeres, oxygen enrichment, and hot blast have improved the productivity and flexibility of the process. However, these improvements are often the result of long trial-and-error procedures; experiments were often initiated based largely on intuition. They were frequently high-risk ventures because of the potential consequences in product quality and productivity.

Today there is a great deal of pressure to further improve the cupola melting operation, especially in the areas of energy consumption and materials usage. Rising prices for gas and coke provide the incentive for such efforts. The main obstacle to reducing energy usage is that most cupolas were designed and built at a time when energy was inexpensive and they do not have the flexibility to easily adapt to the changing energy situation. For example, the use of water-cooled furnace shells using no refractory lining became prevalent as higher-temperature hot blast systems became available. If, for energy usage reasons, the hot blast temperatures must be significantly reduced, use of the water-cooled shell may cause technical difficulties.

The foundryman is in the position of having to be innovative to accomplish his energy goals, but the economic risks of process changes are high. In order to reduce these risks, a means of evaluating proposed process changes without actually implementing them would be desirable. The use of process modeling provides this opportunity.

The model described in this paper is a mathematical description of the cupola combustion and melting processes. It uses both fundamental thermochemical relationships, such as reaction rate equations, and empirical expressions developed from experimental data.[1-3] It combines many of the characteristics of earlier cupola models[4-6] and introduces several innovations in the energy balance considerations.

As with any model, the cupola model was designed with specific objectives: to estimate the effect of operating parameters on fuel usage and productivity and to determine a least-cost charge material mix to give a specified product. While it does provide additional useful information, this information is generally a secondary result of the model's primary objectives.

This paper discusses the capabilities and limitations of the model, presents examples of model results, and describes the general architecture and approach of the algorithm itself.

The Model's View of the Cupola Melting Process

The cupola belongs to the family of vertical shaft furnaces, which also includes blast furnaces and certain types of coal gasifiers. All operate on the principle of continuous countercurrent flow. In the cupola, solid melting stock and coke are charged at the top of the stack. Combustion of the coke occurs near the bottom. As the coke burns, the hot gases rise through the full stack, heating and melting the metal. The molten metal descends through a coke bed and is tapped out at the bottom of the furnace, just below the combustion region. Figure 1a illustrates this process.

In the development of the cupola model, the primary functions of the cupola were defined as:

- melting of metallic materials to provide a specified iron chemistry;
- superheating the iron to a specified tap temperature; and,
- transfer of carbon to the iron from the coke.

The typical charge makeup for a cupola is determined to a great degree by material availability, cost, and product chemistry desired. In general, the metallic portion of the charge will consist of cast iron returns, such as gates and runners, as well as scrap castings, hot or cold pressed iron briquettes, and steel scrap in the form of flashings, bushelings and fragmentized steel sheet. In some instances borings and turnings from machining operations are included. Product chemistry adjustments are generally made by additions of ferroalloys, such as ferrosilicon and ferromanganese, in lump or briquette form.

During the melting process, various oxides and other nonmetallics accumulate and form a slag, which floats on the molten iron and is removed by various mechanical separation devices. The majority of the slag originates from ash present in the coke. To facilitate both removal of ash from the coke surfaces and separation of slag from the iron, various fluxes, such as limestone, are added in the charge.

The basic fuel of a cupola is foundry coke, which provides the necessary combustion material and a source of carbon for the metal. The amount of coke added in the charge can vary a great deal, depending on the furnace operating parameters,

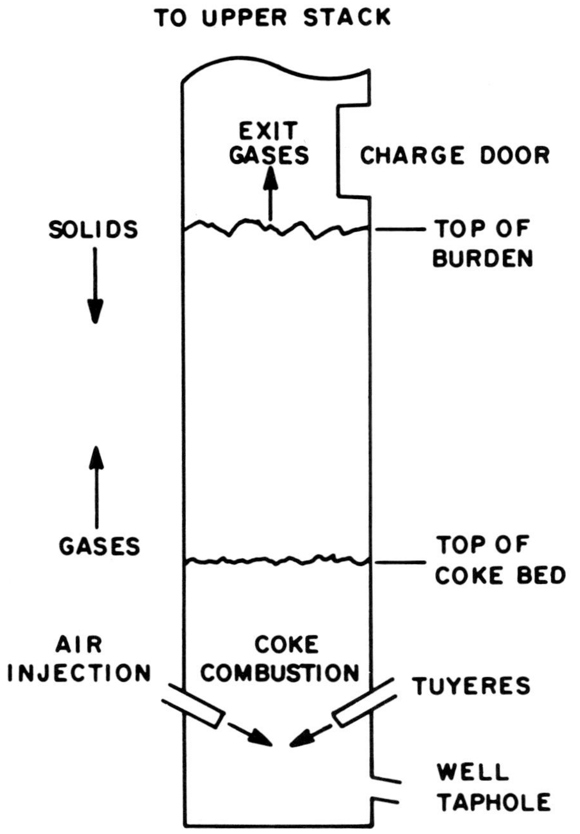

Fig. 1a. Basic components of a cupola melting furnace.

combustion efficiency, and carbon requirements of the metal. Typically, the coke addition is 13 to 18 percent of the metal addition.

The energy efficiency of the cupola is largely determined in the lower regions where the gas-carbon reactions occur. Air is injected through tuyeres and reacts with incandescent coke. The air is often preheated and may vary in chemical composition depending on ambient humidity and the use of oxygen enrichment. The furnace atmosphere around the tuyeres is highly oxidizing and can cause significant loss of easily oxidized metals such as silicon.

The primary gaseous reactions occurring in this region are:

(a) $C_{coke} + O_2 \longrightarrow CO_2 + 96{,}428$ cal

(b) $C_{coke} + H_2O \longrightarrow CO + H_2 - 29{,}000$ cal

Reaction (a) is the primary source of heat in the process, typically supplying 80-90 percent of the total heat input. Reaction (b) absorbs heat and also consumes coke which would otherwise be used in combustion. High humidity can thus adversely affect cupola performance.

As the hot gases rise, the furnace atmosphere is altered due to heat transferred to the descending charge and by the reaction of carbon dioxide to carbon monoxide:

(c) $C_{coke} + CO_2 \longrightarrow 2CO - 38{,}840$ cal

Reaction (c) is again endothermic and consumes additional coke. As a result, the gases become cooler and less oxidizing as they rise. Eventually, one of the reacting components is depleted or the temperature slows the reaction rates to negligible levels. Once this condition has been reached, no further coke is consumed. The various cupola regions mentioned above are illustrated in Fig. 1b.

As gases continue to rise up the stack, they transfer sensible heat to the cooler charge materials and to the furnace walls. The major additional compositional change results from the calcination of limestone, which occurs when the limestone reaches about 816C (1500F).

(d) $CaCO_3 \longrightarrow CaO + CO_2 - 42{,}500$ cal

The gases eventually reach the top of the charge burden and their temperature at that point is defined as the "exit gas temperature."

The composition of the exit gas, especially the CO and CO_2 levels, provides a good indicator of the combustion efficiency in the coke bed, where direct measurements are not generally feasible. Based on past experience, most cupola operators prefer a CO level of about 13 percent, though the gas composition is usually not measured in production.

The iron also undergoes both thermal and composition changes. When metallics are charged into the upper portions of the cupola, they are preheated by the gases; the rate of preheating is determined primarily by the size of the solid pieces. The use of large, bulky charge materials may hinder the heat transfer, causing the material to melt far down in the stack and in extreme cases may even prevent complete melting. Charge material size is therefore kept within certain limits so that all metal is in the liquid state when it enters the coke bed.

Once in the coke bed, the metal drips down and makes contact with the coke. Some of the carbon in the coke dissolves in the iron. The exact amount of dissolved carbon varies with temperature, coke quality, amount of steel vs. cast iron in the charge, and several other factors. The amount of carbon transferred to the iron is normally on the order of 10 percent of the total coke addition and can significantly affect product quality and coke usage. The carbon transfer reaction is shown below:

(e) $C_{coke} \longrightarrow C - 5400$ cal

As the iron passes down through the coke bed, it is also heated to well above its melting point by the hot gases. This superheating is not efficiently done in a cupola because of the high gas temperatures required. The metal then collects at the bottom of the stack and flows out through a refractory tap hole.

Basic Structure and Use of the Model

The primary objective of the model is to estimate the coke rate and melt rate for specified operating conditions and product requirements. In addition, the model provides other information, such as exit gas composition and slag volume, and recommends a charge mix which minimizes total material cost. The information content of the model results is given in more detail below.

A diagram depicting the model structure is given in Fig. 2. The objectives are accomplished within this structure in several stages:

1) Input information is conditioned (converted to proper units, sortec, etc.)
2) A multizone heat balance is performed over the entire stack, beginning at the tuyeres. The main results of this steady state simulation are estimates of production rate and the amount of carbon used in the combustion reactions.
3) A matrix of linear equations is constructed which performs a mass balance on the system. All elemental

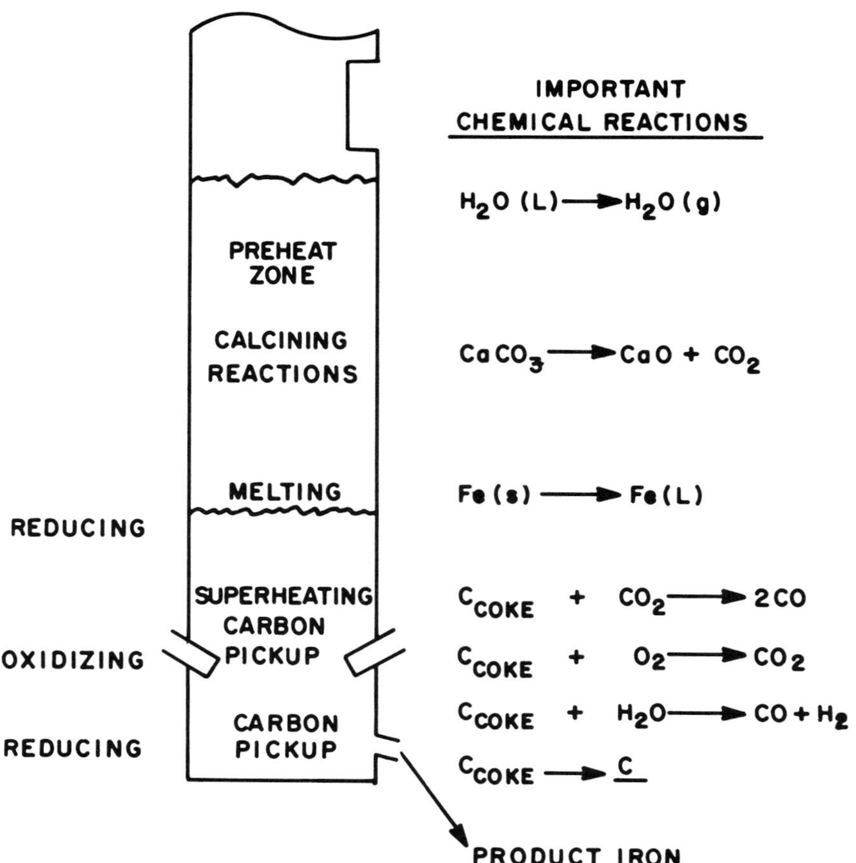

Fig. 1b. Basic reactions in the cupola melting process.

components which contribute to the iron and slag output of the cupola are considered. The matrix is solved by a linear programming technique to determine the charge mix which gives the required product at least cost.

Information on all cupolas which are to be simulated is stored in data files. This information includes physical data, such as shell diameter, water jacket dimensions and maximum charge height. Also included are a stored set of operating variables; these are displayed to the user during operation of the model. The user can revise these values and the new values will then be stored. If no changes are made, the stored values are automatically inserted into the model.

A similar arrangement is used to specify product requirements. Maximum and minimum limits can be set for major chemical constituents (C, Si, Mn, S and CE) and maximum values for relevant trace elements. Iron temperature at the cupola spout is also specified.

The final portion of the input information consists of a list of charge materials (and complete chemical analyses) which the model uses in charge mix determination. It is unlikely that every material (20 maximum) will be used. The model is constrained to provide a product iron whose composition is in a range specified by the user. It must provide this product by manipulation of charge materials and consideration of elemental recovery factors calculated internally.

The current model output consists of the following:

- recommended charge mix
- charge material total cost/ton of iron tapped, based on recommended charge mix
- final iron chemistry

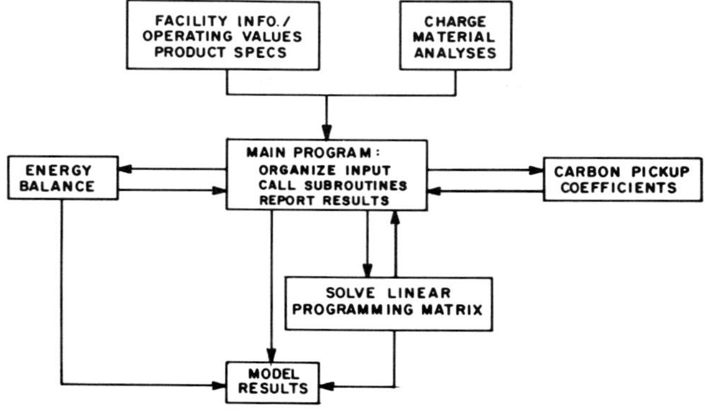

Fig. 2. Schematic diagram of cupola model.

- slag composition and volume
- melt rate and coke rate
- gas requirements to preheat blast air
- heat distribution among furnace products
- stack gas composition and temperature.

Heat Balance Procedure

The model performs the heat balance in the cupola by dividing the stack into disc-shaped zones and establishing steady state conditions within each zone. The entire heat balance is rate-based, balancing energy and mass fluxes per minute. This time unit was chosen because normal input parameters (e.g., air volume) are given in units per minute. The procedure begins at tuyere level.

**HEAT DISTRIBUTION
LEAVING ZONE**

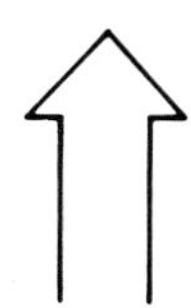

**FIND NET CHANGE IN HEAT
CONTENTS OF GAS, METAL FOR ZONE**

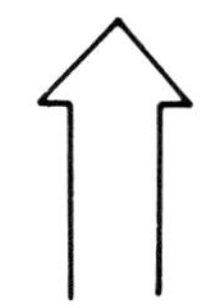

**HEAT DISTRIBUTION
ENTERING ZONE**

Fig. 3. General zone heat balance procedure.

The method, as illustrated in Fig. 3, involves the following steps:

1) Determine the total sensible heat input into the zone.
2) Calculate the net sensible heat evolved or absorbed within the zone.
3) Determine the total sensible heat out of the zone.

Coke Bed Zones

Steps 1 and 3 are straightforward and involve simply an efficient bookkeeping system. Determination of the change in sensible heat within the zone requires that the contributions of the various exothermic and endothermic processes be known.

The model considers the following quantities:

1) heat evolved by reaction (a) as a function of the amount of oxygen which reacts in the zone;
2) heat absorbed by reaction (c) as determined by temperature/gas composition conditions within that zone;
3) heat absorbed by the metal and slag during preheating, melting, and superheating;
4) heat lost through furnace walls and tuyeres;
5) heat absorbed by reaction (b); the consumption of the total moisture in the air blast is considered to take place linearly over the first two feet of the stack;
6) heat absorbed by reaction (e); total carbon pickup is distributed over the height of the coke bed;
7) heat evolved by metal oxidation reactions.

The model calculates these heat quantities (as functions of temperature when applicable) and repeats the procedure until it determines the zone temperatures at which steady state conditions exist.

The specific expressions used to calculate these values are listed below:

A. Amount of oxygen reacting in zone.

$$O_2^r = O_2^o [1 - \exp(-vt)]$$

where

O_2^r = amount reacted (moles/min)

O_2^o = amount available (moles/min)

v = reaction velocity (sec^{-1}) - 0.35 (v/d)

V = average gas velocity (in./sec)

d = coke diameter (in.)

t = residence time of gas in zone (sec).

The gas velocity and residence time are determined by blast rate, temperature, coke size, and zone volume.

B. Amount of CO formed in zone.

$$CO^r = C^o \cdot R_{CO}$$

where

CO^r = moles of CO formed per minute

C^o = moles of carbon available in zone

R_{CO} = moles of carbon reacted to CO per mole available per minute = $aP_{CO_2}/(1 + b\,P_{CO} + c\,P_{CO_2})$

$a = 3.72 \times 10^8 \exp(-61000/RT)$

$b = 4.00 \times 10^{-4} \exp(40300/RT)$

$c = 3.16 \times 10^{-2} \exp(6100/RT)$.

C. Heat absorbed by metal in zone (coke bed zones only).

$$\Delta H = \int_{T_{in}}^{T_{out}} C_p \, dT$$

The value of T_{out} is given by the boundary conditions at the bottom of each zone, as determined by the results in the previous zone.

The value of T_{in} is calculated by the expression:

$$T_{in} = T_{gas} + (T_{out} - T_{gas})/(1.02 \times 0.525^t)$$

where

T_{gas} = temperature of stack gases in zone (F)

t = residence time of metal in zone (min).

D. Heat loss through walls and tuyeres.

The total heat loss to tuyere cooling water is estimated by using operating data from a cupola similar to that being simulated. The heat quantity is determined from the cooling water flow rate and temperature change. This heat loss is then distributed by the model over the first two zones of the stack, about 0.8 feet.

The total losses through the cupola shell are likewise determined. However, little is known about the distribution of this loss and estimates are therefore used. The model was run with no losses to the shell considered, the internal gas temperature profile was determined, and the wall losses were then distributed in proportion to the temperature difference across the shell for each zone.

E. Other heat balance effects.

The remaining heat balance items listed above are fixed for a given zone and therefore do not change as the zone temperature converges on steady state conditions. They are simply included in the heat balance expression as constants.

Figure 4 shows a typical zone in the coke bed portion of the model and illustrates the approach taken to balance the heat contributions of the various sources and sinks. Figure 5 is a flow diagram which outlines the method used in the model to determine steady state conditions for this zone.

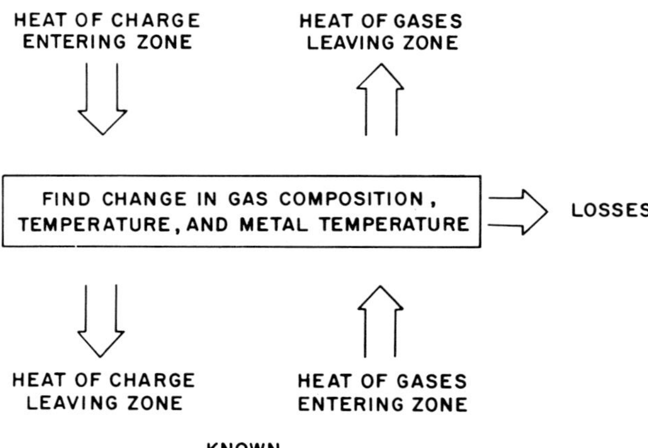

Fig. 4. Heat balance procedure for coke bed zone.

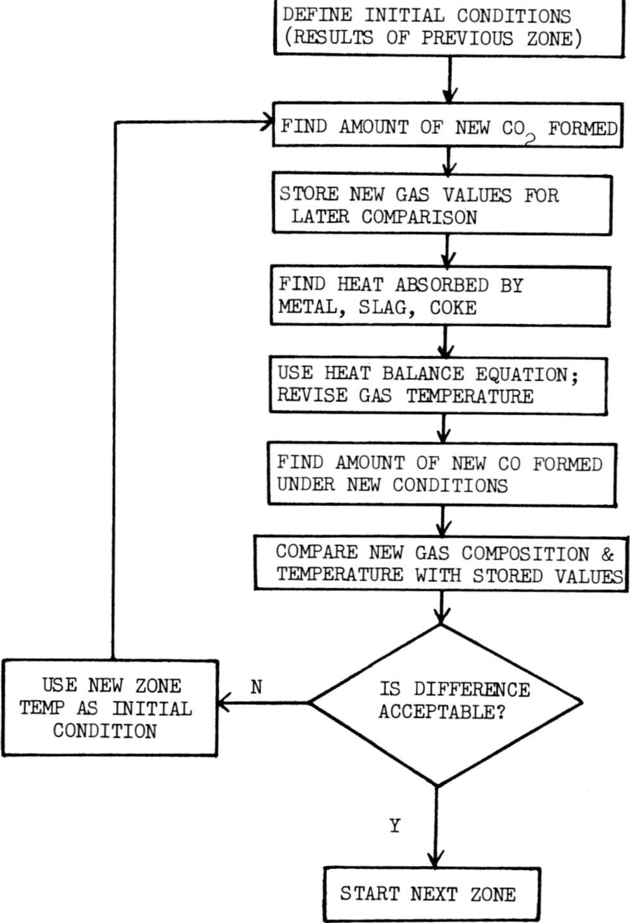

Fig. 5. General procedure for coke bed zone calculations.

The first zone, at tuyere level, is a special case. Since there are no previous zones to define the input conditions (i.e. at the bottom of the zone), these values are determined from user input to the model. The initial gas composition and temperature correspond to the blast air composition and temperature, and

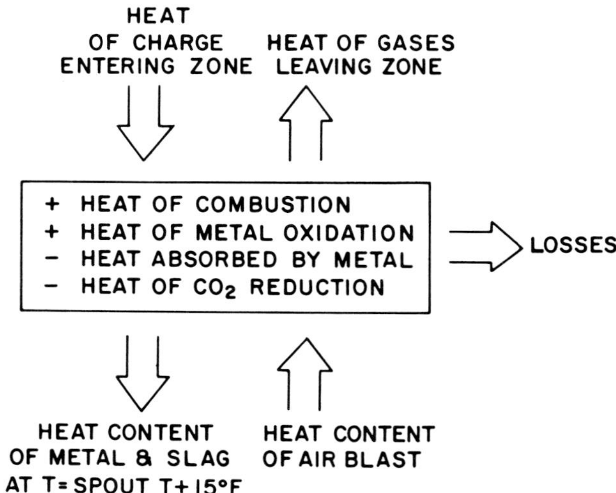

Fig. 6. Heat balance in tuyere zone calculations.

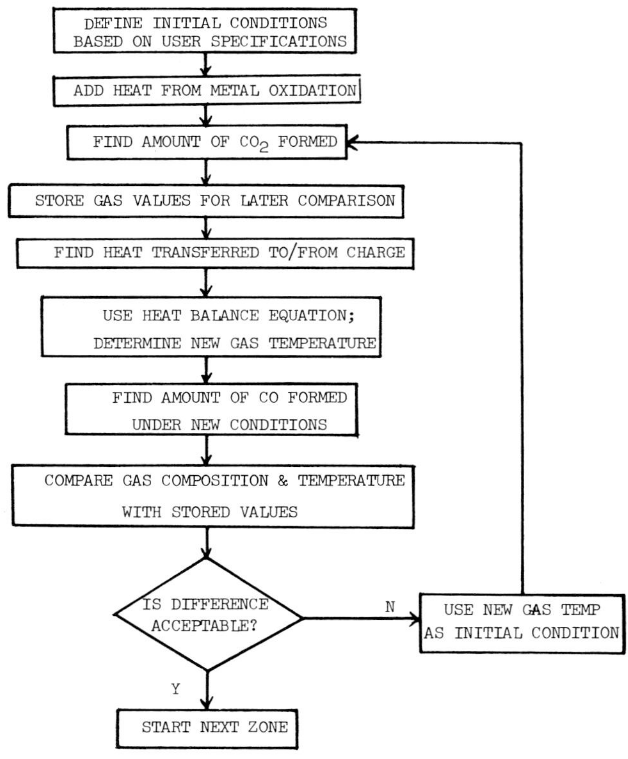

Fig. 7. General procedure for tuyere zone calculations.

the value for T_{out} of the metal and slag is given by the specified tap temperature (minus an assumed 8°C (15°F) loss in the well).

The heat effect of the metal oxidation reactions is also considered in this zone. Using various metal loss factors, the rate of heat generation is calculated and added to the total heat in the first zone. This zone is illustrated in Fig. 6; the procedure for performing the heat balance is given in Fig. 7.

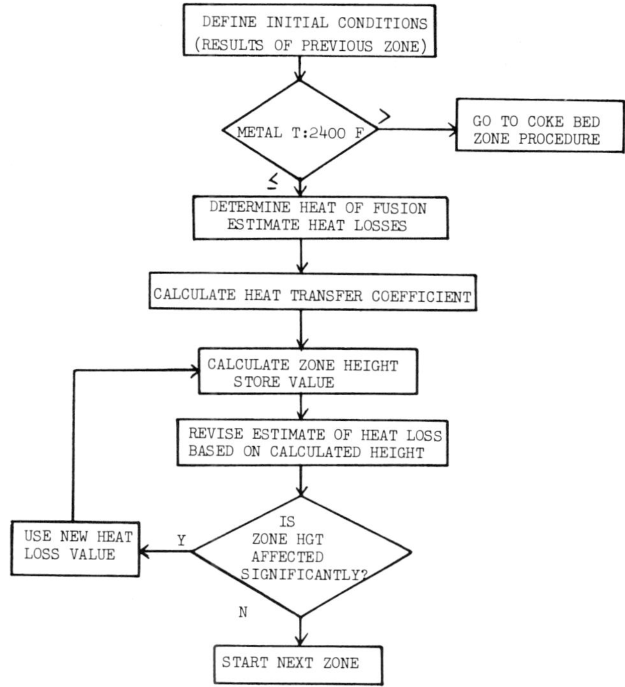

Fig. 8. Heat balance in melt zone calculations.

Fig. 9. General procedures for melt zone calculations.

Melt Zone

The model identifies the melt zone as the zone where the value of T_{out} of the metal drops below 1316C (2400F). While this temperature value is somewhat arbitrary, it is approximately the liquidus temperature of a typical cupola charge mix.

In the melt zone, the model transfers a known heat quantity, equal to the heat of fusion, and determines the height of the zone required to transfer that amount of heat. The procedure uses convection heat transfer relationships which are summarized by the expression:

$$ZHT = H_f/(h \cdot A \cdot f_A \cdot T_L)$$

where

ZHT = melt zone height (ft)

h = convection heat transfer coefficient

A = cupola cross-sectional area at melt zone (ft^2)

f_A = surface area/volume ratio of charge mix

T_L = log mean temperature

$$= (T_{GI} - T_{GO})/\log [(T_{GI} - T_{out})/(T_{GO} - T_{in})]$$

T_{GI} = gas temperature entering zone (F)

T_{GO} = gas temperature leaving zone (F).

The value of h is calculated by the model using a procedure outside the scope of this report.

The model defines the top of this melt zone as the top of the coke bed. No gas-carbon reactions which would alter the stack gas composition are involved in the melt zone or in the zones above it. Figure 8 illustrates the melt zone and Fig. 9 presents the heat balance procedure.

Stack Conditions Above the Coke Bed

By the time the model moves on to consider the portion of the cupola stack above the coke bed, the melt rate, coke rate, and combustion ratio have all been determined. The amount of heat transfer to be considered, therefore, is simply that required to heat the charge from ambient temperature to the temperature at the melt zone. The model assumes that the rate at which this heat is transferred is always sufficient to satisfy the melt rate determined in the combustion region. That is, the rate of heating of the solid charge is not the limiting factor in determining melt rate. If conditions in actual operation were such that large charge pieces were used or if the cupola stack were short relative to its diameter, the melt rate of the operation may be significantly lower than model estimates. Experience with monitoring actual operations indicates that the model gives satisfactory melt rate estimates if

$$\frac{\text{tuyere-to-charge door distance}}{\text{diameter at tuyere}} \geq 2.25$$

The remaining heat balance contributors considered by the model are:

- Heat lost to cupola walls above the coke bed
- Heat absorbed by limestone/dolomite calcination
- Heat absorbed by evaporation of surface water on the charge materials.

Mass Balance Procedure

When the heat balance has been completed, a series of linear equations is formulated to do a mass balance on the system. These equations balance the total input contribution for each element with the total output. Figure 10 shows the general form of the matrix, illustrating the various input and output classifications. The model solves these equations with a linear programming technique, using minimum charge material cost as the objective function. Once it has been determined that the product specifications can be met using the given charge material list (solution is feasible), the model will calculate the most economic mix.

A detailed discussion of the formulation of the equations themselves is outside the scope of this paper. However, the physical significance of the equations is described in Table 1. Each equation represents a separate row of the matrix in Fig. 10.

Similarly, the variables used in these expressions are represented by the individual columns of the matrix and the physical significance of each column is given in Table 2.

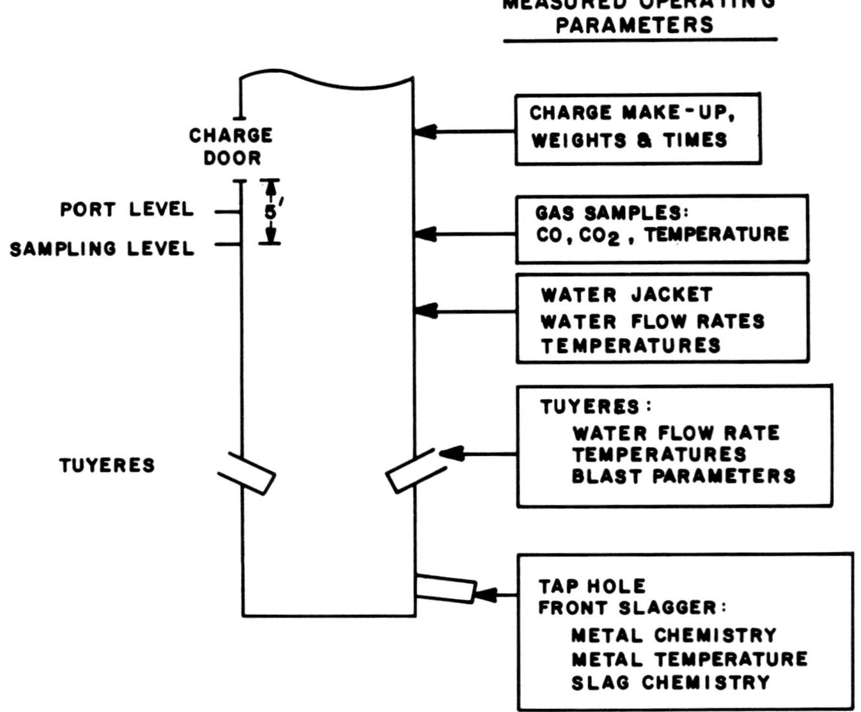

Fig. 10. General layout of linear programming matrix defining system mass balance.

Fig. 11. Sampling locations used during cupola operation monitoring.

Both the heat balance and mass balance are accomplished internally and require no user intervention or input.

Cupola Monitoring and Model Validation

In order to validate the mathematical model, a monitoring program was designed to provide data from operating cupolas. This information was then compared to the model output to ensure that the model gives an acceptable representation of actual conditions. Both the monitoring and validation procedures are described below.

Cupola Monitoring

The monitoring program was designed to provide detailed information on both the input and output variables. The parameters which were measured are described below:

1) *Metal temperature* was monitored continuously by an infrared pyrometer aimed at the metal stream. The temperature was recorded on magnetic tape at the rate of one reading per minute.

2) *Metal composition* samples were taken by hand ladle once per hour. Two samples were taken in rapid succession and combined in order to minimize the effect of short-term fluctuations of cupola iron composition. These samples were later analyzed for percent carbon, silicon, manganese, phosphorous, sulfur, chromium, and aluminum.

3) *Slag composition* samples were also taken by hand ladle once an hour at the front slagger and were analyzed for percentages of SiO_2, Al_2O_3, CaO, MgO, CaS, and MgS.

4) *Cooling water* outlet temperatures were measured with immersion thermocouples at the tuyere, lower shell, and upper shell (when applicable) outlet pipes. Data were recorded electronically at a rate of one measurement per minute for each parameter. Inlet temperatures were checked periodically with a thermocouple, but remained essentially constant during the monitoring period. Water flow rates were recorded manually from flow meter readings.

5) *Stack gas temperature and composition data* were accumulated using two probes inserted through cupola ports at diametrically opposite locations, and an additional movable probe inserted through the charge door. These locations are shown schematically in Fig. 11. The probes consisted of L-shaped pipe to allow extension (3-6 feet) below the charge door lip. This extension minimized mixing of the stack gases with air drawn through the charge door and also reduced the effects of the CO combustion in the afterburner region.

Table 1. Matrix Equations: Physical Significance

Row		Function
1	Charge	**Material Cost Summation**
2	Fe	Input Summation
3	C	
4	Si	
5	Mn	
6	S	
7	P	
8	Cr	
9	Mo	
10	Cu	
11	Al	
12	Ni	
13	Sn	
14	Pb	
15	Fe	Input/Output Relation
16	MgO	Input Summation to Slag
17	CaO	
18	SiO_2	
19	Al_2O_3	
20	FeO	
21	MnO	
22	Cr_2O_3	
23	S	
24		Definition of Basicity Ratio
25		Slag Summation
26		Sulfur Input to Stack Gases
27	C	Input/Output Relation
28	Si	
29	Mn	
30	S	
31	P	
32	Cr	
33	Mo	
34	Cu	
35	Al	
36	Ni	
37	Sn	
38	Pb	
39	C	Specification for Product - max
40	C	" " " - min
41	Si	" " " - max
42	Si	" " " - min
43	CE	" " " - max
44	CE	" " " - min
45	Mn	" " " - max
46	Mn	" " " - min
47	S	" " " - max
48	S	" " " - min
49	P	Specification for Product - max
50	Cr	
51	Mo	
52	Cu	
53	Al	
54	Ni	
55	Sn	
56	Pb	
57		Total Carbon Balance - Fuel + Metallics
58		Product Quantity Definition = 1000 pounds
59		BHN Definition
60		Product BHN Specification - max
61		Product BHN Specification - min
62		Carbon Equivalent (CE) Definition

Table 2. Matrix Variables: Physical Significance

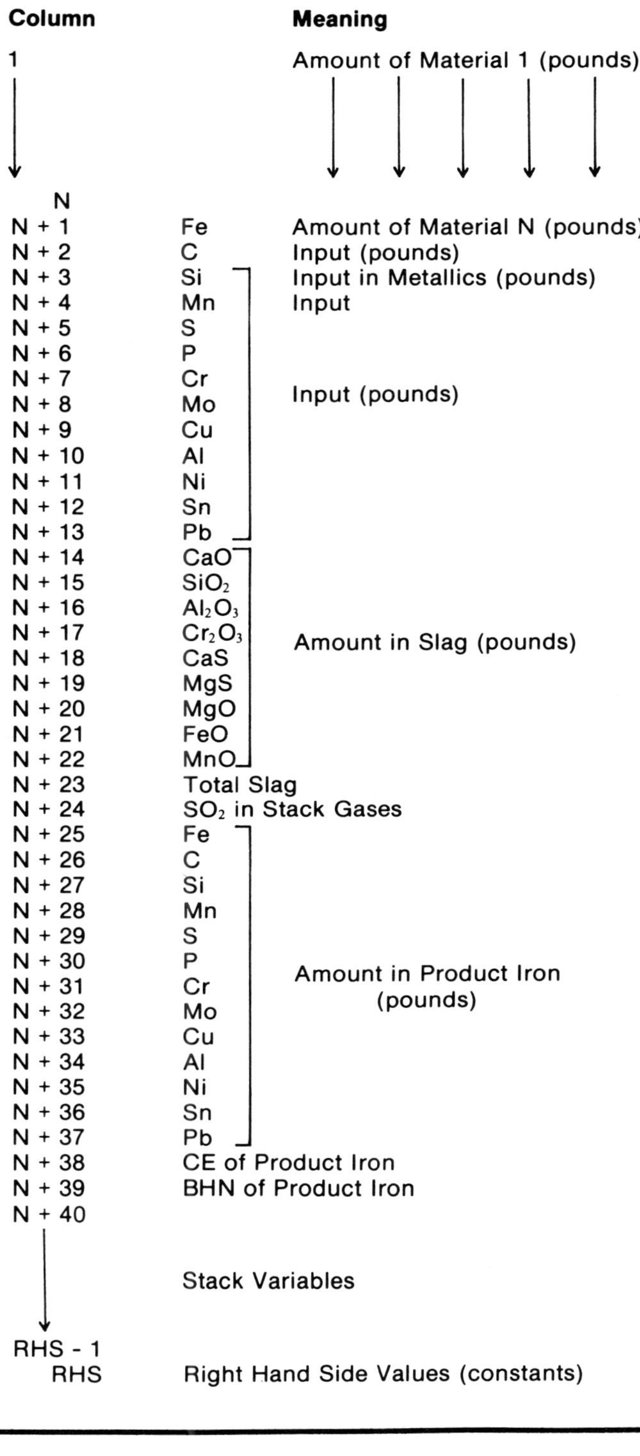

Column		Meaning
1		Amount of Material 1 (pounds)
N		
N + 1	Fe	Amount of Material N (pounds)
N + 2	C	Input (pounds)
N + 3	Si	Input in Metallics (pounds)
N + 4	Mn	Input
N + 5	S	
N + 6	P	
N + 7	Cr	Input (pounds)
N + 8	Mo	
N + 9	Cu	
N + 10	Al	
N + 11	Ni	
N + 12	Sn	
N + 13	Pb	
N + 14	CaO	Amount in Slag (pounds)
N + 15	SiO_2	
N + 16	Al_2O_3	
N + 17	Cr_2O_3	
N + 18	CaS	
N + 19	MgS	
N + 20	MgO	
N + 21	FeO	
N + 22	MnO	
N + 23		Total Slag
N + 24		SO_2 in Stack Gases
N + 25	Fe	Amount in Product Iron (pounds)
N + 26	C	
N + 27	Si	
N + 28	Mn	
N + 29	S	
N + 30	P	
N + 31	Cr	
N + 32	Mo	
N + 33	Cu	
N + 34	Al	
N + 35	Ni	
N + 36	Sn	
N + 37	Pb	
N + 38		CE of Product Iron
N + 39		BHN of Product Iron
N + 40		Stack Variables
RHS - 1		
RHS		Right Hand Side Values (constants)

Table 3. Cupolas Used in Model Verification

Parameter	Cupola A	B	C	D
Diameter (in.)	138	108	96	34
Blast Rate (SCFM)	40000	18000	16000	3600
Water Jacket	Yes	Yes	Yes	No
Blast Temp (F)	1000	900	900	70
No. of Tuyeres	10	8	6	4

A K-type thermocouple was inserted in each of the probes to measure gas temperature. Data were recorded on magnetic tape at a rate of one reading per minute. The large number of measurements facilitated identification of extraneous data points; such readings occurred when the thermocouple tip touched the solid charge or furnace shell. One of the probes contained, in addition to a thermocouple, stainless steel tubing for sampling stack gas. Stainless tubing was used to prevent contamination of the gas stream by chemical reactions. Glass wool filters were used to remove large particulates and oil from the gas stream and molecular sieve desiccant was used to remove water vapor. The tubing was then connected to a pump to provide a continuous gas stream for sampling, thereby eliminating the need for purging.

The tubing was subsequently connected to two nondispersive infrared analyzers (NDIR), one calibrated for percent CO and one for percent CO_2. The gas stream was sampled continuously and the output was recorded on magnetic tape at a rate of one reading per minute.

Occasional cleaning of the apparatus and filter changes were required but the probes and infrared apparatus held up well in the casting plant environment over a week long period. The NDIR analyzers were calibrated daily.

6) *Melt rate and coke rate* were determined by monitoring the number and makeup of the charges. The time of actual entry into the cupola was also recorded for each charge.

7) *Air blast parameters,* which include blast rate and temperature, were taken directly from the blast charts located in the control panels. Local conditions of ambient temperature and relative humidity were used.

All data were later digitized and placed in a master file containing other monitored data. The file was organized to provide minute-by-minute operational data. Computer software was designed to calculate information such as weighted averages of blast parameters, wind-on time, and melt rate per unit of air blast.

Model Validation

In order to validate the model, the input parameters of the actual operations were used to input to the model. These values included:

- blast rate, temperature, and composition;
- ambient temperature and relative humidity;
- cooling water flow rates and temperatures;
- coke analysis;
- desired product chemistry ranges and temperature;
- desired slag basicity ratio.

The model was then used to simulate the cupola for the specified operating conditions and the model output was compared to the output of the actual installation with regard to:

1) coke rate (metal-to-coke ratio);
2) melt rate;
3) exit gas temperature and composition;
4) final product chemistry.

This comparison provided a basis for adjustment of several assumptions and expressions in the model in order to achieve close approximation to actual data in the relevant range. The details of this calibration are not presented in this report.

In order to calibrate the model effectively, a wide range of operating conditions is desirable. However, since each individual production cupola uses a rather narrowly-defined set of operating parameter values (largely dictated by physical

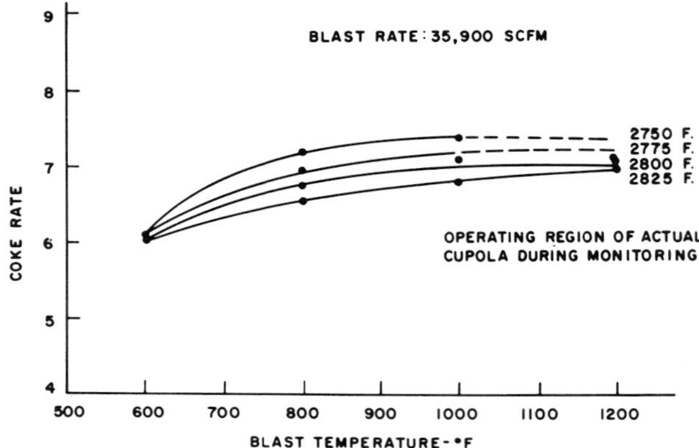

Fig. 12. Effect of blast temperature and metal temperature on coke rate for a 138-inch water-cooled cupola.

size), data were obtained for several markedly different facilities (Table 3).

Examples of Model Results

Two examples illustrating model results are presented below. In the first example, the effect of hot blast temperature on coke rate is investigated for a 138-inch water-cooled cupola. In the second example, the effect of oxygen enrichment on coke rate is studied for a 108-inch water-cooled cupola.

Effect of Hot Blast Temperature on Coke Rate

Figure 12 presents model results illustrating the change in coke rate with changes in hot blast temperature over the range of 316 to 649C (600 to 1200F). For each of the four curves indicated, all input conditions except hot blast temperature remain constant. The effect of metal temperature is also indicated; each of the curves represents a different iron temperature over the range of 1510 to 1552C (2750 to 2825F).

For a given iron temperature, 1552C (2825F) for example, the model results indicate a gradual deterioration of coke rate as hot blast temperature is lowered. As the blast temperature is reduced from 649 to 538C (1200 to 1000F), the coke rate decreases from 6.95 to 6.78. The effect on coke rate becomes more significant in lower blast temperature ranges. Reduction of blast temperature from 427 to 316C (800 to 600F) lowers the coke rate from 6.55 to 6.05. The model can thus be used to estimate the coke penalties involved with a reduction in natural gas available to the air blast preheater.

Figure 12 also indicates that a reduction in metal temperature can result in a significant improvement in coke rate. For example, at a hot blast temperature of 538C (1000F), reducing the specified iron temperature from 1552 to 1510C (2825 to 2750F) improves the coke rate from 6.78 to 7.34. Use of the model in this way can aid in establishing the most energy efficient means of attaining a given metal temperature in a duplexing (cupola/holding furnace) operation.

The operating conditions encountered during monitoring of the actual facility are approximated by the shaded area in Fig. 12. It is apparent that these data represent a very small range of potential operating conditions. Investigation of conditions significantly different than the usual operating region involves considerable risk if done with a production cupola, whereas the model can provide estimates of these effects with no risk.

Effect of Oxygen Enrichment on Coke Rate

The use of oxygen enrichment of the air blast has been shown to improve productivity and energy efficiency in many installations.[7-9] In considering the application of oxygen enrichment, use of the model can provide estimates of potential benefits for a specified cupola. Figure 13 illustrates model results for simulation of a 108-inch water-cooled cupola. The level of oxygen enrichment varies from 0 to 2%. The effects of oxygen were investigated at three specified iron temperatures ranging from 1510 to 1552C (2750 to 2850F) and these results are presented in the three curves in Fig. 13.

At a given metal temperature, use of oxygen improves coke efficiency significantly. For example, at a metal temperature of 1538C (2800F), increasing the oxygen enrichment level from 0 to 2% improves the coke rate from 6.20 to 7.20. This relative effect is consistent for all the metal temperatures studied, though the magnitude of improvement is somewhat greater at higher metal temperatures.

Figure 13 also shows that specifying a high metal temperature is deleterious to coke efficiency regardless of the level of oxygen enrichment. These results are consistent with the previous example.

The cupola model can be used in such studies for any of the process and material variables listed earlier. The programming features are designed to facilitate the desired changes in input parameters and the use of multiple runs.

Concluding Remarks

A process model of cupola melting has been developed to analyze the effects of changes in operating parameters on coke consumption and charge material costs.

Variables which have been studied include:

last rate and temperature;
- coke quality (size and composition);
- ambient temperature and humidity;
- oxygen enrichment;
- product chemistry and temperature requirements.

The model gives quantitative estimates of the effects of changes in these variables and provides this information with no risk to the actual facility.

In order to use the model effectively, several general points should be remembered. The model adjusts its estimates of cupola productivity in discrete steps when attempting to balance the energy in the system; care should therefore be taken in examining the results for a single run. It is more applicable to establishing trend data, such as the rate of change in coke usage over a range of conditions, rather than to providing single point values.

Also, the model functions most efficiently and most reliably under relatively moderate cupola operating conditions. It is our

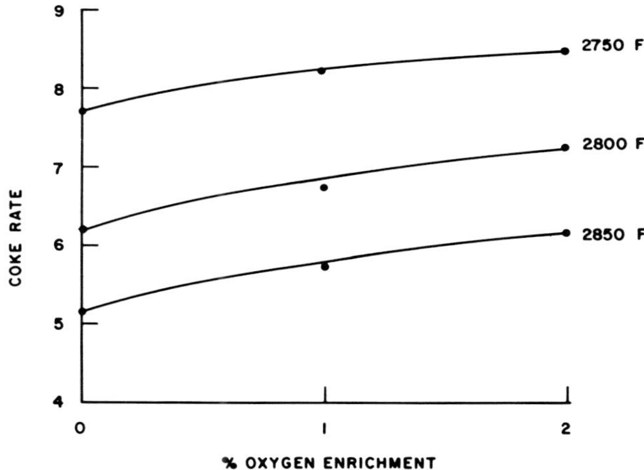

Fig. 13. Effect of oxygen enrichment on coke rate for a 108-inch water-cooled cupola.

experience that the model provides satisfactory results over any range of conditions likely to be encountered in a production gray iron cupola. However, it can also be used to investigate some rather extreme conditions (e.g., the use of coke with 20% ash). It will provide results for such conditions, but the user should anticipate some loss in the accuracy of the estimates.

References

1. *Cupola Handbook,* American Foundrymen's Society, 4th ed., 1975.
2. H. J. Leyshon, "General Principles of Cupola Operation," *BCIRA Journal,* vol 7, pp 224-248 (August 1957).
3. W. W. Holden, "How Coke Size, Screening, and Handling Affect Cupola Melting," *American Foundryman,* pp 38-40 (November 1954).
4. A. B. Draper and R. C. Creese, "Cupola Charging Model to Yield a Gray Iron of the Required Physical Properties at Minimum Cost," U.S. Steel Corporation unpublished report (November 1968).
5. P. E. DeGarmo, R. C. Creese, and S. A. Rao, "Determining Lowest-Cost Charge Mixes," *Foundry,* pp 63-65 (August 1966).
6. N. Meysson, "Mathematical Model Simulating Cupola Operation," *Fonderie,* pp 165-173 (May 1975).
7. J. V. Harding and J. A. Charles, "The Use of Oxygen in Cupolas," *The British Foundryman,* pp 365-370 (August 1961).
8. F. Carl, "Costs vs. Benefits of Oxygen-Enriched Cupola Melt," *American Foundryman,* pp 28-29 (February 1951).
9. J. Goswami and A. B. Chatterjea, "Melting Iron in the Cupola with Oxygen-Enriched Blast," *Foundry Trade Journal,* pp 255-261 (April 1, 1971).
10. G. A. K. Jungbluth, "Chemical Reactions in the Cupola," *Foundry Trade Journal,* pp 377-387, 405-411 (October 1955).

Production, properties, and industrial uses of magnesium and its alloys

by P. A. Fisher

Developments in the magnesium industry during the past decade are reviewed. Usage for aerospace applications has declined and consumption for nuclear power is static. Specialized alloy development for these markets has continued but at a slower pace than that of post-war years. Consumption for general engineering structural applications has increased slowly, mainly as pressure die castings. Developments include fluxless melting, hot-chamber casting, and alloy development for improved creep resistance, castability, and electroplating characteristics. The most notable increases in consumption have been for alloying into aluminium and for uses in metallurgical treatment such as production of nodular iron. The restricted use as an engineering material is ascribed to unfavourable economics rather than technical limitations. The most significant developments are therefore in extraction technology where new processes should reduce extraction costs and where the overall trend is towards larger units giving an economic advantage of scale.

During the decade covered by the review, magnesium achieved its 50th year as an engineering metal. Although it had been used on a very limited scale for military purposes during World War I, its industrial use dates from the mid 1920s. During much of the time the developments in its metallurgy and technology were spurred by its importance as a strategic material, as an essential material in post World War II aerospace developments and, finally, in its uses in the nuclear energy programme.

A characteristic of the past decade has been the absence of significant development pressures of this type. Development has been more concerned with consolidation and stabilization into a natural order dictated by normal commercial pressures. The industry is comparatively small, total Western world production in 1976 was of the order of 200000 t, but its growth rate has been good, averaging 5.8% annual increase in consumption over the years 1964—74. Extraction units tend to be small compared with those of the aluminium industry. Details of Western world plants are given in Table

Mr Fisher was with Magnesium Elektron Ltd, Manchester, until his recent retirement.

1 (compiled from data published by US Bureau of Mines).

Production in the USSR is estimated to be 75000 t/year. Little information is published on applications of magnesium in the Eastern bloc but development of various aspects of technology is reported.

Emley's treatise 'Principles of magnesium technology'[1] remains the authoritative work and is used in this review as the base for subsequent developments. The alloy nomenclature normally used in the Western world (primarily the ASTM system) has been used in this review, with the nominal content of the major alloying elements included in parentheses, where appropriate, for those not familiar with the system. The 'Handbook of Soviet alloy compositions'[2] includes the designations and compositions of magnesium alloys used in the USSR.

PRIMARY EXTRACTION

About 80% of Western world production is by electrolysis, the remainder being by thermal reduction. Production in the USSR is understood to be entirely by electrolysis.

Table 1 Western world magnesium extraction plants

Location	Process	Capacity, t
Canada		
Chromasco	Thermal	12000
France		
Sofrem	Thermal	11000
Italy		
SIIM	Thermal	12000
Japan		
Furukawa	Thermal	7000
Ube	Thermal	5500
Showa Denko	Thermal	250
Norway		
Norsk Hydro	Electrolytic	44000
USA		
American Magnesium	Electrolytic	10000
Dow Chemical	Electrolytic	120000
NL Industries	Electrolytic	45000
North West Alloys	Thermal	24000

Electrolysis

The I. G. and Dow processes, described by Emley, are still used for the majority of electrolytic production. It will be recalled that, although magnesium is readily available as a chloride solution (e.g. in seawater and natural brines) the direct production of anhydrous magnesium chloride from these sources was not commercially possible owing to hydrolysis, particularly beyond the dihydrate. The Dow process therefore uses a feedstock containing about 1.7 molecules of water, the cell discharging hydrochloric acid which is re-used to prepare the stock. In the I. G. process anhydrous magnesium chloride is prepared by chlorinating the oxide in the presence of carbon, the cell providing chlorine for re-use in the chlorinating stage. Extensive treatment of electrolytic technology is given in the 1972 Russian textbook by Strelets of which an English translation is now available.[3] Hoy-Petersen[4] has reviewed the important variables of the I. G. process. The most significant development of the past decade has been the work to establish a technology to produce anhydrous chloride from natural brine. The bonus is of course a theoretical 2.9 kg of saleable chlorine for each kilogram of magnesium. Fougner[5] has reviewed the probable forms of its implementation.

Solar evaporation to concentrate the magnesium chloride and to precipitate sodium and potassium salts is believed to be followed by spray drying to a water and MgO content of about 5% each. Melting of the spray-dried product followed by chlorination to remove the remaining water and to convert the oxide impurity to chloride is seen as the final step. For the operation in Norway, where solar evaporation would be impracticable, the magnesium chloride concentrate is a byproduct of the German potash industry.

In practical terms, new facilities to use this technology by N. L. Industries Inc. and American Magnesium Co. have both suffered extensive delays due to technical problems although both facilities are now in production. Dow have announced that they are developing new electrolytic technology which will result in a significant reduction of energy consumption; it seems almost certain that this will include the new chloride technology. Finally, Norsk Hydro have announced a specific programme involving construction of a 15000 t/year prototype plant, which, if successful, will lead to a new 100000 t/year plant in the 1980s. Apart from the new chloride technology it would be expected that these developments will include such features as larger cells, requirements to conform with more stringent environmental legislation, developments to improve labour productivity, etc. No major new magnesium electrolytic facility was built during the period 1943–70 and new facilities must presumably now include a considerable forward leap in technology.

The literature published in the USSR has been more prolific than that in the West and covers a number of specific aspects of magnesium electrolysis.[6-10] Some indication of the scope and direction of Russian development is given in a number of articles.[11-17] A recent paper by Petrunko *et al.*[18] indicates that dehydrated carnallite is generally used in the USSR. Anhydrous $MgCl_2$ arising from production of titanium is also electrolysed. The article also gives a general description of developments in cell technology.

One novel process has been reported.[19] This involves electrolysis of magnesium sulphide dissolved in a molten bath of magnesium chloride and sodium fluoride.

Thermal reduction

The Pidgeon process is now obsolescent, the plants formerly operated by Alabama Metallurgical Co. and Magnesium Elektron Ltd both having closed. The original plant at Haley, Ontario, continues to operate, although with an output much below its original capacity of 12000 t/year.

The Magnetherm process[20] has continued its industrial development. Apart from the original plant at Marignac, France, a 25000 t/year plant built under licence by North West Alloys Inc. (a subsidiary of Alcoa) commenced production in 1976, and a licence was negotiated with Magnohrom for a 5000 t/year plant to be built in Yugoslavia. Avery[21] has patented a number of improvements to this process, including the ability to operate at atmospheric pressure and means for continuous, as opposed to batch, operation. As far as is known the Avery developments are not being operated commercially. No technical information has been published on the Amati thermal process, operated at Bolzano, Italy. Plans are being considered for an 18000 t/year plant in Canada, using this process.[22]

Purity of primary magnesium

In Western world production little or no refining of metallic impurities is needed. The 99.95% purity produced by the Pidgeon process at Haley, Ontario, meets special requirements such as those for nuclear energy, other supplies fall within 99.5% purity, which is satisfactory for all other requirements. An examination of purification by vacuum distillation[23] has been made, the interest probably being to provide a second source to the Haley supplies. In the USSR the concern with purity appears to be related to the magnesium used for production of titanium. A number of articles[24-28] deal with refining methods based on coprecipitation as intermetallics in the liquid. A recent article[29] surveys the purity of electrolytic magnesium made in various countries and gives details of a three-layer electrorefining technique. This would evidently be similar to that used for production of super-purity aluminium.

SECONDARY PRODUCTION

Western world production of magnesium secondary is reported[30] as 21000 t in 1976, much of this being produced in the USA. A substantial part of the scrap recovered consisted of Volkswagen engine castings. Volkswagen were believed to consume some 40000 t/year of magnesium alloy during peak production of the 'Beetle' in the 1960s. Consumption of magnesium is now at a much lower level.

Japanese tonnages reported as secondary are not included in the above estimate since they consist of electrolytic magnesium recycled from the chloride arising from the production of titanium. It has been estimated[31] that, of the recoverable primary magnesium sold in the USA, only about 6% is recovered. Processes, almost invariably involving use of a chloride refining flux, and permitting recovery of nearly all forms of magnesium scrap are in regular use. The majority of scrap is of various magnesium—aluminium alloys which can be reprocessed to a specified composition without much difficulty. A significant volume of magnesium alloys containing zirconium, rare-earth metals, etc. requires physical separation and separate treatment since its admixture with magnesium—aluminium alloys degrades both alloy types. The use by Volkswagen of a new alloy containing silicon is expected to complicate secondary production in future years.

Lockwood[32] has reviewed the recovery of automotive scrap in the USA. Separation of magnesium alloy by liquation melting is not a useful technique. Crushing followed by selective flotation in an appropriate heavy media is used to a limited extent in Europe.

PRODUCTION OF ENGINEERING COMPONENTS

Although the use of magnesium for 'exotic' applications in aerospace, nuclear energy, etc. represents a declining fraction of total consumption (probably now less than 10% of consumption in the Western world) the nature of the requirements has assured continued development in technology, although at a slower pace than that in the years following World War II. Developments for general engineering applications are strictly conditioned by the dictates of economics and have tended to follow a separate pattern, although some overlap has occurred.

Castings

An overall review of developments in industrial practice for the production of sand castings (mainly

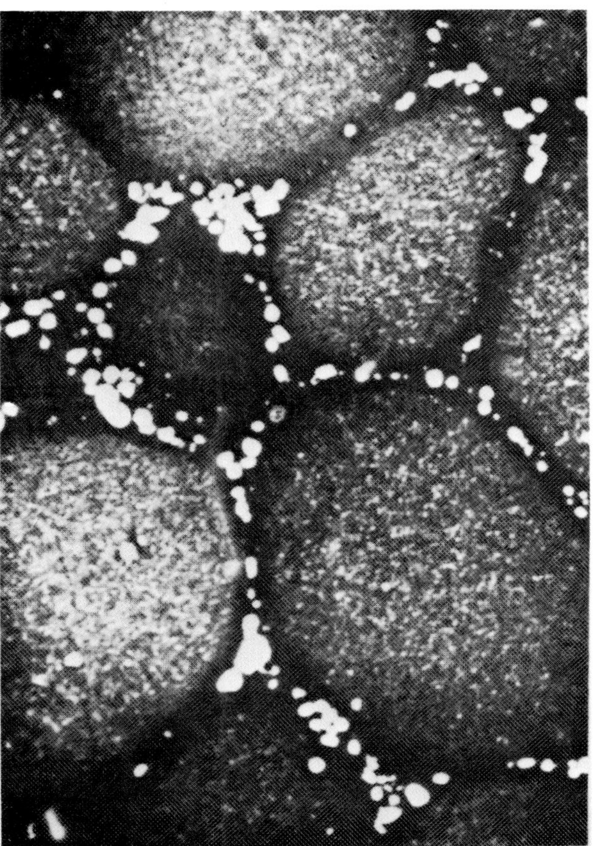

1   **ZE63 alloy heat treated with conventional SO₂ protection**   ×750

2   **ZE63 alloy heat treated in hydrogen**   ×750

for aerospace applications) has been given.[33] The development of fluxless melting by Hanawalt and his co-workers[34] has been applied to aerospace castings in one known instance. The technique involves use of small additions of SF_6 to air or CO_2 but for aerospace castings the gas mixture has been more frequently applied to replace SO_2 while pouring and during heat treatment. Most of the other developments in foundry technology represent adaptions to magnesium practice of general foundry developments designed for such requirements as greater dimensional accuracy, production of thinner sections, more effective use of skilled labour, etc.

Alloy development has included further work on the novel conception involved with ZE63 alloy.[35] This alloy depends on diffusion of hydrogen during heat treatment to achieve a metallurgical trans-formation and permits use of rare-earth metals to achieve structural soundness, while avoiding the embrittling effects normally present. Figure 1 shows the microstructure after a conventional heat treatment at 753K with SO_2 protection. In Fig. 2 the alloy has received the same time—temperature cycle but in a hydrogen atmosphere. Although industrial use has confirmed the advantages in strength and uniformity of properties expected of the alloy, such use has been limited by the time, cost, and restriction of section thickness involved in heat treatment. It has been found that condition-ing of the metal surface by control of humidity of the heat-treatment atmosphere, during the initial stages of heat treatment, is beneficial in speeding up hydrogen penetration.[36] A new casting alloy, termed QH21, has been introduced. This combines the use of thorium with neodymium-rich rare-earth metals and silver effectively to extend the creep strength of previous alloys from 473 to 523K.[37]

The valuable work carried out at the Canadian Bureau of Mines in filling in the background to magnesium sand-casting technology has continued. Couture and Meier[38] have investigated the effect of testbar variables on tensile properties of a number of alloys. Lagowski[39] has described the segregation and inclusion effects in magnesium—zirconium alloys and discussed[40] the proposed additions to the standard ASTM radiographs, cover-ing these alloys. Lagowski and Meier[41] have listed the improvements to the strength of magnesium—aluminium alloys obtainable by controlled solidifi-cation, addition of silver, and grain refinement by copper, and have described[42] the effects of cold work on the tensile properties of a number of alloys. Crawley and Lagowski[43] have investigated the effect of a two-step aging treatment on the precipitate in AZ91 alloy and the benefits of intro-ducing a cold-working step in the heat treatment of AZ92.[44]

Although, as with most metals, magnesium casting alloys benefit from application of cold work, their tolerance for such work is very limited and the cracking which results if the tolerance is exceeded can be disastrous to subsequent perfor-mance.

Other useful articles which may be characteri-zed as filling in gaps in casting technology include investigations of the relationship between micro-porosity and tensile properties.[45-46] Little work of Russian origin appears to have been published but Greshishchev[47] has investigated the design of feeding systems for an AZ81 type alloy.

Work on casting technology for general engineering purposes has been predominantly concerned with the pressure die casting process. Street[48] has pointed out that metal loss may ex-ceed 15% in small-quantity batch production, re-ducing to 5—7% for continuous operation. (These figures will presumably relate to overall loss after repeated recycling of scrap.) Hanawalt's work with SF_6 atmospheres is very relevant, not only to a reduction of these losses but also towards avoiding the cost and nuisance involved in the use of chloride melting fluxes. The SF_6 technique appears successful for use with clean pre-alloyed ingot but is understood to be less successful in handling scrap, particularly of the large surface-area type arising from the pressure die casting process. A method enabling such scrap to be re-charged direct to a bale-out furnace is described by Kittilsen.[49] A crucible having two compart-ments is used, the molten metal passing through a flux layer into the bale-out section.

The melting of magnesium alloys is still mainly carried out in gas- or oil-fired crucible furnaces, with some use, particularly in Germany, of mains-frequency induction melting. Calamari[50] has included details of magnesium melting in a review of low-frequency induction melting practices.

Automatic metering of molten alloy to cold-chamber machines has been applied in the larger installations for many years and there has been little indication of further developments in the past decade. A unit has been marketed in Europe under the trade mark 'Normagic'[51] and Russian interest in this technique is indicated in an article by Gluschchenko et al.[52]

The most significant development has been the more general use of hot-chamber machines. Al-though used in Germany for many years and inten-sively developed by Dow Chemical Co. in the 1950s and 1960s such machines remained in limited use, possibly owing to engineering maintenance problems. The introduction by Frech of machines up to about 100 t locking force and adapted for magnesium from the standard zinc machine was paralleled by other manufacturers with machines up to 300 t capacity. The higher shot rate obtain-able, as compared with cold-chamber equipment, made magnesium competitive for a wider range of applications and many of these machines have been sold to established foundries, particularly in Europe. An interesting development was the evolution of substantial new capacity based on magnesium hot-chamber practice.[53-54] Some typical hot-chamber castings are shown in Fig. 3.

3 Typical hot-chamber pressure die castings

Bauer[55] has reviewed the various systems available for magnesium. Nelson[56] has studied the heat balance in a die when casting aluminium and magnesium alloys and concluded that magnesium could be cast at a rate 75% faster than aluminium. The majority of magnesium pressure die castings are made in alloys containing 8—9%Al. Alloy development to meet specific requirements has included use of AM60 (6%Al) to provide the better impact strength required for automotive wheel castings,[57] and AS41[58] (4%Al, 0.7%Si) for the improved creep strength required for later versions of the Volkswagen air-cooled engine.

AS21 (2%Al, 0.7%Si) provides still better creep strength but suffers, as do all these alloys with aluminium contents below about 8%, from reduced castability. Alternative alloys, involving the addition of calcium or rare-earth metals to the general purpose aluminium-containing alloys[59] do not appear to have been adopted commercially, again probably owing to poor castability. Foerster[60] has discussed the factors affecting the design of magnesium alloys for pressure die casting. A new range of alloys involving higher zinc contents than had previously been considered desirable was announced by Foerster.[61] Typically, ZA124 contains 4% aluminium and 12% zinc. Over the range of alloys Foerster claims improvements in castability, creep strength, and plating characteristics. The characteristics of the main pressure die casting alloys are given in Table 2. Very little detailed information has been published on the technology of magnesium pressure die casting, other than that given in Emley's 'Principles of magnesium technology'[1] and Beck's earlier 'Technology of magnesium and its alloys'. A sponsored project has been carried out by the BNF Metals Technology Centre

| Designation | Nominal composition, wt-% | | | | Tensile properties | | Elongation, % | Characteristics |
	Al	Zn	Mn	Si	0.2% proof stress, MNm^{-2}	Ultimate tensile strength, MNm^{-2}		
AZ91	9.0	1.0	0.2	...	160	225	1.5	General purpose
AZ81	8.0	0.7	0.2	...	150	220	2.0	General purpose
AZ61	6.0	0.6	0.2	...	145	220	4.5	Preferred for electroplating
AM60	6.0	...	0.2	...	135	260	6.0	Improved impact strength
AS41	4.5	...	0.3	0.7	135	235	4.5	Improved creep strength

to identify and measure the important variables and the results are now generally available.[62] Russian experience has been that hot cracking is one of the main reasons for high reject rates.[63] The addition of bismuth is claimed to alleviate hot cracking.[64]

Magnesium-alloy castings are produced in limited quantities by the gravity die process but no significant developments are reported over the past decade. The low-pressure process is used, notably in Italy for the production of automotive wheels.[65] An account[66] has been given of a novel process designed to meet a specific need. The requirement was for small cylindrical pieces of a magnesium alloy for administration to cattle. The density required was higher than that available from conventional magnesium-base alloys and the chemical and physical nature of the articles was constrained by possible physiological effects in the cattle. The problem was solved by heating a magnesium—aluminium alloy to a temperature between its solidus and liquidus points, stirring in iron shot to give the required density and then moulding the mixture in a multicavity die by a simple pressing operation.

This approach to metals handling has, of course, been developed by Mehrabian and Flemings, and their co-workers at the Massachusetts Institute of Technology (MIT). Work on magnesium alloys, using the principles of 'rheocasting' and 'thixocasting' is reported by Bennett *et al.*[67] AZ91B alloy test panels were less porous and had slightly better mechanical properties compared with those made by conventional pressure die casting. A substantial reduction in weight loss during a wear test was achieved by incorporation of silicon carbide in the alloy. Addition of lubricants such as molybdenum disulphide and carbon black were less effective in reducing wear.

Wrought forms

Most of the stock for subsequent working is produced by the direct chill process but, compared with advances in other metals, the process has seen negligible development for magnesium in the past decade. Products are limited to cylindrical stock for extrusion, forging, conversion to powder, etc., and slab for rolling. As far as is known all products are made by the vertical drop technique. Melting is normally carried out in gas- or oil-fired steel crucibles and the use of pumps or siphon devices to transfer metal to the mould is common.[68] A number of references from the USSR indicate an interest in the application of electro-magnetic induction to the handling of molten magnesium.[69-72] A review of the development of electromagnetic moulds in the USSR[73] indicates that these have been applied to magnesium but, as with their application to aluminium, suggests that further development is required before the process can be justified economically. The possibility of casting magnesium by the horizontal process has been considered by Vyatkin and Chukhrov,[74] who point out that the asymmetry of the thermal balance may lead to anisotropic mechanical properties.

Preheating for hot working is normally carried out in muffle furnaces or by induction. An examination of the use of modern gas-fired direct-flame-impingement furnaces for preheating extrusion stock suggested that these might be satisfactory for Magnox alloys, having a margin of about 100K between preheat temperature and solidus. For AZ61 alloy, where the margin was about 30K the risk of fire was considered too high. No developments in extrusion practice for magnesium have been reported although the use of modern presses and techniques is undoubtedly applied. Forged magnesium components are almost exclusively confined to use by the aerospace and nuclear industries and no significant developments in techniques have been reported.

A little experimental work has been carried out with the pendulum mill which, like the planetary mill, seems ideally suited for rolling magnesium alloys. Although further development would have been required, the production of strip at 0.75 mm thickness appeared feasible. Indeed, the remarkable feat of converting 300 mm cylindrical billet into 230 × 0.75 mm strip in only two operations (one extrusion and one rolling) was demonstrated. The development of wrought alloys, as with those for castings, has tended to be specifically for aerospace or general engineering use, with a small degree of overlap. For aerospace applications the potential of magnesium—yttrium alloys was examined by London et al.[75] Work on direct production of magnesium—yttrium master alloys by electrolysis is reported by Aamland et al.[76] No further work on development of magnesium—lithium alloys has been reported in Western world literature and the final status of these alloys in the USA has been summarized by Jackson and Frost.[77]

Evidence of interest in the USSR is given by publications of Vyatkin et al.[78] and Dritz et al.,[79] who were concerned with a complex alloy (Mg—8Li—5Al—4Cd—1Zn—0.5Mn) termed 1MV2. Borisov et al.[80] claim that the corrosion resistance of this alloy in boiling water is superior to that of the American alloy LA141 (Mg—14Li—1Al). Magnesium alloys with scandium and yttrium were discussed by Busk,[81] and Dritz et al.[82] have studied the effects of other additions to an alloy containing nominally 11%Sc and 0.8%Mn. The interest in these alloys is largely concerned with development of high-temperature properties but, as far as is known, the alloys have not been adopted commercially. Significant improvements in room temperature strength are claimed by Stratford[83] for ZM61 alloy (Mg—6Zn—1Mn). This alloy benefits from use of a stepped aging treatment and is capable of giving a 0.1% proof strength of 376 MNm^{-2}.

For general engineering purposes alloy development has been primarily directed towards improvement in working characteristics with retention of medium strength. With ZM21 alloy[84] (Mg—2Zn—1Mn) formation of the zinc-rich beta phase is suppressed by the addition of manganese and the resultant increase in solidus temperature enables hot deformation to be carried out significantly faster than, for example, with the more generally used AZ31 alloy. The improvement also extends to rolled forms where ZM21 alloy is less prone to edge cracking. Similar advantages are claimed for magnesium—zinc alloys containing calcium,[85] rare-earth metals,[86] and silicon.[87]

The need to use elevated temperatures for deformation of magnesium has lead in the past to the production of complex pressed shapes reminiscent of those obtainable in superplastic alloys. The very fine structure obtainable by addition of zirconium has also appeared a good base for the superplastic phenomenon. Although no specific magnesium alloys have been developed for this purpose Attwood and Hazzledine[88] have studied superplasticity in ZK60 alloy (Mg—6Zn—0.5Zr). Valiev and Kaibyshev[89] report on microstructural changes during superplastic deformation of an alloy containing 1.5%Mn and 0.3%Ce.

Powder-compacted and fibre-strengthened products

Emley describes the American work on atomized powder-compacted alloys which resulted in the extrusion alloy ZK60B (Mg—6Zn—0.6Zr). The suppression of grain growth, ascribed partly to the cored structure of the powder and partly to the oxide coating on the particles, leads to high tensile strength coupled with high compressive strength, a combination difficult to achieve with more conventional wrought magnesium alloys. Isserow and Rizzitano[90] have repeated this work using powder produced by the rotating electrode process and claim tensile properties substantially higher than

Table 3 Properties of dispersion-hardened magnesium extrusions

| Composition | Fabrication data | Extrusion conditions | | Elongation, % | Yield strength, MNm^{-2} | | Tensile strength MNm^{-2} |
		Temp., K	Speed, mms^{-1}		Tension	Compression	
Pure Mg	<1 μm Powder	866	<5	1	241	248	275
ZK10 + 2MgO	Screw extruded	711	20	7	241	151	289
Mg—1Zn —1.6Si	Atomized 35—65 mesh	589	25	3	296	258	344
Mg—1Zn —1.6Si	Atomized 35—65 mesh	589	500	7	234	151	289
1Zn—1.6Si	Atomized 100 mesh	589	500	4	310	199	337
AK11	Interference hardened	589	500	8	303	310	351
AK11	Interference hardened	755	350	10	330	248	351

those quoted by Emley. This may be associated with a faster quenching rate when producing the powder or, perhaps more probable, with the low (295K maximum) extrusion temperature. Compressive properties are not quoted. Foerster[91] has reviewed a number of aspects of extruded powder-compacted alloys and gives the results quoted in Table 3. It is interesting to note the high level of properties obtained in the dilute AK11 alloy. This comprised pellets of ZK10 alloy (Mg—1Zn—0.6Zr) coated with aluminium which, during extrusion, diffuses to precipitate an insoluble aluminium—zirconium compound. The effect was first explored by Busk and Leontis.[92] None of these techniques are effective in improving the elastic modulus of magnesium alloys and such improvements can only be derived from composite structures. Huseby et al.[93] describe the preparation of extrusions from a mixture of magnesium and 25 vol.-% boron powders. Young's modulus was increased from 44.1 to 77.9 GNm^{-2} at room temperature and retained a value around 76 GNm^{-2} at 673K. It is noted, however, that the hardness of the composite precluded machining, the specimens being prepared by grinding.

Graphite fibre reinforced magnesium composites containing 11 vol.-% graphite were prepared by Levitt et al.[94] Magnesium will not wet graphite and the fibres were coated with titanium, increasing the effective volume fraction to 31%. The tensile strength was not outstanding, values around 220 MNm^{-2} being reported. The value for Young's modulus was, however, nearly twice that for pure magnesium. In a subsequent paper Richman et al.[95] have studied the causes of exfoliation and disintegration in such composites. This is an effect of corrosion associated with localized porosity and lack of wetting. These authors found that methane was evolved during corrosion when the aluminium-containing alloy AZ91 was used as the matrix. They deduce that aluminium carbide was formed during preparation of the composite. With a pure magnesium matrix exfoliation occurs, without gas evolution, by the disruptive effect of oxide formation. It is claimed that composites with good wetting and bonding characteristics have adequate corrosion resistance. However, in his review of carbon fibre reinforced metals, Baker,[96] who does not mention magnesium, comments that swelling of carbon—aluminium composites may be due to rapid corrosion along the fibre/matrix interface. Reinforcement with boron and tantalum fibres has been studied by Ahmed and Barranco.[97] The strength of the boron fibres used was reduced by about 50% on heating to temperatures above 773K. The strength of magnesium composites was that expected from heat-affected fibres, reaching a longitudinal value around 480 MNm^{-2} with about 0.30 volume fraction of fibre. The elastic modulus was about 138 GNm^{-2}. The incorporation of relatively low strength but ductile tantalum fibres improved the ductility of magnesium and produced modulus values around 90 GNm^{-2} at 0.3—0.4 volume fraction.

SURFACE PROTECTION PROCESSES

All of the processes in general use are described in Emley's 'Principles of magnesium technology'.[1] For aerospace applications the 'Surface sealing' process forms the basis of British Specification DTD 911C and is used for specific applications in other countries. The Dow 17 and HAE anodizing processes are widely used. Adamson et al.[98] have evaluated the Dow 17 process. Resin-based paints are, of course, now generally used and have undoubtedly contributed significantly to improved protection.

An approach to simplified procedures by attempting to incorporate an organic protectant into a fluoride film during the anodizing process is described by Yauw and Schlick.[99] The possibility of impregnating anodic films was examined by King et al.[100] Neither approach has resulted in schemes giving protection as good as that available from older methods. A study of protective coatings resistant to elevated temperatures has been made by Adamson et al.[101]

A new anodizing process, claimed to be more suitable for high volume production was announced by Kotler et al.[102] Termed the MGZ process, its development stems from an original concept of Dr Evangelides, the inventor of the HAE process. A complex film of oxides and fluorides of magnesium, phosphorus, vanadium, and chromium is built up on the magnesium surface. Corrosion resistance better than that provided by the Dow 17 and HAE films, coupled with excellent wear resistance against rubber, are claimed.

Simple chromate conversion films, used without further protection or as a base for paint schemes, are normally used for applications in general engineering. The problems of complying with environmental legislation concerning disposal of chromate waste liquors can be met but there will almost certainly be a future interest in the development of chromate-free systems.

Electroplated finishes have been available for many years but have not been widely applied. Their use on magnesium pressure die castings to compete with zinc for decorative applications is the present dominant interest. Spencer[103] reviewed the technology in 1970 and quoted an estimate that bright chromium plating of a magnesium component would cost 15—25% more than an identical zinc-base part. Olsen[104] has reviewed present trends in Europe, pointing out that some routine production is now carried out in Germany. For plated pressure die castings AZ61 alloy (Mg—6Al—1Zn) is preferred to the more commonly used AZ91 composition since aluminium-rich segregation in the latter alloy leads to defective local areas in the plated coating. Safranek et al.[105] report on a comparison of USA and European plating systems for magnesium, including corrosion testing of typical components.

In the USSR Roikh et al.[106] have studied the protection provided by vacuum aluminized coatings.

At a thickness of 20 μm corrosion of the underlying magnesium continued through pores in the aluminium film but at 40 μm thickness the film was self sealing. In Japan, Yamada *et al.*[107] have studied the pretreatment of magnesium for enamelling. Poor adhesion was ascribed to the formation of metallic lead when firing the low-temperature frits employed. This was overcome by pretreatment in sodium silicate solution and prefiring.

WELDING AND JOINING

Welding of magnesium is now almost entirely by inert-gas shielded arc. Apart from its use for structural joints it is also generally used for repair of defects in castings. Spot welding is employed to a limited extent, notably in the production of nuclear fuel elements.

Various aspects of welding technology are described in publications from the USSR. Ryazantsev and Fedoseev[108] have examined the use of filler rod containing zinc and rare-earth metals for joining wrought alloys and Ivanov *et al.*[109] have studied the effects of cleanliness of the base material on its weldability. Osokin and Umnaev[110] describe procedures for repairing defects in large thin-walled castings of various alloys. In India, Dutta *et al.*[111] have used chloride flux protection when pressure welding gas-bottle halves. No mention is made of problems in ensuring freedom from chloride inclusions. Experimental work with electron beam welding of magnesium is described by Cull[112] and Ivanov and Zhulin.[113] Weld strengths better than 90% of that of the parent metal are obtainable.

There appears to have been no significant development of other joining techniques although adhesive bonding has been used in the production of an automotive wheel.[114]

ENERGY CONSIDERATIONS

As with many other energy-intensive materials, the justification, in terms of energy balance, for using magnesium as a structural material has been examined. The most comprehensive treatment is given in the report of a conference organized by the MIT,[115] in which Kenney and Clark[116] argue that, with current technology applied to automotive applications, magnesium provides a 78% better energy return than aluminium, considering both metals as substitutes for cast iron. This ratio should increase as new extraction technology is applied, reaching 90% when the availability of magnesium scrap permits use of 50% secondary metal. The Kenney and Clark study is based on more modern data than that used in an earlier review by Bauer,[117] which was less favourable to magnesium.

The MIT report contains much useful data on energy consumption, and prospects for its reduction, by both the magnesium and aluminium industries. For production of primary metal consumption of energy by existing electrolytic processes is about 45% greater per kilogram for magnesium than for aluminium. Credits for by-products bring the energy requirements for the Magnetherm process close to those for aluminium. Taking known developments into account the energy requirements for both metals should be approximately equivalent by 1990. Energy requirements for conversion to semi-finished products are generally greater for magnesium than those for aluminium, an exception being pressure die casting where the hot-chamber process is advantageous to magnesium. The lower energy efficiencies of other magnesium processes are, in part, a reflection of limited scale of operation and lack of development but, particularly in rolling technology, also reflect the limited capacity for cold work characteristic of most magnesium alloys.

APPLICATIONS

The estimated Western world consumption of primary magnesium for 1975 and 1976 by geographical area and broad category of end use is given in Table 4, reproduced from a paper by

Table 4 Estimated consumption of primary magnesium distribution by market, t (Ref. 118).

Country	Year	Aluminium alloys	Chemical metallurgical	Structural	Total
Americas	1975	48000	31500	21500	101000
	1976	51000	35000	21000	107000
Japan	1975	8100	3000	...	11000
	1976	10600	3600	...	14200
Europe	1975	15500	10000	22000	47500
	1976	21500	11500	23000	56000
Others	1975	4000	...	...	4000
	1976	4000	...	...	4000
Total	1975	75600	44500	43500	163500
	1976	87100	49100	44000	181200

Table 5 **Consumption of primary magnesium in the USA, (short tons $\times 10^3$)**

Use	1955	1965	1974
Structural			
Pressure die castings	...	5.45	11.80
Gravity die castings	10.37	0.97	1.00
Sand castings	...	2.95	1.37
Extrusions	4.41*	5.99	7.32
Sheet, plate	6.42	4.94	6.03*
Total	21.20	20.30	25.52
Alloying			
Aluminium alloys	11.10	26.27	62.15
Other	0.36	0.14	0.06
Total	11.46	26.41	62.21
Chemical/Metallurgical			
Cathodic protection	3.94	4.60	10.44
Chemicals	0.12	3.81	9.20
Nodular iron	†	†	10.60
Reducing agent for Ti, Zr, etc.	8.06	8.47	7.57
Other, including powder	1.79	3.83	2.50
Total	13.79	20.71	40.31
Total consumption	46.45	67.42	130.04

* Includes forgings
† Not estimated separately.

Campbell.[118] Of the total consumption about 60% is marketed in the Americas (mainly in the USA) for which the end-use distribution is very similar to that of the average for the Western world. European consumption is more heavily slanted to structural uses, mainly by the continuing demand for pressure die castings in Germany, including substantial use by Volkswagen. Conversely, Japanese consumption for structural purposes is insignificant. More detailed statistics of end-use consumption in the USA are published annually by the US Bureau of Mines. Table 5, compiled from the Bureau's 'Minerals Yearbooks'[119] shows how market distribution in the USA has changed in two decades.

Structural engineering applications

The variety of applications of magnesium in aerospace engineering has declined during the past decade for a number of reasons. In airframe construction the possible effects of corrosion, or the unacceptable cost of preventive maintenance, have generally limited use of magnesium to components which are readily accessible for inspection. The increased power available from modern engines has, of course, reduced the need for weight reduction to some degree. For engine components such as compressor housings the development of increased power has resulted in operating temperatures beyond the capability of light alloys. Magnesium alloys are, however, still used extensively for the cooler engine components and notably for the massive gear-box castings required by helicopters. Many other structural components, including landing wheels, are still in regular production.

Few totally new components types have been introduced. One exception is a thrust reverser casting (Fig. 4) of which several designs are used to make up an engine set. A typical casting measures about 750 × 400 × 50 mm, comprising a 'waffle' type section with wall thickness around 2.5 mm. High mechanical strength and close dimensional accuracy are required.[33]

In military engineering magnesium alloys have been used in a French tank for the main gear box,

4 **Casting for jet engine thrust reverser unit**

in gun stabilizer castings, and for track idler sprockets. The last application is a further example of a successful use in an environment which might be thought inherently too hostile for magnesium. In military ballistics the 'Sabot', a type of driving band for high-velocity shells, is used by many countries.

In nuclear engineering the use of magnesium cans in 'Magnox' stations has been eminently successful. Harris[120] has reported on studies of irradiated cans. Unfortunately magnesium is not used in any other type of reactor.

Wrought magnesium has been more widely applied to general engineering applications in the USA than in Europe and typical American applications include items such as step ladders, sack trucks, bakery delivery racks, etc., which have few European counterparts. Conversely, the application in Europe of an extruded alloy for production of pencil sharpeners, machinability being of prime importance, does not appear to have been duplicated in the USA. Scherer[121] has reviewed American applications of wrought magnesium.

More interest has focused on development of cast products, particularly those made by pressure die casting, notably in Germany and the USA. Bolstad[122] has reviewed some of the European applications and, as would be expected, the majority are used in transport or where portability is important. Magnesium wheels have been standard equipment on racing cars for many years but are now also used as standard or optional equipment by the car enthusiast and in more expensive models. Production of pressure die cast wheels[57] has been established and many are produced by the low-pressure process.[65] The total usage by Volkswagen has declined as their air-cooled engine has been phased out but this company still remains the largest producer and user of magnesium structural castings, notably for gear boxes. Magnesium castings are used extensively in chain saws and some European manufacturers use them for do-it-yourself equipment such as electric drills.

In the USA the applications are generally similar with the exception that large-scale sales to the automotive industry have not yet developed although much promotional effort has been made. In 1973 the production of one model of fuel-pump housing accounted for nearly 13% of the total use of magnesium pressure die castings in the USA. Total sales to the USA automotive industry amounted to 1820 t, compared with the use by that industry of 300000 t of aluminium castings.[117] One application specific to the USA is the large deck casting for rotary lawnmowers. Some models have this casting in aluminium alloy and presumably the magnesium component commands a price premium. Magnesium baseball bats have been cleverly designed to reproduce the performance of the traditional wooden type, and are, of course, a development unique to North America.

Many current applications have been produced

in magnesium for over 40 years and their continued use is confirmation of the technical adequacy of magnesium alloys for the service conditions involved. An application first made nearly 50 years ago and only recently revived is described by James.[123] This is the use of magnesium castings for a rigid diving suit which, in 1974, held the world record of 300 m depth. Such suits are in regular use for marine salvage operations and provide an interesting example of the degree to which magnesium components can be made to withstand a corrosive environment.

Aluminium alloying

This now accounts for nearly half of total Western world consumption. Redhair[124] has reviewed the consumption of magnesium by the US aluminium industry, pointing out that the average content of magnesium was 1.06% in 1975. Notable uses include the easy-open beverage-can lid and plate for cryogenic gas storage, both alloys containing about 4.5% magnesium. Lugagne[125] states that the average magnesium content of aluminium alloys in Europe is 0.67% but that this is increasing.

An interesting facet of this market for magnesium is that most of the metal is eventually lost when the alloy is demagged for subsequent use as secondary alloy. It has been estimated[31] that only about 4% of the consumption by the aluminium industry is recycled. The move in the USA to recover used aluminium beverage cans and to re-use this scrap for the same product is expected to improve the percentage recovered.

Products and applications in chemical and metallurgical treatment

This broad category covers a variety of applications where magnesium is used as a chemical (e.g. in Grignard reactions, production of titanium, zirconium, etc., as flares for production of light, etc.), or for its electrochemical characteristics (e.g. cathodic protection, primary batteries), or for metallurgical treatments such as in nodular iron, steel desulphurization, etc. Hock[126] has reviewed the uses of magnesium in this category. A number of specific products and compositions have been developed for these applications.

Powder and granulated forms

The majority of magnesium powder is produced by mechanical comminution, typically with a milling cutter or by abrasion with a steel rasp. The initial product may be milled to produce a finer or more rounded particle. Magnesium powder is also produced by gas jet or centrifugal disintegration of molten metal. In the latter cases an atmosphere inert to molten magnesium must normally be used. Fulford[127] has reviewed the safety aspects in handling magnesium powder, and Popov[128] reports on conditions required for spontaneous combustion. Developments in production of comminuted powder have been mainly concerned with improvements in productivity or the production of special shapes for specific applications. For granulated forms produced by centrifugal disintegration Barannik *et al.*[129] describe the inclusion of a small quantity of molten chlorides which coats the magnesium granule and avoids the need for a protective atmosphere. In principle, such powder could be produced direct from the electrolytic reduction cell and hence its production cost should be low. The chlorides are, of course, deliquescent and storage and transport of such granules requires special precautions. The development is specifically for use in desulphurizing blast furnace iron and Voronova[130] reports that it is now a routine process in the USSR. Studies of magnesium powder produced by condensation in vacuum or inert gas are reported by Kvater and Frishberg[131] and by Granqvist and Buhrman.[132]

The potential use of magnesium for storage of hydrogen has been examined. Magnesium—10Al granules are reported to be resistant to fragmentation during the hydrogenation—dehydrogenation cycle and have an attractive ratio of H_2 capacity to density. However, the temperature required for the hydrogen cycle is around 573K, somewhat too high to permit heating by the exhaust gas of a hydrogen-fuelled vehicle. The addition of Cu or Ni has an advantangeous catalytic effect but a Mg—25Ni alloy is more prone to fragmentation. Much useful data has been presented in a number of reports from Brookhaven National Laboratory[133-135] and, in examining the use of magnesium hydride as an automotive fuel, Douglass[136] has included a study of microstructural effects.

Additives for treatment of ferrous metals

The addition of magnesium to molten cast iron, to produce nodular iron, has been a routine practice for over 25 years. The majority of such iron is made by treatment with ferrosilicon containing 5 or 10% magnesium. Within the past decade considerable interest has developed in the use of magnesium for desulphurizing blast furnace iron. The increased scale of operation and limitations on the composition of the additive have required special developments. The problems and possible application of the techniques developed for nodular iron have been summarized.[137] The most popular magnesium desulphurizing practice has been the use of Magcoke by plunging. Magcoke is a composite containing about 45% magnesium and is made by quenching heated coke in molten magnesium. Results of its industrial use in North America, given in a number of papers, have been summarized[138] and an account of European experience has been given.[139] Other forms of composites, with either a ferrous base or using refractories, have been developed.

The use of gas injection techniques for addition of desulphurizing additives is becoming more popular and Koros *et al.*[140] describe one such technique for administering a mixture of lime and magnesium powder. Tokuda and Lu[141] have pre-

sented a bibliography on desulphurization external to the blast furnace.

Primary battery anodes

Magnesium alloys are used in seawater-activated primary batteries, usually with silver chloride cathodes. This system was developed to use the magnesium alloy AZ61 (Mg—6Al—1Zn). Although the theoretical open circuit emf for this type of cell is 2.6 V practical working voltages are of the order of 1.1 V. King[142] has summarized the technical problems inherent in this application and given details of alloys containing lead[143] and thallium[144] which provide improved operating voltages.

Williams[145] and King[142] have described various primary magnesium-containing batteries activated by seawater. Many of these require hard alloys in the form of sheet at 0.25 mm thickness, a difficult rolling operation. Magnesium—carbon dry cells have had considerable military use in the USA and are reported[146] to avoid the need for the refrigerated storage applied to zinc cells in tropical conditions. Efforts to develop civilian applications do not appear to have been successful.

Uses in cathodic protection

Cast magnesium anodes are usually supplied in AZ63 (Mg—6Al—3Zn) and extruded shapes in AZ61 (Mg—6Al—1Zn) alloys. AM503 (1.25% Mn) is used for higher performance requirements. The original work by Robinson and George,[147] reported by Emley, was well founded since no improved compositions have appeared. In the USSR, Kotik and Kirina[148] have repeated this work and come to the same conclusions. Lyublinskii[149] has studied the effect of seawater salinity on the performance of magnesium anodes.

The use of magnesium for cathodic protection has developed to the greatest extent in North America, being widely used in ceramic-lined steel hot-water heaters. Legislation requiring mandatory protection of gas pipelines has also resulted in a significant demand. Scherer[150] has described the advantages of magnesium in this application. Lennox[151] has compared the electrochemical properties of magnesium with those of aluminium and zinc and Peart and Steinicke[152] describe some applications. Woody[153] outlines experience in the use of various protective systems for pipeline distribution. Boening[154] discusses both technical and economic factors in the protection of offshore structures; magnesium gives good protection but is more expensive than aluminium.

Market developments

The use in Grignard reactions is mainly concerned with production of lead additives for petrol and is thus threatened by the trend to lower-lead or lead-free fuel. Some companies prefer to use sodium for this application. The use for production of zirconium and titanium is increasing in line with the growing demand for these metals but much of the magnesium is recycled by electrolysis of the chloride byproduct. For production of nodular iron 12864 t of primary magnesium was consumed in the USA in 1975, about 13% of total consumption. This use has grown rapidly over the past 20 years but indications of market saturation[141] suggest that future growth may be more modest. The use for desulphurization of blast furnace iron is expected to be a major growth area, depending on future economics, and probably requiring a granulated form of magnesium. The main competitive material is calcium carbide.

A number of applications fall marginally between the categories of 'chemical' and 'structural' or may be judged as belonging to both. Substantial quantities of sheet (estimated at 60% of total USA sheet production in 1972) and extrusions are produced, mainly in the USA, for use in printing. The rapid etching characteristics and freedom from environmentally objectionable byproducts enabled the capture of some of the market served by zinc. Both aluminium and plastics are of course also used now. One of the early applications of magnesium, the 2 kg incendiary bomb, made efficient use of the material, first as a container and subsequently for generation of heat. As far as is known this application has now been entirely superseded. Continued ingenuity in the application of specific characteristics of magnesium is illustrated in two proposals for its use as a corrodable link. In one proposal the marker buoy for a lobster pot is submerged with the pot and floats to the surface after an interval dependant on the time taken for the magnesium link to corrode. Submersion of the buoy is a deterrent to poaching and avoids damage from marine traffic. In the other suggestion a spring-loaded door to a lobster pot is held open by a magnesium pin until corrosion fracture permits the door to close. In addition to simplifying construction of the pot this device was claimed to trap lobsters which would escape from conventional pots. Neither suggestion has yet resulted in significant sales of magnesium. It is of course a characteristic of applications in this general area of chemical type uses that no scrap is returned to serve a future market for secondary alloys. The applications for printing are of course salvaged.

ECONOMIC CONSIDERATIONS

Although the consumption of magnesium increased at an average annual rate of 5.8% over the years 1964—74, growth in its use as a structural engineering material has been disappointingly slow. The alloys have been shown to provide satisfactory technical performance in many applications but usage has generally been restricted to those instances where a specific property, usually low density, is of particular importance. Bearing in mind the large-scale usage by Volkswagen, magnesium appears to have been close to more general acceptance as an engineering material for many years, but has never quite managed to achieve such acceptance.

Although considerable effort has been made in promoting wrought forms of magnesium, particularly in the USA, its use in pressure die castings is generally believed to offer the most immediate potential for growth. In this form competitive materials include plastics and zinc, but aluminium dominates the applications most widely suited to magnesium. Hence the economic acceptability of magnesium components for general engineering purposes is related to their price ratio with aluminium. Since material costs represent a high proportion of total costs for pressure die castings, the price ratio of the primary materials is significant. However, while aluminium pressure die castings are mainly produced from secondary alloy, for magnesium the volume of secondary is insignificant in relation to the potential size of the market. The price ratio is therefore essentially between primary magnesium and secondary aluminium.

A qualitative estimate of the relationship between price ratio and development of automotive markets has been given,[155] and is reproduced in Table 6. This estimate has been included in the survey by the American National Materials Advisory Board (NMAB)[31] which points out that the consumption of magnesium pressure die castings in the USA appears to have been stagnant at about 9 Mt/year over a period when the price ratio was around 1.36.

The consumer is, of course, concerned with the cost of the finished component and the cost of conversion from the primary material is important. Campbell[156] has estimated that, for a large volume production, using cold-chamber technology, magnesium castings can be produced at 84% of the cost of their aluminium equivalents when the primary material price ratio is 1.23. In this estimate, lower machining costs for magnesium contribute significantly to the cost advantage. Campbell also points out that the relatively more difficult technology involved with magnesium makes small volume production more expensive than aluminium even at primary material price ratios where high volume production would be more economic. Development of a broad technological base from a multitude of small users is therefore inhibited.

Developments in conversion technology, notably fluxless melting and hot-chamber casting, have been estimated[31] as potentially equivalent to a movement in price ratio of 0.20—0.25, making a significant penetration of the automotive market possible at a ratio of 1.35.

The future needs of the automotive industry for weight saving are likely to be more onerous than those in the past. Holland[115] has estimated that consumption could increase from a present average of 0.11 kg per car in USA, to a value of 4.54—9.08 kg per car by 1985 at a price ratio of 1.5. If the ratio fell to 1.3 consumption could rise to 9.08—18.16 kg per car.

Table 6 Effect of metal price ratio on automotive markets[31]

Magnesium/aluminium price ratio	Effect on magnesium usage
1.1 : 1	Major penetration of automotive market possible
1.25 : 1	Slow penetration of structural markets typical of past 20 years
1.5 : 1	Little acquisition of new markets. Existing applications would decline by product obsolescence, and hence total market would probably decline slowly
2.0 : 1	Loss of existing applications to be feared, with the extent depending on the length of time that this ratio was expected to persist

In the light of the above, one of the most significant events of the past decade has been the increase in the price of magnesium over the period 1973—74 to lead to magnesium/aluminium price ratios of around 2.0. The higher price for magnesium is said to be needed to justify reinvestment in extraction capacity. For the important German pressure die casting market the ratio has been held to around 1.7. As a measure intended to preserve such markets as a base for future expansion the market impact of these ratios seems very much in line with the opinion summarized in Table 6.

There has naturally been much speculation about the cost benefits which will result from current developments in extraction technology. In the USA the NMAB survey[31] concludes that it seems unlikely that future costs will be significantly lower than the traditional factor of 25—40% more than those of aluminium. This conclusion seems to be based more on opinion than on the background data included in the report. A reasonable analysis of the data suggests a future ratio of about 1.1. Stein[115] has attempted a breakdown of extraction costs for magnesium and aluminium to arrive at the selling prices of 96 and 55 cents per lb* said to be needed as reinvestment levels for each metal. With the new magnesium technology he forecasts that the price ratio could drop to between 1.25 and 1.02 in 1985 depending upon unpredictable movements in the cost of alumina. An interesting facet of Stein's estimate is his assumed values for below the line costs (i.e. depreciation, debt servicing, taxes, profits, etc.). For magnesium he assumes these to total 59 cents per lb and for aluminium

---

*1 cent per lb = 0.25p per kg for an exchange rate £1 = \$1.76.

22 cents per lb. The NMAB report suggests that the capital cost for magnesium extraction is not very different from that for aluminium. It is therefore of interest to speculate to what degree the values assumed by Stein are related to the fact that his hypothetical aluminium plants was of 150000 t/year capacity while that for magnesium was of only 20000 t/year capacity

Apart from the market for pressure die castings, public discussion of the economic value of magnesium in specific applications appears to have been confined to the potential market for desulphurizing blast furnace iron. The primary competitive material is calcium carbide and it has been pointed out[137] that consumption of carbide is 7—10 times that of magnesium. The price of primary magnesium is presently near or slightly above the upper limit to be inferred from these consumption factors. The potential new market available for competitively priced magnesium is estimated to be second only to the automotive market. Other surveys of the economic performance of the magnesium industry[157,158] are now mainly of historical interest although they include much useful data.

FUTURE DEVELOPMENTS

The present structure of the magnesium industry is more akin to that of tin than it is to that of aluminium. As with tin, production tends to be in relatively small units (with one exception) and the markets concerned with ancillary metallurgy or chemistry rather than primary engineering. Magnesium differs, however, in that it has significant engineering applications, that it has demonstrated its suitability for use in a wide range of further applications, and that it is now developing technology to make its use economic in such applications. A further significant difference from the tin industry is of course the fact that magnesium is a structural metal for which most countries could be self-dependent, given the requisite energy.

The automotive market clearly has the greatest potential for magnesium as an engineering material. However, its general use, even to the limited extent of 4.5—9 kg per car would imply a growth for the magnesium industry, both in size and type, more closely akin to that of aluminium than that of tin.

It seems most probable that it is this type of development and growth which will become apparent in the next decade. It has been pointed out that the growth in magnesium consumption resembles that of aluminium, with a 40 year time lag. It is perhaps the exponential nature of the early growth years of aluminium which is the most significant aspect of this comparison. No firm data is available concerning the economics of the new chloride process and comparatively little experience of its industrial application has yet been generated. Neither has there been any public discussion of possible changes in marketing philosophy resulting from adoption of a technology which simultaneously generates large quantities of chlorine. Continued development in the use of the 'Magnetherm' process seems probable. Indeed, all the factors which led Campbell to entitle his paper[156] 'An exciting future for magnesium?' are still present, although some seven years have elapsed without obvious impact on the market. It now seems probable that the future will involve much restructuring of primary extraction facilities and long-term development of automotive (and other) usage. The time scale for major developments in the industry is, therefore, probably of the order of 10—20 years, a period in which other, more fundamental, changes to industrial society seem almost certain to supervene.

ACKNOWLEDGMENTS

The author's thanks are due to the Directors of Magnesium Elektron Ltd for permission to publish this review. Figure 3 is reproduced by kind permission of Tonsberg Presstoperi and Norsk Hydro a.s.

REFERENCES

Many important papers on magnesium technology and business development are presented at the annual conferences of the International Magnesium Association. In referring to such papers the name of the Association has been abbreviated to 'IMA' and the year of the conference at which the paper was presented is given. Copies of the Proceedings of IMA conferences may be purchased from:

The Secretaries,
International Magnesium Association,
1406 Third National Building,
Dayton,
Ohio 45402, USA.

1. E. F. Emley: 'Principles of magnesium technology'; 1966, London, Pergamon Press.
2. M. J. Wahll and R. F. Frontani: Battelle Columbus Laboratories, Report No. MCIC-HB-05, February 1975.
3. Kh. L. Strelets: 'Electrolytic production of magnesium'; 1972, (English translation 1977), Springfield, Va, US Department of Commerce, National Technical Information Service.
4. N. Hoy-Petersen: *J. Met.*, 1969, **21**, (4), 43.
5. S. Fougner: 'Light metals 1974', Vol. 2, 515; 1974, AIME.
6. B. A. Overin and B. A. Grubiyan: *Sov. J. Non-Ferrous Met.*, 1975, **16**, (5), 55.
7. N. M. Zuev *et al.*: *Tsvetn. Met.*, 1975, (8), 47.
8. M. L. Rudnitskii *et al.*: *ibid.*, 1976, (1), 53.
9. N. M. Zuev *et al.*: *ibid.*, 1976, (4), 48.
10. M. L. Rudnitskii *et al.*: *ibid.*, 1977, (1), 51.
11. O. A. Lebedev *et al.*: *Sov. J. Non-Ferrous Met.*, 1971, **12**, (6), 44.
12. N. M. Zuev *et al.*: *ibid.*, 1971, **12**, (9), 39.
13. N. A. Frantasev *et al.*: *ibid.*, 1972, **13**, (7), 50.
14. P. A. Donskikh and V. A. Kolesnikov: *ibid.*, 1975, **16**, (4), 61.
15. Yu. V. Dobrunov *et al.*: *Fiz.-Khim. Mekh. Mater.*, 1976, **12**, (3), 98.
16. K. D. Muzhzhavlev *et al.*: *Tsvetn. Met.*, 1976, (5), 48.
17. P. A. Donskikh: *ibid.*, 1977, (1), 49.
18. A. N. Petrunko *et al.*: *Proc. IMA Conf.*, 1977, 53.
19. H. Winterhager *et al.*: *Metall*, 1976, **30**, (6), 547.
20. F. Trocmé: 'Advances in extractive metallurgy and refining', (ed. M. J. Jones), 517; 1972, London, The Institution of Mining and Metallurgy.
21. US Patent Nos. 3579326, 3658509, 3681053, 3698888.
22. *Am. Met. Market*, 16 March 1977.
23. G. Revel *et al.*: *CR Hebd. Séances Acad. Sci.*, 1975, (c), **281**, (24), 1065.
24. N. A. Frantas'ev: *Sov. J. Non-Ferrous Met.*, 1964, **5**, (5), 48.
25. A. E. Andreev *et al.*: *ibid.*, 1964, 5, (10), 47.
26. I. P. Vyatkin and V. A. Kechin: *ibid.*, 1972, **13**, (3), 40.
27. V. A. Kechin *et al.*: *Tsvetn. Met.*, 1976, (4), 57.
28. O. A. Putina *et al.*: *ibid.*, 1976, (10), 43.
29. O. A. Lebedev *et al.*: *ibid.*, 1977, (1), 51.
30. G. Campbell: *Proc. IMA Conf.*, 1977, 1.
31. 'Trends in usage of magnesium', National Materials Advisory Board, Washington DC, 1975.
32. L. F. Lockwood: *Light Met. Age*, 1975, **33**, (5, 6), 18.
33. P. A. Fisher: *Proc. IMA Conf.*, 1976, 40.
34. J. W. Fruehling and J. D. Hanawalt: *Trans. AFS*, 1969, **77**, 159.
35. P. A. Fisher *et al.*: *Foundry*, 1967, **95**, (8), 68.
36. British Patent No. 1465687.
37. W. Unsworth *et al.*: 'QH21 a high performance magnesium casting alloy for aerospace applications', Paper presented at 106th AIME Annual Meeting, Atlanta, Ga, 1977.
38. A. Couture and J. W. Meier: *J. Mater.*, 1966, **1**, (4), 837.
39. B. Lagowski: *Mod. Cast.*, 1967, **52**, (2), 87.
40. B. Lagowski: *J. Test. Eval.*, 1974, **2**, (4), 221.
41. B. Lagowski and J. W. Meier: *Mod. Cast.*, 1968, **53**, (5), 83.
42. B. Lagowski and J. W. Meier: *ibid.*, 1968, **53**, (6), 150.
43. A. F. Crawley and B. Lagowski: *Metall. Trans.*, 1974, **5**, (4), 949.
44. B. Lagowski and A. F. Crawley: *ibid.*, 1976, **7A**, (5), 773.
45. M. Robba: *Alluminio*, 1966, **35**, (2), 59.
46. G. A. Fowler: *Mod. Cast.*, 1967, **51**, (3), 89.
47. B. A. Greshishchev: *Russ. Cast. Prod.*, 1975, **8**, 331.
48. A. Street: *Int. Metall. Rev.*, 1975, **20**, 121.
49. B. Kittilsen: *Giesserei*, 1969, **56**, (20), 596.
50. E. Calamari: *Fonderia Ital.*, 1975, **24**, (11), 353.
51. Trade mark of Norsk Hydro a.s.
52. I. N. Glushchenko *et al.*: *Russ. Cast. Prod.*, Nov. 1971, 468.

53. H. Mezger: *Proc. IMA Conf.*, 1974, 53.
54. A. Hofmaier: *Proc. IMA Conf.*, 1976, 32.
55. A. F. Bauer: *Mod. Met.*, 1973, **29**, (7), 29.
56. C. W. Nelson: *Diecasting Engr.*, 1970, **14**, (4), 12.
57. C. Pinamonti and A. Fiorelli: *ibid.*, 1967, **11**, (6), 20.
58. British Patent No. 1216377.
59. W. Schmidt: *Giesserei*, 1970, **57**, (17), 535.
60. G. Foerster: *Met. Eng. Q.*, 1973, **13**, (1), 19.
61. G. Foerster: *Proc. IMA Conf.*, 1976, 35.
62. 'BNF guide to better magnesium diecasting', 1976, Wantage, Oxon., BNF Metals Technology Centre.
63. I. F. Boltenkov *et al.*: *Russ. Cast. Prod.*, 1976, **3**, 126.
64. British Patent No. 1291553.
65. V. Campagnolo: *Proc. IMA Conf.*, 1975, 45.
66. 'Press cast process for magnesium', *Machinery (Lond.)*, 1970, **116**, 301.
67. F. C. Bennett *et al.*: *Proc. IMA Conf.*, 1977, 23.
68. P. A. Fisher: *Met. Mater.*, 1972, **6**, (2), 88.
69. V. P. Polishchuk *et al.*: *Russ. Cast. Prod.*, 1968, **12**, 536.
70. I. P. Vyatkin *et al.*: *Sov. J. Non-Ferrous Met.*, 1970, **11**, (4), 57.
71. I. P. Vyatkin *et al.*: *ibid.*, 1970, **11**, (10), 53.
72. B. I. Bondarev and V. D. Mishchenko: *ibid.*, 1972, **13**, (4), 72.
73. Z. N. Getselev: *Tsvetn. Met.*, 1976, (10), 56.
74. I. P. Vyatkin and M. V. Chukhrov: *ibid.*, 1976, (12), 43.
75. R. V. London *et al.*: *ASM Trans. Q.*, 1966, **59**, (2), 250.
76. E. Aamland *et al.*: US Bureau of Mines, Report of Investigation RI 7722, 1973.
77. R. J. Jackson and P. D. Frost: 'Properties and current applications of magnesium—lithium alloys', NASA SP—5068, 1967. Supt. of Documents, US Govt Printing Office, Washington DC.
78. I. P. Vyatkin *et al.*: *Sov. J. Non-Ferrous Met.*, 1972, **13**, (6), 48.
79. M. E. Dritz *et al.*: *Russ. Metall.*, 1974, (5), 174.
80. V. V. Borisov *et al.*: *Tekh. Legk. Splavov. Nauchno-Tekh. Byul. Vilso*, 1975, (11), 76.
81. R. S. Busk: *Mod. Met.*, 1968, **24**, (6), 43.
82. M. E. Dritz *et al.*: *Russ. Metall.*, 1971, (6), 138.
83. D. J. Stratford: *J. Inst. Met.*, 1968, **96**, (3), 87.
84. British Patent No. 1078629.
85. British Patent No. 1063276.
86. British Patent No. 1094237.
87. British Patent No. 1137613.
88. D. G. Attwood and P. M. Hazzledine: Fourth Int. Conf. Strength of Metals and Alloys, Nancy, France. 1976, i, 413. Laboratoire de Physique du Solide, Parc de Saurupt, 54042, Nancy.
89. R. Z. Valiev and O. A. Kaibyshev: *Izv. Akad. Nauk SSSR*, 1976, (1), 82.
90. S. Isserow and F. J. Rizzitano: *Int. J. Powder Metall. Powder Technol.*, 1974, **10**, (3), 217.
91. G. S. Foerster: *Met. Eng. Q.*, 1972, **12**, (1), 22.
92. R. S. Busk and T. E. Leontis: *Trans. AIME*, 1950, **188**, 297.
93. I. C. Huseby *et al.*: *Int. J. Powder Metall.*, 1973, **9**, (2), 91.
94. A. P. Levitt *et al.*: *Metall. Trans.*, 1972, **3**, (9), 2455.
95. M. H. Richman *et al.*: *Metallography*, 1973, **6**, (6), 497.
96. A. A. Baker: *Mater. Sci. Eng.*, 1975, **17**, (2), 177.
97. I. Ahmed and J. M. Barranco: *Metall. Trans.*, 1973, **4**, (3), 793.
98. K. G. Adamson *et al.*: D. Mat. Report No. 192, 1973, Procurement Executive, Ministry of Defence, London.
99. A. E. Yauw and H. Schlick: *Plating*, 1968, **55**, (12), 1295.
100. J. F. King *et al.*: D. Mat. Report No. 193, 1973, Procurement Executive, Ministry of Defence, London.
101. K. G. Adamson *et al.*: D. Mat. Report No. 196, 1973, Procurement Executive, Ministry of Defence, London.
102. G. R. Kotler *et al.*: *Proc. IMA Conf.*, 1976, 45.
103. L. F. Spencer:, *Met. Finish.*, Part 1, 1970, **68**, (12), 32; Part 2, 1971, **69**, (2), 43.
104. A. L. Olsen: *Proc. IMA Conf.*, 1977, 36.
105. W. H. Safranek *et al.*: *ibid.*, 39.
106. I. L. Roikh *et al.*: *Prot. Met. (USSR)*, 1972, **8**, (1), 97.
107. T. Yamada *et al.*: *J. Met. Finish. Soc. Jpn*, 1976, **27**, (2), 85.
108. V. I. Ryazantsev and V. A. Fedoseev: *Avtom. Svarka.*, 1975, (3), 18.
109. V. Ya. Ivanov *et al.*: *Weld. Prod. (USSR)*, 1975, **22**, (7), 33.
110. A. A. Osokin and V. Ya. Umnaev: *Svar. Proizvod.*, 1976, **5**, 48.
111. S. K. Dutta *et al.*: Proc. Int. Conf. on 'Welding of castings', Bradford, England, 1976, Vol 1, 129; 1977, Abington, The Welding Institute.
112. G. M. Cull: *Met. Constr.*, 1969, **1**, (8), 378.
113. V. Ya. Ivanov and A. I. Zhulin: *Svar. Proizvod.*, 1976, **1**, 15.
114. British Patent No. 1331332.
115. Int. Conf. on 'Energy conservation in production and utilisation of magnesium', 1977. Massachusetts Institute of Technology, Cambridge, Mass.
116. G. B. Kenney and J. P. Clark: *ibid.*, 132.
117. A. Bauer: *Proc. IMA Conf.*, 1974, 37.
118. G. Campbell: *Proc. IMA Conf.*, 1977, 1.
119. 'Minerals Yearbook'; 1955, 1965, 1974; Washington DC, US Department of Interior, Bureau of Mines.
120. J. E. Harris: *Microstructural Sci.*, 1976, **4**, 45.
121. J. Scherer: *Proc. IMA Conf.*, 1974, 26.
122. J. A. Bolstad: *Proc. IMA Conf.*, 1975, 40.
123. D. James: *Proc. IMA Conf.*, 1974, 23.
124. M. L. Redhair: *Proc. IMA Conf.*, 1977, 13.
125. P. Lugagne: *ibid.*, 7.
126. A. L. Hock: *Chem. Ind.*, 1971, **3**, 78.
127. B. B. Fulford: *Powder. Metall.*, 1976, **19**, (2), 73.

128. E. I. Popov: *Sov. Powder. Metall. Ceram.*, 1974, **13**, (7), 594.

129. I. A. Barranik *et al.*: *Tsvetn. Met.*, 1976, (4), 66.

130. N. A. Voronova: *Proc. IMA Conf.*, 1977, 44.

131. L. I. Kvater and I. V. Frishberg: *Russ. Metall.*, 1973, (6), 54.

132. C. G. Granqvist and R. A. Buhrman: *J. Appl. Phys.*, 1976, **47**, (5), 2200.

133. J. J. Reilly and R. H. Wiswall: *Inorg. Chem.*, 1968, **7**, 2254.

134. J. J. Reilly and R. H. Wiswall: *ibid.*, 1970, **9**, 1678.

135. J. J. Reilly and R. H. Wiswall: *ibid.*, 1974, **13**, 1.

136. D. L. Douglass: *Metall. Trans.*, 1975, **6A**, (12), 2179.

137. P. A. Fisher: *Proc. IMA Conf.*, 1974, 60.

138. P. A. Fisher: *Met. Mater.*, 1973, **7**, (11), 501.

139. H. Jaunich: *Proc. IMA Conf.*, 1974, 45.

140. P. J. Koros *et al.*: *Proc. IMA Conf.*, 1976, 53.

141. M. Tokuda and W.-K. Lu: Symposium on 'External desulphurisation of hot metal', 1975, McMaster University, Hamilton, Ontario.

142. J. King: *Proc. IMA Conf.*, 1973, 12.

143. US Patent No. 3288649.

144. British Patent No. 1241223.

145. J. Williams: *Engineering*, 1970, **209**, 60.

146. R. F. Udell: *Proc. IMA Conf.*, 1973, 7.

147. H. A. Robinson and P. F. George: *Corrosion*, 1954, 10.

148. V. G. Kotik and L. F. Kirina: *Tr. Vses. Nauchno—Issled. Inst. Stroit. Magistral'n Truboprovod.*, 1974, **30**, 159.

149. E. Ya. Lyublinskii: *Prot. Met.* (*USSR*), 1973, **9**, (2), 194.

150. J. G. Scherer: *Proc. IMA Conf.*, 1973, 4.

151. T. J. Lennox: Proc. Third Int. Congress on 'Marine corrosion and fouling', 1973, 176; Evanston, Ill., Northwestern University Press.

152. J. F. Peart and A. R. Steinicke: *Aust. Corros. Eng.*, 1973, **17**, (8), 17.

153. C. L. Woody: Corrosion 76. Int. Corrosion Forum, National Association of Corrosion Engineers, Houston, Texas, 1976, Paper No. 160.

154. D. E. Boening: 'Proc. 8th offshore technology conf. 1976', Vol. 3, 101; Houston, Texas, AIME.

155. G. Campbell and P. A. Fisher: *Eng. Min. J.*, March 1975, 184.

156. G. Campbell: *Met. Mater.*, 1971, **5**, (5), 169.

157. 'Economic analysis of the magnesium industry'; Charles River Associates Inc., Report PB 176473, 1967, Clearinghouse, US Dept of Commerce, Springfield, Va.

158. 'Roskill report on magnesium'; 1973, London, Roskill Information Services Ltd.

SOME THERMAL AND MECHANICAL PROPERTIES
OF COMPACTED GRAPHITE IRON

by

R. W. Monroe
C. E. Bates

of

Southern Research Institute
Birmingham, Alabama

ABSTRACT

Compacted graphite cast irons have been recognized for twenty years or more but have received considerably more attention in the past five years because of the special combination of properties they possess. This paper summarizes the available property data and presents the results of some recent research on both thermal and mechanical properties of these irons in relation to the properties of conventional high strength gray iron and unalloyed ductile iron.

It was found that the tensile strength and modulus of CG/V irons are intermediate to gray and ductile irons with the same alloy content. Higher nodularity values increase the modulus and tensile strength. The thermal expansion, heat capacity, and enthalpy of all the near-eutectic cast irons were independent of graphite type. The thermal conductivity was dependent on the graphite shape.

INTRODUCTION

Compacted graphite iron has received considerable attention
as an engineering material in the past few years,even though
it has been recognized for 20 years or more.(1,2)* Compacted graphite irons are irons that exhibit a graphite morphology intermediate to the interconnected, flake graphite in
gray cast iron and the spheroidal graphite in ductile cast
iron. Compacted graphite, when observed in two dimensions
under an optical microscope, appears as thickened flakes
with rounded ends. Three-dimensional views of compacted
graphite, as revealed by SEM examination of deeply etched
samples, reveal interconnected graphite similar to flakes
but more compacted in nature. The terms "vermicular graphite" or "compacted graphite" have been generally accepted
for the intermediate graphite shown in Figure 1(b).(3)

LITERATURE REVIEW

Compacted graphite irons may be produced by several modification processes, including -

1. desulfurization of base metal to low values of around
    0.002%, followed by rapid solidification,

2. treatment of iron with controlled amounts of elements
    which promote graphite spheroidization, e.g., Mg or Ce,
    and

3. treatment of iron with both spheroidizing and anti-
    spheroidizing elements, e.g., magnesium, or magnesium
    and cerium, with titanium.(4)

The influence of various elements on graphite morphology in
the Fe-C-Si system is quite diverse with the absolute influence of any element being dependent on the presence of other
elements and the cooling rate of the iron.

GRAPHITE MORPHOLOGY AND TERMINOLOGY

Several efforts have been made to provide classifications
for the diverse forms of graphite found in castings. ASTM
has one classification in which vermicular or compacted
shapes are designated as ASTM P-graphite.(5) Another classification proposed by L. Sofroni et al distinguishes intermediate forms of graphite based on the length, thickness,
and length-to-thickness ratio of the graphite.(4) Still
another classification, proposed by the Ductile Iron Society

* Numbers in parentheses refer to items in "References."

includes nine types of graphite.(6)   Most of the published
properties of compacted graphite irons appear to have contained Sofroni Type III graphite, which is similar to ASTM
Type P and DIS Type VII.

MECHANICAL PROPERTIES

Typical properties of several quasi-flake irons produced by
Morrogh are given in Table 1.   The tensile strength ranged
from about 55,000 to 76,000 psi at hardness values of 179-
238 BHN.(1,2)

Experimental heats of iron produced by Sofroni were treated
with an iron-silicon-calcium-aluminum-titanium-magnesium
alloy, and Table 2 contains some property data.   The tensile strength and elongation properties of these irons approach those obtainable in high strength gray irons possessing acicular matrix structures.   The better irons exhibited
tensile strengths of 51-73,000 psi at hardness of 150-240 BHN.

Additional property data on CG/V irons at two carbon equivalent values cast in section thicknesses ranging from 30 to
200 mm (1.2 to 7.9 in.) is shown in Table 3.(7)   A single
treatment alloy containing magnesium, titanium, and cerium
was used to prepare these irons.   The nodularity was about
10% and the fracture stress ranged from about 46,000 psi in
the thickest sections to 65,000 psi in a 1.2 in. diameter
bar at a C.E. of 4.0-4.3%.

Property data on compacted graphite irons produced with a
FeSiMgTi alloy and cast into 1.2-in. dia round bars are
presented in Table 4. (7)   Test specimens were machined
from the 1.2 in. dia bars, which had been cast into sand
molds at various time intervals after treatment.   These
irons exhibited tensile strength values of 58,000-69,000 psi
at Brinell hardness values of 180-200.   The higher strengths
were associated with higher nodularity values in the irons.

Table 5 presents some European data on 25 mm dia bars from
irons with flake, compacted, and nodular graphite.(8)   These
irons had approximately the same composition but ranged in
strength from about 16,000 to 63,000 psi, depending on the
graphite shape.   The elongation value similarly increased
with nodularity from nil in gray iron to about 25% in the
spheroidal graphitic iron.

MODULUS OF ELASTICITY

The modulus of elasticity of CG/V iron in tension and compression is compared to the modulus of nodular and flake

graphite iron in Figure 2. The modulus was slightly affected by section size and carbon equivalent, with lower values
being found in larger sections at higher carbon equivalent
values. Compacted graphite and nodular irons maintain their
modulus to higher tensile stresses compared to flake graphite iron.(9)

Table 6 presents additional modulus data from several
sources for flake, compacted, and nodular graphite irons in
compression.(3,7,9-11)

THERMAL CONDUCTIVITY

The thermal conductivity of all cast irons appears to be
controlled by the form, amount, and distribution of graphite.
The interconnected nature of the graphite contained in compacted graphite iron enhances its thermal conductivity when
compared with ductile iron.

Table 7 presents some published thermal conductivity data
for cast irons with different graphite structures.(3,7,10-
13) The conductivity of compacted graphite iron was generally lower than that of flake graphite iron. However, for
near eutectic compositions the conductivity was comparable
to that of low carbon equivalent, high strength gray cast
irons. Increased carbon equivalent, flake length, or ferrite content of the matrix in compacted graphite irons with
a given nodularity increased the thermal conductivity of the
metal.

While some property data are available in the literature on
compacted graphite irons, a program to provide a set of
thermal and mechanical properties to define CG/V iron properties in relation to gray and ductile iron properties was
conducted. The experimental portion of this program was
undertaken to produce and evaluate several irons.

PROCEDURES

The CG/V iron production procedure consisted of tapping iron
from an induction furnace onto ladle additions of FeSiMgTi
alloy. Test bars of one and two-inch diameters were poured
to provide metal stock for subsequent mechanical and thermal
property determinations. The bars were cast in no-bake
bonded molds.

MECHANICAL AND PHYSICAL PROPERTIES

A series of cast irons with a range of graphite structures
were produced and evaluated to determine the mechanical,
thermal, and microstructural properties. The mechanical

properties evaluated included modulus of elasticity, yield strength, ultimate tensile strength, elongation, reduction in area, and Brinell hardness (converted from Rockwell B and C). Tensile properties were determined at four temperatures: room temperature (RT), 425°F(218°C), 850°F(454°C) and 1300°F (704°C).

Tables 8, 9 and 10 list the composition and mechanical properties for eight cast irons possessing a variety of graphite structures. At room temperature, tensile strengths ranged from 33,000 to 58,000 psi for gray cast irons with Brinell hardnesses of 195 to 245 and elastic moduli of 11 to 15 x 10^6 psi. Compacted graphite irons with nodularities ranging from 14 to 51% exhibited tensile strengths of 50,000 to 87,000 psi, Brinell hardnesses of 170 to 205, and elastic moduli of 17 to 23 x 10^6 psi. A slightly alloyed nodular iron exhibited a tensile strength of 100,000, a Brinell hardness of 220, and an elastic modulus of about 22 x 10^6 psi. The effect of nodularity on ultimate tensile strength at room temperature is illustrated in Figure 3. Ultimate tensile strength, elongation and tensile reduction of area increased as the nodularity increased.

The tensile strengths of the irons decreased at elevated temperatures, but the prior order of strengths was maintained until near 1300°F, where strength values converge to the range of 10-14,000 psi. The ranking of modulus values among the irons generally followed the same order as observed for tensile strength.

THERMAL CONDUCTIVITY, EXPANSION,
ENTHALPY AND HEAT CAPACITY

Thermal conductivity, thermal expansion, enthalpy, and heat capacity were measured for each iron over a range of temperatures from approximately 200°F(90°C) to 1400°F(760°C). No significant difference in thermal expansion, enthalpy, or heat capacity was found among the different irons across the entire temperature range examined. This is a reasonable result since there was no large variation in chemical composition. The single curve in Figure 4 represents the enthalpy of all irons investigated as a function of temperature. The equation describing the curve was analytically determined to be -

$$H = 0.5721 \times 10^{-1} \, T + 0.4444 \times 10^{-4} \, T^2 + 0.4643 \times 10^4 \, T^{-1} - 0.5211 \times 10^2$$

where -

"H" is the enthalpy above the reference temperature of 85°F, and "T" is the temperature of interest, in degrees Rankine. The heat capacity or rate of change in enthalpy with temperature is derived from the enthalpy. The heat capacity of these irons at temperatures from 200°f to 1033°F is described by -

$$C_p = 0.5721 \times 10^{-1} + 0.8888 \times 10^{-4}\ T - 0.6215 \times 10^{4}\ T^{-2}$$

where -

"C_p" is the heat capacity at constant pressure for the temperature of interest and "T" is expressed in degrees Rankine.

The thermal expansion data for all irons is shown in Figure 5. The curve represents the total expansion above room temperature as a function of temperature. The rate of thermal expansion at any particular temperature can be obtained from the slope of a tangent line to the curve at the point.

Thermal conductivity, unlike the other thermal properties, is structure sensitive and varies with the matrix and graphite form. Ferritic matrices impart higher thermal conductivities than pearlitic. The most dramatic influence results from·the shape of the graphite; the thermal conductivity of the iron increases as the graphite becomes less nodular and more elongated. The effect of nodularity on thermal conductivity at 200°F is illustrated in Figure 6. There was a smooth decrease in conductivity as the nodularity increased.

CONCLUSIONS

This project was undertaken to compile available data on compacted graphite iron production and conduct an experimental program to develop a set of thermal and mechanical and properties of CG/V irons in relation to competitive ductile and gray cast irons. The conclusions drawn from this study are:

1.   The tensile strength and modulus of CG/V irons are intermediate to gray and ductile irons containing approximately the same concentrations of alloy elements. The ultimate tensile strength and modulus of CG/V irons increase with nodularity.

2.   The thermal expansion, heat capacity, and enthalpy of all the gray, ductile, and CG/V irons included in this study were essentially equal.

3.   The thermal conductivity of CG/V irons decreased as the nodularity increased. The 200°F thermal

conductivity of near eutectic CG/V irons was approximately 35% higher than near eutectic ductile irons. At temperatures approaching 1300°F, the conductivities of all irons examined converged to a value of about 210 Btu in./hr ft^2 °F.

REFERENCES

1.   H. Morrogh and W. J. Williams, "The Production of Nodular Graphite Structures in Cast Iron," <u>Journal of the Iron and Steel Institute</u>, March 1948, pp 306-322.

2.   H. Morrogh, "Production of Nodular Graphite Structures in Gray Cast Irons," <u>AFS Transactions</u>, <u>56</u>, 1948, pp 72-90.

3.   J. Sissener, W. Thury, R. Hummer, E. Nechtelberger, "Der Einsatz des Werkstoffes Gußeisen mit Vermiculargraphit aus technischer und wirtschaftlicher Sicht," Giesserei-Praxis Nr. 22/1972, pp 396-404./"Cast Iron With Vermicular Graphite," <u>AFS International Cast Metals Journal</u>, <u>8</u>, December 1974, pp 178-181.

4.   L. Sofroni, I. Riposan and I. Chira, "Some Considerations on the Crystallisation Features of Cast Irons with Intermediary-Shaped Graphite (Vermicular Type)," The Metallurgy of Cast Iron, Georgi Publishing Company, St. Saphorin, Switzerland, 1975, pp 179-195.

5.   W. Rhury, "Effect of Small Ti and Cu Additions on Mg-Free Nodular Graphite Cast Iron," <u>AFS Cast Metals Research Journal</u>, December 1970, pp 163-166.

6.   C. K. Donoho, "Classification of Ductile Iron, <u>Modern Castings</u>, July 1961, pp 65-71.

7.   G. F. Sergeant and E. R. Evans, "The Production and Properties of Compacted Graphite Irons," <u>BCIRA Journal</u>, GFS/BER/I.45/4, March 6, 1978.

8.   N. P. Lillybeck, M. G. McKimpson, R. T. Wimber, and D. W. Donis, "An Evaluation of Graphite-Shape Consistency in Compact-Flake-Graphite Iron from Mechanical Property Data," <u>AFS Transactions</u>, <u>85</u>, 1977, pp 129-132.

9.   M. J. Lalich and J. J. LaPresta, "Progress in the Use of Compacted Graphite Cast Irons for Engineering Applications," Foote Mineral Company, Exton, Pa.

10.   R. D. Schelleng, "Effect of Certain Elements on the Form of Graphite in Cast Iron," International Nickel Company Technical Paper 486C dated March 29, 1966.

11.   K. R. Hiemer, "Gu eisen mit vermiculargraphit und seine V Verabeitung zu Zylinderdeckeln Fur Hochleistungs-Dieselmotoren," _Geisserei_ 63 (1967) Nr. 10, pp 285-291.

12.   E. R. Evans, J. V. Dawson, and M. J. Lalich, "Compacted Graphite Cast Irons and Their Production by a Single Alloy Addition," _AFS Transactions_, _84_, 1976, pp 215-220.

13.   M. J. Lalich, "Effects of Rare Earths on Structure and Properties of Cast Iron," _Foundry Management and Technology_, March 1978, pp 118-119.

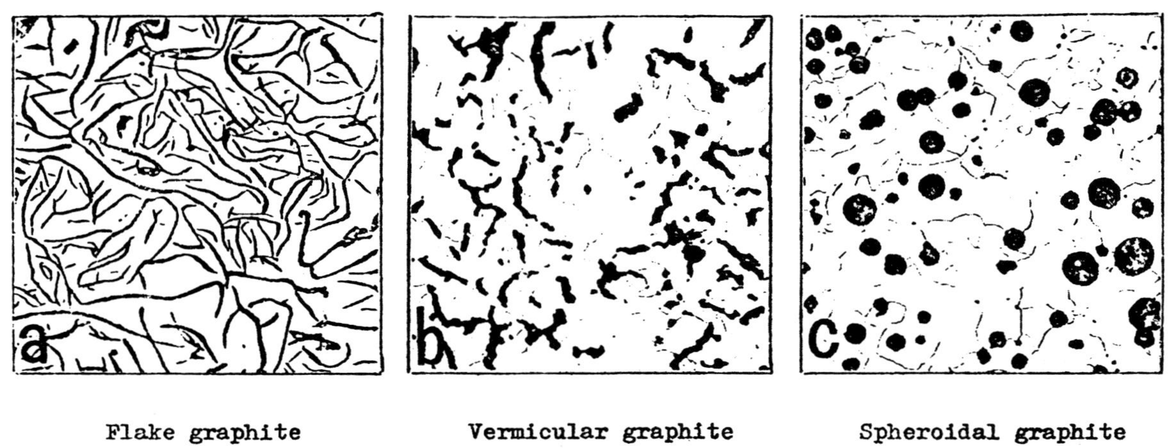

Figure 1.   Typical graphite shapes in gray, compacted and
ductile irons.(3)

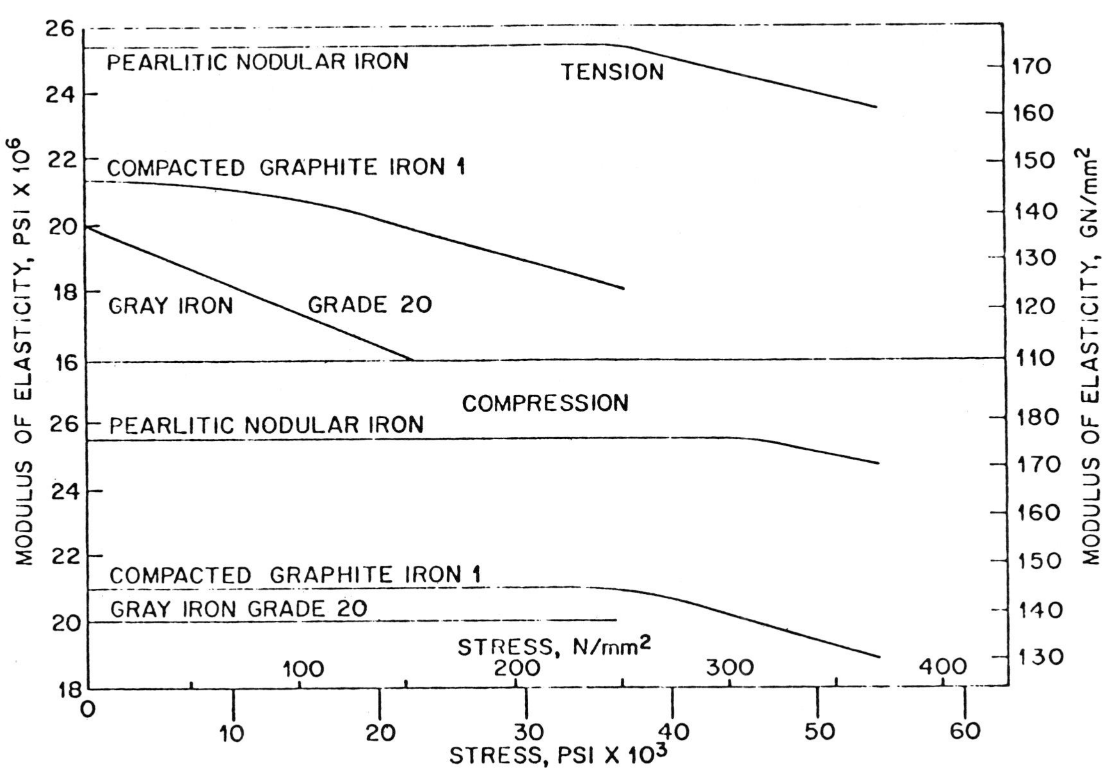

Figure 2. Modulus of elasticity for cast irons as a function of
stress in compression and tension. (9)

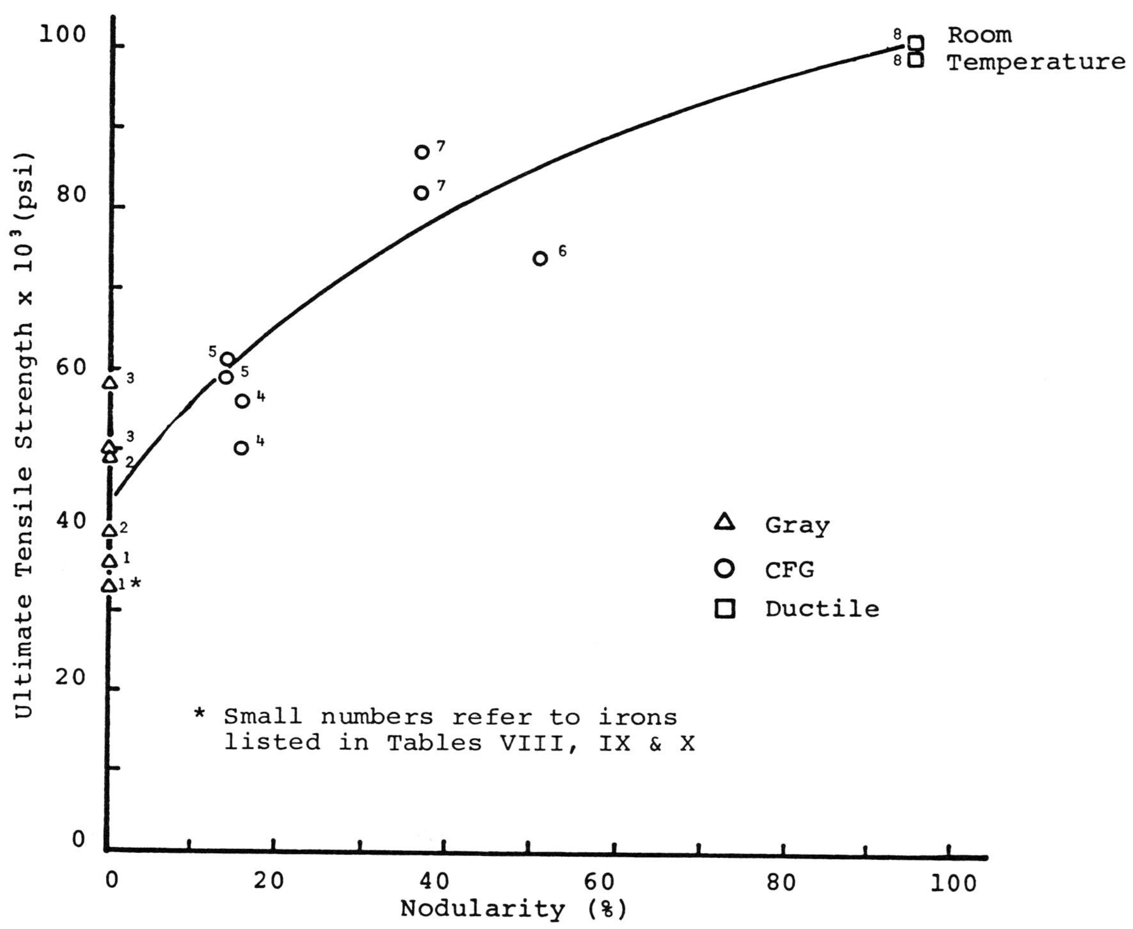

Figure 3. Effect of nodularity on the ultimate strength
of experimental irons.

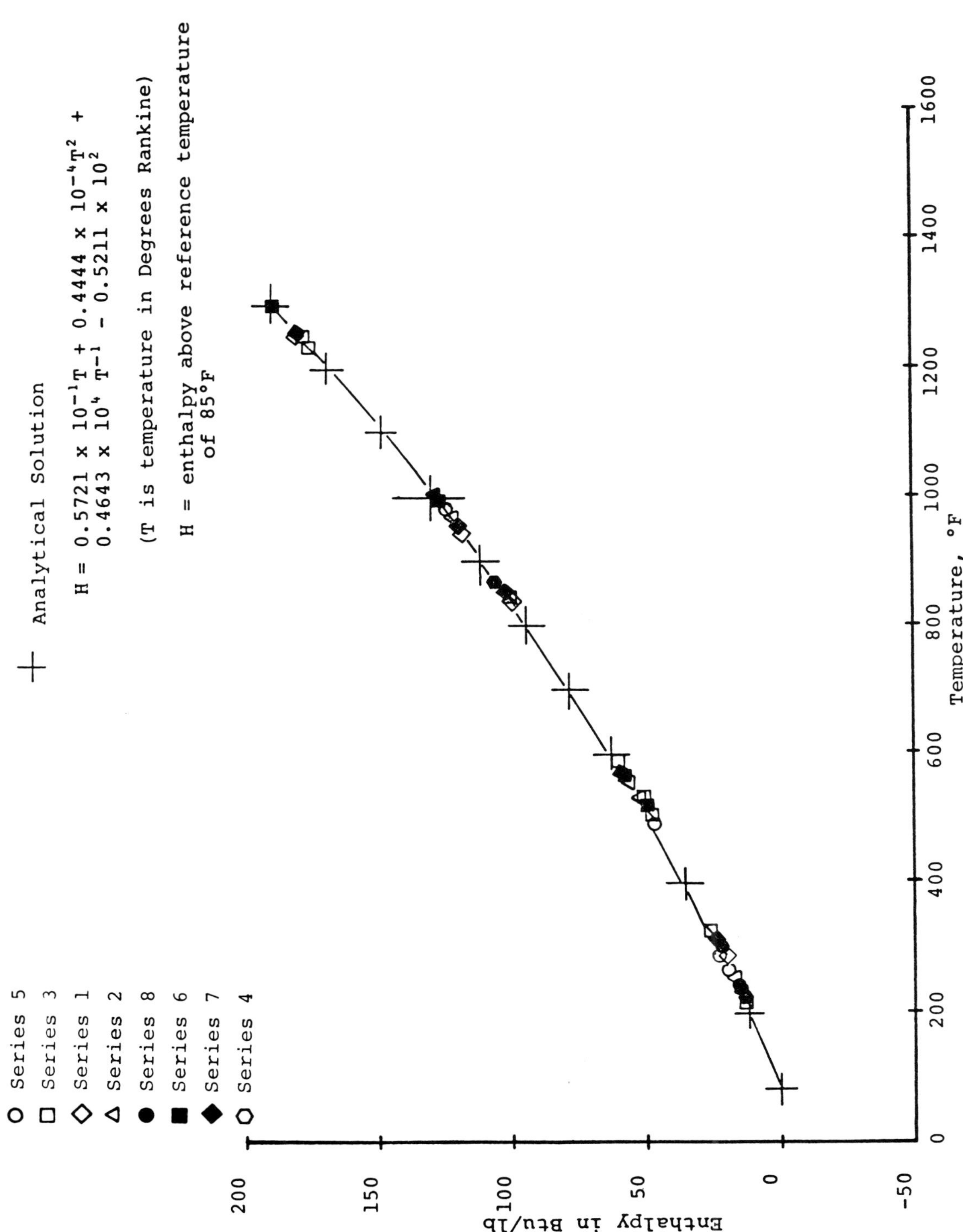

Figure 4. Enthalpy of compacted, gray and ductile irons.

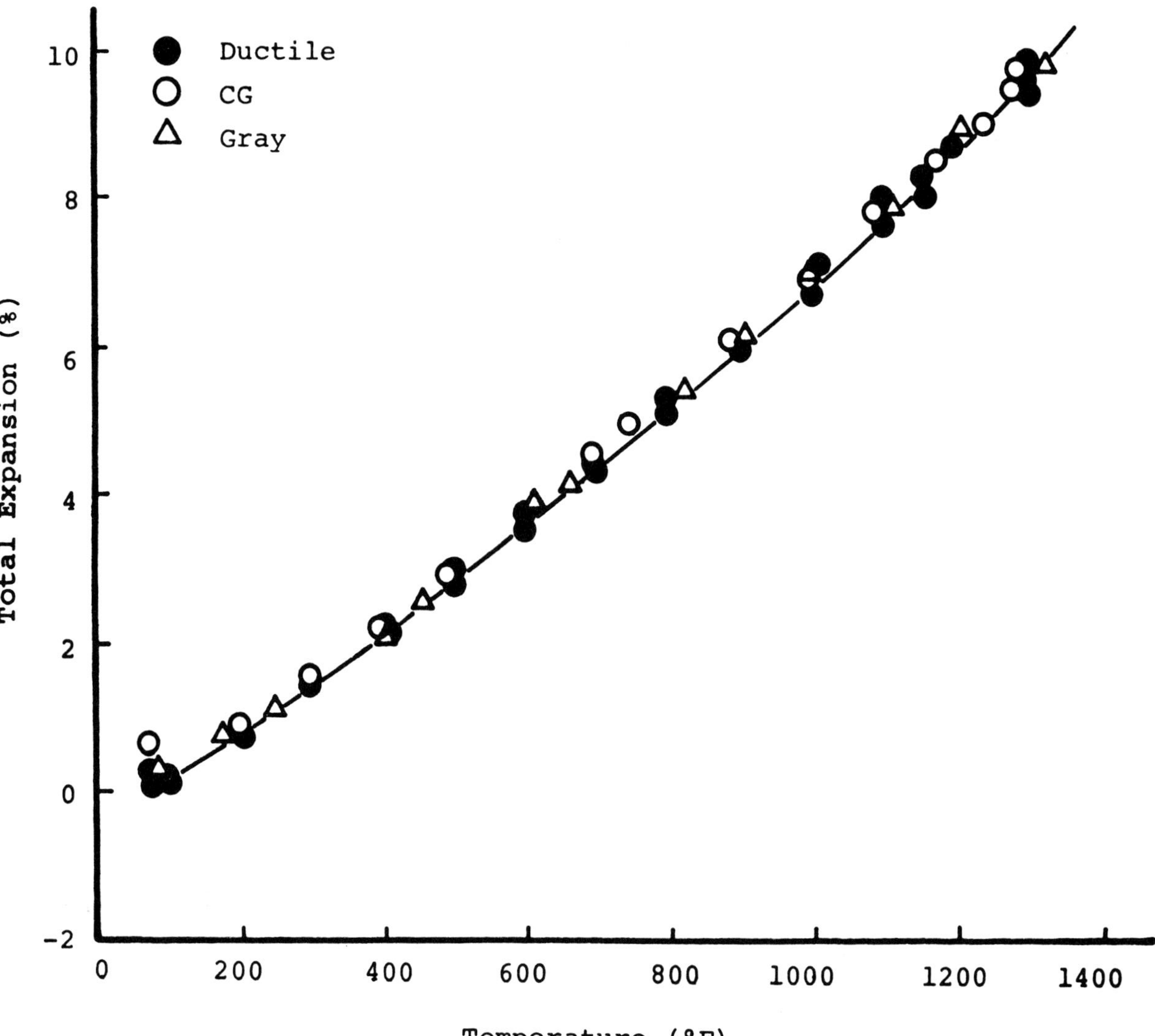

Figure 5. Thermal expansion of experimental cast irons.

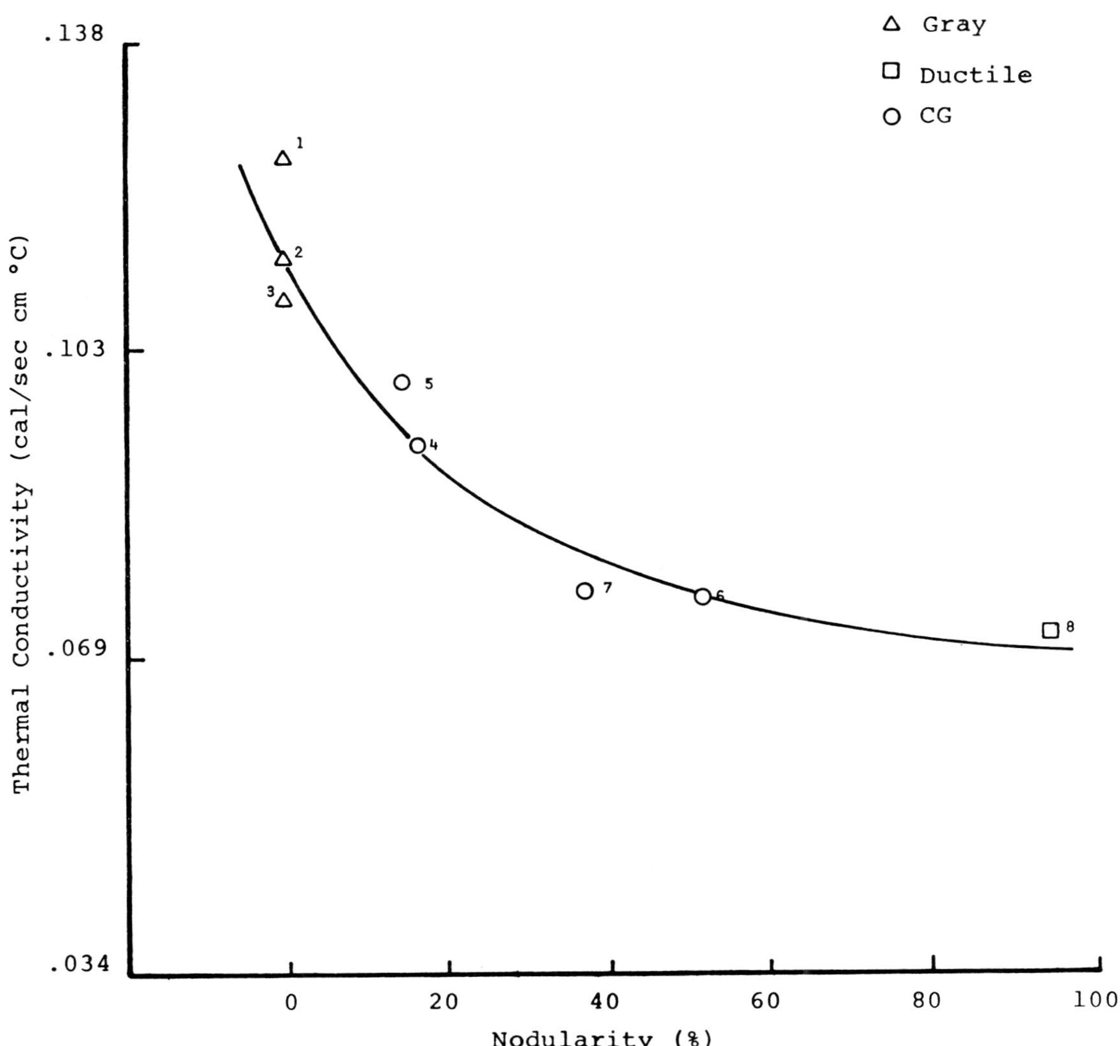

Figure 6. Effect of nodularity on 200°F thermal conductivity of compacted, gray, and ductile irons.

Table 1

Properties of Several "Quasi-flake" Irons
Produced Using Mischmetal (1, 2)

Casting Set No.	Bar Size, In.	Tensile Strength, psi	Izod Impact Strength, ft-lb	Deflection* In.	Brinell Hardness Number	Ce, %	Casting Set No.	Iron Chemistry				
								C	Si	Mn	S	P
1	1.6	54,800	32	0.51	179		1	3.90	2.96	0.51	0.006	0.024
	1.2	62,700	46	0.68	189	0.016						
	0.875	65,100	76	0.39	213							
	0.6	75,900	–	0.35	239							
2	1.2	58,200	54	0.60	221	0.040	2	3.67	2.67	0.86	0.005	0.051
3	3	40,100	–	0.21	167	0.040	3	3.79	2.80	0.53	0.028	0.015
		50,200	–	0.35	176	0.053						
3		53,800	–	0.34	181	0.072						

* Tested on 18-in. centers.

Table 2

Characteristic Types of Vermicular Graphite After Sofroni (4)

Type of Vermicular Graphite	Maximal Lgth, μm*	Maximal Thickness, μm	Length-to-Thickness Ratio	Tensile Strength 1000 lb/in^2	Elongation (%)	Brinell Hardness Number
I	20	10	2-4	43-65	2-5	150-240
II	150	50	2-5	51-73	3-9	150-240
III	150	20	3-10	43-65	1.0-3.5	150-250

* μm = 10^{-6} meter

Table 3

Properties of Compacted Graphite Irons (7)

Carbon Equivalent	Matrix Structure	Cast Section Size, In.			
		1.2	2.1	1.75 Keel	7.9
		Tensile Strength, 1000 psi			
4.3	Ferrite	52.9	47.1	44.9	40.6
4.0		58.0	50.7	47.8	43.5
4.3	Pearlite	63.8	53.6	52.2	46.4
4.0		66.7	56.5	55.8	49.3
		0.2% Yield Strength, 1000 psi			
4.3	Ferrite	33.7	33.3	30.4	27.5
4.0		41.3	36.2	34.1	31.9
4.3	Pearlite	44.2	39.1	34.8	30.4
4.0		49.3	40.6	38.4	33.3
		Elongation, %			
4.3	Ferrite	4.5	4.5	5.5	4.5
4.0		2.0	2.5	3.0	3.0
4.3	Pearlite	1.5	1.0	2.0	2.0
4.0		1.0	1.0	2.0	1.5
		Hardness, BHN			
4.3	Ferrite	140-155	135-150	120-130	120-130
4.0		180-205	170-180	135-145	130-140
4.3	Pearlite	225-245	175-245	195-205	160-180
4.0		210-260	175-240	195-215	160-190
		Modulus, 10^6 psi			
4.3	Ferrite	23.0	23.0	23.0	23.0
4.0		24.5	23.9	23.0	23.9
4.3	Pearlite	23.9	23.1	23.9	22.5
4.0		23.9	-	23.0	22.5

Table 4

Some Properties of Compacted Graphite Iron(8)

Test Bar	Pour Time, min[a]	Nodularity[b] (%)	Pearlite[b] (%)	Carbon[c] Equivalent (%)	Element (%)				Tensile Strength (psi)	BHN
					Mg	Ti	C	Si		
8	3½	10	40	3.85	0.012	0.06	3.18	2.68	60,871	194
9	3½	10	40	3.85	0.012	0.06	3.18	2.68	58,070	191
10	7	10	30	3.90	0.011	0.09	3.22	2.70	58,812	194
11	10	5	40	3.88	0.011	0.07	3.21	2.68	58,899	184
12	10	5	40	3.88	0.011	0.07	3.21	2.68	58,812	193
13	2	50	30	3.94	0.023	0.10	3.21	2.94	69,266	186
14	2	50	30	3.94	0.023	0.10	3.21	2.94	69,121	187
15	6	15	40	3.91	0.021	0.11	3.17	2.96	66,076	200
16	6	15	40	3.91	0.021	0.11	3.17	2.96	65,105	198
17	10	5	30	3.93	0.022	0.12	3.21	2.89	63,017	202
18	10	5	30	3.93	0.022	0.12	3.21	2.89	63,539	198

a Time bar poured after treatment
b Visual estimates from center of test bar
c CE = %C + (%Si)/4 + (%P)/2

Table 5

Comparison of the Properties of Cast Irons
with Flake, Compacted and Nodular Graphite
(As-cast 25 mm dia bars) (3)

	Flake	Compacted	Nodular
Matrix (%)			
Ferrite	0	>95	100
Pearlite	100	< 5	0
Tensile Strength (psi)	15,900	48,800	63,600
0.2% Yield Str (psi)	n.d.	37,300	41,400
Elongation, %	n.d.	6.7	25.3
Hardness, BHN	156	150	159
Modulus of Elasticity x 10^- psi	14,060	22,960	25,370
Composition, %			
C	3.61	3.61	3.56
Si	2.49	2.54	2.72
Mn	0.05	0.05	0.05
C.E.	4.44	4.46	4.47
Graphite Type, %			
K	0	<5	80
L	0	<5	20
P	0	95	0
A	100	0	0

Table 6

Elastic Properties of CG/V Iron

Reference	Modulus of Elasticity, $\times 10^6$ psi	
10	20.0-21.4	Tension
9	21	Tension
3	22.9	Tension
11	20.3-23.2	Tension
7	22.5-24.6	Resonant frequency

Table 7

Thermal Conductivity Properties for Various Cast Irons

Ref.	Graphite Structure	C.E.	Cast Section Size	Thermal Conductivity, cal/sec cm °C					
				R.T.	100	200	300	400	500
10	Compacted	4.09	–	∿0.10	–	–	–	–	–
	Nodular	4.16	–	∿0.089	–	–	–	–	–
	Flake	4.33	–	∿0.115	–	–	–	–	–
7	Flake	3.8	–	–	0.120	0.117	0.108	0.100	0.092
	Flake	4.0	–	–	0.128	0.121	0.113	0.103	0.092
	Nodular	4.2	–	–	0.077	0.083	0.079	0.075	0.070
	Compacted	3.9	–	–	0.091	0.098	0.094	0.089	0.084
	Compacted	4.1	–	–	0.104	0.103	0.096	0.009	0.084
	Compacted	4.2	–	–	0.098	0.104	0.098	0.092	0.086
11	Flake, DIN 1691	–	–	0.12	0.16	–	–	–	–
	Nodular, DIN 1693	–	–	0.06	0.10	–	–	–	–
	Compacted, GGV30	–	–	0.10	0.12	–	–	–	–
3	Flake	–	30 mm	0.100	–	–	–	–	–
	Flake	–	25 mm	0.101	–	–	–	–	–
	Compacted	–	25 mm	0.085	–	–	–	–	–
	Nodular	–	25 mm	0.078	–	–	–	–	–
12	ASTM A48-74 Flake Class 25	4.4	Medium-Light	0.125	–	–	–	–	–
		–	Heavy	0.120-0.135	–	–	–	–	–
	ASTM A48-74 Flake Class 45	3.6	Medium-Light	0.113	–	–	–	–	–
		–	Heavy	–	–	–	–	–	–
	Nodular	4.2	Medium-Light	0.080	–	–	–	–	–
		–	Heavy	0.090	–	–	–	–	–
	Compacted	4.2	Medium-Light	0.118	–	–	–	–	–
		–	Heavy	0.105-0.120	–	–	–	–	–
13	Flake	4.5	As-cast 1" thick	0.101	–	–	–	–	–
	Compacted	4.5	ditto	0.085	–	–	–	–	–
	Nodular	4.5	ditto	0.078	–	–	–	–	–

Table 8

Composition of Cast Irons With Different Graphite Forms

No.	C.E.	C	Si	Mn	S	P	Mg	Ti	Cu	Cr	Ni	Mo	Sn
1	4.33	3.51	2.45	0.56	0.078	0.043	–	0.005	0.17	0.13	0.23	0.10	–
2	3.86	3.10	2.29	0.76	0.053	0.035	–	0.01	0.24	0.15	0.11	0.01	–
3	4.20	3.40	2.41	0.55	0.086	0.043	–	0.004	0.45	0.35	0.76	0.59	–
4	4.30	3.47	2.50	0.49	0.030	0.014	–	0.068	0.029	–	–	–	–
5	–	–	2.42	0.48	0.014	0.023	0.025	0.056	0.15	0.09	0.21	0.02	–
6	4.47	3.63	2.52	0.39	0.016	0.01	0.021	0.039	–	–	–	–	–
7	4.45	3.61	2.51	0.42	0.017	0.01	–	0.033	–	–	–	–	0.08
8	4.42	3.64	2.35	0.48	0.018	0.03	0.042	0.01	0.06	0.06	0.99	0.01	–

Table 9

Mechanical Properties and Thermal Conductivity
of Cast Irons with Different Graphite Forms

Iron No.	C.E. (%)	Nodularity	Matrix Structure	Tens Str, (ksi)	Yield Strength (.2% Offset) (ksi)	Modulus of Elasticity ($\times 10^6$ psi)	Elong, (%)	Red. in Area (%)	BHN[c]	Thermal Conductivity at 200°F	
										BTU in ($\frac{\text{BTU in}}{\text{hr ft}^2\ °F}$)	Cal ($\frac{\text{Cal}}{\text{sec cm }°C}$)
1	4.33	Flake, Cl. 30	100% pearlite	36 33	–	14 13	–	0.3 0.3	195	360	.124
2	3.86	Flake, Cl. 40	100% pearlite	40 49	–	11 15	–	0.1 0.1	215	330	.114
3[a]	4.20	Flake, Cl. 50	100% pearlite	50 58	–	15 11	–	0.1 0.1	245	315	.109
4	4.30	16%	27% pearlite	56 50	44 44	19 17	4.3 1.2	2.5 1.5	170	270	.093
5	–	14%	48% pearlite	59 61	45 45	18 20	3.0 2.8	2.4 2.1	185	290	.100
6	4.47	51%	68% pearlite	74 –	49 –	19 21	5.8 –	3.2 –	205	215	.074
7	4.45	37%	75% pearlite	82 87	56 57	22 23	3.9 4.3	2.0 2.4	200	220	.076
8[b]	4.42	95%	68% pearlite	100 101	61 62	25 19	9.9 9.7	7.2 6.7	220	210	.072

a Iron #3 contained 0.45% Cu, 0.35% Cr, 0.59% Mo, 0.76% Ni
b Iron #8 contained 1.0% Ni
c BHN was converted from Rockwell

Table 10

Elevated Temperature Mechanical Properties of
Cast Irons with Different Graphite Forms

Iron No.	Nodularity	Tensile Strength (ksi)				Elastic Modulus (x 10^6 psi)			
		R.T.	425°F	850°F	1300°F	R.T.	425°F	850°F	1300°F
1	flake	36	29	30	7	14	11	6	8
		33	46	25	10	13	13	9	5
2	flake	40	31	38	10	11	10	9	8
		49	43	29	–	15	10	–	–
3	flake	50	54	46	7	15	12	8	3
		58	48	44	10	11	14	8	–
4	16%	56	48	41	11	19	16	21	6
		50	49	41	11	17	20	21	6
5	14%	59	50	49	11	18	22	16	5
		61	52	47	11	20	19	14	5
6	51%	–	79	58	12	19	16	16	–
		74	75	61	11	21	21	15	3
7	37%	82	79	65	11	22	26	18	11
		87	81	66	14	23	21	17	8
8	.95%	100	89	59	13	25	20	19	5
		101	90	61	11	19	18	22	4

Fifty Years of Progress in Cast Iron Inoculation

USA OFFICIAL INTERNATIONAL EXCHANGE PAPER

V. H. Patterson, M. J. Lalich
Foote Mineral Co
Exton, Pennsylvania

Introduction

Inoculation is a means of controlling the structure and properties of gray and ductile cast irons by minimizing undercooling. Through proper use of inoculants in gray iron, for example, graphite in the castings will tend to precipitate during solidification in the form of small type-A flakes uniformly distributed throughout the matrix. If, on the other hand, no inoculation is used in gray cast iron, there will be a tendency for undercooling to take place. This can cause formation of types B and D graphite flakes in the castings, resulting in lower mechanical properties than might be expected for a given cast iron. In cases of severe undercooling, excessive chill, or areas of cementite will be present. Such conditions result in poor machinability as well as poor mechanical properties.

Before discussing the progress which has been made in the inoculation of gray cast iron during the last 50 years, a brief review of the history of gray cast iron will be presented. This review will trace some of the significant events which led to inoculation of cast iron.

Early Progress in Iron Melting

Cast iron was first produced successfully by the Chinese between 800-700 BC.[1] Even though iron was produced many centuries earlier, it apparently could not be cast because the furnaces were incapable of producing the required temperatures. However, as pointed out by Simpson,[1] the Chinese "had developed melting equipment capable of producing greater draft than hitherto had been possible."

Another reason for the success of the Chinese in being able to produce cast iron, as mentioned by Simpson,[1] was that they reduced iron oxide by heating in the presence of an excess amount of carbon, apparently in the form of charcoal. This procedure resulted in a soft, pure iron with a melting point of 1530C (2786F). The iron was then carburized, reducing its melting point to about 1170C (2138F) thereby making it easier to melt in their high-draft furnaces.

Additional references indicate that the Chinese used some high-phosphorus coal along with high-phosphorus iron ore as charge materials.[1,2] By lowering melting temperatures, these materials reduced the amount of blast needed to melt the iron.

From these early beginnings, interest in cast iron continued to grow. Many applications for this new cast metal were made possible by improvements in melting equipment and techniques as well as great progress in the art of molding. Several engineering applications used cast iron from time to time, including iron chain suspension bridges, the first of which were constructed by the Chinese in 56 AD.[2] However, iron was not generally cast in substantial quantities in Europe until the 14th century AD.[1]

Development of the Blast Furnace

Although the early furnaces for melting iron were probably a very crude form of blast furnace, the development of the Catalan forge in Spain in the 8th century AD was probably the forerunner of the blast furnace.[1] In the Catalan forge, iron ore and charcoal were charged vertically in the top, resulting in a loupe or ball of iron which was "hooked out and hammered into a bloom."[1] By modifying this simple furnace, the Swiss made an improved melting unit which was vertical, aboveground and charged with alternate layers of ore and charcoal. Improvements leading to the development of the true blast furnace were made by German and Swedish craftsmen in about 1000 AD.[1]

During the next 300 years these early blast furnaces were improved and made larger. In 1325 AD water-driven bellows which delivered sufficient draft to make hot molten metal directly from the blast furnace were introduced. Development of these bellows led to production of substantial amounts of pig iron in Europe by 1400 AD[1] and marked the beginning of modern iron foundry practice.

Early Improvements in Cast Iron Quality Through Use of Fluxes

When the Spanish Armada attempted to invade England in the 16th century, an important step in improving cast iron quality was discovered.[3] In his historical book *Full Fathom Five* about an expedition organized to recover the buried wrecks of the "invincible Armada" off the coast of England, C. Martin[3] indicates that the cast iron cannons, shot and anchors of the Spanish fleet were inferior to those used by the British. Martin cites this as an important reason the British were able to defeat the Spanish, thus preventing the conquest of England.

Even though the Spaniards possessed a good-quality hematite ore, they produced poor-quality iron guns, anchors and shot due to their lack of knowledge of cast iron behavior. Historical evidence indicates that in smelting and fluxing of the ore, refining after smelting and in molding and casting techniques, the Spanish were years behind the English. Practically all their iron castings contained slag.

The inferior quality and brittle nature of the shot coupled with the explosive force of the potent black powder caused the shot to crack and partially disintegrate prior to hitting its target. Similarly, many of the cast iron guns exploded during the firing, indicating poor strength and poor ability to absorb shock and vibration. For the same reasons, Spanish anchors broke under the stresses of heavy seas and were the cause of many shipwrecks.

What were the reasons for superiority of the English cast iron, which was the envy of their continental competitors? Martin points out that no magic formula existed. All the practices of the 16th-century founders of the Weald of Sussex — seat of the English iron founding industry at that time — are known to us. Practices included weathering the ore for several months to wash out many impurities. The ore was then crushed and washed again. Fossilized gray shells inherent in the ore resulted in a high degree of fluxing during smelting, which greatly reduced slag in the iron. After smelting, the iron was further refined several times by remelting, allowing removal of surface dross and other impurities.

The advanced knowledge of the British founders during this period is demonstrated by the fact that the gun and shot molds were dried and warmed prior to casting of the iron. The metal, in turn, was poured each time at as even a temperature as possible.

This practice minimized undercooling and established close to equilibrium conditions of solidification.

After pouring, the castings were allowed to cool gradually in the molds to room temperature. The procedure minimized stresses in the finished castings. The Spanish, on the other hand, as pointed out by Martin, cooled the castings as quickly as possible to expedite production. Their practice often involved water-quenching the castings, which contributed to stresses and cracking.

For several years after the defeat of the Spanish Armada, iron founders on the continent attempted to determine the reasons for the better quality of the British castings. In 1619, a Dutchman, Jan Andries Moerbeck, proved that he was onto something new and revolutionary in the art of iron founding by applying for and obtaining a 12-year patent involving the use of iron ore from the Weald of Sussex. By comparing the English ore having the built-in flux with their flux-free but otherwise good-quality hematite ore, the Dutch developed the use of limestone for fluxing. This new technique spread rapidly across the continent to Germany, France and eventually to Spain and was a major contribution to the development of engineering cast irons.

Refinements in the Process of Making Cast Iron

The next significant development, credited to an English ironfounder named Darby in 1730, was the discovery and production of coke which lowered the cost of producing cast iron. This development encouraged foundrymen to experiment for better-quality cast iron with improved mechanical properties. As a result, French founders tried remelting pig iron in separate smaller furnaces. This type of refining resulted in more uniform iron with respect to chemistry and was another big step toward the development of engineering-grade cast iron. Until this time, apparently most iron castings were poured using iron directly from the blast furnace.

The improved quality iron produced by remelting pig iron in separate furnaces made it possible for James Watt to build the first steam engine in 1765. Watt's steam engine was in turn, used to provide air blast for operating the first cupola bult in 1794 by John Wilkinson.[1] Controlled air blast plus higher melting temperatures in the cupola further improved cast iron quality. Designers, engineers, builders and others thereafter became more interested in cast iron as an engineering material.

Applications for the steam engine in such fields as land and sea transportation, agricultural equipment and, later, electrical power generation, created a demand for large quantities of high-quality gray cast iron. As this demand grew, so did the need for higher-strength and better-quality iron requiring more efficient melting equipment, improved charge materials and closer control of the melting operations.

Early Use of Ferrosilicon in Cast Iron

About 1810, Bergelius — a Swedish chemist — and Stromeyer — a German physicist — operating independently, produced ferrosilicon.[1] A mixture of silica, carbon and iron filings was melted in a sealed crucible. Stromeyer produced several grades of ferrosilicon by this method.

Although there appears to be no record as to how the ferrosilicon was used, it was probably added to the melting furnace. The iron founders probably became interested in a source of silicon because of differences in silicon content in various pig irons produced by different furnaces, differences due to varying silica content in the iron ores used. Advantages of higher silicon in making softer and less brittle irons were obvious. By adding silicon to the furnaces along with charge materials consisting of scrap and pig iron, foundrymen were

able to make consistently good-quality cast iron. They soon learned that it was advantageous to have silicon low in thick-section castings and high in thin sections. It is not known whether ferrosilicon was added to the ladles in the early to middle 19th century. In 1885, Turner[4] ran a number of experiments in which ferrosilicon was added to white iron to produce high-quality gray iron castings. It is reasonable to assume that the ferrosilicon was added to the iron in the ladle. If so, this would be an indication that some of the early investigators recognized the chill-reducing potential of adding ferrosilicon to the ladle. In 1920, G. Schury[5] discussed the use of ferrosilicon briquets in the cupola. A discussor of the paper indicated that he had added ferrosilicon to molten iron as early as 1890 for improving cast iron properties. However, no mention is made of ferrosilicon grade or addition method used.

At the time of the discovery of ferrosilicon and possibly for over a hundred years thereafter, furnace charges for making cast iron consisted of pig iron and cast iron scrap. Furnace and tapping temperatures were much lower than they are today. Consequently, iron temperature was not high enough to completely dissolve all the graphite in the charge. Undissolved graphite particles may have acted as nuclei for precipitation of more graphite as the liquid iron temperature dropped, thus minimizing undercooling. Under such conditions of solidification, inoculants might well be made necessary. Clark[6] refers to this possibility in a recent review covering the inoculation of cast iron. Clark states that, "while there are differences of opinion whether or not graphite can persist in molten cast iron, it does appear that something persists which serves to nucleate graphite during the solidification of the iron. In fact, it seems likely that graphite was the first inoculant commonly used in cast iron, originally introduced by the melting stock, or later by the intentional addition of electric furnace, or Mexican graphite."

Development of Ladle Treatments for Control of Graphite Structure in Cast Iron

Possibly, many iron foundrymen in the early part of the 20th century observed benefits resulting from the addition of various materials to molten iron in the ladle. However, there was very little exchange of foundry technical information in those days and certain practices were consequently kept secret. For example, in investigating and collecting facts for his "History Cast in Metal," Sanders found some very interesting information.[7] At the time of the first World War, many old German iron founders reportedly used silicon and iron dust from the rafters in the melting area of the foundry as ladle additions for stoveplate iron. They found that this very fine dust not only minimized chilled edges on the stoveplate but also produced a tighter structure in the casting.

Sanders refers to other practices used by European foundrymen for improving the properties of gray cast iron, such as the addition of small wafers of beeswax, pattern wax, candle ends and other similar volatile materials to the ladles of molten iron. Another practice was "stirring of the metal with green limbs from trees in order to blend and purify the metal." Still another method of "calming the metal and making it more machinable" was to stir the molten iron with charred sticks. All these techniques probably provided enough nucleation to prevent excessive undercooling of the iron during solidification.

After World War I, many European foundrymen emigrated to the US and brought these trade secrets with them. This knowledge served as an impetus for further investigation by American foundrymen into ways of improving gray cast iron through the use of ladle additives. Concurrently, development work continued in other countries throughout the world. As a result, the process of inoculation for control of gray cast iron structure and properties really started to gain momentum shortly after 1920.

Probably some of the first work done in the US on ladle additions to control gray cast iron structure was carried out by Crosby from 1922-28 at the Studebaker foundry in South Bend, Indiana. Timmons,[8] a former associate of Crosby, reports that in 1922 or 1923 Crosby used mixtures of graphite and ferrosilicon as ladle additives to gray cast iron. Actually, Crosby attempted to add graphite to ladles of cast iron during tapping and found the practice difficult because of the high silicon content in the base iron. For this reason, a low-silicon base iron was produced in the cupola and was found to be more compatible with the graphite addition. Silicon was then added, following the graphite addition, to achieve the desired final silicon level. Microscopic examination of this iron revealed a uniform type-A graphite structure in a predominantly pearlitic matrix with mechanical properties superior to gray cast irons made by other methods. As a result, for several years engineering gray iron castings produced at Studebaker were made from cupola-melted iron treated in this manner.

While Crosby was treating gray irons by the above techniques, other investigations were also being carried out. One such investigation was made by Meehan,[9] who in 1922 applied for a patent covering a new method for making gray cast iron. Meehan found that uniform and high-strength gray iron castings could be produced by adding calcium silicide or magnesium silicide, each of which is a strong reducing agent, to molten white iron. He stated in his patent that the above additives acted to rid the molten white iron of occluded gases and to precipitate the carbon content of the white iron bath as graphite particles during solidification. By thus converting white iron into gray iron he claimed to "obtain substantial uniformity in the product which is of greater strength than gray iron produced by other methods, while not too hard to be readily machined." Meehan's preferred way of making the calcium or magnesium silicide addition was to the ladle while it was being filled with molten iron.

In later patents, Meehan discussed various additives to control the graphite structure of cast irons which, because of their compositions, would normally be gray in character.[10] He also mentioned uses of calcium metal and ferrosilicon as graphitizing agents added to the ladles of molten iron, recognizing that ferrosilicon had been the material generally used for this purpose.

Clark[6] states that in the 1920s the use of small ladle additions of aluminum was not uncommon in the foundry industry for minimizing oxidation of the iron. In fact, in the late 1920s the practice at the former Cadillac foundry was to add small quantities of aluminum to the ladles of gray cast iron for improving casting strength and minimizing porosity and shrinkage in the castings. According to Clark, as early as 1921, Moldenke reported that ladle additions of up to 0.15% zirconium to cast iron as silicon-zirconium resulted in improved transverse strength. Moldenke credited this improvement in cast iron properties to deoxidation.

Development of High-Strength Cast Irons

In the 1920s, as pointed out by Clark,[6] there was considerable interest in high-strength gray cast iron for engineering applications, especially those of an automotive nature. However, iron foundrymen recognized and were concerned about the weakening effect of the graphite phase in cast iron. They were aware of the advantages of small randomly distributed graphite flakes.[11] Much effort was therefore spent in finding ways to control graphite type and size. Although ladle additions of ferrosilicon and other ferroalloys were used to adjust composition and chilling tendencies of the cast iron, levels of aluminum and calcium in the ferrosilicons were not well

controlled at that time because the nucleating importance of these elements was not realized. This might account for the early success of Meehan in achieving consistent results through the addition of calcium silicide, which he insisted be made and controlled to close chemical limits. As we now recognize, controlled amounts of calcium are very effective in reducing chill and promoting uniformly dispersed type-A graphite flakes in gray iron castings.

Other work in the late 1920s indicates that iron foundrymen at that time realized the importance of ladle additions of ferrosilicon in controlling the structure of gray cast iron for achieving maximum mechanical properties. Clark refers to the findings of Bornstein[12] in producing 65,000 psi gray cast iron using ladle additions of 0.75% nickel plus 75% ferrosilicon to a high-steel-charge base iron. Such high strength without the late silicon addition would have been difficult to achieve.

During the late 1920s and early 1930s investigations continued on ladle additions to gray cast irons. However, the inoculating grades of alloys as we know them today did not exist. Thus, control of the important nucleating elements calcium and aluminum was poor. In fact, it probably wasn't until the work of Lorig, Kinnear and Barlow in 1938 that the importance of these two elements was realized.[13] They reported that the tensile properties of gray cast irons were decreased by 5000-10,000 psi when poured from highly superheated base iron. However, by reducing base silicon in the furnace about 0.75% and adding it to the ladle during tapping, tensile properties of the superheated iron were restored. The authors speculated that the calcium and aluminum present as impurities in commercial ferrosilicon could have been the effective part of the addition.

In the 1920s, Smalley[14] recognized the value of ladle additives to gray cast iron for controlling its structure and properties. He controlled chill by adding steel or ferrosilicon to the ladles of iron. If the iron was hard and showed excessive chill, ferrosilicon was added to the ladle. Smalley states that the value of silicides, silicon alloys and other alloys having high affinity for oxygen as graphitizing agents for cast iron had long been established. He credits the first use of calcium silicide in cast iron to two German foundrymen, Wohler and Goldschmidt. In about 1923 or 1924, Smalley[14] evaluated the relative influence of ferrosilicon, calcium silicide, magnesium silicide and zirconium silicon alloys on the hardness, tensile strength and graphitization of white cast iron. Composition of the magnesium alloy was 12% magnesium, 40% manganese, 40% silicon, balance iron. The ferrosilicon was the 50% grade and probably contained only minor amounts of calcium and aluminum. The zirconium silicide alloy contained 35% zirconium, 45% silicon and 20% iron. His tests indicated that calcium silicide was the best of the graphitizers tested.

Development of Improved Proprietary Inoculants

In the work done prior to 1930 on ladle additives for cast iron, no reference is made to inoculation. Just when the word inoculation was first used to describe the addition of materials to the ladle for controlling cast iron structure is not known. However, it was probably in the 1930s. In the Howe Memorial Lecture in 1937 Merica[15] states that "the graphite residue theory was also probably at the bottom of the reasoning which led to the development of the so-called 'inoculation' or ladle graphitizing methods developed at the Ross Meehan foundries." In this same paper he referred to a statement made by Strauss of Vanadium Corp to the effect that as early as 1920 ferrosilicon was used by Outerbridge in Philadelphia in inoculating ladle iron for production of lathe bed castings. During the 1930s rapid strides were made in developing new and improved inoculants due to the increased production of high-strength cast irons. Merica[15] refers to the "new attitude toward cast iron" by prominent companies such as Ford Motor Co; Caterpillar Tractor Co; Campbell, Wyant and Cannon Foundry Co and

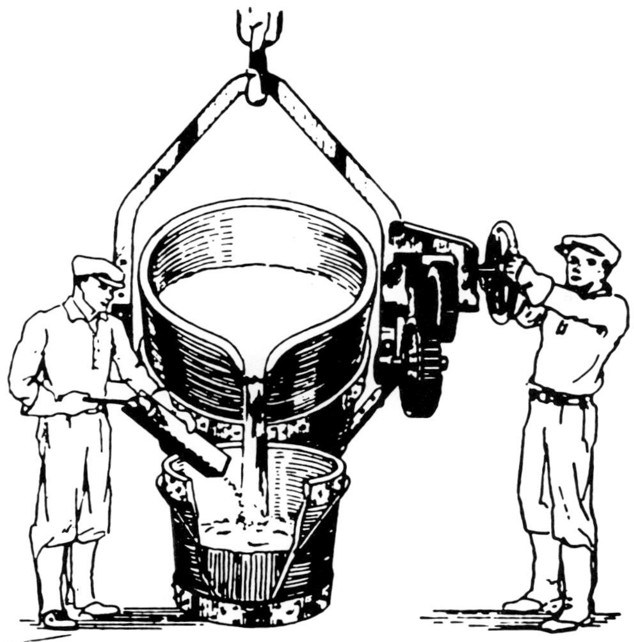

Fig. 1. Method of adding alloys to ladle of molten iron.

other leading manufacturers of automotive and diesel engines, as being responsible for interest in methods of controlling cast iron structure. He states that, in 1936, 10-15% of gray iron castings produced in the US were high-strength grades. Since that time, production of engineering grades of gray cast iron throughout the world has increased manyfold.

Further stimulus was given to the interest in gray iron inoculation by the work of Eash[16] of International Nickel Co, probably carried out in the late 1930s and reported in 1941. Using an inoculant which added 0.7% nickel and 0.35% silicon,

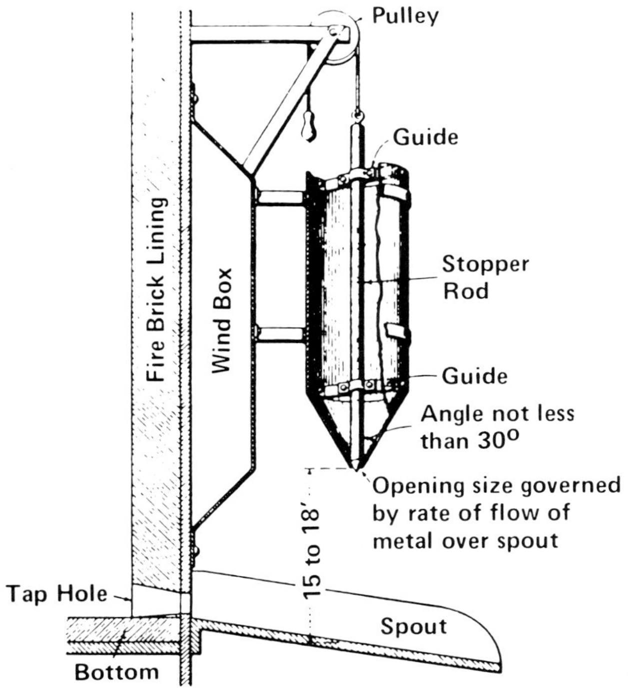

Fig. 2. Method of adding alloys to molten iron in furnace spout.

Eash reports tensile strength increases of more than 20,000 psi for inoculated gray cast irons of 2.25% C and 2.25% Si versus uninoculated irons of the same composition. However, in irons with higher carbon equivalents, improvement in strengths of the inoculated irons was less. This is because in comparable gray iron castings under similar casting conditions, the lower the carbon equivalent, the greater is the tendency for undercooling.

Burgess of the Union Carbide and Carbon Research Laboratories,[17] in pointing out that Eash had limited his paper to a discussion of the effect of inoculating the iron with nickel plus ferrosilicon, mentions that satisfactory results can often be obtained with straight ferrosilicon additions. Burgess also indicated that a given amount of a complex additive not only was more effective than a comparable amount of 75% ferrosilicon but that the inoculating effect lasted longer. Due to these advantages, a considerable number of special graphitizing inoculants were developed. Burgess mentioned the silicon-manganese-zirconium alloy as a typical proprietary complex graphitizing inoculant. He further pointed out that by the combined use of chromium additions with a graphitizing inoculant added to the ladle of molten iron, high-strength irons with unusually high impact values and low chill could be obtained. Burgess was apparently referring to his 1936 patent covering the use of ladle additions of silicon to low-silicon base irons for controlling chill depth and section sensitivity of cast irons alloyed with chromium. This work led to the introduction in about 1939 of a stabilizing inoculant by the Vanadium Corp of America which proved to be very effective. This alloy contained about 40% chromium, 10% manganese, 16% silicon, 1% each of calcium and titanium, balance iron and was designed to add chromium to produce pearlitic gray cast irons with little or no chill.

During the 1930s the major research in the US leading to the development of the more complex graphitizing inoculants was carried out in the laboratories of two leading ferroalloy producers — Union Carbide and Carbon Corp and the Vanadium Corp of America (now Foote Mineral Co). This research resulted in improvements in standard ferrosilicons used for inoculation plus the development of seven proprietary inoculants shown in Table 1, taken from the 2nd edition of the Alloy Cast Irons Handbook published in 1944 by the American Foundrymen's Association. Other inoculants in use at that time are also listed in Table 1. Of the 18 shown, inoculants 2, 3, 8, 9 and 15 are still in use. Inoculant 15, however, has been modified to include some barium and much less calcium than that shown.

Development of Inoculating Methods

During the 1940s more effort was spent in determining the proper methods of inoculation rather than in developing new inoculants. However, considerable advancements were made in the methods of adding inoculants to molten iron. Two methods which resulted from that work are shown in Fig. 1 and 2 taken from the Alloy Cast Irons Handbook, 2nd edition.

Figure 1 shows the inoculant being added to the molten iron at the point where the stream from the transfer ladle hits the metal in the pouring ladle. This method is an effective way of adding inoculants in the ladle since it insures good mixing action as well as uniform distribution of the inoculant throughout the iron. The method of inoculation shown in Fig. 2, while more costly than the technique used in Fig. 1 since it involves more specialized equipment, insures good mixing action. Experience with this type of addition has demonstrated that finer particles in the inoculant will float on the liquid surface and lose effectiveness through oxidation. In addition, uniform flow of inoculant through the reservoir opening is difficult to maintain without some type of vibrating device.

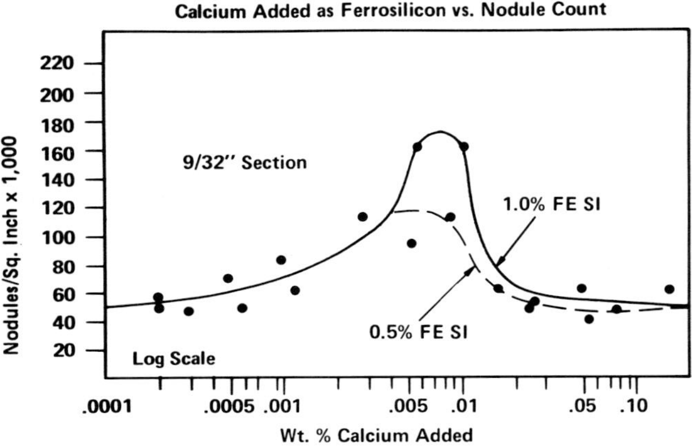

Fig. 3. Effect of calcium added with ferrosilicon on nodule count in ductile cast iron.

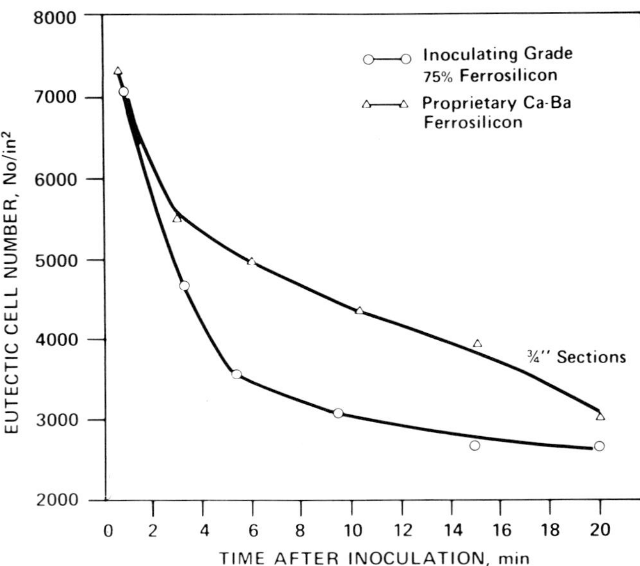

Fig. 4. Example of beneficial effect of proprietary inoculant on nucleation and fade of cast irons.

Impact of Ductile Iron on Inoculation

With the advent of ductile cast iron in 1948, investigations of inoculants — often referred to as post inoculants — for controlling distribution of the graphite nodules were begun. Ferrosilicon was considered to be the best available at that time. In comparing 75 and 85% ferrosilicons, there appeared to be very little difference between the two. Aluminum contents of 1.00-1.50% and calcium contents of 0.50-1.10% in the ferrosilicons were found to be important in obtaining maximum nodule counts and minimizing carbides in the as-cast irons, Fig. 3. For inoculating thin-section ductile iron castings, a 1% addition of either the 75 or 85% ferrosilicon became common practice but for heavier section castings less ferrosilicon was required. From this time on, ferroalloy producers established and maintained close control over calcium and aluminum contents of the ferrosilicons.

Development of Inoculants and Inoculating Methods During the 1950s

In the 1950s, many articles were published on the inoculation of gray and ductile cast irons. Most articles described improved properties obtained from inoculating irons. Although only a few patents were obtained involving new inoculant compositions, one in particular — awarded to Kessler — was significant.[18] It was a graphitizing inoculant having 72% silicon, 20% manganese, 1% calcium, 1% aluminum, 2% barium and 4% iron. It was claimed that the barium addition caused iron to solidify with a microstructure of short thick graphite particles uniformly distributed throughout a pearlitic matrix.

Other articles described techniques for introducing alloys and inoculants, including magnesium ferrosilicon, below the molten iron surface. Inert gases such as nitrogen were used to either convey granular additives through graphite — or refractory tubes immersed in the molten iron — or to agitate the liquid metal while the inoculants were being added. Many of these methods were unsatisfactory because of temperature losses in the liquid irons caused by injection of carrier gases. In other instances, improperly sized inoculants with excessive fines clogged the tubes resulting in poor metal treatment.

During this period the porous plug method of introducing inert gases into ladles of molten iron was developed in France. Bucket-shaped ladles were fitted with silicon carbide plugs through which the gases were blown to achieve active and methodic agitation of molten metal. Properly sized additive materials, thrown on top of the agitated molten metal bath, resulted in uniform dispersion of inoculant and effective nucleation of iron.[19]

Other methods used during the 1950s involved conical refractory nozzles specifically designed to uniformly add inoculants and alloying elements to molten cast iron during furnace tapping. Electronic devices were developed for automatically adding inoculants to molten cast iron. Magnesium wire injection of molten iron through small openings in the lower portion of the ladles was patented in East Germany.[20] Graphitizing agents were added under gas pressure to molten irons through special conduits from a number of reservoirs. These conduits were equipped with special measuring devices capable of introducing controlled amounts of inoculants and other alloys into the iron.[21]

Even though many inoculating techniques of the 1950s are not in general use today, they added to the general overall knowledge of inoculation and helped to develop new ideas and understanding concerning structural control of cast iron.

Inoculant Developments During the 1960s

During the 1960s many developments in inoculation of irons occurred due primarily to changes in melting methods. Electric

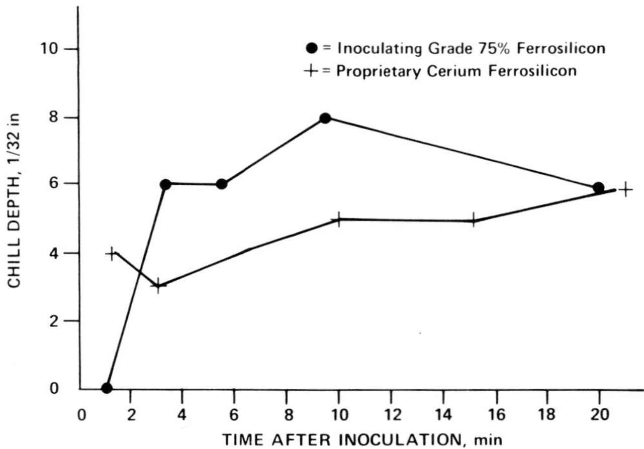

Fig. 5. Beneficial effect of proprietary inoculant on chill control with respect to time in gray cast iron.

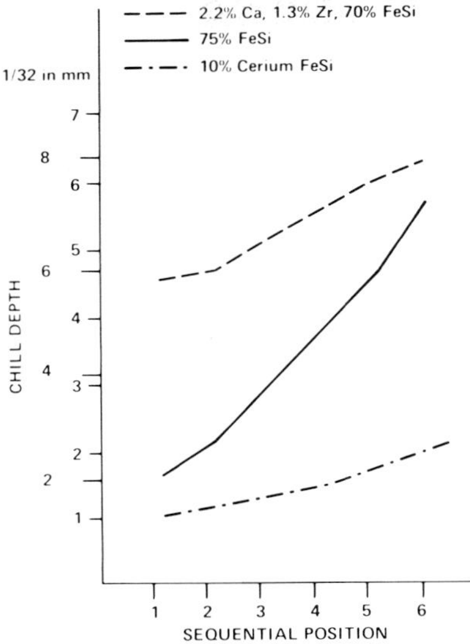

Fig. 6. Example of chill variation in castings made in multiple-cavity mold using granular inoculants in the mold.

furnaces — induction and arc — were replacing many cupolas because of air pollution control requirements. When cupolas were retained and equipped with pollution control systems, hot-blast and oxygen-enrichment equipment was often installed. Steel and purchased scrap were being used in the charges to replace some or all of the pig iron previously used and melting temperatures were increased resulting in more oxidation of the metal. All these factors made it necessary to develop more effective graphitizing additives. As a result, new complex proprietary inoculants designed to improve the nucleating and fading characteristics of the older products became available (Fig. 4 and 5). Some were suitable for both gray and ductile cast irons. Most of these complex inoculants are still in use today. As conditions vary, inoculants exhibit different performance characteristics. Thus for a given application some inoculants are better than others. For this reason foundrymen prefer to select one or more of the available proprietary inoculants according to their compatibility with certain melt conditions.

Inoculants containing barium, strontium, cerium and magnesium were developed in the 1960s. A US patent was granted in 1962 for an inoculant containing barium.[22] In 1963, a British patent was obtained on the use of strontium in ferrosilicon for inoculating purposes.[23] Prior to that time, Dawson had also recognized the benefits of cerium as an inoculant.[24] However, the first commercially available inoculant containing cerium was developed and reported by Mickelson in 1967.[25] An exothermic ferrosilicon inoculant containing about 2.00% magnesium was produced in 1960[26] and has been used to some extent in gray and ductile cast irons.

Inmold inoculation[27] and instantaneous ladle inoculation[28] were probably the most significant and novel inoculating techniques developed in the 1960s. In the former method, fine-mesh graphitizing inoculant is placed either in the base of the sprue or in a suitable chamber in the runner system of the mold. The mold inoculant compensates for fading of the primary inoculant which may have taken place prior to pouring and prevents undercooling and chill in thin-section castings. The process works well but to insure uniform inoculation, alloy solution rate and gating design must be carefully controlled. For example, in a multiple-cavity mold, some castings may be properly inoculated while others may be inadequately treated (Fig. 6).

This problem led to the development of solid or compact inoculants for insertion in the pouring basins, bases of the sprues or runner systems of the molds.[29] These inoculants were in the form of lumps, pellets or briquets, and helped to eliminate inconsistent results obtained using the loose fine-mesh materials. The solid-insert inoculants were later improved by casting them to the proper shape and size for the quantity of metal treated.[30]

A tablet type of mold inoculant was also introduced in the 1960s, consisting of a powdered graphitizing inoculant bonded by a wax or stearic acid substance.[31] Depending on mold size, one or more of these tablets can be placed in the mold. Success of inmold inoculation probably led to the development of the inmold nodulizing process.

The instantaneous ladle inoculation process consists of a thin-walled steel tube filled with fine-mesh inoculant or a solid bar of the inoculant held by a guide so that its end is kept in contact with the pouring lip of the ladle. As the molten iron flows from the ladle into the mold it dissolves that portion of the rod with which it is in contact. This method was highly effective. However, its use was discontinued after attempts to mechanically and uniformly feed the inoculating rod into the molten metal stream failed.

Recent Innovations in Inoculation

During the 1970s there has been increasing interest in cast iron inoculation, including mold and ladle techniques. In addition, some new inoculants have been developed. The high degree of interest has been due in part to increased use of automatic pouring in combination with automatic molding.

In 1971, Moore and Kessler[32] devised a gating system which included a treatment basin in the mold for late inoculation, nodulization or addition of alloying elements to iron. In this system molten iron enters the treatment basin from the downsprue. It then passes through a dam-gate into the mold cavity. The dam-gate controls the time during which the molten metal is in contact with the additive in the treatment basin and also provides cleaning of the molten iron before it enters the mold cavity. Developments such as this are important for successful use of small-mesh-size granular additives in the mold.

A recent development in ladle additives is custom-blended inoculants for gray cast iron. These inoculants are mixtures of ferrosilicon, graphite, silicides and fluxing agents which control inoculant solution rate. The idea is based on the assumption that all foundries have different operating techniques. Therefore, there is an inoculant which will work best for each foundry, providing the foundry practice is controlled within reasonable limits. Once the proper tailor-made inoculating mixture is found, consistent results should be obtained indefinitely.[33]

Probably the newest and most interesting process developed in the 1970s for controlled inoculation of gray and ductile cast iron is the patented controlled quality (CQ) inoculation process.[34] Essentially, the CQ process uses a special steel wire having a core composed of the inoculant, Fig. 7. Figure 8 shows the inoculation wire with the inoculant which is sized 40 x 140 mesh. The wire is fed directly into the metal stream at a controlled rate while pouring. As the wire melts in the molten metal, the inoculant is thoroughly and uniformly mixed with the metal, Fig. 9. Since metal immediately enters the mold after inoculation, no inoculant fading takes place. The usual amount of inoculant added is 0.02-0.03%. In the operation of this process, an electrical mechanism automatically lowers the inoculant feeder head so that the tip of the wire is in the pouring basin after the mold is in position for pouring. As the molten metal starts entering the pouring basin, an electric eye starts the wire feeder. When pouring is completed, either sensing or timing devices stop the wire feeder and raise the feeder head so that the next mold can move into the pouring position. The CQ process

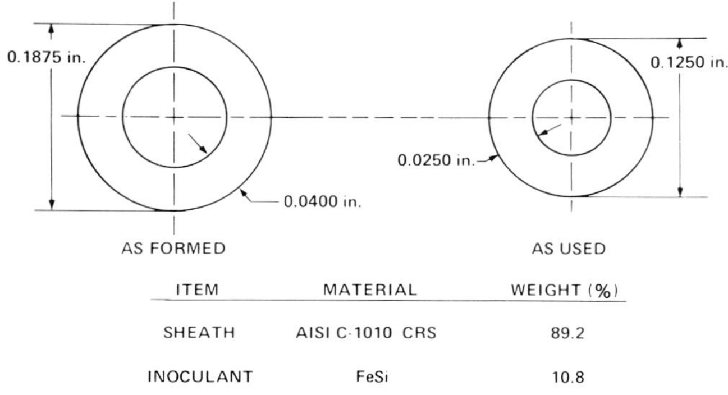

Fig. 7. Cross section of inoculating wire used in the controlled quality (CQ) inoculation process.

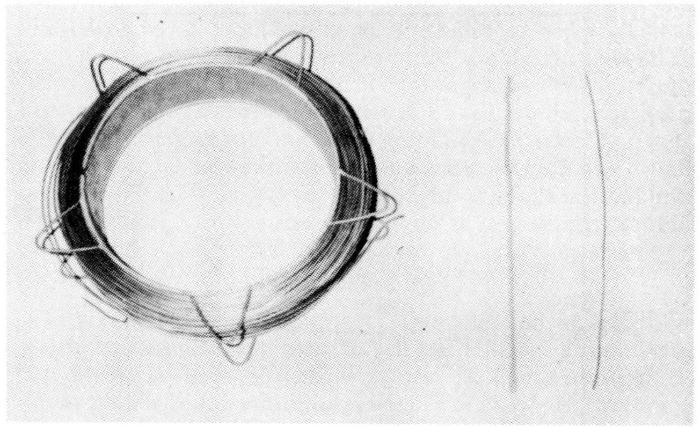

Fig. 8. Inoculating wire coil and small section showing fine-mesh inoculant.

thus provides the potential of computerized control of inoculating procedures which can be used advantageously with automatic molding and pouring.

Current Inoculants Used as Ladle Additives

Some inoculants in use from 1940 through the early 1960s are listed in Table 1 for comparison with some graphitizing inoculants currently in use and shown in Table 2.

Alloy 1 in Table 2 is the most popular and widely used inoculant for gray and ductile iron and has been referred to as inoculating grade 75% ferrosilicon throughout this text. The next two are modifications of this grade. Inoculant 2 has low aluminum and low calcium and, although sometimes used, is generally considered a poor inoculant. Because of its higher calcium content, inoculant 3 is thought to resist fading to a greater extent than inoculant 1, although this has not been proven. Inoculant 4 is 50% ferrosilicon modified with more aluminum and calcium than the standard 50% grade to achieve a lower-cost substitute for 75% ferrosilicon. Inoculant 5 is the same as inoculant 4 but contains a small amount of magnesium. This relatively new inoculant was developed for restoring some of the magnesium lost to fading during holding of treated ductile iron. Inoculant 6, an exothermic ferrosilicon containing a small amount of magnesium, can be used for the same purposes as inoculant 5.

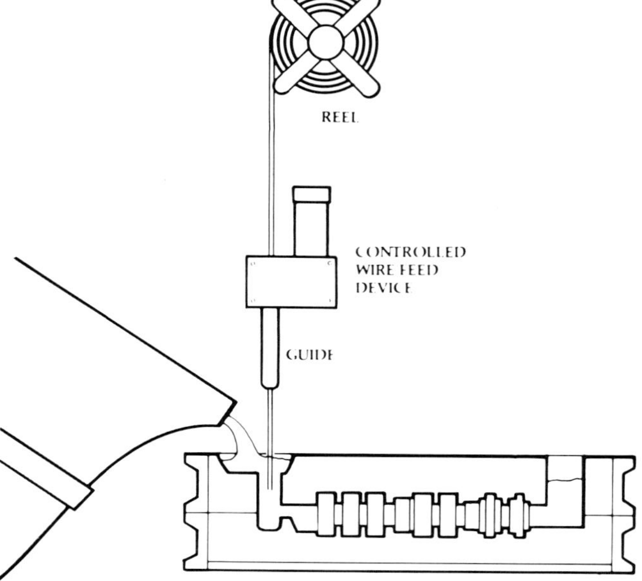

Fig. 9. Diagrammatic sketch illustrating automatic operation of CQ process.

Table 1. Typical Analyses of Some Ladle Inoculants Used in 1944

No.	INOCULANT	COMPOSITION — PERCENT								
		C	Ca	Cr	Mn	Si	Ti	Zr	Al	Fe
1.	Ca-Metal		100							
2.	Ca-Si		30-35			60-65				
3.	Ca-Si-Ti		5-8			45-50	9-11		1.50	bal.
4.	Cr-Si-Mn-Ti-Ca	3.0	1*	38-42	8-11	14-16	1*		1*	bal.
5.	Cr-Si-Mn-Ti-Ca	3.0	1*	28-32	14-16	15-21	1*		1*	bal.
6.	Cr-Si-Mn-Zr (3 grades)			30-52	5-10	14-35		1-6		bal.
7.	FeSi		0.5-0.8			80-90			1.25*	bal.
8.	FeSi		0.5-0.8			74-78			1.25*	bal.
9.	Graphite	90-100								
10.	Mo-Si**					30*				
11.	Ni-Si***					30				bal.
12.	Si-C	28-46				45-56				
13.	Si-C	50				42*				
14.	Si-Mn				20-25	47-54				bal.
15.	Si-Mn-Zr		2.5		5-7	60-65		5-7	1.75	
16.	Si-Ti					20-25	20-27			
17.	Si-Zr					47-52		35-40		bal.
18.	Si-Zr					39-43		12-15		bal.

* Approximate
** 60% Mo
*** 60% Ni

Inoculant 7, standard calcium silicide, was an effective inoculant in the early days of inoculation. However, this inoculant, finds little use today because the high calcium content makes it very reactive and produces considerable dross.

Inoculant 8 is a popular graphitizer for gray cast iron and is a very effective chill reducer in thin-section castings. The high titanium content helps prevent pinholing caused by nitrogen picked up from charge materials or molding sand additives.

Inoculants 9 and 10 are used in both gray and ductile cast irons. Manganese in these alloys lowers their melting point, thereby increasing their rate of solubility in irons. Barium and calcium, being strong nucleating agents, help prevent chill in castings and the high barium concentration makes the alloy more resistant to fading.

Inoculant 11, an effective inoculant for preventing chill in gray cast iron is becoming increasingly popular, particularly for inoculating electric furnace irons. This inoculant is effective for combating the carbide-forming tendencies of small amounts of residual elements such as chromium often present in gray cast iron. Its low aluminum content helps prevent pinhole porosity in thin-section, high-carbon-equivalent gray iron castings. While inoculant 11 is not used in ductile iron strictly as an inoculant, it is added in controlled amounts, in combination with cerium-free magnesium alloys, to condition the melt for high graphite

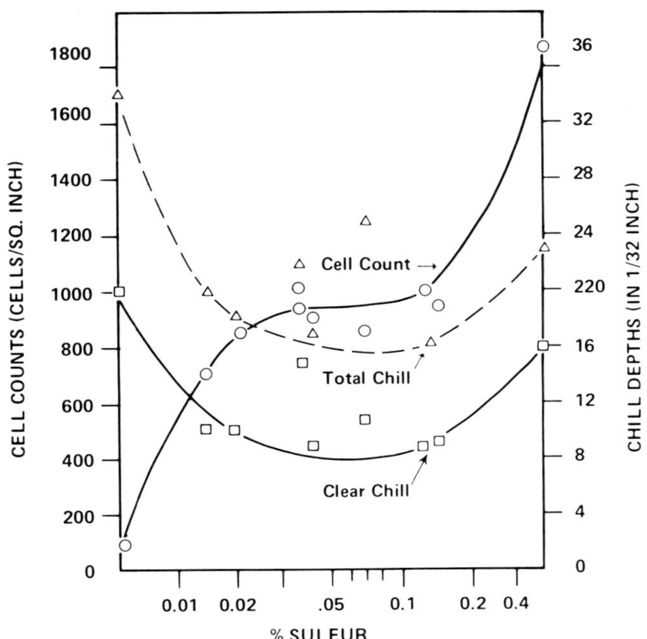

Fig. 10. **Relationship of sulfur content in gray irons to cell counts and chill depths.**

Table 2. Current Inoculants for Gray and Ductile Iron

COMPOSITION — PERCENT*

INOCULANT NUMBER	SI	AL	CA	SR	CE	MN	BA	MG	ZR	Other	FE
1	74/79	0.60/1.25	0.50/1.00								BAL
2	74/79	0.40/0.50	0.10/0.20								..
3	74/79	0.60/1.10	1.0/2.0								..
4	46/50	1.20 Max.	0.60/0.90								..
5	46/50	1.25 Max.	0.60/0.90					1.0/1.50		NaNo$_3$	..
6	58/61	0.90/1.20	0.50/0.70					2.0/2.50		10.0	..
7	60/65	0.90/1.10	28/32							Ti	..
8	50/55	1.0/1.30	5.0/7.0							9/11	..
9	60/65	1.0/1.50	1.5/3.0			9.0/11.0	4.0/6.0				..
10	60/65	0.75/1.25	0.60/0.90			5.0/7.0	0.60/0.90		5.0/7.0	Total	..
11	36/40	0.50 Max.	0.50 Max.		9.0/11.0					R.E. 11·15	..
12	73/78	0.50 Max.	0.10 Max.	0.60/1.00							..
13	46/50	0.50 Max.	0.10 Max.	0.60/1.00							..
14	78/82	1.0/3.0	2.25/2.5						1.25/1.75		..
15	74/79	3.0/4.0	0.50/0.80					0.50/1.00			..

*There are other inoculants of a proprietary nature available for inoculating cast iron. However, their analyses are confidential. Most of them are blends of crushed ferrosilicon, calcium silicon, magnesium ferrosilicon, rare earth compounds, graphite and fluxes.

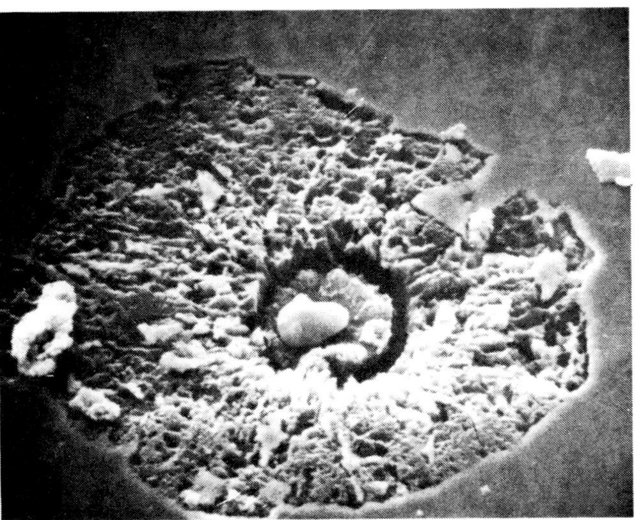

Fig. 11. **Magnesium-calcium sulfide nucleus in graphite nodules: a) X2500 and b) X10,000. (plasma etch)**

nodule counts and low susceptibility to chill after normal post-inoculation. This inoculant will also neutralize the deleterious effects of tramp elements such as bismuth and lead in gray and ductile cast irons.

Inoculant 12 is an effective graphitizing inoculant which can be used for both gray and ductile cast iron. Designed to minimize susceptibility to shrink in gray iron castings, this inoculant, reduces chill without decreasing eutectic cell size as much as other inoculants. Inoculant 13 is similar to inoculant 12 except that it is made from lower-cost 50% ferrosilicon.

Like the other proprietary inoculants, inoculant 14, is usually more effective than the inoculating grades of 75% ferrosilicon. It can therefore be used in smaller quantities. This product is unique in that it is supplied with a standard aluminum content for green sand applications and a higher aluminum content for effective inoculation of castings made in dry sand molds.

The last inoculant is produced in various cast shapes and sizes, providing a flexible system for inmold inoculation.

In summarizing Table 2, numerous graphitizing inoculants are currently available for controlling the structure of gray and ductile cast irons. Each has its own individual characteristics which can only be highlighted in a paper of this nature. Actual performance of an inoculant depends on charge materials, type of furnace, melt conditions, temperature, etc. Carefully controlled trials should thus be made to evaluate inoculant performance in actual foundry practice.

Understanding Mechanisms of Inoculation

Very little investigation was carried out during the early years of inoculation to determine why the practice worked. However as new inoculants were developed, investigators became increasingly curious about the mechanisms of inoculation which control cast iron structure. Improved understanding of nucleation and growth phenomena, associated with the inoculation practice, gradually resulted. Even today these mechanisms of inoculation are not well understood and numerous theories which are the subject of continuing discussion exist.

Theories such as the graphite nuclei theory, the gas theory of inoculation and the silicate slime theory were discussed in detail by Boyles[35] as early as the mid-1930s. He concluded from his discussion that "both chemical and physical factors were probably involved in the action of inoculants on the graphite structure in cast iron."

Inoculants were recognized as excellent deoxidizers which could as such react with oxygen and possibly other gases in the iron, gases such as nitrogen and hydrogen. The tendency of some elements in inoculants to form compounds with sulfur in the molten iron was also known. In fact, Boyles found through extensive experiments that sulfur was an essential ingredient in gray cast iron for favorable response to inoculation. This concept, Fig. 10, was later confirmed and developed further by Wallace and Muzumdar.[36]

Early theories discussed the alternatives of homogeneous and heterogeneous nucleation of graphite. For example, Gulliver[37] stated that "the nucleus is most effective when it is itself of the same substance as that about to crystallize." Boyles pointed out that this statement contains the essence of the graphite nuclei theory of inoculation.

In a more recent investigation concerning the nature of nuclei in gray cast iron, Lux[38] discussed both homogeneous and heterogeneous nucleation of graphite. He concluded that the elements calcium, strontium and barium can readily form salt-like carbides of the CaC_2 type. Carbides of these elements exhibit ionic bonding between the metal and the carbon and are "thus expected to be insoluble in an iron, carbon, silicon melt." For this reason he concluded that salt-like carbides may act as effective heterogeneous nuclei for graphite formation. In regular gray cast iron, however, sulfur and some oxygen are usually present. Thus it is presently thought that at normal melt temperatures these elements make calcium carbide thermodynamically unstable because of the strong affinity of calcium for sulfur and oxygen.

In a recent investigation covering nucleation of spheroidal graphite in ductile cast irons, Lalich and Hitchings[39] confirmed the findings of other investigators by demonstrating the importance of nonmetallic inclusions. For example, they found that compounds of magnesium calcium sulfide acted as heterogeneous nuclei for graphite nodules in ductile cast irons made from magnesium ferrosilicon treatment alloys (Fig. 11). They concluded from the evidence obtained in their research that the majority of nodules in ductile cast iron is associated with nonmetallic inclusions and that graphite growth in some instances is related to the shape and distribution of the nonmetallic inclusions.

Numerous investigators have postulated that modification of the amounts of surface-active elements such as oxygen and sulfur can affect graphite growth characteristics. Bates, McSwain and Scott,[40] for example, conclude from their work that interfacial energies between the liquid and precipitating graphite are altered by adsorbed surface-active elements. Thus they postulate that the preferred direction of graphite growth is a function of amounts of the surface-active elements present.

Summary

When it was realized that gray cast iron could be used as an engineering material, the need for structural control became evident. Use of ladle additives gradually evolved, along with development of better charge materials and melting techniques, to achieve the structural control required.

Today's inoculating techniques and materials are a tribute to 50 years of progress in the understanding of cast iron metallurgy. Rapidly developing foundry technology will lead to further changes and improvements in the nature of inoculants and inoculating methods.

Acknowledgements

The authors gratefully acknowledge the following people for providing information used in the preparation of this paper:

Mr. Clyde A. Sanders, American Colloid Co, who provided the facts concerning ladle additives discovered accidentally by German foundrymen during World War I.

Mr. Ralph A. Clark, Consultant, Union Carbide Corp, for permission to use some of his unpublished work on early developments in the inoculation of cast iron.

Mr. G. A. Timmons, Climax Molybdenum Co, for providing information concerning the early use of graphite additives in cast iron.

Messrs. James Vanick and Keith Millis, Ductile Iron Society, for furnishing historical evidence relating to the inoculation of gray and ductile iron.

J. R. Nieman and his associates, Caterpillar Tractor Co, for the information concerning the patented controlled quality (CQ) inoculation process.

Management and staff members of Foote Mineral Co for their support and assistance.

References

1. B. L. Simpson, History of the Metal Casting Industry.
2. C. A. Sanders, D. C. Gould, History Cast in Metal.
3. C. Martin, Full Fathom Five, p 247-261.
4. T. Turner, Metallurgy of Iron, London, Griffin (1895).
5. G. Schury, Giesserei, vol 7, p 241-244 (1920).
6. R. A. Clark, "Inoculation of Cast Iron," (unpublished) (Dec 9, 1976).
7. C. A. Sanders, personal correspondence, (Jan 4, 1977).
8. G. A. Timmons, Climax Molybdenum Co, Ann Arbor, Michigan, personal conversation (Feb 1, 1977).
9. US Patent #1,499,068, Making Gray Iron, A. F. Meehan (Jun 24, 1924).
10. US Patents #1,683,086 and #1,683,087, "Cast Iron and the Method of Making Same," A. F. Meehan (Sep 4, 1928).
11. J. W. Bolton, "Graphite in Cast Iron," AFA Transactions, vol 31, p 68-81 (1923).
12. H. Bornstein, "Discussion - Cast Iron Developments," AFA Transactions, vol 36, p 835 (1928).
13. C. H. Lorig, H. B. Kinnear, T. Barlow, "Summary Report on Electric Furnace Melting of Cast Iron," Battelle Memorial Institute (Feb 10, 1938).
14. O. Smalley, Proceedings of the Institute of British Foundrymen, vol 16, p 495 (1922) and AFA Transactions vol 34, p 881-895 (Oct 1926).
15. P. D. Merica, "Progress in Improvement of Cast Iron and Use of Alloys in Iron," AIME Transactions, vol 125, p 13-45 (1937).
16. J. T. Eash, "Effect of Ladle Inoculation on the Solidification of Gray Cast Iron," AFA Transactions, vol 49, p 887-910 (1941).
17. C. O. Burgess, Union Carbide and Carbon Research Laboratories, Inc, Niagara Falls, NY.
18. H. H. Kessler, "Alloy for Addition to Molten Cast Iron," US Patent #2,810,639 (application date Aug 10, 1956).
19. "Ladle and Method for the Treatment of Molten Metals," L'air Liquide SA, British Patent #760,561 (application date Nov 15, 1954).
20. R. Radtke, East German Patent #2339 (18b, 1/03) (May 21, 1952).
21. Bergen and Babcock British Patent #808,293 (applied for on May 29, 1956).
22. Vanadium Corporation and R. L. Mickelson, "Inoculating Alloy," US Patent #3,137,570 (Aug 10, 1962).
23. BCIRA and J. V. Dawson, "Improvements in the Manufacture of Cast Iron," British Patent #1,002,107 (filed Aug 19, 1963).
24. J. V. Dawson, "Factors Influencing the Inoculation of Cast Iron," BCIRA Journal, vol 9, no. 2, p 199-236 (Mar 1961).
25. R. L. Mickelson, "Cerium-Silicon Alloy Reduces Chill in Cast Iron," Foundry, vol 95, p 145-146, 149 (Jun 1967).
26. Unpublished report, Foote Mineral Co technical files.
27. W. J. Dell, R. J. Christ, AFS Transactions, vol 72, p 408-416 (1964).
28. S. I. Karsay, "Method and Apparatus for Treating Molten Metals," US Patent #3,367,395 (filed May 12, 1965).
29. H. Trager, A. Kaune, "Process for Inoculating Cast Iron," German Patent #OLS1,458,899 (applied for Nov 17, 1965).
30. G. Hillner, K. H. Kleeman, "Mold Inoculation of Gray and Ductile Cast Iron, New Solutions to Old Problems," AFS Transactions, vol 83 (1975).
31. C. Hall, Foseco International, Ltd, "Production of Cast Iron," British Patent #1,105,028 (filed Jun 27, 1966).
32. W. H. Moore, H. H. Kessler, US Patent #3,746,078 (filed Feb 4, 1971).
33. M. D. Bryant, J. Briggs, R. Neuman, "The Case for Inoculants Containing Graphite in the Production of Gray Iron Castings," 81st AFS Casting Congress, Cincinnati, Ohio (Apr 25-29, 1977).
34. J. R. Nieman, L. W. McFarland, "Method and Apparatus for the Introduction of Additives into a Casting Mold," US Patent #3, 991,808 (Nov 16, 1976).
35. A. Boyles, "The Structure of Cast Iron," American Society for Metals (1947).
36. K. M. Muzumdar, J. F. Wallace, "Inoculation-Sulfur Relationship in Cast Iron," AFS Transactions, vol 80, (1972).
37. B. Lux, "Hypothesis on the Nucleation of Eutectic Graphite in Inoculated Gray Iron by Salt-Like Carbides," AFS Transactions, vol 72 (1964).
38. J. H. Gulliver, Metallic Alloys, Charles Griffen Co, 4th ed (1921).
39. M. J. Lalich, J. R. Hitchings, "Characterization of Inclusions as Nuclei for Spheroidal Graphite in Ductile Cast Irons," AFS Transactions, vol 84 (1976).
40. R. H. McSwain, C. E. Bates, W. D. Scott, "Iron-Graphite Surface Phenomena and Their Effects on Iron Solidification," AFS Transactions, vol 82 (1974).

SECTION VII
Cleaning, Inspection, Quality Control

Latest developments and possible trends in fettling

W. McCormack, BCIRA

Introduction

Fettling accounts for 15 – 25 per cent of the labour cost of a casting in the UK, and as much as 35 per cent in Sweden. The cost will rise as working conditions are improved, during our progress into the '80s; therefore the amount of fettling required to bring a casting to an acceptable state for selling should be reduced as much as possible.

There are several possible ways of approaching the problem, all of which should be used. The first step is to use properly designed and maintained pattern equipment and core boxes to produce the best moulds and cores. There must be good control over moulding, melting and casting operations to give the best castings. Even with the best castings that can be made, the cooperation of the customer should be sought to examine jointly the possibility of reducing the fettling by accepting flash, etc., which can instead be removed by subsequent machining. In many cases it may be possible to transfer much of the fettling now carried out in the foundry, to the subsequent machining of the casting – with a resultant saving to both foundry and customer. It may also be possible to agree on a reduced amount of fettling; since much of this may be for cosmetic purposes only, with little effect on the functional value of the castings.

With fettling reduced to a minimum the job, unless using an automatic machine, is still arduous, dusty and noisy. Working conditions need to be improved – both to meet better standards and to attract the required labour.

CURRENT DEVELOPMENTS AND THEIR EXTENSIONS TO FUTURE USE IN FETTLING

Manual fettling

In many foundries, because of the variety of castings produced and the limited number of each, it is often not practical to consider any other way of fettling than manual fettling methods.

<u>Pedestal grinders</u>   The smallest castings are often hand-held against a pedestal grinder. Such fettling may be improved in two ways. First, improvements in automatic machines are required, making shorter runs economical; second, improving the dust control of pedestal grinders and their method of use is vital. The dust control of pedestal grinders can be adequate when correctly adjusted, but the control deteriorates as the wheel is worn unless proper maintenance and adjustment are carried out. Abrasive-belt grinders of the kind[1,2] shown in Fig. 1 offer an alternative to grinding-wheel machines in many cases, and I believe that the '80s will see a much wider application of these. The newer machines are powerful enough to remove 320 mm^3 of cast iron per mm of belt width per second (0.5 in^3/in belt width). With this rate of metal removal some assistance is required for pushing the workpiece against the belt.

Paper presented at BCIRA Conference "Foundry Technology for the 80's" held at the University of Warwick, England, April 1979. The bound volume of papers presented at the conference is available through BCIRA.

305

The abrasive-belt grinder has a number of advantages over the conventional
bonded abrasive-wheel grinding machine. The abrasive particles are oriented
before being bonded to the belt so that the action of bending the belt round
the work pulley exposes a multitude of minute cutting edges, which remove
material in small cuts rather than by the abrasive action of a conventional
grinding wheel, thereby producing less heat. This, together with easier heat
dissipation because of the length of belt, results in cooler abrasive and
workpiece. The hazard of a bursting abrasive wheel is not present with an
abrasive-belt grinder. The abrasive wheel develops an uneven working-edge,
with wear, and continued grinding operations reduces its diameter - both of
which make continual maintenance and adjustment of dust control equipment
necessary for efficient dust control without excessive air extraction. The
abrasive belt, however, runs over a hard-rubber wheel at the point of contact
with the workpiece, and a true contour is maintained which remains in the same
place relative to the workrest and dust control equipment. This means that,
provided the dust control equipment is properly adjusted initially, its
efficiency will not fall as grinding progresses. A stable grinding-point also
permits a much wider application of simple fixtures, such as the one
illustrated in Fig. 2. A casting is placed in the fixture and oscillated against
the belt. The second fixture, Fig. 3, is basically a chuck which holds a
quick-change mandrel. The workpiece is placed on the mandrel, pushed against
the belt (to a stop if need be) and rotated fully or partially - depending on
the application. Such an approach is 'easier on the hands' of the fettler and
it gives a better, more accurate finish to the casting than a part ground
free-hand. The higher cost of abrasive-belt grinding can be justified provided
use is made of the potentially higher metal-removal rate - which needs sufficient
power for the belt drive and may need assisted pressure on the workpiece.

<u>Fettling-benches</u>   Castings of somewhat larger size will often be fettled on
a fettling-bench. In many cases the castings to be fettled are difficult to
support and hold whilst fettling, and many unsafe and otherwise undesirable
improvisations are resorted to. New techniques are therefore being developed
using rapidly adjustable holding or packing devices. One such development is
the Fetlok casting support system, shown in Fig. 4. This is available in one
size at present, but the range will be extended. Basically, the unit consists
of a table covered with pins upon which the casting is placed, depressing the
pins, Fig. 5. Movement of a lever then locks the pins, which hold the casting
where it has been placed, rigidly enough for fettling to be carried out. Dust
control is taken care of by an extraction hood at the back of the bench. The
Fetlok is suitable for hand-grinding and also chipping.

A casting of the next larger size will probably be fettled on a
fettling-stool in front of a fettling-booth. Such castings will be too heavy
for easy manual lifting. The temptation for easy working is to move round the
casting so that the operator comes between the workpiece and the dust
extraction, which is unsatisfactory from a dust control point of view. There
are two ways of overcoming the difficulty. The first is to put the casting on
a turn-table, which requires the operator to move the casting periodically;
the second alternative is the BCIRA fettling-hood, as seen in Fig. 6. This
hood allows the operator to work round the casting in the natural manner, using
his shoulder to push the hood round as he moves. A further advantage of the
fettling-hood is that the hood itself controls the air flow past the casting,
giving more efficient dust control for lower air extraction than with a
fettling-booth. This gives a lower running cost, because of the reduced power
requirement and the reduced heating required for the lower volume of make-up
air necessary.

<u>Air-extraction zones</u>   Some castings fall between the two size ranges just
discussed. They may, for instance, be too heavy to lift onto a fettling-bench
until several castings have been cut from a spray or the runners and risers
have been removed. BCIRA has developed a combined fettling-booth/fettling-bench,

Fig. 7, where initial fettling takes place on the floor, Fig. 8, in front of
the lower ventilation slot, and final fettling takes place on the bench, Fig. 9.
The air flow is diverted from one position to the other by means of a simple
lever.

Large castings are currently fettled in a very large booth or a down-draft
fettling area, both of which use very large air volumes. It is desirable to
apply dust control to the fettling-tool and thereby use less air, and the LVHV
system was designed for this purpose. It has three main drawbacks: the hose for
the extraction is an added burden to the operator, efficient extraction depends
on the operator's using the tool in the correct manner (which may entail his
using only a small area of the grinding-wheel), and the extraction hoods require
adjustment as the wheel wears.

<u>The BCIRA suspension system for portable grinders</u>   These shortcomings are
overcome by a new BCIRA development, shown in Fig. 10. The object was to make
lighter work of fettling by supporting the weight of the grinder with a balancer,
but as soon as this was done it became clear that here was an opportunity to
incorporate a better dust control than LVHV. The support system allows the use
of a larger-bore exhaust pipe, which allows a higher air-extraction rate using
a lower-power simple fan - compared to the expensive high-power exhauster
required for LVHV. The increased extraction gives better dust control over the
full grinding area of the wheel. For instance, the 230 mm (9 in) disc-grinder,
shown in Fig. 11, can be used over the whole exposed segment with good dust
control. The cup-grinder, Fig. 12, can be used over its full face, from a new
wheel until it is worn out, without the need for adjustment of the extraction
hood. The supporting jib is designed so that it cannot be over-loaded or
damaged by hook-up with the main foundry crane. The system, as currently
developed, is not suitable for internal work on a casting, but in the 1980s we
can expect a combination of this development with a small swing-frame grinder
which will extend the work range which can be covered.

For external fettling the present system is easy to use, and offers good
dust control on the highest-power hand-grinders available. The next stage is
clearly to use a higher-powered tool, which will probably be heavier. During
fettling it is necessary to move the grinder about the casting, and the cable
coupling between the grinder and the balancer allows the beam assembly to swing
about the grinding position, pulling on the grinder and tending to make control
difficult. This effect, for a light-weight beam and balancer designed for
normal hand tools, is not disturbing to the operator. However, in order to
support a heavier grinder the balancer and beam must be larger, and the increased
inertia of these units will make control of the grinding operation more
difficult.

Preliminary tests have indicated that a different kind of balancer, Fig.
13, which eliminates the cable and moves the weight of the actual balancer
nearer to the wall or pivot, reduces the problem to an acceptable level. The
balancer is shown with a normal hand tool, but is capable of supporting a much
heavier grinder. The unit also has the advantage that a simple hand control
can be used to give weight to the tool and assist grinding when necessary. The
balancer can also be used as a light crane, as seen in Fig. 14, to manoeuvre
castings. The operator can, therefore, remove castings from one stack, fettle
them and re-stack without any effort and without waiting for the main shop crane.

There have been other advances in hand-fettling which apply to a range of
casting sizes.

<u>Chipping hammers</u>   A new generation of reduced-vibration chipping hammers[3] is
now available, Fig. 15. These hammers use a cushion of air to isolate the
impact from the hammer itself so that less vibration is transmitted to the
operator. The tool can be seen in action in Fig. 16.

<u>Impact hammers</u>   Hammers are the most common impact tools for removing runners, risers and thin flash. These hammers, particularly the larger ones, are heavy to use and entail a large swing - which can be dangerous. Alternatives are now available. There is the modified humane killer seen in Fig. 17, a modified rifle, and the BCIRA pneumatic impact cylinder. Most of these tools are suspended in use by a balancer system, and are less tiring and safer to use than normal hammers.

<u>Friction band-sawing</u>   Only limited application of high-speed friction band-sawing to the fettling of iron castings has yet taken place, but within its scope it is the fastest method of cutting available. Basically, a friction band-saw[4] is similar to a normal toothed band-saw, but the blade travels at a much higher speed - of up to 75 m/s (15 000 ft/min). A commercially available friction band-saw can be seen in Fig. 18. The workpiece is pressed against the blade, where the friction between the blade and the workpiece rapidly generates heat which softens the material so that it is easily swept away by the blade. The length of the blade ensures that any heat generated in it is readily dissipated. The speed of cut will decrease with increasing material thickness, since the specific pressure applied to the blade will fall and the heat generated will be spread over a wider cut. Because of this, the process is limited to a material thickness of 25 mm. However, a friction band-saw with a blade speed of 50 m/s (10 000 ft/min will cut most cast irons at a rate of 2 mm/s at a thickness of 25 mm. If the thickness of the cast iron is reduced to 12 mm the cutting rate rises to 10 mm/s. If the runners and risers of a casting are within its scope a friction band-saw can be used to cut off runners, risers and flash rapidly, and since the cut can be contoured the subsequent fettling will be greatly reduced or even eliminated. Friction band-saws seem to be less hazardous to use, and to produce less dust, than the circular cut-off wheel. The latter, however, is available with a two-station jig system, where a casting is loaded onto a jig whilst a previously loaded casting is being cut on a second jig. With suitable guarding, this arrangement overcomes the hazards of the cut-off wheel and could equally be applied to the friction band-saw - although the cut would probably be limited to a straight line. It seems to follow logically that in the '80s a flexible jig system, such as Fetlok, will be made available on a two station cut-off machine, either friction band-saw or slitting-wheel, to produce a very versatile piece of equipment. It would be a relatively simple matter to add a light-beam device to the loading station to indicate the cutting-line, allowing precise location and cutting of the workpiece.

<u>Cryogenic fettling</u>

Certain cast irons become embrittled at very low temperatures, and this can be used to simplify removal of runners and risers. A machine[5] has been constructed in which casting sprays are placed in baskets which travel into a tunnel containing a pre-cooling zone and then a chilling-tank. Runners and risers are then removed, in most cases, by dropping the sprays 20 cm onto a slab. In the few cases where this impact is insufficient, a single blow with a small 1 kg hammer completes the operation. Liquid nitrogen is used as the cooling medium. Efficient design of the apparatus is said to reduce the consumption of liquid nitrogen to 0.40 litres per kilogramme of mass cooled, provided that the immersion time is carefully controlled so that the thinner neck sections have cooled sufficiently for brittle fracture to occur but energy has not been wasted cooling the bulk of the casting excessively. The system uses a well-known phenomenon which has not been widely applied because the motivation has not been sufficient to overcome the expense but some castings, particularly nodular iron castings, require many hammer blows for removal of runners, and its advantages will lead to more widespread application in the '80s. It may be possible to increase the range of casting size which can be accommodated, and to reduce the nitrogen consumption by cooling only the neck of the runner before impact, but this will need special appliances for directing the liquid nitrogen. For long production runs and for castings of a similar size and layout, it may be possible to construct a readily adjustable simple frame-type apparatus but, for full versatility with economy, a robot may be required.

Automatic fettling

Smaller castings, in long runs, can often be fettled on an automatic machine
of the types seen in Figs. 19 & 20, which are readily available and have been
fully described at an earlier conference. These machines are limited in the
range of castings which any one of them can handle, and require a set of jigs
for each casting. To be economical, therefore, there must be enough work for
each machine and sufficient numbers of each casting to justify the jigs, and
sufficient castings in a batch to justify fitting the jigs. It is generally
accepted that a machine is not worth setting up for less than a two-hour run.
The machines can fettle from 200 to 1200 castings per hour. The limitations of
the machines are being tackled in a variety of ways, and we can look forward to
more versatile automatic fettling-machines in the future.

At BCIRA we are developing a simple grinding machine, Fig. 21, for small
round parts such as valve seats, burner caps, etc. The machine uses a removable
plate with cut-outs mounted on an indexing table to transport the castings to a
position between a single, simple, replaceable jig and a support table in front
of and slightly above a grinding-wheel. The jig is lowered to clamp the
casting, and the clamping action continues – bringing the flash-line onto the
centre line of the wheel, then the casting is rotated to remove flash and ingate.
Indexing of the table replaces the ground casting and ejection is automatic.
Casting feed can be manual or automatic. 40mm-diameter castings can be fettled
at the rate of 1000 per hour. Replacement of the jig enables a limited range
of castings to be handled, and replacement of the jig and table plate greatly
extends the range. The first machine of this type will handle castings of 25 mm
to 100 mm diameter.

An alternative to fully automatic fettling which may well prove to be the
most economical and productive is a combined man/machine operation where a man
inspects a casting and performs the variable fettling of the flash and sand
burn-on, etc., where necessary, and an automatic machine then performs the
fixed fettling of stubs from runners and risers. The automatic machine may be
in the foundry, or it could be the first operation in the subsequent machining
of the casting.

An alternative to grinding which can be used for automatically fettling
malleable iron castings or other castings which have been annealed is the broach
method. This is illustrated in Fig. 22, which shows a typical broach and
casting. Such equipment has been extended in range, and is now capable of
fettling crankshafts at the rate of 600 per hour. The system is restricted to
castings which have been annealed, since any chilled flash is likely to damage
the dies. The process does offer a quick, quieter form of fettling with less
dust than grinding, and is more readily automated, but long production runs are
necessary before the cost of the dies can be justified; then the costs can be
significantly lower than other fettling methods.

A similar principle can be applied to more complex castings, where the
casting is held and the flash and stubs cut off by means of press-operated
trimming tools which clip the excess material from the casting. These can
approach the casting from different directions. This technique, again, is
limited to long runs, but given these can be very competitive with other methods.

Water-jetting and laser-cutting

A method of cutting which may be useful for fettling is water-jetting.[6] This is
essentially a high-pressure, narrow jet of water which may contain an abrasive
and is capable of cutting metal. Due to the high velocities and pressures of
up to 340 MPa (50 000 lbf/in^2) the wear on nozzles and pumps is high, making
running costs excessive at present. The process will be useful, however, for

internal fettling, and equipment to fettle automatically using water-jetting
will be developed in the next few years.

The laser, a powerful light beam capable of cutting metal, will also need
automatic equipment before it becomes a useful tool for fettling.

Because, for safety, both laser and water cutting need automatic equipment,
and would therefore have to fettle all faces of a casting that might need
fettling, they may be too slow to become competitive with a manual fettler.
Since either method has the potential for automatically fettling internal
cavities beyond the capability of existing automatic machinery, they have
applications in this field - leaving the external fettling to other automatic
equipment, or a fettler, or a combination of the two.

Servo-arms

For some years now the GABLIN grinder[7] has been available: this is essentially,
a grinder on the end of an adjustable boom mounted on a movable carriage, used
to grind ingot moulds. The unit incorporates a basic principle which many will
be endeavouring to apply to fettling in the '80s and will continue to do so
until quieter, less dusty means of fettling are discovered. The principle is
to remove the operator from the source of noise and dust and let him perform
his tasks from a distance, often seated in an air-conditioned, acoustically
treated cab. Several servo-arm manipulators are available, such as those seen
in Fig. 23.

One of these servo-arms is controlled by manipulating a smaller arm or
master unit,the movements of which the larger working arm or slave unit follows,
Fig. 24. This unit is easier to control than the more usual lever-operated
manipulators, particularly for intricate operations. The units can be
servo-operated with force feed-back so that the pressure applied to the work
can be gauged. This makes the unit suitable for fitting to a grinder or a
chipping-hammer, and one is shown in Fig. 25. This application will
undoubtedly be used in the '80s but, at present, the one servo-arm known to
have suitable control capability for this type of application has not been
proved under foundry conditions.

Another method that is being studied for large castings is the use of a
modified gantry-type manipulator for grinding.* Most of these systems require
a fair degree of skill and judgement, and use flexibly mounted grinders to
allow contour-following and avoid damage. As development proceeds it will
probably be found that increased flexibility demands better vision, and
closed circuit television could become necessary for the best performance. In
the long term, control of these manipulators will probably use
micro-processers and mini-computers, to the stage at which they become robots.
It may then be possible for the operator to grind one casting and produce a
programme which can be stored and selected so that the arm grinds an identical
casting automatically from then on. It then becomes a simple matter to store
a range of programs from which the appropriate one for the casting is selected
- either automatically or manually. [8]

Robots

Castings made in sand moulds inevitably vary. The position of flash, burn-on
and defects is often different, from one casting to another. This problem
presents the greatest barrier to automatic fettling by robots and, as a result,
the castings require inspection by a man before being offered to a robot.
This limits the application of robots to fettling, since their cost can rarely
be justified on the basis of the complete elimination of fettling labour
because a man is still needed. They may be justified, however, for the
improved working conditions and productivity they can bring about.

* Work carried out by the Association of Swedish Foundries in collaboration
  with some Swedish foundries.

It is doubtful whether the problem of casting variation can be solved in the near future, and almost certain that it will not be solved by a device capable of operating (or being maintained) within a foundry fettling-shop this century. The simpler problems of casting orientation and identification of shape are not insurmountable, and as a first step several casting-recognition systems are being developed. Some of these are capable of selecting up to fifty different castings and rejecting others which do not conform to its recognizable shapes. The system can then orient the castings, which includes turning them over so that they can be picked up by a robot coupled to the same control system.[9]

Using existing technology, a robot[8] has been applied to the grinding of gearboxes, Fig. 26.* Here, as in most applications to date, the operator inspects, then places, the casting to be fettled onto a jig on one edge of a turn-table. While he is doing this the robot, beyond a protective screen, is fettling the previous casting according to its pre-recorded program. The turn-table is rotated, changing the castings, and the operator inspects the finished casting as he removes it from the jig.

The robot can accommodate a range of programs, and it is said to improve production; it clearly improves the work of the operator. Another application is shown in Figs. 27-29.* Here the operator inspects castings then loads them onto a jig and the robot, carrying a slitting-wheel, removes large runners and risers, again according to a pre-recorded program.

A development now almost complete will enable robots to use oxy-acetylene torches to cut runners and risers off steel castings.[10]

In one particular application a robot has been used to remove sprues from castings and thus replace several operators. The casting spray is made on a DISA line, and is designed to engage and ride on two vibrating arms which penetrate the moulds at the knock-out. After travelling along the arms to remove the major part of the sand from the castings, the spray is picked up by the robot and placed between the jaws of a clipping-press, which removes the castings from the spray. The sprue is then discarded into a bin by the robot.

A more advanced robot system is being developed, which has a manipulator table coupled to the same computer, Fig. 30. With this system it will be possible to combine the actions of the robot and the manipulator table to improve the versatility of the robot, its speed and range of operation and, probably, the resulting finish on the casting - if the robot can be adapted for grinding castings.

The technology exists to present a robot system with a stillage of pre-inspected castings which it can pick out and orient on a transfer mechanism, then selecting a suitable program to manipulate a fettling-tool over all the areas of the casting that could need fettling. The robot could probably never be specific in the selection of areas which actually need fettling on the individual casting it is handling. A simple system could direct fettled castings to the correct bin or, if required, the castings could be automatically sealed, painted, boxed and placed on pallets.

<u>CONCLUDING REMARKS</u>

As I see the future, the amount of fettling will be reduced. Where possible this will initially come about by better design and maintenance of equipment and control of production, and no matter how good or how efficient fettling becomes, these factors should never be overlooked. Some fettling will inevitably remain.

* Work carried out by the Association of Swedish Foundries in collaboration with some Swedish foundries.

Small castings of up to 2 kg will probably be fettled on a specialized
machine, as in Figs. 19, 20 & 21, where medium to long production runs of
identical or similar castings are produced. For medium to long production runs
of medium-size castings (2 - 20 kg) the fettling will be divided into variable
and fixed fettling. The variable fettling will be the flash and sand burn-on,
etc., which a man will continue to fettle, since a machine can only do this work
by grinding every surface which may exhibit the defect - and is therefore
inefficient and time wasting. The man will leave the fixed fettling, such as
stub removal, to machining; his job should therefore become lighter and more
interesting, since it is more variable. The machine might be in the foundry,
or it could be the first machine in subsequent manufacture. Where production
runs are shorter and more varied, robots will come into their own, but limited
to performing the fixed fettling if maximum productivity is to be achieved.
The large castings will continue to be fettled by operators, but using grinders
or 'chippers' mounted on servo-arms which could be fitted with programming units.

We are looking to a future, therefore, of more mechanization in the
fettling-shop, which will be less of a bottle-neck than at present. The
fettling-shop will be a better place to work in - which is just as well,
because the fettler will be with us for many years to come.

REFERENCES

1.   Van Cleave, D.A.
         High-pressure grinding speeds abrasive machining.
         Iron Age, 1976, vol. 218, No. 2, 50-52.

2.   Feinberg, B.
         New credibility for coated-abrasive machining.
         Manufacturing Engineering and Management, March, 1975, vol. 74, 28-30.

3.   Lindqvist, B.
         Vibration control of chipping hammers.
         In: BCIRA. The working environment in ironfoundries. University of
         Warwick, 22-24 March 1977, Birmingham, BCIRA, 1977.

4.   Wilkie Brothers Foundation
         Fundamentals of band-machining, New York, Delmar Publishers Inc.,
         1964, pp. 105-125.

5.   A cryogenic method for removing feeder heads.
         Fondeur d'Aujourd'hui, 1977, No. 285, 25-27.

6.   Butler, P.
         Pressures favouring water-jets build up.
         Engineer, 1978, vol. 246, No. 6371, 22-27.

7.   Simonis, W.
         The machines GAB 400 and GAB 120 for dressing and finishing operations
         of castings.
         In: United Nations Economic Commission for Europe, Seminar on engineering
         equipment for foundries, Geneva, 28 November - 2 December, 1977,
         Oxford, Pergamon Press, 1979.

8.   Svensson, I.
         Manipulators and industrial robots in sand foundries.
         AFS International Cast Metals Journal, 1977, vol. 2, No. 3, 13-19.

9. (a)   Agin, G.J.
          An experimental vision system for industrial applications.
          In: Proceedings of 5th international symposium on industrial
          robots, Chicago, September, 1975. Bedford, International
          Fluidics Services Ltd, 1975.

9. (b)   Abraham, R.G., Stewart, R.J.S. & Shum, L.Y.
          State-of-the-art in adaptable-programmable assembly systems,
          Pittsburgh, Pa., Westinghouse R&D, 1977.

10.   Suschil, T.
          Automation - the robots are coming.
          Modern Casting, 1978, vol. 68, No. 10, 46-47.

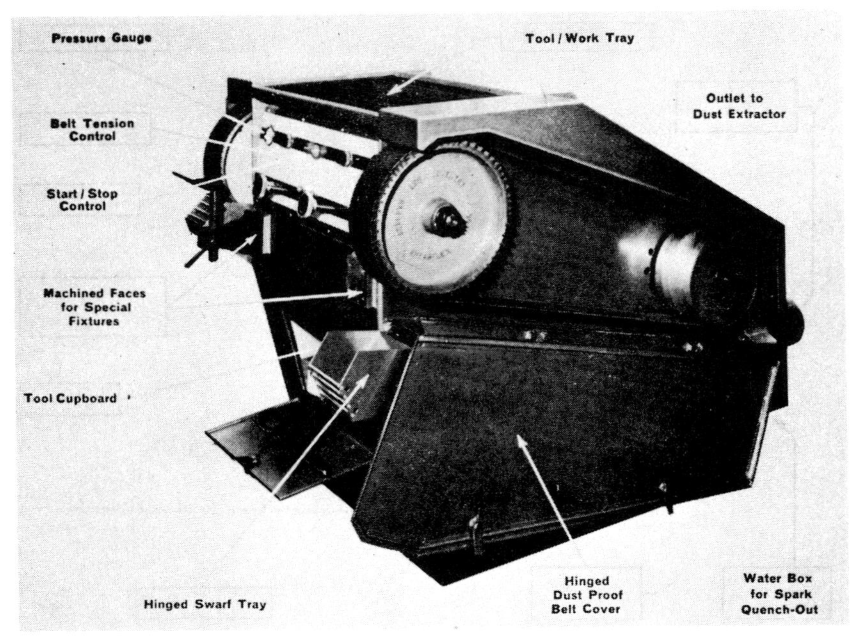

Fig.1.   Typical abrasive-belt grinder.

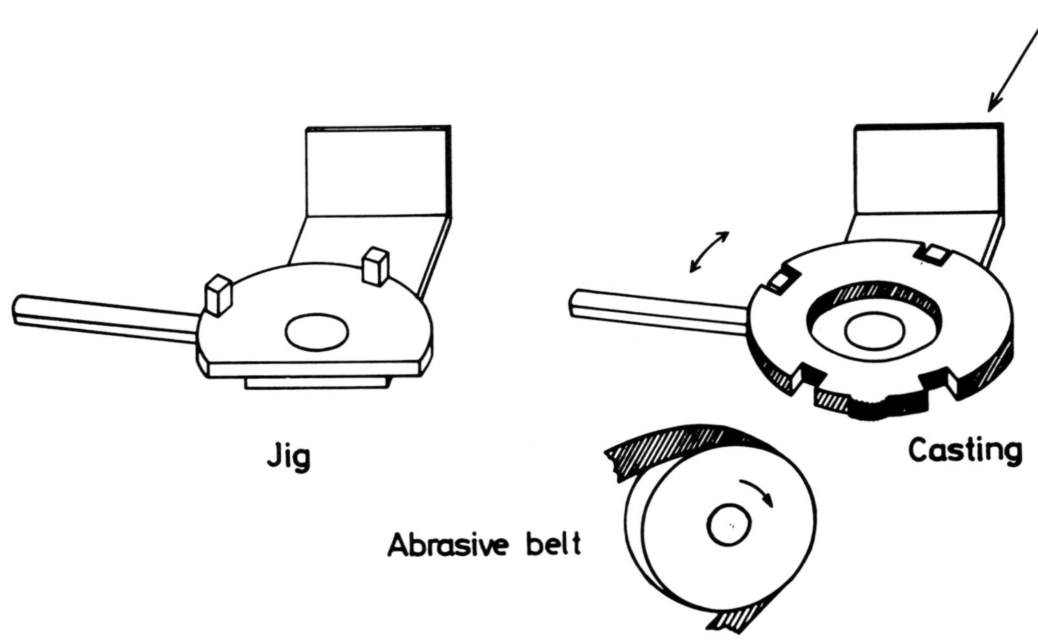

Fig.2.   Jig to aid fettling on an abrasive-belt grinder.

Fig.3.   Rapidly changeable jig for fettling.

Fig.4.   The 'Fetlok' casting support system.

Fig.5.   The 'Fetlok' casting support system in use.

Fig.6.   BCIRA designed fettling-hood.

Fig.7.   BCIRA combined fettling-bench/booth.

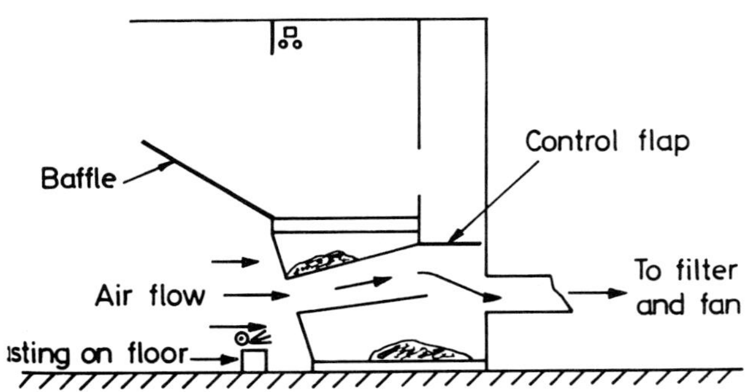

Fig.8.   BCIRA combined unit used as a
          fettling-booth.

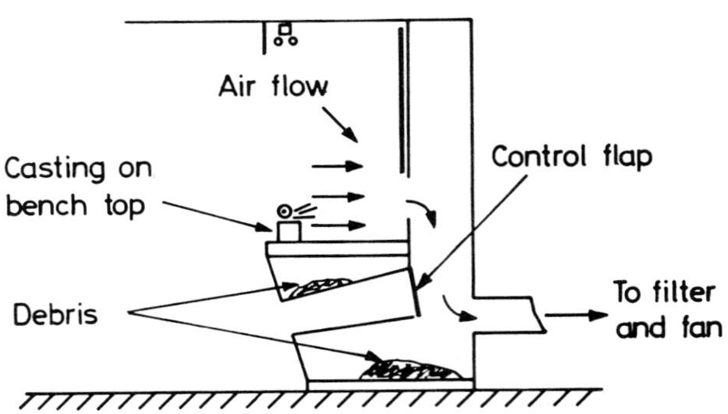

Fig.9.   BCIRA combined unit used as a
          fettling-bench.

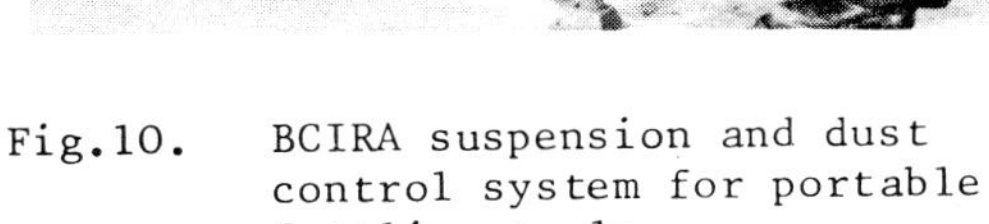

Fig.10.   BCIRA suspension and dust control system for portable fettling tools.

Fig.11.   Disc grinder fitted with extraction hood for BCIRA suspension and dust control system.

Fig.12.   Cup-stone grinder fitted with extraction hood for BCIRA suspension and dust control system.

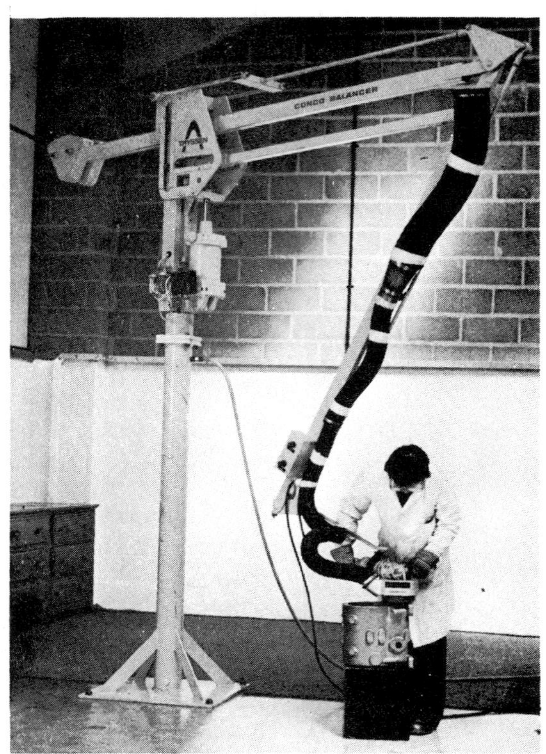

Fig. 13   A commercial balancer used with
           the BCIRA dust control system.

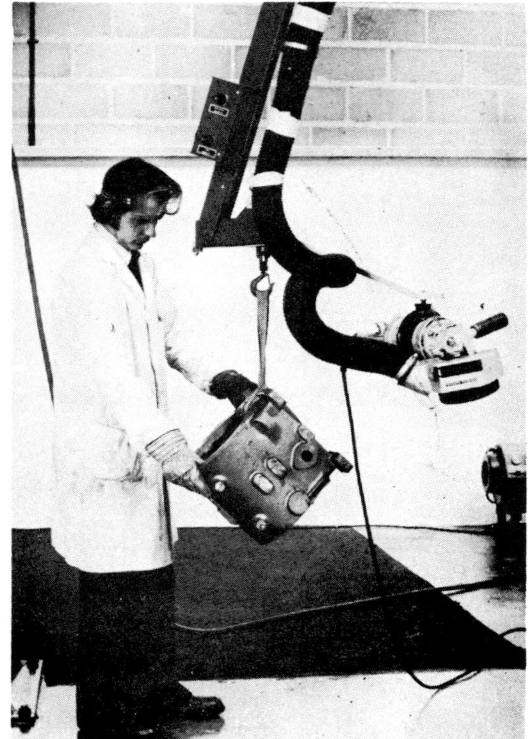

Fig. 14   A commercial balancer,
           illustrating how it can be used
           to lift castings as well as
           support the fettling tool.

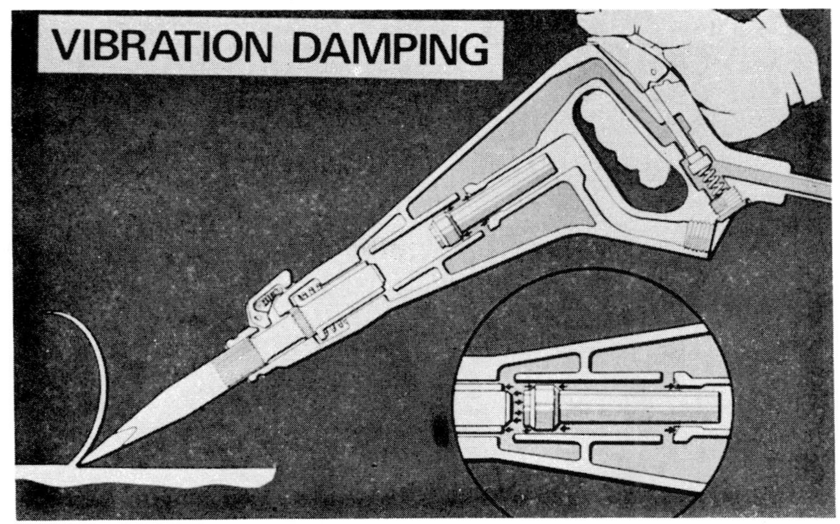

Fig.15.   Reduced-vibration chipping hammer.

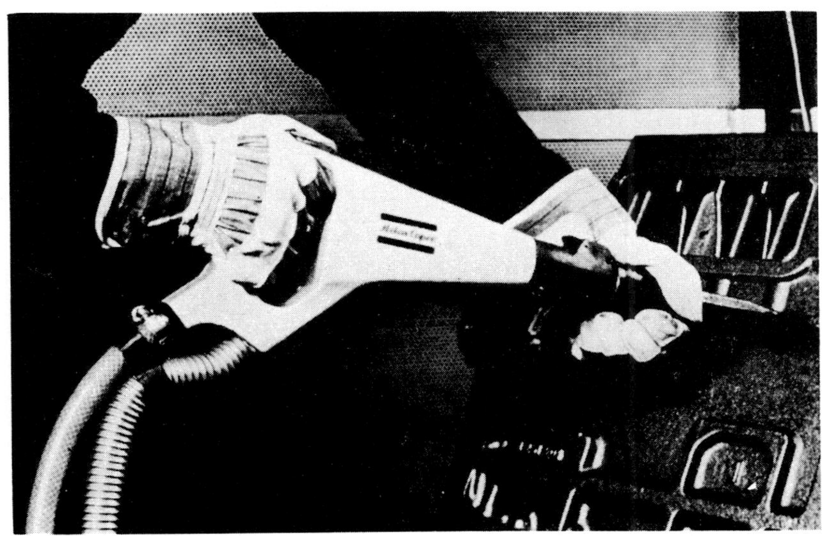

Fig.16.   Reduced-vibration chipping hammer in action.

Fig.17.   Commercial impact hammer using cartridges.

Source: *Foundry Technology for the 80's*, 1979

Fig.18.   Friction band-saw fettling castings.

Fig.19.   Automatic fettling machine.

Fig.20.   Automatic fettling machine.

Fig.21. BCIRA designed machine for automatically
         fettling small circular castings.

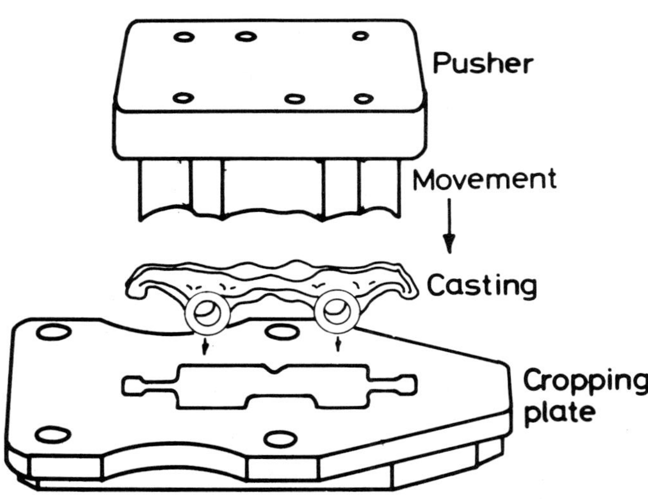

Fig.22. Casting with broach for clipping.

Fig.23. Manually controlled hydraulic
         servo-arm.

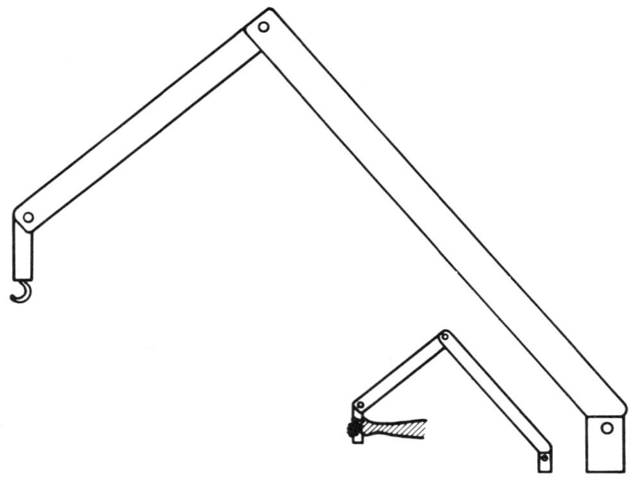

Fig.24.   Servo-arm control using a master/slave system
          with spatial coordination.

Fig.25.   Grinder mounted on a servo-arm.

Fig.26.   Grinding a gear-box with a robot.

Fig.27.   Fettling a casting with a robot-mounted slitting wheel.

Fig.28.   Fettling a casting with a robot-mounted slitting wheel.

Fig.29.   Fettling a casting with a robot-mounted slitting wheel.

Fig.30.   A robot with a coupled, programmable, manipulating table.

The New Family of Zirconia Abrasives

by L. E. Erickson, Product Manager, and
J. J. Paterno, Asst. Dir. Res. & Dev.,
Grinding Wheel Division
P. A. Henderson, Supvr., Prod. Eng.,
Abrasives Marketing Group,
J. E. Patchett, Asst. Dir. of Research,
Norton Research Corp. (Canada) Ltd., and
R. W. Story, Supvr., Manufacturing Eng.,
Coated Abrasive Divison,
The Norton Company,
Worcester, Massachusettes

ABSTRACT

The zirconia alumina family of abrasive materials, first introduced in the 1960s, can truly be termed the most significant, far-reaching development in abrasive technology in decades. Due to the unique characteristics of these materials, zirconia alumina has reduced the cost of removing stock from castings, enhanced the quality of cleaning operations and dramatically increased cleaning room productivity. These combined achievements have provided time and dollar savings to foundrymen. Today, zirconia alumina abrasives surpass previously used aluminum oxide products for virtually every abrasive operation performed in the foundry cleaning room. Continued research and development should make possible an even greater variety of applications for these new abrasives in the future.

History

Prior to the turn of the century, naturally occurring emery and corundum were the primary abrasive materials used in the manufacture of grinding wheels. The abrasive capabilities of these two minerals were recognized by Greek tradesmen as early as 500 B.C. We know that their cutting efficiency was based on a relatively high content of aluminum oxide.

These emery and corundum grinding wheels, however, proved untrustworthy for modern industrial applications. Because these minerals were used in their natural state, they were characterized by unpredictability and could not be controlled. As grinding processes became more sophisticated, uses for emery and corundum grinding wheels gradually disappeared.

The greatest development of the 1800's in abrasive technology was the successful fusing of bauxite, a mineral also known to have a high content of aluminum oxide. For the first time bauxite was fused in an electrical arc furnace where temperatures of over 2000C (3632F) were used to melt the material and reduce its level of impurities. This fusing process transformed one of nature's softest materials into one having unprecedented hardness. By 1904, the production of fused aluminum oxide abrasive had become a commercial success. In time, the abrasive industry developed a rapidly cooled aluminum oxide by means of a casting process. This enabled control of the product's crystalline structure, thereby further enhancing its cutting ability.

After almost 60 years of improving on the basic alumina formulas, these original aluminous abrasives have now been surpassed by alloy abrasives; specifically, the family of zirconia alumina compositions. The purpose of this paper is to explain the magnitude of this long-awaited development.

It seems appropriate to discuss zirconia alumina abrasives at this time because the availability problems that have concerned users of these products over the past two years are being solved through expanded production operations that soon will assure capability of sample supply.

Characteristics

The nature of zirconia alumina abrasive is much easier to explain to those in the foundry industry than to many other industrial groups. In simple terms, one can say that zirconia alumina abrasives are true alloys with properties very similar to those which foundrymen incorporate into martensitic iron, steel and other metal castings.

Zirconia abrasive is characterized by extreme strength, hardness and sharpness. These properties allow zirconia

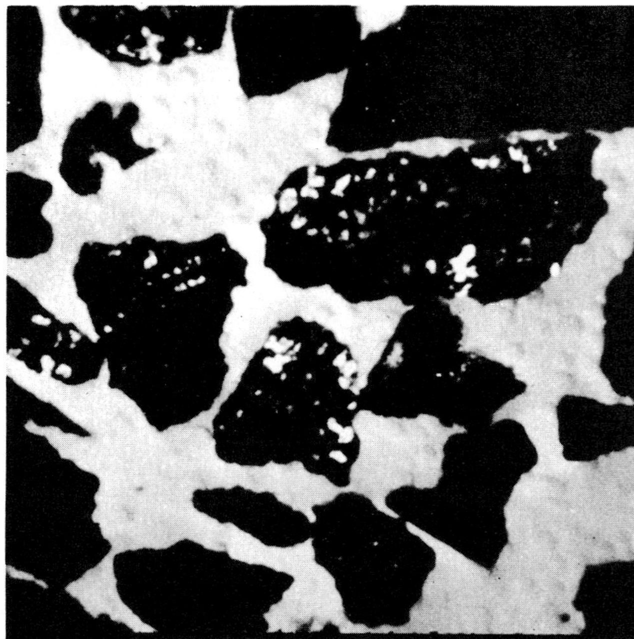
Fig. 1. Crystal size of fine crystalline alumina.

alumina abrasive to stay sharp for a much longer time and thus produce a high rate of stock removal as compared to conventional aluminum oxide abrasives that wear down and dull much more quickly. The standard 25% zirconia abrasive has a remarkably low dulling rate while the newer 40% zirconia abrasive has a unique microstructure that allows for a self-sharpening feature. Individual abrasive grains of the newer 40% zirconia abrasive, rather than being ejected from the grinding wheel or coated belt when dull, as in conventional aluminum

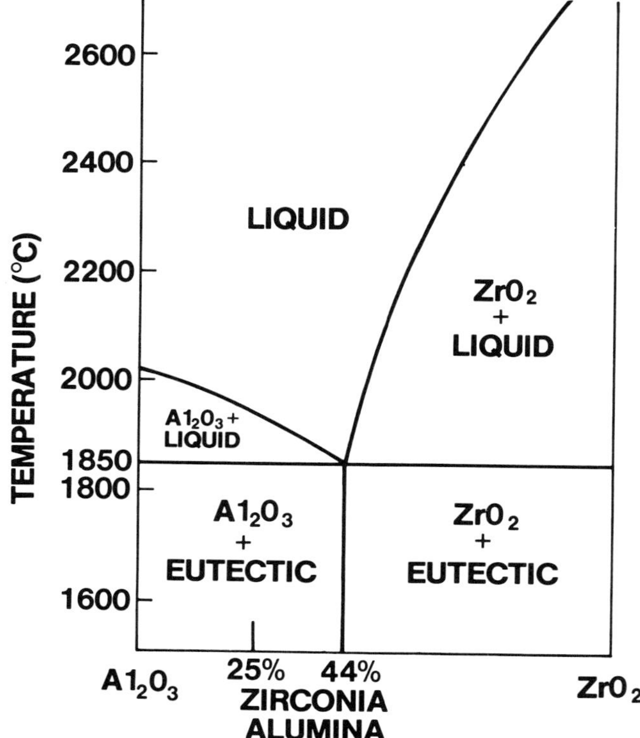

Fig. 2. Zirconia alumina phase diagram.

oxide abrasives, are retained and continually fractured, exposing sharp new edges to the workpiece.

A visual comparison of the two zirconia alumina compositions is given in Figs. 1-4. Figure 1 shows the crystal size which is typical for standard, fine crystalline alumina, which averages about 400 microns. Figure 2 is a phase diagram of the zirconia alumina system, including the two alloy compositions, 25% and 40% zirconia. Figure 3 shows a polished section of the standard 25% zirconia abrasive and the dentritic or branchlike crystalline structure of the alumina in this abrasive. Figure 4 depicts a polished section of 40% zirconia composition and shows the extremely fine microstructure of the alumina zirconia eutectic at 5000X magnification.

Impact

Although it may be obvious, it is important to point out that major new developments in abrasives, such as zirconia alumina, occur very infrequently. In the interim period between these developments, it falls upon the abrasive industry to constantly seek improvements in bonding formulations and manufacturing techniques. For zirconia alumina related operations, the cumulative effect of wheel improvements, plus corresponding equipment improvements, is tabulated in Fig. 5.

As these illustrations indicate, improvements in all aspects of the grinding operation machine variables as well as wheel variables have been realized. In many cases these changes have been interdependent. That is, an improved abrasive would not achieve its maximum benefits without higher horsepower, wheel speeds and head pressures to utilize the abrasive to its maximum potential. This interdependence will be given more detailed treatment a little later.

Bonded Abrasives

The real measure of an abrasive development is how production costs can drop with a new abrasive material or composition. In a recent study of floorstand grinding,* grinding costs using standard aluminum oxide and zirconia alumina abrasives were discussed. The data, summarized in Fig. 6, compares grinding cost as a function of the type of abrasive used. As the table shows, the 25% zirconia alumina composition operated at the lowest cost level of all the abrasive types tested. Production rates from foundry users of zirconia alumina grinding wheels support this generalized data. It's really like comparing the single engine biplane of World War I vintage (aluminum oxide) to modern jet aircraft (zirconia alumina) and it has taken the abrasives industry about as long to develop the technology.

Similar information relating to cost has been developed for the steel industry. As shown in Fig. 7, the conditioning cost in steel mills has been significantly reduced with the development of zirconia alumina abrasives.

The cost reductions that zirconia alumina abrasives have brought to the foundry and steel industries are just two examples of the emphasis that has been placed on developing new and better products for the primary metals industry. In the face of ever-increasing worldwide inflation, these figures constitute a truly significant achievement.

Achievements of this nature are not without their price however. To the grinding wheel manufacturer, this price is the maintenance of an aggressive, technically superior research and development organization that can take a promising development, optimize the bond that constitute the grinding sheel and then ensure that the product is feasible for manufacture. Also required is a marketing organization capable of exploring the full range of application for the product, for many a promising

---

*Modern Casting, Feb. 1973, June 1973.

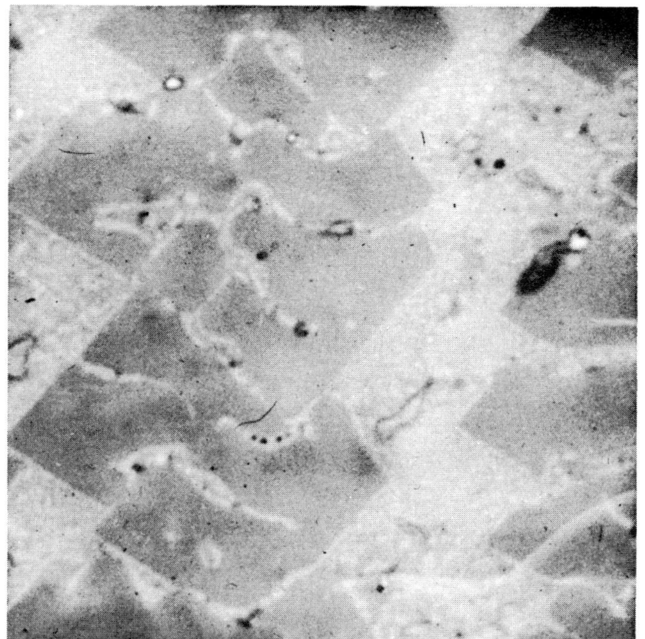

Fig. 3. Polished section of 25% zirconia alumina.

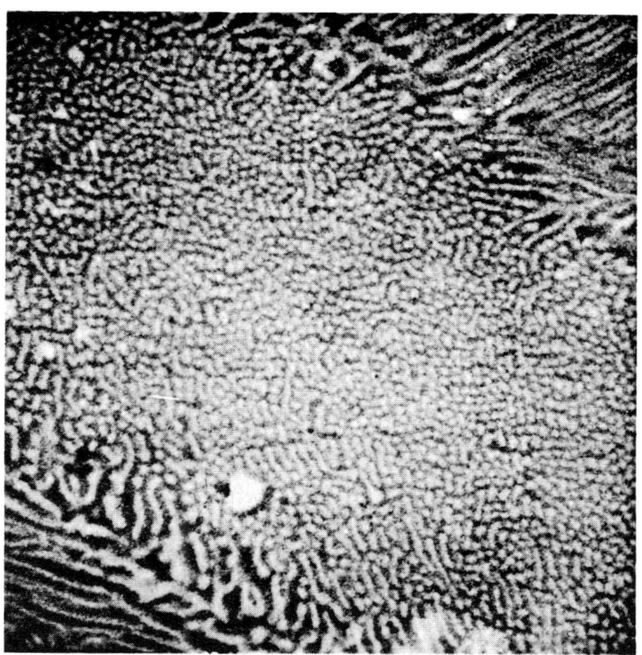

Fig. 4. Polished section of 40% zirconia alumina.

Fig. 5. Floorstand grinding, improvements in product and process.

FACTOR	1930s	1940s-1950s	early 1960s	TODAY
Abrasive	Al_2O_3	Al_2O_3	Improved AL_2O_3	Al/Zr
Horsepower	5-10	10-15	10-20	20-30
Speed	6500 rpm	9500 rpm	9500 rpm	12500 rpm
Bond	Vitrified	Resinoid	Resinoid	Safety Web Resinoid

Fig. 6. Floorstand grinding, performance and cost summary nodular iron grinding.

		AL_2O_3	Sintered AL_2O_3	25% ZrO_2/ 75% AL_2O_3
a).	Stock removed/ cycle, lb.	1.14	2.46	17.8
b).	Grinding time/ cycle, min	0.43	0.56	4.45
c).	Wheel loss/ cycle, in.^{3}	2.75	4.55	1.65
d).	Abrasive cost, in.^{3}	7	8	10

Other inputs:

    a. Dressing time- 5 sec/cycle
    b. Labor and overhead costs - $15/hr, or 25¢/min.

		AL_2O_3	Sintered AL_2O_3	25% ZrO_2/ 75% AL_2O_3
e).	Grinding time cost/cycle, ¢	10.8	14	111
f).	Dressing time cost/cycle, ¢	2.1	2.1	2.1
g).	Abrasive cost/cycle, ¢	19.3	36.4	16.5
	Total e&f&g	32.2¢	52.5¢	129.6
h).	Unit cost of stock removed/cycle, ¢/lb.	28	21	7

Fig. 7. Progress in steel conditioning.

DATE	LABOR AND OVERHEAD RATE PER HOUR	COST PER LB METAL REMOVED	METAL REMOVED LBS. PER HOUR
1948	$5	22¢	75
1957	5	17.3	150
1958	10	15	200
1959	10	12	260
1965	15	7.5	300
1966	15	5.5	400
1969	20	5.0	650
1972	30	2.5	1000

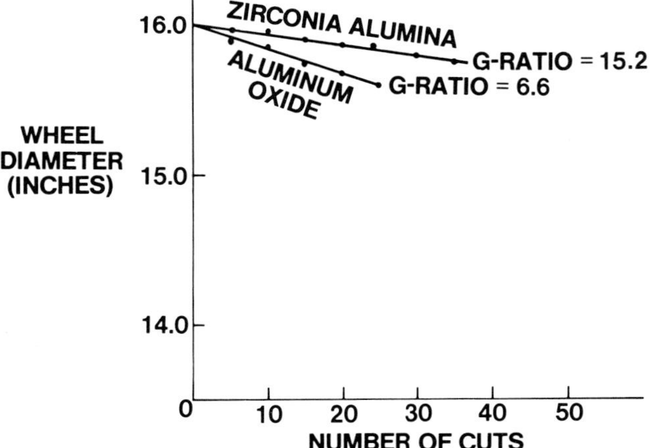

Fig. 8. Cut-off test.

development has died an early and perhaps unnecessary death because of misapplication early in its life cycle.

The producer must also establish lines of communication with customers so that they are aware of the potential for improving their process. He must ensure that adequate supplies of the product are available when and where customers need them. For zirconia alumina abrasives, the adequacy of supplies is partially

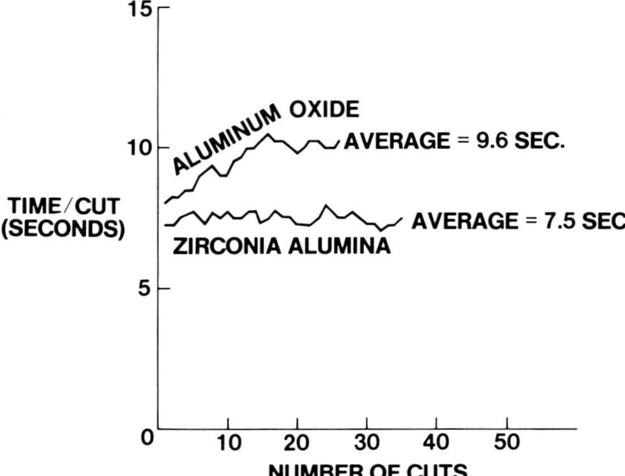

Fig. 9. Cut-off test.

ensured by the product itself. Because of their longer life and better operating efficiencies, zirconia alumina products provide four times the capability of conventional products. Simply put, this means one zirconia alumina grinding wheel is equal, in terms of production, to four conventional grinding wheels.

The successful application of zirconia alumina products in industry has created shortages in the supply of this abrasive during 1974. This shortage was due to limited fusion furnace capacities and a shortage of raw materials, particularly zircon sand. Today, through the opening of new mines in Australia, South Africa and other locations throughout the world, the supply of zircon sand has eased. Also, production facilities for furnacing zirconia grain are being expanded to meet the greater demand. Facilities for the manufacture of zirconia alumina grinding wheels are being expanded. The fact is that over the past three years, such expansion projects and the increased availability of zircon sand have substantially increased the abrasive industry's zirconia alumina product capacity. Projects currently underway will provide even further increases in the near future.

At least one manufacturer is paying the price by investing in projects that will increase their product capacity to assure the foundry industry that zirconia alumina abrasive products are and will be available when they need them. There is a further price, however, that foundrymen must pay. As we continue to push the efficiency of cleaning room grinding operations higher and higher, and associated costs become lower and lower, the nature of the interaction between the wheel and the machine environment becomes quite critical. Wheel speed, horsepower, machine rigidity and maintenance all assume far greater import than in the past.

For example, with the advent of the latest zirconia alumina abrasive developed for foundry cut-off, a general rule which applies is that the machine must be capable of delivering to the wheel one horsepower per inch of wheel diameter. In other words it takes a 25 hp machine to fully utilize a 24-in. diameter zirconia alumina cut-off wheel. The result is that foundrymen may have to make some rather large initial investments which will amortize through greater productivity.

Once this price is paid, some tremendous strides in productivity, cost reduction and grinding quality are possible. Figures 8-10 give comparisons based on several measures of grinding performance between a conventional aluminum oxide wheel, commonly used in foundry cut-off operations, and a zirconia alumina cut-off wheel. Both had internal reinforcing and were operated at 3,500 rpm. The machine was a 20 hp chop stroke type, and the material cut was 2-in. diameter, 304 stainless bars. All cuts were made under a constant force of 143 lb.

Figure 8 shows a plot of wheel diameter vs number of cuts.

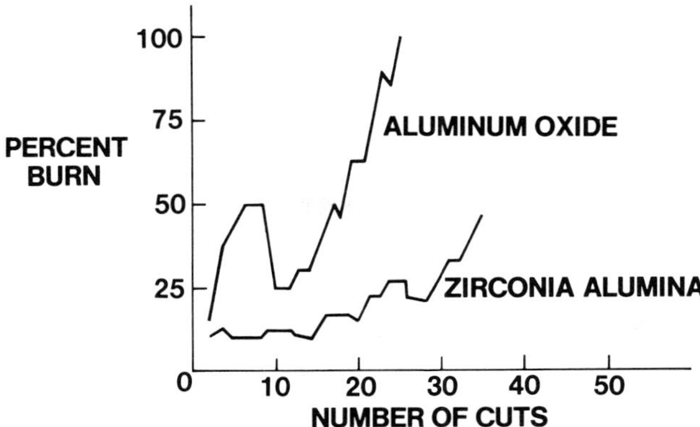

Fig. 10. Cut-off test.

The conventional wheel wore at a much faster rate than the zirconia alumina wheel. This observation is verified by calculated G-ratios, or ratios of metal removed to wheel wear. For the conventional aluminum oxide wheel, this ratio was 6.6 against a ratio of 15.2 for the zirconia alumina wheel. This indicates a performance edge greater than a factor of two for the zirconia alumina wheel.

G-ratio alone cannot tell the entire story, however. High G-ratios may sometimes be achieved by hard wheels, which last quite long, but at a sacrifice of cut rate. Figure 9 shows that this is not a factor in this comparison. Note that over the 25 cuts made with the aluminum oxide wheel, the average time per cut was 9.6 sec, while with the zirconia alumina wheel, the average time per cut for 35 cuts was 7.5 sec, a reduction of 22%.

Further, the quality of the cut was greatly improved. Under grinding test conditions, workpieces cut with conventional aluminum oxide abrasives displayed a high degree of burn. Those cut with zirconia alumina abrasives, however, were relatively burn free. Figure 10 plots the percentage of visual burns observed on each cut surface. Note that after the 25th cut with the conventional wheel, the burn level had reached 100%.

The zirconia alumina family finds utility in other operations such as portable snagging using raised hub wheels, cup wheels or straight wheels, on air or on electric powered portable grinders. Operator opinion constitutes a major part of the evaluation of grinding wheels of this type, as actual performance is quite difficult to measure. Some laboratory evaluations have been developed, however, which simulate actual operating conditions and correlate quite closely with user reports.

Figure 11 illustrates one such evaluation. This graph plots wheel wear expressed as a percentage of the usable wheel life against material removal in grams. The significance of the two lines may be seen by drawing a vertical line at the 15% wheel wear position. At this level of wheel wear, the conventional wheel had removed 610 grams of metal, while the zirconia wheel had removed 1,000 grams. The lines continue to diverge as well, meaning that as the wheels continue to wear, the difference in stock removal becomes greater and greater.

Coated Abrasives

The success of zirconia alumina abrasives in bonded applications led to attempts to extend their use to coated abrasives. As a first step in this development, the same standard 25% zirconia composition used in grinding wheels was incorporated in a heavy duty coated abrasive product design. Grinding tests, however, quickly established that this product would have very limited application in the field. In fact, tests showed no advantage over aluminum oxide abrasives except at very high pressures. The successful adaptation of zirconia

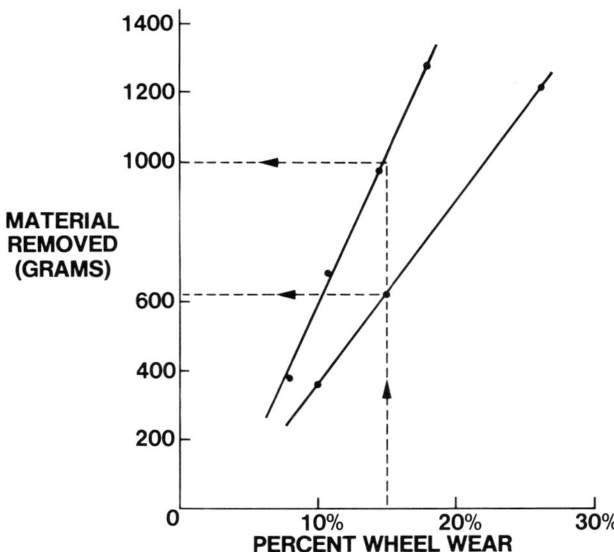

Fig. 11. Portable wheel grinding test.

alumina to coated abrasives had to await the development of an entirely new zirconia alumina composition.

Zirconia alumina abrasives are, as a class, quite resistant to wear. However, as generally formulated they had a characteristic tendency to dull quite quickly under low pressure grinding conditions through the formation of flats on the cutting surface. The key to the successful development of a zirconia alumina line of coated abrasives then was in the modification of the abrasive itself, to incorporate a unique microstructure.

The plot of typical grinding test results, measured in laboratories shown in Fig. 12, gives some idea of the relative performance of aluminum oxide, the standard 25% zirconia composition and a 40% zirconia alumina composition over a range of grinding pressures. The aluminum oxide has a very definite optimum pressure range. The standard 25% zirconia alumina composition works poorly at low pressure, better at high pressure. The 40% zirconia alumina composition on the other hand, has a very broad range over which grinding performance is fairly uniform.

In addition, the 40% zirconia alumina abrasive generally maintains a much higher rate of cut than comparable aluminum oxide products under the same grinding conditions (Fig. 13).

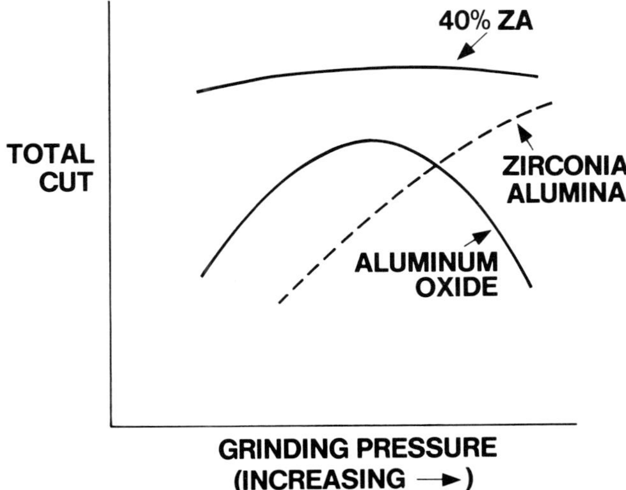

Fig. 12. Coated abrasive test.

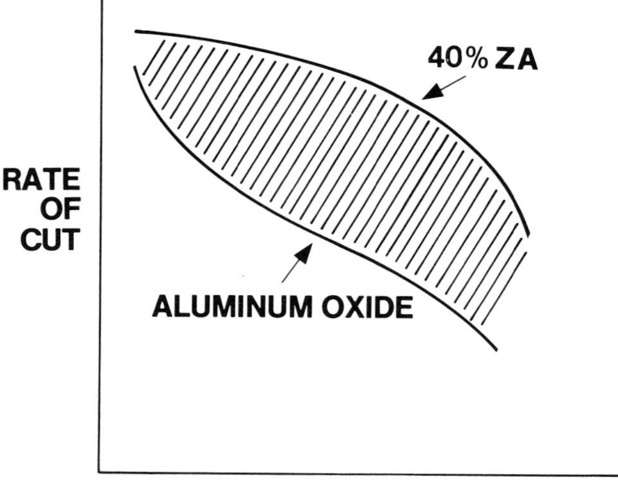

Fig. 13. Coated abrasive test.

By grit size	
Grit	Success Ratio
80X	59%
60X	80%
50X	85%
40X	82%
36X	85%
24X	79%

By type of metal	
Aluminum	76%
Brass and bronze	71%
Stainless & high alloy steel	77%
Tool steel	86%
Mild steel	88%
All others	89%

Fig. 14. Overall reorder success.

The overall performance advantage of the 40% zirconia alumina abrasive is proportional to the shaded area between the two curves. As far as materials to be ground are concerned, it was again found that the 40% composition had very broad application. Although the advantage of the 40% zirconia alumina abrasive does vary with different materials, there are very few on which it shows no advantage at all.

Research and development test data predicted a significant performance advantage for the 40% zirconia composition in commercial coated abrasive applications. This has been confirmed by the product's acceptance by the foundry industry.

One of the measures of a new product's success is the "recorder success ratio", which is simply the number of customers who place orders after trying the product, as against the total number of trials. As shown in Fig. 14, the overall reorder success ratio for zirconia alumina products has been more than 80%. The foundry industry in particular has had a success ratio at least as good as shown by the overall results and perhaps a bit better.

The wide operating range of the 40% zirconia alumina abrasive is of particular value in foundry applications because of the range of workpieces involved and the resulting variation in unit pressures. The overall success ratio of better than 80% speaks for itself but perhaps a few specific examples of the production increases possible in foundry applications will help to dramatize the point.

A large manufacturer of ductile iron castings was using a standard type 25% zirconia alumina belt to grind castings on a floorstand grinder with a 40 hp drive at 8,250 sfpm. Three gates were being removed from a caterpillar casting using power assist. Average production was 25 castings per belt which was considered quite satisfactory. The belts made from the 40% zirconia composition averaged 75 castings, an increase of 200%.

Another example: at another foundry using the same type of equipment and type of abrasive belt, average production was 57 malleable iron castings per belt. The 40% zirconia alumina abrasive averaged over 100 pieces — a 100% increase in production.

Perhaps the best success story has been in nonferrous foundry applications because in this case possible advantages could not be predicted. This turned out to be one of the most productive applications. These examples show that applications available for 40% zirconia alumina abrasives can result in significant process performance and cost improvements for foundrymen.

Summary

Several years of research and experience in the marketplace have shown that wherever there is an application for abrasives in the foundry, there is an application for zirconia alumina abrasives. Some of the zirconia alumina products, such as floorstand snagging wheels, are relatively mature, with well-defined product and application guidelines. Others such as cut-off abrasives, are in their infancy, being less than a year old with wide possibilities yet to be explored. Future zirconia aluminia applications, such as precision cylindrical grinding, are now under investigation. In total, the development of this new family of abrasives demonstrates continuing commitment to foundrymen in the furthering of improved grinding for the foundry industry.

The Effect Upon Mechanical Properties of Variation in Graphite Form in Irons Having Varying Amounts of Carbide in the Matrix Structure and the Use of Nondestructive Tests in the Assessment of the Mechanical Properties of Such Irons

G. F. Sergeant
A. G. Fuller
BCIRA, Birmingham, England

Introduction

The work described in this report is a continuation of that previously carried out on behalf of the Ductile Iron Quality Control Committee of the American Foundrymen's Society. One of the important objectives of the work has been to evaluate alternative methods for the nondestructive assessment of the effect of variations in graphite form and in matrix structure upon the mechanical properties of the irons produced, so that recommendations can be made to foundries concerning the most suitable nondestructive tests which might be applied. The application of these tests should enable foundries to guarantee that all castings supplied to their customers have at least the minimum properties specified.

Earlier reports[1-3] were concerned with irons in which the matrix structure was either ferritic or pearlitic and only the graphite form was varied. It was shown that both measurements of resonant frequency by sonic testing and of ultrasonic velocity were very sensitive to the changes in mechanical properties caused by the presence of non-nodular graphite, either in a compacted (ASTM type IV) or in a "spiky" ASTM type III form. Provided that the matrix structure was known and remained consistent, measurements made of resonant frequency or ultrasonic velocity provided good estimates of the mechanical properties of the castings, and either measurement could be used to ensure that their properties conformed with users' requirements.

In a second phase of this work,[4] both matrix structure and graphite form were varied simultaneously to give a series of irons having various amounts of ferrite and pearlite in the matrix and containing increasing amounts of non-nodular graphite in the compacted (ASTM type IV) form. In addition to the use of resonant frequency or ultrasonic velocity measurements to assess the influence of graphite form, eddy current and coercive force testing were used to detect and measure the effect on properties of changes in matrix structure. It was shown that by combining either resonant frequency or ultrasonic velocity measurement with either coercive force or eddy current testing, it was possible to indirectly assess the

mechanical properties of the castings, and hence to guarantee the engineering properties upon which they might be sold.

An assessment was also made of the possibility of using metallographic techniques including the use of image analysis as a means of assessing the properties of castings. However, the conclusion was reached that the metallographic methods currently available were not a satisfactory alternative to the other methods of nondestructive testing.

The present work investigates the use of sonic or ultrasonic testing in combination with eddy current or coercive force testing to evaluate the properties of ductile irons having various amounts of carbide and non-nodular graphite in otherwise pearlitic matrices. Since the work as a whole has now covered the principal structural variations likely to be encountered in the production of ductile iron castings, a short supplementary report has also been prepared. This takes into account results from all phases of the work carried out on behalf of AFS, and re-examines them to provide, in a single document, a statement concerning the use of nondestructive tests in ductile iron quality control for the purpose of evaluating the mechanical properties of castings.

Investigation Program

In accordance with Contract Proposal SP.874/2, it was agreed that a series of nine irons would be produced in which the following variations in structure would be obtained in the as-cast condition: 1) proportion of graphite present in nodular form — greater than 90%, near to 70%, near to 50% and 2) proportion (excluding area occupied by the graphite) of uniformly distributed carbides in an otherwise pearlitic matrix — 2-5%, 7-12% and 17-22%.

Variation in the amount of non-nodular graphite would be produced by control of the addition of the nodularizing alloy to reduce residual magnesium content. The non-nodular graphite would be in a compacted (ASTM type IV) form. Small additions of tin would be made to promote pearlite within the matrix. Initial tests showed that by the addition of ferrochromium alone it was not possible to obtain uniformly distributed carbides at the higher levels required. After consultation with AFS, in accordance with the contract proposal, it was agreed that the required variation in the amount of carbides would be obtained by constant small additions of ferrochromium to the furnace charge and a varying but small addition of ferroboron to the metal in the ladle prior to casting.

Due to the marked inoculating effect of magnesium ferrosilicon, it was agreed that ladle inoculation with ferrosilicon might be reduced or omitted. The latter was found to be necessary in order to obtain the higher amounts of carbide. Therefore, in order to maintain consistent treatment procedures throughout the series, inoculation was omitted from all melts and the silicon content of the furnace charge was increased in order to maintain cast metal compositions which were similar to those in the earlier phases of the work.

From each of the nine treatments, two pairs of keel blocks would be cast and these would be used to provide test specimens for:

1) Determination of tensile strength, yield strength at 0.1, 0.2 and 0.5 percent offset, elongation, hardness, un-notched impact values, and impact transition temperatures.

2) Measurement of the velocity of ultrasonic energy, relative attenuation, resonant frequency, damping capacity (by both band width and decay methods), determination of modulus of elasticity, and measurements using coercive

Table 1. Furnace Charge Materials and Ladle Additions

BCIRA Melt Ref. No.	Iron No.	Furnace Charge, kg						Ladle Additions, kg	
		OBV Pig Iron	Swedish Iron	Ferrosilicon	Ferro-manganese	Tin	Ferro-chromium	MgFeSi	Ferro-boron
X507	1	76.850	0.100	2.135	0.460	0.080	0.375	0.960	0.045
X523	2	76.850	0.100	2.135	0.460	0.080	0.375	0.960	0.090
X545	3	76.850	0.180	2.030	0.460	0.080	0.400	1.008	0.180
X669*	4	74.280	2.665	2.115	0.460	0.080	0.400	1.000	0.030
X525	5	76.850	0.100	2.135	0.460	0.080	0.375	1.008	0.090
X570	6	76.850	0.180	2.030	0.460	0.080	0.400	1.008	0.180
X710*	7	74.280	3.320	1.460	0.460	0.080	0.400	2.000	0.030
X514	8	76.850	0.235	2.080	0.460	0.080	0.375	1.056	0.090
X613*	9	74.280	3.360	1.420	0.460	0.080	0.400	2.000	0.200

Table 2. Compositions of Charge Materials and Ladle Additions

Material	C %	Si %	Mn %	S %	P %	Ni %	Cr %	Mo %	Cu %	Ti %	V %	Al %	Co %	Sn %	As %	Pb %	Sb %	B %	Ca %	Mg %	Ce %
OBV 548	4.06	0.06	<0.005	0.008	0.02	<0.005	0.01	<0.025	0.01	<0.01	0.02	<0.005	<0.03	<0.01	<0.01	<0.0002	<0.001	<0.001			
OBV 549	4.20	0.06	0.02	0.014	0.015	0.03	<0.01	<0.025	0.01	<0.01	0.03	<0.005	<0.03	<0.01	<0.01	<0.0002	<0.001	<0.001			
Swedish Iron	<0.02	0.01	<0.12	0.02	0.02	0.02	<0.01	<0.02	0.02	<0.02	<0.02	<0.02	<0.03	<0.02	<0.01	<0.0002	<0.001	<0.001			
Lumpy FeSi		76.82										1.31							1.04		
FeMn	0.09	0.84	88.6																		
MgFeSi		41.6										0.99							0.96	6.48	0.60
FeCr	0.03	0.70					66.4														
FeB	0.14		2.48				0.64					0.36						19.2			

force and eddy current testing on as-cast and on fully machined specimens.

3) Assessment of graphite structure and of the amounts of carbide in the matrix both by visual assessment and by measurement using image analyzing microscopes.

Experimental Procedures

Production of Materials

The series of irons were produced to have a nominal composition of C, 3.6%, Si, 2.4% (after magnesium treatment), Mn, 0.4%, S, 0.02% (sulfur content prior to magnesium treatment) and P, 0.029% max.

The charge for each melt consisted of oxygen blown Vantit pig iron together with small additions of Swedish iron, ferrosilicon and ferromanganese, tin to promote the formation of pearlite, and ferrochromium to aid the formation of carbides in the as-cast material. The weight of ferrosilicon added to the furnace charge was adjusted according to the weight of the magnesium ferrosilicon nodularizing alloy to be added, so that after treatment all irons had a similar silicon content. The makeup of the charge for each 80 kg melt is shown in Table 1. The composition of the charge materials is shown in Table 2.

Melting was carried out in a high frequency induction furnace. Metal was superheated to 1500C (2730F), held for 5 minutes, and then allowed to cool to 1470C (2680F). It was then tapped from the furnace and poured into a ladle which had at its base the required weight of crushed (½-⅜ in. mesh) magnesium ferrosilicon alloy. The weights of the magnesium ferrosilicon alloy additions were calculated to provide the necessary variation in residual magnesium content to give the required

Iron No.	COMPOSITION, %								
	C	Si	Mn	S	P	Mg	Sn	Cr	B
1	3.47	2.47	0.44	0.016	0.010	0.027	0.091	0.27	0.0079
	3.51	2.44	0.44	0.016	0.009	0.025	0.089	0.27	0.0074
2	3.64	2.59	0.45	0.018	0.010	0.025	0.082	0.25	0.013
	3.64	2.58	0.46	0.017	0.010	0.025	0.082	0.25	0.012
3	3.73	2.29	0.44	0.010	0.007	0.026	0.081	0.29	0.027
	3.69	2.32	0.44	0.011	0.008	0.026	0.082	0.29	0.028
4	3.66	2.62	0.48	0.014	<0.010	0.027	0.080	0.29	0.0049
	3.63	2.63	0.49	0.014	<0.010	0.026	0.081	0.29	0.0052
5	3.62	2.53	0.47	0.016	0.009	0.025	0.078	0.26	0.012
	3.63	2.58	0.48	0.016	0.009	0.026	0.078	0.26	0.012
6	3.58	2.51	0.48	0.013	0.009	0.026	0.084	0.27	0.028
	3.58	2.47	0.48	0.013	0.008	0.025	0.082	0.28	0.026
7	3.51	2.46	0.45	0.012	0.010	0.044	0.082	0.29	0.0073
	3.51	2.41	0.45	<0.010	0.010	0.043	0.081	0.29	0.0065
8	3.52	2.56	0.45	0.013	0.008	0.028	0.079	0.26	0.013
	3.48	2.55	0.45	0.014	0.008	0.029	0.079	0.26	0.013
9	3.66	2.30	0.37	0.012	0.010	0.028	0.086	0.28	0.032
	3.58	2.24	0.37	0.012	0.010	0.028	0.082	0.28	0.031

NOTE: Two samples were cast, one with each pair of keel blocks.

variations in graphite structure. Ferroboron, necessary to obtain the required amount of carbides, was added to the ladle after the magnesium reaction had subsided. The additions made are shown in Table 1. Pouring began when the metal temperature had fallen to 1380C (2515F). From each melt two pairs of keel blocks, 1¾ in. wide by 1¾ in. deep, were cast into greensand molds. One chill cast sample for spectrographic analysis was poured with each of the pairs of keel blocks.

Analysis

The chill cast spectrographic samples were analyzed using a direct reading vacuum spectrometer, to determine the silicon, manganese, phosphorus, magnesium, tin, chromium and boron contents. Carbon and sulfur contents were determined, on solid samples cut from the spectrographic test pieces, using a carbon and sulfur determinator. The compositions of the irons are shown in Table 3. They were similar to those produced in the earlier work carried out on behalf of AFS.

Sectioning of Keel Block Castings

From each melt, four 1¾ in. keel blocks were available, produced as two pairs. The first pair were identified by the letters A and B and the second pair by C and D. The machining program for each melt was as follows:

Block A: Sonic test bar and one tensile bar, cut side by side from bottom section of keel (Fig. 1). Top sections were retained as spare material.

Block B: Nine un-notched impact test pieces from a slab cut from the bottom of the keel and a further nine test pieces from a slab immediately above (Fig. 2). The orientation of the test pieces was recorded so that

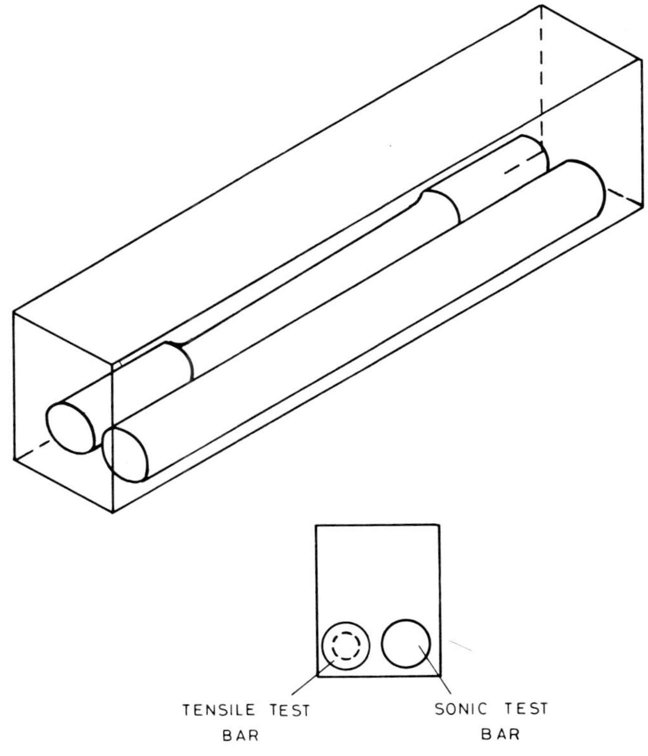

Fig. 1. Positions of tensile and sonic test bars in keel blocks A and C.

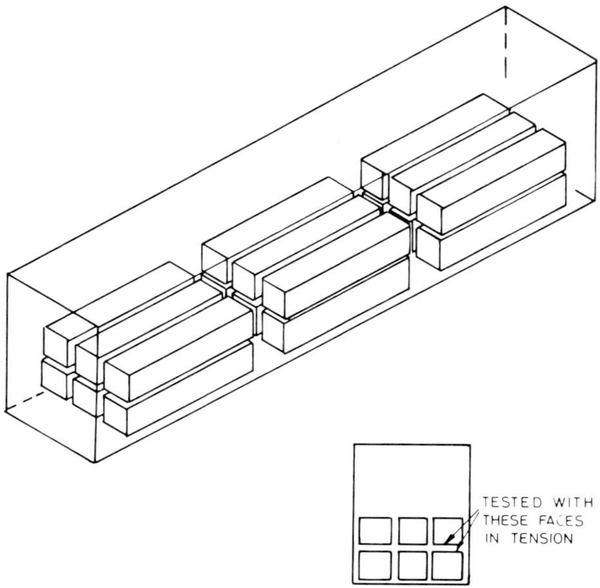

Fig. 2. Position of un-notched impact test pieces in keel block B.

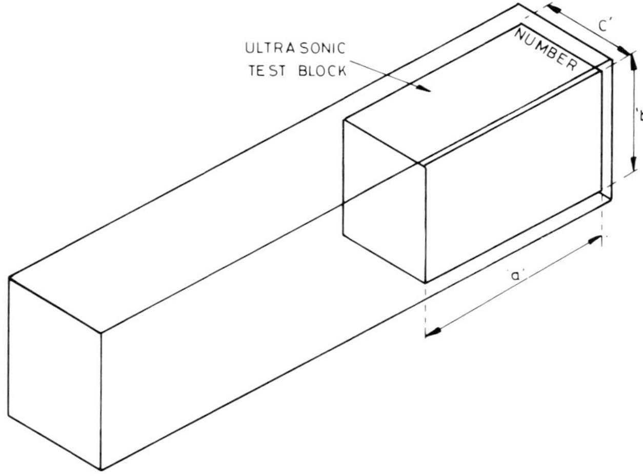

Fig. 3. Position of ultrasonic test block in keel block D.

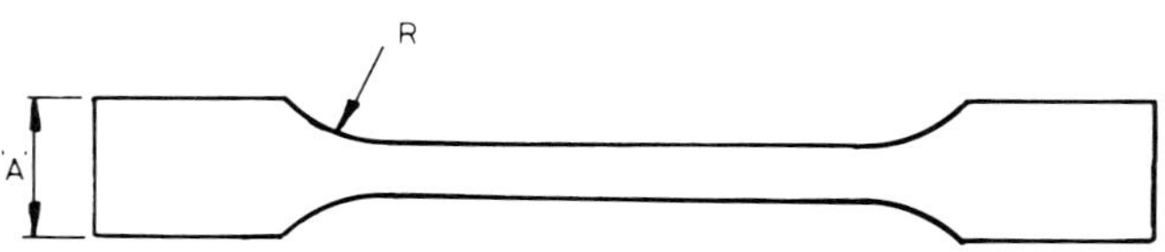

DIAMETER OF GAUGE LENGTH – 0.564 in
GAUGE LENGTH 2.8 in
RADIUS R – 1 in
DIAMETER 'A' 0.875 in   THREADED $^{13}/_{16}$" B.S.F.

Fig. 4. Tensile test piece.

during impact testing the faces in tension would correspond to the common cut face between the two slabs.

Block C: One unmachined test piece for eddy current testing, obtained by cutting the "keel" section off. This section was shaped on its cut face to provide a test piece 9 in. long by 1.5 in. high with five of its six faces having an as-cast surface. After eddy current testing had been completed, the upper part of the keel was cut off and the bottom section was machined to produce one tensile bar and one sonic bar as shown in Fig. 1. This sonic test bar was also used to make measurements of eddy currents on a test piece having machined surfaces.

Block D: One ultrasonic test block (Fig. 3). The remainder of the block was retained as spare material.

The sonic test bars, which were approximately 7.5 in. long, were centerless, ground to a diameter of 0.75 in. The results of resonant frequency measurement were corrected to take account of variations from the standard length of 7.5 in. Tensile test pieces were prepared in accordance with British Standard Specifications BS.18/1962 and BS.1452/1961 (Fig. 4). These tests meet the requirements of ASTM. They were 0.564 in. in diameter, with a minimum gauge length of 2.5 in., and had threaded shoulders. Un-notched impact test pieces were of 10 mm square cross section and had a length of 55 mm.

The rectangular test blocks for ultrasonic testing were approximately 65 x 40 x 40 mm, and were surface ground on all faces. The number of the block was stamped on the face originally adjacent to the end of the keel block so that the direction of measurement could be identified.

Tensile Tests

Tensile tests were carried out on a 50 ton tensile testing machine with facilities for the autographic recording of the stress-strain curve, to allow the 0.1, 0.2 and 0.5 percent offset yield strengths to be determined. Extensometers of high sensitivity and accuracy were used to obtain the extension values. The tensile strength and elongation at failure were determined. Hardness was measured on the shoulders of the broken tensile test pieces.

Impact Tests

Impact tests were carried out on a standard Charpy impact test machine with a striking energy of 120 ft lb. Specimens which were 55 mm long, were broken over a 40 mm span. Tests were carried out in the temperature range -75C (-100F) to +440C (+825F).

Sonic Tests

The longitudinal resonant frequency of the sonic test pieces was measured using the BCIRA Sonic Testing equipment to an accuracy of ±0.02 percent. The dimensions and weights of each bar were measured to allow the modulus of elasticity to be calculated from the following expression:[5]

$$E = 16\ Wlf/\pi d^2 g$$

where

W is the weight of the bar,
l is the length,
d is the diameter,
f is the resonant frequency and
g is the acceleration due to gravity.

The ratio of the diameter to the length of the bars was in all cases less than 0.15:1.0, so that correction for finite diameter was unnecessary.

Damping Capacity

Band Width — This was determined by the standard procedure of measuring the band width of the resonant frequency curve, at the point at which the amplitude of vibration was $1/\sqrt{2}$ of the maximum at resonance. In order to determine this as precisely

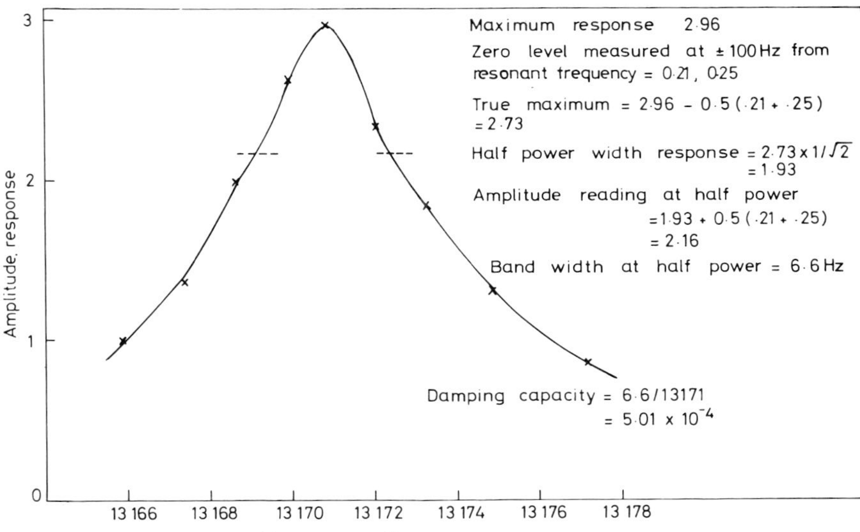

Fig. 5. Amplitude/frequency response curve for sonic test bar showing calculation of damping capacity.

as possible, the amplitude response of the resonant frequency test bar, at a series of frequencies slightly above and below the true resonant frequency, was accurately determined. The measurements obtained were plotted graphically as illustrated by Fig. 5, and from these graphs the band widths were read off. The damping capacity of each material was calculated as:

$$\text{Damping capacity} = \frac{\text{Band width}}{\text{Resonant frequency}}$$

Decay — This was determined by exciting a sonic test bar at its longitudinal resonant frequency using the BCIRA Sonic Testing unit, together with specially constructed equipment which, after the power to the drive transducer had been switched off, automatically counted the number of cycles for the amplitude of vibration to fall between two fixed levels set in the ratio of 2:1. The damping capacity was calculated from this count "n" using the expression

$$\text{Damping capacity} = \ln\left[(0.5)^{-4n}\right] = 0.693/n$$

Ultrasonic Velocity Measurements

The ultrasonic velocity measurements at nominal frequencies of 1.25, 2.5, 5.0 and 10.0 MHz were made by measuring the transmission time for ultrasonic waves along the three mutually perpendicular directions of the ultrasonic test blocks. The directions referred to as a, b and c are shown in Fig. 3. Ultrasonic reflections were obtained using a standard ultrasonic test set running at a pulse repetition frequency of about 500 MHz. The detected pulses were fed to an accurately calibrated oscilloscope and the time delay between successive multiple reflections measured. This technique eliminated any error due to the finite time required for the ultrasonic energy to pass through the ultrasonic probe. When the nominal test frequency was 10 MHz, the attenuation was such that velocity measurements could only be made in direction "c."

Attenuation of Ultrasonic Energy

Absolute values of attenuation cannot be measured due to variations in probe characteristics. Relative attenuation however was measured using 20 mm square twin probes, the material from melt No. 9 being chosen as the standard. The method used was based on the measurement of the decrease in back wall echo shown by multiple reflections. A calibrated attenuator was used to measure the additional attenuation required to reduce the amplitude of unrectified responses shown on the screen of an oscilloscope to comparable levels for all materials. At 1.25 MHz three or four reflections were used. At 2.5 and 5.0 MHz one or two reflections were used. At 10 MHz no measurements were possible.

The measurements of relative attenuation made with this series of irons should not be compared with those made on the previous ferritic/pearlitic series of irons[4] due to changes in probe characteristics. Measurements which can be compared are shown in the supplementary report.

Coercive Force Measurement

Measurements of coercive force were made using the BCIRA equipment. The readings recorded were an uncalibrated measurement of the reverse field voltage across the coil to reduce the remanent magnetism to zero. These measurements may be compared directly with those made in the previous report.[4]

Eddy Current Measurement

These measurements were made by an instrument operating at a test frequency of 5 Hz. The larger unmachined blocks were tested in a 6 in. coil system, whereas the machined sonic test bars were tested in a 2 in. internal diameter coil. The reference material for the tests on unmachined pieces was a block of the same size, but having a different composition. For the tests on machined materials, a sonic test bar giving an indication in the middle of the range was used. Tests were carried out when the current in the magnetizing coil was 0.2 and 1.0A. For convenience in reporting, these are referred to as low and high current, respectively. The results reported are measured amplitudes on the screen of the cathode ray tube at the fundamental frequency.

Eddy current measurements depend upon both the reference material chosen and the setting of the controls of the instrument. The measurements recorded in this report on both machined and unmachined test pieces were made with the equipment adjusted to give maximum discrimination for the series of irons being tested. The results shown in the tables and figures of this report should not be compared with those shown in the previous report which were made using different reference materials and different instrumental control settings. However in the supplementary report, the irons from all the work carried out on behalf of AFS are compared under common test conditions.

Metallographic Examination

Samples for metallographic examination were cut from the shoulder of one of the broken tensile test bars from each melt.

Table 4. Results of Tensile Tests

Iron No.	Yield Strength						Tensile Strength		Elongation	Hardness
	0.1% offset		0.2% offset		0.5% offset					
	$tonf/in^2$	psi	$tonf/in^2$	psi	$tonf/in^2$	psi	$tonf/in^2$	psi	%	HB 10/3000
1	23.9	53500	25.9	58000	29.2	65400	30.7	68800	1.0	244, 241
	22.5	50400	24.8	55600	27.6	61800	30.8	69000	0.5	240, 244
2	24.3	54400	26.4	59100	29.6	66300	29.9	67000	<1.0	257, 260
	24.5	54900	26.6	59600	29.2	65400	30.8	69000	<1.0	250, 244
3	24.8	55600	26.8	60000	29.4	65900	29.6	66300	2.0	256, 253
	24.4	54700	26.4	59100	29.0	65000	29.1	65200	<1.0	266, 274
4	26.1	58500	28.0	62700	30.8	69000	36.3	81300	1.0	256, 257
	25.9	58000	27.8	62300	30.6	68500	37.0	82900	2.0	253, 256
5	25.5	57100	27.4	61400	30.3	67900	32.9	73700	1.0	255, 260
	25.6	57300	27.5	61600	30.4	68100	32.8	73500	1.0	262, 257
6	26.3	58900	28.4	63600	30.6	68500	32.2	72100	2.0	257, 259
	26.2	58700	28.2	63200	31.2	69900	31.5	70600	<1.0	269, 269
7	26.4	59100	28.2	63200	31.3	70100	38.4	86000	2.0	255, 256
	26.1	58500	27.9	62500	31.2	69900	42.9	96100	1.5	272, 277
8	27.0	60500	28.8	64500	31.9	71500	35.0	78400	1.0	257, 263
	27.1	60700	29.0	65000	32.0	71700	35.6	79700	1.5	257, 259
9	26.7	59800	28.7	64300	32.2	72100	33.8	75700	0.5	285, 288
	26.8	60000	28.7	64300	32.3	72400	35.2	78800	1.5	285, 286

There were no significant variations in either graphite or matrix structure in specimens cut from other sections of keel blocks produced in the same melt. Detailed assessment of the structures was made by the following means:

1) Using a projection microscope, the percentage of graphite in nodular form was estimated by two experienced metallurgists making independent assessments.

2) The amount of carbide was estimated on the basis of the area which it occupied within the matrix after the graphite had been excluded. This was done by projecting the image of several fields of each specimen onto the screen of a microscope and comparing these with photographs of structures in which the amount of carbide was stated.[6]

3) By etching in a 10% solution of ammonium persulfate to darken all matrix constituents except the carbides, the areas occupied by carbides were measured using an image analyzing microscope. After exclusion of the area occupied by the graphite, the amount of carbide was calculated as a proportion of the remaining matrix.

4) A detailed assessment of graphite structure in the specimens cut from the broken tensile test bars was made using an image analyzing microscope. The total area of graphite, the sum of the projected heights of the graphite, the number of particles and the number of ends of particles were counted. Each of these measurements was made on 40 different fields on each sample, each approximately 0.01 in. x 0.007 in. The measurements were made of only graphite particles with widths greater than 12, 61, 122, 244 and 488 micro-inches respectively. Average values of the measurements are reported together with maximum and minimum values, and standard deviations.

Results

The results of the tensile tests shown in Table 4 include the 0.1, 0.2 and 0.5 percent offset yield strengths, tensile strength, elongation at failure, and the hardness measured on the shoulders of the broken tensile test pieces.

Table 5 summarizes the results of the un-notched impact tests. The full results are given in Table 6 and the impact transition curves are shown in Fig. 6.

Measurements based on sonic testing, including resonant frequencies corrected to a length of 7.500 in., modulus of elasticity, and damping capacity measured by both decay and band width methods, are shown in Table 7 together with the densities of the materials. The detailed results of measurements of ultrasonic velocity are shown in Table 8. These are summarized in Table 9 to show average values of velocity for each of the irons produced, and the attenuation measured at the various frequencies.

Table 10 shows the results of estimates of nodularity made during examination of the microstructure, together with estimates of the amount of carbide made by visual comparison with photographs of structures containing stated amounts of carbide. It also includes measurements of the amount of carbide obtained using the image analyzing microscope. Table 11 shows the average values of the graphite parameters measured using the image analyzing microscope. Tables 12a-d show for each

Table 5. Summary of Impact Test Results

UN-NOTCHED TESTS

Iron No.	Impact Value in Ductile Range, ft-lb	Transition Temperature, °C
1	10.0	200
2	9.2	200
3	8.0	200
4	16.8	208
5	11.6	200
6	10.8	250
7	25.2	217
8	15.6	221
9	10.8	225

Equivalent Temperatures

°C	°F	°C	°F
-75	-103	175	347
-40	-40	200	392
-20	-4	250	482
RT	65	270	518
60	140	300	572
100	212	350	662
125	257	380	716
150	302	440	824

graphite particle width, the maximum, minimum and mean values of measurements made of graphite parameters together with the standard deviation of the observations. Table 13 summarizes the results of the statistical examination of the relations which exist between the graphite parameters measured using the image analyzing microscope, yield strength, tensile strength, resonant frequency, ultrasonic velocity at a nominal frequency of 2.5 MHz and estimated percentage nodularity.

Micrographs representative of the structure of sections cut from the tensile test pieces, machined from the first pair of keel blocks to be cast, are shown in Fig. 8-16. These were the sections on which measurements were made using the image analyzing microscope.

The results of measurements of coercive force and of eddy current taken on test pieces having as-cast surfaces are shown in Table 14, and are repeated for eddy current tests on machined test bars in Table 15. Table 16 provides a summary of a statistical evaluation of the relations between mechanical properties and resonant frequency, damping capacity, coercive force measurement and eddy current readings on machined test pieces. A similar evaluation in which measured values of resonant frequency and damping capacity were replaced by those of ultrasonic velocity and attenuation at a nominal frequency of 2.5 MHz is given in Table 17. The latter analysis was repeated also for nominal ultrasonic test frequencies of 1.5 and 5.0 MHz. Results similar to those shown in Table 17 were obtained and therefore have not been included in this report.

Discussion of Results

Relation between Mechanical Properties, Graphite Form and Matrix Structure

The irons numbered for reference purposes as 1 to 9, were grouped in series of three. Irons 1 to 3 had between 40 and 50 percent of the graphite in nodular form, irons 4 to 6 between 65 and 75 percent and irons 7 to 9 were all classified as fully nodular. Within each group of three, the amount of carbide present in the matrix structure increased from the iron having the lowest number to the one having the highest number. On the basis of visual comparison with photographs of structures having stated amounts of carbide in them, the amounts of carbide present in irons 1, 4 and 7 were between 1.0 and 2.5 percent, in irons 2, 5 and 8 between 7.5 and 8.0 percent, and there was 15.0 percent of carbide in irons 3, 6 and 9. Measurement of the area of carbide made using the image analyzing microscope showed that the amount present had been underestimated at the lower levels when using the comparative method, but this method was reasonably accurate when carbide was present in greater quantities. Thus the series of irons produced, which had structures shown in the following chart, had a range of variation which was in accordance with that agreed to be necessary for the work to be carried out. The graphite present in a non-nodular form was compacted and similar to the ASTM type IV.

Iron Reference Number	Nodularity %	Carbide Content %		
1 – 3	40 – 50	3.5	8.4	13.9
4 – 6	65 – 75	1.9	8.7	13.4
7 – 9	100	1.9	7.1	14.4

Table 6. Un-notched Impact Results

Iron No.	Charpy Impact Values, ft lb															
	Temperature, °C															
	-75	-40	-20	Room Temp	60	100	125	150	175	200	250	270	300	350	380	440
1	2.5			3.0	4.0	5.0		9.0		10.0	8.0		8.5		10.0	10.5
2	2.0			3.5	4.0	5.0		8.0		7.5	8.0		7.0		10.0	10.5
3	2.0			3.0	4.0	4.0		6.5		8.0	9.0		8.0		8.0	7.5
4	2.0	3.0	4.0	4.0	5.5	6.0	8.5	13.0	14.0	16.0		17.0		16.5		
5	2.0			3.0	5.0	5.0		8.5		11.5	11.0		9.0		12.0	12.0
6	2.0			3.5	4.0	4.0		7.0		10.0	7.5		9.5		12.0	11.0
7	2.5	5.0	6.0	5.5	9.0	16.5	18.0	22.5	21.0	21.0		26.0		24.5		
8	2.0			3.0	5.0	6.0		10.0		14.0	15.5		16.5		15.0	14.5
9	2.0	2.0	2.0	3.0	3.5	5.5	6.0	10.0	9.0	9.5		12.5		10.0		

Equivalent Temperatures

°C	°F	°C	°F
-75	-103	175	347
-40	-40	200	392
-20	-4	250	482
RT	65	270	518
60	140	300	572
100	212	350	662
125	257	380	716
150	302	440	824

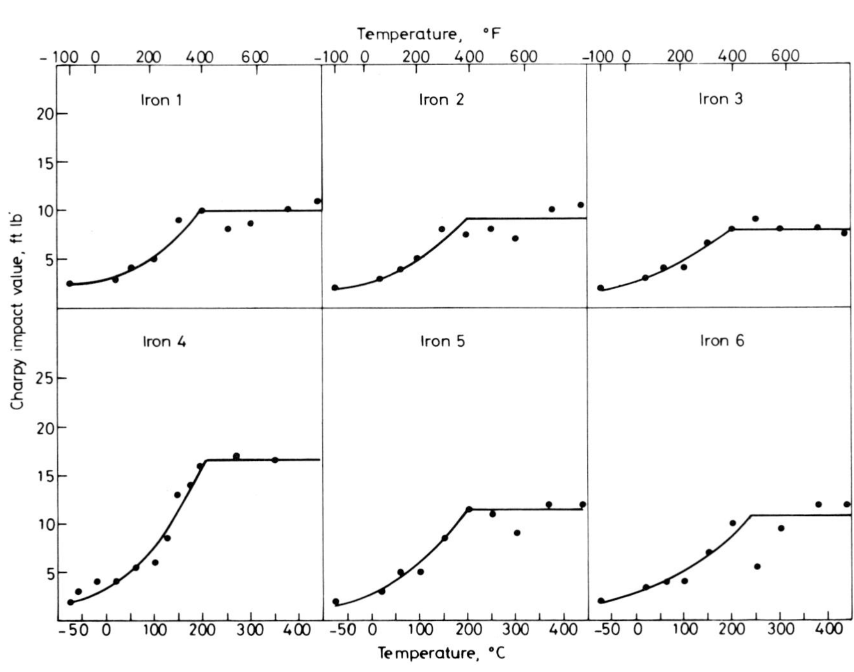

Fig. 6. Un-notched impact values, irons 1-9.

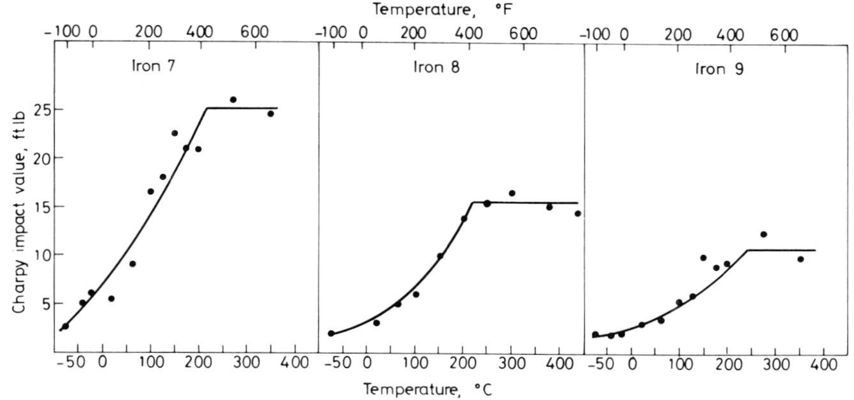

Fig. 6. (continued)

Table 7. Sonic Testing Results

Iron No.	Resonant Frequency, Hz (Corrected to 7.5 in long bar)	Modulus of Elasticity X 10^6 psi	Density		Damping Capacity	
			kg/m^3	lb/ft^3	Decay x 10^{-4}	Band Width x 10^{-4}
1	12710	24.3	7160	447.0	10.3	4.0
	12655	24.2	7170	447.6		
2	12756	24.4	7150	446.3	9.7	4.3
	12710	24.3	7160	447.0		
3	12776	24.7	7180	448.2	9.0	3.9
	12771	24.7	7190	448.8		
4	12864	25.3	7150	446.3	7.8	4.1
	12971	25.3	7140	445.7		
5	12908	25.1	7170	447.6	8.9	3.5
	12858	24.9	7170	447.6		
6	12897	25.2	7190	448.8	8.3	3.1
	12895	25.2	7210	450.1		
7	12998	26.0	7200	449.5	7.3	4.6
	13094	26.0	7200	449.5		
8	13025	25.6	7190	448.8	7.7	3.5
	12983	25.5	7180	448.2		
9	12944	25.5	7220	450.7	7.3	4.3
	12921	25.4	7220	450.7		

Table 8. Ultrasonic Velocity Measurements

Iron No.	Ultrasonic Velocity, $\times 10^5$ in/s											
	1.25 MHz			2.5 MHz			5.0 MHz			10.0 MHz		
	a	b	c	a	b	c	a	b	c	a	b	c
1	2.183	2.195	2.197	2.187	2.199	2.205	2.188	2.196	2.205	-	-	2.208
2	2.198	2.210	2.209	2.201	2.216	2.220	2.201	2.219	2.220	-	-	2.226
3	2.221	2.240	2.234	2.225	2.246	2.242	2.224	2.246	2.249	-	-	2.249
4	2.258	2.268	2.266	2.259	2.272	2.278	2.261	2.272	2.285	-	-	2.281
5	2.234	2.240	2.250	2.238	2.244	2.249	2.236	2.244	2.255	-	-	2.262
6	2.231	2.250	2.253	2.240	2.259	2.255	2.236	2.259	2.258	-	-	2.275
7	2.279	2.287	2.289	2.283	2.296	2.298	2.282	2.293	2.305	-	-	2.308
8	2.252	2.263	2.266	2.262	2.267	2.275	2.261	2.272	2.271	-	-	2.281
9	2.271	2.303	2.293	2.280	2.307	2.295	2.279	2.312	2.301	-	-	2.311

Conversion 1×10^5 in/s $\equiv$ 2.54 km/s

Table 9. Average Ultrasonic Velocity and Attenuation Results

Iron No.	Average Ultrasonic Velocity, $\times 10^5$ in/s				Average Attenuation, dB/in			
	1.25 MHz	2.5 MHz	5.0 MHz	10.0 MHz	1.25 MHz	2.5 MHz	5.0 MHz	10.0 MHz
1	2.192	2.197	2.196	2.208	0.762	0.330	-0.965	-
2	2.206	2.212	2.213	2.226	0.305	0.152	-1.270	-
3	2.232	2.238	2.240	2.249	0.152	-0.330	-1.120	-
4	2.264	2.270	2.273	2.281	0	0.152	0.635	-
5	2.241	2.244	2.245	2.262	0.762	0	-1.118	-
6	2.245	2.251	2.251	2.275	0	-0.152	-1.422	-
7	2.285	2.292	2.293	2.308	0.610	0.483	0.330	-
8	2.260	2.268	2.268	2.281	0.610	0.483	0	-
9	2.289	2.294	2.297	2.311	Standard			

Table 10. Assessment of Graphite and Matrix Structures

Iron No.	Carbide Content, %		Nodular Graphite, %
	Visual	Quantimet	Visual
1	2.5	3.5	40
2	8.0	8.4	40
3	15.0	13.9	50
4	1.0	1.9	75
5	7.5	8.7	65
6	15.0	13.4	70
7	1.0	1.9	100
8	8.0	7.1	100
9	15.0	14.4	100

Table 11. Graphite Parameters Determined Using An Image Analyzing Microscope

Graphite Parameter	Graphite width, $\times 10^{-6}$ in	Iron Number								
		1	2	3	4	5	6	7	8	9
Area, %	>12	9.6	8.7	9.6	9.5	9.2	9.2	8.9	10.6	9.5
	>61	8.5	7.6	8.8	9.0	8.5	8.6	8.5	10.0	9.0
	>122	7.2	6.5	8.0	8.4	7.7	8.0	7.9	9.5	8.5
	>244	5.4	4.8	6.4	7.4	6.4	6.9	7.0	8.4	7.6
	>488	2.6	2.8	4.2	5.5	4.4	5.1	5.3	6.6	6.1
Particle Count, $\times 10^3/\text{in}^2$	>12	183	168	160	129	229	122	139	179	138
	>61	175	152	145	114	184	107	106	108	123
	>122	163	145	122	88	131	88	88	85	86
	>244	119	97	81	48	72	45	50	42	38
	>488	72	73	62	39	56	36	37	34	30
Projected Height, in/in^2	>12	186	159	144	100	129	105	91	104	86
	>61	171	149	136	89	126	98	85	92	78
	>122	156	137	127	81	112	92	79	84	72
	>244	116	99	106	70	92	79	69	73	61
	>488	55	56	64	60	65	61	59	64	53
End Count $\times 10^3/\text{in}^2$	>12	610	531	515	434	581	432	317	516	375
	>61	373	347	304	229	337	233	189	215	200
	>122	309	289	233	147	217	165	127	154	125
	>244	277	238	183	86	139	90	79	75	58
	>488	188	183	143	78	125	75	57	67	47

Table 12(a). Maximum and Minimum Values, Means and Standard Deviations of Image Analyzing Microscope Measurements of Graphite Areas as Percentages of the Total Structure

Iron No.	Graphite Size, $\times 10^{-6}$ in																			
	>12				>61				>122				>244				>488			
	Max.	Min.	Mean	Std. Dev.	Max.	Min.	Mean	Std. Dev.	Max.	Min.	Mean	Std. Dev.	Max.	Min.	Mean	Std. Dev.	Max.	Min.	Mean	Std. Dev.
1	13.7	1.0	9.6	3.9	12.0	0.9	8.5	3.4	10.3	0.8	7.2	2.8	7.4	0.7	5.4	1.9	4.8	0.5	2.6	1.0
2	12.8	1.0	8.7	3.3	12.0	0.9	7.6	3.0	11.0	0.8	6.5	2.6	9.4	0.6	4.8	2.2	7.0	0.3	2.8	1.9
3	20.3	1.3	9.6	4.2	19.0	1.1	8.8	3.9	18.0	0.8	8.0	3.6	15.0	0.5	6.4	3.1	10.4	0.3	4.2	2.5
4	19.3	3.6	9.5	4.0	18.2	3.3	9.0	3.8	17.3	2.9	8.4	3.7	15.5	2.4	7.4	3.4	12.4	1.4	5.5	2.8
5	17.0	1.5	9.2	4.0	16.0	1.3	8.5	3.8	14.4	1.2	7.7	3.5	12.4	1.0	6.4	3.1	10.4	0.8	4.4	2.5
6	18.4	1.7	9.2	4.1	17.7	1.5	8.6	3.9	17.0	1.3	8.0	3.7	15.6	1.1	6.9	3.4	13.1	1.0	5.1	2.8
7	18.1	3.8	8.9	3.2	17.3	3.5	8.5	3.0	16.4	3.3	7.9	2.9	14.8	2.8	7.0	2.7	12.7	2.2	5.3	2.3
8	20.1	4.2	10.6	3.9	19.2	3.8	10.0	3.8	18.1	3.5	9.5	3.7	16.3	3.1	8.4	3.4	12.9	2.3	6.6	3.0
9	21.0	3.8	9.5	3.8	20.3	3.5	9.0	3.6	19.4	3.2	8.5	3.5	17.9	2.8	7.6	3.3	15.0	2.1	6.1	2.8

Table 12(b). Maximum and Minimum Values, Means and Standard Deviations of Image Analyzing Microscope Measurements of Particle Counts ($\times 10^3 / in.^2$)

Iron No.	Graphite Size, $\times 10^{-6}$ in																			
	>12				>61				>122				>244				>488			
	Max.	Min.	Mean	Std. Dev.	Max.	Min.	Mean	Std. Dev.	Max.	Min.	Mean	Std. Dev.	Max.	Min.	Mean	Std. Dev.	Max.	Min.	Mean	Std. Dev.
1	457	70	183	117	374	47	175	103	373	35	163	96	302	12	119	78	163	12	72	41
2	387	116	168	101	377	81	152	94	361	70	145	87	186	47	97	64	116	47	73	41
3	364	128	160	81	327	70	145	69	268	47	122	53	233	12	31	47	174	12	62	37
4	271	93	129	74	187	58	114	47	175	47	88	33	81	35	48	18	58	23	39	17
5	352	120	229	59	304	128	184	53	268	81	131	48	221	35	72	45	116	35	56	35
6	213	116	122	62	165	47	107	36	163	47	88	25	152	23	45	26	105	23	36	21
7	282	105	139	59	211	93	106	36	175	58	88	31	93	23	50	19	70	23	37	15
8	503	73	170	91	176	93	108	44	129	70	85	30	81	35	42	17	81	23	34	15
9	271	82	138	61	269	10	123	45	175	70	86	27	70	35	38	16	35	12	30	16

Results of the un-notched impact tests shown in Tables 5 and 6 and also in Fig. 6 have been used to produce Fig. 7 which shows more clearly the effect of increasing the amount of carbide and non-nodular graphite upon the impact values in the ductile range, and also upon the temperature at the start of the ductile to brittle transformation. This temperature, which was 230C (430F) for irons in which all the graphite was in a nodular form, was reduced slightly as the amount of non-nodular graphite was increased. However it was unaffected by the amount of carbide present in the matrix.

A similar trend for the impact transition temperature to decrease as the amount of non-nodular graphite increased was found in the previous work. [1,2,4] Impact values in the ductile range decreased as the amount of carbide in the matrix structure increased. This effect was most pronounced in the irons having the highest proportion of nodular graphite present. At any given carbide content, the impact value also decreased as the amount of non-nodular graphite in the structure increased. This decrease was greatest in the irons containing the smaller amounts of carbide.

The un-notched impact values of a ductile iron are a measure of the work done to produce failure. This is also represented by the area under the tensile stress strain curve. Since, as discussed later, increasing both the amounts of carbide and non-nodular graphite caused reductions in tensile strength and elongation, the effect of these structural variations upon un-notched impact values might have been anticipated. However Fig. 7 shows that the weakening effect of carbides and of non-nodular graphite is not directly additive. If one cause of weakening is already present, then the effect of the second is reduced.

From Fig. 6 it is possible to predict the un-notched impact properties of the iron at a temperature of 25C (75F) which is within the transition range. As shown in the following chart, un-notched impact values at this temperature were affected by changes in the amount of carbide and non-nodular graphite, in the same way as that which occurred in the ductile range.

At the low temperature of -75°C (-100°F) all irons had similar un-notched impact values of approximately 2 ft lbs, resulting from brittle impact failure.

Table 12(c). Maximum and Minimum Values, Means and Standard Deviations of Image Analyzing Microscope Measurements of Projected Heights (in./in.^{2})

Iron No.	Graphite Size, x 10^{-6} in																			
	>12				>61				>122				>244				>488			
	Max.	Min.	Mean	Std. Dev.	Max.	Min.	Mean	Std. Dev.	Max.	Min.	Mean	Std. Dev.	Max.	Min.	Mean	Std. Dev.	Max.	Min.	Mean	Std. Dev.
1	395	21	186	111	311	19	171	100	284	17	156	91	208	15	116	61	95	14	55	25
2	375	23	159	96	356	19	149	91	323	16	137	83	180	10	99	47	109	8	56	23
3	296	52	144	62	279	48	136	59	262	36	127	55	190	11	106	43	99	8	64	28
4	172	44	100	30	159	29	89	28	144	24	81	25	128	21	70	25	105	19	60	22
5	279	56	129	65	257	49	126	61	237	40	112	58	206	37	92	48	132	33	65	28
6	246	47	105	47	227	43	98	44	214	33	92	43	188	20	79	41	101	16	61	26
7	164	61	91	27	150	57	85	24	139	51	79	23	129	41	69	21	117	29	59	20
8	216	62	104	32	191	52	92	27	175	39	84	26	157	30	73	25	133	26	64	22
9	148	58	86	24	130	50	78	21	117	38	72	20	112	22	61	22	107	18	53	21

Table 12(d). Maximum and Minimum Values, Means and Standard Deviations of Image Analyzing Microscope Measurements of End Counts (x10^3/in.^{2})

Iron No.	Graphite Size, x10^{-6} in																			
	>12				>61				>122				>244				>488			
	Max.	Min.	Mean	Std. Dev.	Max.	Min.	Mean	Std. Dev.	Max.	Min.	Mean	Std. Dev.	Max.	Min.	Mean	Std. Dev.	Max.	Min.	Mean	Std. Dev.
1	1445	81	610	274	874	58	373	190	768	47	309	198	640	12	277	199	337	12	188	113
2	1445	151	531	239	932	93	347	188	768	81	289	199	663	12	238	201	291	23	183	133
3	1096	154	515	129	606	72	304	93	431	70	233	90	291	23	183	173	291	47	143	93
4	735	213	434	35	304	60	229	38	256	59	147	39	186	35	86	36	93	35	78	46
5	1061	247	581	92	606	130	337	86	442	94	217	82	430	35	139	108	337	24	125	102
6	910	154	432	74	420	153	233	58	268	105	165	67	198	0	90	64	163	0	75	63
7	666	85	317	39	293	107	189	25	187	70	127	33	81	70	79	33	93	47	57	26
8	1061	270	516	99	374	153	215	28	349	82	154	42	163	12	75	37	186	12	67	41
9	759	154	375	15	304	152	200	17	210	81	125	26	128	12	58	30	105	12	47	29

Nodularity %	Impact value, ft lbs		
	Carbide content, % of matrix		
	2.0-3.5	7.0-8.5	13.5-14.5
40-50	3.2	3.0	3.0
65-75	4.0	3.6	3.1
100	8.8	4.0	3.0

It should be noted that in comparison with a fully pearlitic iron having a fully nodular graphite structure but free from carbides, the presence of even the smallest quantity of this constituent had a large effect on reducing un-notched impact properties in the ductile range. In the carbide-free iron the impact value was 45 ft lbs, but this was reduced to 25 ft lbs in the presence of 2 percent of carbide. However at room temperature the change in properties due to the presence of carbide was less marked.

The results of measurements of the 0.1, 0.2 and 0.5% offset yield stress, tensile strength, elongation and hardness are shown in Table 4. Values of the modulus of elasticity determined by resonant frequency measurements are shown in Table 7. The effects which various amounts of carbides and non-nodular graphite had upon offset yield stress and tensile strength were easily detected from changes which they caused in the shape of the tensile stress-strain curve.

With increasing amounts of carbide in the structure, the onset of plastic deformation was delayed, and in the plastic region the stress-strain curves were raised, requiring a higher level of stress to produce a given strain. However elongation and tensile strength fell. Thus carbide had its expected effects of, in the first instance, stiffening the matrix and raising the limit of proportionality and the yield stress, and then, due to its brittleness, fracturing under higher tensile loads to produce a severe notch effect and reduce ultimate properties. On the other

Graphite Parameter	Graphite width in x 10^6	0.2% offset yield strength	Tensile Strength	Resonant Frequency	Ultrasonic Velocity	Nodularity
Area %	>12					
	>61					
	>122	+		+	+	++
	>244	++		++	++	++
	>488	++		++	++	++
Particle Count thousands/in^2	>12					
	>61	-		-	-	-
	>122	--	-	--	--	--
	>244	--		--	--	--
	>488	--	-	--	--	--
Projected Height in/in^2	>12	--	-	--	--	--
	>61	--	-	--	--	--
	>122	--	-	--	--	--
	>244	--	-	--	--	--
	>488					
End Count thousands/in^2	>12		-	-	--	-
	>61	--	-	--	--	--
	>122	--	-	--	--	--
	>244	--	-	--	--	--
	>488	--	-	--	--	--

+ Indicates increase as graphite parameter increases
- Indicates decrease as graphite parameter increases

-)
or) significant at 95% level
+)

--)
or) significant at 99% level
++)

Table 13. Significance of Correlations between Graphite Parameters Measured Using the Image Analyzing Microscope, Mechanical Properties, Resonant Frequency and Ultrasonic Velocity

Table 14. Coercive Force and Eddy Current Readings on Unmachined Keel Blocks

Iron No.	Coercive* Force	Eddy Current*	
		High Current	Low Current
1	154	-2.0	-1.5
2	165	+1.6	+1.3
3	174	+0.3	-0.2
4	157	-1.8	-1.0
5	163	-2.0	-1.5
6	167	-1.0	-0.5
7	215	-5.0	-4.6
8	154	+3.1	+2.4
9	185	-5.9	-4.2

* Arbitrary readings

Table 15. Eddy Current Readings on Machined Test Bars

Iron No.	Eddy Current*	
	High Current	Low Current
1	1.7	3.8
2	1.3	2.5
3	-1.0	0.5
4	2.1	3.0
5	2.1	2.3
6	1.8	1.5
7	-3.4	-5.0
8	4.8	3.5
9	1.4	-0.7

* Arbitrary Readings

Table 16. Results of Multiple Regression Analysis between 0.2% Offset Yield Stress, Tensile Strength and Resonant Frequency in Combination with Other Measurements Made on Irons Containing Various Amounts of Carbide

Measurement combined with Resonant frequency	Significant variables	
	0.2% offset yield stress	Tensile strength
Damping capacity decay	$(Decay)^2$	$(Resonant\ Frequency)^2$
Damping capacity band width	Resonant Frequency	$(Resonant\ Frequency)^2$ $(Band\ width)^2$
Coercive force	Resonant Frequency	$(Resonant\ Frequency)^2$
Eddy Current (High Current)	Resonant Frequency	$(Resonant\ Frequency)^2$
Eddy Current (Low Current)	Resonant Frequency $(Eddy\ Current)^2$	$(Resonant\ Frequency)^2$ $(Eddy\ Current)^2$
Hardness	Resonant Frequency Hardness	$(Resonant\ Frequency)^2$

Measurement combined with Ultrasonic velocity	Significant variables	
	0.2% offset yield stress	Tensile strength
Relative attenuation	Ultrasonic velocity	$(Ultrasonic\ velocity)^2$ Relative attenuation
Coercive force	Ultrasonic velocity	$(Ultrasonic\ velocity)^2$
Eddy Current (High current)	Ultrasonic velocity Eddy current	$(Ultrasonic\ velocity)^2$
Eddy Current (Low current)	Ultrasonic velocity	$(Ultrasonic\ velocity)^2$ $(Eddy\ current)^2$
Hardness	Ultrasonic velocity	$(Ultrasonic\ velocity)^2$ Hardness

Table 17. Results of Multiple Regression Analysis between 0.2% Offset Yield Stress, Tensile Strength and Ultrasonic Velocity at 2.5 MHz in Combination with Other Measurements Made on Irons Containing Various Amounts of Carbide

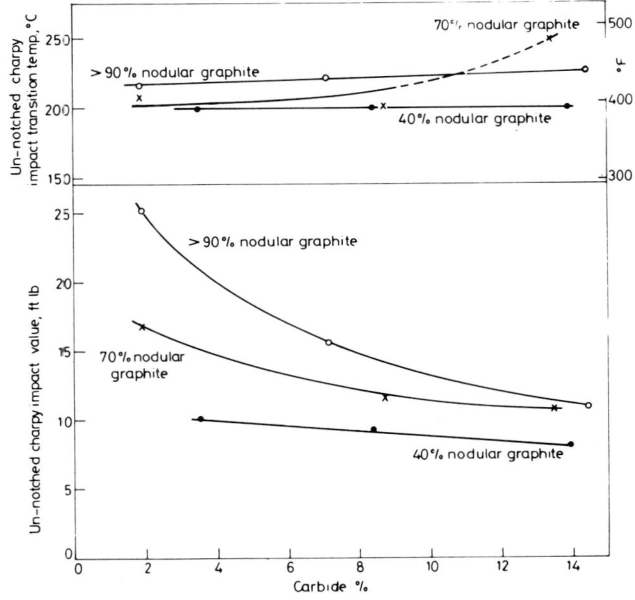

Fig. 7. Relation between un-notched impact properties and amount of carbide in irons having varying amounts of graphite in nodular form.

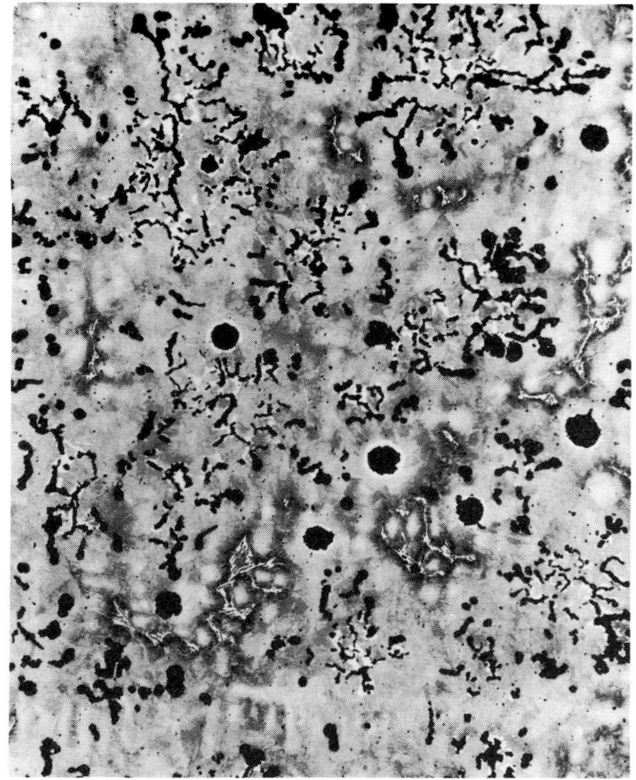

Fig. 8. Iron 1 50X, etched 4% picral.

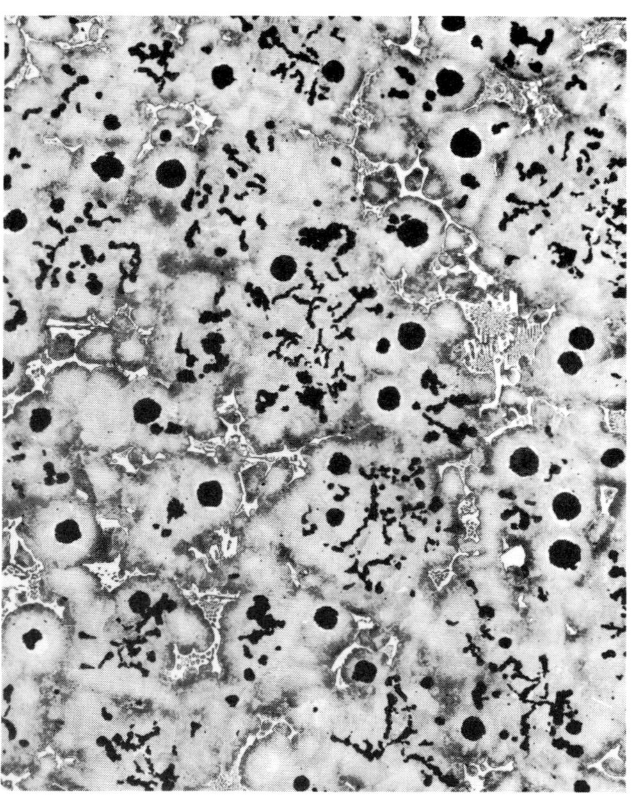

Fig. 10. Iron 3 50X, etched 4% picral.

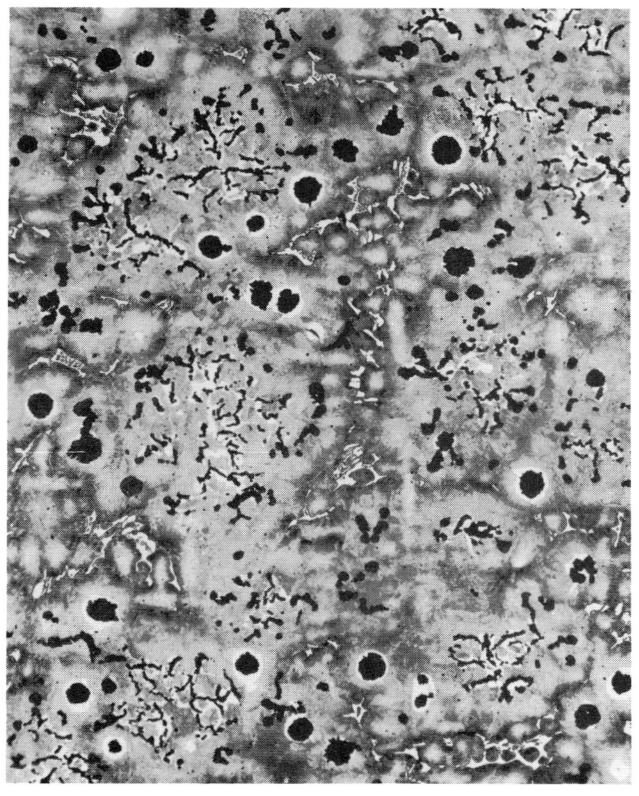

Fig. 9. Iron 2 50X, etched 4% picral.

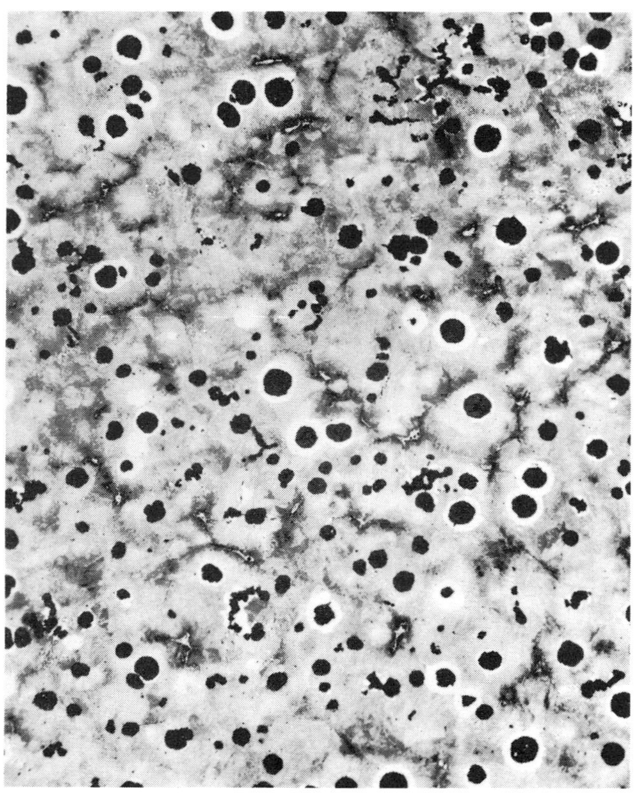

Fig. 11. Iron 4 50X, etched 4% picral.

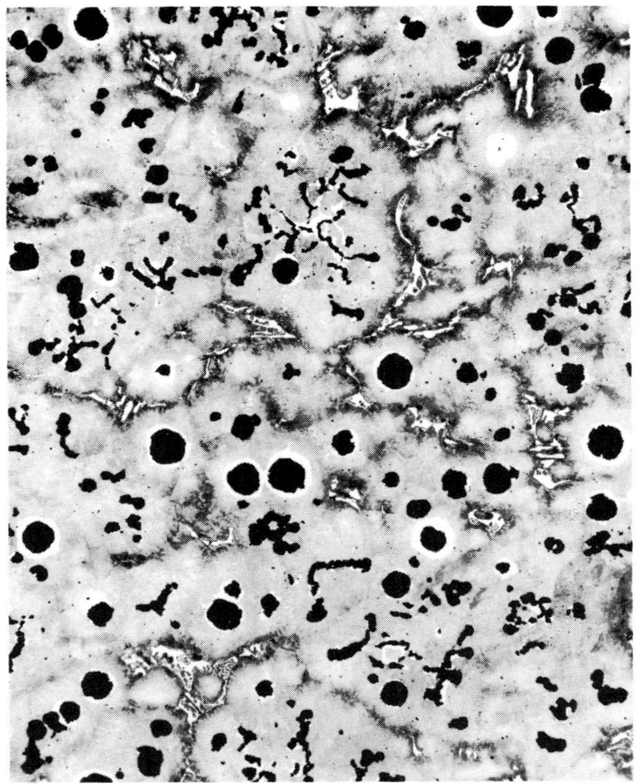

Fig. 12. Iron 5 50X, etched 4% picral.

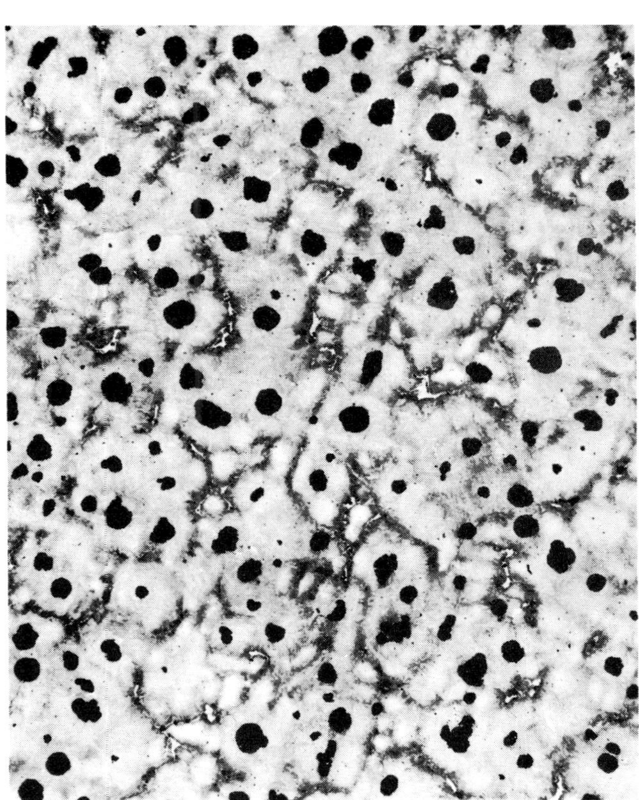

Fig. 14. Iron 7 50X, etched 4% picral.

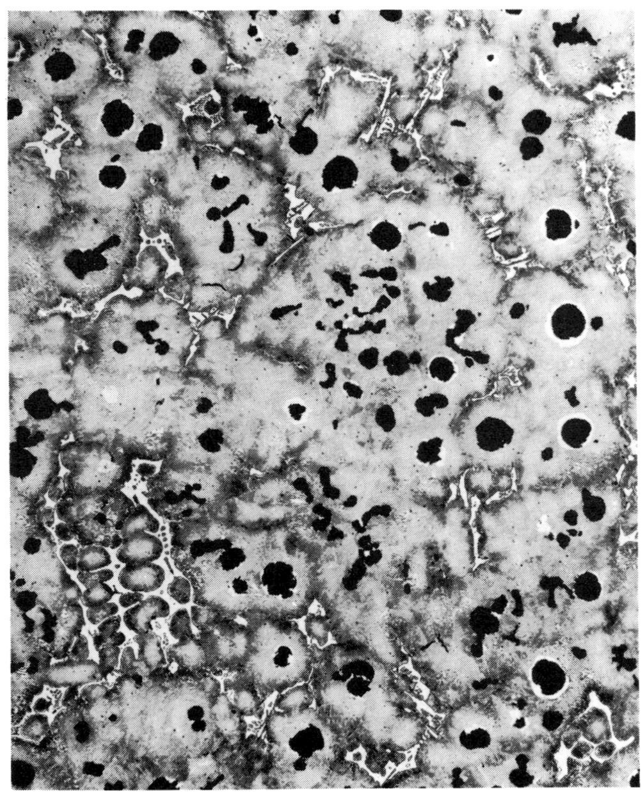

Fig. 13. Iron 6 50X, etched 4% picral.

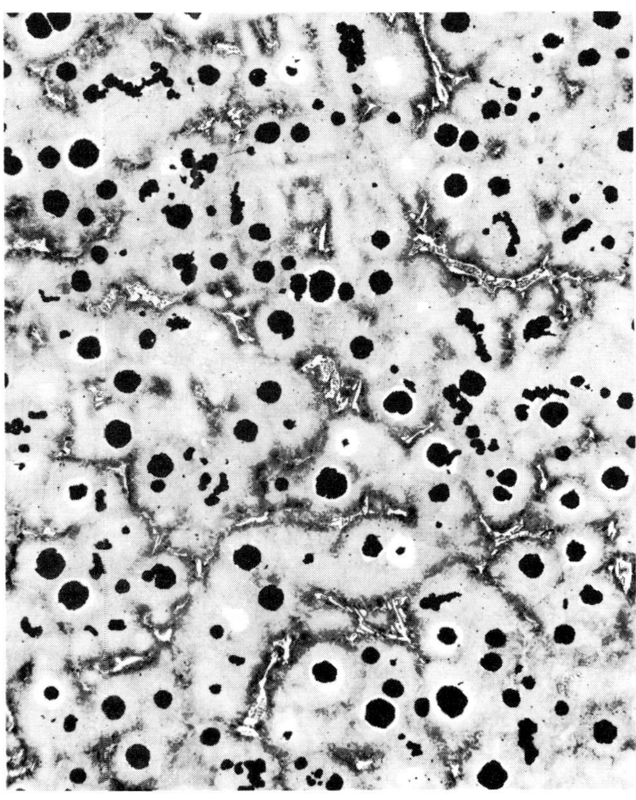

Fig. 15. Iron 8 50X, etched 4% picral.

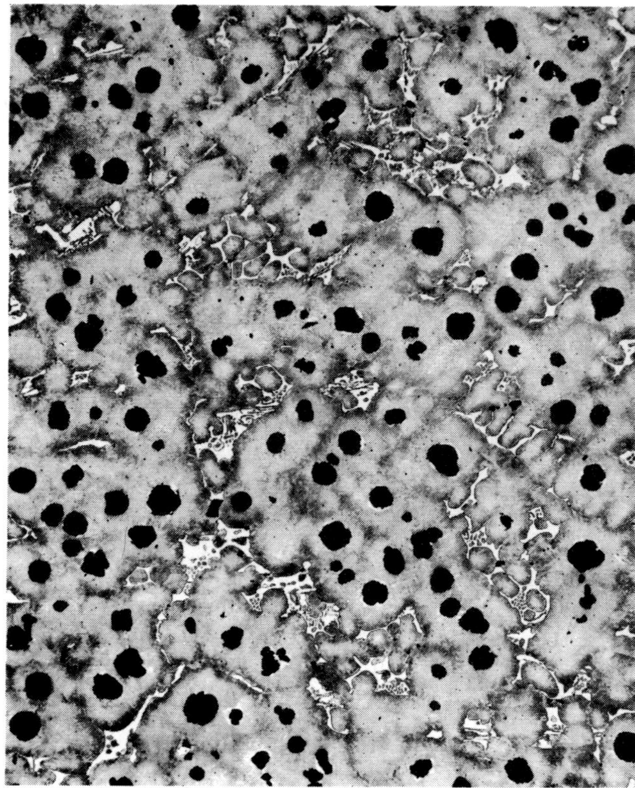

Fig. 16. Iron 9 50X, etched 4% picral.

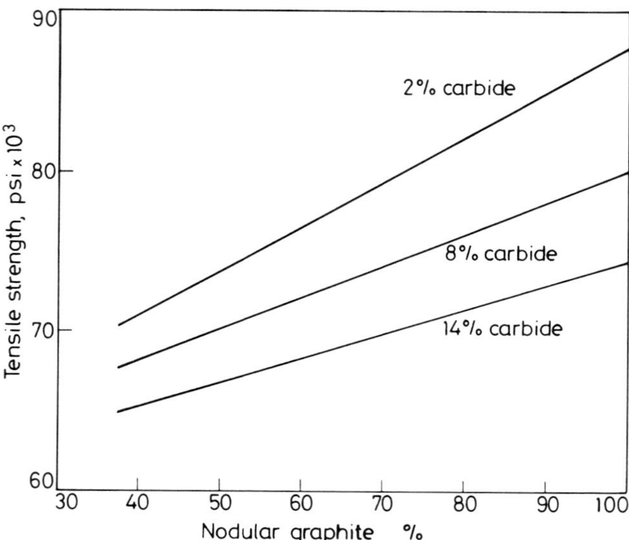

Fig. 17. Relation between tensile strength and amount of graphite in nodular form in irons having varying proportions of carbide (extrapolated values for irons having as-cast structures).

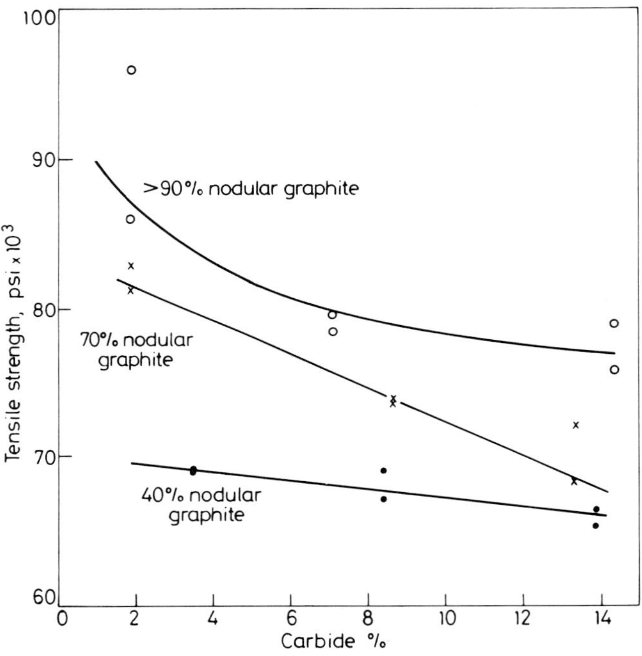

Fig. 18. Relation between tensile strength and amount of carbide in irons having varying proportions of graphite in nodular form.

hand the \presence of increasing amounts of non-nodular graphite led to the onset of plastic deformation at lower stress levels and lowered the stress-strain curve in the plastic region. Thus its effects are to reduce the limits of proportionality and the yield stress values.

As in the previous work, tensile strength and elongation decreased as the amount of non-nodular graphite increased. These effects may be attributed to the local stress raising effects of the non-nodular graphite, which might have been more severe had it been present in a "spiky" rather than in a compacted form.

Figures 17 and 18 further illustrate the effects of these structural variations upon tensile strength. As previously found, if the matrix structure of the iron does not change, then tensile strength decreases linearly with increases in the amount of non-nodular graphite. The effect of increasing amounts of carbide in decreasing strength was greatest in irons having fully nodular structures and decreased as the amount of non-nodular graphite increased. In irons of high nodularity, tensile strength decreased rapidly at first and then more slowly as the amount of carbide increased. The chart on the next page summarizes the effect of increasing amounts of carbide and of non-nodular graphite upon tensile strength. It includes results previously obtained for an iron having a fully pearlitic matrix. It also illustrates that the weakening effect of non-nodular graphite is reduced in irons in which weakening is already induced by the brittleness of carbide in the matrix.

Effect of Carbides and Non-Nodular Graphite on Tensile Strength

The effects of various amounts of carbide and non-nodular graphite upon the 0.2% offset yield stress are shown in Fig. 19 and 20. These were similar to the effects upon the 0.1% and the

0.5% offset yield stress which are therefore not illustrated. When comparing the effects of these structural variations upon the 0.2% offset yield stress, the vertical scales of Fig. 19 and 20, which show the strength, are twice as large as those of Fig. 17 and 18. The effect of increasing carbide content upon the yield strength was to cause a relatively rapid increase initially, which then continued at a slower rate as the amount of carbide further increased. This pattern was similar irrespective of the amount of non-nodular graphite present. Increasing amounts of non-nodular graphite reduced the yield stress. However, as in previous work, the presence of a small amount of non-nodular graphite in the structure has relatively little effect upon yield

Nodularity %	Tensile strength, psi x 10³			
	Carbide content, % of matrix			
	0	2	8	14
100	132	92	80	72
50	93	73	70	65
Percentage decrease	30	15	13	10

stress although it had a greater effect on tensile strength. This evidence, supported by work at BCIRA,[7] provides further confirmation that the presence of small amounts of non-nodular graphite in a compacted form do not affect the yield strength. This is important since this is the property upon which most castings are designed.

A summary of the effect of the structural variation upon the 0.2% offset yield stress is shown below.

Nodularity	0.2% offset yield stress, psi x 10³		
	Carbide content, % of matrix		
%	2	8	14
100	62.8	64.2	64.7
50	58.6	61.0	61.8
% decrease	6.7	5.2	4.7

As a consequence of the effects of carbides in increasing yield stress, but of reducing tensile strength, the ratio of the yield stress to tensile strength increased as their amount increased. Also, as in previous work this ratio increased as the amount of non-nodular graphite increased (Fig. 21). The chart below shows the extent of these changes. Typical values of this ratio in a carbide-free fully nodular iron having a pearlitic matrix are between 0.60 and 0.65.

Effect of Carbides and Non-Nodular Graphite on the Ratio of Yield Stress to Tensile Strength

Nodularity	0.2% offset yield stress / tensile strength		
	% carbide in matrix		
%	2	8	14
100	0.73	0.80	0.90
50	0.80	0.87	0.95

Hardness increased with an increase in the amount of carbide in the matrix and decreased slightly with an increasing amount of non-nodular graphite. Measured elongations were low even

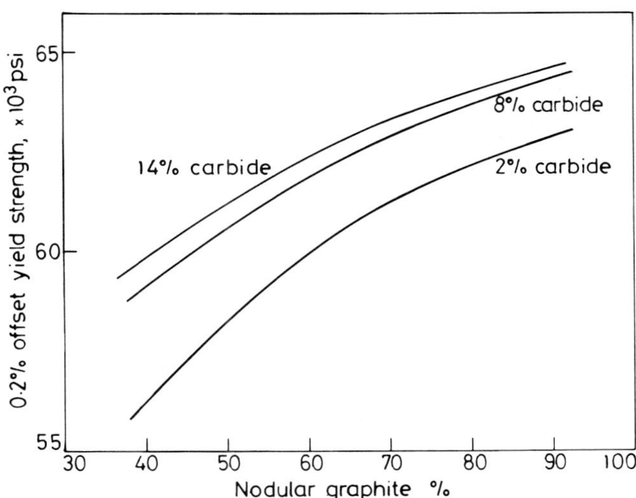

Fig. 19. Relation between 0.2% offset yield strength and amount of graphite in nodular form in irons having varying amounts of carbide (extrapolated values for irons having as-cast structures).

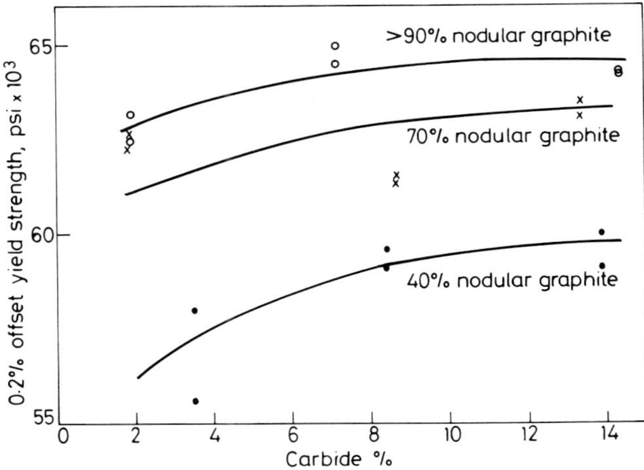

Fig. 20. Relation between 0.2% offset yield strength and the amount of carbide in irons having varying proportions of graphite in a nodular form.

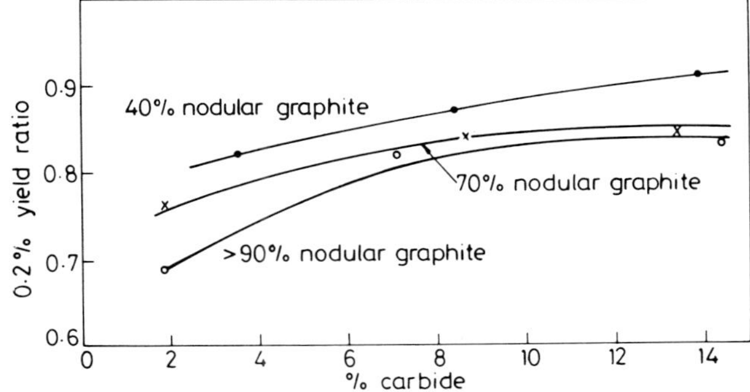

Fig. 21. The relation between 0.2% yield ratio and carbide content in irons containing varying amounts of nodular graphite.

in fully nodular irons containing the least amount of carbide. They decreased further with increases in the amount of carbide and non-nodular graphite. The modulus of elasticity fell with increasing amounts of non-nodular graphite but was not greatly affected by increases in the amount of carbide.

The effects on properties due to changes in the amounts of carbide and non-nodular graphite were as expected. However this evaluation was only one of the objectives of this work. A second was to evaluate the use of nondestructive testing methods to predict the mechanical properties of the irons produced. As in the previous work it was found that provided the matrix structure was known, and did not vary, a visual estimate of the amount of graphite present in nodular form provided a reasonable means of estimating the likely properties. This however assumes that the form of the non-nodular graphite does not change, and in all the results illustrated in this report, it was in a compacted form.

Although in practice this form of non-nodular graphite is likely to occur most frequently, it can be present in other forms which, if it is in similar quantities to compacted graphite, may cause a much greater reduction in mechanical properties.[1] Thus quality control procedures based upon estimates of nodularity may be very misleading unless the form of the non-nodular graphite is taken into account. This makes the form of quality control more difficult to successfully apply, as it becomes increasingly dependent upon the skill and judgment of the operators to ensure that they are to recognize and take into account subtle variations in both the graphite form and matrix structure. While this form of assessment may continue to be useful for the purposes of process control within the foundry, it is not a substitute for other methods of testing when the foundry has to guarantee the quality of castings supplied to its customers.

Measurement of Graphite Parameters Using the Image Analyzing Microscope and Their Relation to Mechanical Properties

Measurements of the area of graphite, the number of particles, their projected height and the number of ends, were made using an image analzying microscope. The results shown in Tables 11 and 12 may be compared with those obtained in the previous work. As previously found, there were no consistent trends between the measured graphite parameters and the variations in the matrix structure, even though with an increase in the amount of carbide, a small reduction in the area occupied by the graphite might have been expected. As the amount of non-nodular graphite increased, particle counts, projected heights and end counts increased and the area occupied by larger graphite particles decreased. End counts were higher than in previous work due to the absence of inoculation.

Table 13 shows the results of a statistical evaluation of the significance of relations between graphite parameters measured using the image analyzing microscope, and mechanical properties, resonant frequency and visual estimates of nodularity. The significance is shown at two levels, better than 95% but less than 99% indicated by a single plus or minus sign, and better than 99% indicated by two plus or two minus signs. The plus signs also indicate that the value of a given measurement increased as the graphite parameter increased. Conversely a minus sign indicates that the value decreased as the value of measured graphite parameter increased.

In comparison with the previous work,[1,2,4] a larger number of relations were found to be significant at the better than 99% level. With measurements of graphite area significant relations existed only when fine particles were excluded. Then as area increased, nodularity, 0.2 percent offset yield strength, resonant frequency and ultrasonic velocity increased. With increases in the amount of non-nodular graphite the number of graphite particles increased, their elongated shape resulted in an increase in projected height, and their less regular outline caused the image analyzing microscope to count a greater number of particle ends. Thus, as values of particle count, projected height and end count increased, the nodularity, 0.2 percent offset yield stress, resonant frequency and ultrasonic velocity decreased. Correlations with tensile strength were at a lower level of significance. This may be explained by the fact that while increasing amounts of carbides had a relatively small effect upon 0.2 percent yield stress, resonant frequency and ultrasonic velocity, their effect upon tensile strength was much greater.

In addition to the statistical evaluation described above, graphs were prepared relating measured graphite parameters with mechanical properties. Although the degree of correlation between graphite parameters and properties was statistically high, the slopes of the relationships were steep, with only a small variation in measurement corresponding to a relatively large change in properties. It was concluded therefore that the measurements that can be made using the image analyzing microscope are unsuitable for casting quality control, even if it was conceivable that specimens of the required quality could be obtained representing the casting produced. In the future, the attachment of microcomputers to image analyzing microscopes might enable a much wider range of detailed measurement to be made of both graphite and matrix constituents. When the use of such equipment becomes established, it might be appropriate to re-examine the use of these instruments in quality control, but even then there would be the practical difficulty of the preparation of a casting for examination.

Relations between Mechanical Properties and Measurement of Resonant Frequency, Ultrasonic Velocity and Attenuation, Damping Capacity, Coercive Force and Eddy Currents

The relation between resonant frequency and 0.2 percent offset yield stress is shown in Fig. 22. It can be represented by a single line providing a reasonably good estimate of yield stress, irrespective of the amount of carbide present. This reflects the relatively small effect which changes in the amount of carbide had upon yield stress, and is in contrast to the previous work[4] where separate lines defined the relation according to the amount of ferrite and pearlite in the matrix structure.

The relation between resonant frequency and tensile strength has also been represented by a single line (Fig. 23) but in this instance, the scatter in the prediction of tensile strength from resonant frequency is much greater. There is again no clear separation of points on this graph according to the amount of carbide present in the matrix structure. The reason for this is apparent from Table 7. Measurements of resonant frequency were insensitive to variations in the amounts of carbide, even though these had considerable effect upon tensile strength. In the other work carried out for AFS,[1-4] it has been shown that provided the matrix structure is known and remains consistent, resonant frequency measurements can be used to provide a very good estimate of tensile strength.

When using ultrasonic velocity measurements to assess properties, the relationship with yield stress was again best represented by a single line irrespective of the amount of carbide present (Fig. 24). However with tensile strength, three separate lines were drawn in order to define its relationship with ultrasonic velocity at each of the levels of carbide (Fig. 25). In consequence, the prediction of tensile strength from ultrasonic velocity measurement was more accurate provided that the amount of carbide in the matrix structure was known. The reasons for this difference in the relationships between resonant frequency and ultrasonic velocity with tensile strength are not altogether clear. Table 8 does show however that measurements of ultrasonic velocity were far more sensitive to changes in carbide content than those of resonant frequency.

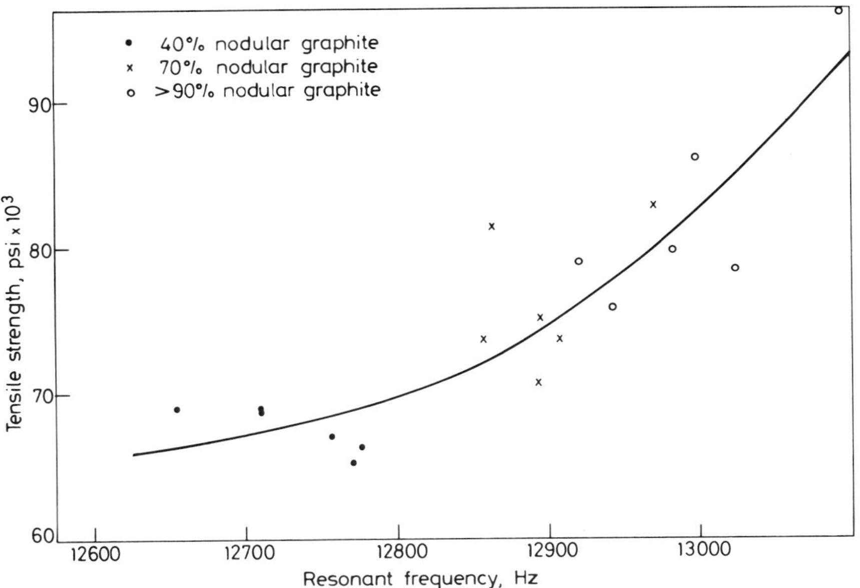

Fig. 22. Relation between 0.2% offset yield strength and resonant frequency in irons containing varying amounts of carbide and nodular graphite.

Fig. 23. Relation between tensile strength and resonant frequency in irons containing varying amounts of carbide and nodular graphite.

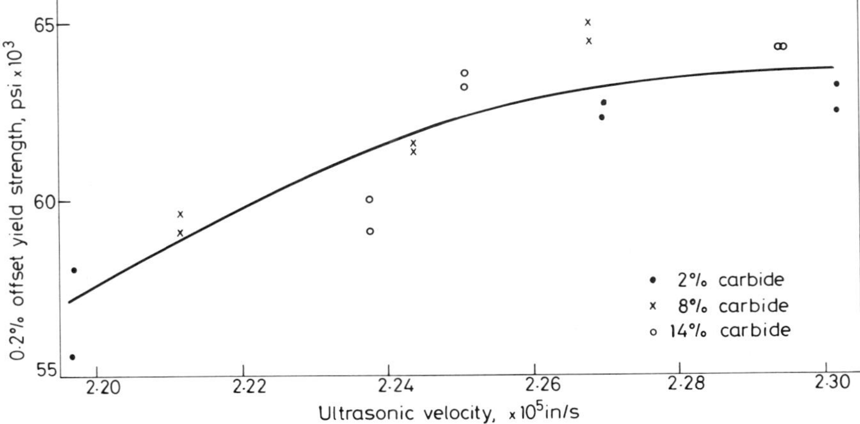

Fig. 24. Relation between 0.2% offset yield strength and ultrasonic velocity in irons containing varying amounts of carbide.

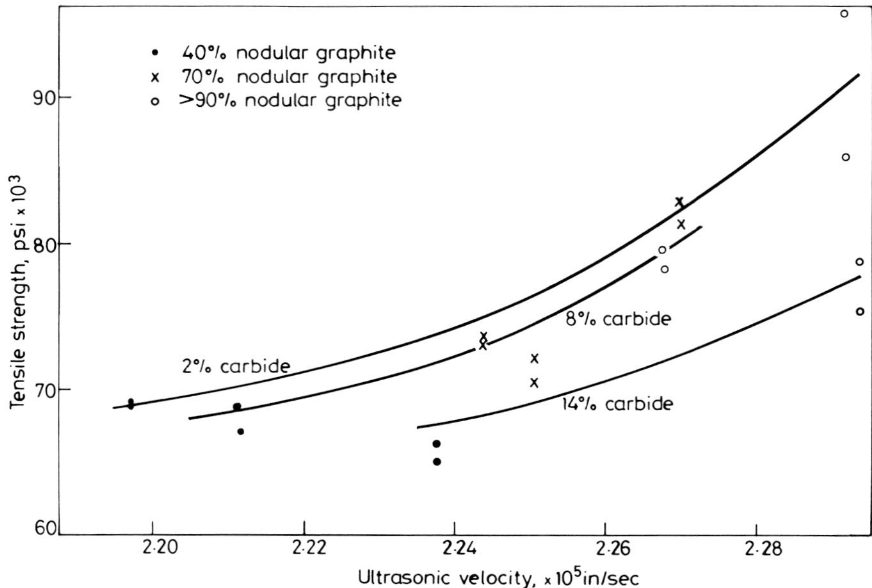

Fig. 25. Relation between tensile strength and ultrasonic velocity in irons containing varying amounts of carbide.

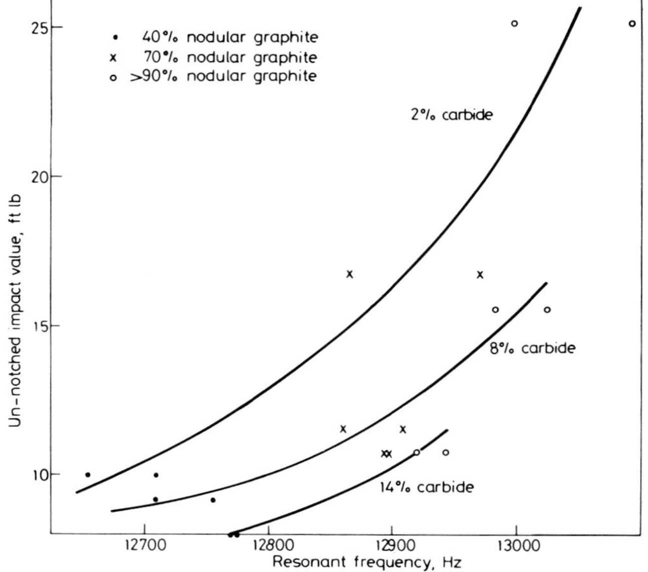

Fig. 26. Relation between un-notched impact values and resonant frequency of irons containing varying amounts of nodular graphite.

Figures 26 and 27 show that provided the matrix structure was known, there were good relationships between un-notched impact values and resonant frequency or ultrasonic velocity measurements. Figure 28 illustrates the relationship between resonant frequency and ultrasonic velocity. Compared with previous work the line drawn has been slightly displaced so that, for a given value of resonant frequency, the ultrasonic velocity in the carbide containing irons was slightly higher. This is in keeping with the observations made that changes in the amount of carbide appeared to have a greater effect upon ultrasonic velocity than upon resonant frequency.

In each part of the work so far reported, values of resonant frequency measured on 7.5 in. long bars and of ultrasonic velocity have been quoted, which have to be exceeded if the good mechanical properties associated with fully nodular graphite structures are to be obtained. In the case of heat treated ferritic or pearlitic irons the minimum value of resonant frequency was 12850 Hz. In as-cast irons, the limit was between 12950 and 13050 Hz according to the amount of pearlite in the matrix. With carbides present, it is recommended that the resonant frequency should exceed 12975 Hz. With ultrasonic velocity measurements the lower limits for acceptance which were recommended were 2.19×10^5 in./sec for heat treated irons, and between 2.22 and 2.24×10^5 in./sec depending on the amount of pearlite in the as-cast condition.

When carbides are present it is recommended that the ultrasonic velocity should exceed 2.27×10^5 in./sec if it is to be guaranteed that irons have good nodular structures. If it is known by other means that irons have good nodular graphite structures, measured values of ultrasonic velocity in excess of 2.26×10^5 in./sec might be used as a means of indicating that some carbides are present in the structure. Thus such high measurements might indicate the need to heat treat to remove carbides.

The comments made in the previous paragraph serve to underline the difficulty of guaranteeing casting quality based

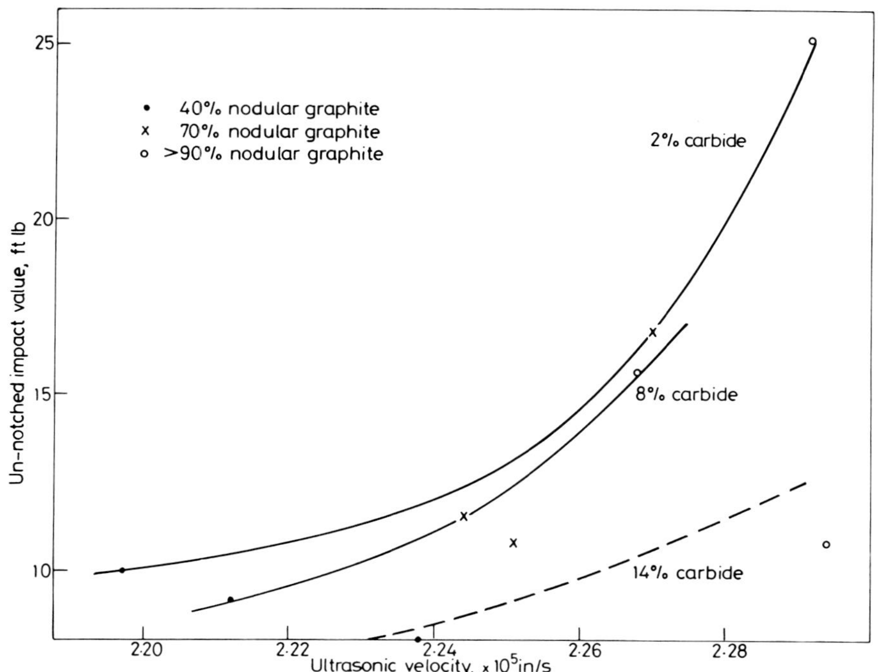

Fig. 27. Relation between impact value and ultrasonic velocity for irons containing varying amounts of nodular graphite.

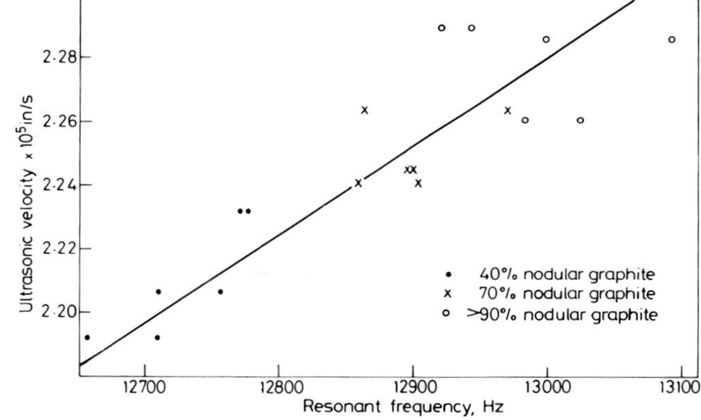

Fig. 28. Relation between ultrasonic velocity and resonant frequency for irons containing varying amounts of carbide and nodular graphite.

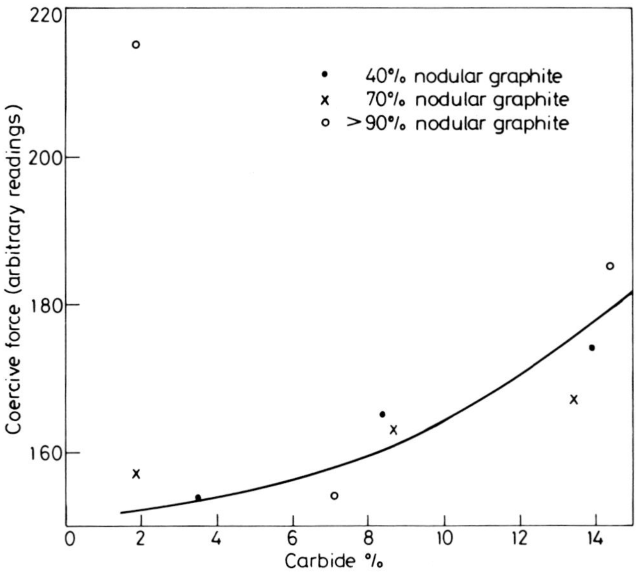

Fig. 29. Relation between coercive force and carbide content for irons with varying amounts of nodular graphite.

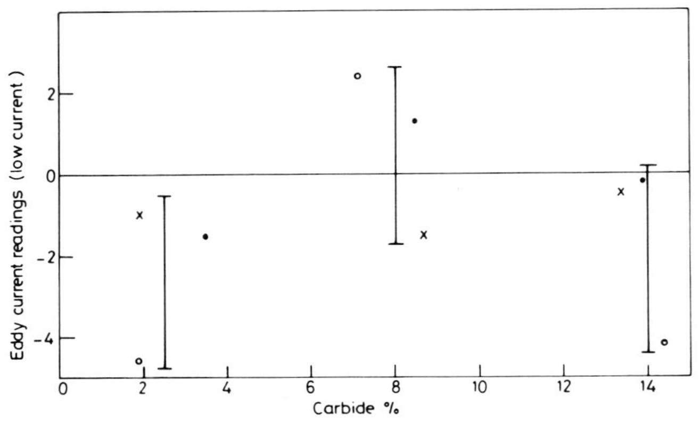

Fig. 30. Relation between eddy current readings (low current) and carbide content in irons with varying amounts of nodular graphite — unmachined test castings.

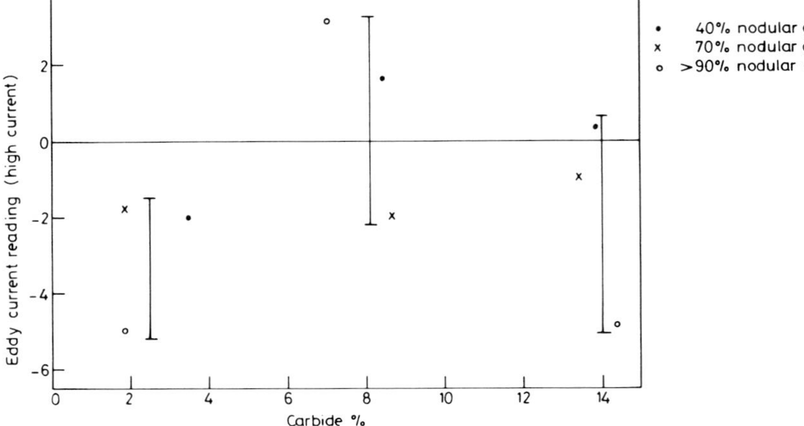

Fig. 31. Relation between eddy current (high current) and carbide content in irons containing varying amounts of nodular graphite — unmachined test castings.

upon measurements of resonant frequency or ultrasonic velocity alone, unless control within the foundry is such that the production of castings having consistent, known matrix structures can always be achieved. If castings having a range of matrix structures are produced, or if there is a range of variation which may occur without causing the properties of castings to fall below the minimum specified values, it is necessary for resonant frequency or ultrasonic velocity measurements to be combined with a second nondestructive test, sensitive to matrix variations, to provide the standards of inspection expected of a rigorous casting quality control system.

There are two alternative approaches. One is to obtain additional information from resonant frequency or ultrasonic testing by adding to these tests, measurements of damping capacity or relative attenuation. The other is to use a separate test such as eddy current or coercive force measurement, to assess matrix structure and to combine these with either resonant frequency or ultrasonic velocity measurement.

The first approach has the attraction that only a single test would be required to make an assessment of casting structure and hence properties. However it is known that the extent to which either damping capacity or relative attenuation is affected by changes in matrix structure is small,[8,9] and is overshadowed by the effects of changes in graphite form. Nevertheless, if small consistent changes of damping capacity or relative attenuation do occur with changes in matrix structure, mathematical methods of analysis could separate the effects due to graphite form and matrix structure and thus provide a means of estimating properties. The possibility of these being successful does not appear likely from the present work. Measurements of damping capacity either by the decay or band width methods (or of relative attenuation) did not differentiate between irons having 2, 8 and 14 percent carbides in the structure. These

methods did however separate irons into groups according to the amount of non-nodular graphite they contained. It must be concluded therefore that at the present time, there is no alternative to the use of a second test which is much more sensitive to matrix variations.

There was a good relation between coercive force reading obtained using the BCIRA equipment and the amount of carbide present (Fig. 29). Measurements were not affected by variations in the amount of non-nodular graphite present, and were similar irrespective of the size or shape of the test piece. They did not alter with change from an as-cast to a machined surface. Eddy current readings, at either high or low current, made upon unmachined test pieces failed to differentiate between irons having various amounts of carbide (Figs. 30 and 31). This was an unexpected result. However as shown in Figs. 32 and 33, when the eddy current tests were repeated upon machined test pieces, reasonably good correlations existed between the reading made and the carbide content of the iron. In the previous work,[4] eddy current testing was equally successful with both as-cast and machined test pieces, although readings varied with both the size of the test piece and surface condition.

A statistical examination of the information available was made in order to assess how the various nondestructive tests might best be combined to predict either the tensile strength or the 0.2 percent offset yield stress of castings. For this purpose, it was assumed that resonant frequency measurements might be combined with measurements of either damping capacity, hardness, coercive force or eddy currents and that ultrasonic velocity measurements might be combined with measurements of either relative attenuation hardness, coercive force or eddy currents. Since the relationship between the results of any of the nondestructive test measurements and the casting properties to

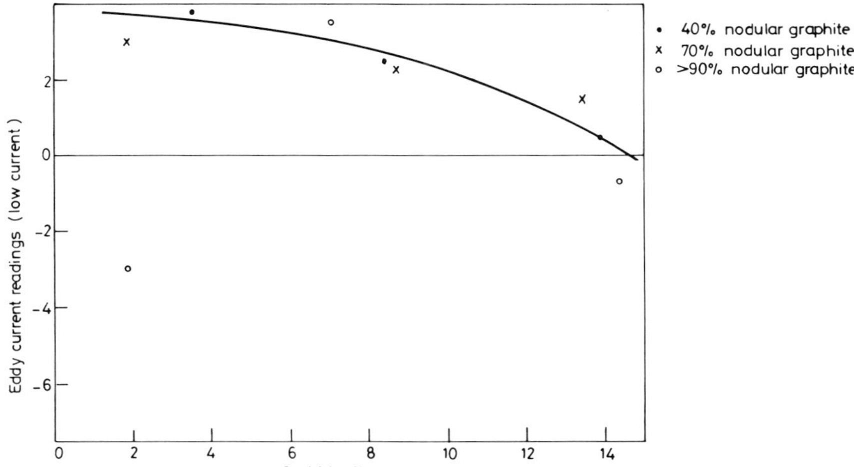

Fig. 32. Relation between eddy current reading (low current) and carbide content in irons containing varying amounts of nodular graphite — machined test castings.

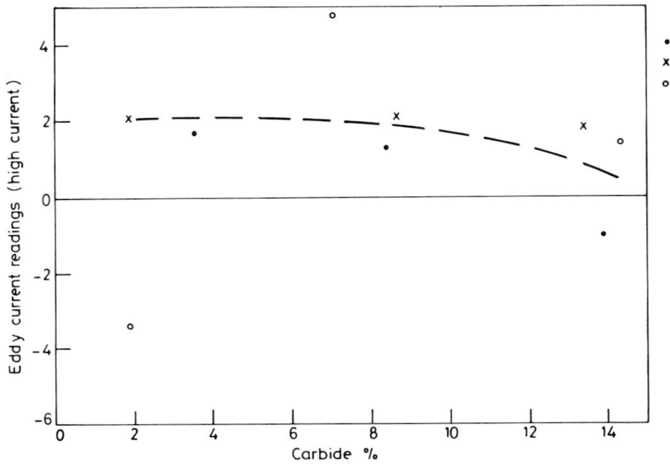

Fig. 33. Relation between eddy current readings (high current) and carbide content in irons containing varying amounts of nodular graphite — machined test castings.

be predicted might not necessarily be a straight line, the statistical analysis also included values of measurements from resonant frequency and the other nondestructive tests which had been raised to the power of 2. The statistical technique employed was stepwise multiple regression in which the resonant frequency or the ultrasonic velocity measurement and the value of these measurements raised to the power of 2 were combined in turn with the results of each of the other relevant nondestructive tests having values which were those measured and also raised to the power of 2. These groups of measurements were then used as independent variables in the prediction of casting properties.

In the stepwise multiple regression technique, the term such as resonant frequency or the square of the resonant frequency which has the greatest importance in the accurate prediction of properties is obtained first. This is followed by the other variables which have a significant role in the prediction. The variables are listed in the stepwise regression technique in their order of decreasing importance and those which have no significant benefit when used for the prediction of properties are omitted.

As a measure of the accuracy with which a prediction of properties can be made using all the variables of significance in a particular analysis, the regression is completed by the calculation of a standard error. Ninety-five percent of all further prediction of properties using the nondestructive testing methods to which the analysis refers should lie within $\pm$ 2 standard errors. The smaller the standard error, the more accurate is the prediction of properties. The results of the stepwise regression analysis are shown in Tables 16 and 17 and are commented upon in the following paragraphs.

For example, Table 16 shows that for the prediction of tensile strength, the most important measurement is the resonant frequency raised to the power of 2 but consideration of the standard errors in the paragraph below shows that the accuracy of prediction is enhanced when this measurement is combined with the squared values of measurement of damping capacity using the band width method or of eddy current measurements at a low current. A similar method of interpretation should be employed when using resonant frequency measurements in combination with other tests to predict the offset yield stress and also in Table 17 when ultrasonic velocity measurements are combined with the other tests.

In the analysis based on the use of Sonic Testing (Table 16), with one exception, the first and therefore the most important variable entering the regression equation for the prediction of the 0.2 percent offset yield strength was the resonant frequency. This would be expected because the property was affected mainly by the amount of non-nodular graphite present and only to a smaller extent by changes in the amount of carbide. The standard error of prediction of 0.2 percent offset yield strength from a measurement of resonant frequency was 1220 lb/in.^{2}.

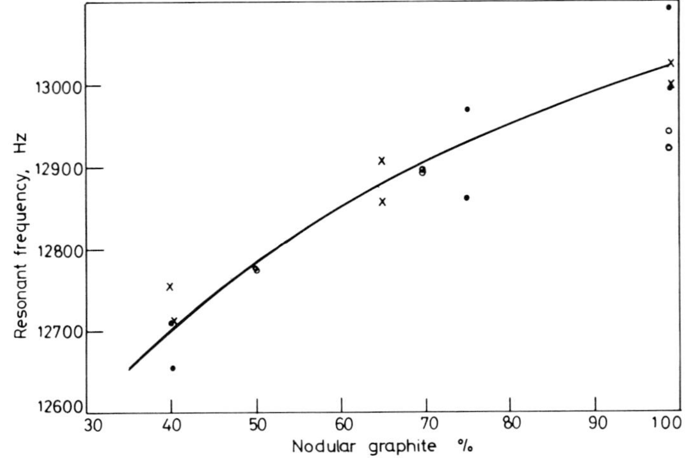

Fig. 34. Relation between resonant frequency and nodularity in irons containing varying amounts of carbide.

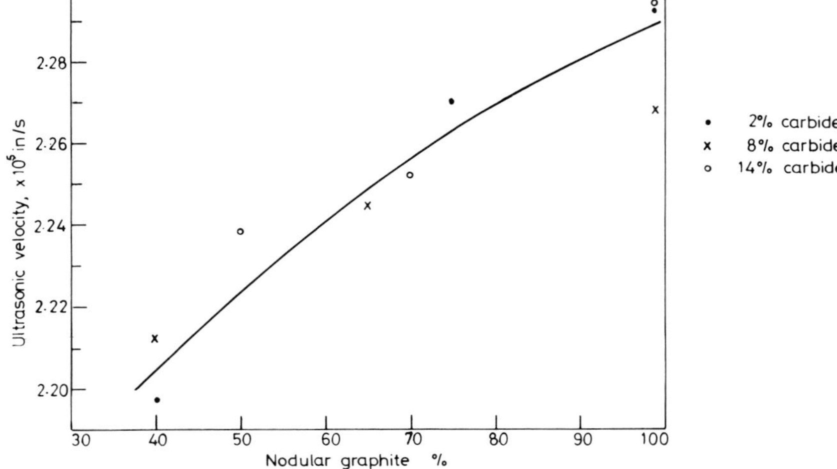

Fig. 35. Relation between ultrasonic velocity and nodularity in irons containing varying amounts of carbide.

When measurements of hardness were combined with resonant frequency, the standard error of prediction was reduced to 850 lb/in.[2] The smallest standard error of 680 lb/in.[2] was obtained when eddy current readings made at low current on machined specimens were combined with resonant frequency. Predictions of tensile strength based upon resonant frequency measurements alone were less precise. The standard error of prediction was 4380 lb/in.[2]. The most accurate predictions were made by combining results of resonant frequency and eddy current testing made at a low current on machined test pieces. Then the standard error of prediction was reduced to 2910 lb/in.[2]. This standard error indicates that 95% of the tensile strengths predicted on the basis of these measurements would lie within ±6 percent of the strengths measured by conventional testing methods.

In a similar analysis based on measurements of ultrasonic velocity rather than on resonant frequency, the results, in Table 17, show that ultrasonic velocity was the first variable entering the regression equation for the prediction of 0.2 percent offset yield stress. The standard error of prediction was 1350 lb/in.[2]. This was slightly greater than that when predictions were based on resonant frequency measurement. When eddy current readings made at high current were combined with ultrasonic velocity the standard error of prediction was 930 lb/in.[2] which again was slightly higher than when the same measurement was combined with resonant frequency.

In the prediction of tensile strength the most important factor in the regression equation was again the ultrasonic velocity. Based on this measurement alone standard error of prediction of tensile strength was 5220 lb/in.[2]. This error was slightly greater than when predictions were based upon resonant frequency measurement. When ultrasonic velocity measurement was combined with measurement of hardness, or eddy current readings made at low current on machined specimens, standard errors of prediction of tensile strength of 2300 and 1900 lb/in.[2] respectively were obtained. The accuracy of these predictions was slightly better than those based on resonant frequency and are such that 95% of the predicted values of tensile strength made by use of these nondestructive tests should lie within ±5 percent of the strength determined by conventional methods.

Thus it has again been shown that resonant frequency or ultrasonic velocity measurements, combined with a second non-destructive test sensitive to changes in matrix structure, provide a reliable means of predicting the principal mechanical properties of castings. As in previous work, it was found that the agreement between actual and predicted values of the 0.2 percent offset yield strength was much better than with tensile strength. Without attempting to detract from the advantages to foundries of being able to use these tests to guarantee to the user that castings have the tensile strengths expected, consideration might also be given to the benefits which could arise from promoting the use of the same measurements to predict, within relatively narrow limits, the values of offset yield strengths. In the longer term, this approach may prove to be the most

356

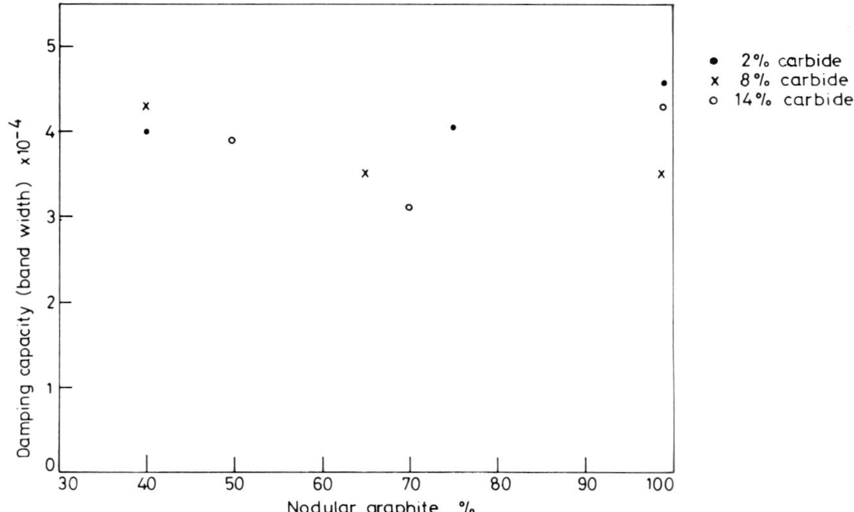

Fig. 36. *Relation between damping capacity (band width) and nodularity in irons of varying carbide content.*

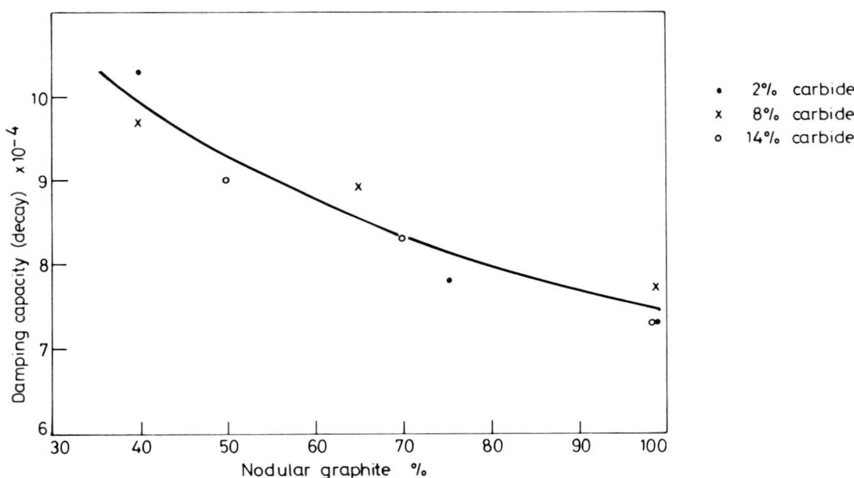

Fig. 37. *Relation between damping capacity (decay) and nodularity for irons containing varying amounts of carbide.*

successful for promoting the use of ductile iron castings, since yield strengths are the principal properties considered in design.

Relationships between Sonic and Ultrasonic Measurements and Graphite Structure

Figures 34 and 35 show the relations between resonant frequency or ultrasonic velocity measurements and estimates of nodularity. These have been represented by a single line, the position of which is independent of the amount of carbides in the matrix structure. These graphs show that the resonant frequency of a 7.5 in. long bar should exceed 12975 Hz and the velocity of ultrasonic energy should exceed 2.27×10^5 in. /sec if irons are to contain almost all of the graphite in a fully nodular form. These are the same values as those quoted in earlier sections of this report, as those which should be exceeded if optimum mechanical properties are to be obtained.

The correlations shown in Table 13 between estimates of nodularity and the graphite parameters, measured using the image analyzing microscope were similar to those found in previous work. Measurements of damping capacity made by the band width method showed that the results all fell within a narrow range of values. It was not possible to use this measurement to estimate percentage nodularity (Fig. 36). On the other hand, the greater discrimination possible with measurement of damping capacity using the decay method enabled a line to be drawn in Fig. 37 to show an increase in damping capacity with increase in the amount of graphite present in a non-nodular form. This relationship was not affected by the amount of carbide present in the matrix structure, although, as in previous work, it had been found that it varied according to the amount of pearlite present.

General Comment

BCIRA staff believe that the AFS Ductile Iron Quality Control Sub-Committee, by their continuing sponsorship of this work, have emphasized the importance which they attach to the promotion within the industry of methods which enable the producer of ductile iron castings to guarantee that each casting supplied has at least the minimum combination of properties expected by the user. The results of this work have successfully shown that this objective may be achieved by the application of simple, established nondestructive testing methods which can be applied to irons having a very wide range of variation in both matrix structure and in graphite form.

There is a natural tendency to think of these tests being applied just for the prediction of the tensile strength of castings, since this is a property which appears in most specifications. However it has been shown in all the work done on behalf of AFS that offset yield strengths may be estimated using the same nondestructive testing methods with precision greater than that obtained in the estimation of tensile strength. Design stress recommendations for ductile irons are based upon offset yield strength properties.

When promoting the use of ductile iron castings in new applications, including those in which the properties of the casting may be critical to the safety of the equipment of which they form a part, it may be in the interests of the industry to ensure that designers and engineers are fully aware that the property on which design is based can also be guaranteed. In parallel with this concept of promoting the ability of foundries to guarantee the actual properties expected within castings, it would also be advantageous to discourage the idea that the specification of a minimum level of nodularity alone provides the safeguards which might be required. As has been shown throughout the work, this is a subjective assessment, very dependent upon the form of any non-nodular graphite present, and which relates only very approximately with either tensile strength or offset yield strength.

There can be little doubt that within the next few years engineers will require the quality of many of the components, which they will be using, to be guaranteed. If the foundry industry as a whole is to retain or to expand its markets, further consideration will have to be given as to how the results of this research may be translated into easily applied quality control procedures, which form an integral part of the casting production process. Essentially this raises two questions. The first concerns the methods of testing which should be chosen, and the second whether or not suitable equipment for these tests is currently available. However in considering the decisions which might have to be made, it is helpful to first specify the principal requirements which should be met. These are:

1) The nondestructive testing methods chosen should be as sensitive as possible to graphite form and matrix structure, so that the estimates made of the properties of castings have accuracy acceptable to the user.

2) The properties determined should be representative of the bulk of the casting and should not be merely those which occur at its surface.

3) Little or no preparation of the castings should be required before testing.

4) Minimum dependence should be placed upon the skill of operators.

5) Methods of testing should be such that they may be automated to provide facilities for testing large numbers of castings at high speeds.

At the moment, the best which can be achieved is a combination of resonant frequency or ultrasonic velocity measurement to estimate the effects of graphite form and amount, with either eddy current, coercive force or hardness measurement to represent the effect of matrix structure. These measurements when used in combination, provide a very satisfactory means of assessment of tensile strength and yield strength values of castings, and within the next few years, are likely to become the principal methods used in casting quality control. Thus it is with these methods of testing that the producer of ductile iron castings will be able to guarantee to the user that all castings sent to them by the foundry have at least the minimum agreed properties, and that any castings which have inferior properties will have been removed. Furthermore, these tests are able to meet the principal requirements listed above, which will enable them to be engineered into an inspection line which would make the verification of casting properties an integral part of the casting production process.

Metallographic techniques are clearly unsuitable. Ultrasonic velocity or resonant frequency measurements are already well established as good methods for the evaluation of the effect of graphite form and amount on mechanical properties. The choice of which is to be used will to an extent be determined by the number of castings of similar type to be inspected, but will also be influenced by development in equipment and techniques and the availability of automated inspection procedures. An example of what might be achieved is shown in Fig. 38 which illustrates the automatic sonic testing of automobile crankshafts. There is undoubtedly scope for further development of both sonic and ultrasonic equipment.

For the assessment of matrix structure the choice is likely to lie between the use of eddy current testing or coercive force measurement. Although testing rates with the former can be very high, calibrations have to be established for each type of casting to be tested. Measurements are susceptible to changes in casting dimensions and may be greatly influenced by variations in structure at cast surfaces. Developments in eddy current testing procedures, such as multifrequency testing, may help to overcome some of the limitations, but the extent to which they might be successful in casting quality control have yet to be demonstrated and evaluated.

On the other hand there are many advantages to the use of coercive force measurement for the assessment of matrix structure. Although rates of testing are lower than those which can be obtained when using eddy currents, separate calibrations do not have to be provided for each type of casting, and measurements are not susceptible to changes in dimensions or to local variations in structure at cast surfaces. The main limitation is the lack of commercially available equipment.

In the longer term there are new developments taking place in the field of nondestructive testing such as acoustic emission or optical holography.[10] At present the application of these techniques is restricted mainly to the aircraft industry where relatively high inspection costs, due to complexity of nondestructive testing equipment, low speeds of operation and considerable dependence on the skill of the operators can be accepted. However, developments in nondestructive testing do take place very rapidly and these tests have the particular advantage that they do measure the performance of a component under an applied stress. Thus they are sensitive to the effects of discontinuities such as porosity as well as to variation in material properties. It is recommended that those concerned with casting quality control should keep abreast of these developments.

BCIRA believes that with the completion of this work, the AFS Ductile Iron Quality Control Committee has established the testing methods which foundries can use to ensure that castings are sold on the basis of their properties on which they are designed. However, as has been discussed in this section of the report, the large scale application of these quality control methods in foundries may at the present time be hampered by lack of commercially available equipment to enable the full

Fig. 38. Automatic sonic testing of nodular iron crankshafts.

potential of the tests to be realized. It is in this field that further development may be required.

Conclusions

A series of magnesium-treated irons were produced having pearlitic matrix structures, and containing various amounts of uniformly distributed carbides. The graphite form ranged from fully nodular to a mixture of nodular and compacted (ASTM type IV) graphite. From assessment of both graphite and matrix structure by metallographic techniques, by measurement of resonant frequency, damping capacity, ultrasonic velocity and relative attenuation, by coercive force and eddy current testing, and from measurement of the mechanical properties of the irons, the following conclusions can be drawn.

Relationships between Mechanical Properties, Graphite Form and Matrix Structure

1) Un-notched impact values decrease as the amount of carbide and non-nodular graphite present in the structure increases. Changes in the amount of carbide have the most effect upon the properties of irons having fully nodular graphite structures, and the effect of increasing amounts of non-nodular graphite is greatest in irons containing the smallest amounts of carbide.

2) Impact transition temperatures are reduced slightly as the amount of non-nodular graphite increases, but are not affected by the amount of carbide present.

3) Tensile strength is reduced by the presence of both non-nodular graphite and carbides. The effect of the former is greatest in irons which are free from carbides, while the effect of the latter is greatest in irons having fully nodular graphite structures.

4) As the amount of carbides in the iron increases, the yield strength shows a rapid increase initially, but this then continues at a slower rate. Small amounts of non-nodular graphite have very little effect upon yield strength, but subsequent increases cause relatively large reductions.

5) Since the 0.2 percent offset yield strength is the property upon which most castings are designed, it is probable that small amounts of non-nodular graphite may be tolerated in castings without seriously affecting their performance.

6) Hardness measurements increase with an increase in the amount of carbide present, elongations at failure fall, but the modulus of elasticity does not change appreciably. Increasing amounts of non-nodular graphite reduce hardness, elongation and modulus of elasticity.

Relationships between Mechanical Properties and Graphite Structure Using the Image Analyzing Microscope

1) Although the mechanical properties of the irons correlate well with some of the graphite parameters measured using the image analyzing microscope, the slopes of these relationships are very steep. Consequently, small variations in measurement made correspond to large changes in properties, making these relationships unsuitable for use in casting quality control.

2) Accurate measurement of graphite and the matrix structure depends on the preparation of good, representative samples from castings, and this would be difficult to achieve for the routine assessment and control of casting quality.

3) In series of irons in which the matrix structure is consistent, mechanical properties relate reasonably well with visual estimates of nodularity.

Assessment of Mechanical Properties by Nondestructive Testing Methods

1) Ultrasonic velocity and resonant frequency decrease as tensile strength and yield strength decrease. For the assessment of yield strength it is not necessary to know the amount of carbide in the matrix and the accuracy of prediction is similar using either resonant frequency or ultrasonic velocity measurement. If tensile strengths are to be predicted as accurately as possible, the matrix structure must be known and the use of ultrasonic velocity measurement is preferred.

2) The amount of carbide in the matrix structure can be assessed by coercive force measurement and by eddy current testing carried out on machined test pieces. The latter test method is most satisfactory with materials having as-cast surfaces.

3) In the presence of carbides, the mechanical properties associated with fully nodular graphite structure are obtained when the resonant frequency of a 7.5 in. long bar exceeds 12975 Hz and when the ultrasonic velocity exceeds 2.27×10^5 in./sec. These values differ from those reported in previous work.

4) Ultrasonic velocities higher than 2.24 x 10^5 in./sec may indicate the presence of carbides in fully nodular irons.

5) On machined test pieces the use of eddy current tests in combination with either resonant frequency or ultrasonic velocity measurement enables properties to be predicted more accurately than when either of the two latter test methods are used alone.

6) Un-notched impact values in the ductile range can be predicted reasonably accurately from measurements of resonant frequency or ultrasonic velocity, provided that the amount of carbide in the matrix is known.

Relationships between Graphite Structure and Measurements Made by Nondestructive Tests

1) The resonant frequency of a standard length bar, and the ultrasonic velocity increase as the amount of nodular graphite increases.

2) Ultrasonic attenuation increases as the amount of non-nodular graphite increases, but varies only to a small extent with changes in the amount of carbide in the matrix.

3) Damping capacity, measured by the decay method, decreases as the amount of nodular graphite increases, but is not sensitive to variations in the amount of carbide in the matrix. Band width measurements are not sensitive to changes either in the amount of non-nodular graphite or of carbides in the matrix.

4) Relationships exist between resonant frequency, ultrasonic velocity and graphite parameters measured using an image analyzing microscope, when only coarser graphite particles are considered.

General

1) To ensure that the properties of ductile iron castings conform to the users' requirements, nondestructive testing of castings by resonant frequency or ultrasonic velocity measurements should be used to assess the effect of graphite form upon mechanical properties.

2) These tests may be used alone if matrix structure is known and does not vary, but if irons are produced which contain various amounts of carbide, then they should be used in conjunction with eddy current testing.

3) Visual assessment of nodularity is unlikely to be a satisfactory quality control procedure since non-nodular graphite can occur in more than one form. The use of image analyzing microscopes requires good specimen preparation, and at the present stage of their development, are also unsatisfactory except possibly in some limited and specialized circumstances.

4) The adoption by foundries of the recommended methods of nondestructive testing to indirectly assess the mechanical properties of castings enables them to guarantee the engineering qualities upon which castings are sold and to ensure that the customers' expectations of the mechanical properties of the cast materials are fulfilled.

References

1. P. J. Emerson and W. Simmons, "Final report on the evaluation of the graphite form in ferritic ductile iron by ultrasonic and sonic testing, and on the effect of graphite form on mechanical properties," Report SP.536 to AFS Ductile Iron Quality Control Committee 12E, Sept 1975, *AFS Transactions,* vol 84, p 109-128, (1976).
2. A. G. Fuller, "Evaluation of the graphite form in pearlitic ductile iron by ultrasonic and sonic testing, and the effect of graphite form on mechanical properties," Report SP.689 to AFS Ductile Iron Quality Control Committee 12E, Jan 1977, *AFS Transactions,* vol 85, p 509-526, (1977).
3. A. G. Fuller, "Effect of graphite form on fatigue properties of pearlitic ductile irons," Report SP.689/1 to AFS Ductile Iron Quality Control Committee 12E, March, 1977, *AFS Transactions,* vol 85, p 527-536, (1977).
4. A. G. Fuller, P. J. Emerson, and G. F. Sergeant, "Report on the effect upon mechanical properties of variation in graphite form in irons having varying amounts of ferrite and pearlite in the matrix structure, and the use of nondestructive tests in the assessments of mechanical properties of such irons," Report SP.874/1 to AFS Ductile Iron Quality Control Committee 12E, April, 1979, AFS 1980 Casting Congress, Paper 80-09.
5. S. Spinner and W. E. Tefft, "A method for determining mechanical resonance frequencies and for calculating elastic moduli from these frequencies," Proceedings ASTM, vol 61, p 1221-1238, (1961).
6. E. F. Ryntz Jr., "Reference microstructures for visual estimation of iron carbide content in nodular iron," *AFS Transactions,* vol 82, p 551-554, (1974).
7. G.N.J. Gilbert, Unpublished work at BCIRA.
8. B. V. Kovacs and G. S. Cole, "On the interaction of acoustic waves with SG iron castings," *AFS Transactions,* vol 83, p 497-502, (1975).
9. A. G. Fuller, "Ultrasonic testing of cast iron," *BCIRA Journal,* vol 10, No. 3, p 339-365, (1962) BCIRA report 649.
10. A. G. Fuller, "Casting quality control," BCIRA, *Foundry Technology for the 80's,* (Birmingham, BCIRA: University of Warwick), 3-5 April 1979.

Casting quality control

A. G. Fuller, BCIRA

<u>THE CHANGING ATTITUDE</u>

Previous papers presented to this Conference have described the technical changes
and advances which may be expected in the '80s. These changes should improve
both manufacturing efficiency and working conditions, enabling the industry to
continue to produce castings at costs which make their use economic in
competition with other materials and methods of manufacture. What has not so
far been considered is whether, solely as a result of the implementation of such
changes, the ironfounding industry will be able to retain or preferably increase
its share of the market in which castings may be used as engineering components.
The answer is probably 'No'. Improved technical efficiency alone will not be
sufficient unless it is matched by comparable advances by the industry in
marketing its products as engineering components of guaranteed quality.

Modern design places a considerable premium upon weight saving. To achieve
this, the engineer will strive to use to the full the properties of the available
materials. The liability of a producer for the performance of his products
requires the maintenance of quality standards in engineering components with
greater consistency and to higher levels than those previously demanded. The
designer in selecting materials for components may be as much concerned that they
will have the expected properties and freedom from imperfections likely to cause
failure in service, as he will be with the cost of their manufacture.

The foundry industry in general is at a disadvantage. Although some
companies have taken steps to promote the use of castings as engineering
components of consistent quality, the industry as a whole has not yet done
sufficient actively to challenge the outdated image of iron castings as cheap
components of variable quality over which little control is exercised. In the
'80s the industry will have to meet increasing demands for guarantees of the
quality of castings in terms of:

   Freedom from defects affecting performance

   Properties

   Dimensional accuracy

The future of the industry depends upon its ability to motivate the required
changes in attitude to casting quality control in order to give the guarantees
demanded. This will require assurances that castings have not only been produced
from metal having the appropriate properties, but also that each individual
casting has the properties claimed for it. There are still those who are so bold
as to believe that separately cast test bars can be used to indicate the quality
of batches of castings and that by this means the need to measure the properties
of individual castings can be avoided. At best, test bars are a crude guide to
the grade of metal used to produce the castings and reveal nothing of their other

Paper presented at BCIRA Conference "Foundry Technology for the 80's" held
at the University of Warwick, England, April 1979. The bound volume of papers presented at the conference is available through BCIRA.

attributes such as soundness.

Non-destructive testing, which is necessary if the quality of each casting
is to be examined and its quality guaranteed, does not have a uniformly good
reputation in the foundry industry. Regular users are usually well satisfied
with the results obtained. Others who use such tests only occasionally, as for
example in the case of a dispute with the customer concerning casting quality,
may be much less satisfied. The reason for this is that the ad hoc use of non-
destructive testing is not practicable. The techniques are not always easy to
apply and there is no opportunity to develop the skills needed in the interpretation of results. The lack of confidence in non-destructive testing has been
one of the factors which may have held back the development of casting quality
control systems in which each casting is inspected. A second reason is the doubt
which has existed concerning the willingness of the customer to pay the higher
prices required for castings of guaranteed quality. It would be incorrect to
infer that the industry has been passive in its approach to quality and that
little has been done to improve the control of the quality of its products.
There have been considerable improvements in process control, including control
of metal composition by means of techniques such as rapid shop-floor thermal
analysis for determining carbon and silicon contents. The belief that process
control can be an alternative to casting quality control has been strengthened
in the last decade by the attention which some important users of castings have
given to different foundries' process control systems when rating them as
prospective suitable suppliers. However, process control—like the use of test
bars—does not guarantee that castings are to the standards expected by the user.
It is internal to the foundry, its value lying in reducing scrap through
obtaining consistent process operation; especially the scrap found at final
inspection.

The demand which will be made to an increasing extent throughout the '80s
for foundries to supply castings of guaranteed quality means that the time for
debate as to whether non-destructive testing of castings is preferable or
desirable and who should pay has passed. The answer to Who pays? is known - it
will be the customer. There may be a difficult period in which the customer is
torn between the alternatives of quality and low price, but at the conclusion it
can be expected that there will be many more foundries whose existence, prosperity
and reputation depend upon the guarantees of quality given with each casting.
In the next decade there will be improvements in many of the existing methods of
non-destructive testing, making their application and use easier. Other methods
at present not used for the inspection of iron castings will be developed,
probably replacing some of those which are currently used. Towards the end of
the decade it might be expected that attitudes of engineers to non-destructive
testing will also have changed. It may be foreseen that from then on towards
the turn of the century, emphasis will be placed more and more upon quality
control tests in which castings are tested and their quality guaranteed under
conditions which simulate those to which they are subjected in service. The
challenge to the industry is to decide how such tests are to be used so that
during inspection every casting is examined, the only ones despatched being those
known to meet at least the minimum standards in the specification agreed with
the customer.

DEVELOPMENTS IN TESTING PROCEDURES

Without exception, the non-destructive testing methods in use today will
continue throughout the '80s to be the principal ones used for inspection of
castings. Rapid development is taking place in all aspects of these tests.
Existing methods will be improved. New techniques, which a few years ago were
regarded as suitable only for the laboratory, are now finding industrial
application and some may find their place in casting quality control within the
decade. Much of the development in non-destructive testing is stimulated by the
high standards of inspection required in aeronautical and space engineering,

where the value of the products justifies the development costs incurred. There
has been little incentive for the manufacturers of non-destructive testing
equipment to extend new developments to the foundry industry, which has long
been regarded·as an area in which the potential sales are low, although as the
non-destructive examination of castings increases this attitude may change. The
industry will, at the start of the decade, have to depend upon its own skills
and those of research establishments, in order to achieve improvement in testing
techniques. In assessing developments in techniques which may occur within the
decade a cautious approach has been adopted. Table 1 summarizes the tests
currently available.

The alternatives are those which are likely to become available and be used
during the '80s. The developments for the future stem from a change of attitude
to the role of non-destructive testing, in which methods are developed to test
the performance of components under service conditions, such as under stress.
These new methods are already used in some inspection activities, but it is
uncertain whether or not they will be used in casting quality control within the
decade.

TABLE 1

Quality to be measured	Available now	Alternatives during the '80s	Future developments
Internal soundness	Weighing and measuring; ultrasonics; onto film X-ray	X-ray with TV-screen presentation Thermography	Acoustic emission Holography
Sub-surface defects	No generally applicable test	No real breakthrough in sight	
Surface cracks	Magnetic crack detection	Dye penetrants	
Structure and indirect assessment of properties, particularly strength	Graphite form and amount: Sonic testing and ultrasonic velocity measurement Matrix structure: Coercive force and eddy-current testing	Unnecessary	Methods for detecting metallurgical variations such as chill or primary mottle
Dimensional accuracy	Gauges, acceptance fixtures and jigs, non-contact sensing, optical projection	Digitizers	?

It is beyond the scope of this paper to consider in detail the methods of
operation, advantages and disadvantages of the use of established testing methods;
its main purpose is to suggest the developments likely to occur. These may be
divided into three groups:

Improvement in the methods

Greater automation in testing procedures

The use of microprocessors and data processing

<u>Improved methods</u>

1.   Internal defects   Ultrasonic inspection[1] will be the most widely used
non-destructive testing technique for the detection of defects in castings. It
will be used in preference to assessment of soundness by weighing and measuring,
as these methods do not locate the position of defects. It will also be
preferred to conventional onto-film X-ray methods which, because they are slow,
are only suitable for use with small quantities of prototype or highly priced
large special-purpose castings. At the present time ultrasonic testing suffers
from two difficulties when used in manual inspection systems. It is not easy to
obtain reproducible testing conditions when working with as-cast surfaces because
it is difficult to achieve consistent acoustic coupling between the probes and
the casting. Although, as in automatic testing, this may be overcome by
immersing the casting in a bath of water, this is not a convenient way of
carrying out inspection manually. The second disadvantage of conventional
ultrasonic testing procedures is that the operator has to interpret a pattern of
signals on the screen of a cathode-ray tube, which may be very complex. To
detect a defect he may have to find an echo in a pattern containing a multiplicity
of other echoes resulting from the reflections of energy from surfaces within
the casting and coarse graphite in the structure. Successful use of ultrasonics
may depend greatly upon the skill and experience of the operator in making such
assessments, and if this is not well developed the reliability of testing may be
seriously affected.

Already new types of ultrasonic probes exist which can give improved coupling,
and some examples are shown in Fig. 1. They include hard-faced probes for use
on areas of the casting prepared by grinding, and also probes which have flexible
faces or which are mounted onto rubber rollers to enable them to be squeezed
firmly against a cast surface to improve contact. Although each type of probe
provides some benefit, the advantage of better coupling is in part offset by
loss of freedom to move such probes in all directions within small areas of the
casting surface. This freedom of movement is essential when searching for and
locating defects. A better prospect lies in the development of non-contact
probes. These are already used for ultrasonic thickness measurement where the
sensitivity required is relatively low. It is expected that within a short time
ways will be found of obtaining the increased sensitivity required for defect
detection, and these probes can be expected progressively to replace those in use
today.
Dependence on the skill of the operator to interpret signals will also
decrease rapidly. Electronic logic circuits capable of making complex decisions
are already built into some modern ultrasonic equipment. These replace the very
simple logic circuits of earlier equipment, which were limited to deciding
whether or not a signal was present in a preset position. It can be expected
that microprocessors will be incorporated in the next generation of ultrasonic
equipment to provide even greater facilities for automatic decision-making.
Although such developments may make the setting-up of testing parameters a highly
skilled operation, the actual testing will become relatively unskilled.

The pre-eminence of ultrasonic inspection for detection of internal defects
may be challenged by the development, for the examination of ferrous components,

of X-ray systems with TV-screen presentation.[2] This method is already used for
the inspection of non-ferrous castings, and similar units are used in the
examination of luggage at many airports. Fig. 2 shows a system in operation.
The component to be examined is placed on a conveyor belt which carries it through
the X-ray unit, which operates continuously. The results are shown on a TV
screen which is watched by the operator, who then decides whether ot not the
material is of acceptable quality. These X-ray units operate at very low power
because of the high level of intensification of X-ray images which can be
obtained using modern electronic circuits, thus they avoid the need for heavy
screening required with higher-power X-rays and make this method of testing
safe for shop-floor quality control. At the moment they cannot be used for the
inspection of ferrous materials because TV screens suitable for use with the
higher-energy X-rays required are unavailable. However, such screens are being
developed for steelworks application and as they become generally available low-
power X-ray examination could find extensive use in casting quality control.

Another method of detecting internal defects which has been suggested as
having considerable potential is thermography.[3] It consists of heating either
the whole of the component to be tested or its surface layers. The pattern of
energy emitted is examined using an infrared detector, or photographically using
a film sensitive to infrared radiation. As shown in Fig. 3, the presence of
defects can be clearly shown. Thermography has been applied for the detection of
defects in iron castings, but the results have been disappointing owing to
variable infrared emission from cast surfaces. It is therefore unlikely that it
will find application in casting quality control, but it might be used in the
examination of machined castings.

2.   Sub-surface defects   Sub-surface pinholes and fine porosity are very
difficult to detect in iron castings and no generally applicable method of
non-destructive testing for such defects is available. In steels sub-surface
defects may be detected by eddy-current testing, but the structural variations
which occur at cast surfaces of iron castings have a considerable influence upon
the eddy currents generated and this prevents the application of this method of
testing. Some success has been achieved using ultrasonic testing methods, but
the techniques which have to be employed are more suited to automatic inspection
methods and in any case have to be set up separately for each type of casting
inspected. The detection of this type of defect in iron castings is a difficult
problem and no real breakthrough is in sight. There is a need for considerable
research and development in this field, since it is most desirable that such
defects should be found before castings are machined.

3.   Crack detection   Magnetic-particle inspection[4] is technically the most
suitable method. Its drawbacks are that it is difficult to automate even when
large numbers of castings have to be inspected, and the use of iron dust
suspended in paraffin used to show the cracks makes working conditions dirty.
An earlier limitation, that testing had to be carried out in a darkened room, has
been overcome by the availability of high-intensity ultraviolet lights which
enable inspection to take place in a dim light. The alternative is the use of
dye penetrants.[5] These have been much improved to give better penetration into
cracks, more rapid development of crack indications and easier removal of excess
fluids from surfaces. Working conditions are cleaner and, although each casting
has to be individually inspected, the procedures for revealing the cracks can be
automated. Fig. 4 shows an installation for the treatment of batches of
components in which they are transferred from tank to tank until the process of
delineating cracks has been completed. In one important respect dye penetrants
are less satisfactory than magnetic-particle inspection, in that they will only
reveal cracks which have an open mouth at the casting surface, whereas magnetic-
particle inspection will often detect cracks which are sealed at the surface.
Dye-penetrant testing is most likely to come into wide use if there is considerable
pressure to improve working conditions.

4. Structure and properties There are no non-destructive tests which provide a direct measurement of the strength of castings. These properties depend upon graphite form and amount, and matrix structure. However, non-destructive tests are available which provide an indirect indication of both of these factors and thus provide a means by which the properties of castings may be guaranteed.

The resonant frequency of a casting measured by sonic testing,[6] and the velocity of ultrasonic energy[7] measured by the time required for the transmission of energy through a section of known thickness, are both determined by the modulus of elasticity of the iron. The modulus varies with graphite form and amount but is substantially independent of the matrix structure. If the matrix structure remains consistent, graphite form and amount determine tensile strength so that in these conditions, using either of the two methods, an indirect indication of the strength of castings may be obtained. They are applicable to the control of quality of irons having nodular, flake and compacted graphite structures. Fig. 5 shows the relation between resonant frequency and tensile strength for a nodular iron crankshaft. A calibration of this kind has to be set up for each type of casting to be tested. Fig. 6 shows the relation between ultrasonic velocity and tensile strength for flake graphite irons. This relation between strength and ultrasonic velocity is independent of the type of casting on which the measurement is made.

Ultrasonic velocity measurement or sonic testing can be operated manually or used in automatic testing systems. Where the quantities of castings of similar type are large, to justify the setting up of the necessary calibration curves, and the shape of the casting is other than in the form of hollow cylinders, or chunky, sonic testing is the best method. Although in automatic testing systems methods have been devised for determining ultrasonic velocity in which the thickness does not have to be directly measured, in manual testing methods thickness must either be measured separately or the equipment used must include thickness-sensing devices such as callipers. Manual testing therefore requires access to both faces of the casting. New probe systems (Fig. 7) are being developed for manual testing, which will overcome the need to measure thickness and provide a direct reading of ultrasonic velocity with access only to one face. The availability of such probes will make ultrasonic velocity measurements easier to carry out.

If matrix structure varies as well as graphite structure, resonant frequency or ultrasonic velocity measurements must be combined with a second non-destructive test sensitive to changes in the matrix for an indication of casting properties to be obtained. Fig. 8 shows how resonant frequency and hardness measurements may be combined to estimate the tensile strength of castings having a wide range of matrix structures. The choice for this second test was between eddy-current,[8] coercive-force[9] and hardness measurement. In the USA it has been claimed that damping-capacity measurement[10] may also be used to assess matrix structure, but this has not been confirmed by BCIRA, other than to differentiate between castings which have and have not been heat-treated.

For assessing matrix structure eddy-current testing will be the first choice of most foundries. This is because it is fast in operation, the equipment is inexpensive, and the methods are easily incorporated into automatic testing systems. Fig. 9 shows an automatic eddy-current testing unit used for the assessment of the structure of small castings. It incorporates an electronically controlled gate separating acceptable castings from those which are unsatisfactory.

Considerable development is taking place in eddy-current testing procedures. Long testing coils have been developed through which castings can be passed, thus eliminating the present need to position them accurately within the coils during testing. The eddy currents generated depend upon the magnetic and electrical properties of the areas being examined and are very sensitive to variations in structure at the casting surface. New eddy-current testing equipment makes

simultaneous use of several different testing frequencies to allow penetration
of the effects to different depths. By mathematical procedures carried out
within the eddy-current testing equipment, the effects of unrepresentative surface
layers can be eliminated and a more accurate assessment of general matrix
structure obtained. The main limitations to eddy-current testing are the need
for the test to be calibrated for each type of casting, and its sensitivity to
small changes in casting size. The alternative is coercive-force testing which,
although not sensitive to variations in casting size and therefore not needing
to be calibrated for each type of casting, is much slower in operation. It is
unlikely, however, that this method of testing would be used in preference to
eddy-current testing.

Hardness testing is not always a satisfactory alternative to either eddy-
current or coercive-force testing, since it provides only a local evaluation of
structure.

In this field the need to combine several measurements to provide an
assessment of structure is most likely to make use of the power of microprocessors
to calculate from several sets of data the casting structure and express it in
terms of a numerical read-out, or to initiate sorting gates to separate
unsatisfactory or doubtful castings from the batch being examined.

5.   Dimensional accuracy   Acceptance fixtures and gauges are the simplest
means for checking dimensions. Fixtures such as that shown in Fig. 10, which is
used to check that the casting for an air-braking system will locate correctly
into an automatic machine, and that the bores are concentric and the sump studs
correctly positioned, are already used by a number of foundries, and it is
forecast that many more will use such devices in the future. Where the quantity
of castings produced of similar type is not sufficient to justify the costs of
acceptance fixtures, simpler go/not-go gauges will need to be used. Measurement
by micrometer, calliper or dial-gauge is slow and cumbersome, and is suited only
for checking the dimensions of prototype castings or of very small quantities.

Acceptance fixtures and gauges have a disadvantage: they are subject to
wear as a result of continual contact with cast surfaces, and with time their
accuracy is lost. However, there is no need for contact with the casting even
though the objective is to measure its dimensions, and there are two ways in
which this can be done. Firstly an image of the casting is projected onto a
screen and its dimensions can be compared with those of a projected master
layout. These optical methods[11] (Fig. 11) are widely used for engineering
applications such as the checking of threads, but since only a flat image is
produced their use in quality control will be limited to those applications
where casting size and shape are critical in two dimensions only. The second
method of non-contact measurement is applicable to castings of any shape and
relies upon the use of non-contact proximity sensors.[12] These are electric or
pneumatic devices which, when placed close to an object, determine the distance
by which they are separated by measuring the pressure drop or electrical
capacitance across the gap. To use such sensors for checking the dimensions of
castings, they would be held in jigs similar to acceptance fixtures into which
the castings would be placed. Multiple sensors placed around the castings,
positioned so as to measure the critical dimensions, would sense the gaps and the
measurements made by each would be fed to a small data-processing unit which
would decide whether or not the casting was acceptable. Although the preceding
description of use has been written conditionally, such methods are used in
engineering, but as far as is known have yet to be used with castings. Their
extension for this application is only a matter of time.

In the '80s an alternative method of checking dimensions will become
available, based on the use of digitizers. A digitizer consists of an arm which
can be moved to any position, connected to a sensitive servo-unit which measures
the spatial position of the end of the arm. Servo-units of this type are used to

control the movement of tools in numerically controlled machines. Their use to
check dimensions is the reverse of this application. A digitizer is connected
to a small computer which is used to record and display the position. The use
of ditigizers to check dimensions may become very important. They offer many
advantages, including speed, and flexibility, for inspecting both small and
large quantities of castings, with very little jigging of castings being required.
As with the use of non-contact proximity sensors, decisions concerning the
acceptability of castings may be made automatically. However, with a computer
available facilities could also be provided to retain records of casting
dimensions, so that trends towards loss of dimensional accuracy can be identified
and corrected before castings fall outside specifications.

6.   <u>Testing under service conditions</u>   The developments in non-destructive
testing described in the preceding sections have all been concerned with those
which make their use easier or which extend the range of application of well-
established procedures. Such tests confirm that the properties of castings are
in accordance with agreed specifications which are assumed to be, at least,
sufficient to obtain satisfactory service performance of the component. However,
in engineering a significant change in attitude to non-destructive testing is
taking place, in that increasing emphasis is being placed upon the use of methods
of testing in which the component is, for example, stressed to the levels which
it will encounter in service to prove that it has the required combination of
properties and freedom from defects. Methods such as optical holography and
acoustic emission, which until recently have been used mainly in the laboratory,
are now being used in engineering inspection, especially in the aircraft and
space industries. The methods are considerably more advanced than any likely
to be used in casting quality control in the early part of the '80s, but as time
progresses and attitudes within engineering change these important advances in
non-destructive testing techniques may become a part of foundry inspection for
safety-critical components.

Holography[13] is an optical method of measuring in three dimensions the small
changes in shape of a component subjected to stress. The extent to which
dimensional changes occur is a function of design, properties of the material,
and freedom from defects. The first use of holography in connection with iron
castings may be by engineers seeking to improve use of materials by optimum
design. The principles of holography are shown in Fig. 12. Visible light of
fixed single wave-length from a laser is split into two beams, one of which is
directed onto the component under test, which is subjected to stress. Light
reflected by the component is directed towards a TV scanner where it is mixed
with the second beam, which has been reflected from a mirror, to produce a
hologram. This hologram is viewed on a TV screen simultaneously with one
previously produced for an unstressed component. The two combine to produce a
three dimensional image of the component with interference fringes, the spacing
of which indicates the movements which have occurred due to stressing. Well-
spaced fringes show little movement, close spacing shows higher movement. An
example reproduced from the ASM handbook on non-destructive testing and quality
control, Fig. 13, shows the result of testing a stressed honeycomb-core sandwich
panel. The close spacing of the interference fringes in the area marked by ink
was due to poor bonding in that area causing greater strain. Optical holography
is a method of testing which is selective. It is sensitive to defects in
critical locations which will affect the service performance of components, but
it will not lead to components being rejected because of defects in locations
where their occurrence is unimportant.

Acoustic emission[14] has similar applications in that it is also used to
evaluate the performance of a component under stress. Its principles are simple.
When a stress is applied to a component, noise of ultrasonic frequency is
emitted due to the release of strain energy during plastic deformation, slip or
reorientation of grain boundaries. This noise is detected by sensors placed at
various points around the component and subsequent mathematical analysis of

amplitude enables the location and significance of structural variations and discontinuities to be assessed. At the present time, its use as an inspection method is limited by the difficulty in analysing the results, but as experience is gained this should be simplified.

The industry would be well advised to keep itself informed on developments in both optical holography and acoustic emission. Research into the application of these techniques in quality control is being carried out by major engineering companies, including users of iron castings. They may become standard casting-quality control tests of the future as the requirements move from compliance with specification to tests of service performance.

The desirability of automation

Relatively small amounts of non-destructive testing are carried out by foundries. Too frequently the present image of non-destructive testing is that of a specialist situated in a corner of a foundry or carrying equipment around the shop floor, painstakingly examining piles of castings, and not always succeeding in separating good from bad. Fortunately for the future of casting quality control, this is a false impression.

If success is to be achieved, as much care and attention must be paid to the methods and facilities for carrying out tests, to the training of operators and the prevention of fatigue, as is given to any other aspect of foundry operations. Castings have to be handled, tests have to be carried out reproducibly and reliably, results have to be interpreted, decisions made and good castings separated from those which are unsatisfactory. The only way in which the required standards can be achieved and maintained is to make testing as simple as practicable and its dependence upon special skills of the operator as low as possible. In the ideal situation all tests would be carried out automatically.

However, for a variety of reasons the ideal will be beyond the capability of many foundries. Quantities of castings for inspection may be too low to justify the high capital cost of automatic inspection units or suitable methods may be unavailable. But even if these limitations exist, much can be done to improve the way in which testing is carried out.

The simplest approach is to provide jigs to position castings whilst testing using manual methods, and Fig. 14 shows their use in the inspection of automobile brake components. Use of jigs makes consistency in testing easier to achieve, and the ultrasonic equipment has built-in logic circuits to determine whether or not castings are acceptable. In the '80s many foundries will have to provide similar facilities for casting quality control. If more than one type of test is required, the various tests will need to be positioned in a line along which the castings will pass, their quality being verified at each stage.

Between such simple methods and fully automatic systems, there is considerable opportunity to make testing easier and more reliable by making at least part of the operation automatic. Fig. 15 shows an inspection unit installed for the detection of defects in brake discs, by ultrasonic methods. Testing, as in other methods of automatic ultrasonic inspection, is carried out by immersing the castings in a tank of water in which they are rotated beneath the ultrasonic probe so that the whole of the casting is inspected. Results are shown on a chart recorder and, although an operator is required to place the castings in position and to interpret the results, this unit has been very successful. It has been installed in a smaller foundry producing relatively small quantities of the type of casting to be inspected. It is the type of purpose-built unit which is relatively inexpensive and which will become much more widely used as the demand for casting quality control increases.

Fully automatic testing systems will be used by the foundries where the
production of large quantities of castings of similar type justifies the
considerable expense of purpose-built testing systems. Fig. 16 shows an
automatic unit for the measurement of the properties of nodular iron crankshafts
by means of sonic testing. It incorporates a conveyor system to carry the
castings to the testing unit, automatic positioning of the casting during test,
measurement of resonant frequency, and marking of unsatisfactory castings.

The golden rule in casting quality control is ease of use of the tests. It
should be the object to achieve this by introducing as much mechanization into
the testing methods as possible.

<u>Microprocessors and data processing</u>   Ideally, non-destructive testing should
be simplified to the extent that its operation is no more difficult than using
a go/not go gauge. In the previous section consideration has been given to the
desirability of improving the handling of components during testing, and
reference has been made in automatic sonic testing of crankshafts to the
equipment containing the logic system necessary to decide whether or not
castings are acceptable. Such facilities are a common feature of most fully
automatic testing systems.

However, in the majority of cases where manual methods of testing are used
only relatively simple decision-making facilities are available. In medical
applications of non-destructive tests, such as ultrasonic scanners and X-ray
equipment, very sophisticated data-processing facilities are available, and in
many cases this technology could be used in metallurgical applications. The
dramatic fall in the cost of data processing, due to the development of
microprocessors, should remove the cost barrier. Initially, microprocessors,
will find application by being included in non-destructive testing equipment for
the purpose of taking away from the operator the need to interpret complex test
results. At first sight this may appear to be a rather mundane application, but
the difficulty which operators face in making such decisions concerning casting
quality should not be underestimated. Their developments are expected to take
place rapidly, and will have a considerable impact during the '80s since the
foundry industry will benefit from a general requirement that these improvements
be made.

A second area in which data processing and microprocessors will be applied
during the '80s is to combine the results of several measurements each of which
assesses a particular aspect of casting quality. Reference has been made to
combination of measurements from several non-contact sensors to determine
whether overall casting dimensions are within specification. A particular case
where the use of microprocessors would be advantageous lies in the assessment of
the properties of castings. These have to be indirectly assessed from measurement
of graphite form and amount and matrix structure. The way in which such results
have to be combined is relatively complex (Fig. 8), and although, with the aid
of graphs, it is not beyond the capability of an operator, such duties are
unnecessary. They could easily be replaced by a microprocessor, and would be a
further step towards making casting quality control a simple operation.

A feature common to most non-destructive tests is that they provide more
information about the structure and properties of material tested than can be
easily interpreted. In the laboratory where time is available this information
can be used, but to make non-destructive tests usable in production they are set
up to accentuate their sensitivity to the aspect of quality to be measured. In
consequence, much useful information is lost. For example, when ultrasonic energy
is transmitted through a casting, its velocity is related to graphite form and
amount and is affected by some microstructural variations. The signal is
attenuated, by the graphite, by variations in matrix structure, and by the
presence of defects. In practice, ultrasonics are used in separate tests to find
defects, or to measure graphite form, but never used to evaluate matrix structure.

The availability of microprocessors should result in a significant change in the
future, making each non-destructive test much more searching than it is at the
moment. For example, with suitable development it would be possible to make a
single ultrasonic test check soundness, properties, and freedom from unwanted
microstructural variation such as the presence of dispersed or centre-line
carbides. This is a much more complex use of microprocessors, but such
developments are likely to be achieved within the decade. Their contribution
will be to make possible a much more thorough examination of casting quality
without increasing the cost or number of tests to be carried out.

In the field of the use of data processing in casting quality control, there
is little previous experience. If past patterns are repeated the priority given
by the equipment manufacturers to the foundry industry will be low, even though
the application of microprocessors may be of greater significance than any
improvement in testing methods or development in methods of automatic inspection.

THE STEPS TO BE TAKEN

At the forefront of changes to be made there must be a concerted effort by the
industry, in its marketing and promotional activities, to establish a new image
as a supplier of engineering components which it guarantees as having the
combination of freedom from defects, properties and dimensional accuracy
expected by the user. Relationships with customers will have to be established
in which matters concerning casting quality are of equal importance with those
concerning delivery and price.

The specific steps which will have to be taken are:

a.   Agreed quality standards   Casting quality control must be carried out to
standards agreed by the foundry and the user. To achieve this the industry will
need to develop a much greater understanding of the properties of the materials
it produces and, in co-operation with the users, apply this knowledge to the
intended use of the castings. Foundries must discuss with users the quality
standards required, and resist the imposition of any which are unreasonable.
Issues which have been hidden need to be openly discussed. It may be possible
but unnecessary to produce all castings to such high degrees of soundness that
no defects can be detected using the most sensitive testing methods, since small
imperfections do not necessarily affect proof-stress values[15] and may be
unimportant to performance in fatigue loading.[16]

b.   Knowledge of quality which can be maintained   Individual foundries must
know the quality standards which they can offer and maintain. The production of
castings to standards of quality for which processes within the foundry are
unsuitable, or for which methods of control are inadequate, should not be
undertaken. The highest standards of quality might be offered by specialized
foundries providing a service paid for by an additional premium on the price of
castings.

c.   Design for inspection   The complexity of shape of castings may make the
application of non-destructive inspection difficult. When agreeing quality
standards, the foundry should ask for minor changes in casting design which can
make testing easier. The designer may not appreciate the requirements of
inspection unless the foundry is able to discuss them with him. Fig. 17 shows
how a simple modification to design, incorporating pads on the castings, has made
inspection easier to carry out.

d.   Choice of tests   Each foundry will have to establish for itself the
methods of non-destructive testing most suited to the castings which it produces.
Once established, methods of testing should be recorded so that they can be
reproduced whenever castings are made from the same pattern. It is important that
records of the tests carried out should be kept as part of the assurance given

by the foundry.

e.   <u>Expertise in techniques used</u>   Most foundries encourage the development of
expertise in casting production methods but, in the future, promotion of similar
proficiency in casting quality control will be required. Successful use of
non-destructive testing depends upon the technical understanding of the methods
employed and the transference of the necessary knowledge and skill to those
carrying out the work. The services of a specialist will be required to develop
the most suitable methods of testing, but for those carrying out the tests most
training could be carried out in the foundry, with the possibility that their
understanding of the methods could be increased by attendance at short courses
widely available.

f.   <u>Skilled inspectors of high integrity</u>   The inspector carrying out tests
to guarantee the quality of castings has considerable responsibility for the
reputation of the foundry. Such persons must be trained to develop the skills
required and be of the highest integrity. Inspection should be free from
pressure from any source to release castings whose quality cannot be fully tested
or guaranteed.

g.   <u>Flexible approach</u>   There may be more than one way of providing to the
customer castings of the required quality. Throughout this paper the emphasis
has been on checking castings. For example, if close dimensional tolerance is
required it might be better for the foundry and for the customer if, instead of
trying to provide this by control and selection of foundry processes, it was
achieved by the foundry carrying out the first machining operation, thereby
adding to the value of the products. Alternative ways of achieving the required
quality should always be sought.

    The theme of this paper has been that castings are engineering components
and that the quality of each must be guaranteed if the industry is to retain its
markets and to meet its responsibilities to the user of its products. Inspection
will become a major tool in influencing the image and status of iron castings.
The attention paid by foundries to casting quality control will have a growing
influence upon their reputations and their ability to obtain markets for their
products. There is no room for complacency. The industry faces a challenge from
traditional users of its products: to react by guaranteeing their quality, or
to face a continually diminishing market for castings.

<u>REFERENCES</u>

1.   Emerson, P.J. Some practical observations on ultrasonic testing of iron
      castings. British Foundryman, 1973, vol. 66 No. 10, 282-287.

2.   Hall, H.T., Lavender, J.D. & Ball, J.I. Recent developments in the
      application of the techniques of X-ray television fluoroscopy to foundry
      problems. British Foundryman, 1969, vol. 62 No. 8, 296-308.

3.   Lawson, W.D. & Sabey, J.H. Infrared techniques. In: Sharpe, R.S. (ed.)
      Research techniques in non-destructive testing. Vol. 1. London, Academic
      Press, 1970, 443-478.

4.   American Society for Metals. Metals handbook, 8th ed. Vol. 11.
      Non-destructive inspection and quality control. Metals Park, Ohio, ASM,
      1975, 44-74 Magnetic particle inspection.

5.   Magistrali, G. Liquid penetrant inspection. In: Bolis, E. (ed.)
      Non-destructive inspection practices. Vol. 1. Neuilly sur Seine, France,
      North Atlantic Treaty Organization, Advisory Group for Aerospace Research
      and Development, 1975, 169-180. (AGARD AG 201).

6. Fuller, A.G., Emerson, P.J. & Rew, R. Sonic testing: a simple
    non-destructive test for verifying casting quality. BCIRA Journal, 1963,
    vol. 11 No. 3, 358-375. BCIRA Report 696.

7. Fuller, A.G. Evaluation of the graphite form in pearlitic ductile iron by
    ultrasonic and sonic testing and the effect of graphite form on mechanical
    properties. Transactions of the American Foundrymen's Society, 1977,
    vol. 85, 509-526.

8. Emerson, P.J. Unpublished work at BCIRA.

9. Emerson, P.J. Unpublished work at BCIRA.

10. Kovacs, B.V. Quality control and assurance by sonic resonance in ductile
    iron castings. Transactions of the American Foundrymen's Society, 1977,
    vol. 85, 499-508.

11. McMaster, R.C. (ed.) Non-destructive testing handbook, Vol. 1. New York,
    Ronald Press Co., 1959, 12.1-12.37 Optical projectors and comparators.

12. Abbe, R.C. & O'Brien, M. A brief report on non-contact gaging. Quality
    Management & Engineering, May, 1972, 16-19.

13. American Society for Metals. Metals handbook, 8th ed. vol. 11.
    Non-destructive inspection and quality control. Metals Park, Ohio, ASM,
    1975, 198-233 Inspection by optical holography.

14. Hutton, P.H. & Ord, R.W. Acoustic emission. In: Sharpe, R.S. (ed.).
    Research techniques in non-destructive testing. London, Academic Press,
    1970, 1-30.

15. Gilbert, G.N.J. Unpublished work carried out at BCIRA.

16. Palmer, K.B. Unpublished work carried out at BCIRA.

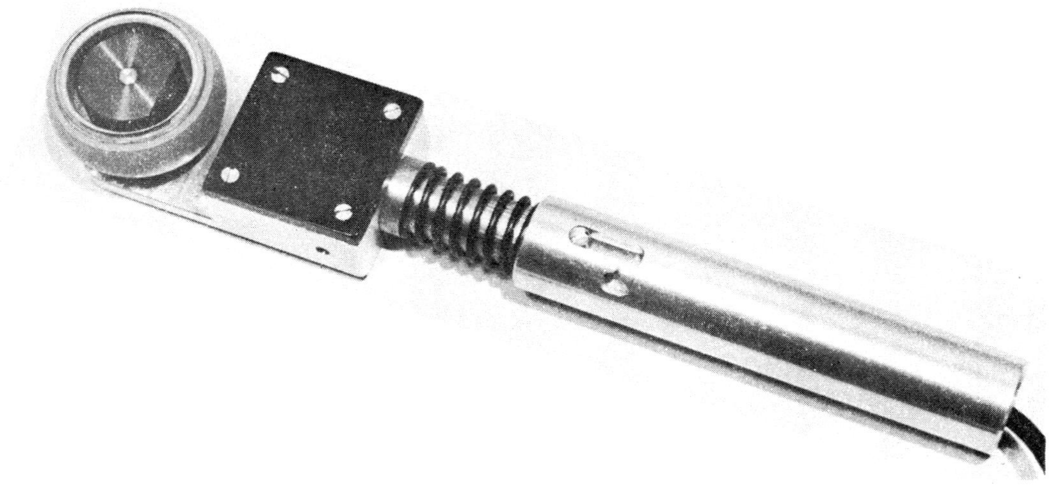

Roller probe

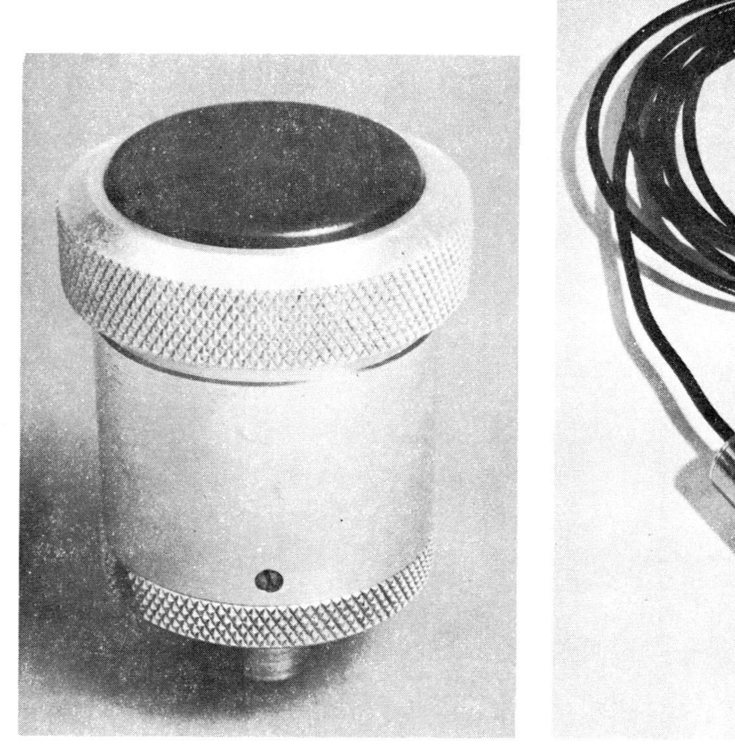

Soft faced probe

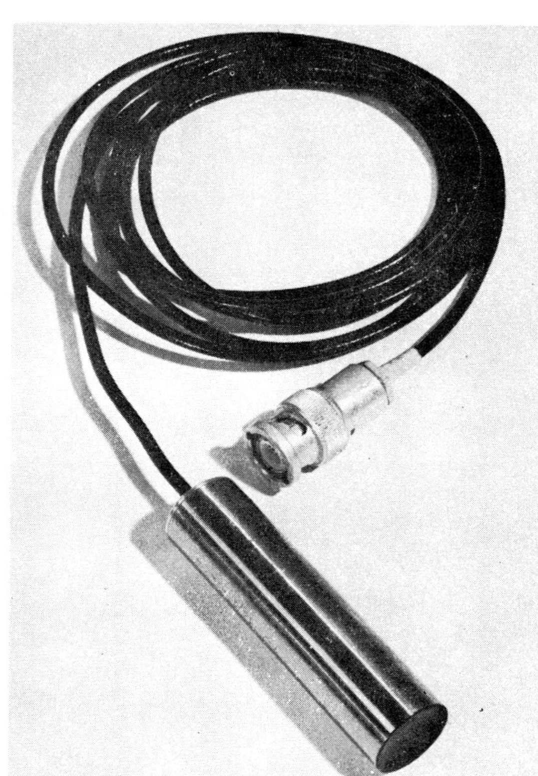

Hard faced probe

Fig. 1 Alternative types of ultrasonic probes.

Fig. 2 Automatic X-ray inspection unit for detecting
        defects. (Courtesy Non-destructive testing)

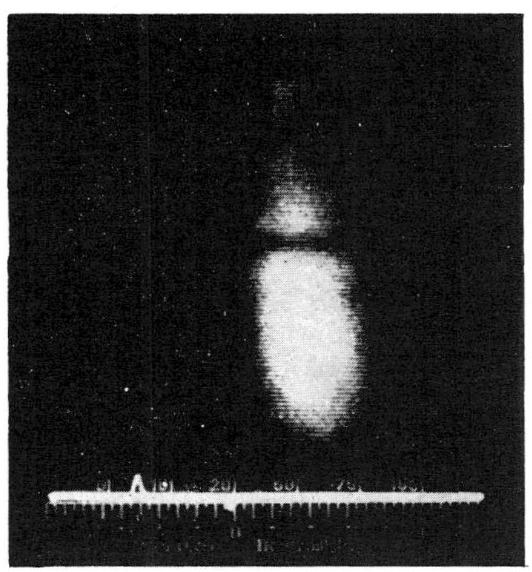

Fig. 3 Thermograph of a plate
        showing distortion
        of surface temperature
        distribution due to
        sub-surface voids.
        (Courtesy Academic
        Press)

Fig. 4 Automatic unit for dye penetrant
        inspection. Components are
        transferred from tank to tank to
        complete development of crack
        indications.
        (Courtesy of Magnaflux
        Corporation)

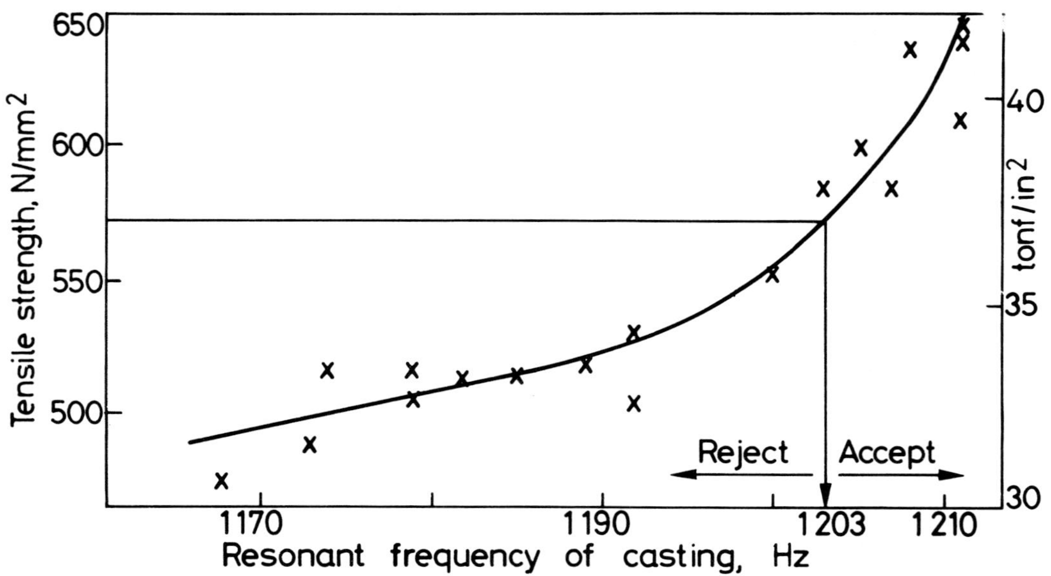

Fig. 5 Relationship between resonant frequency and
tensile strength of pearlitic nodular iron
crankshafts.

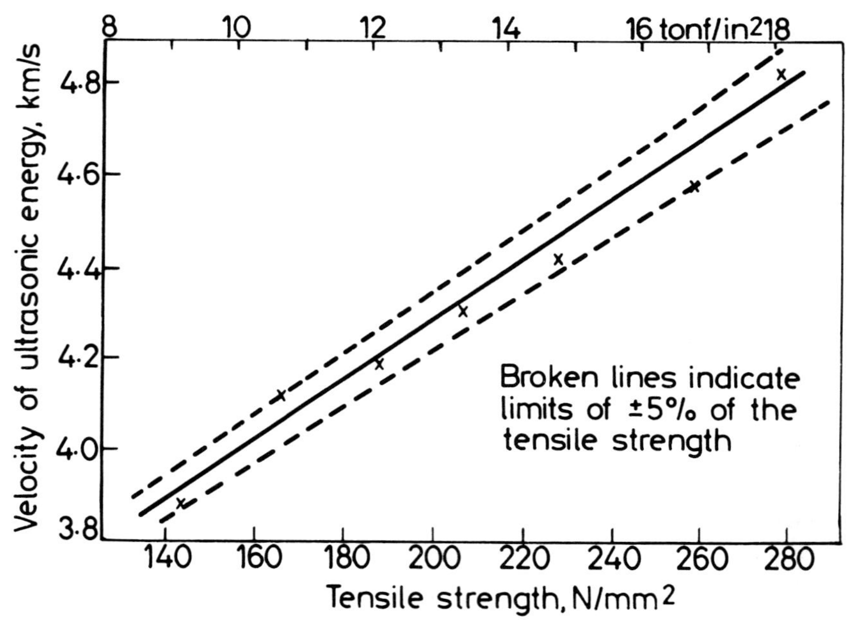

Fig. 6 Relationship between ultrasonic velocity and
tensile strength of flake graphite cast irons.

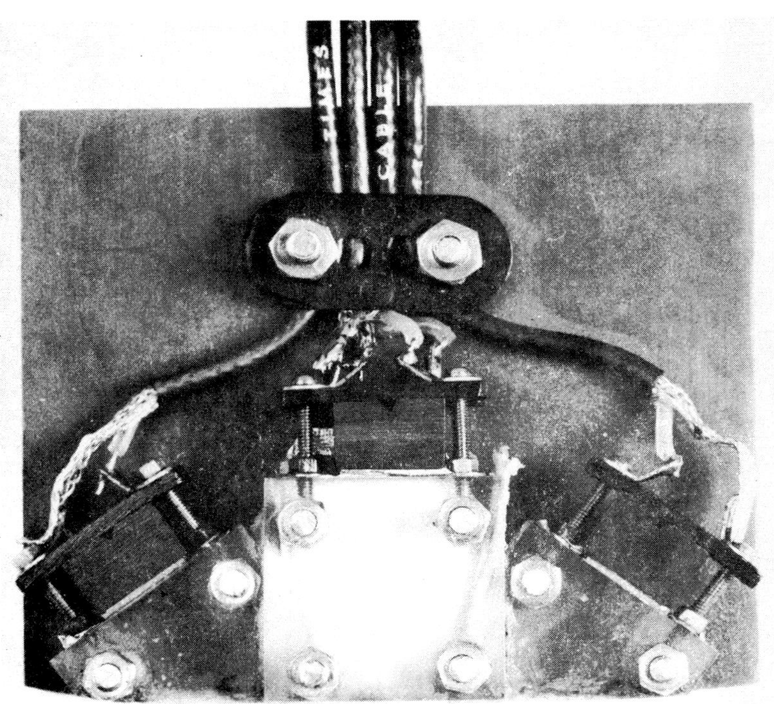

Fig. 7   Ultrasonic probe system developed by
         BCIRA to allow measurement of
         ultrasonic velocity to be made from
         one face of a casting without
         requiring measurement of thickness.

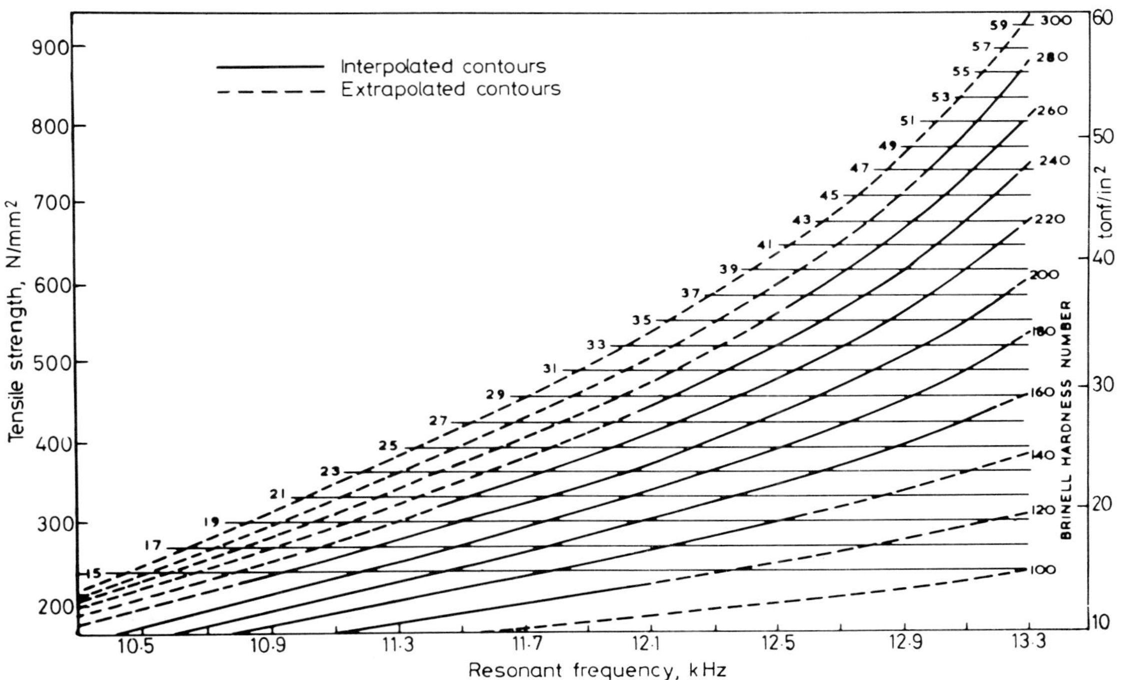

Fig. 8   Relationships between hardness, resonant frequency
         and tensile strength of irons having varying matrix
         and graphite structures cast as bars having a length
         of 7.5 in.

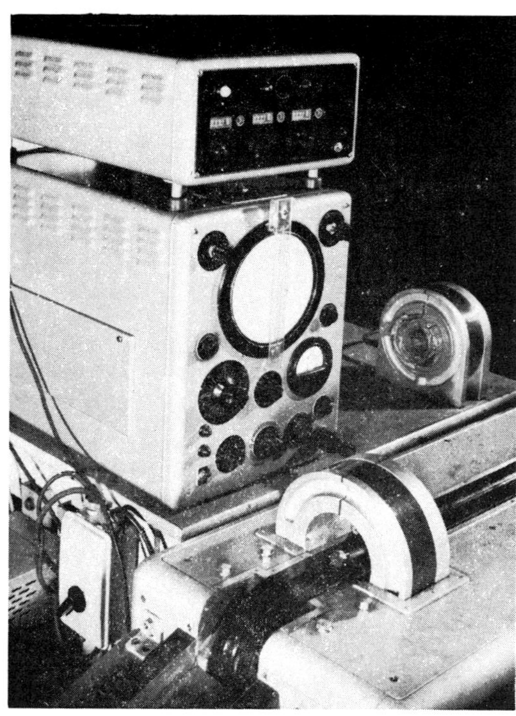

Fig. 9 Eddy current testing apparatus with monitoring system for the automatic sorting of materials of acceptable and unacceptable quality.

Fig. 10 Arrangement of acceptance fixture for checking dimensions of cylinder used in an air braking system for commercial vehicles.

Fig. 11 Equipment for optical projection
method for checking dimensions.
(Courtesy of Non-destructive
testing)

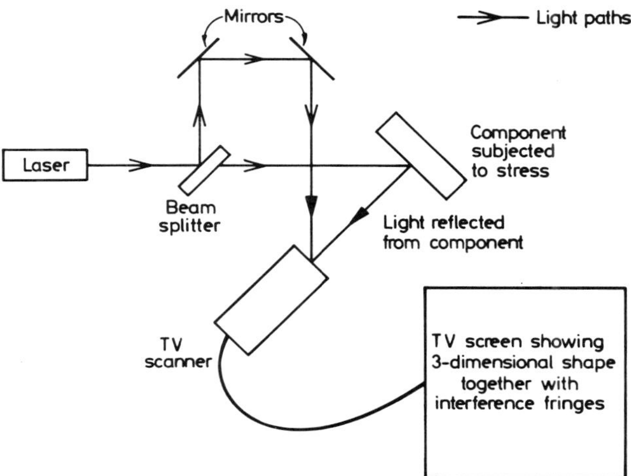

Fig. 12 Principles of holography.

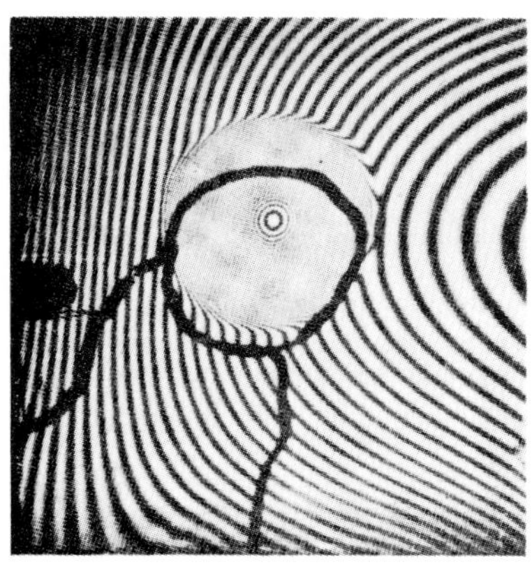

Fig. 13 Optical holograph
of a stressed
honeycomb panel
showing close
spacing of
interference
fringes in a
poorly bonded
area. (Courtesy
ASM)

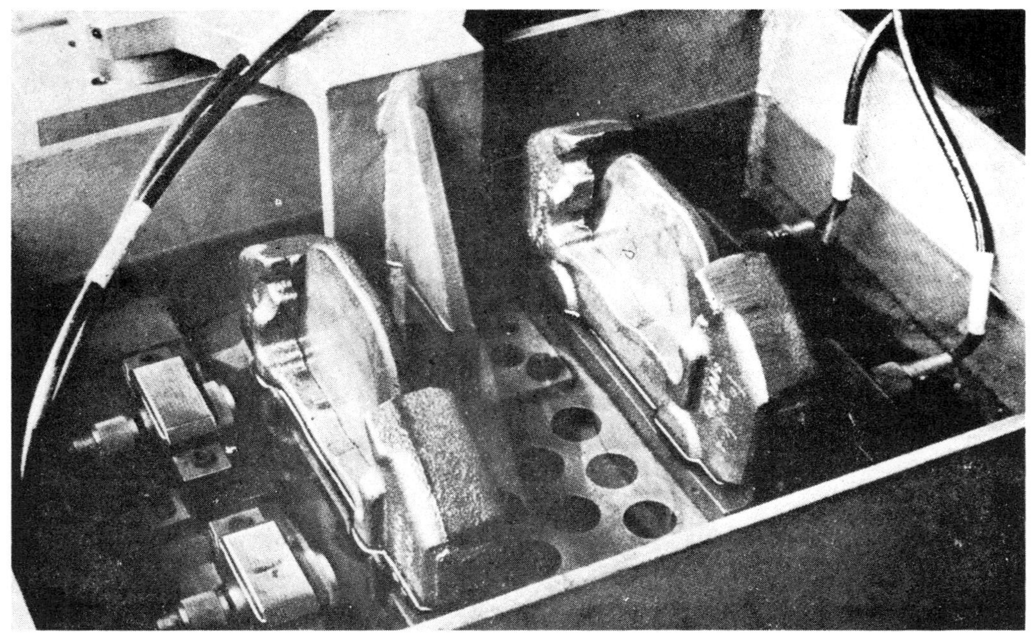

Fig. 14   Jig to hold brake calipers during ultrasonic testing to assess
properties (Courtesy Non-destructive testing)

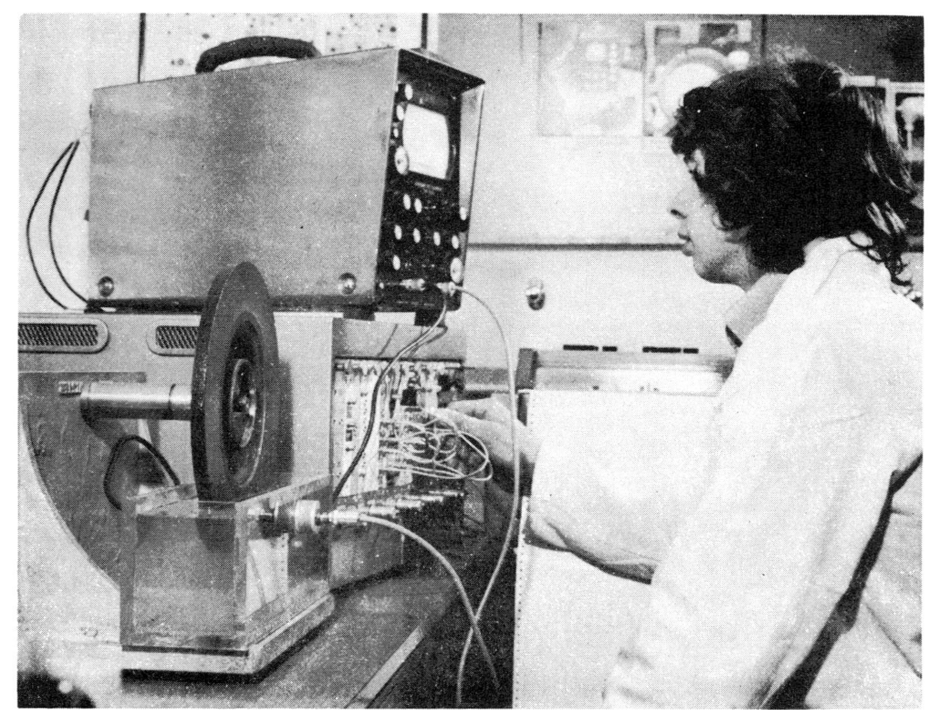

Fig. 15   Semi-automatic unit for ultrasonic examination
of discs used in braking systems.

Fig. 16   Automatic sonic testing of nodular iron crankshaft.

Fig. 17   Modification of casting design to make inspection easier.
Cross section has been changed to provide 2 parallel surfaces
from which ultrasonic energy may be transmitted and reflected.

Materials, Equipment, Auxiliary Processes

The Economics of Sand Reclamation as Opposed to Disposal

F. W. Rohr
Perkins & Will, Chicago, Illinois

ABSTRACT

Increasing costs and new methods of operation have resulted in the consideration of methods to reclaim sand in many foundries. Before a decision can be made regarding sand reclamation, a detailed economic evaluation of existing sand handling practice in the foundry as well as reclaiming methods should be made.

This discussion includes information concerning evaluation of sand handling and a review of reclamation methods and applications. Information is also included to indicate how accumulated cost data can be used for analysis of capital investment projects such as sand reclamation. Various criteria of capital investment decision making are outlined including risk factors and return on investment. Methods of capital investment analysis normally used for evaluation of alternative industrial projects are illustrated by sample problems, including payback period, return on investment and discounted cash flow.

Introduction

Sand is seemingly among the most abundant materials on earth, yet like many other things that used to be inexhaustible, it is beginning to have an increasing value. Sand is not just sand to the foundryman but a specialized material that must be at the right place at the right time. This degree of speciality, plus material handling, the timing factor and the cost to discard it are factors that give foundry sand a meaning and a cost.

Reclaiming sand in foundries is not a new subject. It has been done in many foundries for a number of years. Those foundries who have been active in this field have already developed considerable expertise, sometimes after experimentation to fit a process to an individual need. Others may also have tried it and discarded the idea as being not workable at the time. Progressive foundries today know the requirements for physical properties of the different types of sand that they use plus availability of various binders, material handling and reclamation equipment and the cost to dispose of the discarded material. This discussion is not intended to be a technical review of the subject. Various case studies have already been presented by foundrymen and manufacturers can be contacted for specific information. This discussion is intended to review the economic factors and methods that can be used by foundries to determine the feasibility of various sand reclamation systems.

Factors in Economic Considerations

The decision to make a change in foundry operation such as methods and equipment for sand reclamation must be preceded by an objective approach. A step by step program should be carefully planned over a period of time, evaluation of equipment and methods, space utilization, personnel training and financial arrangement.

Since major changes in methods, equipment and finances will have a lasting effect on the future success of the firm, a consideration should also include the market for the castings to be produced. Equipment purchased for a specific need could be an undesirable investment if the market for the product manufactured changes because of competition, material availability, technological changes or environmental requirements. For example, future emphasis on lighter automobiles may result in a greater substitution of aluminum for cast iron parts. Another possibility is that the complete upgrading and modernizing of a plant including the possibility of sand reclaiming could result in opening up new markets for the foundry. The commitment to invest in equipment for a specialized product of a particular size may also be more of a deterent than planning for diversified sizes and weights.

The basic steps in developing a program of improvement for an existing foundry, including sand reclamation, is to determine the cost of existing practices. This information can be obtained from plant accounting records. If a new plant operation is being planned, costs must be estimated for operations with and without sand reclaiming. These factors generally include but are not limited to the following:

1) Purchased sand
2) Shipping
3) Unloading
4) Storage of new sand
5) Moving or handling of sand
6) Binders or other additives
7) Handling and disposition of dust from collectors
8) Expense of rehandling or reusing if performed
9) Handling and conveying of used sand to storage
10) Storage of used sand
11) Loading, transport and disposal of used sand.

Some items may be identified from records while others may be less obvious. For example, purchased sand is generally paid from an invoice. The price may or may not include shipping or unloading. Unloading may be simply a dumping operation from a truck or may involve movement of sand by means of plant personnel and a vehicle. In any event, it can be an expense item, and should not be regarded as part of overhead if a cost analysis of the operation is to be made.

Storage of sand may or may not be considered an expense depending on the firm's accounting methods. The space or location of storage may be treated as an allocation because of land cost, taxes or cost per square foot of storage space. This space allocation could change if less inventory is needed when sand reclamation methods are used.

Moving and handling sand from storage areas could be a significant factor in some foundries especially if sand is used only once. Larger inventories of sand would mean that a greater amount would be handled or transported over a longer distance by a vehicle or conveying means. Binders and additives should be identified as a cost item since these types of materials may change if sand reclamation methods are used.

The costs associated with dust or sludges from collectors should be estimated not only from a handling standpoint but also because of the cost of collection. Baghouse maintenance, cost of bag replacement and fan horsepower for example, are all a part of the cost of collector operation. Dust or sludges collected may also be considered a hazardous waste and require special treatment and disposal methods. Dusts and sludges from sand used in systems designed for reclaiming may result in even more of a problem since they may contain greater amounts of chemicals. These materials are being considered a hazardous material in more and more areas today and disposal costs may be higher than other foundry wastes, due to the necessary preparation of landfill sites.

In studies of operations of various foundries, molding sand is used anywhere from one to twenty times before it is discarded. From a material handling standpoint, handling cost and molding sand are probably the largest expense items in the foundry. The expense factors may differ greatly between various foundries but will include labor, owning and operating cost of equipment, cost of tests to monitor quality and finally the cost of disposal.

The cost of storage may not be significant in many foundries but, as in the case of storage of new materials, metal for melting and finished castings requires space and an allocation of cost for land, taxes or other accounting factors.

Consideration of Sand Reclamation Methods

Sand reclaiming methods have been used with some degree of success for many years. In some cases, the systems developed were accepted as satisfactory for a particular foundry. But in general, reclamation has not been adapted by the industry as a whole because of various technical, operating and economic problems. Wet and thermal systems, although more effective in many cases and the most capable of restoring the sand to its original specifications, are more costly to operate especially since the rapid rise of fuel costs. Dry systems, although not as capable of refinishing the sand grains, have proven adequate for many purposes.

In the evaluation of sand reclaiming processes consideration must be given to the degree to which sand should be processed before it can be renewed or blended with new sand. The quality of reclaimed sand can vary with different systems from lumpbreaking and screening to near new condition. Specialized systems may also be used to recover and reuse zircon and chromite sands.

Economic evaluation of sand reclamation processes will require a new look at the entire foundry practice. It is possible that sand reclamation would not be feasible if system sand is simply diverted at some point in the material flow through the plant and processed. Reclaiming may only prove to be a valuable asset if all factors in the foundry operation, including the product, are considered. These factors should include but

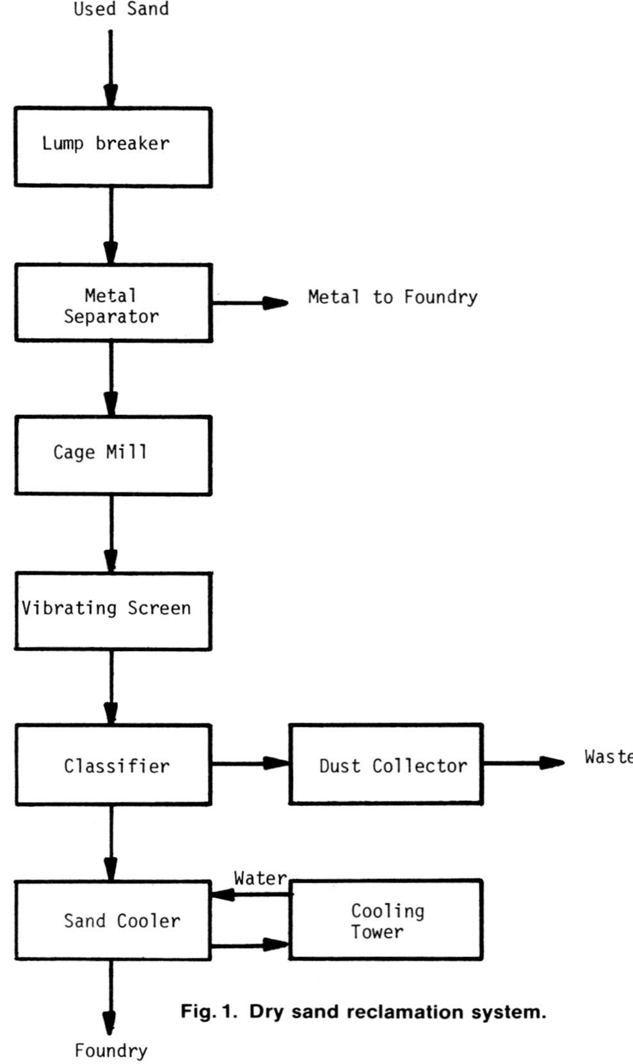

Fig. 1. Dry sand reclamation system.

not be limited to:

1) Purchased sand
2) Shipping
3) Unloading
4) Storage of new sand
5) Moving or handling of sand
6) Binders or other additives
7) Handling and disposition of dust from collectors
8) Reclamation equipment
9) Fuel, electric power and water
10) Labor for sand reclaiming system
11) Spare parts and maintenance
12) Debt service on capital cost for system
13) Handling, storage and blending of used and new sand
14) Loading, transport and disposal of used sand.

Although these factors appear similar to those included for existing sand handling practice, there are significant differences. For example, there would be less expense for purchased sand, shipping, unloading, storage and moving of sand. Bonding materials may be similar however, the obvious increased expense will include handling and disposition of chemical dusts or sludges and cost of equipment. More expense for utilities will also be required along with labor, maintenance and payments for the installed equipment. An additional factor to offset the

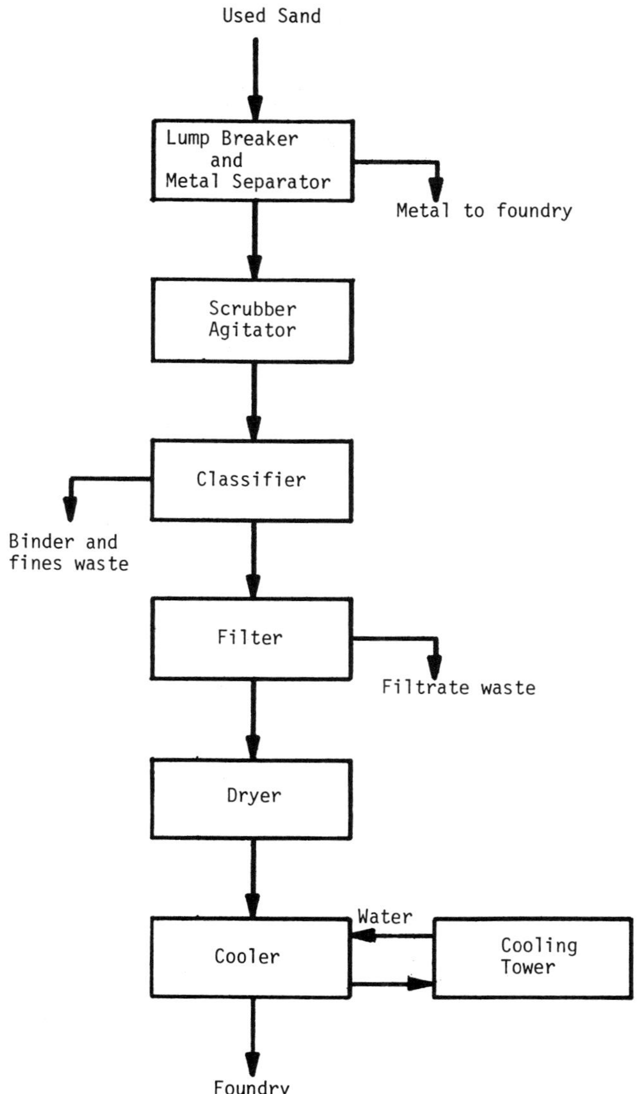

Fig. 2. Wet scrubbing sand reclamation system.

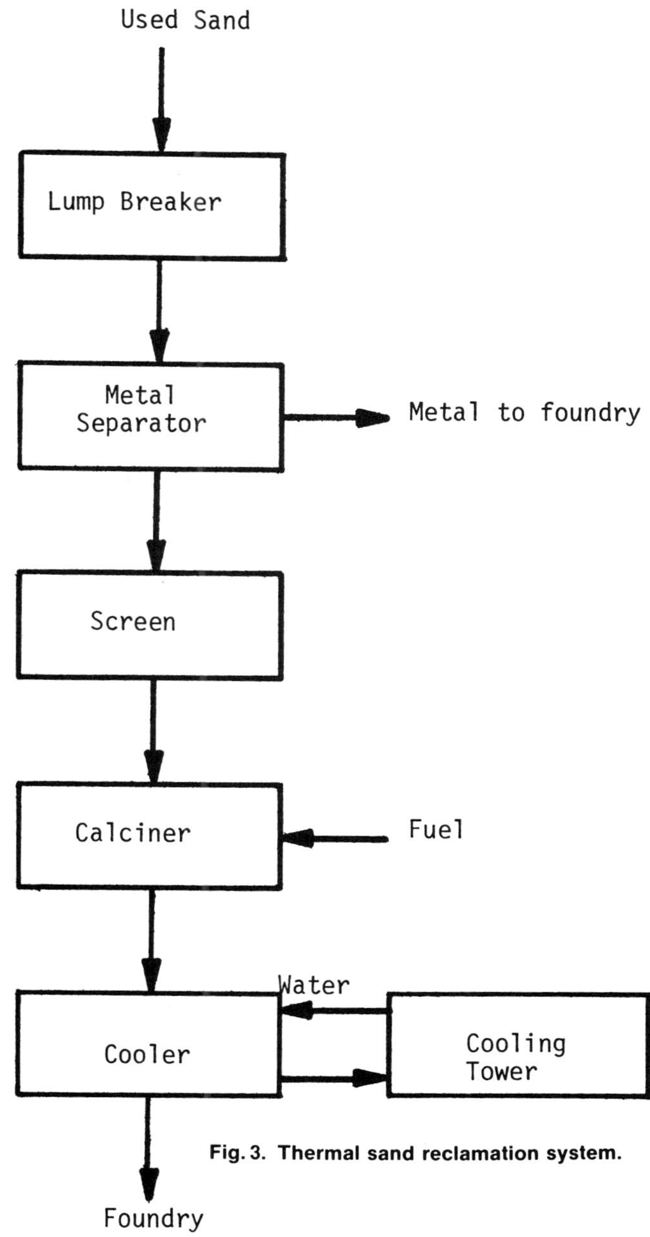

Fig. 3. Thermal sand reclamation system.

expense of equipment is the decreased amount of core and molding sand for disposal.

The decision to use a wet, dry or thermal method of sand reclamation depends on the types of binders used, sand, metals cast and the degree to which sand is to be cleaned of residue. These systems are not interchangeable in terms of effectiveness and will not reclaim 100% of the sand in use. Where these systems are economically feasible, however, the process can result in an 80 to 90% recovery rate, more consistent sand quality, more control of binder usage and possibly better castings.

Dry Systems

Dry systems have the lowest first cost although this may be partly offset by maintenance due to abrasion of equipment and high sand loss due to greater fines generation. The basic process includes:

1) Lumpbreaking and screening
2) Metal separation
3) Classification
4) Fines removal
5) Cooling.

Figure 1 shows a schematic of this type of system. The lumpbreaking and screening process is intended to break down the build-up of carbonaceous coatings on sand particles that gradually prevent adequate bonding. The process will remove fines, clays, other binders and coatings to some extent but gradually the build-up of coatings will require replacing all or part of the sand. Dry systems can be used for reclaiming nobake sands. New methods of dry sand reclamation are a direct outgrowth of nobake binders used in molding. Various binders react differently in the sand reclamation process. Furan resin binders are hard and brittle and tend to break loose from sand grains. Other binders such as oil nobake types have a softer film and do not break off as readily, while silicates become hard glass coatings and are the most difficult to remove. The type of sand used also influences the yield of reclaimed sand since coatings tend to separate from round grains more readily than angular shapes. Claims have been made as high as 90% for dry reclaimed nobake sands.

Wet Systems

Wet scrubbing systems remove clays, fines and a greater amount of carbonaceous coatings on sand grains than dry methods. These systems also cost more than dry types and have greater operating cost.

Wet reclamation systems can vary considerably in their degree of sand treatment and effectiveness. The simplest type of system includes wet clarification and drying where only fines are removed and grains are washed to remove superficial soluable portions of binders. Other more complex systems remove fines and include scrubbers which agitate sand grains in a slurry, tend to break up agglomerated sand grains and scrub off binder deposits. A flow sheet of this type of system is shown in Fig. 2.

Thermal Systems

Thermal or calcining systems remove carbonaceous coatings and organic binders by subjecting sand up to 1500F (815.56C) in the presence of oxygen. This type of system results in the most effective removal of organic materials from sand grains but also requires the use of fuel unless methods are found to use waste heat from other sources at an elevated temperature. Figure 3 shows the processes in a thermal system.

Evaluation of Economic Alternatives

The primary purpose for development of economic information about the cost of foundry operations is to evaluate engineering and operating alternatives. The problem of economically appraising costs of projects is faced by all firms and the success of a company is frequently determined by its effectiveness in selecting the most profitable alternative.

The investment of funds in machinery, buildings or other property with a useful life of more than one year is called a capital investment. These funds must be recovered by the company through its operations and the company is accountable to itself, its investors or for loans necessary for the investment. When evaluating a capital investment project, management wants to know what the profit will be, the payback period or the return on investment. This sounds like a simple question, therefore, management would like a simple answer so that it can decide what to do.

As a result of competition in world markets, increased pressure by overhead and wage increases, profit margins are smaller than ever before. Business today faces the difficult situation of deciding where and how to invest its limited capital in the most effective projects in order to survive. Therefore, evaluation of capital projects is not a simple problem and involves far more than engineering decisions.

The engineer planning the details of a capital investment is generally asked for information regarding a proposed project. This information would normally include estimates of the cost of installation, labor required, operating cost and expected rates of production. The information is important for a management decision to undertake a capital project because it has an effect on operating cost, overhead and sales. The engineer should recognize the importance of this because if the project costs are higher than expected, or the rate of production is lower than expected, the entire financial program of the company may be altered. If a major capital program is involved, such as sand reclamation, any drastic changes in the financial program may result in jobs being in jeopardy and a bank loan to cover the deficiency.

Since preparation of project planning information can be of such importance, it is desirable for engineers and those who help plan important projects to become knowledgeable about the "business end of the business." This knowledge will help them to plan better and more profitable projects for the company and will help them sell the projects to management.

The actual financial evaluation of various capital projects requires comparison to certain criteria which financial planners try to make as objective as possible but in many cases is subjective and a matter of judgment. These criteria are risk and economic return.

Risk Factors

Every firm takes risks in day to day business. If it elects not to spend money but invest it in government bonds or other securities considered safe, it may find someday that it is out of business because its business was lost to competition who spent their money and assumed risks to improve operations.

The prudent thing for a firm to do is to carefully consider its options which all involve risks. Evaluation of risk is one of the basic reasons large firms do market research to plan for new products, new plants and methods. This research is for the purpose of becoming more objective in financial planning, but many forces today react in the market to create dynamic unpredictable situations. Despite the best long range plans, risk factors can develop quickly in political conditions, rate of inflation, technological advance, international markets and errors in initial planning which have great impact on the duration of capital improvement projects. The assumption that the market will remain stabilized for a planned rate of sales is not always a safe assumption. Because of various risk factors, it may be a wonder sometimes that large capital investments are made at all. But, it must be done if the firm is to remain in business.

Rate of Return

Because of various risk factors, firms normally seek a greater rate of return on capital projects than for funds which it can invest elsewhere. For example, if it can invest its funds in bonds for a 6% return, it certainly would expect to receive a return on investment greater than that for its own capital projects. Usually in a well-run firm, the risk factors are not as important as the rate of return which must be expected before management will fund a project. The relative rates of return for various capital projects within a firm result in competition between projects for funding.

Therefore when management considers a project as important as sand reclamation in competition with other projects, it is important that engineers or planners who develop the project also develop realistic financial information which will help the firm make the decision to proceed as planned or modify with options for phased development.

Measurement of Rate of Return on Investment

Engineers who plan capital projects and the financial analysts of a firm both have a responsibility to work together as partners for greater profitability. Engineers would do well to study methods of investment analysis and financial experts would profit by learning a little about engineering.

Since money for investment is limited, firms strive to reinvest earnings at the greatest return on investment and recycle those earnings as much as possible. For this reason the terms "return on investment" and "payout time" are heard often. In the objective analysis of investments, distinct criteria are available to compare the relative merits of various projects. These criteria are: payment time, rate of return on investment and rate of return on discounted cash flow.

Table 1. Discounted Cash Flow Analysis

Year	0	1	2	3	4	5	6	7	8	9	10
Capital Investment	1,000,000										
Terminal Value	100,000										
Investment Tax Credit		50,000	50,000								
Cash Flow											
Operating Cost Saving		200,000	200,000	200,000	200,000	200,000	200,000	200,000	200,000	200,000	200,000
Depreciation		90,000	90,000	90,000	90,000	90,000	90,000	90,000	90,000	90,000	90,000
Taxable Income	50,000	110,000	110,000	110,000	110,000	110,000	110,000	110,000	110,000	110,000	110,000
Income Tax 50%		55,000	55,000	55,000	55,000	55,000	55,000	55,000	55,000	55,000	55,000
After Tax Saving		55,000	55,000	55,000	55,000	55,000	55,000	55,000	55,000	55,000	55,000
Depreciation		90,000	90,000	90,000	90,000	90,000	90,000	90,000	90,000	90,000	90,000
Annual Cash Flow	850,000	195,000	195,000	145,000	145,000	145,000	145,000	145,000	145,000	145,000	145,000

The term cash flow is the common denominator used in analyzing capital expenditures. Cash flow is not the same as profit, earnings or income. Profit is the amount of money left after taxes, operating costs, payments on debts and depreciation. Cash flow, however, is the after-tax profit plus depreciation. Cash flow is therefore spendable cash because depreciation is charged off as a tax shield but the company doesn't actually spend it during that particular year. For example, if $100,000 is charged to depreciation, that amount of earnings is shielded from being taxed as shown in the following example.

Project Estimated Costs:

Capital investment	$1,000,000
Annual depreciation	100,000
Annual savings	350,000
Annual operating cost	150,000

Cash Out:		**Cash In:**	
Operating cost	$150,000	Savings	$350,000
Depreciation	100,000		
Total	$250,000	Total	$350,000

Cash Flow:

Change in flow	= $350,000 - 250,000 = $100,000
Income tax 50%	= $ 50,000
After-tax saving	= $ 50,000
Cash flow	= After tax saving + depreciation
	= $50,000 + $100,000
	= $150,000

Payment Time

Evaluating a project by quickly estimating a payback period is popular because of its ease of determination and understanding. It is commonly determined by dividing the estimated capital cost of the project by the annual cash flow.

$$\text{Payback time} = \frac{\$1,000,000}{\$150,000} = 6\text{-}2/3 \text{ years}$$

Engineers sometimes use a shortened method by which the capital investment is divided by the net annual savings.

$$\text{Payback time} = \frac{\$1,000,000}{\$350,000 - 150,000} = 5 \text{ years}$$

Either method is a measure of the length of time required to recover the initial investment. The methods, however, do not include the economic value of time or effect on taxes and tend to favor projects which have high cash flow regardless of what may happen to the flow of cash after the payback period is over.

Return on Investment

The return on investment, or ROI, found by an abbreviated method also uses the same values as used to determine payback period.

$$\text{ROI} = \frac{\text{Cash flow}}{\text{Capital investment}} \times 100\%$$

$$= \frac{\$150,000}{\$1,000,000} \times 100\% = 15\%$$

This method also sometimes uses variations in the net yearly return to determine the cash flow value.

The method, however, does not include changes in the value of money, taxes or changes in cash flow over periods of time. Both payback and ROI are useful for preliminary sorting of alternatives or comparisons where all the alternatives have the same life and cost factors.

Discounted Cash Flow Rate of Return

The discounted cash flow method of analysis (DCF) has become one of the best methods for the analysis of projects because it includes provisions for income taxes, inflation, depreciation, investment tax credit and salvage value. As in the case of other analysis procedures, this method also uses cash flow as a basis.

The discounted cash flow procedure determines the discount rate that must be applied to annual cash flow to make the sum of the annual cash flow over the given depreciation period equal to the amount of the initial investment. For example, in the following project, the initial investment is similar to the previous examples for payback and ROI.

Project Estimate Costs:

Capital investment	$1,000,000
Service life	10 years
Terminal value	$100,000
Annual savings	350,000
Annual operating cost	150,000
Operating cost saving	200,000

In the DCF method used in Table 1, depreciation and operating cost are shown as straight line functions. The development of actual cases, however, provides an opportunity to include the projected effects of inflation. Operating cost savings will increase with time for example, in a sand reclamation system, if it can be shown that the rate of purchased materials and services such as sand, disposal costs and fuel will rise faster than plant labor. Various forms of depreciation evaluation can also be used which have a more beneficial effect than the straight line method.

Table 2. Discounted Cash Flow Percentage Determination

		Annual After Tax Cash Flow	Trial at 15%		Trial at 12%	
			Discount Factor(1)	Present Value	Discount Factor (1)	Present Value
Year	0	-$850,000	1.000	-$850,000	1.000	-$850,000
	1	195,000	.8607	167,836	.8869	172,946
	2	195,000	.7408	114,456	.7866	153,387
	3	145,000	.6376	92,452	.6477	101,166
	4	145,000	.5488	79,576	.6188	89,725
	5	145,000	.4724	68,498	.5488	79,576
	6	145,000	.4066	58,957	.4868	70,586
	7	145,000	.3499	50,736	.4317	62,597
	8	145,000	.3012	43,674	.3879	56,246
	9	145,000	.2592	37,584	.3396	49,242
	10	145,000	.2231	32,350	.3012	43,674
				$776,119		$879,145

(1) Discount factors from Table of present value of $1 received at end of year (n)

Calculation of the discounted flow rate is a trial and error solution. The expenditure for the project at year zero is $850,000 and the sum of the discounted cash flow for each year for trials of 12 and 15% are shown in Table 2. Since the discounted cash flow must equal the initial expenditure to determine the proper discount, the cash flow trial of 15% is less than $850,000 and 12% is slightly more. The discounted cash flow rate is therefore between 12 and 15%. If discounted cash flow analysis is performed frequently it can be done more quickly using a computer program. Programmable pocket calculators can also be obtained which use financial models for these calculations.

Methods of treatment of the various elements of a discounted cash flow analysis vary from firm to firm and the example shown is for illustration only. Actual application can be complex and it is suggested that project planners coordinate their effort with their decision making financial counterparts. The methods shown are simplified for illustration but include some of the basic elements used.

Application of Financial Planning to Sand Reclamation

The application of financial planning methods to sand reclamation as a capital project involves collection of all of the various pieces of information so they can be analyzed collectively. For example, all items noted earlier concerning the economic evaluation of existing sand handling practice must be obtained first. For a comparison to determine if a capital project to reclaim sand will result in an adequate return on investment, the differences in existing costs must be compared with the various alternatives of costs for proposed reclaiming methods and systems by one of the available financial investment analysis methods. The methods of analysis for capital investments are also simplified for illustration and variations used in return on investment and discount cash value should be studied before making major decisions.

Thus far in the industry, the divergence of risk factors, capital investment, technological state of the art and return on investment levels have made acceptances of sand reclaiming a matter of individual plant preference. An objective analysis of all of the technical and financial factors is the only practical method for a foundry to make that decision.

References

1. F. Dettore and J. Parish, "Reclamation of No-Bake Sands," Modern Casting, p 102-104 (Apr 1972).
2. K. J. Smith, "Foundry Sand Reclamation Techniques," Foundry M & T, p 46-52 (Nov 1976).
3. R. M. MacDonald, "The Rebonding of Reclaimed Silicate Sands," Modern Casting, p 66-67 (June 1976).
4. A. J. Wagner, "Sand Reclamation System at Frederick Iron & Steel," Modern Casting, p 66-67 (Apr 1973).
5. T. Harty, "New No-Bake Sand Reclaimer at Rodney Hunt Co.," Modern Casting, p 48-49 (Feb 1972).
6. G. X. Diamond, "How Delray Foundry Recirculates No-Bakes," Modern Casting, p 34-35 (July 1972).
7. E. V. Akerlow, "Conservation of Foundry Wastes," Modern Casting, p 38-39 (Aug 1975).
8. K. Lowe and D. S. MacKinnon, "A Short Course In The Fundamentals Of Capital Investments," Pulp and Paper, Editorial Dept. 500 Howard St., San Francisco, CA 94105 (1976).
9. G. M. Hollander, "Cost Studies and Engineering Economy: A Review Of Terms," Specifying Engineers, p 79-81 (Sept 1976).
10. E. L. Grant and W. G. Ireson, "Engineering Economy," Fifth Edition, Ronald Press Co., New York, NY (1970).

Application of Computer-Aided Design to a Steel Wheel Casting

AFS RESEARCH

A. Jeyarajan, *Graduate Student*
R. D. Pehlke, *Professor and Chairman*
University of Michigan, Ann Arbor, Michigan

ABSTRACT

A computer program developed in earlier research to simulate solidification was used to study computer-aided design of a commercial casting. Solidification of a steel rail wheel casting was numerically simulated. The wheel had a flanged rim connected to a hub through a web. The casting was fed by a single riser attached to the top of the hub. Various casting designs were investigated with respect to avoiding shrinkage unsoundness in all parts of the casting. As part of the research, a computer program was developed to estimate the thermal properties of exothermic materials that are commonly employed as risering aids in the applicable foundry practices.

Effects of riser size and pouring temperature on the time-temperature profiles during solidification were studied. The computed solidification sequence for the rail wheel indicated that the web would solidify first, thereby preventing the rim from being fed. The computer predicted that chilling the rim with chromite sand facing would not be sufficient to alter the solidification sequence and that either an increase in web thickness or exothermic padding over the web would be necessary. The thicknesses of metal and exothermic padding needed to make the web solidify after the rim were determined by computer calculations. The castings with designs involving chromite sand facing around the rim and an increase in web thickness were experimentally cast and sectioned to verify the computer predictions.

A computer program was developed to calculate the shape and dimensions of shrinkage cavity in the riser from the computer temperature distributions present during solidification. Cases involving sand risers and insulated risers were investigated with respect to sufficient riser size to provide required feed metal.

Introduction

A major application of the computer in the foundry industry could be the design of castings and prediction of their integrity. With this objective, the American Foundrymen's Society has sponsored a series of computer-oriented projects on heat transfer and solidification. This research has been monitored by the AFS Heat Transfer Committee and carried out at The University of Michigan.

In the initial phases of the AFS-sponsored research, solidification of several casting geometries was simulated using the digital computer[1-3] and calculations were compared with experimental results available in the literature.[4,5] Computer techniques were subsequently developed for determining mold material thermal properties as functions of temperature,[6] and for employing these properties in the simulation of solidification. Computer calculations were compared with laboratory thermocouple measurements of a 3.5 in. square bar casting poured in commercially-pure aluminum and silicon brass. It was established that computer simulation could produce accurate time-temperature profiles of solidification.[7]

In the next phase of AFS-sponsored research, using laboratory aluminum castings poured in dry sand molds, it was demonstrated that computer simulation could predict sound and unsound casting designs.[8] The casting shape employed was axi-symmetric but with the complexities of corners, curved surfaces and a riser neck. Later this work was extended to castings with graphite end chills.[9] The computer was used to predict casting designs with sufficient and insufficient chilling. Casting soundness and prediction of time-temperature profiles were verified experimentally. The salient point in the work with chills was that air gap formation across the chill-metal interface was included in the simulation.

Several other studies of computer simulation of casting solidification have been reported, including the work of Henzel and Keverian[10] who simulated solidification of steel castings, Berry and co-workers[11] who investigated freezing against chills, Erickson and Houghton[12] who studied the solidification of pure lead in graphite molds and Weatherwax and Riegger[13] who simulated the solidification of an aluminum alloy in a steel die. However, no work has been reported employing computer simulation to evaluate casting designs for shrinkage unsoundness and to assess alternative casting designs and foundry practices. Ruddle[14] and Umble[15] have studied the use of insulated risers with the aid of empirical formulas. Their work suggests that computer simulation can be utilized to evaluate foundry practices involving risering aids.

The casting employed in this study is shown in Fig. 1. The wheel has a central hub and a flanged rim which is attached to the hub through a web. The entire casting was fed by a single cylindrical riser attached to the top of the hub. The casting was bottom gated through the hub. This arrangement preserved the

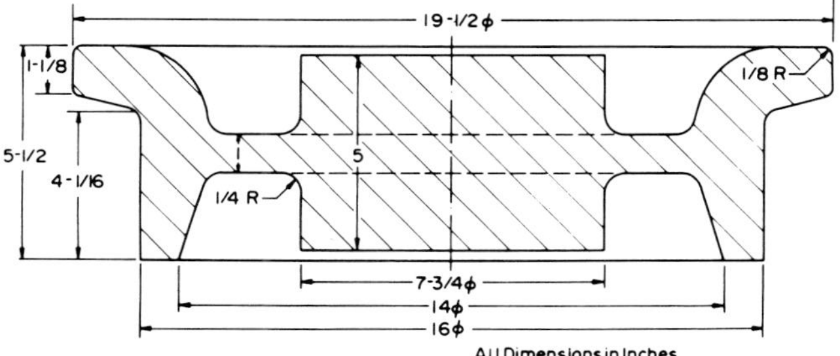

Fig. 1. Cross sectional view of rail wheel casting.

Fig. 2. Photograph of the pattern.

radial symmetry of the casting and facilitated use of a two dimensional simulation to represent three dimensional heat transfer and solidification.

Experimental Procedure

The castings were poured in killed steel with a specified composition of .20-.28% C, .45-.85% Mn. The final silicon compositions ranged from .30-.61%. The steel was melted in a 5 ton arc furnace and poured into green silica sand molds. Molds had about a 5 in. facing of sand mix containing 3 screen AFS 50 mesh New Jersey silica sand bonded with 5% western bentonite, 1% cereal binder and 3.5% moisture. In experiments involving chilling of the rim, a 3 in. thick chromite sand facing was employed around the rim and the flange. The chromite sand used was AFS 55 mesh sand bonded with approximately 2.5% western bentonite, 1% cereal and about 2.5% moisture.

Facing sands were applied to the pattern in desired thicknesses and the remainder of the flask was filled with backing consisting of recycled system sand. Molds were then butt rammed with pneumatic equipment. Mold hardness ranged from 80 to 85 and the green compression strength of the molding sand mix was 5-6. Figure 2 presents a photograph of the soft pine wood pattern employed. After filling the mold with steel through a bottom horn gate, the riser was covered with 1 in. of proprietary exothermic topping material. The experimental castings were radiographed and sectioned to study internal soundness and obtain riser shrinkage cavity profiles.

Computer Programs

Numerical Simulation

The computer program generated in earlier research work involving laboratory aluminum castings[8,9] was modified to handle cases where selective portions of the casting were mildly chilled using molding materials such as chromite sand, or insulated using materials with an exothermic action to obtain a desired solidification profile. The program segment that inputs casting geometry and dimensions into the simulation also was modified so that it could treat any axi-symmetric geometry as long as all cylindrical surfaces of the casting are parallel to each other and adjacent surfaces are at right angles to each other.

Inclined or curved surfaces as encountered in the rim portion of the rail wheel casting can be treated by approximating those boundaries as a jagged series of steps constructed from the rectangular grid employed in the simulation. Thus the final version of the simulation program was versatile enough to be applied to solidification of any axi-symmetric casting with chills, exothermic padding and riser insulating materials. Air gap formation, in cases involving chills, also can be simulated.

Input into the computer consists of values of the coordinates of the corner points of the casting and other pertinent variables including the number of nodes in the axial and radial directions, as well as initial metal and mold temperatures. The program sets up the casting geometry in terms of nodal coordinates and, for each node, automatically uses the proper finite difference approximation describing heat flow at that location and the applicable thermal properties. Even though considerable mathematical derivation and programming expertise was necessary to develop the program, the user need not have any programming knowledge.

Output from the simulation is in the form of printouts of temperatures at each nodal point. Also, the axial temperature distribution can be either plotted using a computer driven plotter or viewed on a graphics terminal with a cathode ray tube display.

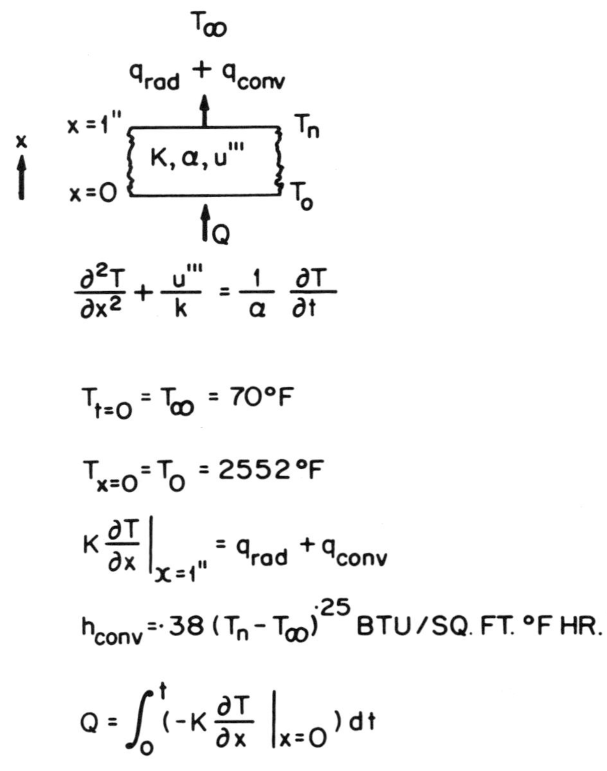

$$\frac{\partial^2 T}{\partial x^2} + \frac{u'''}{k} = \frac{1}{\alpha} \frac{\partial T}{\partial t}$$

$$T_{t=0} = T_\infty = 70°F$$

$$T_{x=0} = T_0 = 2552°F$$

$$K \frac{\partial T}{\partial x} \bigg|_{x=1''} = q_{rad} + q_{conv}$$

$$h_{conv} = \cdot 38 (T_n - T_\infty)^{.25} \text{ BTU/SQ. FT. °F HR.}$$

$$Q = \int_0^t \left(-K \frac{\partial T}{\partial x} \bigg|_{x=0} \right) dt$$

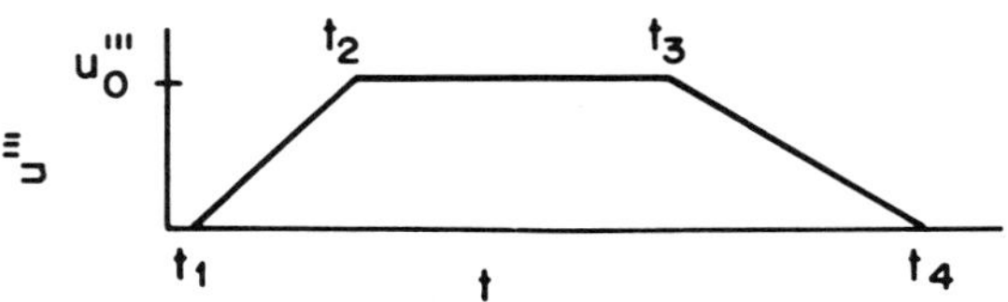

Fig. 3. Heat transfer conditions employed in the thermal property evaluation of exothermic materials.

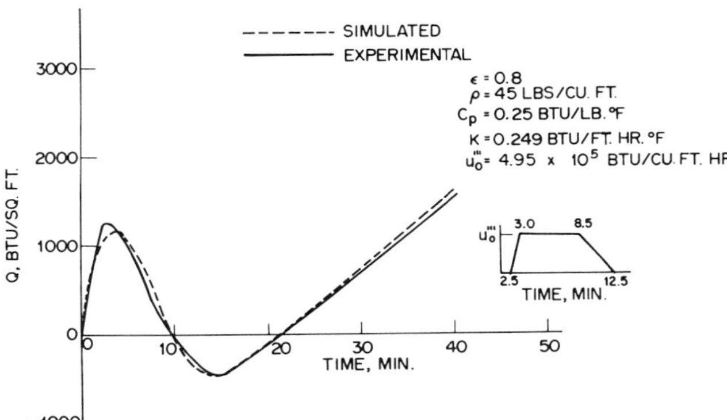

Fig. 4. Computed and experimental net heat transfer into 1-in. thick specimen of riser sleeve material.

Riser Shrinkage Cavity Profile

A risering system is considered sufficient if there is directional solidification toward the riser and, the shrinkage cavity is confined to the risering system and does not extend into the casting. In computer-aided design of castings one must check the solidification sequence inside the casting and also assess the shape of the riser shrinkage cavity to decide whether the casting under consideration is sound or not. A computer program was developed to calculate the shrinkage cavity in the riser from the simulated temperature distribution.

The first step in computing the shrinkage cavity profile was to calculate locations of liquidus and solidus isotherms at each time step by interpolation. The volumes swept by these two fronts between two successive time steps multiplied by the corresponding shrinkage factors represent the liquid state shrinkage due to superheat and liquid-solid shrinkage due to solidification. By subtracting the total shrinkage from the amount of liquid in the riser, a new liquid level in the riser was calculated from the computed liquid pool diameter in the riser. By repeating this procedure over the entire solidification period, the complete profile of the shrinkage cavity in the riser was obtained.

It must be remembered that in the numerical simulation of solidification the riser was assumed to be full at all times during solidification and metal flow was neglected. This results in somewhat higher temperatures being predicted in the riser, and hence may lead to small errors in the shrinkage cavity profile computation. This effect would manifest itself as a slightly shallower cavity being predicted in the riser.

Thermal Properties of Materials

To incorporate the presence of exothermic materials into the simulation, the thermal properties of these materials have to be known. The thermal properties of these materials were available in the form of measurements of net heat transfer into a 1 in. thick specimen as a function of time.[16] A computer program was developed to simulate heat transfer in a specimen with exothermic action and compute the net heat transfer. Computed and experimental net heat transfer rates were compared and the thermal properties obtained by trial and error. The initial and boundary conditions and the equations necessary to calculate the net heat transfer are presented in Fig. 3. Figure 4 presents the computed and experimental heat transfer curves for the riser insulating material employed in this study, along with the set of data that produced the computed results. Similar results are presented in Fig. 5 for the substance which was assumed as exothermic padding material for the web of the casting in this study. The relatively simple technique developed in this research to obtain apparent thermal properties of materials can be employed for a wide variety of materials, including those with no exothermic action.

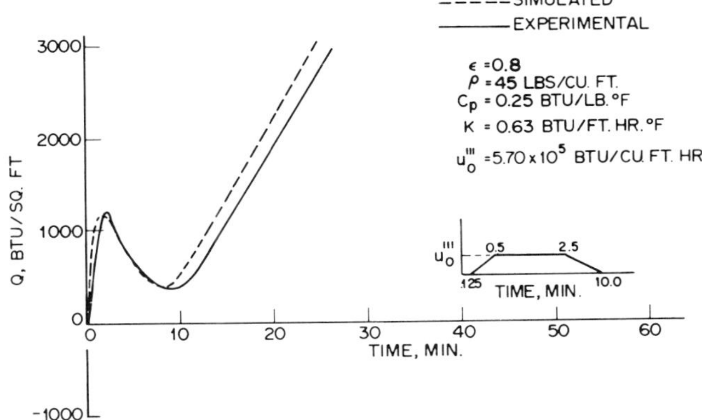

Fig. 5. Computed and experimental net heat transfer into 1-in. thick specimen of exothermic padding material.

Table 1. Details of Test Wheel Castings

Casting	Riser Dia.(in.)	Riser Height(in.)	Tap Temp.(°F)	Special Features
A	7.5	6.5	3050	
B	8.5	7.0	3050	
C	7.5	9.0	2930	
D	6.0	9.0	2930	1-in. thick riser sleeve
E	7.5	9.0	2930	3-in. thick chromite sand facing around rim and flange
F	7.5	9.0	2920	Metal padding-web thickness = 2-in.
G	7.5	7.5	2930	1-in. thick exothermic padding over the web

Thermal properties of the green silica sand were computed from the dry sand properties that were evaluated in earlier research.[6] The latent heat of evaporation of water was taken into account by appropriately increasing the heat capacity values of dry sand in the temperature range 192-232F (88.9-111.1C). The increased thermal conductivity of the sand due to the presence of moisture was calculated by multiplying the dry sand conductivities by a factor suggested by Paschkis.[17] The density of the chromite sand was assumed to be 185 lb/cu ft and the heat capacity to be 0.27 BTU/lb-F. The thermal conductivity was computed from the observation that solidification times in silica sand molds are 1.46 times the solidification times in chromite sand molds[18] and that the solidification times are inversely proportional to the heat diffusivities.[19] The properties of steel employed had been evaluated in earlier research.[1]

Experimental and Computed Results

Table 1 presents details of the castings employed in this study. Castings A-F were both experimentally cast and numerically simulated, whereas casting G was only simulated. Figure 6

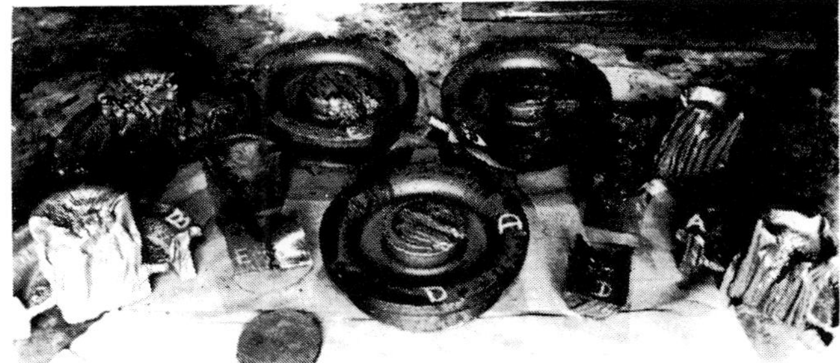

Fig. 6. View of several of the experimental castings.

Source: *Transactions of the American Foundrymen's Society*, Vol 86, 1978, 457-464

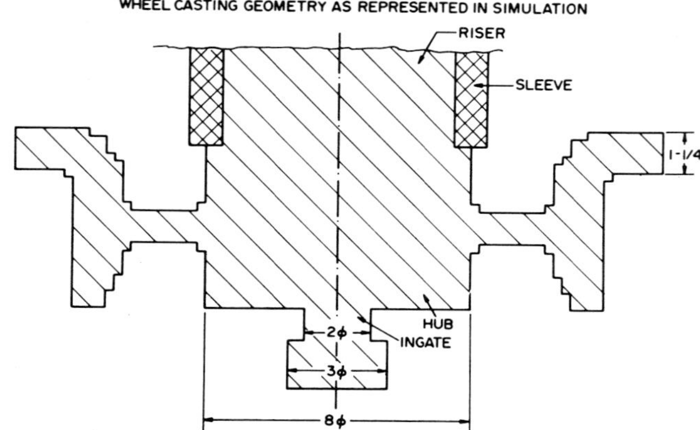

Fig. 7. Geometry of the casting as represented in the simulation.

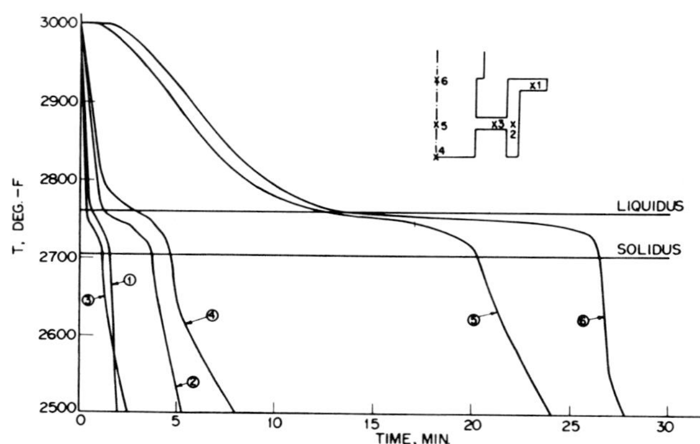

Fig. 8. Computed time-temperature profiles for several locations in Casting A.

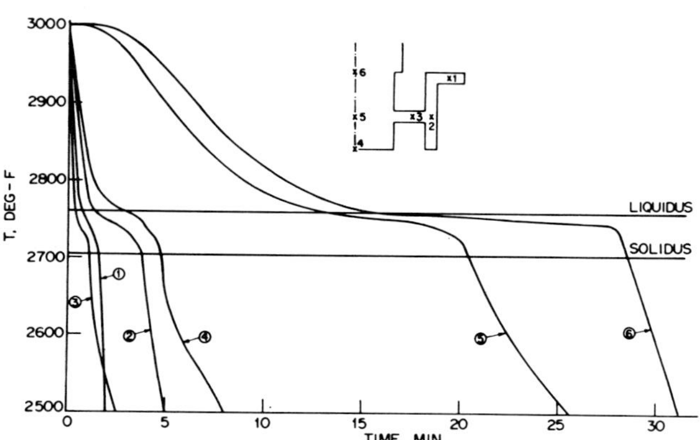

Fig. 9. Computed time-temperature profiles for several locations in Casting B.

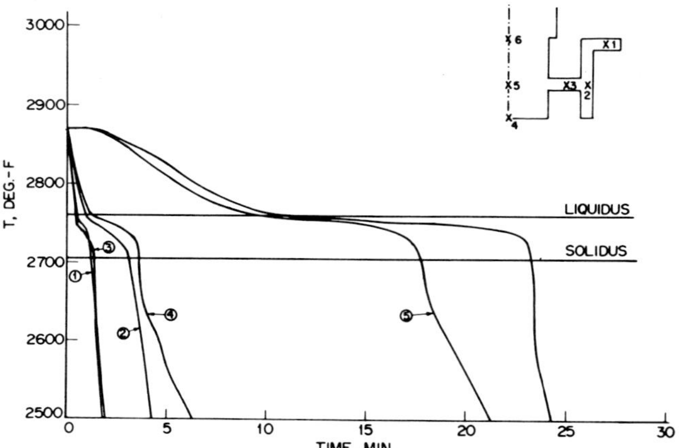

Fig. 10. Computed time-temperature profiles for several locations in Casting C.

presents some of the castings poured. The casting geometry, as represented in the numerical simulation, is shown in Fig. 7.

Figures 8, 9 and 10 present computed time-temperature profiles in castings A, B and C, respectively. A general picture of solidification inside the rail wheel casting can be obtained from these figures. Changes in riser diameter have noticeable effects only on the cooling curves of the locations near the riser (Fig. 8 and 9). Decreasing the pouring temperature from 3050F (1676.7C) to 2930F (1610C) decreases solidification-end time for location 6 from about 27 min 30 sec to about 24 min (Fig. 8 and 10). With regard to riser height, a computer simulation was carried out for a casting with riser height, H = 7 in. but in all other respects identical to casting A. No appreciable differences from casting A were noted in the simulated results.

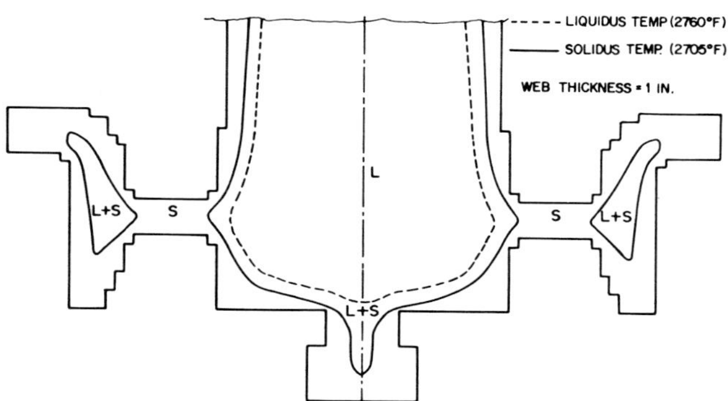

Fig. 11. Computed isotherms of liquidus and solidus temperatures inside Casting C at time = 2 min.

Directional Solidification

To present the solidification patterns inside the casting, isotherms of liquidus and solidus temperatures shortly after the web had solidified, were plotted. Such a plot for casting C is presented in Fig. 11. It can be observed that the web solidifies first and prevents the rim from receiving feed metal in its last stages of solidification. This solidification pattern was obtained in castings A-D which all had identical casting designs except for riser dimensions and pouring temperatures. All these castings were radiographed in a direction parallel to the casting axis at two exposures, as presented in Fig. 12 and 13. Shrinkage defects are indicated by the radiographs. However, because the thicknesses of the shrinkage cavities in the direction of the x-ray were very small compared to the rim height, it was not possible to identify positively and specifically shrinkage defects in the radiographs. Consequently, the castings were sectioned to study internal soundness. Figure 14 presents the photograph of a portion near the rim of the cross section of casting A. Shrinkage defects can be seen in the area where the web joins the rim.

The possible means of eliminating the shrinkage defect include chilling the rim, so that it would solidify before the web, and delaying the solidification of the web by exothermic padding or by increasing the web thickness (metal padding). Casting E was made with 3 in. thick chromite sand facing all around the rim and flange. The computed isotherms of liquidus and solidus temperatures, at time = 2 minutes for casting E are plotted in Fig. 15. It is observed that chromite sand does not have enough chilling power to remove heat from the rim to an extent to freeze it before the web. Figure 16 presents a similar plot for casting F which had an increased web thickness, 1/2 inch of metal padding on each side of the web (i.e. the total web

Fig. 12. Low exposure radiograph (positive) of Casting C showing details in the flange and web portions.

Fig. 13. High exposure radiograph (positive) of Casting C showing details in the rim portion.

Fig. 14. Cross sectional view of the web-rim-flange portion of Casting A.

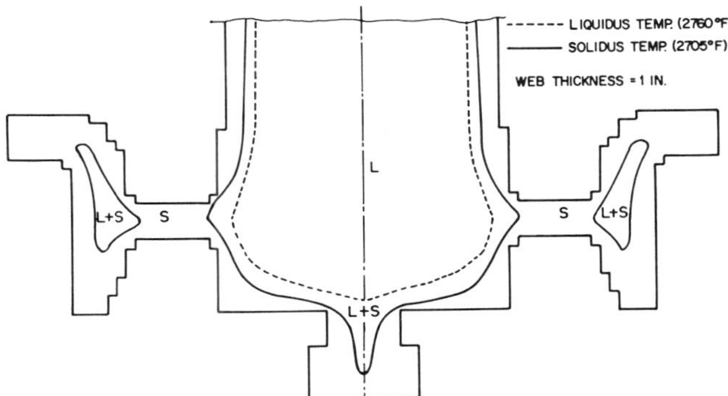

Fig. 15. Computed isotherms of solidus and liquidus temperatures inside Casting E at time = 2 min.

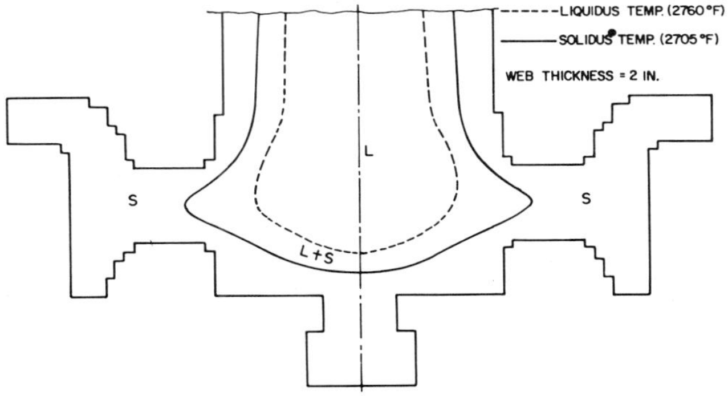

Fig. 16. Computed isotherms of solidus and liquidus temperatures inside Casting F at time = 2 min.

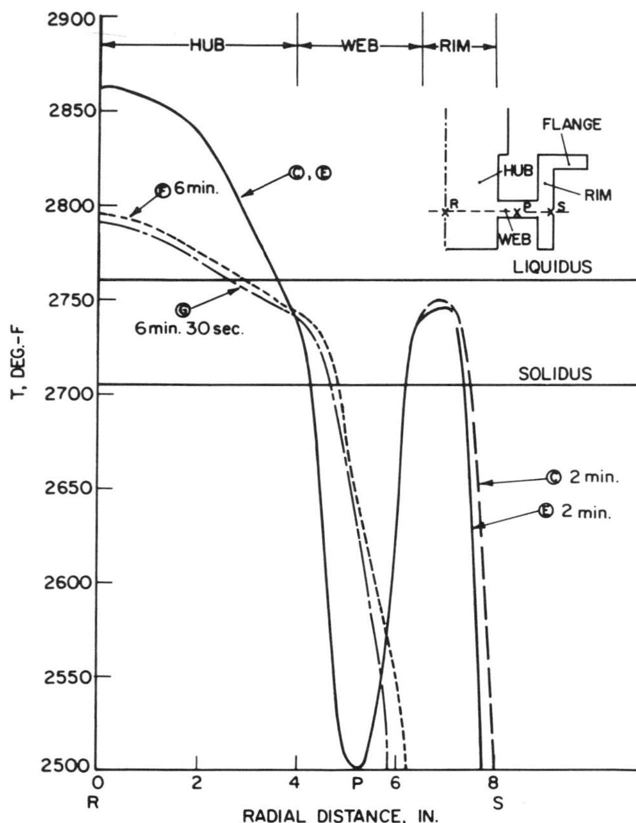

Fig. 17. Computed temperature distribution along the line RPS in Castings C, E, F and G.

Fig. 18. Cross sectional view of the web-rim-flange portion of Casting E.

and E. Casting F was chosen as a test casting which should have a sound rim and was experimentally cast. Photographs of cross sections of castings E and F are presented in Fig. 18 and 19, respectively. Casting E has a shrinkage defect while casting F is sound, as predicted by the computer. Radiographs of casting F are shown in Fig. 20 and 21 and indicate a sound casting.

thickness was 2 in.). The web solidifies after the rim, resulting in directional solidification from the rim to the riser. Casting G which was only simulated had 1/2 inch of exothermic padding on each side of the web. Figure 17 presents the computed temperature distributions along the center line (line RPS) of the web shortly after point P located in the web has solidified, for castings, C, E, F and G. Castings F and G are shown to solidify progressively from the rim toward the hub, unlike castings C

Fig. 19. Cross sectional view of the web-rim-flange portion of Casting F.

Fig. 20. Low exposure radiograph (positive) of Casting F showing details in the flange and web portions.

396

Fig. 21. High exposure radiograph (positive) of Casting F showing details in the rim portion.

Riser Design

To study the aspects of riser design by computer, the shrinkage cavity profiles of castings A, B, D and E were calculated and these predicted results were compared with experimentally observed shrinkage cavity profiles (see Fig. 6). Approximate values for shrinkages occurring during solidification of steel castings have been indicated as a function of composition by Wlodawer.[20] In practice, inaccuracies can be introduced by mold wall movements during solidification. For liquid-solid shrinkage occurring during phase transformation, two values, 3.5% and 4.7% were used, and the liquid state shrinkage taking place during dissipation of superheat was assumed to be 7.5×10^{-5} cu in./cu in.-F.

Figure 22 presents computed and experimental riser shrinkage cavity profiles for casting B. Excellent agreement is observed between the computed and experimental profiles.

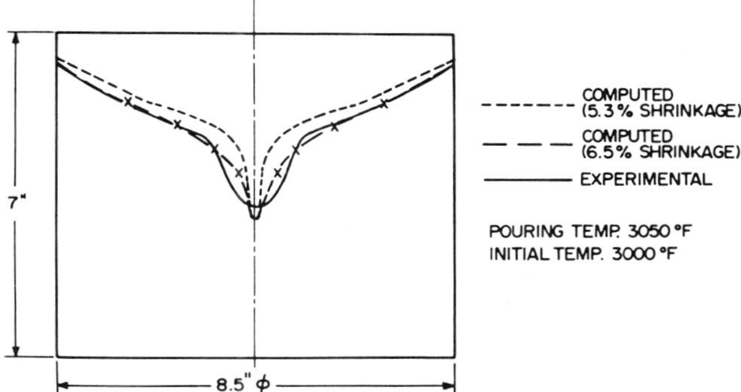

Fig. 22. Computed and experimental riser shrinkage cavity profiles of Casting B.

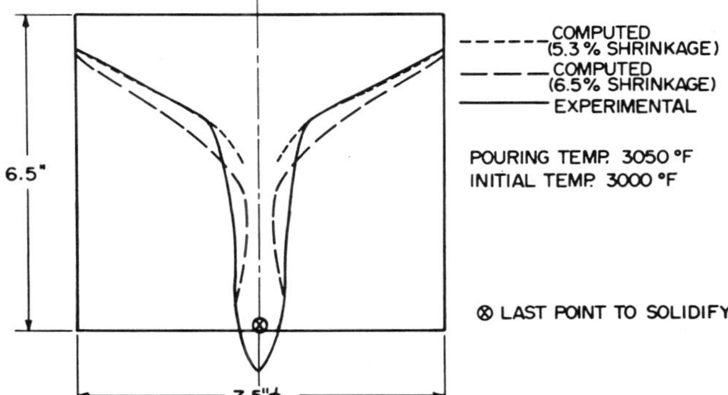

Fig. 23. Computed and experimental riser shrinkage cavity profiles of Casting A.

Similar observations can be made for castings A and E whose riser shrinkage cavity profiles are presented in Fig. 23 and 24, respectively. From computed profiles, one could conclude that a 7.5 in. dia. riser is inadequate and an 8.5 in. dia. riser is sufficient. These conclusions are supported by the experimental results.

Empirical calculations based on Wlodawer's modulus method[20] indicate that an H=D=10 in. height-diameter sand riser would be sufficient. This riser would result in a cast metal yield of about 56%, excluding the gating system. From computer simulation it was predicted that a riser with H=7 in. and D=8.5 in. would be sufficient and this riser results in about a 68% yield. In addition to the improved casting yield, the smaller diameter riser would simplify riser removal and reduce cost in cleaning and finishing operations. Computer costs are very small compared to these savings and a greater profit would

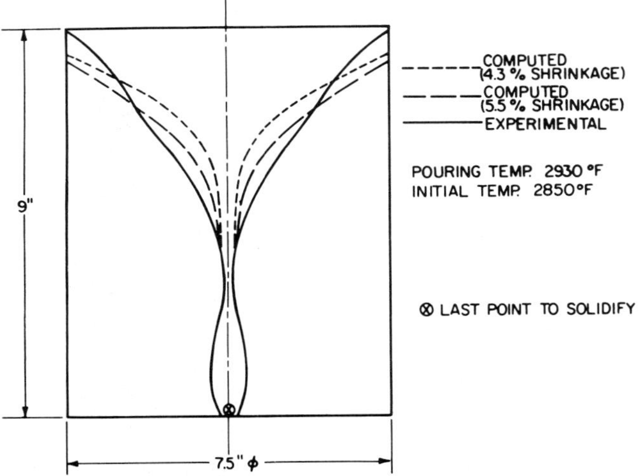

Fig. 24. Computed and experimental riser shrinkage cavity profiles of Casting E.

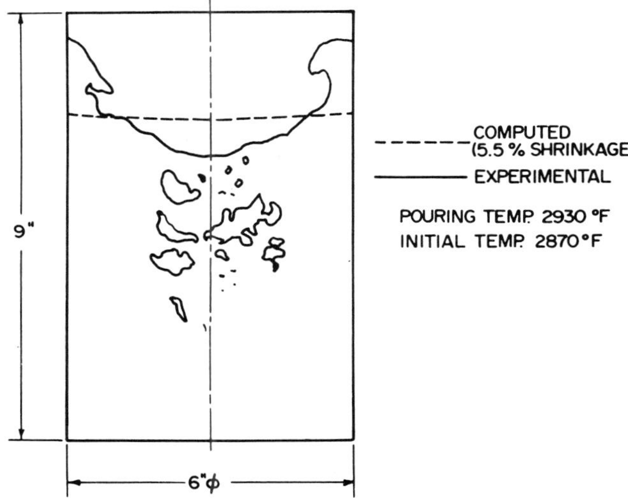

Fig. 25. Computed and experimental riser shrinkage cavity profiles of Casting D.

result as the number of pieces to be made from the computer-predicted casting design increases.

Riser insulating sleeves are very commonly employed in steel foundries to improve casting yield. Casting D was made with a 1 in. thick riser insulating sleeve around a 6 in. dia riser. Figure 25 presents the computed and experimental shrinkage cavity profiles in the riser. Good agreement is observed between the computed and experimental results. The slight differences between the predicted and measured profiles are due mainly to the fact that the exothermic material applied on the top of the riser breaks up as solidification proceeds. This exposes the liquid metal surface and results in increased heat losses through the upper surface. The effects of such heat losses through the top of the riser become more apparent in cases of sleeved risers due to the fact that the heat loss through the insulated side walls is quite small. In the case of sand risers, the same amount of radiation loss would not be very significant compared to the heat loss through uninsulated side walls. To improve the accuracy of the computed shrinkage cavity profiles for sleeved risers, these heat losses through the top of the riser should be taken into account.

Conclusions

Computer-aided design was demonstrated with a steel wheel casting showing that castings can be designed more economically with the aid of the computer. Numerical simulation of solidification predicted that a typical design for this type of wheel casting would produce some shrinkage in the rim. Alternative casting designs involving metal and exothermic padding which would eliminate the shrinkage were evaluated using the computer. The computer predictions for one of these casting designs were then experimentally verified.

In addition, a computerized technique was developed to determine the thermal properties of materials with exothermic action from straightforward heat transfer measurements that conventionally are used to describe the relative effectiveness of these materials.

It was also demonstrated that the shape of the riser shrinkage cavity can be accurately calculated using the computer and that the computer-predicted shape can be used to determine the sufficiency of a risering system. Casting yield can be greatly increased by computer-aided design.

Acknowledgements

The authors wish to express their appreciation to David B. Hess, Foundry Engineer at the Steelton, Pennsylvania plant of Bethlehem Steel Corporation for his assistance in formulating and supervising the experimental phase of this research. We also thank Bethlehem Steel Corporation for providing the facilities and materials for the experimental work, and the Lebanon Steel Foundry, Lebanon, Pennsylvania, for radiographing the castings. Finally, we also gratefully acknowledge the invaluable encouragement, guidance and technical assistance of the American Foundrymen's Society Heat Transfer Committee, currently chaired by W. C. Erickson. Partial financial support of the research was provided by the American Foundrymen's Society.

The casting designs studied and test castings poured at the Steelton plant of Bethlehem Steel Corporation were part of the investigation of steel casting solidification. These castings are not current products, and the casting arrangements and production procedures should not be taken as representative of present operating practice.

References

1. R. E. Marrone, J. O. Wilkes and R. D. Pehlke, "Numerical Simulation of Solidification, Part I: Low Carbon Steel Castings - 'T' Shape," Cast Metals Research Journal, p 184 (Dec 1970).
2. R. E. Marrone, J. O. Wilkes and R. D. Pehlke, "Numerical Simulation of Solidification, Part II: Low Carbon Steel Castings 'L' Shape," Cast Metals Research Journal, p 188 (Dec 1970).
3. R. E. Marrone, J. O. Wilkes and R. D. Pehlke, "Numerical Simulation of Solidification of a Copper-base Alloy Casting-Flanged-Barrel Shape," AFS Research Report (1972). Abstracted in Cast Metals Research Journal, p 94 (June 1972).
4. F. A. Brandt, H. F. Bishop and W. S. Pellini, "Solidification of Corner and Core Positions," AFS Transactions, vol 61, p 451 (1953).
5. M. J. Weins, P. K. Trojan and R. A. Flinn, "Application of Basic Solidification Phenomena to Mold Design for Copper-base Castings," AFS Transactions, vol 71, p 656 (1963).
6. M. J. Kirt and R. D. Pehlke, "Determination of Material Thermal Properties Using Computer Techniques," Cast Metals Research Journal, p 117 (Sept 1973).
7. D. J. Cook, M. J. Kirt, R. E. Marrone and R. D. Pehlke, "Numerical Simulation of Casting Solidification," Cast Metals Research Journal, p 49 (June 1973).
8. A. Jeyarajan and R. D. Pehlke, "Casting Design by Computer," AFS Transactions, vol 83, p 405-412 (1975).
9. A. Jeyarajan and R. D. Pehlke, "Computer Simulation of Solidification of a Casting with a Chill," AFS Transactions, vol 84, p 647-652 (1976).
10. J. G. Henzel and J. Keverian, "Comparison of Calculated and Measured Solidification Patterns in a Variety of Steel Castings," Cast Metals Research Journal, vol 1, no. 1 (June 1965).
11. D. R. Durham, D. Verma and J. T. Berry, "Some Further Observations on Freezing from Chills," AFS Transactions, vol 84, p 787-792 (1976).
12. W. C. Erickson and A. V. Houghton, "Simulating the Effect of Thermocouple Assemblies on Temperature Measurements," AFS Transactions, vol 85, p 59-64 (1977).
13. R. B. Weatherwax and O. K. Riegger, "Computer-aided Solidification Study of a Die-cast Aluminum Piston," AFS Transactions, vol 85, p 317-322 (1977).
14. R. W. Ruddle, "Risering Aids in Steel Foundry Practice," AFS Transactions, vol 83, p 577-584 (1975).
15. A. E. Umble, "Insulated Risers for Large Steel Castings," AFS Transactions, vol 84, p 49-54 (1976).
16. R. W. Ruddle, Foundry Products Div., FOSECO MINSEP INC., Private Communications (July 29, 1977).
17. V. Paschkis, "Heat Flow in Moist Sand," AFS Transactions, vol 59, p 389-391 (1951).
18. C. Locke, C. W. Briggs and R. L. Ashbrook, "Heat Transfer of Various Molding Materials for Steel Castings," AFS Transactions, vol 62, p 599 (1954).
19. R. W. Ruddle, The Solidification of Castings, The Institute of Metals, London (1957).
20. R. Wlodawer, Directional Solidification of Steel Castings, English Translation, Pergamon Press, Oxford (1966).

Practical Application of Infrared Thermographic Inspection Techniques

A. Ward
Ward Associates
Stillwater, Oklahoma
D. R. Ferrell, *Vice President*
Energy Services Engineering, Inc.
Stillwater, Oklahoma

Introduction

Infrared radiation is that portion of the electromagnetic spectrum between visible light and microwaves. Infrared exists just beyond the red end of the visible light spectrum at wavelengths from 0.75 to 1000 microns (μ).

No one particular person or organization is credited with the derivation of the word infrared. It is assumed to be derived from the Latin word "infra," meaning below or beneath, and when joined to the word red, the meaning becomes below, beneath or beyond red. It was referred to as ultra-red in English scientific literature of the 1880s during a period in which significant advances in the science took place.

Infrared radiation can essentially be divided into four separate regions. See Fig. 1.

A measure of the response of infrared can be measured in a quantity called spectral radiant admittance. This factors peaks in the middle infrared range and most commercial detectors operate in this region. The earth's atmosphere is essentially transparent to infrared in the first three regions. Extreme infrared is opaque in the atmosphere and its applications are limited to scientific and special military purposes.

History[1]

The history of infrared dates from the year 1800 when Sir William Herschel conducted experiments in the visible light spectrum. Using thermometers placed along the length of a beam of light split by a prism into its constituent colors, Herschel discovered that heat was being generated just beyond the red end of the light spectrum in spite of the fact that no visible light existed there. He initially disregarded this discovery and made only brief mention of it at the time and in the papers subsequent to the discovery in the year 1801. Other researchers had conducted similar experiments in the latter 1700s, but no one had ever explored the maximum range of temperature which Herschel was able to guess as existing beyond the end of the visible light spectrum.

The balance of the 19th century, as far as infrared is concerned, was devoted to the discovery and assembling of bits and pieces of information related to infrared. Very few practical applications, however, were discovered for the information. Research was instrumental, though, in prodding the development of devices which were much more accurate and responsive to measuring temperature.

One outcome of the development of devices which could measure temperature more accurately was the development of instruments which would respond to infrared radiation in unique ways. In 1850, John Herschel, son of Sir William Herschel, was instrumental in the development of a radiation detection process that used an evaporation principle of a thin film of oil to form a "heat picture." In 1843, it was discovered that certain materials phosphoresce when exposed to infrared radiation. In 1870, the first photoconductive detector was developed that used the direct interaction between photons of infrared energy and the electronic structure of the detector material as a method of determining the presence of infrared radiation. This development was further enhanced during World War I by the Germans who discovered that the sensitivity of a detector could be improved dramatically by cooling.

Starting in the early 1900s an increased interest was placed on the application of infrared to solve numerous problems. National standards for thermal radiation were developed and the detection and measurement of atmospheric and stellar temperatures, the application of infrared for medical and therapeutic services and the development of heat absorbing glasses for protecting the eyes were examples of developments during this period.

This period between the World Wars is marked by the development of image converters and detectors and the use of infrared spectroscopy as a primary tool for chemists. An image converter was developed just prior to World War II that allowed the seeing of human images in the dark.

Germany made the most significant strides in the development of infrared during World War II based upon the false assumption that the United States and its allies were pursuing infrared at a rapid pace. Their assumption was incorrectly based upon the significant development the United States and England made in the employment of radar.

The period following World War II is marked by its significant advances in the detection and conversion of infrared images into visible and usable information. Most notable is the

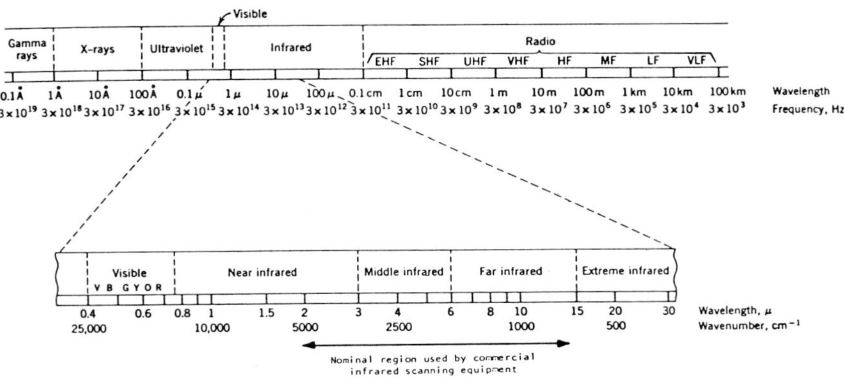

Fig. 1. Electromagnetic spectrum.

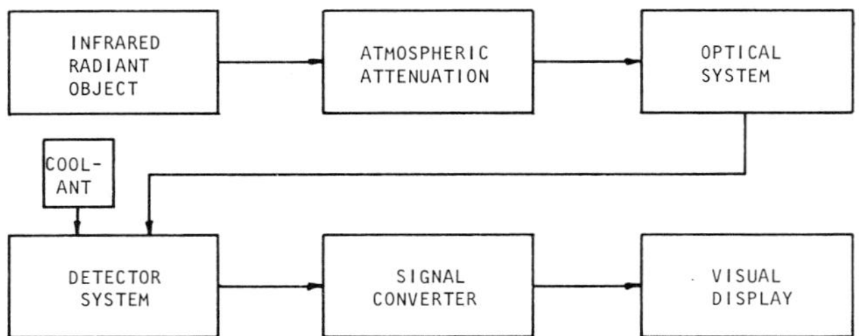

Fig. 2. Block diagram of a typical infrared detecting system.

fact that this image conversion was done in a passive manner. In other words, no transmission of infrared energy was required in order to produce a reflected image. The passive infrared devices used the natural emitted radiation of an object to form an image.

Further refinements and improvements in resolution have resulted in the use of infrared for nondestructive testing in industrial equipment as well as the detection of cancer in medical patients. Most recently, infrared has been developed as a tool for maintenance and the improvement of quality control and the detection of overheated mechanical, electrical and heat process equipment to determine if there are problems that need to be corrected before actual downtime occurs.

The greatest majority of infrared developments, of course, have occurred in the military. As late as the latter 1960s, it was estimated that the total of budgets devoted to infrared devices included approximately three-fourths to military applications. The other one-fourth of market applications, devoted to the civilian sector, included equipment and research in infrared spectroscopy, process control, intrusion detection, fire monitor warning, medical diagnosis and industrial infrared surveys for maintenance and quality control.

Infrared Detector Systems[1]

A block diagram of the essentials of a typical infrared detecting system is shown in Fig. 2. These elements include the infrared radiating object, the atmosphere and its limiting factors, the optical receiver, the infrared detector, plus any cooling system required by the particular detector, a signal converter and the display system.

Infrared Radiant Objects

In very basic terms, all objects, animate or inanimate, radiate infrared energy transferred by electromagnetic waves. In order to detect the amount of infrared radiation being transmitted by a particular object, it is necessary to formulate a system for measuring its efficiency as a radiator. The convention developed for this particular type of measurement includes radiant flux density, the amount of energy per unit area of a source compared to a hypothetical perfect radiator. Common terminology today refers to a perfect radiator as a "black body," also termed a perfect absorber of radiant heat energy. Radiant flux density is referred to as radiant emittance. When the radiant emittance of an object is compared to that of a perfect thermal absorber, the resultant term, emissivity, is used to rate the thermal efficiency of a particular object. A perfect energy absorber, or "black body," would have an emissivity of 1.0 which, as mentioned before, is hypothetical only and does not exist in nature. It can be approximated only through experimental methods.

The human body has an emissivity factor that approaches 0.99 wavelengths greater than 4 μ. This factor is not dependent upon skin color. This is one of the factors that makes the use of infrared in medical applications so appropriate. Objects exhibiting a high emissivity factor include some oxidized metal surfaces, painted surfaces, asbestos, glass paint, numerous alloys and so on. See Table 1.

Most organic materials, such as wood, plastics, plants, animals, tile and painted metal have emissivities approaching unity. Oxidized metals have emissivity factors ranging anywhere from 0.1 to 0.8. Only those factors allowing an emissivity factor near unity would be useful in determining the true thermal characteristics of an object when viewed by an infrared detector.

Atmospheric Attenuation

The transmittance of infrared radiation from a particular object to an infrared detector will be affected by numerous factors. The primary factor to consider is the absorption of infrared radiation by the gaseous molecules of the atmosphere. Scattering of infrared radiation can be accounted for by the molecules themselves, as well as any moisture in the atmosphere such as haze and fog. The absorption and scattering of infrared radiation is also variable along the range of infrared rdiation from near infrared to extreme infrared and has been plotted graphically from experimental methods. For example, the atmosphere is essentially opaque to the extreme infrared region from 15 to 1000 μ. The attenuation of infrared radiation in the first three bands shows a widely varying range of transmittance characteristics in the near and middle infrared regions, with a gap in transmittance capabilities from approximately 5-1/2 to 7-1/2 μ. The far infrared region includes an extensive solid range of transmittance characteristics favorable to detection. The absorption and detection of infrared will also vary with the density of the atmosphere, a factor encountered with a change in altitude. That particular subject is beyond the scope of this paper and will not be treated at this point.

One other factor which may affect the transmittance of infrared radiation includes scintillation. Scintillation is essentially the property that affects a beam of electromagnetic energy as it passes through regions of the atmosphere that vary in

Table 1.[1] Emissivity (Total Normal) of Various Common Materials

Material	Temperature (°C)	Emissivity
Metals and their Oxides		
Aluminium:		
polished sheet	100	0.05
sheet as received	100	0.09
anodized sheet, chromic acid process	100	0.55
vacuum deposited	20	0.04
Brass:		
highly polished	100	0.03
rubbed with 80-grit emery	20	0.20
oxidized	100	0.61
Copper:		
polished	100	0.05
heavily oxidized	20	0.78
Gold: highly polished	100	0.02
Iron:		
cast, polished	40	0.21
cast, oxidized	100	0.64
sheet, heavily rusted	20	0.69
Magnesium: polished	20	0.07
Nickel:		
electroplated, polished	20	0.05
electroplated, no polish	20	0.11
oxidized	200	0.37
Silver: polished	100	0.03
Stainless Steel:		
type 18-8, buffed	20	0.16
type 18-8, oxidized at 800°C	60	0.85
Steel:		
polished	100	0.07
oxidized	200	0.79
Tin: commercial tin-plated sheet iron	100	0.07
Other Materials		
Brick: red common	20	0.93
Carbon:		
candle soot	20	0.95
graphite, filed surface	20	0.98
Concrete	20	0.92
Glass: polished plate	20	0.94
Lacquer:		
white	100	0.92
matte black	100	0.97

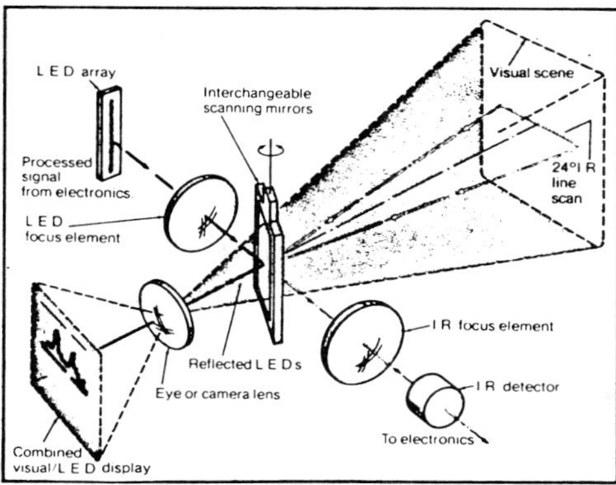

Fig. 3. Typical line scanning mechanism.

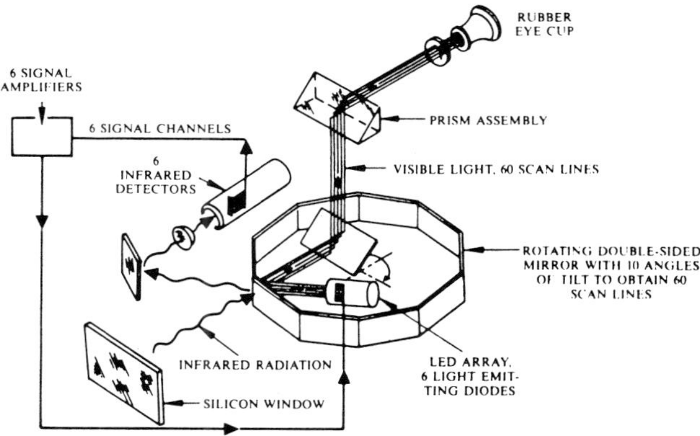

Fig. 4. Typical raster scanning mechanism.

temperature, such as the air above a heated road or surface. The effect is a rapid change in the apparent intensity of radiated energy. Obviously, heated air creates an unstable situation and a beam of infrared energy can be scattered at a random time varying rate.

Optical Systems

During the early development of optical systems for infrared detectors, it was necessary to choose materials which possessed the most favorable characteristics for passing or reflecting infrared energy. Since the availability of materials transparent to infrared was very limited, the choice in the majority of cases was for reflective optics. Due to extensive research conducted by the military and industrial concerns, materials transparent to infrared materials are becoming more numerous in the market today, allowing designers greater variety of choice in the development of infrared detectors. In choosing materials, it is necessary to consider factors such as the ability of that material to transmit infrared radiation and how it is affected by temperature, the refraction of the material and its variation with temperature, the hardness of the material and its ability to withstand scratches and physical attacks, the resistance of the material to corrosive atmospheres or liquids, density, thermal conductivity, thermal expansion, specific heat, the elastic deformation of the material, its softening or melting temperatures and its radio frequency properties. Many of these characteristics are of primary concern to military and sophisticated research applications. Some of the other characteristics are not as critical in industrial applications but are a consideration nonetheless. Examples of materials which are transparent to infrared and are used extensively include germanium (Ge), silicon (Si), arsenic trisulfide, strontium titanate, sapphire, calcium aluminate and fused silica.

Detector Systems

Once the infrared radiation has been passed through infrared transparent materials, it is necessary to devise a scanning system so that a thermal image can be formed.

The purpose of most scanning systems is to cover a field of view in a rectangular pattern. This is done by a combination of methods that involve optical as well as opto-mechanical scanning techniques. Basically, the infrared image is focused by various optical devices on an infrared detector material (example: indium-antimonide). A raster is then formed by rotating a scan mirror within the search field to form an image. Typically, a method of scanning could conceivably be accomplished by rotating a mirror in two different axes in front of a detector. In other words, a mirror could be rotated about one axis and "wobbled" about the other axis to form a horizontal and vertical field of view.

One typical method of creating a scan pattern in use today is to focus the infrared image onto a rotating mirror of multiple sides. Each of the mirror faces is tilted slightly with respect to the preceding mirror to form a multi-line variation in scan. From there the image is focused onto several detectors which convert the image into electrical signals for reprocessing through a light emitting diode array to a viewer raster.

Other methods of scanning involved the use of devices similar to television cameras which can detect infrared radiation. The use of this type of device has the advantage of increasing resolution while using no cryogenic coolant such as argon or nitrogen.

Figures 3[2] and 4[3] illustrate examples of opto-mechanical scanning methods used in two commercially available infrared viewers.

Infrared Limitations

Limitations to consider when using infrared in any type of application, especially those found in the foundry industry, include solar gain, the emissivity of the radiating object and air movement.

Solar gain is very simply that factor caused by the reflection of sunlight off objects placed outside. For instance, electrical distribution systems include large insulators, transformers and capacitors placed at electrical substations. These objects, when viewed during daylight, appear to have numerous bright spots merely due to the reflection of sunlight off their surfaces. Even though those surfaces normally exhibit a high capability to transmit thermal energy properly, the reflection of sunlight will cause false images to appear in the infrared detector. Infrared scanning systems which use the far infrared region of the spectrum (see Fig. 1) can eliminate this problem.

Emissivity, as mentioned before, is a limiting factor that needs to be given more attention than it is normally given. Only those objects which possess an emissivity approaching unity will exhibit true thermal characteristics. Those with a low emissivity such as polished metal surfaces or brightly plated metal, will give a false image. For example, the emissivity of human skin is 0.99 but the emissivity of a polished metal watchband on a person's arm is very low. When viewed with an infrared detector, the watchband will appear to be at a very much different temperature even though it is obvious that the watch will approximate the same temperature as the surface of the skin. Emissivities of materials commonly used in foundries may range from 0.20 to 0.98. For example, cast iron or steel with a rusted surface may exhibit an emissivity factor of 0.69, or as a gray surface may be 0.31 and when oxidized at 593C (1100F) may be 0.79.

Table 1 lists the emissivity of various materials found in industrial facilities. The important factor to keep in mind when conducting an infrared scan is to disregard or take into account objects with a low emissivity factor and either disregard or adjust the data obtained.

The next limiting factor which needs to be taken into account is air movement. Specifically, the rapid movement of air or wind across a surface, especially those surfaces viewed outdoors, will not exhibit true thermal characteristics because the heat is being conducted off the surface rapidly. This factor could serve to distort very important thermal information especially when scanning is being conducted for maintenance purposes.

Types of Commercially Available Infrared Detectors

The types of infrared detection and scanning equipment which may be obtained in the civilian market can be classed into three broad headings:

1) Pyrometers
2) Line Scanners
3) Thermal Imaging Equipment.

Pyrometers include those devices which use a very simple, single element infrared detector to determine the temperature at a specific point on a surface. The pyrometer's main attraction is its relatively low cost compared to other thermal imaging devices as well as the ability to obtain temperature by remote sensing. However, when the pyrometers are compared to thermal imaging devices their limitations become quite evident. First of all, they are able to detect temperature only at a single point. Thermal imaging either in a qualitative or quantitative sense must be accomplished by manually moving the pyrometer over a surface and noting the temperatures at a specific set of points in a grid system. This procedure can be time-consuming which makes the information available limited in value.

The second type of infrared imaging unit is line scanners. Line scanners use the same principle of detection as pyrometers, but an opto-mechanical scanning method can be used to process thermal information received from a particular line in the field of view. Thermal data is converted and processed at a time varying rate, then reproduced in the visual image with an amplitude readout to show the variations in temperature along a particular line in the field of view. Figure 3 is an illustration of this principle. Line scanners are a definite improvement in technology over pyrometers, but, as with any improvement, the cost is proportionately higher. Spatial resolution available with this type of detector is very good and the information presented is quantitative in nature and improves the ability of the operator to make a sound decision on the subject being viewed.

The third class of infrared equipment available is thermal imaging devices. These devices take full advantage of the range of opto-mechanical and electronic scanning techniques in order to produce a circular or rectangular raster which actually produces a thermal image. Depending upon the amount of technology involved and the price that the user wishes to pay, the image formed is either strictly qualitative in nature or a combination of qualitative and quantitative information.

Further refinements include the capability to read out specific temperature ranges of the object being scanned. This type of information is quite useful in research where specific thermal mapping of a device is required. Even finer improvements can be accomplished by using electronic circuitry that will "assign" different colors to various thermal images. Most thermal images are comprised of a black through white scale or black and red images, the brighter color representing higher intensities of temperature. The human eye is generally only able to detect as many as eight different shades of gray in this region. Electronic circuitry, however, can define a number of gray scale ranges and assign colors to those ranges and integrate them with the thermal image to allow the user greater flexibility in defining temperature loss.

The latest developments in thermal imaging devices include equipment which uses strictly an electronic scanning method similar to that used in television cameras. These devices require no cooling and the resolution obtained approaches that of standard television circuitry. Their advantage lies in improved flexibility by eliminating the cumbersome cooling system required in other systems. Their main limitation, however, is the high cost and sensitive electronics. The purchasers of these devices, generally, have been large firms which are able to absorb the initial cost in either research related projects or specific maintenance uses.

Types of Infrared Survey Work Available

The public has become acquainted with the use of infrared by utility companies who have used aerial infrared thermographers to produce thermal images of cities, towns and industrial plants. The use of infrared viewing in this manner was derived directly from military applications and its usefulness is apparent.

The next classification of infrared survey work available includes the use of infrared scanning techniques similar to those used in an aircraft, but applied to a moving vehicle such as a van or truck. In this method, a single line scan is used with the second dimension of scanning supplied by the motion of the vehicle itself. Such images are then transferred to photographic film and displayed on a single roll. This method is very useful in slow "drive-bys" of buildings to determine if there are any flaws in construction or thermal characteristics.

Both aerial and moving vehicle thermography are beyond the scope of this paper and will not be discussed further. However, aerial infrared surveys can be used to find and identify high energy sources and may be used some day to isolate excessive heat losses in manufacturing operations without entering the facility property.

The third type of infrared survey work includes infrared thermography used at ground level with a hand-held infrared imaging device such as those described in the previous section. While the use of such a device is limited generally to a local area, the usefulness of the information obtained can be of far greater value. Using the proper combination of equipment, a thermal imaging device can be carried into a plant and used to observe any abnormal thermal characteristics of mechanical devices, electrical distribution systems or heat process equipment within that plant.

By conducting a formal infrared survey, a foundry or plant can prevent many unscheduled shutdowns by detecting failure conditions before they cause production downtime. Infrared inspection can be performed while the equipment is under full load. Infrared inspections should be conducted before plant shutdowns or turnaround to aid in ordering materials and scheduling maintenance. By pinpointing areas of heat loss in industrial equipment, the decision process is quickened by making it possible to further define problems either known or unknown and base maintenance and quality control decisions on that information. In addition, infrared inspection and surveillance can be used to monitor the start-up of furnaces, molding equipment and other processing equipment. Overheating wires or connectors indicate improper installation or failure. Cracks in furnace linings can be detected, overheating hydraulic motors can be found and clogged cooling lines detected and corrected. Overheating motors, pumps and capacitors can be found and corrected before a costly fire occurs, or downtime occurs due to failure of components.

Practical Applications

The practical applications of infrared scanning techniques in the foundry industry become quite evident when we examine the results of those surveys. Infrared surveys can be conducted with either in-house equipment purchased by the company or by services contracted from a private firm whose primary interest is in serving the client. For the detection of actual maintenance problems, the second method is probably the best, because the consulting firm which has been hired to conduct the infrared survey is not adversely influenced by political and personal considerations within the company. In other words, the problem areas found are reported without regard to external influences. The ownership of equipment in-house is quite useful for emergency purposes and on-call uses where it is not feasible to wait for a consulting firm to conduct the survey. Ownership of in-house equipment carries the added financial burden of requiring the use of personnel on the maintenance staff devoted to the use of the survey equipment. But assuming that only one or two persons within a plant devote at least one-third of their time to the use of infrared and the conducting of periodic surveys, it can be seen that the cost of hiring a consulting firm to conduct the survey on a periodic basis can be justified.

In order to gain maximum use of infrared equipment and personnel, it is necessary to establish a pattern by which infrared surveys are conducted. It is necessary to plan and discuss with the maintenance, operations or engineering people who will be involved in the survey exactly which areas need to be scanned. This can generally be best accomplished by reviewing a plant layout diagram and marking the machinery and equipment along the route that will be observed. Once the route is established, the equipment must be activated and the survey started. The infrared survey is recorded methodically to assure that a correct analysis is made. A typical form is shown in Fig. 5.[6] It is also possible to videotape-record and take photographs and thermograms during the survey. In subsequent paragraphs, a typical survey will be followed showing equipment viewed along the route to give examples of the type of work which infrared analysis can perform in a foundry.

The first item observed is a blast furnace in the external areas of the plant. In this particular case we are assuming regularly programmed inspections are conducted in order to achieve two objectives:

1) Immediate detection of trouble areas on the furnace and its related equipment in order to conduct required maintenance.
2) The accumulation of baseline thermal data for comparison at later dates to detect any deterioration of the furnace itself.

In Fig. 6[4] we see a composite of the blast furnace itself and the related thermograms of that particular unit. The equipment used was a liquid cooled infrared scanner. Subsequent thermograms in this report were obtained using an argon gas cooled infrared viewer interfaced with an audio-video recorder. These two particular models are very typical of the type of equipment in common use today.

By close examination of the thermograms in Fig. 6 it is possible to achieve the two objectives just stated. Part a of the figure shows a possible scab buildup inside the furnace, indicated by the large cool area in the upper right portion of the furnace stack. It will be necessary to inspect this furnace more frequently in order to track the course of the scab buildup. Part b of Fig. 6 shows an indication of possible furnace channeling where hot gases progress unevenly up through the furnace by different temperatures in the gas pipes. The hot spot in the center of part c of Fig. 6 indicates a possibly inoperative cooling plate. Part d shows a buildup of matter in the bottom of gas pipes as indicated by the cooler temperatures in this particular area.

PAGE___ OF___

INFRARED SURVEY TYPE: DATE:

LOCATION:

WEATHER CONDITIONS:

COMPANY REPRESENTATIVE: OPERATOR:

TIME	TAPE SEQUENCE	SUBJECT RECORD	(°F) (°C)

Signed:

Fig. 5. Infrared survey form for collecting data.

The thermogram of a lime kiln, Fig. 7,[4] is a composite of several different thermograms taken and assembled into a mosaic. Due to the variation of the thermal pattern along the length of this kiln, it was necessary to observe more closely what the temperatures were in specific areas. These temperatures are indicated on the thermogram. By observing differences in temperature, plant personnel can forecast refractory replacement and assist in preventing unscheduled kiln downtime, thereby insuring a more dependable flow of product to the steel plant.

Steam lines or hot air ducts which have been insulated often show the same pattern indicated in Fig. 7, where insulation has failed and high heat loss results. Steam leaks which occur are clearly detectable using infrared viewing. Process performance is reduced and the cost of operations increased as a result of these losses.

Downtime costs are elusive and difficult to calculate, even when a specific example is examined in a broad industry such as the foundry industry. One foundry may assign its cost differently than another, so that a specific downtime cost is not interchangeable. However, taking a simple approach in an area where infrared inspection can easily pinpoint electrical efficiency or defective refractories and potential failure, consider the melting operation.

Cupola lining failure, holding ladle refractory failure, electric furnace lining or electrical component failure can cause direct production downtime. At certain melting rates, lost primary melting time, based on a selling price of $850 per ton and 60% yield, can result in the following lost production based on selling price:

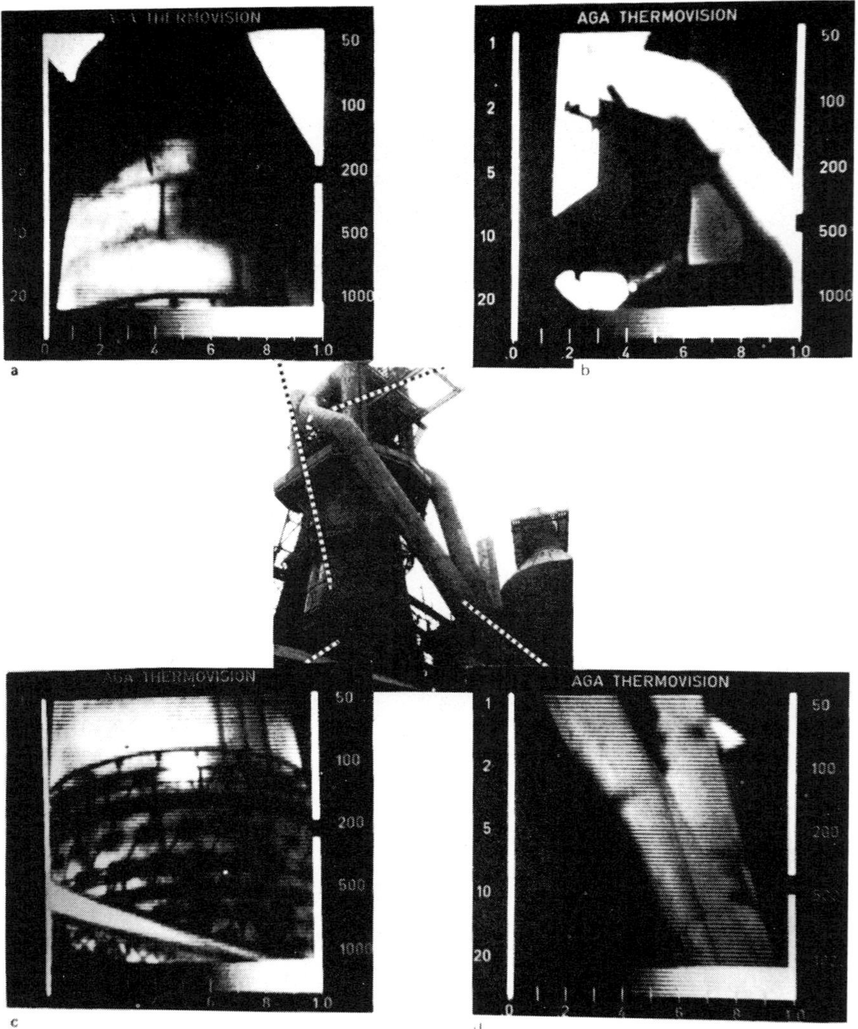

Fig. 6. Composite of a blast furnace and thermograms.

Melting Rate Tons per hour	Lost Production per hour
1	$ 510
3	$1,530
5	$2,550
10	$5,100

Improved efficiency as a result of correcting inefficient capacitors, loose or corroded contactors and wiring in the control cabinet or buss systems is more difficult to quantify in monetary value. However, it is clear that more efficient electrical power input will result in more melting power delivered to the furnace.

Examples of potential failures are shown in the following figures.

Figure 8 shows refractory erosion in a railroad molten steel transfer car. The lining was inspected, found to be near failure and corrected.

Figure 9[6] shows overheating of insulators and of the power components in a substation, indicating possible power failure and energy loss.

Figure 10[6] shows an overheating electric motor which was driving a gear reducer, showing near failure and excessive energy use.

Figure 11[6] shows an overheating electrical connector, which was tightened before a short occurred.

Figure 12[6] shows a comparison between the cylinder heads of two air compressors used in a foundry. The one on the left is normal. On the right, notice the oil line which appears cooler than the head. This indicates an overheating which was confirmed by an on-site check. The cool oil line was on a compressor which had just been overhauled.

Figure 13[6] shows an overheating conduit which indicated the insulation was faulty on the electrical wiring.

Figure 14[6] shows loose or corroded connections at a relay panel.

Cost savings and reduced energy use, such as a reduction in gas consumption, electrical power, steam use and other heat energy, can also be elusive when the actual location of loss cannot be pinpointed. Infrared viewing inspection allows specific areas to be located. Once temperature and heat transfer factors are determined, then engineering calculations can be determined.

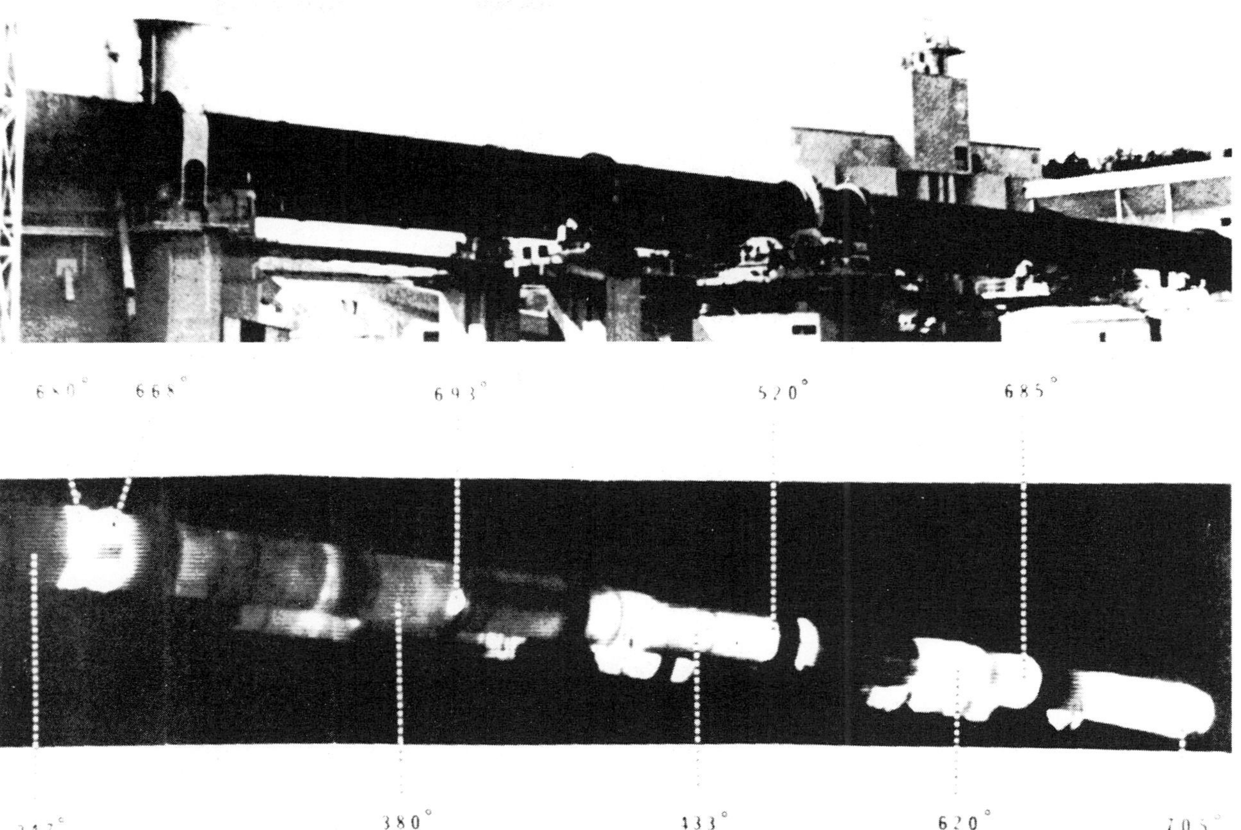

Fig. 7. Lime kiln with composite thermogram.

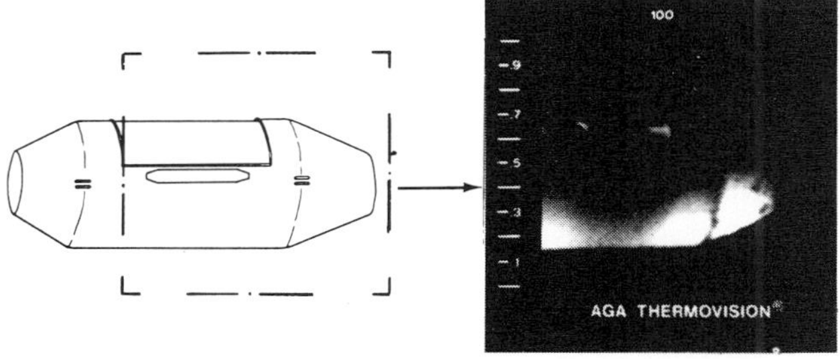

Fig. 8. Railroad molten steel transfer car. Thermogram shows areas of heat loss indicating refractory failure in regions indicated on diagram.

Fig. 9. View of electrical components in a power substation. The thermogram to the right showed a set of overheated insulators; below, an overheated circuit breaker.

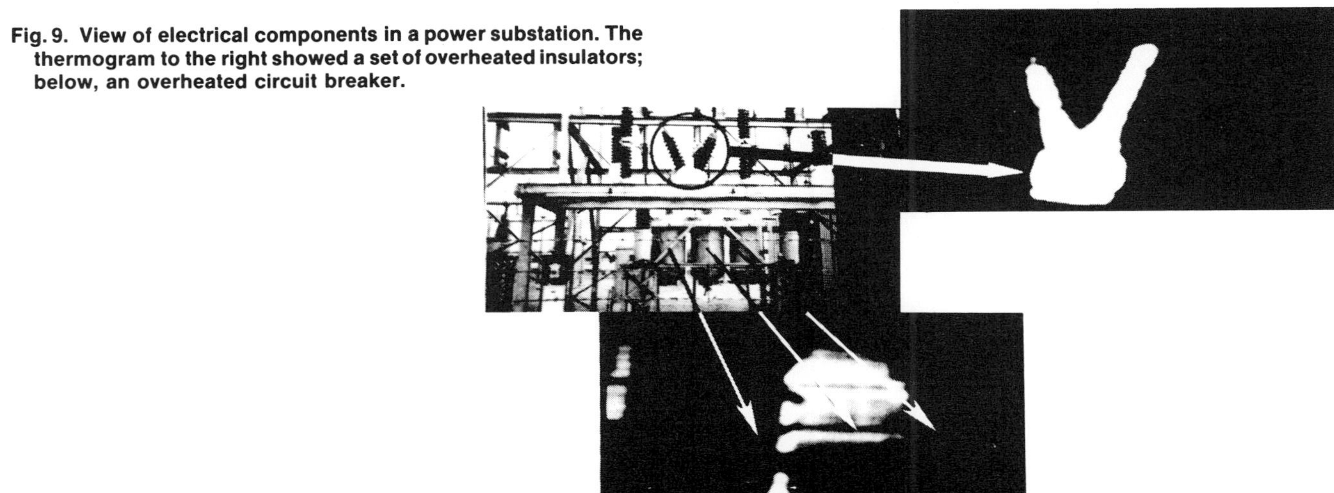

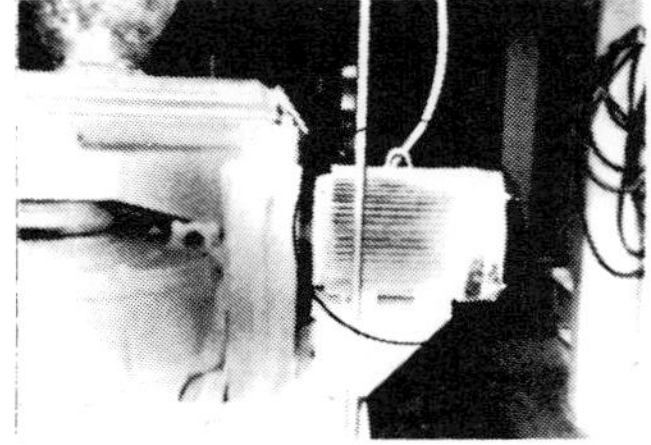
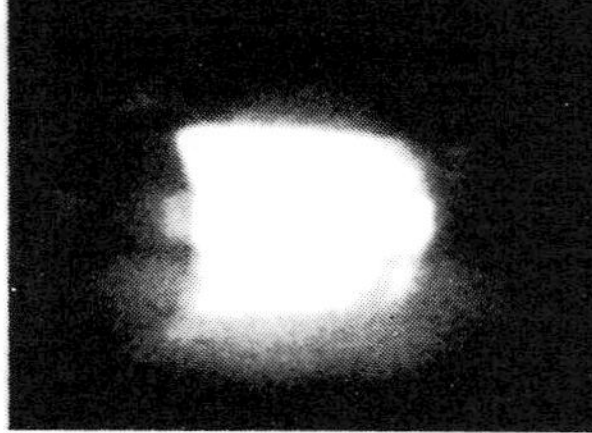

Fig. 10. Overheating electric motor showing near failure and excessive energy use. (b)

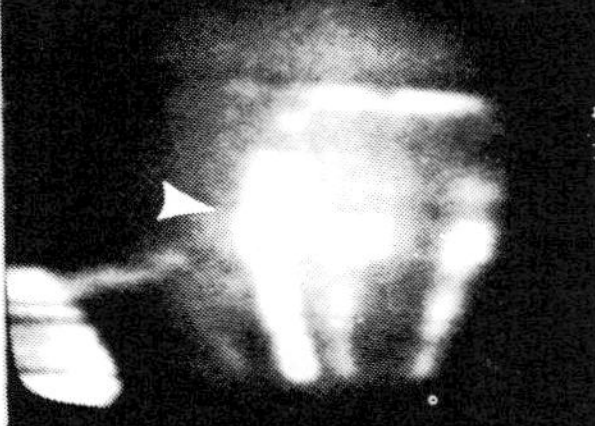

Fig. 11. Three phase circuit breaker panel; the left-hand lead shows up hotter than the other two indicating a possible overloading (see arrows).

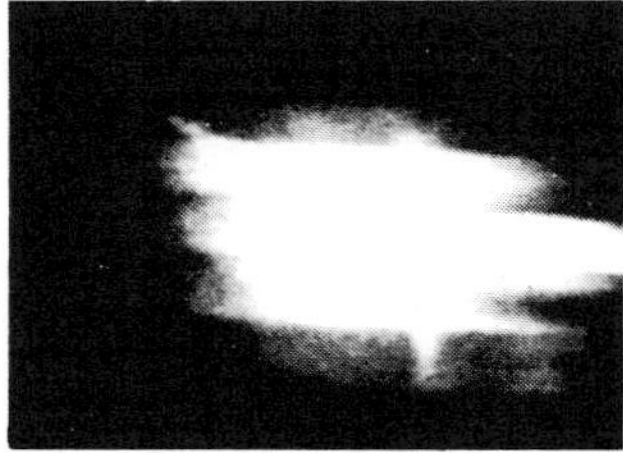

Fig. 12. Comparison of two air compressor heads. The one on the left is normal, but on the right, the oil line across the head appears to be cooler. An on-site inspection revealed that the head on the right was only running at 50% capacity and was actually overheated.

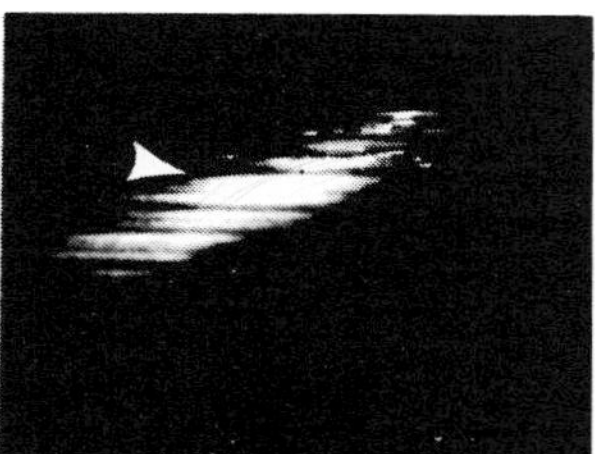

Fig. 13. A scan of multiple conduit runs allowed maintenance personnel to find this overheated conduit (see arrows).

Fig. 14. The connectors at this relay panel indicate a loose or corroded connection due to their heat pattern.

Furnaces present a natural heat loss potential. Erratic heat patterns in refractory linings, excessive heat loss or exhaust control areas, can show areas where energy costs can be reduced.

Figure 15[5] shows the exhaust system in a gas furnace. Figure 15 a shows the actual exhaust system. Figure 15 b is the thermogram showing the excessive heat loss pattern. Further investigation showed that the exhaust was 1260C (2300F) and the molten metal was at 538C (1000F) in a holding furnace. The exhaust damper was made automatic and the energy costs reduced 55% for the furnace operation.

Figure 16[5] shows a variation in heat loss from a furnace. The loss was causing cracks in the castings. Refractory material had fallen, causing increased energy costs and a defective product. The furnace was repaired and normal operation resulted.

Figure 17[5] shows the extreme heat loss occurring in a car bottom furnace operation. The loss was not visible to the unaided eye. Replacement of door seals effectively reduced the energy loss.

Figure 18[5] shows excessive heat loss at the access door on an indirect heating oven. Correct sealing and insulation would correct this problem.

Figure 19[5] shows high temperature and energy loss in the upper part of the thermogram and heat loss at the valve which is not insulated.

Figure 20[6] is a thermogram showing the operation on a good steam trap on the right and the heat loss through a faulty steam trap on the left.

To show just how practical the proper use of infrared inspection can be, consider Fig. 21[6] which shows an overheating insulator in the thermogram. After a power failure, it was discovered that a cracked insulator was the cause. Figure 22[6] shows an overheating section on a heat treatment furnace wall, which was due to refractory failure. Heat loss was also occurring which would increase energy usage and costs. Figure 23[2] shows overheating, detected with a line scanner, in electrical power insulators.

Once an area of apparent excessive heat loss is found, accurate analysis is needed in order to make good management decisions. A company may choose to develop their own in-house method of energy calculations based on estimates and averages. The use of infrared viewing and having an actual heat loss pattern allows more accurate data and resulting energy calculations. Companies that have been using infrared inspection and accompanying thermograms claim that they have more than paid for infrared viewing equipment costing over $60,000 in less than three years of use.[7]

Using hand calculators, infrared viewing inspection, thermograms, temperature measurement, emissivity and other applicable data, heat loss can be determined. As a result, insulation and refractories can be replaced in order to reduce heat loss and related energy costs. Once the heat loss is known in BTU, then the cost can be determined using the fuel cost per unit BTU. The cost of repair or replacement of the excessive heat loss area can also be estimated. The difference in the heat loss before and after the proposed correction can provide a simple return-on-investment estimate. For example, if the estimated energy loss before correction is $15,000 per year, and estimated energy loss after correction is $5,000 per year, then a $10,000 savings is estimated. If the correction cost is $10,000, then a one year payback is expected. The corrective action may be justified in order to reduce energy consumption, even when payback or return-on-investment is not high.

Although in the past energy calculations have been on a dollars per square foot per year ($/ft^2/yr) basis, energy costs

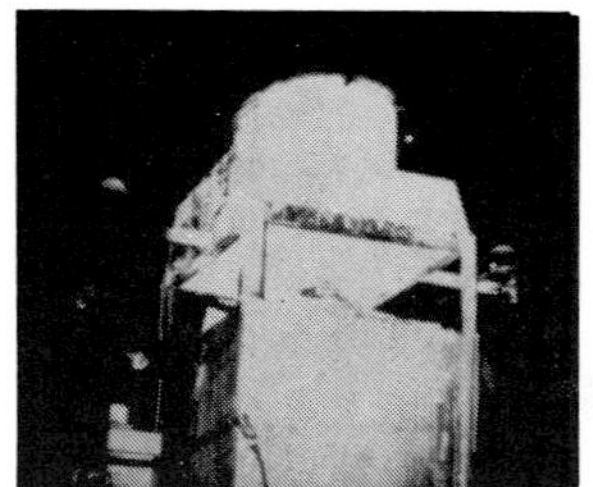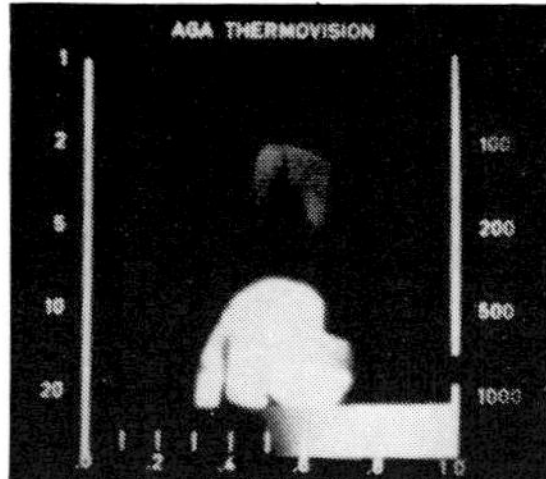

Fig. 15. Exhaust system in a gas furnace.

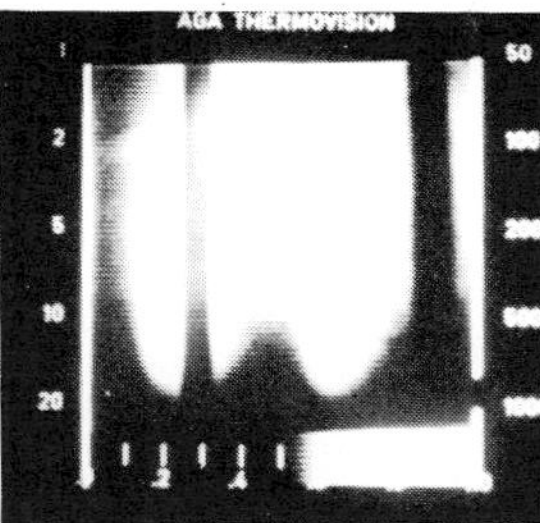

Fig. 16. Heat treatment furnace showing variation in temperature pattern caused by fallen refractory.

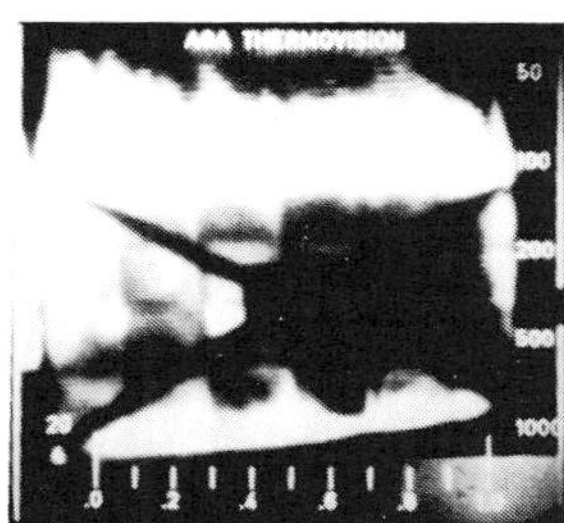

Fig. 17. Heat loss caused by faulty seals in a car bottom furnace.

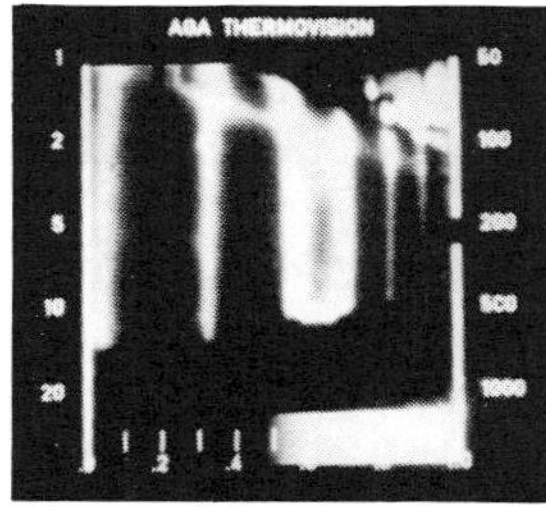

Fig. 18. Excessive heat loss at the access door on an indirect heating oven.

Fig. 19. Heat loss caused by an uninsulated valve.

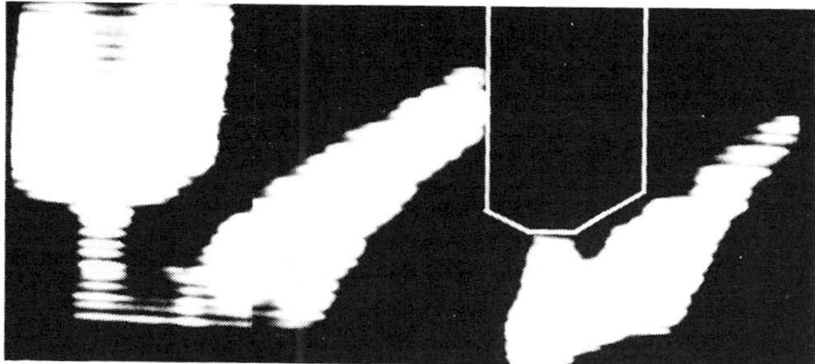

Fig. 20. Composite of two steam traps; the left one operating normally, the right one inoperative (outline).

Fig. 21. Overheated insulator on a power line pole.

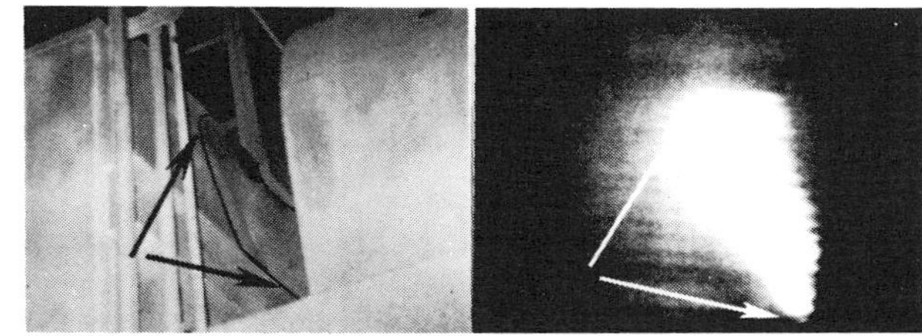

Fig. 22. Overheating section on a heat treatment furnace.

may require that a dollars per square foot per hour ($/ft^2/hr) base be used in the future, due to higher energy costs.

In order to obtain an estimate of the surface heat loss of a furnace wall, hot air duct, insulated vessel or similar item, the radiant heat loss and convective heat transmission loss must be considered. A mathematical expression of these losses would be:

$$Q_T = Q_R + Q_{CV} \text{ in BTU/ft}^2\text{/hr}$$

where

Q_T = total heat loss
Q_R = radiant heat loss
Q_{CV} = convection heat loss

from the area under analysis.

Fig. 23. Overheating in electrical power insulators detected with a line scanner.

The Stefan-Boltzman Law is used for Q_R

$$Q_R = 0.174E \left[\left(\frac{T_s}{100}\right)^4 - \left(\frac{T_a}{100}\right)^4 \right]$$

where

T_s = surface temperature in degrees Rankin (°R)
T_a = ambient temperature in (°R)
E   = emissivity
°R   = T °F + 460 (or 459.69)

or

°R = T °C + 273 (or 273.15), where C is Celsius.

Langmuir's Equation provides Q_{CV},

$$Q_{CV} = 0.296 \, (t_s\text{-}t_a) \, 5/4 \, \frac{V + 68.9}{68.9}$$

where

t_s = surface temperature °F
t_a = ambient temperature °F
V = wind velocity (feet per minute).

Therefore, the total heat loss can be quickly determined by estimating emissivity, measuring wind velocity and ambient temperature and obtaining the surface temperature. The correct emissivity is important for accurate calculations.

Summary

The applications of infrared inspection in the foundry industry are limited only by the extent to which management desires to examine a new way to approach old problems. Advantages of infrared scanning are evident: remote sensing of high energy process equipment under load (mechanical, thermal or electrical) and the resultant elimination of guesswork in making key decisions.

Cost savings are realized in the correction of maintenance problems which could cause downtime. This downtime can be used as the justification for the use of infrared surveys. The payback is very quick even if only one hour of downtime is eliminated. Payback in any case may depend on the accounting methods used by the foundry.

Infrared scanning services can be obtained either by in-house ownership of equipment or by contracting with private consulting firms. The first method requires a longer payback period, depending on the cost of equipment purchased and the time spent by plant personnel. The use of consulting firms will be more economical to the foundry for work performed on a periodic basis. Costs are straightforward and easily accounted for.

The type of infrared imagery available depends on the type of equipment used and the amount of information needed by the user. Generally, qualitative data is all that is needed. Qualitative infrared data is very useful in pinpointing locations of energy loss. Further refinements to aid in thermal mapping can be made by adding color assignments to various temperature gradations in the black through white scale. More sophisticated equipment will pinpoint temperature differentials. These advanced techniques have their place in research and specialty applications, but most users will find the simpler qualitative approach more to their advantage.

Recommendations

The high cost of infrared equipment has slowed the introduction of infrared scanning to the foundry industry for many years. Recently, however, many contractors have assumed the capital expenditure for infrared equipment and have offered their services to industry at costs commensurate with need.

Infrared scanning is based on a highly advanced technology, but its use in the foundry is relatively simple. When contracting to have infrared surveys performed, it is important that at least one person in the plant be appointed an expert: someone who has been briefed on infrared capabilities and can coordinate the use of survey results between management, engineering, operations and maintenance. Also, during an infrared survey it is important that maintenance personnel accompany the surveyors, not only to assist in access to restricted areas, but to confirm suspicions about potential trouble areas and to note previously unsuspected problems. Infrared surveys performed on a periodic basis are perhaps most useful in maintaining an ongoing record of maintenance and quality control problems and can point out where progress is being made in correcting problems.

Infrared detection allows for the checking of a far greater number of points in an electrical system in less time than virtually any other system. For those surveys devoted strictly to electrical systems, an itinerary should be planned ahead of time to reduce delays and to obtain the best results. A brief summary of items which can be surveyed include high-voltage fuse connections, knife switches, disconnect and transfer switches, terminations and transformers.

Mechanical components which must be checked include power transmission couplings, bearings, gear boxes and reducers. Heat process applications, especially important in the foundry, can be checked to determine the condition of mud rings in kilns, compressed air leaks, heat exchanger operation, cupola wall hot spots, refractory breakdown, burner operation, ladle lining, cooling lines, steam lines, etc.

Other unusual applications include checking for buried line leaks, hydraulic fluid flow, determining tank levels and looking for more moisture leaks in flat roofs.

References

1. R. D. Hudson, Jr., "Infrared System Engineering," John Wiley & Sons, Inc., New York (1969).
2. Courtesy, Barnes Engineering Company, Stamford, Connecticut (1978).
3. Courtesy, Hughes Aircraft Company, Industrial Products Division, Carlsbad, California (1977).
4. J. D. Fleischer, "Practical Aids to Increase the Utility of Thermography in Steel Plant Operations," Energy Conservation and Plant Maintenance, Proceedings of the Third Biennial Infrared Information Exchange, IRIE '76, p 69-70.
5. C. W. Hurley and K. G. Kreider, "Applications of Thermography in Industry," Energy Conservation and Plant Maintenance, Proceedings of the Third Biennial Infrared Information Exchange, IRIE '76, p 53-59.
6. Courtesy, Ward Associates and Energy Services Engineering, Inc., Stillwater, Oklahoma (1978-79).
7. L. J. Anderson, "Energy Conservation with Thermography," Energy Conservation and Plant Maintenance, Proceedings of the Third Biennial Infrared Information Exchange, IRIE '76, p 61-62.

Production Welding of Cast Irons

S. D. Kiser

Technical Service Specialist
Huntington Alloys, Inc
Huntington, West Virginia

ABSTRACT

The use of cast irons for welded components has been severely restricted by the lack of a welding wire that would produce high-quality welded joints in the materials by an automatic process. Merely because of welding requirements, forged or cast steel has been used for many components that otherwise could have been made of a cast iron. The use of economical cast irons for such parts could substantially reduce production costs for many types of assembled equipment. A recently developed flux-cored wire now makes it possible to extend the benefits of cast irons to welded structures.

Many attempts have been made to develop a welding wire for cast irons but until recently efforts were thwarted by the low ductility of heat-affected zones in welded cast irons. Shrinkage stresses during cooling of the weld metal often caused heat-affected-zone cracking. It has long been known however that cast irons could be successfully welded with nickel-alloy weld metal having a high (about 1%) carbon content. The low solubility of carbon in nickel results in precipitation of graphite during cooling of the weld metal. The resulting volume increase holds shrinkage stresses to a low level and prevents heat-affected-zone cracking. The graphite-rejection principle is the major reason for the success of nickel-alloy coated electrodes (AWS classes ENi-Cl and ENiFe-CI) for manual welding of cast irons.

With coated electrodes, carbon can be added to the weld metal through the flux coating. A conventional bare wire for automatic welding cannot feasibly be manufactured with the required high carbon content. With flux-cored wire however the graphite-rejection principle can be applied to automatic welding through carbon added to the flux core. The flux-cored wire for cast irons produces high-quality welded joints having favorable heat-affected zones. The wire can be operated at travel speeds up to 100 in./min (254 cm/min) and provides deposition rates of up to about 18 lb/hr (9.2 kg/hr).

Introduction

Cast ductile, malleable and gray irons are economical, highly serviceable materials used for numerous components of assembled equipment such as highway vehicles, farm machinery, construction equipment and machine tools. Throughout the years, however, the use of cast irons has been impeded by the lack of welding products capable of creating high-quality welded joints in the materials. The problem has become especially acute with the development of modern manufacturing technology in which high-speed, automatic welding is integrated with the assembly sequence. Because of the welding requirements, designers are often forced to specify expensive forged or cast steel for parts that would deliver equal performance in cast iron. The substitution of cast iron for such parts would save literally millions of dollars in many industries. The opportunity to gain those savings is now presented by a new flux-cored welding wire.[1]

Several factors influence the achievement of satisfactory welds in cast irons. The deposited weld metal must have high strength and ductility. Because of the comparatively low ductility of most cast irons, the shrinkage stresses developed during cooling of the weld deposit can cause the base material to crack in the heat-affected zone (HAZ). The cracking problem is compounded by metallurgical changes in the HAZ that further decrease ductility in that area.

Most of the problems with manual welding of cast irons were solved over twenty years ago with the introduction of high-nickel (Ni)-coated electrodes especially formulated for cast irons.

The electrodes, AWS classes ENi-CI and ENiFe-CI, deposit either Ni or Ni-iron weld metal with a high (about 1.5%) carbon (C) content. The high-Ni weld metal provides the necessary strength and ductility. The high C content, which is added to the deposit through the flux coating, serves to decrease shrinkage stresses in the welded joint. Carbon has a low solid solubility in Ni and any excess C present in the molten weld metal will be rejected as graphite during solidification. The graphite increases the volume of weld metal and thereby decreases the amount of shrinkage and accompanying stress.

The Ni-alloy-coated electrodes that incorporate the graphite-rejection principle produce high-quality welds in cast irons. However, the low deposition rates and manual operation restrict use of the coated electrodes to repair welding and small-scale assembly operations. Large-scale production-line welding requires a filler wire that can be deposited at high rates by automatic processes.

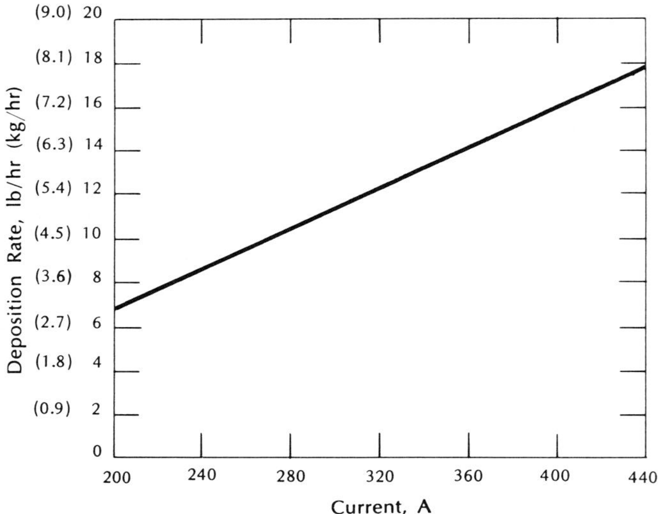

Fig. 1. Effect of welding current on deposition rate of flux-cored wire.

Welding Procedures

The new flux-cored wire is specifically designed for high-speed production welding of cast irons. The product consists of a Ni-iron tubular wire filled with C, slagging ingredients and deoxidizers. It is produced as standard 50-lb (23 kg) coils in diameters of 0.093 and 0.078 in. (2.4 and 2.0 mm). The nominal chemical composition of weld metal deposited by this flux-cored wire is 50% Ni, 1% C, 0.6% silicon (Si), 4.2% manganese (Mn) and 44% iron (Fe).

Welding with the cored wire is performed with standard equipment for flux-cored arc welding, usually the same equipment used for gas-metal-arc welding or submerged-arc welding. Direct-current power sources with constant potential and reverse polarity generally give the best results. A pulsing-arc power source has also been used successfully.

The core of the flux-cored wire contains deoxidizers and is designed to be operated without shielding gas. If welding conditions require greater protection, a shielding gas can be used with no effect on mechanical properties of the joint. Carbon dioxide is the most commonly used gas. The cored wire can also be operated with submerged-arc flux.

A wide range of welding travel speeds can be used with the flux-cored wire. High-quality welds have been produced at travel speeds from 10-100 in./min (25-254 cm/min).

The high deposition rates typical of automatic welding processes are readily achieved with this flux-cored wire. As shown in Fig. 1, deposition rates vary with welding current. Rates of 13-18 lb/hr (5.9-9.2 kg/hr) are obtained at normal welding currents. Such deposition rates are up to nine times greater than those obtainable with coated electrodes.

Preheat is not required for joints in ductile and malleable iron. For heavy-section, fully restrained joints in gray iron, preheating may be advantageous. None of the cast irons requires a postweld heat treatment after being welded with this wire.

Although a need has long existed for a filler wire that would be equivalent to the coated electrodes, the very feature that accounts for the success of the electrodes prevents the manufacture of a conventional bare wire. A Ni or Ni-iron composition having the required high C content would be impossible to process by known methods.

The advent of flux-cored wire, however, presented the opportunity to combine the features of the coated electrodes with the high-speed welding of a bare wire. After much development work, a flux-cored wire having a Ni-iron sheath and a high-C core was successfully produced. Like the coated electrodes, the cored wire deposits sufficient C to promote graphite precipitation and lessen shrinkage stresses. In addition, the new wire produces a much more favorable HAZ which further decreases the possibility of base-metal cracking. The product is now in commercial use.[1] For the first time, high-quality cast-iron welds are being produced on the assembly line.

Mechanical Properties

Joints in cast irons welded with this Flux-cored wire exhibit high mechanical strength. Table 1 lists typical mechanical properties of all-weld-metal samples and transverse samples from joints in

Table 1. Mechanical Properties of Joints Welded with the Flux-Cored Wire

Specimen	Shielding	Yield Strength (0.2% Offset)		Tensile Strength		Elongation,	Reduction of Area,	Hardness,
		ksi	MPa	ksi	MPa	%	%	Rb
All-Weld Metal	None	45.0	310	69.0	476	15.5	14.5	81
All-Weld Metal	CO₂	45.5	314	72.0	496	21.0	18.8	80
All-Weld Metal	Sub. Arc Flux	49.0	338	74.0	510	18.5	20.6	86
Transverse	None	43.5	300	66.0	455	-	-	-
Transverse	CO₂	44.0	303	66.0	455	-	-	-
Transverse	Sub. Arc Flux	45.0	310	64.0	441	-	-	-
All-Weld Metal*	CO₂	44.0	303	68.0	468	15.0	16.2	80
Transverse*	CO₂	43.5	300	67.7	467	-	-	-

*Pulsing-arc power source.

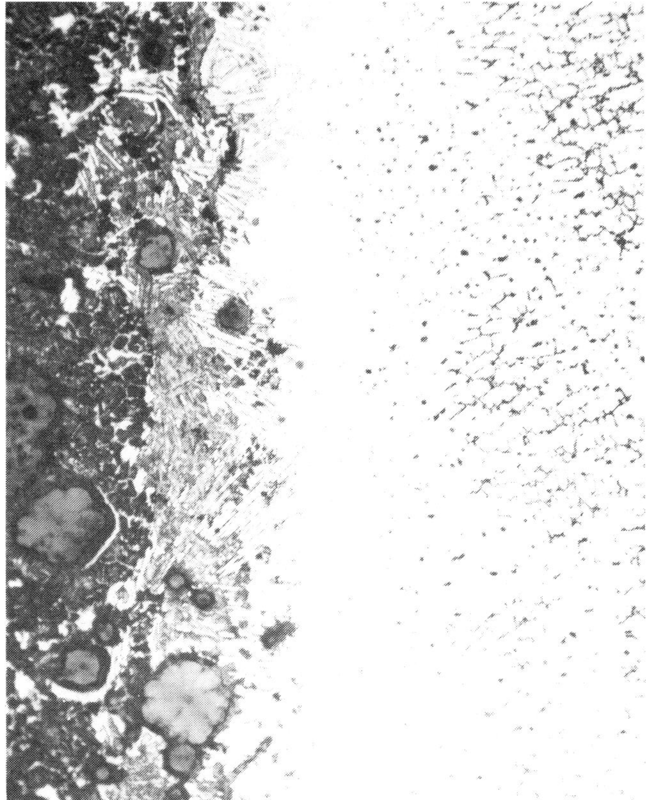

ductile iron. As indicated by the data, the tensile strength of weld metal deposited by the wire typically ranges from 68,000 to 74,000 psi (469-510 MPa), well above the 65,000 psi (448 MPa) minimum tensile strength of the most frequently used ductile iron (SAE grade D4512). In addition to their high strength, the weld deposits exhibit good ductility.

Heat-Affected Zone

The heat-affected zones of joints in cast irons have been the subject of much research and discussion. At one time, any welded joint in cast iron was suspect because of the formation of brittle carbides in the HAZ. The brittle carbides, coupled with the heavy shrinkage stresses of low-C welding products, often led to cracking problems. Now, however, metallurgists and design engineers accept the fact that all metals undergo metallurgical changes in their HAZs during fusion welding. The current approach for all materials, including cast iron, is to experimentally evaluate the effects of welding on service performance. For example, welded joints are often subjected to accelerated simulations of service conditions.

One of the advantages of flux-cored arc welding with the cored wire is an improved HAZ structure. Figure 2 compares the HAZs in joints welded with ENiFe-CI coated electrode and the fux-cored wire. The microstructure of the joint welded with the coated electrode contains a dense, continuous network of carbides along the HAZ. In contrast, the microstructure of the joint welded with the cored wire contains only sparse, discontinuous carbides in the HAZ, eliminating the possibility of a continuous band of brittle structure. The condition is further enhanced by the very irregular fusion line, which results from the high current density of flux-cored arc welding.

Mechanical tests confirm the absence of continuous brittle structures in HAZs of welds made with the cored wire. Figure 3 shows failure locations in transverse fatigue-test specimens. In none of the specimens does the path of failure follow the HAZ. Tensile tests produce similar results.

The effects of welding on service life should be evaluated for any welded structure. For most applications, however, the metallurgical changes in cast-iron joints welded with the cored wire are no more detrimental than those in welded joints in

Fig. 2. Heat-affected zones from joints in ductile iron welded with, top) ENiFe-CI coated electrode and bottom) the flux-cored wire.

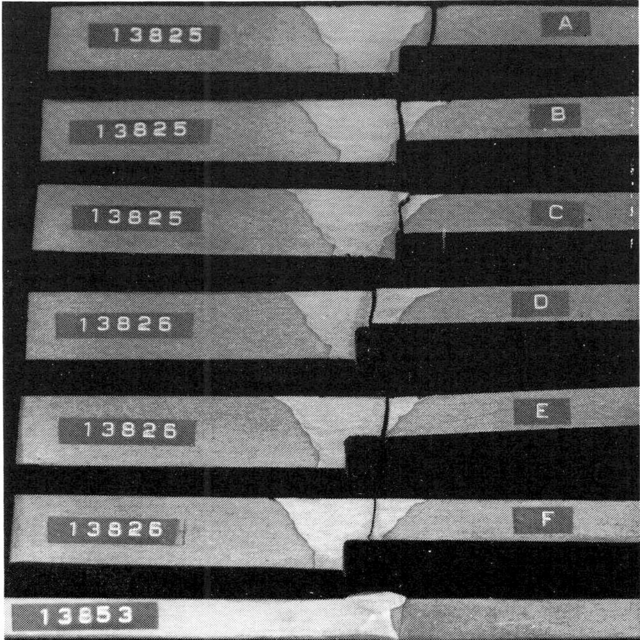

Fig. 3. Fatigue-test specimens from joints in ductile iron welded with the flux-cored wire.

Fig. 4. Conventional universal joint composed of a cast iron yoke on the differential unit and a forged steel yoke welded to the drive shaft.

many low-alloy steels that are routinely used for welded construction.

Applications

Several manufacturers have begun programs to take advantage of the cost reductions available with the use of welded cast-iron components. The examples that follow are typical of the applications in which the cored wire is being used or evaluated.

Universal Joints

A typical universal joint in an automobile drive-shaft/differential assembly is shown in Fig. 4. The universal joint is composed of two different materials. The drive-shaft yoke, which must be welded to the carbon-steel drive shaft, is forged steel. The differential yoke, however, requires no welding and is a cast iron, in this case malleable iron. Cast iron obviously meets the requirements of power transmission since it is presently used for half of the universal joint. Steel is used for the welded half of the unit only because of the need to produce high-quality welds at high production speeds.

An extensive study conducted by one manufacturer has demonstrated the suitability of universal joints made entirely of ductile iron and welded with the cored wire. Figure 5 shows the welded joint in a prototype drive-shaft yoke of ductile iron. Prototype units have passed all life-cycle tests, including requirements for torsional fatigue strength and ultimate torsional strength. The redesigned unit can be produced at lower cost even though the ductile iron yoke weighs only 3-4 lb (1.4-1.8 kg).

Wheel Lock Bar

Figure 6 illustrates a lock bar used to lock the railroad wheels in position on a dual-purpose highway/railroad vehicle. The lock

Fig. 6. Lock bar for the wheels of a highway/railroad vehicle. The cast ductile-iron lugs are welded to the steel tube with the flux-cored wire.

Fig. 5. Ductile iron universal-joint yoke welded to steel drive shaft with the flux-cored wire.

bar consists of two square-cornered locking lugs welded to the ends of a carbon steel tube. Cast steel has been used for the lugs because of the welding requirements. Prototype units having lugs of a ferritic ductile iron and welded with the cored wire are now being field tested. The use of ductile iron instead of steel would significantly reduce the cost of the units.

Hydraulic Cylinders

Figure 7 shows a section of a prototype hydraulic cylinder for farm machinery. The cylinder incorporates a cast head of ductile iron welded with the cored wire. The previous standard material for the head was cast steel. Substitution of ductile iron for steel required only a slight modification in joint design. Cylinders with welded cast-iron heads have met all test requirements and are scheduled to become standard production items. Cost studies project a substantial savings.

Figure 8 illustrates another hydraulic cylinder in which the cast steel head was redesigned in malleable iron and welded with the cored wire. The cylinder, which has a diameter of 3 in. (7.6 cm), is now in commercial production. The cast-iron heads are being welded with the cored wire at the same speed used for welding the cast steel with mild-steel wire. The malleable iron heads are being produced at a net savings of 55%, including cost difference for the welding products.

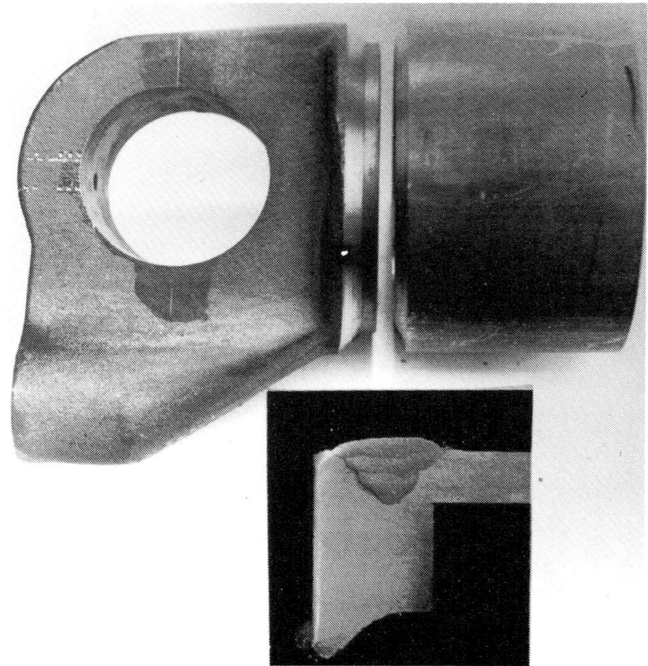

Fig. 7. Cast ductile iron head with a portion of the steel hydraulic cylinder to which it will be welded. Inset shows cross section of a finished weld bead.

Fig. 8. Welded joint between the malleable iron head and carbon steel tube of a mass-produced hydraulic cylinder.

Fig. 9. Wheel-spindle assembly for large off-road vehicle produced by welding the steel spindle to its ductile iron housing.

Wheel Spindles

Flux-cored arc welding with the cored wire has been adopted as the standard production method for the heavy-equipment wheel spindle shown in Fig. 9. A single-pass fillet weld is used to join the steel spindle to the ductile iron backing plate into which the spindle is pressed. The welded joint meets all the requirements imposed by the vehicle's strenuous off-road service.

Other Applications

The most apparent advantage offered by the new flux-cored wire is the opportunity to reduce materials cost through the use of cast irons instead of steel. In many other instances, however, the use of the cored wire can improve existing cast-iron assemblies. In some applications, cast irons are now being considered for design reasons in addition to cost.

Figure 10 illustrates an example of the use of the cored wire to improve an existing cast-iron component. The hollow, ductile iron tail shaft for an automobile transmission requires installation of a mild-steel disc to prevent loss of transmission fluid. The disc is rolled into the end of the shaft and the resulting mechanical seal is unreliable in service. The cored wire enables the disc to be welded, providing a permanent seal.

Fig. 10. Ductile iron tailshaft for automobile transmission with unwelded seal plate.

Another application for the cored wire is spot welding of various clips and brackets to cast-iron components. Resistance welding, although widely used to attach such items to steel, produces very brittle welds in cast irons. The high-Ni weld metal provides excellent ductility.

Figure 11 shows a prototype steering-column assembly for heavy off-road equipment. Standard units consist of a stamped steel flange welded to the carbon-steel steering column. Designers are now considering use of the illustrated malleable iron flange because of the ability of cast iron to dampen the engine-compartment noise transmitted to the operator by the steel flange. The machined, cast flange also provides a more consistent mating surface.

For those instances in which manual welding is being used on the production line, substantial cost savings may be possible through substitution of automatic welding with the cored wire. Figure 12 shows an example now in commercial production. The balance weights on brake drums for large trucks were formerly attached by manual shielded-metal-arc welding. The cored wire permits the use of automatic welding to plug-weld the carbon-steel weights on the drum, greatly reducing welding costs. The high current density associated with the flux-cored wire enables fusing through the casting skin without creating hard spots on the inner drum surface.

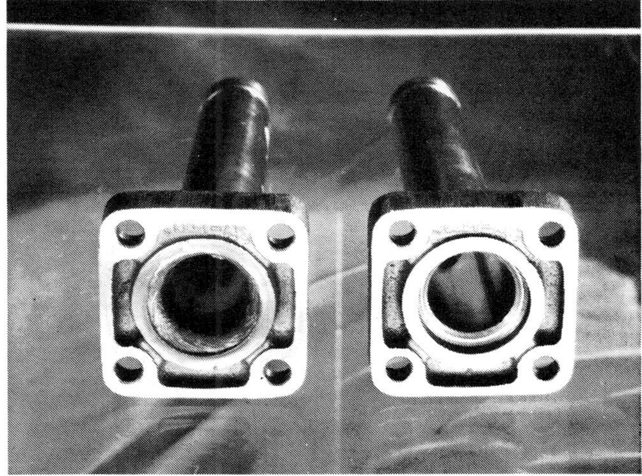

Fig. 11. Malleable iron steering-column flange for off-road equipment. The unit on the left was welded with the flux-cored wire. An unwelded joint is shown on the right.

Fig. 12.   Brake drum with carbon steel balance weights attached with the flux-cored plug welds.

Fig. 13.   Cast iron cylinder liner with carbon-steel water jackets welded to it with the flux-cored wire.

Another example of the use of automatic welding with the cored wire to reduce assembly costs is shown in Fig. 13. Production of a large diesel engine involves joining two carbon-steel water jackets to the illustrated cast cylinder liner of a proprietary gray iron. The standard joining method has been by brazing with an expensive silver brazing alloy. Units welded with the flux-cored wire are now being tested and offer substantial cost savings.

The flux-cored wire can also be used to increase foundry production. The rejection rate for large or complex castings can be reduced by welding together smaller, easier-to-cast components. In one example, a 195-lb (88 kg) transmission case for heavy equipment has been difficult to cast in one piece because of the intricate core configuration. Studies indicated that the shape could readily be cast in two sections and welded with the flux-cored wire.

Summary

The development of a new welding product, a flux-cored wire, has made assembly-line welding of cast irons a commercial reality. High-quality welds are now being produced in ductile, malleable and gray irons by automatic, high-deposition-rate processes.

The availability of an automatic welding method for cast irons gives manufacturers of various types of assembled equipment the opportunity for dramatic reductions in materials costs. Previously, many components that required welding had to be produced in forged or cast steel. Those parts can now be redesigned in economical cast irons and welded on the assembly line at the same production rates and with equivalent joint integrities.

The new flux-cored wire is presently being used or evaluated for numerous applications in highway vehicles, construction equipment, farm machinery and other welded structures. The results of both experimental testing and in-service use clearly indicate that welded cast irons will become important materials of construction.

References

1. NI-ROD (Trademark of Huntington Alloys, Inc) FC 55 cored wire.

Hot Rolling and Forging of Ductile Iron

L. A. Neumeier, Supervisory Metallurgist
B. A. Betts, Metallurgist and
R. L. Crosby, Metallurgist
U.S. Bureau of Mines
Rolla Metallurgy Research Center
Rolla, Missouri, United States

ABSTRACT

The Bureau of Mines investigated hot rolling and forging characteristics of experimental ductile iron castings made with charges ranging up to 70% of foundry pig irons and 95% of a steelmaking pig iron. Both sand and permanent mold castings were evaluated. Between 1550 and 1950F (843 and 1066C), most castings could be rolled to 90% reduction or forged to 70% without serious cracking. Some were deformed further. Charge and composition have less bearing on workability than on subsequent properties. Nickel produced some adverse effect, as did poor nodularity. Permanent mold castings, which had finer nodules and primary carbides, could be worked as readily as sand castings at 1750 and 1950F (954 and 1066C). Plasticity improved with temperature. Small billets were also forged cold to 50% reduction without cracking.

With equivalent nodularity, composition affects properties of wrought materials by altering matrix structure and strength. Properties vary with reduction and improve with working temperature. Rolled material has high strength and anisotropy, and low ductility, particulary in the transverse direction. Annealing reduces strength and improves ductility, but anisotropy persists. At 70% reduction, impact resistance is about twice as good in the longitudinal direction as in the transverse. Annealing about doubles impact resistance. Impact properties decreased somewhat for materials made with steelmaking pig iron, which had more phosphorus. Cross rolling can reduce anisotropy by equalizing directional nodule deformation.

Although workability and ductility are inferior to steel, the results demonstrate that more advantage could be taken of the plasticity of ductile iron to work rough shapes to final dimensions.

Introduction

Ductile iron, often called nodular cast iron, was discovered some 25 years ago. An historial review has been given recently.[1] Its discovery made it possible for the first time to form graphite nodules directly during solidification. Mechanisms remain controversial, but processes are well established. Basically, low-S melts with appropriate C, Si and Mn are nodularized shortly before casting by treating with Mg, usually in conjunction with some rare earth metals, and are graphitized with agents such as ferrosilicon. Production with a variety of techniques[2] has grown rapidly, to about 2.5 million tons annually in this country alone.

Unlike gray cast iron, which contains embrittling graphite flakes, the nodules of ductile iron are dispersed in a continuous matrix, which is similar to steel. This structure imparts toughness and ductility and permits plastic deformation.

Ductile iron is actually a family of materials depending on composition, heat treatment and other factors. It can probably be processed to the widest range of properties of any material.[3] Compared with steel, it has generally inferior ductility, weldability, and impact and fatigue resistance, but superior castability (high C and Si), machinability, wear and corrosion resistance, and damping capacity.

Since its inception, there has been scattered interest in assessing the workability of ductile iron, much of it abroad. In 1951, Perry[4] (Canada) reported that ductile iron could be hot rolled and forged at 1550 to 1950F (843 to 1066C), but that edge cracking limited yield. In the same period, Wakamoto[5] (Japan) reported that ductile iron could be hot worked to 50% reduction, and Unksow[6] (Russia) reported that rolling could be carried to 50% reduction and forging to 60%. Chang Tso-mei[7] (China) concluded that maximum deformation was related to reduction rate. Sheffler[8] in 1968 related property anisotropy to the extent of nodule deformation. Okabayshi[9] (Japan) reported that a high-Si ductile iron could be worked at 1000C (1832F). More recently, Dragos[10] (Rumania) measured hardness of ductile iron quenched and tempered after hot pressing to 40% reduction at 950 or 1050C (1742 or 1922F).

In a related area, Russian research originating with Ulitovski[11] in the 1930's has led to a process[12] for making cast iron (not ductile iron) sheet. The iron solidifies rapidly as white cast iron in contact with casting rolls, although the sheet center exits still molten. After passing through work rolls, it is given a graphitize-ferritize anneal to impart some ductility. Such sheet, resistant to atmospheric corrosion, can reportedly be made more cheaply than steel sheet.

Although normally unforgeable, preforms of gray cast iron have been finish forged into flywheels by high-energy-rate forging.[13]

The present research was undertaken as part of the Bureau of Mines program to efficiently utilize the mineral resources of the United States and to stimulate improved technology. Objectives are to clarify conditions under which ductile iron can be hot worked, to measure properties of worked and heat-treated materials, and to determine if various charges and different types of molds can be used without adversely affecting the workability and properties. For example, if ductile iron made by treating molten pig iron could be cast and worked into certain shapes, it would not be necessary to refine out the excess C and Si in steelmaking, and energy use and cost could be lowered for such products.

Results show that ductile iron made with high percentages of certain pig irons, and cast in sand or permanent molds, can be hot rolled and upset forged between 1550 and 1950F (843 and 1066C). Effects on workability of composition, structure, reduction and temperature are discussed. Properties depend on factors such as composition, type of mold, reduction and heat treatment. Heavily worked material has relatively high hardness and strength, but low ductility and impact resistance, particularly in the transverse direction of sheet. Ferritize annealing reduces strength and improves ductility and impact resistance, but substantial anisotropy remains. Cross rolling can reduce anisotropy.

Experimental Procedure

Heats

Heats were melted from four charge types (Table 1). The A-type consisted wholly of electrolytic iron, carbon powder, and ferroalloys, with no pig iron. The B-type contained 70 wt% pig iron, half a special low-P pig iron[14] and half a low-P (LP pig iron). The C-type contained 65% of another foundry pig iron, a bessemer grade. The D-type contained 70% of a basic grade of steelmaking pig iron (Table 2), and the E- and F-types were made with 85 and 95% of the basic grade of pig iron, respectively. Heats were constituted to have nominally 2.2 to 2.6% Si, carbon equivalent (CE) of 4.1 to 4.5 (eutectic composition is 4.3), 0.3 or 0.6% Mn (higher for 95% basic pig iron), with 0.4% Ni in some heats. These are common ranges for ductile iron.

The 115- to 120-lb heats were induction melted in MgO crucibles. Temperature was monitored with Pt, Pt-Rh thermocouples in protection tubes, and CE with a eutectometer apparatus. Manganese and Si were added as ferroalloys. Most heats were nodularized by tapping over Mg-ferrosilicon in a preheated ladle. Nickel was added as Ni-Mg nodularizing alloy. All but one heat made with the basic pig iron (heat 82-D) was nodularized by plunging an Mg-impregnated foundry coke (45% Mg) just before tapping. A small amount of Ce or mischmetal was added to all heats to control trace element effects on nodularization (residual 0.006-0.012% Ce). All heats were ladle inoculated with ferrosilicon; castings were mold inoculated with a few grams of powdered ferrosilicon.

When melting heats 97-E and A02-F, ingredients (2.5% of charge weight) for an oxidizing slag, consisting of a -40 mesh mixture of FeO (3 parts), CaO (1 part) and CaF_2 (1 part), were stirred into the melt while holding at 2325 to 2375F (1274 to 1301C) for 10 min. The slag was removed before heating to the tapping temperature. The additions decreased the Si and Mn substantially (Table 1); the P and S were relatively unchanged.

Heats were tapped near 2820F (1549C) and casting started at about 2595F (1424C).

Table 1. Composition of Heats

Heat No.								CE[1]

Heat No.	C	Si	Mn	Ni	S	P	Mg	CE[1]
No pig iron								
52-A	3.34	2.46	0.58	0.37	0.002	0.015	0.035	4.17
53-A	3.22	2.46	.29	[2]0	.005	.016	.030	4.04
59-A	3.33	2.40	.57	0	.004	.010	.030	4.13
63-A	3.48	2.25	.28	0	.005	.015	.030	4.24
67-A	3.43	2.38	.55	0	.003	--	.035	4.23
68-A	3.44	2.44	.30	.41	.004	.020	.040	4.26
70 pct LP pig iron								
54-B	3.52	2.27	0.28	0	0.009	0.024	0.030	4.29
58-B	3.46	2.72	.55	0	.009	.019	.035	4.37
64-B	3.70	2.09	.26	0	.010	.014	.040	4.40
66-B	3.67	2.16	.55	0	.008	.020	.035	4.40
70-B	3.46	2.50	.25	.40	.007	.017	.050	4.30
70 pct Bessemer pig iron								
62-C	3.55	2.22	0.55	0	0.016	0.035	0.035	4.30
65-C	3.42	2.27	.56	0	.017	--	.035	4.18
70 pct basic pig iron								
81-D	3.58	2.45	0.52	0	0.008	0.074	[3]0.052	4.42
82-D	3.52	2.30	.55	0	.025	.070	.034	4.31
85 pct basic pig iron								
97-E[4]	3.70	1.73	0.48	0	0.008	0.080	[3]0.030	4.31
98-E	3.43	2.22	.60	0	.007	.097	[3] .040	4.20
95 pct basic pig iron								
A01-F	3.54	2.54	0.73	0	0.009	0.10	[3]0.039	4.42
A02-F[4]	3.36	2.16	.56	0	.010	.10	[3] .048	4.11

[1]CE = pct C + 1/3 (pct Si + pct P).
[2]No Ni added.
[3]Mag-Coke nodularization. Others ferrosilicon.
[4]Oxidizing slag added.

Castings

For rolling stock, green-sand slabs and permanent mold ingots were cast. The green-sand slabs were 7 x 6.25 x 1 in. thick. The cropped permanent mold ingots were about 14 x 3.5 x 1.8 in. thick. Each type provided two rolling slabs about 7 in. wide x 3.5 in. long. For hot forging, 3-in.-dia. permanent mold ingots 8 in. tall were cast, which provided machined 2.5-in.-dia. billets 2 in. tall. Small billets for cold forging, 0.75-in-dia. x 0.75-in.-tall, were machined from sand castings. The cast iron permanent molds were coated with silica wash and torched briefly before casting (big end up) to demoisturize. Standard 1-in. keel block molds (core sand) were cast from some heats. Gammagraphs showed that most permanent mold ingots yielded 90% or more below the pipe. Surface defects were ground out before rolling.

Rolling

The slabs were rolled on a two-high mill at 1550, 1750 or 1950F (843, 954 or 1066C) at 175 ft/min, in one direction only, turning over between passes. The roll diameter was 7.75 in. for rolling of the sand castings and 7 in. for rolling of the permanent mold ingots. Reheat time was 30 min to 70% total reduction and 20 min thereafter. At 40 and 70% reduction, the materials were aircooled and specimens sectioned; sheet rolled at 1950F (1066C) was sandblasted to remove encrusted oxide before rolling further. At 90% reduction, the 0.10-in. sheet from sand-cast (SC) slabs and 0.18-in. sheet from permanent mold castings (PMC) was aircooled and sandblasted.

Reduction per pass was varied according to a consistent schedule based on the thickness, temperature, and capacity of the 25-hp mill. To 40% reduction, the reduction per pass was 6 to 10%; from 40 to 90% reduction, it was 10 to 20%. Special rolling beyond 90% reduction, rolling with heavy reduction per pass and cross rolling are described briefly with the results.

Forging

The billets were upset forged at 1550, 1750 or 1950F (843, 954 or 1066C) by striking repeatedly with a 1000-lb pneumatic hammer forge. The usual procedure was to forge from 2 to 1.2 in. thick (40% reduction), reheat 30 min, and forge to 0.6 in. (70%), reheating for 20 or 30 min depending on thickness. One group was forged to 70% at 1750F (954C) without reheating. Selected billets were forged to 80% reduction. Several billets were forged to 0.3 in. (85%) for tensile testing. Completed forgings were aircooled. The small billets were upset forged cold (room temperature) in successive reductions of about 10% of the initial height.

Heat Treatments

Before hot working, the PMC ingots were heated in air for 4 hr at 1750F (954C) to decompose primary cementite (Fe_3C), cooled slowly to ferritize the matrix for easier sectioning, and sand blasted. Specimens for evaluation of properties of wrought materials were ferritize annealed in He by heating 3 hr at 1650F (900C), cooling to 1275F (690C) in 3 hr and holding 6 hr, and furnace cooling. Special heat treatments are described with the results.

Table 2. Pig Iron Analysis

Pig iron grade[1]	C	Si	Mn	S	P
Special low phosphorus	4.3	0.87	0.23	0.016	0.014
Low phosphorus	4.0	2.12	.25	.017	.019
Bessemer	4.1	1.91	.92	.035	[2] .064
Basic	4.2	1.51	.84	.030	.10

[1]First three pig irons, trace Ca, Co, Cr, Cu, Mg, Ni, Ti, and V. Basic grade, also Al, Sn, Zn, and no Co.
[2]Marketed as Bessemer grade, but analyzed less than 0.076 pct P minimum listed in ASTM Spec. A43-67 (4, p. 8).

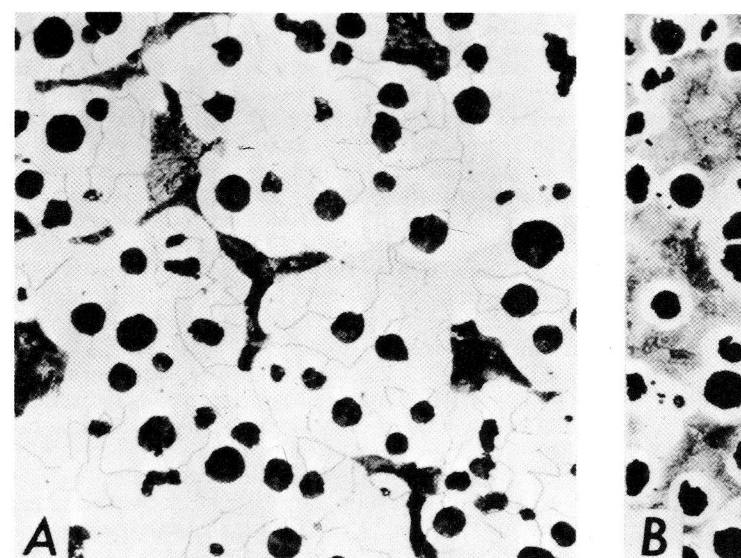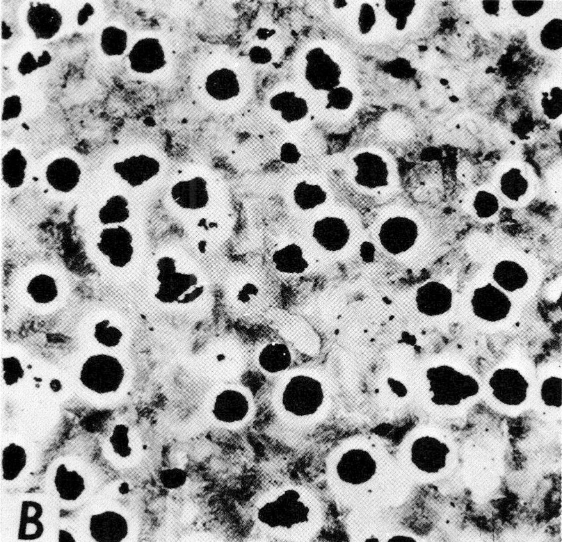

Fig. 1. Microstructures of sand-cast rolling slabs. A) heat 64-B, B) heat 82-D (As-cast, X100).

Hardness

Rockwell and brinell (3000 kg) hardness tests were made on ground surfaces. Average values were calculated for four or more indentations. The rockwell-A test was chosen for worked material because some was too thin to sustain the brinell load; the rockwell-C test was used on specimens of the hard as-cast PMC ingots, some of which cracked under the brinell load.

Tensile Tests

Tensile specimens for wrought materials were sheet-type[14] having a 2-in. gage length. The reduced section width was 1/2 in. for the 0.10-in. sheet and 1/4 in. for thicker materials. The tests were made on a tensile testing machine with a strain rate of 0.02/min to yield and 0.2/min thereafter. Normally, 16 specimens were tested from a sheet, four each longitudinal (L) and transverse (T) for as-rolled, and four each for the ferritize-annealed condition. Keel-block leg specimens (two per casting), standard 1/2-in.-dia.[14] were tested on a second type of tensile testing machine at 4800 lb/min load rate to yield. Tensile strength, 0.2% offset yield strength, and percentage elongation were calculated.

Impact Tests

Impact tests were made on PMC material rolled to 70%

reduction (0.54 in.). From each plate, four each as-rolled L and T specimens were machined and an equal number of ferritize annealed. Specimens were unnotched ASTM type[14] 2.165 in. long and 0.394 in. square. The charpy tests were made at about 75F (24C). Impacts were directed normal to the rolling plane.

Other

The chemical analyses and microscopy were essentially conventional. Carbon was determined from pin castings and other elements from drillings. Microscopy specimens were polished with diamond abrasive and etched in nital. Worked materials were sectioned to examine the region normal to the thickness. For micrographs presented of rolled materials, the rolling direction corresponds to the vertical page direction for L specimens; for T specimens, it is normal to the page.

Results

Microstructures of Castings

The microstructures of two SC rolling slabs are shown in Fig. 1. Because of the higher Mn and P from the basic pig iron, heat 82-D (Fig. 1B) contains more pearlite, but some ferrite (light) remains.

Reflecting the faster solidification, the as-cast structures of

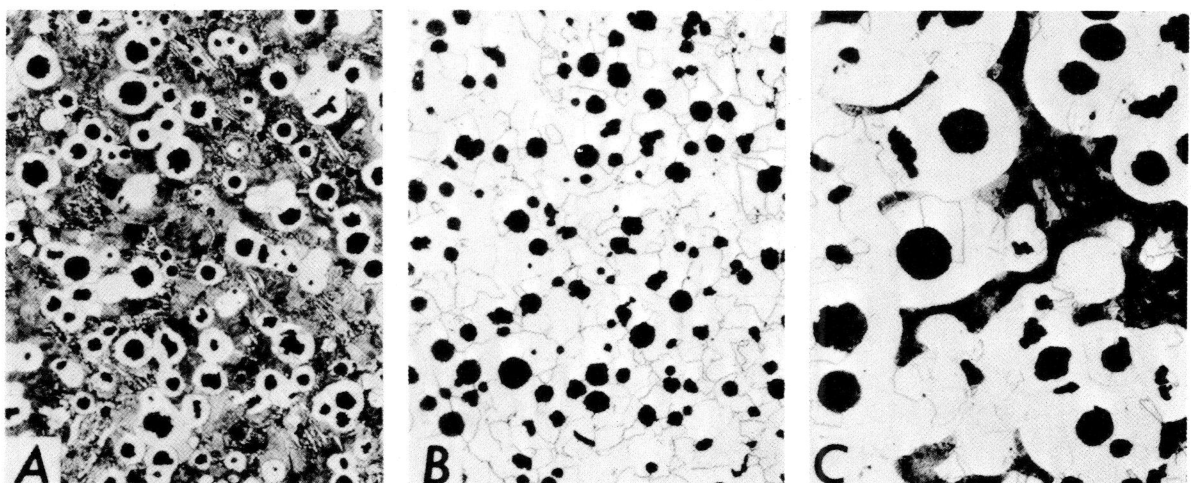

Fig. 2. Microstructures of permanent mold ingot and sand casting from same heat. Heat 54-B. Permanent mold ingot: A) as-cast and B) heated at 1750F (954C) and furnace-cooled, C) sand-cast keel block. (X100)

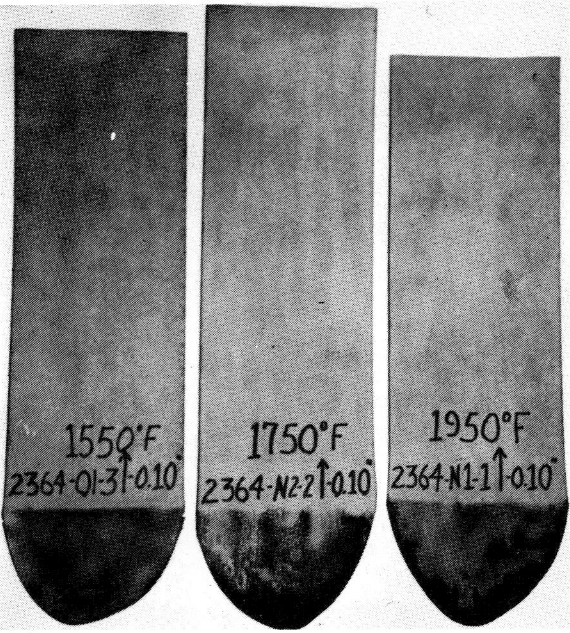

Fig. 3. Sheet rolled from sand-cast slabs to 90% reduction at 1550, 1750 and 1950F (843, 954 and 1066C). Heat 64-B. Sand-blasted except tailing end.

PMC ingots (Fig. 2A) have higher eutectic cell and nodule count, smaller nodules, more pearlite than those of sand castings (Fig. 2C) of similar composition, and also primary cementite (light, acicular). Heating for 4 hr at 1750F (954C) and furnace cooling slowly (Fig. 2B) decomposed the cementite in PMC ingots and ferritized the matrix. Further tests showed that heating for only about 1 hr at 1750F (954C) could decompose the cementite. The period, which depends on section size and composition, can be part of the preheat for working. The nodule size[14] averaged no. 5 (0.08 mm) for SC materials and no. 6 (0.04 mm) for PMC ingots. The nodule count[15] averaged 50 to 75 and 100 to 125 per sq mm, respectively.

Hardness of Castings

Hardness varied substantially depending on type of casting, composition and condition. Sand castings ranged from bhn 149 to 232, with the higher values reflecting factors such as increased pearlite from higher Mn or the presence of Ni. The as-cast PMC ingots with primary cementite ranged between R_c 30 and 47; after heat treatment, the hardness dropped to between bhn 147 and 169; reflecting variable amounts of ferrite hardeners such as Si and Ni.

Rollability

The rollability of SC slabs is coded in Table 3. Rollability was better at 1750 and 1950F (954 and 1066C), although all slabs were rolled to 90% at 1550F (843C). The reason for the severe cracking of the SC slabs from heats 82-D and A01-F during rolling at 1950F (1066C) is not evident from the microstructures or chemical analyses. Sulfur was high in 82-D (trapped sulfides?) but not in A01-F (Table 1). One heat was nodularized with one agent and one with the other. The fact that PMC ingots from the same heats were rolled at 1950F (1066C) without cracking indicates that the behavior is related in some manner to the slower cooling of the sand castings. Precise interpretation must await more detailed studies. At 1750F (954C), only one heat exhibited minor cracking. The two worst results at 1550F (843C), with both edge and surface cracking, occurred for 67-A and 68-A, heats made without pig iron. Although these heats had higher Mn than some, or Ni (Table 1), the increased cracking cannot be attributed directly to the main composition, because similar heats made with pig iron cracked less. Pig irons introduce various trace impurities that can affect factors such as graphite nucleation, cell count, and gas content.[16] In the present study, heats with pig iron generally had higher nodule count and better nodule distribution. An example of sheet rolled from SC slabs at the three temperatures is shown in Fig. 3 (leading end sectioned).

Table 4 lists the rollability coding for PMC slabs. These slabs also rolled without cracking seriously at 1750 and 1950F (954 and 1066C), but they displayed poorer rollability at 1550F (843C) than the SC slabs. There was more edge and surface cracking at 1550F (843C) and most of the initially thicker slabs tended to alligator at the leading end after 70 to 85% reduction. The alligatoring would probably be alleviated with different ratios of roll diameter to sheet thickness, but the most straightforward solution is to roll at higher temperature. The increased cracking at 1550F (843C) in comparison with SC material is evidently related to the lesser fracture path between the smaller and more numerous nodules, coupled with the decreased plasticity at the lower temperature. This indicates that there must be optimum ranges of nodule size and distribution that will produce optimum workability at the lower working temperatures.

The results for both SC and PMC materials show that there is considerable latitude in charge and composition for rolling stock. Some structural and compositional effects will be discussed further with the forgeability results.

In further work PMC slabs from heats 54-B and 58-B were rolled to 96% reduction (to 0.07 in.) at 1950F (1066C) without cracking.

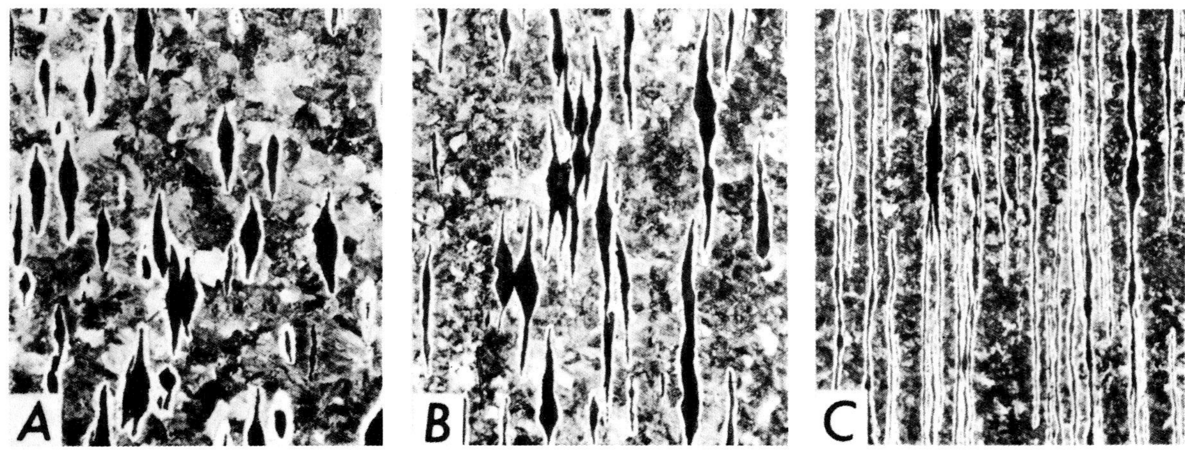

Fig. 4. Microstructures of ductile iron rolled to A) 40%, B) 70% and C) 90% reduction at 1750F (954C). Heat 64-B. As-rolled from sand-cast slabs. (longitudinal, X100)

Table 3. Rollability[1] of Sand Castings — Rolled to 90% Reduction

Heat No.	Rolling temperature, °F		
	1,550	1,750	1,950
63-A	b	a	a
67-A	c,x	a	b
68-A	b,x	b	b
64-B	b	a	a
66-B	a	a	a
70-B	b	a	b
65-C	b	a	a
81-D	b	a	a
82-D	b	a	d,x[2]
97-E	b	b	b
98-E	c	b	a
A01-F	c	b	d,x[2]
A02-F	b	b	b

[1]Rollability code:
  a: essentially no crack.
  b: edge cracks <1/4 inch.
  c: edge cracks 1/4 to 1/2 inch.
  d: edge cracks >1/2 inch.
  x: surface cracks.
[2]Rolling stopped at 40-pct reduction.

Table 4. Rollability[1] of Permanent Mold Castings — Rolled to 90% Reduction

Heat No.	Rolling temperature, °F		
	1,550	1,750	1,950
53-A	c	a	a
59-A	d,x	a	a
54-B	c,x	a	a
58-B	d,x	b	a
62-C	c	a	a
81-D	c,x	b	a
82-D	c	b	a
97-E	c	b	a
98-E	c	b	a
A01-F	c	b	a
A02-F	c	b	a

[1]Rollability code:
  a: essentially no cracks.
  b: edge cracks <1/4 inch.
  c: edge cracks 1/4 to 1/2 inch.
  d: edge cracks >1/2 inch.
  x: surface cracks.

Small SC slabs (2.5 x 2.5 in.) from heat 65-C were cross rolled with 20% reduction per pass to 90% total reduction at 1950F (1066C) and to 80% at 1750F (954C) without cracking. This shows that ductile iron can be cross rolled as a means of reducing the anisotropy resulting from directional nodule deformation.

To check on effects of reduction per pass, small SC slabs (64-B, 66-B) were rolled at 20% or at 40% reduction per pass to 92% total reduction at 1750F (954C) and at 1950F (1066C). With 20% reduction per pass, little cracking occurred at either temperature. However, at 40% reduction per pass, severe cracking started at both temperatures when rolling from 65 to 79% total reduction, showing that reduction rate becomes important at heavier reductions.

A few instances were also noted where surface cracks would develop if initial passes were exceptionally large, especially at 1550F (843C).

Microstructures of Rolled Materials

The microstructures of a SC slab (64-B) as-rolled to various total reductions at 1750F (954C) (Fig. 4) illustrate the nodule deformation as rolling progresses. The nodules are enveloped in ferrite and the remaining matrix is pearlite. The ferrite forms during cooling by diffusion of matrix carbon to the nodules. Close examination reveals that some nodules deform by peeling off outer layers of graphite.

Structures of the same heat (Fig. 5) rolled at 1550 and 1950F (843 and 1066C) show that the ferrite morphology differs. The ferrite envelopes formed after rolling at 1950F (1066C) (Fig. 5B) are similar, although thinner than those for rolling at 1750F (954C) (Fig. 4C). However, for rolling at 1550F (843C) (Fig. 5A), substantial ferrite appears as dispersed particles, many separated from the graphite. This is attributed to the temperature of the thin material cycling below and above the lower critical temperature[3] during final rolling. Below the lower critical, austenite becomes unstable and ferrite can form. When reheated to 1550F (843C), between the lower and upper critical, the ferrite is evidently incompletely redissolved, and is thus

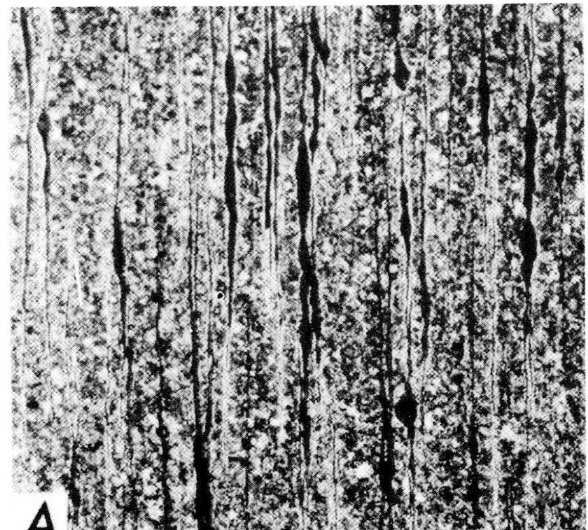
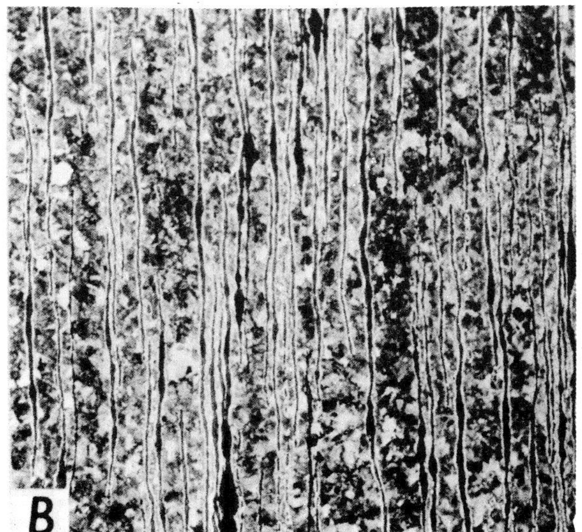

Fig. 5. Microstructures of sheet rolled from sand-cast slabs to 90% reduction at A) 1550F (843C) and B) 1950F (1066C). (longitudinal, X100)

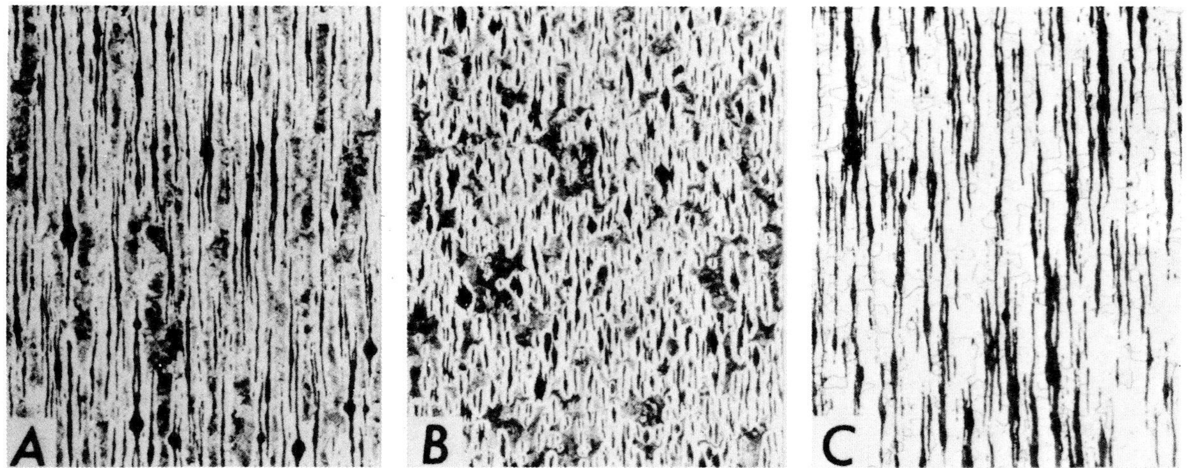

Fig. 6. Microstructures of sheet rolled from permanent mold ingots to 90% reduction at 1750F (954C) Heat 54-B. A) longitudinal and B) transverse; as-rolled, C) longitudinal, ferritize annealed. (X100)

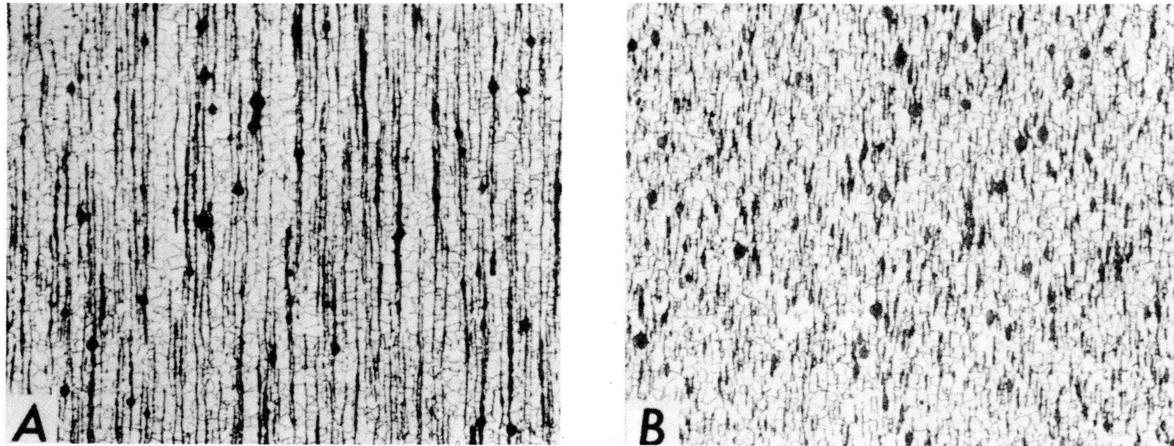

Fig. 7. Microstructures of sheet rolled from a permanent mold ingot to 96% reduction at 1950F (1066C). Heat 54-B. A) longitudinal, B) transverse. (ferritize annealed, X100)

deformed and dispersed in the succeeding pass. This ferrite morphology, not noted in the thicker PMC materials, apparently had no marked effect on rollability, insomuch as the SC sheet could usually be finish rolled at 1550F (843C) with less cracking than the PMC sheet. It is, however, symptomatic of working where lower temperature phases come into play and plasticity is diminished. For both type castings, the amount of ferrite generally decreased with increased rolling temperature.

The microstructures of a PMC slab of similar composition (54-B) rolled to 90% reduction at 1750F (954C) (Fig. 6) illustrate the closer spacing of the smaller rolled-out nodules. As-rolled, more ferrite is present because the thicker PMC material cools slower and also the path is less for matrix carbon diffusion to the more numerous nodules. Comparison of the L and T structures (Fig. 6A and 6B) illustrate the directionally deformed nodules. Figure 6C shows the structure after ferritize annealing.

Fig. 8. Forging sequence. From left: 2-in.-high billet; forged to 40%, forged to 70%, and forged to 80% reduction.

Figure 7 shows the structures (annealed) of sheet from a PMC ingot of the same heat that was rolled to 96% reduction at 1950F (1066C). The rolled-out nodules are closer together and, even with the heavy reduction, some nodules are partially intact.

Microscopic examination of cross-rolled sheet verified that the nodule morphology was essentially the same in L and T directions with respect to the last rolling pass.

Forgeability

The forgeability is coded in Table 5. As for rolling, forgeability improved with temperature. At 1950F (1066C), only one billet cracked somewhat when forged to 70% reduction in height. At 1750F (954C), three forgings cracked, one severely. At 1550F (843C), five of the ten billets forged to 70% reduction cracked extensively, whereas the others had essentially no cracking. Because of more cooling during forging, forgeability was a little poorer when most heats were forged at 1750F (954C) to 70% reduction without reheating (not coded in Table 5). All billets selected for forging to 80% reduction did so without cracking.

The relative ease of forging should relate generally to the rollability, although the nodules are lengthened as well as flattened during rolling. From the standpoint of composition, the three forgings that cracked most severely at 1550F (843C) were from the three heats with the 0.4% Ni addition. The other two that cracked less severely at 1550F (843C), 63-A and 67-A, were from heats made without pig iron. At 1750F (954C), of the three forgings that cracked, two, 68-A (cracked severely) and 70-B, contained Ni; the forging from 68-A was also the only one that cracked at 1950F (1066C). Other than for Ni, which has substantial solubility in austenite, no definite correlation was noted between forgeability and the compositions represented, although the lower Si heats generally produced good results. If present over a wider range than the 1.7 to 2.5% Si in the billets forged, Si, which is soluble in austenite, is a strong graphitizer and raises the lower critical temperature, would be expected to influence workability. Manganese between 0.3 and 0.7% had little effect. All the billets from heats made with basic pig iron, even for 95% of the charge, forged without cracking. The higher P in these heats (0.07 to 0.10%) and high S in 82-D (0.025%) produced no adverse effect. As was the case for rolling, heats with pig iron forged better than those without, and castings with obviously poor nodularity had poor forgeability, although some vermicular graphite had little effect.

The hot forging sequence is illustrated in Fig. 8.

Small billets from heats 64-B, 66-B and 70-B were upset forged cold with successive 10% reductions of the starting height. Three or four billets from each heat were forged in both the as-cast and ferritize annealed conditions. For both conditions, the billets were forged to 50% reduction without cracking. The cracking usually occurred at 60% reduction; but

Table 5. Forgeability[1] of Billets Upset Forged to 70% Reduction

Heat No.	Forging temperature, °F		
	1,550	1,750	1,950
52-A	d	--	a[2]
63-A	c	a	a[2]
67-A	c	b	a
68-A	d	d	b
64-B	a	a[2]	a[2]
66-B	a	a[2]	a
70-B	d	b	a
65-C	a[2]	a[2]	a[2]
81-D	a	a	a
82-D	a	a	a
97-E	a	a	a
98-E	a	a	a
A01-F	a	a	a
A02-F	a	a	a

[1]Forgeability code:
   a: essentially no cracks.
   b: edge cracks <1/4 inch.
   c: edge cracks 1/4 to 1/2 inch.
   d: edge cracks >1/2 inch.
[2]Also forged to 80-pct reduction without cracking.

three did not crack until reduced 70%. Figure 9 shows a starting billet and three billets cold forged to 50% reduction.

Examples of the microstructures of a billet forged to 70% reduction at 1750F (954C) and a billet forged cold to 60% reduction (not annealed before forging) are shown in Fig. 10A and 10B, respectively. The hot forged billet was from PMC material, and the cold forged billet from SC material (larger nodules).

Hardness and Tensile Properties of Wrought Materials

Examples of the hardness of a SC slab and a PMC ingot of similar composition as a function of rolling temperature and percent reduction are shown in Fig. 11. In Fig. 11B, the as-cast hardness is that of the keel-block casting. The general behavior is similar for both, although for a given temperature and reduction the SC material is thinner than the PMC material and therefore cools faster and develops a higher pearlite-to-ferrite ratio. When rolling to 40% reduction, the hardness increases

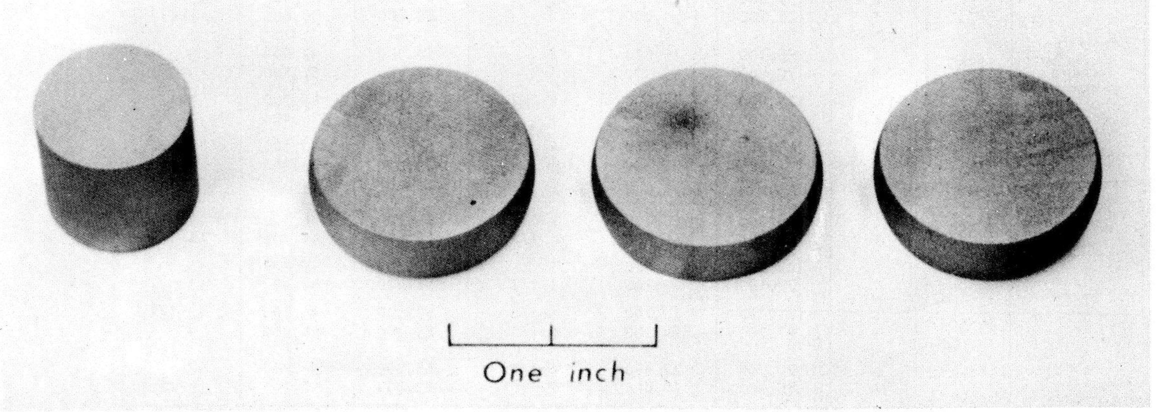

Fig. 9. Ductile iron billets cold forged to 50% reduction. Starting billet on left.

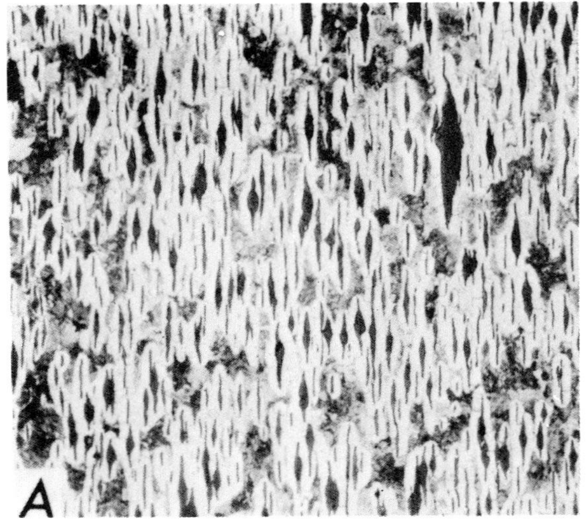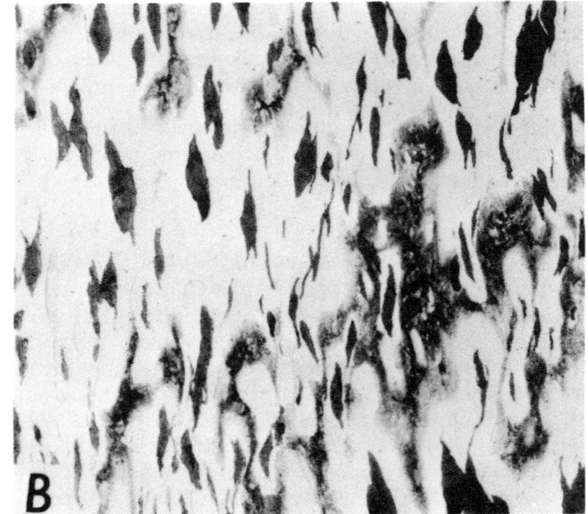

Fig. 10. Microstructures of hot and cold forged billets. Heat 66-B. A) forged to 70% reduction at 1750F (954C), B) forged cold to 60% reduction. (X100)

because cooling is faster than for the castings, permitting more pearlite to form. For rolling to 70 and 90% reduction, hardness becomes very dependent on rolling temperature, resulting from several factors. As rolling temperature decreases, more graphite and ferrite can form because of decreased C solubility in austenite. This, coupled with the lessening graphite spacing and C diffusion path for graphite deposition on the nodules, causes the hardness to tail off at 1750 and 1550F (954 and 843C) at 70 and 90% reduction, even though the sheet becomes thinner and cools faster. Pearlite formed when cooling from higher temperatures is also finer and harder. The hardness for different compositions generally coincided with the relative presence of pearlite stabilizing elements such as Mn, and ferrite strengtheners such as Si and Ni. Other factors equal, the hardness and strength of ductile iron decreases with increased C.

Ferritize annealing significantly reduces hardness. For example, the hardness of the sheet of heat 64-B (see Fig. 11) rolled to 90% reduction between 1550 and 1950F (843 and 1066C)

dropped after annealing to only R_A 30 and 38, respectively. Annealed hardness still reflected the effect of rolling temperature. Hardness data for all rolled materials that were tensile tested are listed in Appendix A.

Selected tensile properties are listed in Table 6 and plotted in Figs. 12 and 13 for sheet rolled from SC and PMC slabs to 90% reduction between 1550 and 1950F (843 and 1066C). A number of factors affect behavior, which may be characterized as follows. For as-rolled sheet, strength is high and elongation low. Strength, and to a lesser extent elongation, increase with rolling temperature. The L specimens exhibit higher strength and elongation than T specimens, because the stress is applied parallel to the graphite "stringers." The L and T tensile strengths differ more than yield strengths. For equivalent conditions, the PMC materials develop less pearlite and lower strength than the SC materials, but elongations are not greatly different. As noted for the hardness, compositional variations that increase pearlite and harden ferrite increase strength. Ferritize annealing

Table 6. Tensile Properties of Sand-Cast and Permanent Mold Materials Rolled to 90% Reduction and Ferritize Annealed

Heat No.	Rolling temp., °F	Orientation	As-rolled			Ferritize annealed		
			Tensile strength, psi	Yield strength, psi	Elongation, pct	Tensile strength, psi	Yield strength, psi	Elongation, pct
				Sand-cast material[1]				
66-B	1,550	L	108,400	84,800	3	53,100	39,300	7-1/2
		T	72,400	67,500	1-1/2	41,400	36,100	2-1/2
	1,750	L	148,300	105,500	4	56,300	37,800	12
		T	88,100	76,600	2	43,400	36,000	3
	1,950	L	157,800	113,100	3-1/2	61,000	41,600	15
		T	118,100	98,000	2	50,600	38,600	4-1/2
81-D	1,550	L	95,700	75,600	3	56,500	42,600	9
		T	70,800	65,200	1	42,400	37,500	3
	1,750	L	125,000	94,900	2-1/2	58,600	43,900	9-1/2
		T	76,500	76,200	1	43,100	38,400	2
	1,950	L	164,300	114,100	4	67,300	48,300	13-1/2
		T	117,500	103,900	1	54,800	43,900	4
				Permanent mold material[2]				
58-B	1,550	L	68,200	60,100	2-1/2	56,000	46,600	3
		T	57,700	56,500	1/2	48,600	45,700	1-1/2
	1,750	L	105,800	74,200	5	69,100	49,200	13
		T	77,100	65,700	1-1/2	56,400	47,400	3
	1,950	L	125,000	78,100	5-1/2	67,500	49,700	13-1/2
		T	97,700	74,100	2-1/2	59,400	48,300	4
81-D	1,750	L	120,700	77,400	6-1/2	63,500	46,200	13
		T	85,900	70,800	2	52,300	42,100	3-1/2

[1]At 90-pct reduction, 0.10 inch thick.
[2]At 90-pct reduction, 0.18 inch thick.

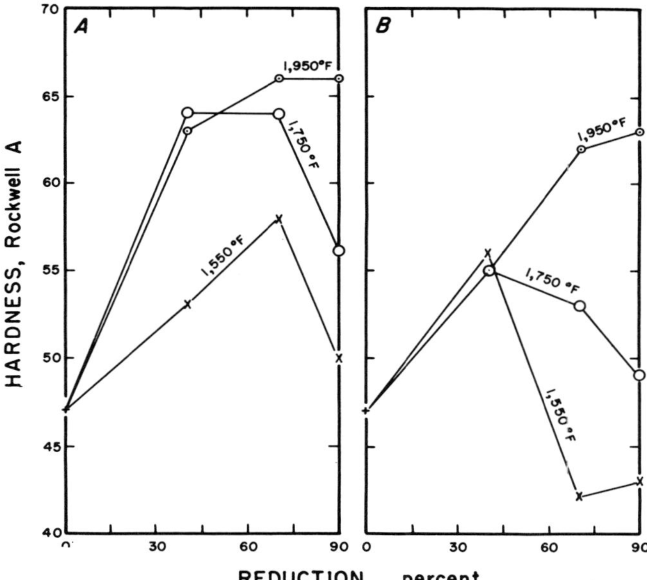

Fig. 11. Hardness as a function of rolling temperature and percent reduction. A) heat 64-B, sand-cast slab, B) heat 54-B, permanent mold ingot.

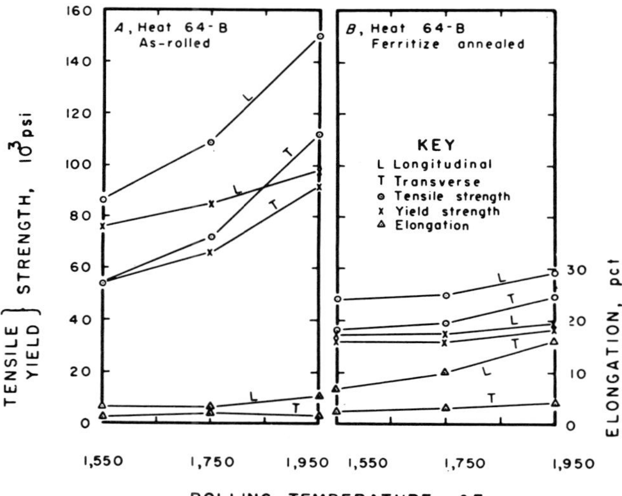

Fig. 12. Tensile properties of sheet rolled from sand-cast slabs to 90% reduction. Heat 64-B. A) as-rolled, B) ferritize annealed.

diminishes anisotropy in strength but anisotropy in elongation remains high. The different graphite distribution leads to higher strengths for the annealed sheet rolled from PMC slabs, in comparison to that rolled from SC slabs, which is converse to the as-rolled behavior where the relative amount of pearlite was a dominant factor. Annealed tensile properties, both strength and elongation, also improved with rolling temperature. As the relative amounts of ferrite and graphite become essentially constant for a given heat, this must result from factors such as small differences in ferrite grain size and secondary graphite deposition from decomposing pearlite.

For comparison with the sheet properties in Fig. 13, the as-cast tensile properties for heat 54-B (keel-block) were 65,200 psi tensile strength, 42,000 psi yield strength and 17.5% elongation. Tensile data for rolled materials not listed in Table 6 are given in Appendix B. The sheet from heat A02-F, which was

made with 95% basic pig iron and had a deoxidizing slag addition, had the best overall tensile properties from the standpoint of yield strength and elongation combination.

Tensile tests on material from heats 64-B and 66-B rolled to only 70% reduction at 1750F (954C) (see Table B3) showed that, as-rolled, anisotropy in strength was less than when rolled to 90% reduction (Fig. 10, Table 6, Table B1). Elongations were relatively unchanged. After annealing, tensile strengths were higher and anisotropy less than for 90%. Yield strengths changed but little. Elongations, however, increased to about double those for the sheet rolled to 90% reduction and annealed, reaching about 20% for L specimens. Nonetheless, anisotropy persisted, because T elongations increased to only 7% for 64-B and 5% for 66-B.

Tensile properties were determined for an annealed sheet that was cross rolled to 90% reduction at 1950F (1066C). The specimens were sub-size (0.25-in.-wide reduced section, 1-in. gage length) so that both L and T sheet directions (with respect to last rolling pass) could be tested from the same sheet. In one direction, tensile strength averaged 56,800 psi, yield strength 43,700 psi and elongation 7%; in the other direction, the values were 54,200 psi, 42,400 psi and 6%, respectively.

For material forged to 85% reduction, the hardness and tensile properties also increased with hot working temperature for both the as-forged and annealed conditions (Appendixes A and B).

Table 7. Impact Energies[1] of Ductile Iron Rolled to 70% Reduction at 1750F (954C) and Ferritize Annealed

Heat No.	Orientation	Energy absorbed, ft-lb	
		As-rolled	Ferritize annealed
53-A	L	57	114
	T	21	41
59-A	L	55	110
	T	18	37
54-B	L	55	100
	T	27	43
58-B	L	54	101
	T	27	47
62-C	L	43	109
	T	18	41
81-D	L	47	106
	T	25	54
82-D	L	45	110
	T	22	44
97-E	L	44	91
	T	25	53
98-E	L	39	93
	T	17	45
A01-F	L	34	92
	T	20	42
A02-F	L	44	104
	T	23	47

[1]Charpy test at room temperature, unnotched specimen.

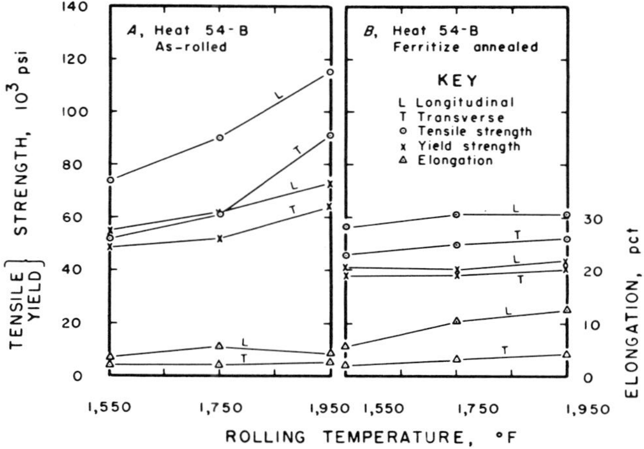

Fig. 13. Tensile properties of sheet rolled from permanent mold ingots to 90% reduction. Heat 54-B. A) as-rolled, B) ferritize annealed.

Impact Properties

Results of impact tests on as-rolled and ferritize annealed specimens rolled from PMC ingots to 70% reduction at 1750F (954C) are given in Table 7. The overall behavior follows a generally consistent pattern. Both as-rolled and annealed, L specimens average roughly double the impact resistance of T specimens. Annealed specimens average about double that of as-rolled specimens. The overall averages were as-rolled, L = 47 ft-lb and T = 22 ft-lb; annealed, L = 103 ft-lb and T = 45 ft-lb. The lowest values occurred for heat A01-F, apparently reflecting the relatively high C, Si and Mn content from this heat made with 95% basic pig iron. However, heat A02-F, made with the same amount of basic pig iron but with the deoxidizing slag addition, had properties consistent with the overall averages. Annealed specimens from heats made with 85 or 95% basic pig iron averaged lower in L strength but higher in T strength. The high P in the heats made with basic pig iron produced no gross embrittlement as often occurs in castings with this much P. Annealed ferritic ductile iron castings tested similarly exhibit about 100 ft-lb for the ductile range and below 20 ft-lb for the brittle range.[17] On this basis, the annealed L specimens represent essentially fully ductile behavior, the as-rolled T specimens are approaching brittle behavior, and the others are intermediate.

Conclusions

From the standpoint of workability, there is considerable latitude for charge materials including various pig irons, although attention must be given to general compositional requirements for good nodularity. For the ranges investigated, the percent C, Si, Mn and the CE did not affect workability noticeably, but factors such as Ni content (soluble in austenite), unusually low nodule count, and obviously poorly formed nodules did, although some vermicular graphite was tolerable. Variations in elements such as Ni and Si deserve further attention.

Although ductile iron is less workable than steel, it has substantial workability between 1550 and 1950F (843 and 1066C), which is lower than the range where steel working is normally started. Under appropriate conditions, it can be rolled to 90% reduction or more and upset forged to beyond 70% without cracking seriously. Rolling reduction rate can affect workability, especially at large total deformation. As noted by Perry,[4] the upper working limit is set by approach of incipient melting (Fe-C eutectic is at 2107F (1153C)), and the lower limit is influenced by the lower critical temperature (increases with Si) where austenite becomes unstable and plasticity diminishes. Workability is better at 1750 and 1950F (954 and 1066C) than at 1550F (843C); oxide scaling is greatest at 1950F (1066C). Cracks can propagate from surface defects, especially at lower working temperatures, where edges of thin materials may also crack more readily because of faster cooling.

The PMC materials, with smaller and more numerous nodules than the SC materials, and with primary cementite because of the faster solidification, could be rolled about as readily as SC materials between 1750 and 1950F (954 and 1066C). At 1550F (843C), the increased cracking of the PMC materials is apparently related to the lesser fracture path between the more numerous nodules. The cementite could be decomposed during heating near 1750F (954C) before working. Somewhat slower solidification such as from preheating or insulating the molds would decrease the cementite. Some cementite should be tolerable during working at the higher temperatures.

For a reason not yet explained, two SC slabs cracked severely during rolling at 1950F (1066C) where workability was normally very good. The PMC ingots from these and the other heats all were rolled at 1950F (1066C) to 90% reduction with virtually no cracking.

Variation in composition affects mechanical properties more than it does workability, by altering pearlite-to-ferrite ratios (such as Mn and Si) and ferrite strength (Ni and Si). Because ductile iron is complex, composition and structure can markedly influence properties even in castings. Properties of sheet from PMC slabs differ from those of sheet from SC slabs because of the different nodule formation.

Heavily rolled material is characterized by high strength and anisotropy, and low ductility particularly for T specimens, which consistently have poorer properties. Ferritize annealing reduces hardness and strength and improves ductility substantially, but anisotropy from the directional nodule deformation persists. Impact resistance of L specimens is about double that of T specimens, and annealed material about double that of material as-rolled to 70% reduction. Annealed L specimens exhibit ductile impact behavior at room temperature. Phosphorus to 0.10% from the basic pig iron charges did not cause any substantial decrease in impact strength. Cross rolling for part of the rolling reduction can reduce anisotropy in structure and properties. Because the interface is weaker, properties and perhaps workability may be affected to some extent by graphite deposition on existing nodules that results from pearlite decomposition or lower working temperatures where austenite carbon solubility decreases. The low ductility of worked material would often dictate annealing, ideally before cooling after working.

Yield strengths are more consistent with those of hot-rolled low-C steel[17] but L elongations with those for high C steel. Some of the better tensile values, such as for sheet rolled to 70% reduction at 1750F (954C) and annealed, had L yield and elongation values comparable to hot-rolled AISI 1030 steel, but the T elongations were much lower. Impact strength of ductile iron castings is only about one-third that of steel.[17] The poorer ductility and impact resistance of worked ductile iron would restrict its applications to where these deficiencies would be tolerable. Its good castability, machinability, damping capacity and wear and corrosion resistance, because of the high C and Si, could be offsetting factors.

Hot working of ductile iron deserves a new look in light of technology advances. With care, large ductile iron castings with good nodularity can be made.[18] Rare earths can alleviate adverse nodularity effects from impurities such as Ti and Pb.[19] Technology is available for minimizing Mg fading during extended holding of melts, permitting many castings from single heats.[20] Ductile iron is being nodularized with unalloyed Mg in converter ladles.[21] Magnesium-impregnated coke developed for ductile iron nodularization now serves to desulfurize pig iron in torpedo cars between blast furnace and steelmaking furnace[22]; additional amounts could feasibly nodularize ductile iron castings for hot working. There is increasing interest in casting ductile iron in permanent molds,[23] nodularizing in the mold[24] and coining of castings.[25] Ductile iron is now being continuous cast through graphite dies.[26]

Several areas may offer potential for hot working of ductile iron. For instance, it may be possible to treat hot metal directly from the blast furnace to desulfurize, nodularize and balance chemistry and cast into ingots for hot working into shapes. Permanent mold castings can develop adequate structures and may offer some economic benefits. More advantage could be taken of the forgeability of ductile iron for certain items, perhaps as forging of cast preforms. Although ductile iron is sensitive to rolling deformation rate at heavy reductions, high-energy-rate forming deserves attention because of the reported success with gray cast iron forging.[13] Items might be cast to rough shape and finished with some rolling or forging, perhaps in conjunction with continuous casting. Innovative thinking should suggest other areas.

References

1. A. P. Gagnebin, K. D. Millis, and H. Morrogh; The First 25 Years of Ductile Iron, Discovery '48, Modern Casting, p. DI 5-DI 16 (May 1973).
2. E. K. Modl, Comparing Processes for Making Ductile Iron, Foundry, Vol. 98, no. 7, p. 43-48 (July 1970).
3. P. S. Cowen, Heat Treatment of Ductile Iron Castings, Part I, Foundry, p. 60-63 (Sept. 1971); Part II, p. 78-81 (Oct. 1971).
4. J. A. Perry and J. E. Rehder, Nodular Iron Hot-Forged and Rolled Experimentally, Iron Age, Vol. 168, no. 14, p. 229-233 (1951).
5. J. A. Wakamoto, J. Saga and T. Tanaka; Hot-Workability of Nodular Graphite Cast Iron, Rept. Osaka Munic. Inst. Ind. Research (Japanese), Vol. 4, no. 2, 18-21 (1952). Abs. in Chem. Abs., Vol. 47, item 11105i (1953).
6. E. P. Unksow and D. I. Berezhkovskii, Forging, Stamping, and Rolling Iron With Spheroidized Graphite, Vestnik Mashinostroeniya, Vol. 33, no. 12, p. 29-35 (1953). Abs. in Chem. Abs., Vol. 48, item 5052e (1954). Review by W. G. Cass, Working Spheroidal Graphite Iron, Russian Experience, Iron and Steel, Vol. 28, p. 319-320 (June 1955).
7. S. Chang Tso-mei and T. Kuoh Sheng-chuen, Study of the Deformability of Foundry Iron With Spheroidal Graphite, Acta Metallurgica Sinica, Vol. 1, no. 2, p. 165-188 (English summary) (1956). Abstract in Abs. J. for Metallurgy, no. 4-5, item 190, p. 96 (1958).
8. K. D. Sheffler and J. F. Libsch, Correlation of Structural and Mechanical Anistropy in Rolled Ductile Iron, Trans. ASM, Vol. 61, no. 2, p. 203-209 (June 1968).
9. K. Okabayashi, Hot Rolling of Nodular Cast Iron, Imono, Vol. 40, no. 3, p. 233-235 (1969). Brutcher Translation No. 7348, Henry Brutcher Tech. Translations, Altadena, Calif.
10. E. Dragos and N. Boer, Influence des Traitements Thermomecaniques a Haute Temperature, sur la Durete et la Structure des Fontes a Graphite Spheroidal (Influence of Hot Working on the Hardness and Structure of Ductile Iron), Fonderie, Vol. 304, p. 297-301 (Aug.-Sept. 1971). Translation by M. T. Rowley, Modern Casting Tech Report 726 (Mar. 1972).
11. A. Ulitovski, Die Forging and Rolling of Cast Iron and Brittle Alloys, Stal, Vol. 7, no. 11, p. 99-111 (1973). Brutcher Translation No. 568, Henry Brutcher Tech. Translations, Altadena, Calif.
12. MPI Digest, Cast Iron Sheets to Replace Steel Strip? Committee of Merchant Pig Iron Producers, AISI, no. 36, p. 3-4 (Apr. 1973).
13. American Society of Metals; Metals Handbook, Forging and Casting, Metals Park, Ohio, Vol. 5, 8th ed., p. 103-104 (1970).
14. American Society for Testing and Materials; 1972 Annual Book of ASTM Standards: Part 2, Ferrous Castings; Ferroalloys, Philadelphia, Pa., p. 8, 150, 214 (1972).
15. S. I. Karsay, Ductile Iron, Part 1, Production, Quebec Iron and Titanium Corp., Sorel, Quebec, Canada, p. 73 (1969).
16. P. F. Wieser, C. E. Bates and J. F. Wallace, Mechanism of Graphite Formation in Iron-Silicon-Carbon Alloys, Malleable Founders Society, Cleveland, Ohio (1967).
17. American Society for Metals; Properties and Selection of Metals, Metals Park, Ohio, Vol. 1, 8th ed., p. 390-188 (1961).
18. S. I. Karsay, Control of Graphite Structure in Heavy Ductile Iron Castings, Modern Casting, p. 85-92 (July 1970).
19. T. Watmough, W. F. Shaw and F. C. Bock, Combined Effects of Selected Elements on the Properties of Ductile Iron, AFS Transactions, Vol. 79, p. 225-246 (1971).
20. R. S. Lee, Extended Holding of Treated Nodular Iron, AFS Transactions, Vol. 79, p. 433-444 (1971).
21. A. Alt, K. Gut, H. Lustenberger and H. G. Trapp, New Method of Treatment With Pure Mg To Produce Nodular Iron, AFS Transactions, Vol. 80, p. 167-172 (1972).
22. Metal Progress, Technology Forecast '73: Magnesium-Infiltrated Coke Makes Low-Cost, Low-Sulfur Steels, p. 48 (Jan. 1973).

Appendix A

Hardness

Hardness data for sheet as-rolled to 90% reduction at various temperatures and ferritize annealed are listed in Tables A1 and A2 for sand-cast and permanent mold materials, respectively.

In Table A3 are hardness values for sand-cast materials as-rolled to 70% reduction at 1750F (954C) and ferritize annealed.

Table A1. Hardness of Sand-Cast Materials Rolled to 90% Reduction[1] and Ferritize Annealed

Heat No.	Rolling temperature, ° F	Hardness, Rockwell A	
		As-rolled	Ferritize annealed
63-A	1,750	56	28
67-A	1.750	61	35
68-A	1,750	63	36
64-B	1,550	50	30
64-B	1,750	56	30
64-B	1,950	66	38
66-B	1,550	55	35
66-B	1,750	58	37
66-B	1,950	69	42
70-B	1,550	56	37
70-B	1,750	62	37
70-B	1,950	67	43
65-C	1,550	55	32
65-C	1,750	57	32
65-C	1,950	69	41
81-D	1,550	56	35
81-D	1,750	61	38
81-D	1,950	68	46
82-D	1,750	56	34
97-E	1,750	60	32
98-E	1,750	61	34
A01-F	1,750	59	38
A02-F	1,550	57	31
A02-F	1,750	62	39
A02-F	1,950	71	48

[1]At 90-pct reduction, 0.10 inch thick.

Table A2. Hardness of Permanent Mold Materials Rolled to 90% Reduction[1] and Ferritize Annealed

Heat No.	Rolling temperature, ° F	Hardness, Rockwell A	
		As-rolled	Ferritize annealed
54-B	1,550	43	41
54-B	1,750	49	40
54-B	1,950	63	42
58-B	1,550	51	44
58-B	1,750	52	44
58-B	1,950	65	45
62-C	1,550	52	38
62-C	1,750	54	41
62-C	1,950	67	45
81-D	1,750	62	43
97-E	1,750	58	40
98-E	1,750	59	45
A01-F	1,750	60	45
A02-F	1,550	56	39
A02-F	1,750	61	44
A02-F	1,950	67	50

[1]At 90-pct reduction, 0.18 inch thick.

Table A3. Hardness of Sand-Cast Materials Rolled to 70% Reduction[1] at 1750F (954C) and Ferritize Annealed

Heat No.	Hardness, Rockwell A	
	As-rolled	Ferritize annealed
64-B	64	41
66-B	61	42

[1]At 70-pct reduction, 0.30 inch thick.

Table A4. Hardness of Billets[1] Forged to 85% Reduction and Ferritize Annealed

Heat No.	Forging temperature, ° F	Hardness, Rockwell A	
		As-forged	Ferritize annealed
66-B	1,750	54	34
66-B	1,950	62	43

[1]Permanent mold castings.

23. A. J. Zuithoff, T. Breedijk and P. M. H. Geelen, The Section Sensitivity of Cast Iron Permanent Mold Castings, AFS Cast Metals Research Journal p. 83-89 (June 1972).

24. G. Mannion, Making Ductile Iron in the Mold, Foundry, p. 80-82 (Jan. 1973).

25. C. A. Sanders, Don't Write Off Green Sand Molding, Foundry, p. 110-113 (Oct. 1971).

26. B. J. Templar, A. Goudie, R. D. Forrest and J. Nichols, Some Experiences in the Continuous Casting of Nodular Iron, Foundry Trade Journal, p. 176-177 (Aug. 17, 1972).

Appendix B

Tensile Properties

In Tables B1 and B2 are listed additional tensile properties of sand-cast and permanent mold materials, respectively, which were not listed in Table 6. The data is for sheet as-rolled to 90% reduction at various temperatures and ferritize annealed.

Tensile data for sand-cast materials as-rolled to 70% reduction at 1750F (954C) and ferritize annealed is given in Table B3.

For the sheet rolled from sand castings, the average standard deviations (four specimens each condition) for tensile strength, yield strength and elongation, respectively, were L (as-rolled) = 4832 psi, 2001 psi, and 0.57%; T (as-rolled) = 4437 psi, 2621 psi and 0.29%: L (annealed) = 1041 psi, 532 psi and 1.02%; and T (annealed) = 1379 psi, 503 psi and 0.60%.

For the sheet rolled from permanent mold ingots, the corresponding values were L (as-rolled) = 2834 psi, 1310 psi and 0.52%; T (as-rolled) = 2663 psi, 1803 psi and 0.40%; L (annealed) = 862 psi, 642 psi and 1.38%; and T (annealed) = 1113 psi, 433 psi and 0.50%.

Table B1. Tensile Properties of Sand-Cast Materials Rolled to 90% Reduction[1] and Ferritize Annealed

Heat No.	Rolling temp. °F	Orientation	As-rolled Tensile strength, psi	As-rolled Yield strength, psi	As-rolled Elongation, pct	Ferritize annealed Tensile strength, psi	Ferritize annealed Yield strength, psi	Ferritize annealed Elongation, pct
63-A	1,750	L	116,600	91,700	3	51,200	35,600	10
		T	60,300	59,700	1	38,900	34,200	3
67-A	1,750	L	134,200	105,200	2-1/2	56,200	40,700	9
		T	58,000	58,000	1-1/2	37,300	36,600	1-1/2
68-A	1,750	L	127,700	97,200	2-1/2	57,700	43,000	8-1/2
		T	74,100	74,100	1	40,700	39,400	1-1/2
64-B	1,550	L	86,200	75,800	3	47,900	36,000	7
		T	54,300	53,900	1	36,400	32,100	2-1/2
64-B	1,750	L	109,100	84,600	3	50,000	35,300	10
		T	72,000	65,900	2	39,000	32,000	3
64-B	1,950	L	150,200	97,600	5	58,400	37,900	16
		T	112,200	91,600	1-1/2	49,200	36,800	4
70-B	1,550	L	91,800	75,700	3	57,500	45,200	7
		T	62,800	62,000	1	45,700	41,400	2
70-B	1,750	L	135,900	98,700	3-1/2	60,700	44,300	13
		T	94,600	85,000	1-1/2	48,200	41,700	2-1/2
70-B	1,950	L	149,100	109,100	1-1/2	64,300	46,100	17
		T	112,200	90,200	1-1/2	54,600	43,900	4-1/2
65-C	1,550	L	86,700	77,800	3	45,900	37,700	5
		T	52,300	58,000	1-1/2	33,100	32,500	1-1/2
65-C	1,750	L	111,700	93,500	2	51,900	38,700	9
		T	72,000	70,000	1	39,300	35,500	2-1/2
65-C	1,950	L	158,900	109,600	4	62,400	42,900	14
		T	117,400	106,000	2	48,000	38,600	4
82-D	1,750	L	114,100	92,900	1-1/2	52,900	39,900	8
		T	64,700	64,700	1/2	36,800	34,700	1-1/2
97-E	1,750	L	133,000	95,200	3	54,600	38,200	12
		T	91,800	82,800	1-1/2	42,100	34,800	4
98-E	1,750	L	132,300	102,100	2-1/2	59,300	43,400	11
		T	91,400	83,200	1-1/2	45,500	39,400	3-1/2
A01-F	1,750	L	137,100	114,200	1-1/2	62,300	46,500	11
		T	85,700	80,700	1-1/2	48,600	42,700	3
A02-F	1,550	L	106,000	82,500	3	54,000	42,000	9
		T	68,500	67,800	1	40,900	37,000	2
A02-F	1,750	L	141,000	96,900	3-1/2	59,600	42,300	14
		T	98,700	87,800	1-1/2	47,000	38,600	3-1/2
A02-F	1,950	L	165,500	119,200	4-1/2	68,800	48,100	17
		T	136,700	117,700	1-1/2	57,100	43,600	5

[1]At 90-pct reduction, 0.10 inch thick.

Table B2. Tensile Properties of Permanent Mold Materials Rolled to 90% Reduction[1] and Ferritize Annealed

Heat No.	Rolling temp. °F	Orientation	As-rolled Tensile strength, psi	As-rolled Yield strength, psi	As-rolled Elongation, pct	Ferritize annealed Tensile strength, psi	Ferritize annealed Yield strength, psi	Ferritize annealed Elongation, pct
54-B	1,550	L	73,500	55,200	3	56,100	41,200	5-1/2
		T	51,500	49,100	2	45,600	38,400	2
54-B	1,750	L	89,900	61,600	5-1/2	61,000	40,200	10-1/2
		T	60,900	52,100	2	49,500	39,100	3
54-B	1,950	L	115,100	73,300	4	61,200	44,000	12-1/2
		T	91,300	64,300	2-1/2	51,900	40,700	4
62-C	1,550	L	90,500	69,900	2-1/2	53,100	42,600	5
		T	59,400	56,200	1	40,600	39,200	1-1/2
62-C	1,750	L	110,300	73,000	4	59,900	43,700	9-1/2
		T	75,500	61,300	1-1/2	49,100	41,200	3
62-C	1,950	L	156,500	104,200	4-1/2	65,300	48,800	14
		T	115,700	93,300	1-1/2	55,700	44,700	4-1/2
97-E	1,750	L	121,900	82,400	3-1/2	59,600	40,600	13-1/2
		T	85,800	64,600	2	49,300	37,900	4-1/2
98-E	1,750	L	127,300	84,500	4	63,500	45,000	13
		T	86,400	68,900	2	52,700	42,400	3-1/2
A01-F	1,750	L	122,100	86,600	3	65,800	47,500	13-1/2
		T	86,800	70,500	2	56,900	44,900	4-1/2
A02-F	1,550	L	120,800	75,400	4-1/2	59,900	44,300	11-1/2
		T	76,300	69,600	1-1/2	46,900	41,600	3
A02-F	1,750	L	134,900	86,000	4-1/2	63,600	45,000	16-1/2
		T	94,500	72,400	2	56,800	41,300	6-1/2
A02-F	1,950	L	152,400	97,700	5	70,300	50,900	19
		T	128,400	91,200	2-1/2	64,800	46,700	9-1/2

[1]At 90-pct reduction, 0.18 inch thick.

Table B3. Tensile Properties of Sand-Cast Materials Rolled to 70% Reduction[1] at 1750F (954C) and Ferritize Annealed

Heat No.	Orientation	As-rolled Tensile strength, psi	As-rolled Yield strength, psi	As-rolled Elongation, pct	Ferritize annealed Tensile strength, psi	Ferritize annealed Yield strength, psi	Ferritize annealed Elongation, pct
64-B	L	126,300	76,900	5	56,800	35,500	19
	T	86,600	69,800	2	46,300	33,400	5
66-B	L	133,400	81,000	5-1/2	60,200	37,200	20-1/2
	T	103,900	70,000	3	52,200	36,400	7

[1]At 70-pct reduction, 0.30 inch thick.

Table B4. Tensile Properties of Billets[1] Forged to 85% Reduction and Ferritize Annealed

Heat No.	Forging temp. °F	As-forged Tensile strength, psi	As-forged Yield strength, psi	As-forged Elongation, pct	Ferritize annealed Tensile strength, psi	Ferritize annealed Yield strength, psi	Ferritize annealed Elongation, pct
66-B	1,750	64,800	64,100	1-1/2	50,200	39,200	4
66-B	1,950	97,800	77,700	2	56,400	41,600	5

[1]Permanent mold castings.

EUROPEAN APPLICATIONS OF
ROBOTS AND MANIPULATORS
IN FOUNDRIES

Norris B. Luther
Managing Associate
Lester B. Knight & Associates
549 W. Randolph Street
Chicago, Illinois 60606

The application of robots and manipulators in European foundries would
appear to be at a level of sophistication that is comparable to applications in the United States. They have similar reasons for adopting the
use of robots, namely:

. Working conditions
. High wages
. Absenteeism
. Desirability of a reliable and consistent work pattern and motion

This report will describe three applications currently being used:

. A cleaning room operation in Sweden
. A core assembly and core setting operation in France
. A block grinding operation in Germany

SWEDEN

The application at the Kohlswa Jernverk in Sweden uses a five-axis robot
in the handling and fettling (grinding) of mill segments. These segments
(see Figure 1) are approximately 10 kg (22 lb) in weight, made of stainless, acid-resisting steel and used in papermills. The robot's task is
to remove ingot remainders (see Figure 2) in the dividing plane around
the entire segment-contour (see Figures 3 and 4). The same robot can in
a similar way fettle pump housings as shown in Figure 5.

To compensate the wear of the grinding disc, the robot is equipped with a
search function followed by coordinate-transformation of the program.
This option enables the robot to first detect the actual size of the
rotating grinding disc and then, upon gaining this information, perform
the fettling (grinding) program. The detection itself is accomplished by
a sensing current in one- or two-axis servo drives.

The workpieces are stored in a magazine with a capacity of 12 pieces,
corresponding to approximately one hour of the robot's work time.

The gripper (not displayed in the photograph in its true version) is
hydraulically driven and clamps the workpiece with a force of approximately 2,000 kp. A check of the clamping force is included to ensure that
the workpiece is properly held in the robot hand.

This robot has been in operation since 1977. Compared with manual fettling, the robot finishes the parts with a 40% shorter cycle time.

FRANCE

Figure 6 shows a robot layout used to assemble and set head cores for an aluminum head at Citroen in France. The layout for this operation consists of the following equipment:

- . 2 furnaces
- . 1 automatic tapping ladle
- . 2 automatic die molding stations
- . 1 index table with four positions
- . 1 lift
- . 1 auxiliary table
- . 1 robot control desk

The robot's electronic control cabinet has been installed in a temperature-controlled room located on a foot bridge 3 meters above the floor.

Prior to the robot installation, the operators were required to perform their tasks in an environment of high temperatures.

Core setting often resulted in core breakage which required the operator to remove the broken core and blow the core pieces and sand from the mold before proceeding with the operation.

Now the cores are handled by the robot using a suspended frame (see Figure 7) equipped with rubber pins (see Figure 8) that are inflated with air to lift the cores from the fixture located on the index table (see Figure 9). Before the cores are set in the mold (see Figure 10) they pass over a mirror to allow the operator to verify core quality. Checking and cleaning of the mold is the operator's responsibility.

After pouring the mold, two operators were required to remove the head and cores from the mold, total weight approximately 40 kg (88 lb), at a temperature exceeding 500 $^{\circ}$C. The total operation required four operators. Now this operation is accomplished with the robot (see Figures 11 and 12) and requires only one operator. His tasks are to place the cores on the large index table (see Figure 13), check and clean the mold after each cycle (see Figure 14), monitor the cleanliness of the cores prior to setting, and observe the total operation. His position is near the robot control station for ease of operation should problems occur.

The cycle for the total operation consists of the following:

- . Take frame core from the index table
- . Place this core in #1 mold
- . Partial closing of #1 mold
- . Take head core from the index table
- . Place this core in #1 mold
- . Close #1 mold
- . Return robot to waiting position
- . Teem aluminum

. Open mold #2 for extra cooling
. Remove head after ejection
. Place head on the lift
. Take frame core from the index table
. Place this core in mold, etc.

The operation can be done with mold #1, mold #2, or simultaneously with the two molds.

The robot has been used on this operation since October, 1978, and is averaging approximately 300 heads per day. It is estimated that 352 heads per day is attainable. Approximately 32,000 cores were handled with only two broken. This foundry now has four robots working on similar production systems, and reports a core scrap reduction from 15% to 2% resulting from improved handling.

This first robot application has operated approximately 18 months, and they report 99.27% up-time. Life of the inflated rubber devices is approximately 12,000 cores.

GERMANY

Figure 15 shows a robot performing the grinding on an engine block main bearing area. This prototype application has ground over 7,000 blocks in a German foundry. The robot has been fitted with both a cup and a cone grinder (see Figure 16). The total production and maintenance data are not available at this time.

There are many other applications of robots throughout Europe, many of them similar to those described in this report. It would appear that European foundries are making excellent technology progress with robot applications.

Figure 1. Mill segment before and after fettling (grinding).
Approximate weight 22 pounds.

Figure 2. Illustrates robot during the operation of removing the risers and
gates from the mill segment.

Figure 3. Shows robot during the operation of removing
the ingot remainders and burrs in the dividing
plane of the mill segment.

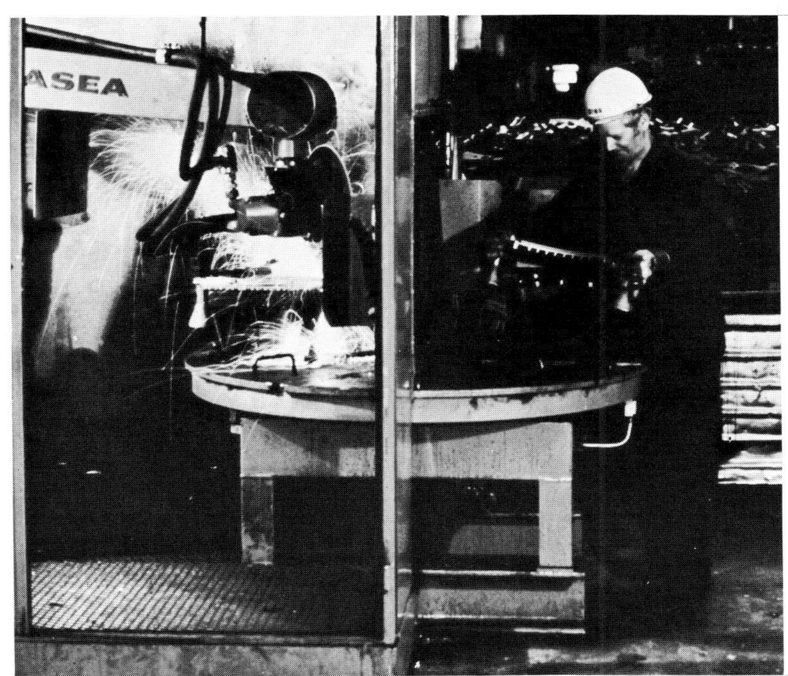

Figure 4. Shows the mill segment being placed in
the clamping device by the operator
prior to indexing the mill segment
into position for the robot.

Figure 5

Illustrates the fettling (grinding)
operation on a pump housing by the
same robot. A variety of other
castings can be fettled in a similar
way.

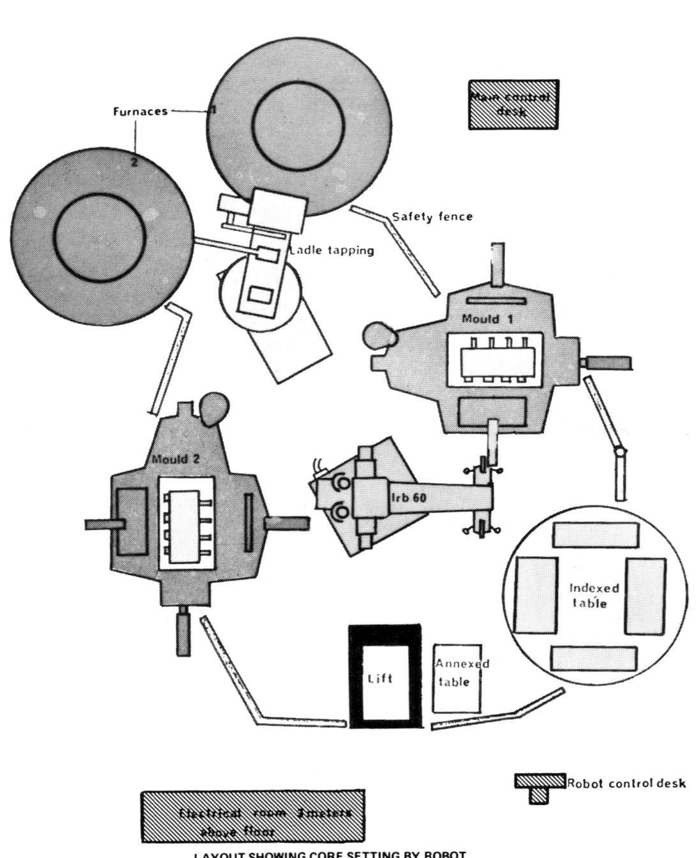

Figure 6.　Layout for an aluminum head process at Citroen in France.
The robot performs the function of setting two cores from
the index table to the molds and removing the hot casting
and cores from the mold stations to the lift table. This
robot removed three operators from a high-temperature
working environment.

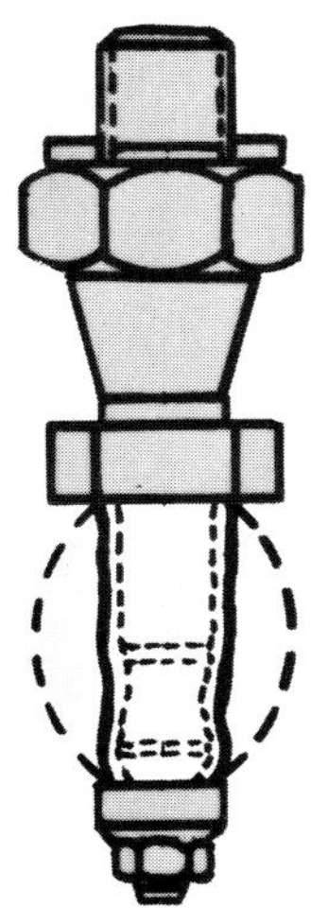

INFLATED RUBBER PINS FOR PICKING UP THE CORES

Figure 8

Illustrates how the rubber pin is inflated.
This principle provides a firm grip on the
core, and with the controlled motion of the
robot is one of the reasons for reduced
core breakage.

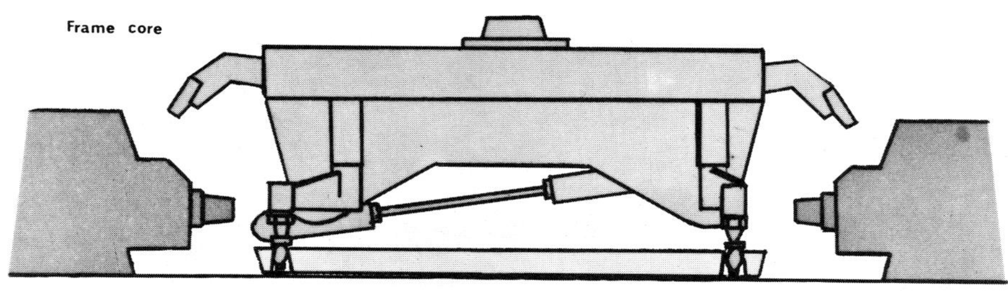

Figure 9.   Illustrates how the frame core is picked up and placed in the mold.

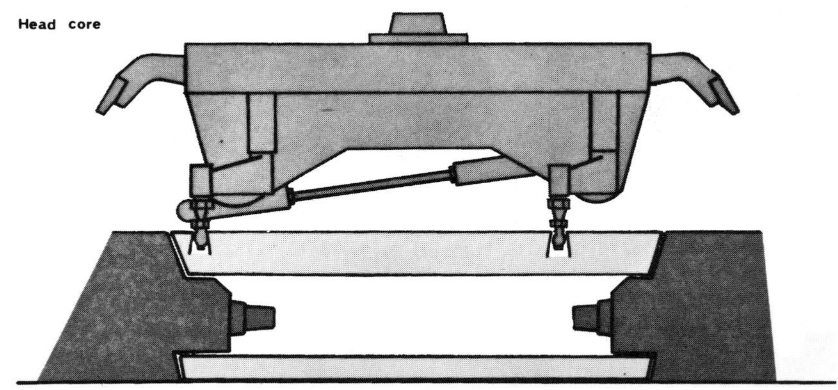

Figure 10.   Illustrates the head core being set into the mold. Note that the mold has been partially closed after setting of the frame core illustrated in Figure 9.

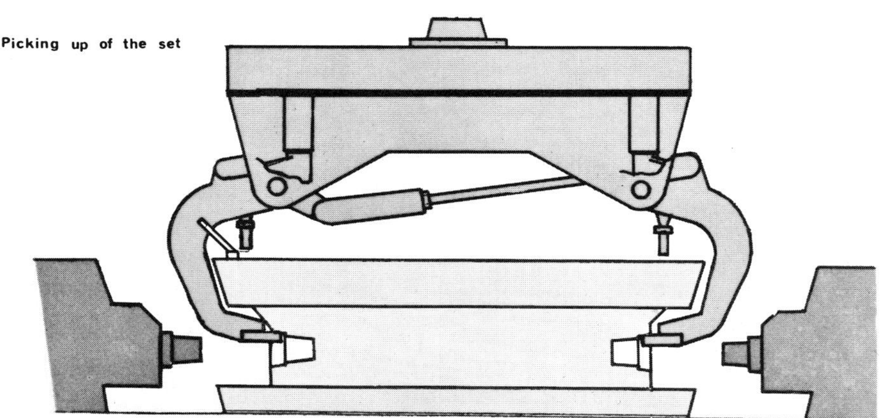

Figure 11.   Illustrates how the frame for core setting is also used to pick up the the casting from the mold. The casting and cores weigh approximately 88 pounds, and the temperature is approximately 500 $^\circ$C. The robot made it possible to remove two operators previously used for this operation.

Figure 12.   Shows the robot with the casting and core sets just after removing from the mold.   Casting is being transferred to the lift table at the lower right of the photograph.

Figure 13.   Shows the operator placing a core set in the fixture located on the index table.   Note that the operator is observing the operation of the mold.

Figure 14.   Shows the operator inspecting the mold.   Note
that the robot is in the wait position prior
to picking up the next core set.

Figure 15.   Shows a robot with a grinder
head used for grinding a four-
cylinder motor block.   This
prototype application has
ground over 7,000 blocks in a
German foundry.

Figure 16.   Shows a robot with a cup
grinding wheel used to
grind on the external surface of a cylinder block.

INDEX